Advances in Experimental Medicine and Biology

Volume 1485

Advances in Experimental Medicine and Biology provides a platform for scientific contributions in the main disciplines of the biomedicine and the life sciences. This series publishes thematic volumes on contemporary research in the areas of microbiology, immunology, neurosciences, biochemistry, biomedical engineering, genetics, physiology, and cancer research. Covering emerging topics and techniques in basic and clinical science, it brings together clinicians and researchers from various fields.

Advances in Experimental Medicine and Biology has been publishing exceptional works in the field for over 40 years, and is indexed in SCOPUS, Medline (PubMed), EMBASE, Reaxys, EMBiology, the Chemical Abstracts Service (CAS), and Pathway Studio.

2023 CiteScore: 5.9

Junjie Xiao
Editor

Advances in Circular RNAs

The Basis and Disease Relevance

Springer

Editor
Junjie Xiao
School of Life Science
Institute of Cardiovascular Sciences
Shanghai University
Shanghai, China

ISSN 0065-2598 ISSN 2214-8019 (electronic)
Advances in Experimental Medicine and Biology
ISBN 978-981-96-9427-3 ISBN 978-981-96-9428-0 (eBook)
https://doi.org/10.1007/978-981-96-9428-0

This Springer imprint is published by the registered company Springer Nature Singapore Pte Ltd.
The registered company address is: 152 Beach Road, #21-01/04 Gateway East, Singapore 189721, Singapore

Preface

Circular RNAs (circRNAs) are a distinct class of non-coding RNAs generated by back-splicing of precursor mRNAs, forming covalently closed loops that confer high stability and diverse regulatory functions. Widely expressed in eukaryotic cells, circRNAs participate in important biological processes—including cell proliferation, differentiation, apoptosis, and stress responses—through mechanisms such as miRNA sponging, transcriptional regulation, and protein interactions. Increasing evidence links circRNAs to the development and progression of various diseases, especially cancer, cardiovascular, and neurodegenerative disorders, underscoring their potential as both diagnostic biomarkers and therapeutic targets. With growing interest and rapid advancements in the field, circRNAs are emerging as a promising frontier in disease diagnosis and precision medicine.

Our research team has made substantial contributions to elucidating the functions of circRNAs, alongside the significant progress achieved by research groups worldwide in recent years. To further promote understanding of circRNA biology and its relevance to human health and disease, this book summarizes the latest advances in circRNA biogenesis, molecular mechanisms, and disease associations. In Part I, we provide an overview of the fundamental concepts and biological significance of circRNAs. In Part II, we focus on bioinformatics tools and computational strategies for circRNA analysis. In Part III, we explore the molecular mechanisms that govern circRNA biogenesis. In Part IV, we examine the diverse ways in which circRNAs regulate gene expression through various molecular pathways. In Part V, we discuss their potential as biomarkers for diseases such as cancer, cardiovascular conditions, and neurological disorders. In Part VI, we address their involvement in the pathogenesis of human diseases, and in Part VII, we broaden the discussion to include circRNAs in archaea. In Part VIII, we present challenges and future prospects in the investigation of circRNAs. We hope this book provides valuable insights for researchers, clinicians, and students in the fields of biochemistry, molecular biology, cell biology, and medicine. We would be honored if it contributes to the development of circRNA-based diagnostic tools and therapeutic strategies, and stimulates future research in this dynamic and rapidly evolving field.

We thank Dr. YANG Tingting for her help in preparing the book.

Shanghai, China Junjie Xiao

Contents

About the Editor

Junjie Xiao is the executive dean of the School of Life Science, Shanghai University, China. In addition, he is the director of the Key Laboratory of Biomedical Materials and Technology for Organ Regeneration of the Ministry of Education, the director of the Shanghai Organ Regeneration Engineering Technology Research Centre, and the head of the Cardiovascular Research Institute of Shanghai University. He is an Associate Editor of *Journal of Molecular and Cellular Cardiology* and *Journal of Cardiovascular Translational Research*, and is on the editorial boards of several other notable journals. He has authored or co-authored in journals such as *Cell Metabolism*, *Circulation*, *Circulation Research*, *Nature Communications*, etc. His research focuses on heart failure intervention and risk prediction, with a particular emphasis on exercise-induced cardiac regeneration and the identification of novel therapeutic targets. He is also interested in epigenetic and metabolic regulations of cardiac function upon exercise and under pathological conditions.

Part I

Overview

An Overview of Circular RNAs

1

Julia Mester-Tonczar and Ena Hasimbegovic

Abstract

Circular RNAs (circRNAs) are a class of endogenous, covalently closed RNA molecules. Unlike linear RNAs, circRNAs are formed through noncanonical splicing, during which a downstream donor site is ligated with an upstream splice acceptor site, building a backsplice junction (BSJ), the distinguishing feature of circRNAs. The inherent feature of circRNAs is their lack of 5′ cap structures and 3′ poly(A) tails, which are typically found in linear RNAs. Due to their resistance to exonucleases, they exhibit increased stability compared to linear RNAs. In the past, circRNAs have been shown to be evolutionarily conserved and possess cell-type and tissue-specific expression patterns. The various important biological functions of circRNAs, including their roles as protein inhibitors (microRNA (miRNAs) sponges) and regulators of RNA-binding proteins (RBPs), have made them interesting biomolecules for the scientific community. However, due to the novelty of this research field, many obstacles are still present, arising from the lack of consensus on the precise methodological standards for reliable identification and nomenclature of newly identified circRNAs, inherent limitations of the applied methodologies, and the open questions regarding the mechanisms of action of this class of RNAs. This chapter will provide a brief summary of the discovery of circRNAs and the early related research. We will outline the importance of RNA sequencing technologies in circRNA research and highlight the more recent findings in this field, with a special focus on the different functions of circRNAs. Finally, we will discuss key challenges associated with circRNA research.

Keywords

CircRNA · CircRNA overview · RNA-seq · CircRNA functions

This chapter briefly summarizes the biology of circular RNAs and the recent research progress in the field of circular RNAs and highlights the breakthroughs as well as the challenges in systematically analyzing novel circular RNAs.

J. Mester-Tonczar (✉) · E. Hasimbegovic
Department of Internal Medicine II, Division of Cardiology, Medical University of Vienna, Vienna, Austria
e-mail: julia.mester-tonczar@meduniwien.ac.at; ena.hasimbegovic@meduniwien.ac.at

J. Xiao (ed.), *Advances in Circular RNAs*, Advances in Experimental Medicine and Biology 1485,
https://doi.org/10.1007/978-981-96-9428-0_1

1 Introduction

1.1 Background: The Beginnings of Circular RNA Research

The initial description of circular RNAs (circRNAs) dates back to 1976, when they were described in viroids, pathogens of higher plants [1]. Another publication in 1976 observed "circular structures" in the Sendai virus containing defective interfering RNAs [2]. Only 3 years later, in 1979, circRNAs were first detected in the cytoplasm of eukaryotic (HeLA) cells using electron microscopy [3]. In 1986, they were identified in the hepatitis delta virus [4]. In humans, the first endogenous spliced circRNA was reported in 1991. In their study, Nigro et al. reported scrambled exons from the *DCC* gene, where an upstream exon was ligated with the last nucleotide of a downstream exon [5]. In 1992, another study uncovered similar findings in the human *ETS-1* gene, where the typical order of exons was also scrambled. The study further observed that this type of splicing only occurred in exons in close proximity to large introns [6].

Despite the abovementioned findings and several publications in the late 1990s and early 2000s, circRNA research progressed slowly because of the widely held belief that the production of molecules was a mere mis-splicing event [7] and the difficulties in their in silico detection.

1.2 The Development of High-Throughput Sequencing and Its Effects on circRNA Research

In general, the early days of sequencing (1970s–1990s) were characterized by expensive experimental approaches, with limited sequencing length and a sole focus on DNA [8–10]. In 1983, Tse-Wen Chang proposed the methodology of antibody matrixes. The concept of these antibody matrixes, often also referred to as antibody microarrays, is the binding of cells to matrixes to simultaneously detect multiple antigens in a single attempt [11].

From 1995 onward, an increase in studies using microarrays allowed researchers to analyze gene expression on a genome-wide level for the first time [12]. In addition, serial analysis of gene expression (SAGE) was developed in 1995 [13].

However, microarrays could only detect already known target genes and information on the exon level was not possible. The major disadvantages of SAGE were its inability to offer information on splice variants and its unreliability in detecting weakly expressed transcripts [14]. This eventually led to the development of exon microarrays [15], which offered additional insights into the field of molecular biology.

However, it was only after the development of massive parallel sequencing techniques in the early 2000s that—for the first time—allowed the generation of large datasets. Massive parallel sequencing also known as high-throughput sequencing or next-generation sequencing (NGS) has the advantage of low sequencing costs and high accuracy with a simultaneous rise in sequencing speed [16, 17]. This technique not only opened the door for studying complex transcriptomes but also created a new scientific field: bioinformatics [10]. Figure 1.1 shows a timeline of selected sequencing technologies from 1977 to 2019.

A timeline of sequencing technologies from 1977 to 2019. Sanger sequencing is considered a first-generation sequencing technique, Illumina sequencing from 2007 and ion torrent from 2011 are commonly categorized as second-generation sequencing also known as next-generation sequencing (NGS), while nanopore sequencing is often referred to as third-generation sequencing (TGS).

The ability to study the whole transcriptome, the entirety of all RNA transcripts contained in individuals or cells, and the ensuing attempts to understand the functional roles and interplay of the detected transcripts are now thought to be the basis of fully understanding the molecular pathophysiology of most conditions [18].

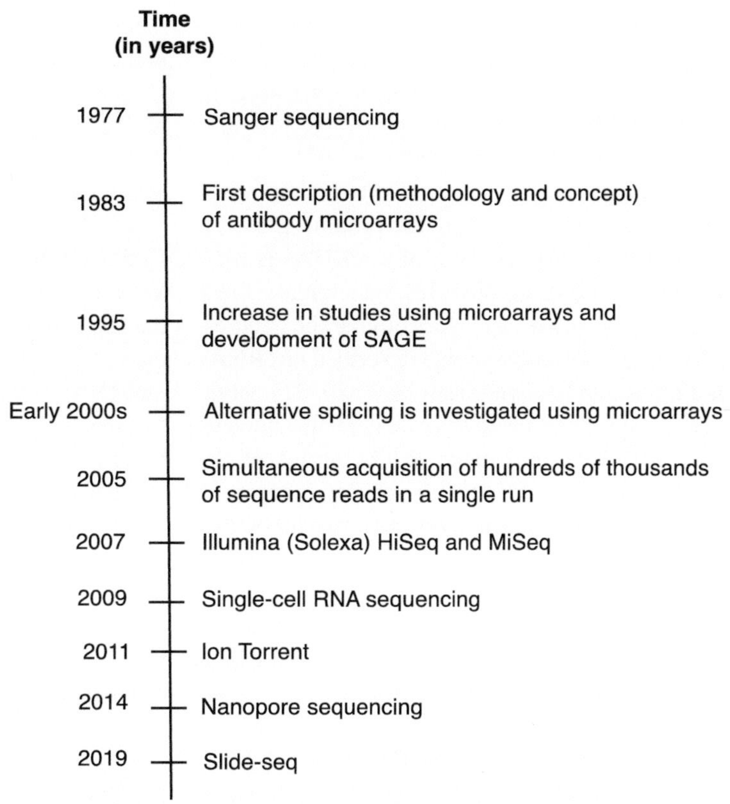

Fig. 1.1 Timeline of selected sequencing technologies

In the early days of transcriptome studies, poly(A) enrichment was performed for most NGS library preparations. Due to their closed-loop structure, circRNAs lack poly(A) tails, which is why these molecules were difficult to profile in early NGS methods [19]..

1.2.1 Detection of circRNAs Using RNA Sequencing (RNA-seq)

The first study to detect circRNAs using NGS was published in 2012. This study used high-throughput sequencing to detect genomic rearrangements in cancers and studied the global expression of circRNAs in poly(A)-depleted human pediatric acute lymphoblastic leukemia samples [20]. Only 1 year later, these findings were extended by other research groups, and, together with the study from 2012, these data represent the beginning of a new era in circRNA research using RNA sequencing (RNA-seq) [20–24].

Simultaneously with the development of high-throughput sequencing, circRNA-specific bioinformatics algorithms allowed the identification of circRNAs in various eukaryotes including plants, fungi, protists, worms, insects, and mammals [25–30]. In addition, new methods for the identification and validation of circRNAs have been proposed, including the analysis of non-polyadenylated libraries combined with a pretreatment of RNase R [21, 22]. As an exonuclease, RNase R mostly degrades linear RNAs and enriches circRNAs. However, several studies have now indicated that some circRNAs can nonetheless be depleted by exonuclease-mediated degradation [19, 31, 32], which is why RNase R treatment is not always recommended when working with these molecules.

1.3 Characteristics of CircRNAs

CircRNAs are single-stranded, endogenous RNA molecules characterized by a covalently closed-loop structure, which lack 5′ cap structures and 3′ poly(A) tails [33]. The size of circRNAs ranges from under 100 nucleotides (nt) to over 4000 nt [34]. Despite their initial classification as a subgroup of long noncoding RNAs (lncRNAs), several studies have since indicated that some circRNAs can be translated into proteins [35–37].

A striking characteristic of circRNAs is their cell and tissue specificity [38], which make them particularly interesting for biomarker studies. Most of the currently known described circRNAs are located in the cytoplasm [3, 7, 20, 21], while only a few have been described in the nucleus [39, 40]. Typically, linear RNAs represent the predominant form of most transcripts. However, the abundance of some circRNAs surpasses that of their linear counterparts [20, 21, 41]. Despite the classification of circRNAs into one larger category based on their circular structure as the unifying trait, the abovementioned variations depict the plethora of forms and functions.

1.3.1 Biogenesis of CircRNAs

While linear RNAs are formed through canonical splicing, where the spliceosome removes introns from pre-messenger RNA (pre-mRNA) and joins exons, circRNAs are—in general—produced via backsplicing, a noncanonical splicing event [42]. The ligation of a downstream donor site with an upstream acceptor site, with the ensuing 3′–5′ phosphodiester bond, forms a backsplice junction (BSJ) [43, 44], a typical feature found in circRNAs. CircRNAs are transcribed by RNA polymerase II, and their transcription is initiated by the spliceosome, with both *cis* and *trans* elements acting as regulators [45–48]. However, the biogenesis of these molecules has not yet been fully studied.

At the present moment, the five known types of circRNAs are exonic and exonic–intronic circRNAs (EIcircRNAs), which display noncolinear 3′–5′ junctions, intergenic circRNAs, which originate from intronic circRNA fragments [49], antisense circRNAs, which are generated through the transcription of the sense strand, and intronic circRNAs (ciRNAs). ciRNAs are the only known type of circRNAs to be formed through a failure of lariat debranching (spliced introns) during canonical splicing [50] and circularized by non-colinear 2′–5′ junctions [51]. A single gene locus can therefore generate multiple types of circRNAs [51]. Figure 1.2 depicts the biogenesis of circRNAs and the currently known types.

CircRNAs represent a subgroup of lncRNAs. Currently, five different types of circRNAs are known: antisense circRNAs (generated through the transcription of the sense strand), exonic circRNAs, exonic–intronic circRNAs, intergenic circRNAs (generated from intronic circRNA fragments), and intronic circRNAs (generated through a failure of debranching of lariats during canonical splicing [52, 53]).

2 Recent Findings in CircRNA Research

The large number of types of circRNAs is a reflection of the number of ways these molecules can regulate gene expression, which have gradually been shown in recent years. This section provides an overview of the recent findings in circRNA research. Figure 1.3 summarizes the versatile functions of circRNAs that are discussed in this section.

Various functions of circRNAs have been described: Translation into proteins in a cap-independent manner, interaction with RNA-binding proteins (RBPs), miRNA sponges, protein sponges, regulation of parental gene expression (i.e., parental gene translation), involvement in alternative splicing, and cell cycle regulation.

2.1 CircRNAs Act as MicroRNA Sponges

MicroRNAs (miRNAs) are small noncoding RNA molecules. They regulate translation by pairing with complementary sequences of messenger RNAs (mRNAs) and hindering mRNA translation [54]. In turn, some circRNAs bind miRNAs and prevent them from interacting with their target mRNAs [22, 24]. As a result,

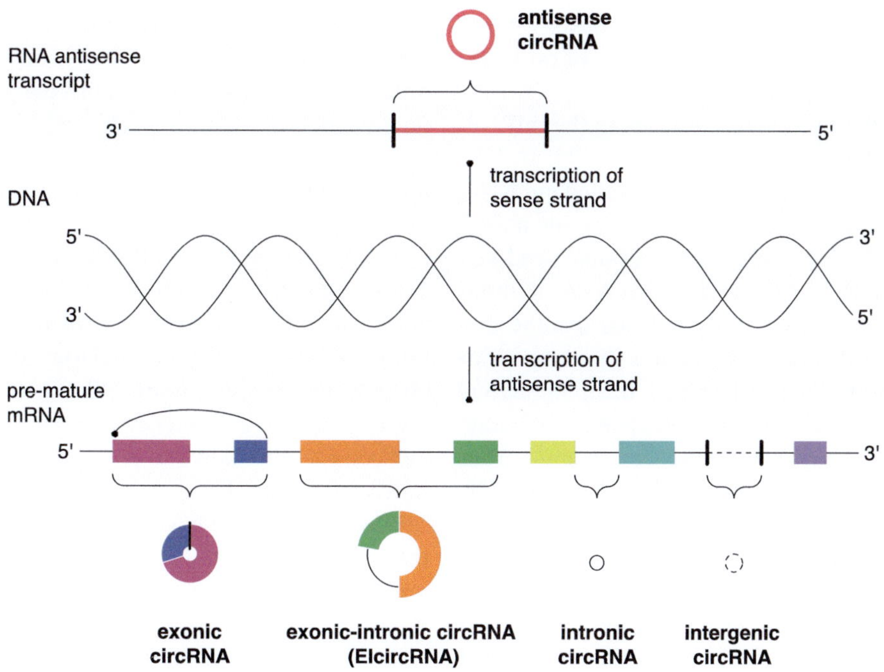

Fig. 1.2 Biogenesis of circRNAs and their different types

Fig. 1.3 Different functions of circRNAs

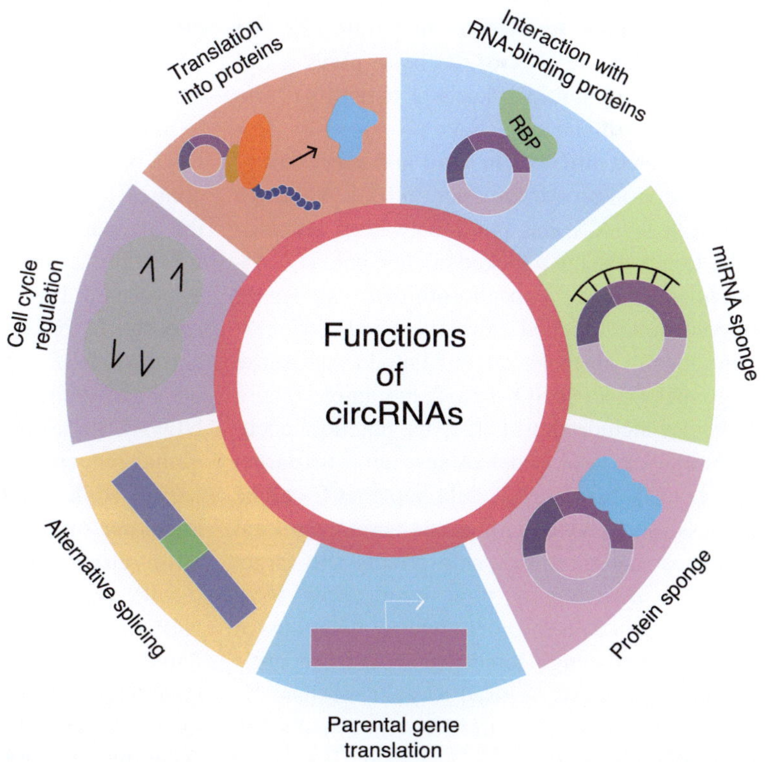

these circRNAs indirectly regulate mRNA translation, serving as an additional layer of posttranscriptional regulation.

In the first two studies describing the miRNA sponging function of circRNAs, published in 2013, two independent groups reported circRNA CDR1as (also known as ciRS-7) as an miR-7 sponge, [22, 24] with 73 conventional seed targets for miR-7 [24]. In addition, one of these studies also showed that the testis-specific circRNA sex-determining region Y (circSRY) sponges miR-138 [24]. Since then, numerous research groups have reported sponge functions of circRNAs in various diseases ranging from cancer [55, 56] to cardiovascular disease (CVD) [57] and neurological disorders [58].

While initial research primarily focused on the investigation of naturally occurring circRNAs as miRNA sponges, recent studies have been examining the potential of artificially engineered circRNAs to target specific miRNAs [59–62]. In 2018, a study attempted to alter the life cycle of hepatitis C virus (HCV) with one such engineered circRNA [60]. The study utilized the previously available knowledge that miR-122, the most abundant miRNA of the liver, plays a significant role in multiple facets of hepatitis C disease. The artificial circRNAs produced by Jost et al. targeted miR-122 in vitro and successfully inhibited viral protein production [60]. Although miR-122 in HCV has previously been successfully targeted using a locked nucleic acid (LNA)-modified DNA phosphorothioate antisense oligonucleotide called Miravirsen [63, 64], certain inherent obstacles are attributed to antisense oligonucleotide (ASO)-based therapies. These obstacles include short half-lives, potential accumulation of LNA nucleotides, and off-target effects [59]. Nowadays, while hepatitis C can be successfully treated with direct-acting antiviral therapies that show few side effects and target proteins involved in viral replication [65], insights gained from preclinical applications of such artificially engineered circRNAs in hepatitis C can nevertheless be used to gain mechanistic insights into how these circRNAs may be used to treat other diseases..

In 2020, artificial circRNAs targeting pro-hypertrophic cardiac miRNAs (miR-132 and miR-212) were applied in vivo in a transverse aortic constriction (TAC) mouse model. Treatment with these engineered circRNA sponges (circmiRs) resulted in an attenuation of pressure overload-induced cardiac hypertrophy [59].

Compared to other miRNA targeting molecules, such as linear miRNA sponges, artificial circRNAs have the advantages of increased stability and the ability to localize to both the nucleus and the cytoplasm [60], all while boasting a comparable efficiency and representing an inexpensive alternative to LNA-containing anti-miRNA drugs [61]..

These recent data have shown the potential for the use of engineered circRNAs as therapeutic miRNA antagonists.

2.2 CircRNAs Interact with RNA-Binding Proteins

In addition to acting as miRNA sponges, circRNAs can also interact with RNA-binding proteins (RBPs). RBPs assemble into ribonucleoprotein complexes [66] and participate in numerous steps of RNA metabolism including RNA splicing, RNA transfer, and translational control [67]. Despite open questions regarding the precise molecular mechanisms of circRNA–RBP interactions, various research groups have established that these molecules regulate one another. In 2021, a study examined two neural-enriched RBPs, NOVA1 and NOVA2, both of which have crucial roles in alternative splicing. A NOVA2 knockout in a mouse model led to a global reduction of circRNAs in the embryonic cortex, while circRNA levels in a NOVA1 mouse knockout model remained unaffected. In addition, cross-linking and immunoprecipitation (CLIP) and reporter analyses indicated that NOVA2 interacts with YCAY clusters in the flanking introns of circEfnb2. This led to the conclusion that NOVA2 represents the first known RBP, which promotes circRNA biogenesis in the brain [68]. Another group observed that several abundant circRNAs are regulated by a mesenchymal splicing factor, the RBP Quaking (QKI) during human epithelial–mesenchymal transition [69]. Furthermore, the modulation of QKI levels affects circRNA abundance via QKI-binding motifs. These additional QKI-binding motifs then lead to de novo circRNA formation, indicating the involvement of QKI in circRNA formation [69]. An additional study also examined the role of QKI and its interaction with circRNAs [70]. In their in vitro model of doxorubicin-mediated cardiotoxicity in human and rodent cardiomyocytes and in vivo in a murine model, *Qki5*, a QKI isoform, decreased the toxic effect of doxorubicin by regulating circRNAs derived from the *Ttn*, *Strn3*, *Arhgap23*, and *Fhod3* genes [70].

In 2019, a research group used RNA-seq to detect a significant upregulation of SCD-circRNA 2 (derived from the stearoyl-CoA desaturase (*SCD*) gene) in hepatocellular carcinoma (HCC). The production of SCD-circRNA 2 was regulated by RNA-binding protein 3 (RBM3), which, in turn, promoted cell proliferation in HCC [67].

2.3 Artificial circRNAs Act as Protein Sponges

In addition to the interaction of circRNAs with miRNAs and RBPs, circRNAs can also be artificially engineered to interact with proteins and can act as protein sponges. Within the field of circRNA research, this approach is relatively novel but a recent study describing the design and application of circRNAs as protein sponges has been published in 2020 [71]. Schreiner et al. published their attempts to design and express artificial circRNAs with specific binding to heterogeneous nuclear ribonucleoprotein (hnRNP) in order to inhibit its function. The designed circRNAs included either CA-repeat elements or CA-rich high-affinity binding sites for hnRNP. The study was able to deliver initial proof of concept for creating circRNA protein sponges, offering a new interference strategy in addition to short interfering RNAs (siRNAs) [71].

2.4 CircRNAs Can Be Translated into Proteins

Despite the common perception of circRNAs as strictly noncoding RNAs, studies have reported that some circRNAs are indeed translated into proteins [35–37]. As circRNAs lack cap structures, translation occurs in a cap-independent manner. Currently, two cap-independent pathways have been described for the translation of circRNAs: a pathway mediated by the internal ribosome entry site (IRES) [36] and a second pathway mediated by N^6-methyladenosine (m^6A), the most abundant RNA base modification in eukaryotes [37]. The IRES-driven translation of circRNAs relies on the start codon of the hosting mRNAs. Increased m^6A-mediated circRNA translation was observed under heat shock conditions [36, 37]. A 2021 study attempted to shed light on the still largely unclear molecular mechanisms underlying circRNA translation by investigating the key elements involved in the process. Chen et al. described that a distinct secondary structure (SuRE) positioned on the IRES in combination with 18S rRNA complementarity are essential factors for the translation of naturally occurring circRNAs [72].

A 2023 study described first efforts to design an approach to create artificial circRNAs for protein production. Chen et al. proposed five elements to increase circRNA translation: vector topology, 5′ and 3′ untranslated regions, IRESs, and synthetic aptamers recruiting translation initiation machinery. The added advantage of the engineered circRNAs compared to mRNA was the larger scale of protein production in vitro and a greater in vitro and in vivo translation [73].

The present circRNA translation research is focused on creating artificial circRNAs, with translational potential. As mentioned above, circRNAs lack essential elements for a cap-dependent translation. However, protein translation can be engineered via IRES or incorporation of the m^6A modification upstream of the open reading frame (ORF).Since the coronavirus pandemic in 2019/2020 and the creation of mRNA vaccines, circRNAs have garnered interest as potential tools for the production of vaccines. Although mRNA vaccines are currently widely used against severe acute respiratory syndrome coronavirus (SARS-CoV-2), new vaccines and technologies are needed to keep pace with fast-spreading variants with immune escape properties like Omicron [74]. One disadvantage of mRNA vaccines is their susceptibility to exonuclease digestion, which can be circumvented using circularized closed molecules like circRNAs. In 2022, Qu et al. successfully created a circRNA vaccine against SARS-CoV-2. The researchers reported robust protection against SARS-CoV-2 in vivo in mice and rhesus macaques. After injection with the vaccine, the animals had potent neutralizing antibodies and T cell responses with no clinical signs of illness [74].

Another study published in 2023 examined the role of engineered circRNAs in therapeutic cancer vaccines and showed that immunization with circRNAs led to an innate activation of dendritic cells and CD8 T cell responses in lymph nodes and tissues in mice [75]. However, further studies are needed to explore the potential and safety of circRNA vaccines in human patients.

These abovementioned findings open the door to the use of circRNAs for vaccines and protein production.

2.5 CircRNAs Are Involved in Parental Gene Expression

CircRNAs are also involved in parental gene expression through different mechanisms of transcriptional and translational regulation. Several reports indicate that some circRNAs, derived from tumor suppressor genes or oncogenes, could regulate transcription of their host genes [76–79]. In 2018, researchers observed that circITGA7 inhibits proliferation and metastasis in colorectal cancer (CRC) through upregulation of its host gene *ITGA7*. The increased expression of *ITGA7* led to the suppression of the transcription factor RREB1 and the subsequent inactivation of the oncogenic RAS pathway, leading them to conclude that circITGA7 could represent a potential target for the treatment of CRC [77]. Another study investigating laryngeal squamous cell carcinoma (LSCC) identified a circRNA that promotes the proliferation of LSCC. Circ-CCD1 is derived from cyclin D1 (CCD1), a proto-oncogene crucial for the cell cycle transition from G1 phase to S phase in cancer. An increased expression of circ-CCD1 was found to be associated with elevated expression of CCD1 and adverse outcomes [79].

2.6 CircRNAs Regulate Alternative Splicing

Another function of circRNAs is the regulation of alternative splicing. Alternative splicing is a process in which a single gene can give rise to multiple proteins [80]. A 2022 study discovered that circRPAP2 is downregulated in tissue samples of patients with breast cancer and in cancerous cell lines, indicating a potential tumor suppressor role for circPAP2. The researchers discovered that circRPAP2 binds to the oncoprotein SRSF1, which, in turn, affected the SRSF1-mediated alternative splicing of PTK2 and resulted in the reduced expression of both PTK2 mRNA and the resulting protein [81].

2.7 CircRNAs Are Involved in Cell Cycle Regulation

The final function of circRNAs discussed in this chapter is their involvement in the regulation of the cell cycle. One of the first studies to show this particular function was published in 2016. In their study, Du et al. investigated circ-Foxo3 [82]. Both linear and circular transcripts of Foxo3 are encoded by *Foxo3*, a known tumor suppressor gene [82–84]. Increased circ-Foxo3 expression inhibited cell cycle progression and cell proliferation. They discovered that the binding of cyclin-dependent kinase 1 (p21) and cyclin-dependent kinase 2 (CDK2) to circ-Foxo3 and the creation of a circ-Foxo3–p21–CDK2 ternary complex halted the cell cycle progression in the G1 phase, which made transition to the S phase impossible [82].

Within the context of cardiac regeneration, a 2022 study investigated circMdc1 and found its expression in cardiomyocytes to be sensitive to oxidative stress. Silencing circMdc1 in vitro resulted in the maintenance of chromosome stability and led to cardiomyocyte cell cycle reentry, hence promoting cardiac repair [85].

Studies on rheumatoid arthritis (RA) have shown that circRNAs are involved in regulating the cell cycle in autoimmune diseases. In their 2020 study, Yang et al. found an increased expression of circRNA 09505 in peripheral blood mononuclear cells (PBMCs) of patients with RA using RNA-seq [86]. Furthermore, they observed that circRNA 09505 acts as an miR-6089 sponge and regulates macrophage inflammation. Upregulation of circRNA 09505 in macrophages led to macrophage proliferation and significantly promoted cell cycle progression, whereas knockdown of circRNA 09505 inhibited cell proliferation and cell cycle progression [86].

In 2022, a research group examined the role of circBNC2 in kidney and liver fibrosis [87]. Fibrosis often occurs after chronic inflammation or repeated injuries as part of the healing response. It is characterized by the excessive deposition of extracellular matrix (ECM) and can affect different organs including the heart, lung, kidney, and liver [88]. Although accumulation of ECM is a typical response after injury, its excessive aggregation can lead to permanent scarring, organ failure, and subsequent death. Wang et al. identified circBNC2 to be enriched in the kidney and liver tissues of mice. This circRNA was abundantly expressed in healthy renal tissue, whereas renal ischemia–reperfusion injury led to a significant downregulation of circBNC2 in a renal ischemia–reperfusion injury mouse model compared to sham-treated mice. In vitro and in vivo gain- and loss-of-function experiments showed that circBNC2 acts as an endogenous negative regulator of cell cycle G2/M arrest. Knockout of circBNC2 in healthy human epithelial cells and hepatocytes resulted in a significant increase of cells arrested in the G2/M phase, thus promoting progression of fibrosis [87].

3 Current Challenges in CircRNA Research

Despite the strides made in both the technologies, which first enabled the effective study of circRNAs, and the discoveries pertaining to the many functions of circRNAs, it is more important than ever to critically assess the ongoing challenges and potential limitations in this new field. The first subsection is dedicated to technical challenges, whereas the second focuses on the obstacles to the research on and use of circRNAs as potential therapeutic or diagnostic targets.

3.1 Technical Challenges

One of the first challenges faced when working with circRNAs is their nomenclature. The pervasiveness of the lack of a standardized nomenclature of circRNAs across publications may result in the effective loss of knowledge or trouble in identifying studies conducted on identical targets as well as in establishing consensus on the role of each circRNA.

The technical issues of isolating and verifying potential novel circRNAs in the laboratory should also not be underestimated. Different isolation protocols have to be applied when working with biological fluids (blood, plasma, or urine) or tissues. Furthermore, the type of starting material often varies between groups. When working with the same body fluid, i.e., whole blood, some groups use ethylenediaminetetraacetic acid (EDTA) tubes, while others prefer PAXgene blood RNA tubes [89, 90], which can tremendously influence the quantity and quality of the detected circRNAs. Another challenge is the use of RNase R to enrich circRNAs. As mentioned in Sect. 1.2.1, some circRNAs are more prone to being degraded when using RNase R, which is why treatment with RNase R should be planned carefully [19]. In addition, there are no standards for the adjustment of RNase R digestion conditions, which can tremendously affect the outcome of digested circRNAs.

The choice of the detection method (quantitative polymerase chain reaction (qPCR), quantitative reverse transcription PCR (qRT-PCR), or digital droplet PCR (ddPCR)) may lead to distinct outcomes of circRNA quantification. This point has to be considered when research groups work with different detection methods but investigate the same circRNAs.

3.2 Medical and Therapeutical Challenges

Dysregulation of circRNAs has been reported in a variety of diseases including CVDs, neurological disorders, and cancer [91–94]. A few years after the initial rediscovery of circRNAs in 2012–2013, first studies came forward reporting the potential of circRNAs to act as biomarkers and therapeutic targets [95–97]. The current strategies to explore the therapeutic potential of circRNAs use gain- and loss-of-function approaches. They include RNA interference (RNAi)-mediated knockdown, circRNA expression vectors, nanoparticle delivery and exosome delivery of circRNA-based therapeutics, and clustered regularly interspaced short palindromic repeats/CRISPR-associated protein 9 (CRISPR/Cas9)-mediated circRNA knockout or knockdown [98–102].

An RNAi knockdown is typically mediated using siRNAs or short hairpin RNAs (shRNAs). As circRNAs are characterized by their unique BSJ, a knockdown of circRNAs usually utilizes this junction, therefore leaving the corresponding linear mRNA unaffected. At the moment, the most convenient methods to perform RNAi-mediated knockdowns in vivo use siRNAs and shRNAs delivered in lipid-based polymers. However, working with RNAi molecules has several limitations including their degradation by nucleases, lack of specificity, and immunogenicity, with these aspects often standing in contrast with one another [97, 103, 104].

The use of circRNA expression vectors is a common method for the overexpression of circRNAs. Different vectors have been used for their overexpression in cell culture and animal models [97, 105]. The most common vectors used at the moment are lentiviral and adenoviral vectors [59, 97, 106].

The delivery of nanoparticles as circRNA-based therapeutics is being investigated. Nanoparticles can carry drugs to therapeutic targets, and their applications range from imaging to the delivery of therapeutic substances [107]. These nanoparticles can vary in size, typically ranging from one-tenth of a nanometer (nm) to a few hundred nanometers and can be made of organic materials like polymers and liposomes or inorganic materials like gold [100, 107]. In contrast to RNAi, the use of nanoparticles as circRNA-based therapeutics has the advantage of being less prone to degradation, a simpler cellular uptake, and the lack of an immune response [100, 108]. However, this method can only be used for the delivery of circRNAs to the cytoplasm as nanoparticles cannot enter the nucleus [100].

Another method for circRNA delivery is based on exosomes. Exosomes are extracellular vesicles with a typical size from 30 to 100 nm in diameter and are found in different biological fluids such as saliva, urine, and blood [100]. In circRNA research, they are currently being investigated as circRNA expression vectors or as vehicles for circRNA targeting agents [100]. As with nanoparticles, exosomes have the advantage of promoting cellular uptake without the risk of immune activation [100].The last strategy to investigate the therapeutic potential of circRNAs discussed in this section is CRISPR/Cas9-mediated circRNA knockout or knockdown. Clustered regularly interspaced short palindromic repeats (CRISPR)-associated protein 9 (Cas9) is a method to edit the genome with high specificity and efficiency [109]. This method knocks out circRNAs through disruption of the pairing introns that flank the circularized exons during circRNA biogenesis [100]. Using CRISPR/Cas9 in circRNA research has the advantage of leaving the linear mRNA unaffected; however, working with CRISPR/Cas9 is more challenging than with other methods because more knowledge and training is required. In summary, the main challenges in circRNA-based treatments is their immunogenicity and off-target effects. While investigating the abovementioned methods in vitro, most of these methods require more research regarding the specificity and efficacy in vivo.

4 Conclusion

The introduction of novel sequencing technologies directly sparked increased interest in circRNAs by revealing key characteristics including their abundance and tissue-specific expression. However, as with most novel molecules, it is essential that the rise in circRNA research is accompanied by appropriate steps to ensure reliable reporting requirements. Despite the numerous advantages conveyed by these novel molecules in diagnostic and therapeutic settings, recent discoveries have also put some of the techniques applied for their study under scrutiny. The vast amount of generated data will also need to be confirmed and put into context of previously available knowledge. Ongoing advances in the study of circRNA biogenesis and function are expected to substantially enhance our understanding and treatment of various diseases.

References

1. Sanger HL, Klotz G, Riesner D, Gross HJ, Kleinschmidt AK (1976) Viroids are single-stranded covalently closed circular RNA molecules existing as highly base-paired rod-like structures. Proc Natl Acad Sci USA 73(11):3852–3856. https://doi.org/10.1073/pnas.73.11.3852
2. Kolakofsky D (1976) Isolation and characterization of Sendai virus DI-RNAs. Cell 8(4):547–555. https://doi.org/10.1016/0092-8674(76)90223-3
3. Hsu MT, Coca-Prados M (1979) Electron microscopic evidence for the circular form of RNA in the cytoplasm of eukaryotic cells. Nature 280(5720):339–340. https://doi.org/10.1038/280339a0
4. Kos A, Dijkema R, Arnberg AC, van der Meide PH, Schellekens H (1986) The hepatitis delta (delta) virus possesses a circular RNA. Nature 323(6088):558–560. https://doi.org/10.1038/323558a0
5. Nigro JM, Cho KR, Fearon ER, Kern SE, Ruppert JM, Oliner JD, Kinzler KW, Vogelstein B (1991) Scrambled exons. Cell 64(3):607–613. https://doi.org/10.1016/0092-8674(91)90244-S
6. Cocquerelle C, Daubersies P, Majérus MA, Kerckaert JP, Bailleul B (1992) Splicing with inverted order of exons occurs proximal to large introns. EMBO J 11(3):1095–1098
7. Cocquerelle C, Mascrez B, Hétuin D, Bailleul B (1993) Mis-splicing yields circular RNA molecules. FASEB J 7(1):155–160. https://doi.org/10.1096/fasebj.7.1.7678559
8. Sanger F, Coulson AR (1975) A rapid method for determining sequences in DNA by primed synthesis with DNA polymerase. J Mol Biol 94(3):441–448. https://doi.org/10.1016/0022-2836(75)90213-2
9. Zimmermann J, Voss H, Schwager C, Stegemann J, Ansorge W (1988) Automated Sanger dideoxy sequencing reaction protocol. FEBS Lett 233(2):432–436. https://doi.org/10.1016/0014-5793(88)80477-0
10. Gondane A, Itkonen HM (2023) Revealing the history and mystery of RNA-Seq. Curr Issues Mol Biol 45(3):1860–1874. https://doi.org/10.3390/cimb45030120
11. Chang TW (1983) Binding of cells to matrixes of distinct antibodies coated on solid surface. J Immunol Methods 65(1–2):217–223. https://doi.org/10.1016/0022-1759(83)90318-6
12. Schena M, Shalon D, Davis RW, Brown PO (1995) Quantitative monitoring of gene expression patterns with a complementary DNA microarray. Science 270(5235):467–470. https://doi.org/10.1126/science.270.5235.467
13. Velculescu VE, Zhang L, Vogelstein B, Kinzler KW (1995) Serial analysis of gene expression. Science 270(5235):484–487. https://doi.org/10.1126/science.270.5235.484
14. Velculescu VE, Zhang L, Zhou W, Vogelstein J, Basrai MA, Bassett DE, Hieter P, Vogelstein B, Kinzler KW (1997) Characterization of the yeast transcriptome. Cell 88(2):243–251. https://doi.org/10.1016/S0092-8674(00)81845-0
15. Clark TA, Schweitzer AC, Chen TX, Staples MK, Lu G, Wang H, Williams A, Blume JE (2007) Discovery of tissue-specific exons using comprehensive human exon microarrays. Genome Biol 8(4):R64. https://doi.org/10.1186/gb-2007-8-4-r64
16. Margulies M, Egholm M, Altman WE, Attiya S, Bader JS, Bemben LA, Berka J, Braverman MS, Chen Y-J, Chen Z, Dewell SB, Du L, Fierro JM, Gomes XV, Godwin BC, He W, Helgesen S, Ho CH, Irzyk GP, Jando SC, Alenquer MLI, Jarvie TP, Jirage KB, Kim J-B, Knight JR, Lanza JR, Leamon JH, Lefkowitz SM, Lei M, Li J, Lohman KL, Lu H, Makhijani VB, McDade KE, McKenna MP, Myers EW, Nickerson E, Nobile JR, Plant R, Puc BP, Ronan MT, Roth GT, Sarkis GJ, Simons JF, Simpson JW, Srinivasan M, Tartaro KR, Tomasz A, Vogt KA, Volkmer GA, Wang SH, Wang Y, Weiner MP, Yu P, Begley RF, Rothberg JM (2005) Genome sequencing in microfabricated high-density picolitre reactors. Nature 437(7057):376–380. https://doi.org/10.1038/nature03959

17. Hong M, Tao S, Zhang L, Diao L-T, Huang X, Huang S, Xie S-J, Xiao Z-D, Zhang H (2020) RNA sequencing: new technologies and applications in cancer research. J Hematol Oncol 13(1):166. https://doi.org/10.1186/s13045-020-01005-x
18. Wang Z, Gerstein M, Snyder M (2009) RNA-Seq: a revolutionary tool for transcriptomics. Nat Rev Genet 10(1):57–63. https://doi.org/10.1038/nrg2484
19. Szabo L, Salzman J (2016) Detecting circular RNAs: bioinformatic and experimental challenges. Nat Rev Genet 17(11):679–692. https://doi.org/10.1038/nrg.2016.114
20. Salzman J, Gawad C, Wang PL, Lacayo N, Brown PO (2012) Circular RNAs are the predominant transcript isoform from hundreds of human genes in diverse cell types. PLoS One 7(2):e30733. https://doi.org/10.1371/journal.pone.0030733
21. Jeck WR, Sorrentino JA, Wang K, Slevin MK, Burd CE, Liu J, Marzluff WF, Sharpless NE (2013) Circular RNAs are abundant, conserved, and associated with ALU repeats. RNA 19(2):141–157. https://doi.org/10.1261/rna.035667.112
22. Memczak S, Jens M, Elefsinioti A, Torti F, Krueger J, Rybak A, Maier L, Mackowiak SD, Gregersen LH, Munschauer M, Loewer A, Ziebold U, Landthaler M, Kocks C, le Noble F, Rajewsky N (2013) Circular RNAs are a large class of animal RNAs with regulatory potency. Nature 495(7441):333–338. https://doi.org/10.1038/nature11928
23. Salzman J, Chen RE, Olsen MN, Wang PL, Brown PO (2013) Cell-type specific features of circular RNA expression. PLoS Genet 9(9):e1003777. https://doi.org/10.1371/journal.pgen.1003777
24. Hansen TB, Jensen TI, Clausen BH, Bramsen JB, Finsen B, Damgaard CK, Kjems J (2013) Natural RNA circles function as efficient microRNA sponges. Nature 495(7441):384–388. https://doi.org/10.1038/nature11993
25. Wang PL, Bao Y, Yee M-C, Barrett SP, Hogan GJ, Olsen MN, Dinneny JR, Brown PO, Salzman J (2014) Circular RNA is expressed across the eukaryotic tree of life. PLoS One 9(6):e90859. https://doi.org/10.1371/journal.pone.0090859
26. Westholm JO, Miura P, Olson S, Shenker S, Joseph B, Sanfilippo P, Celniker SE, Graveley BR, Lai EC (2014) Genome-wide analysis of drosophila circular RNAs reveals their structural and sequence properties and age-dependent neural accumulation. Cell Rep 9(5):1966–1980. https://doi.org/10.1016/j.celrep.2014.10.062
27. Ye C-Y, Zhang X, Chu Q, Liu C, Yu Y, Jiang W, Zhu Q-H, Fan L, Guo L (2017) Full-length sequence assembly reveals circular RNAs with diverse non-GT/AG splicing signals in rice. RNA Biol 14(8):1055–1063. https://doi.org/10.1080/15476286.2016.1245268
28. Zhang G, Diao S, Zhang T, Chen D, He C, Zhang J (2019) Identification and characterization of circular RNAs during the sea buckthorn fruit development. RNA Biol 16(3):354–361. https://doi.org/10.1080/15476286.2019.1574162
29. Cortés-López M, Gruner MR, Cooper DA, Gruner HN, Voda A-I, van der Linden AM, Miura P (2018) Global accumulation of circRNAs during aging in Caenorhabditis elegans. BMC Genomics 19(1):8. https://doi.org/10.1186/s12864-017-4386-y
30. Rybak-Wolf A, Stottmeister C, Glažar P, Jens M, Pino N, Giusti S, Hanan M, Behm M, Bartok O, Ashwal-Fluss R, Herzog M, Schreyer L, Papavasileiou P, Ivanov A, Öhman M, Refojo D, Kadener S, Rajewsky N (2015) Circular RNAs in the mammalian brain are highly abundant, conserved, and dynamically expressed. Mol Cell 58(5):870–885. https://doi.org/10.1016/j.molcel.2015.03.027
31. Gao Y, Wang J, Zhao F (2015) CIRI: an efficient and unbiased algorithm for de novo circular RNA identification. Genome Biol 16(1):4. https://doi.org/10.1186/s13059-014-0571-3
32. Zhang X-O, Wang H-B, Zhang Y, Lu X, Chen L-L, Yang L (2014) Complementary sequence-mediated exon circularization. Cell 159(1):134–147. https://doi.org/10.1016/j.cell.2014.09.001
33. Liu X, Zhang Y, Zhou S, Dain L, Mei L, Zhu G (2022) Circular RNA: An emerging frontier in RNA therapeutic targets, RNA therapeutics, and mRNA vaccines. J Control Release 348:84–94. https://doi.org/10.1016/j.jconrel.2022.05.043
34. Lasda E, Parker R (2014) Circular RNAs: diversity of form and function. RNA 20(12):1829–1842. https://doi.org/10.1261/rna.047126.114
35. Legnini I, Di Timoteo G, Rossi F, Morlando M, Briganti F, Sthandier O, Fatica A, Santini T, Andronache A, Wade M, Laneve P, Rajewsky N, Bozzoni I (2017) Circ-ZNF609 is a circular RNA that can be translated and functions in Myogenesis. Mol Cell 66(1):22–37.e9. https://doi.org/10.1016/j.molcel.2017.02.017
36. Pamudurti NR, Bartok O, Jens M, Ashwal-Fluss R, Stottmeister C, Ruhe L, Hanan M, Wyler E, Perez-Hernandez D, Ramberger E, Shenzis S, Samson M, Dittmar G, Landthaler M, Chekulaeva M, Rajewsky N, Kadener S (2017) Translation of CircRNAs. Mol Cell 66(1):9–21.e7. https://doi.org/10.1016/j.molcel.2017.02.021
37. Yang Y, Fan X, Mao M, Song X, Wu P, Zhang Y, Jin Y, Yang Y, Chen L-L, Wang Y, Wong CC, Xiao X, Wang Z (2017) Extensive translation of circular RNAs driven by N6-methyladenosine. Cell Res 27(5):626–641. https://doi.org/10.1038/cr.2017.31
38. Xia S, Feng J, Lei L, Hu J, Xia L, Wang J, Xiang Y, Liu L, Zhong S, Han L, He C (2017) Comprehensive characterization of tissue-specific circular RNAs in the human and mouse genomes. Brief Bioinform 18(6):984–992. https://doi.org/10.1093/bib/bbw081
39. Li Z, Huang C, Bao C, Chen L, Lin M, Wang X, Zhong G, Yu B, Hu W, Dai L, Zhu P, Chang Z, Wu Q, Zhao Y, Jia Y, Xu P, Liu H, Shan G (2015) Exon-

intron circular RNAs regulate transcription in the nucleus. Nat Struct Mol Biol 22(3):256–264. https://doi.org/10.1038/nsmb.2959
40. Wang L, Long H, Zheng Q, Bo X, Xiao X, Li B (2019) Circular RNA circRHOT1 promotes hepatocellular carcinoma progression by initiation of NR2F6 expression. Mol Cancer 18(1):119. https://doi.org/10.1186/s12943-019-1046-7
41. Liu S, Guo XY, Shang QJ, Gao P (2023) The biogenesis, biological functions and modification of circular RNAs. Exp Mol Pathol 131:104861. https://doi.org/10.1016/j.yexmp.2023.104861
42. Gu A, Jaijyan DK, Yang S, Zeng M, Pei S, Zhu H (2023) Functions of circular RNA in human diseases and illnesses. Noncoding RNA 9(4):38. https://doi.org/10.3390/ncrna9040038
43. Chen L-L, Yang L (2015) Regulation of circRNA biogenesis. RNA Biol 12(4):381–388. https://doi.org/10.1080/15476286.2015.1020271
44. Patop IL, Wüst S, Kadener S (2019) Past, present, and future of circRNAs. EMBO J 38(16):e100836. https://doi.org/10.15252/embj.2018100836
45. Ivanov A, Memczak S, Wyler E, Torti F, Porath HT, Orejuela MR, Piechotta M, Levanon EY, Landthaler M, Dieterich C, Rajewsky N (2015) Analysis of intron sequences reveals hallmarks of circular RNA biogenesis in animals. Cell Rep 10(2):170–177. https://doi.org/10.1016/j.celrep.2014.12.019
46. Starke S, Jost I, Rossbach O, Schneider T, Schreiner S, Hung L-H, Bindereif A (2015) Exon circularization requires canonical splice signals. Cell Rep 10(1):103–111. https://doi.org/10.1016/j.celrep.2014.12.002
47. Kramer MC, Liang D, Tatomer DC, Gold B, March ZM, Cherry S, Wilusz JE (2015) Combinatorial control of drosophila circular RNA expression by intronic repeats, hnRNPs, and SR proteins. Genes Dev 29(20):2168–2182. https://doi.org/10.1101/gad.270421.115
48. Wilusz JE (2015) Repetitive elements regulate circular RNA biogenesis. Mob Genet Elem 5(3):1–7. https://doi.org/10.1080/2159256X.2015.1045682
49. Wang H, Feng C, Wang M, Yang S, Wei F (2019) Circular RNAs: diversity of functions and a regulatory Nova in Oral medicine: a pilot review. Cell Transplant 28(7):819–830. https://doi.org/10.1177/0963689719837917
50. Kristensen LS, Andersen MS, Stagsted LVW, Ebbesen KK, Hansen TB, Kjems J (2019) The biogenesis, biology and characterization of circular RNAs. Nat Rev Genet 20(11):675–691. https://doi.org/10.1038/s41576-019-0158-7
51. Stoll L, Rodríguez-Trejo A, Guay C, Brozzi F, Bayazit MB, Gattesco S, Menoud V, Sobel J, Marques AC, Venø MT, Esguerra JLS, Barghouth M, Suleiman M, Marselli L, Kjems J, Eliasson L, Renström E, Bouzakri K, Pinget M, Marchetti P, Regazzi R (2020) A circular RNA generated from an intron of the insulin gene controls insulin secretion. Nat Commun 11(1):5611. https://doi.org/10.1038/s41467-020-19381-w
52. Mester-Tonczar J (2021) Regulation of circular RNAs CDR1as and novel identified Circ-RCAN2 in a porcine model of myocardial infarction. Doctoral thesis, Medical University of Vienna
53. Liu J, Liu T, Wang X, He A (2017) Circles reshaping the RNA world: from waste to treasure. Mol Cancer 16(1):58. https://doi.org/10.1186/s12943-017-0630-y
54. Bhaskaran M, Mohan M (2014) MicroRNAs: history, biogenesis, and their evolving role in animal development and disease. Vet Pathol 51(4):759–774. https://doi.org/10.1177/0300985813502820
55. Zhang L, Liu Y, Tao H, Zhu H, Pan Y, Li P, Liang H, Zhang B, Song J (2021) Circular RNA circUBE2J2 acts as the sponge of microRNA-370-5P to suppress hepatocellular carcinoma progression. Cell Death Dis 12(11):985. https://doi.org/10.1038/s41419-021-04269-4
56. Liu Z, Zhou Y, Liang G, Ling Y, Tan W, Tan L, Andrews R, Zhong W, Zhang X, Song E, Gong C (2019) Circular RNA hsa_circ_001783 regulates breast cancer progression via sponging miR-200c-3p. Cell Death Dis 10(2):55. https://doi.org/10.1038/s41419-018-1287-1
57. Liu X, Wang M, Li Q, Liu W, Song Q, Jiang H (2022) CircRNA ACAP2 induces myocardial apoptosis after myocardial infarction by sponging miR-29. Minerva Med 113(1):128–134. https://doi.org/10.23736/S0026-4806.20.06600-8
58. Zhou Y, Liu Y, Kang Z, Yao H, Song N, Wang M, Song C, Zhang K, Ding J, Tang J, Hu G, Lu M (2023) CircEPS15, as a sponge of MIR24-3p ameliorates neuronal damage in Parkinson disease through boosting PINK1-PRKN-mediated mitophagy. Autophagy 19(9):2520–2537. https://doi.org/10.1080/15548627.2023.2196889
59. Lavenniah A, Luu TDA, Li YP, Lim TB, Jiang J, Ackers-Johnson M, Foo RS-Y (2020) Engineered circular RNA sponges act as miRNA inhibitors to attenuate pressure overload-induced cardiac hypertrophy. Mol Ther 28(6):1506–1517. https://doi.org/10.1016/j.ymthe.2020.04.006
60. Jost I, Shalamova LA, Gerresheim GK, Niepmann M, Bindereif A, Rossbach O (2018) Functional sequestration of microRNA-122 from hepatitis C virus by circular RNA sponges. RNA Biol 15(8):1032–1039. https://doi.org/10.1080/15476286.2018.1435248
61. Breuer J, Rossbach O (2020) Production and purification of artificial circular RNA sponges for application in molecular biology and medicine. Methods Protoc 3(2):42. https://doi.org/10.3390/mps3020042
62. Müller S, Wedler A, Breuer J, Glaß M, Bley N, Lederer M, Haase J, Misiak C, Fuchs T, Ottmann A, Schmachtel T, Shalamova L, Ewe A, Aigner A, Rossbach O, Hüttelmaier S (2020) Synthetic circular miR-21 RNA decoys enhance tumor suppressor expression and impair tumor growth in mice. NAR. Cancer 2(3):zcaa014. https://doi.org/10.1093/narcan/zcaa014

63. Janssen HLA, Reesink HW, Lawitz EJ, Zeuzem S, Rodriguez-Torres M, Patel K, van der Meer AJ, Patick AK, Chen A, Zhou Y, Persson R, King BD, Kauppinen S, Levin AA, Hodges MR (2013) Treatment of HCV infection by targeting microRNA. N Engl J Med 368(18):1685–1694. https://doi.org/10.1056/NEJMoa1209026
64. Gebert LFR, Rebhan MAE, Crivelli SEM, Denzler R, Stoffel M, Hall J (2014) Miravirsen (SPC3649) can inhibit the biogenesis of miR-122. Nucleic Acids Res 42(1):609–621. https://doi.org/10.1093/nar/gkt852
65. European Association for the Study of the Liver. Electronic address: easloffice@easloffice.eu, Clinical Practice Guidelines Panel: Chair:, EASL Governing Board representative:, Panel members (2020) EASL recommendations on treatment of hepatitis C: final update of the series☆. J Hepatol 73(5):1170–1218. https://doi.org/10.1016/j.jhep.2020.08.018
66. Corley M, Burns MC, Yeo GW (2020) How RNA-binding proteins interact with RNA: molecules and mechanisms. Mol Cell 78(1):9–29. https://doi.org/10.1016/j.molcel.2020.03.011
67. Dong W, Dai Z-H, Liu F-C, Guo X-G, Ge C-M, Ding J, Liu H, Yang F (2019) The RNA-binding protein RBM3 promotes cell proliferation in hepatocellular carcinoma by regulating circular RNA SCD-circRNA 2 production. EBioMedicine 45:155–167. https://doi.org/10.1016/j.ebiom.2019.06.030
68. Knupp D, Cooper DA, Saito Y, Darnell RB, Miura P (2021) NOVA2 regulates neural circRNA biogenesis. Nucleic Acids Res 49(12):6849–6862. https://doi.org/10.1093/nar/gkab523
69. Conn SJ, Pillman KA, Toubia J, Conn VM, Salmanidis M, Phillips CA, Roslan S, Schreiber AW, Gregory PA, Goodall GJ (2015) The RNA binding protein quaking regulates formation of circRNAs. Cell 160(6):1125–1134. https://doi.org/10.1016/j.cell.2015.02.014
70. Gupta SK, Garg A, Bär C, Chatterjee S, Foinquinos A, Milting H, Streckfuß-Bömeke K, Fiedler J, Thum T (2018) Quaking inhibits doxorubicin-mediated cardiotoxicity through regulation of cardiac circular RNA expression. Circ Res 122(2):246–254. https://doi.org/10.1161/CIRCRESAHA.117.311335
71. Schreiner S, Didio A, Hung L-H, Bindereif A (2020) Design and application of circular RNAs with protein-sponge function. Nucleic Acids Res 48(21):12326–12335. https://doi.org/10.1093/nar/gkaa1085
72. Chen C-K, Cheng R, Demeter J, Chen J, Weingarten-Gabbay S, Jiang L, Snyder MP, Weissman JS, Segal E, Jackson PK, Chang HY (2021) Structured elements drive extensive circular RNA translation. Mol Cell 81(20):4300–4318.e13. https://doi.org/10.1016/j.molcel.2021.07.042
73. Chen R, Wang SK, Belk JA, Amaya L, Li Z, Cardenas A, Abe BT, Chen C-K, Wender PA, Chang HY (2023) Engineering circular RNA for enhanced protein production. Nat Biotechnol 41(2):262–272. https://doi.org/10.1038/s41587-022-01393-0
74. Qu L, Yi Z, Shen Y, Lin L, Chen F, Xu Y, Wu Z, Tang H, Zhang X, Tian F, Wang C, Xiao X, Dong X, Guo L, Lu S, Yang C, Tang C, Yang Y, Yu W, Wang J, Zhou Y, Huang Q, Yisimayi A, Liu S, Huang W, Cao Y, Wang Y, Zhou Z, Peng X, Wang J, Xie XS, Wei W (2022) Circular RNA vaccines against SARS-CoV-2 and emerging variants. Cell 185(10):1728–1744.e16. https://doi.org/10.1016/j.cell.2022.03.044
75. Amaya L, Grigoryan L, Li Z, Lee A, Wender PA, Pulendran B, Chang HY (2023) Circular RNA vaccine induces potent T cell responses. Proc Natl Acad Sci USA 120(20):e2302191120. https://doi.org/10.1073/pnas.2302191120
76. Wang H, Gao X, Yu S, Wang W, Liu G, Jiang X, Sun D (2022) Circular RNAs regulate parental gene expression: a new direction for molecular oncology research. Front Oncol 12:947775. https://doi.org/10.3389/fonc.2022.947775
77. Li X, Wang J, Zhang C, Lin C, Zhang J, Zhang W, Zhang W, Lu Y, Zheng L, Li X (2018) Circular RNA circITGA7 inhibits colorectal cancer growth and metastasis by modulating the Ras pathway and upregulating transcription of its host gene ITGA7. J Pathol 246(2):166–179. https://doi.org/10.1002/path.5125
78. Xia B, Hong T, He X, Hu X, Gao Y (2019) A circular RNA derived from MMP9 facilitates oral squamous cell carcinoma metastasis through regulation of MMP9 mRNA stability. Cell Transplant 28(12):1614–1623. https://doi.org/10.1177/0963689719875409
79. Zang Y, Li J, Wan B, Tai Y (2020) circRNA circ-CCND1 promotes the proliferation of laryngeal squamous cell carcinoma through elevating CCND1 expression via interacting with HuR and miR-646. J Cell Mol Med 24(4):2423–2433. https://doi.org/10.1111/jcmm.14925
80. Nilsen TW, Graveley BR (2010) Expansion of the eukaryotic proteome by alternative splicing. Nature 463(7280):457–463. https://doi.org/10.1038/nature08909
81. Yu Y, Fang L (2022) CircRPAP2 regulates the alternative splicing of PTK2 by binding to SRSF1 in breast cancer. Cell Death Discov 8(1):1–12. https://doi.org/10.1038/s41420-022-00965-y
82. Du WW, Yang W, Liu E, Yang Z, Dhaliwal P, Yang BB (2016) Foxo3 circular RNA retards cell cycle progression via forming ternary complexes with p21 and CDK2. Nucleic Acids Res 44(6):2846–2858. https://doi.org/10.1093/nar/gkw027
83. Cho E-C, Kuo M-L, Liu X, Yang L, Hsieh Y-C, Wang J, Cheng Y, Yen Y (2014) Tumor suppressor FOXO3 regulates ribonucleotide reductase subunit RRM2B and impacts on survival of cancer patients. Oncotarget 5(13):4834–4844. https://doi.org/10.18632/oncotarget.2044
84. Myatt SS, Lam EW-F (2007) The emerging roles of forkhead box (fox) proteins in cancer. Nat Rev

Cancer 7(11):847–859. https://doi.org/10.1038/nrc2223
85. Ma W, Wang X, Sun H, Xu B, Song R, Tian Y, Zhao L, Xu Y, Zhao Y, Yang F, Chen H, Gong R, Yu Y, Li X, Li S, Zhang W, Zhang T, Ne J, Cai B (2022) Oxidant stress-sensitive circRNA Mdc1 controls cardiomyocyte chromosome stability and cell cycle re-entry during heart regeneration. Pharmacol Res 184:106422. https://doi.org/10.1016/j.phrs.2022.106422
86. Yang J, Cheng M, Gu B, Wang J, Yan S, Xu D (2020) CircRNA_09505 aggravates inflammation and joint damage in collagen-induced arthritis mice via miR-6089/AKT1/NF-κB axis. Cell Death Dis 11(10):833. https://doi.org/10.1038/s41419-020-03038-z
87. Wang P, Huang Z, Peng Y, Li H, Lin T, Zhao Y, Hu Z, Zhou Z, Zhou W, Liu Y, Hou FF (2022) Circular RNA circBNC2 inhibits epithelial cell G2-M arrest to prevent fibrotic maladaptive repair. Nat Commun 13(1):6502. https://doi.org/10.1038/s41467-022-34287-5
88. Yang H, Cheng H, Dai R, Shang L, Zhang X, Wen H (2023) Macrophage polarization in tissue fibrosis. PeerJ 11:e16092. https://doi.org/10.7717/peerj.16092
89. Yuan Q, Sun Y, Yang F, Yan D, Shen M, Jin Z, Zhan L, Liu G, Yang L, Zhou Q, Yu Z, Zhou X, Yu Y, Xu Y, Wu Q, Luo J, Hu X, Zhang C (2023) CircRNA DICAR as a novel endogenous regulator for diabetic cardiomyopathy and diabetic pyroptosis of cardiomyocytes. Signal Transduct Target Ther 8(1):99. https://doi.org/10.1038/s41392-022-01306-2
90. Wu Y, Zhao T, Deng R, Xia X, Li B, Wang X (2021) A study of differential circRNA and lncRNA expressions in COVID-19-infected peripheral blood. Sci Rep 11(1):7991. https://doi.org/10.1038/s41598-021-86134-0
91. Zhao W, Cui Y, Liu L, Qi X, Liu J, Ma S, Hu X, Zhang Z, Wang Y, Li H, Wang Z, Liu Z, Wu J (2020) Splicing factor derived circular RNA circUHRF1 accelerates oral squamous cell carcinoma tumorigenesis via feedback loop. Cell Death Differ 27(3):919–933. https://doi.org/10.1038/s41418-019-0423-5
92. Ma J, Fang L, Yang Q, Hibberd S, Du WW, Wu N, Yang BB (2019) Posttranscriptional regulation of AKT by circular RNA angiomotin- like 1 mediates chemoresistance against paclitaxel in breast cancer cells. Aging (Albany NY) 11(23):11369. https://doi.org/10.18632/aging.102535
93. Zheng H, Huang S, Wei G, Sun Y, Li C, Si X, Chen Y, Tang Z, Li X, Chen Y, Liao W, Liao Y, Bin J (2022) CircRNA Samd4 induces cardiac repair after myocardial infarction by blocking mitochondria-derived ROS output. Mol Ther 30(11):3477–3498. https://doi.org/10.1016/j.ymthe.2022.06.016
94. Mester-Tonczar J, Einzinger P, Winkler J, Kastner N, Spannbauer A, Zlabinger K, Traxler D, Lukovic D, Hasimbegovic E, Goliasch G, Pavo N, Gyöngyösi M (2021) Novel identified circular transcript of RCAN2, circ-RCAN2, shows deviated expression pattern in pig Reperfused infarcted myocardium and hypoxic porcine cardiac progenitor cells in vitro. Int J Mol Sci 22(3):1390. https://doi.org/10.3390/ijms22031390
95. Salgado-Somoza A, Zhang L, Vausort M, Devaux Y (2017) The circular RNA MICRA for risk stratification after myocardial infarction. Int J Cardiol Heart Vasc 17:33–36. https://doi.org/10.1016/j.ijcha.2017.11.001
96. Li P, Chen S, Chen H, Mo X, Li T, Shao Y, Xiao B, Guo J (2015) Using circular RNA as a novel type of biomarker in the screening of gastric cancer. Clin Chim Acta 444:132–136. https://doi.org/10.1016/j.cca.2015.02.018
97. Holdt LM, Kohlmaier A, Teupser D (2018) Circular RNAs as therapeutic agents and targets. Front Physiol 9:1262. https://doi.org/10.3389/fphys.2018.01262
98. Huang S, Li X, Zheng H, Si X, Li B, Wei G, Li C, Chen Y, Chen Y, Liao W, Liao Y, Bin J (2019) Loss of super-enhancer-regulated circRNA Nfix induces cardiac regeneration after myocardial infarction in adult mice. Circulation 139(25):2857–2876. https://doi.org/10.1161/CIRCULATIONAHA.118.038361
99. Ni H, Li W, Zhuge Y, Xu S, Wang Y, Chen Y, Shen G, Wang F (2019) Inhibition of circHIPK3 prevents angiotensin II-induced cardiac fibrosis by sponging miR-29b-3p. Int J Cardiol 292:188–196. https://doi.org/10.1016/j.ijcard.2019.04.006
100. He AT, Liu J, Li F, Yang BB (2021) Targeting circular RNAs as a therapeutic approach: current strategies and challenges. Signal Transduct Target Ther 6(1):185. https://doi.org/10.1038/s41392-021-00569-5
101. Du WW, Yang W, Li X, Awan FM, Yang Z, Fang L, Lyu J, Li F, Peng C, Krylov SN, Xie Y, Zhang Y, He C, Wu N, Zhang C, Sdiri M, Dong J, Ma J, Gao C, Hibberd S, Yang BB (2018) A circular RNA circ-DNMT1 enhances breast cancer progression by activating autophagy. Oncogene 37(44):5829–5842. https://doi.org/10.1038/s41388-018-0369-y
102. Yang X, Li S, Wu Y, Ge F, Chen Y, Xiong Q (2020) The circular RNA CDR1as regulate cell proliferation via TMED2 and TMED10. BMC Cancer 20(1):312. https://doi.org/10.1186/s12885-020-06794-5
103. Singh S, Narang AS, Mahato RI (2011) Subcellular fate and off-target effects of siRNA, shRNA, and miRNA. Pharm Res 28(12):2996–3015. https://doi.org/10.1007/s11095-011-0608-1
104. Robbins M, Judge A, MacLachlan I (2009) siRNA and innate immunity. Oligonucleotides 19(2):89–102. https://doi.org/10.1089/oli.2009.0180
105. Wei X, Li H, Yang J, Hao D, Dong D, Huang Y, Lan X, Plath M, Lei C, Lin F, Bai Y, Chen H (2017) Circular RNA profiling reveals an abundant circLMO7 that regulates myoblasts differentiation and survival by sponging miR-378a-3p. Cell Death Dis 8(10):e3153. https://doi.org/10.1038/cddis.2017.541
106. Meganck RM, Borchardt EK, Castellanos Rivera RM, Scalabrino ML, Wilusz JE, Marzluff WF,

Asokan A (2018) Tissue-dependent expression and translation of circular RNAs with recombinant AAV vectors in vivo. Mol Ther Nucleic Acids 13:89–98. https://doi.org/10.1016/j.omtn.2018.08.008

107. Zhao X, Zhong Y, Wang X, Shen J, An W (2022) Advances in circular RNA and its applications. Int J Med Sci 19(6):975–985. https://doi.org/10.7150/ijms.71840

108. Yonezawa S, Koide H, Asai T (2020) Recent advances in siRNA delivery mediated by lipid-based nanoparticles. Adv Drug Deliv Rev 154–155:64–78. https://doi.org/10.1016/j.addr.2020.07.022

109. Ran FA, Hsu PD, Wright J, Agarwala V, Scott DA, Zhang F (2013) Genome engineering using the CRISPR-Cas9 system. Nat Protoc 8(11):2281–2308. https://doi.org/10.1038/nprot.2013.143

Guidelines for the Naming of Circular RNAs

2

Pengjie Tao, Wenqin Zhou, Tianzhao Xu,
Christian Bär, Thomas Thum, and Dongchao Lu

Abstract

Circular RNAs (circRNAs) have emerged as pivotal regulators in the posttranscriptional regulatory landscape, exhibiting unique properties such as increased stability and a range of biological functions due to their covalently closed-loop structure. Despite advancements in circRNA research, the field meets a significant challenge in the absence of a standardized nomenclature system, leading to confusion and hindering effective scientific communication. This chapter outlines the historical development, current state, and the necessity for a unified naming system in circRNA research. We highlight the role of databases in shaping circRNA nomenclature and discuss the implications of nomenclature ambiguity, especially with the identification of multiple circRNAs from single gene loci. We propose a systematic approach to naming circRNAs that incorporates the "circ" prefix, references to GENCODE/Ensembl transcripts, and annotations for unique structural features such as retained introns or alternative splicing. The adoption of such a nomenclature system is essential for enhancing clarity, ensuring reproducibility, and advancing our understanding of circRNA investigation.

Keywords

Noncoding RNA · CircRNAs · Naming guidelines · RNA-seq · Database

1 Introduction

Circular RNAs (circRNAs) are a distinctive group of noncoding RNAs with a closed-loop structure that provides them with increased stability and unique regulatory roles compared to their linear RNA relatives [1]. The biogenesis of circRNAs involves a complex interplay between RNA polymerase II activity and the spliceosome, leading to the formation of backspliced products

Pengjie Tao and Wenqin Zhou have equally contributed to this chapter.

P. Tao · D. Lu (✉)
School of Integrative Medicine, Shanghai University of Traditional Chinese Medicine, Shanghai, China
e-mail: lu_dongchao@shutcm.edu.cn

W. Zhou
School of Medicine, Shanghai University, Shanghai, China

T. Xu
Department of Central Laboratory, Zhoupu Hospital, Shanghai University of Medicine and Health, Shanghai, China

C. Bär · T. Thum
Institute of Molecular and Translational Therapeutic Strategies, Hannover Medical School, Hannover, Germany

J. Xiao (ed.), *Advances in Circular RNAs*, Advances in Experimental Medicine and Biology 1485,
https://doi.org/10.1007/978-981-96-9428-0_2

that can evade the typical exonucleolytic degradation pathways [2, 3]. Thus, the lifespan of circRNAs in cells is significantly extended, typically lasting between 18 and 24 hours, which is considerably longer than the 4–8-h lifespan of associated linear RNAs [4, 5]. Importantly, these molecules, characterized by their stable configuration, have demonstrated a broad spectrum of roles in posttranscriptional regulation as well as the potential to translate the novel peptide, positioning them as potential key regulators in a variety of cellular functions [6, 7].

The investigation into circRNAs commenced with a pivotal discovery in 1976 by Sanger and colleagues, who observed a single-stranded circular structure RNA via an electron microscope within viroids [8]. Subsequent research in the 1990s reported the existence of circRNAs in mammalian cells, associated with the tumor suppressor gene *DCC* and the *Sry* gene. These transcripts with a unique structure were initially termed "abnormally spliced or scrambled transcripts" or, more specifically, "circular SRY mRNA transcripts" [9, 10]. Notably, the identification and classification of circRNAs gained momentum in the 2010s, propelled by biochemical strategies and advanced computational methods that could detect these molecules via second-generation RNA sequencing technology [11]. To date, the prediction of over human 315,000 unique circRNAs has been achieved through various circRNA-seq datasets and analytical tools [12]. Due to the growing number of identified molecules and their potential implications in research and clinical applications, the nomenclature of circRNAs has become an important topic for consideration. To further clarify the functions of the newly discovered circRNAs and to make it easier for researchers to communicate, a standardized naming system is necessary.

2 Current State for Naming CircRNAs

As the field of circRNA biology is expanding rapidly, the processes of identifying and characterizing circRNAs have gained notable significance. However, with the growing number of circRNAs being discovered, a considerable challenge has arisen: the absence of a uniform nomenclature system. This section highlights the current dilemmas surrounding the naming of circRNAs, discussing the typical issue for circRNA naming, which can lead to confusion in scientific communication.

2.1 The Role of Databases in CircRNA Naming

Databases play a crucial role in the adoption and maintenance of a standardized nomenclature system. Publicly available databases such as CIRCpedia, circAtlas, circBank, and circBase have been instrumental in archiving circRNAs and providing access to their sequences and annotations (Fig. 2.1). However, the naming strategies used by these databases vary, contributing to the current nomenclature confusion [13–16].

circBase: The Pioneering Database
circBase, established in 2014, was an early database dedicated to circRNA research. It served as a hub for accessing, comparing, and downloading circRNA data, offering a search feature based on sequence data. The goals of circBase were to consolidate evidence of circRNA presence and expression, display circRNAs in the context of the genome, and search circRNA information including sequence, gene, or location. Researchers could also retrieve circRNA data from RNA-seq in FASTA format, including both spliced and genomic sequences. In circBase, circRNAs are named with a "circ" prefix, an underscore, and a number like "circ_000001," a system that is clear but does not provide much biological insight. This database covers circRNAs from humans, mice, and the roundworm *Caenorhabditis elegans*, designated as "hsa_circ_#," "mmu_circ_#," and "cel_circ_#," respectively [9, 13, 17–19].

CIRCpedia2: Expanding the Atlas
CIRCpedia, updated in 2016, is a comprehensive resource that expanded the scope of circRNA research by providing annotations from over 180

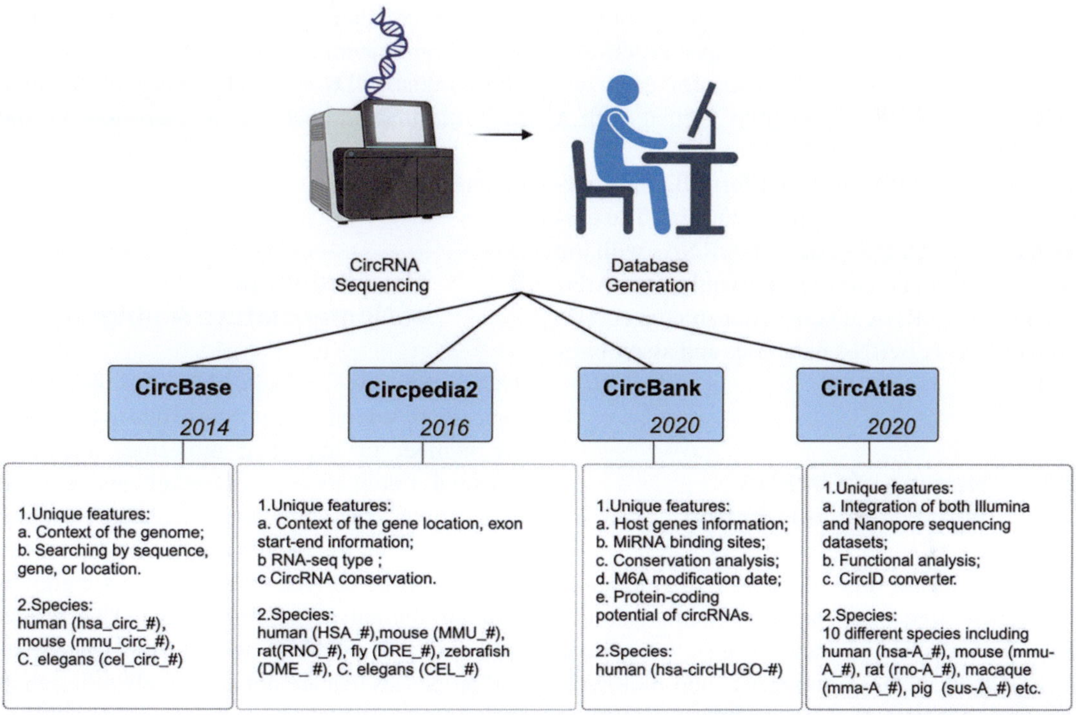

Fig. 2.1 Evolution and distinctive features of circRNA databases. The features were created with BioRender.com.

RNA-seq datasets retrieved from Gene Expression Omnibus (GEO), the Encyclopedia Of DNA Elements (ENCODE) project, and European Bioinformatics Institute (EMBL-EBI). This database's atlas feature enables users to visualize the distribution and characteristics of circRNAs including gene location, exon start and exon end, RNA-seq type, and conservation. CIRCpedia employs a naming format with a "CIRCpedia" prefix, an underscore, and a number like "CIRCpedia_00001." Notably, this database covers more circRNAs from humans, mice, rats, flies, zebrafish, and *C. elegans*, labeled as "HSA_#," "MMU_#," "RNO_#," "DRE_#," "DME_ #," and "CEL_#," respectively [14, 20].

circBank: Enhancing Functional Annotations
circBank, launched in 2020, introduced a novel naming system for circRNAs based on their host genes and includes additional features such as miRNA-binding sites, conservation analysis, m6A modification data, mutation annotations, and the protein-coding potential of circRNAs. This database contains more than 140,000 human-annotated circRNAs from different sources. circBank's naming convention is "hsa-circHUGO-#," where HUGO refers to the host gene's symbol and the number differentiates between circRNAs originating from the same gene. For circRNAs with the same starting site and a different ending site, the more upstream of the ending site, the less the number is. For intergenic regions, the naming convention is "hsa-circChrom#_#," with the first number indicating the chromosome number and the second being the circRNA order number [15].

circAtlas: Integrating CircRNA Database
circAtlas, introduced in 2020, stands out for its integration of both Illumina and Nanopore sequencing datasets, offering full-length sequences and functional analysis for circRNAs. This database employs a standardized nomenclature scheme and provides various search and export functionalities. circAtlas 3.0 comprises a comprehensive compendium of 2609 Illumina

and 65 Nanopore RNA-seq samples from 33 diverse tissues within 10 distinct species. circAtlas provides two types of names for each circRNA: a standardized circAtlas format with a "host gene name," an underscore, and a number like "ABCC2_0015" and a uniform ID like "circABCC2(7 L)" according to the tutorial. The uniform ID can distinguish those circRNAs with the same exon annotation. Importantly, circAtlas provides a circRNA ID converter to convert IDs from different circRNA databases and assemblies [16].

2.2 Naming of CircRNA with Potential Function

In the pioneering time of circRNA exploration, researchers frequently assigned names to circRNAs that reflected their intrinsic properties, such as their mode of biogenesis and biological function. A prime example is fusion circRNAs (f-circRNAs), which originated from cancer-associated chromosomal translocations. These f-circRNAs, exemplified by f-circPR derived from promyelocytic leukemia/retinoic acid receptor-alpha (PML/RARα) fusion and by f-circM9 from mixed lineage leukemia/AF9 (MLL/AF9) fusion, have been demonstrated to propel cellular transformation, reinforce cell survival, and augment resistance to treatment, thereby exhibiting potent tumor-promoting capabilities in in vivo studies [21].

Furthermore, two crucial cardiac circRNAs, identified by their specific circBase IDs (mmu_circ_0000290 and mmu_circ_0000254), have been observed to regulate vital processes like mitochondrial fission, apoptosis, and cardiac hypertrophy by interacting with miR-652-3p and miR-223, respectively. These circRNAs have been named for their functions: mitochondrial fission and apoptosis-related circRNA (MFACR, also known as circ-SMYD4) and heart-related circRNA (HRCR or circ-Pwwp2a), underscoring their potential as therapeutic targets for cardiovascular diseases [22–24].

However, since 2018, there has been a notable shift toward naming circRNAs based on their circBase registry IDs or the host genes from which they originate, reflecting a transition in the nomenclature practices within the scientific community.

3 The Challenge of Nomenclature Ambiguity

The nomenclature of circRNAs has become a topic of considerable importance due to the growing number of identified molecules and their potential implications in research and clinical applications. The ambiguity in circRNA naming originates from the fact that multiple distinct circRNAs can be derived from a single gene locus through alternative splicing events. This results in a variety of circRNAs with different functions and properties that are not adequately reflected in their names [3].

3.1 The Specific CircRNA with a Different Name

The well-known circular RNA, cerebellar degeneration-related protein 1 antisense RNA (CDR1as), exemplifies the common scenario where a circRNA may be identified by multiple names. Hansen and colleagues initially discovered that miR-671 could reduce CDR1 mRNA levels by promoting the cleavage of its natural antisense transcript (NAT) through the action of argonaute protein 2 (AGO-2). Further experiments demonstrated that CDR1 NAT was resistant to enzymes such as nicotinamide pyrophosphatase and terminal 5-phosphate-dependent exonuclease, suggesting that CDR1 NAT is a circular RNA [25]. Subsequent studies revealed that CDR1 NAT possessed a high number of binding sites for miR-7, well over 70, which endowed it with the capacity to act as a sponge for this miRNA, thereby influencing miR-7's regulatory capabilities. Due to this distinctive feature, the circular transcript was also named ciRS-7, indicating its role as a sponge for miR-7 [17].

Nonetheless, other studies in the field have continued to reference this circRNA by its gene locus name, CDR1as, and confirmed it as a key regulator of cell proliferation, differentiation, cancer invasion, and metastasis [26–29]. The variation in naming conventions used across different studies underscores a larger issue of inconsistency, complicating the precise referencing and comparative analysis of circRNAs in scientific literature.

3.2 Different CircRNA Variants with the Same Symbol

Next to the same circRNA having multiple names, nearly identical names with the same index sometimes correspond to different circRNAs. We illustrate this issue using circFNDC3B, a circular RNA derived from the *FNDC3B* gene, which has emerged as a key player in various pathological conditions, including cardiac repair, colorectal cancer (CRC), and Middle East respiratory syndrome coronavirus (MERS-CoV) infection (Fig. 2.2). In the content of cardiac repair, circFNDC3B (mmu_circ_0001113, originating from exons 2 and 3 of the *FNDC3B* gene with 215 nt in length) is downregulated post-myocardial infarction and its overexpression enhances vascular endothelial growth factor-A (VEGF-A) expression, promoting angiogenesis and reducing apoptosis in cardiac endothelial cells [30]. In CRC, circFNDC3B (hsa_circ_0006156, originating from exons 5 and 6 of the *FNDC3B* gene with 603 nt in length) plays the role of a tumor suppressor, mediating the degradation of ASB6 through an RNF41-dependent ubiquitination process [31]. Notably, this circRNA can translate into a novel protein, circFNDC3B-218aa, which inhibits tumor progression and epithelial–mesenchymal transition (EMT) by regulating Snail [32]. In addition, circFNDC3B (hsa_circ_0067985, originating from exons 3 and 6 of the *FNDC3B* gene with 526 nt in length) is identified as a potential miRNA sponge that could influence viral load and related biological pathways in MERS-CoV infection [33]. Although each of the three circRNAs is labeled as circFNDC3B, these circRNAs exhibit distinct functionalities attributed to their unique variants of the circ-*FNDC3B* gene. Therefore, it is necessary to establish a clear and accurate naming system for circRNAs that reflects their functional roles to help avoid misunderstandings.

4 Suggestions for a Common CircRNA Nomenclature

The scientific community has made significant strides in recognizing the importance of a standardized nomenclature for circular RNAs (circRNAs) to ensure clarity and reproducibility in research. In 2023, leading scientists working on circRNA research suggested the scientific community to adopt a standard naming system for circRNAs. By using a common naming system, it becomes easier to clearly communicate and reproduce results [34].

According to the guide of the naming system, the core of the proposed system is the prefix "circ," which unambiguously identifies the molecule as a circular RNA. Next, it is necessary to establish a precise nomenclature for circRNAs, which should include details about the reference linear transcript. This necessity arises since different promoters, splice sites, and polyadenylation signals can result in the generation of multiple unique linear transcripts from a single gene locus. In this context, it is advisable to designate one GENCODE/Ensembl transcript per gene locus as the reference. Ideally, the use of standardized reference transcripts, such as Matched Annotation from NCBI and EMBL-EBI (MANE) transcripts, is recommended [35].

The primary challenge in circRNA nomenclature originates from the generation of multiple circRNAs from a single gene locus with distinct backsplice junctions (BSJs). To clarify these naming protocols in accordance with the guidelines, we offer a series of examples based on circRNAs originating from the *INSR* gene locus. circ-INSR (mmu_circ_0014773) has shown its capacity to prevent cardiomyocyte death and res-

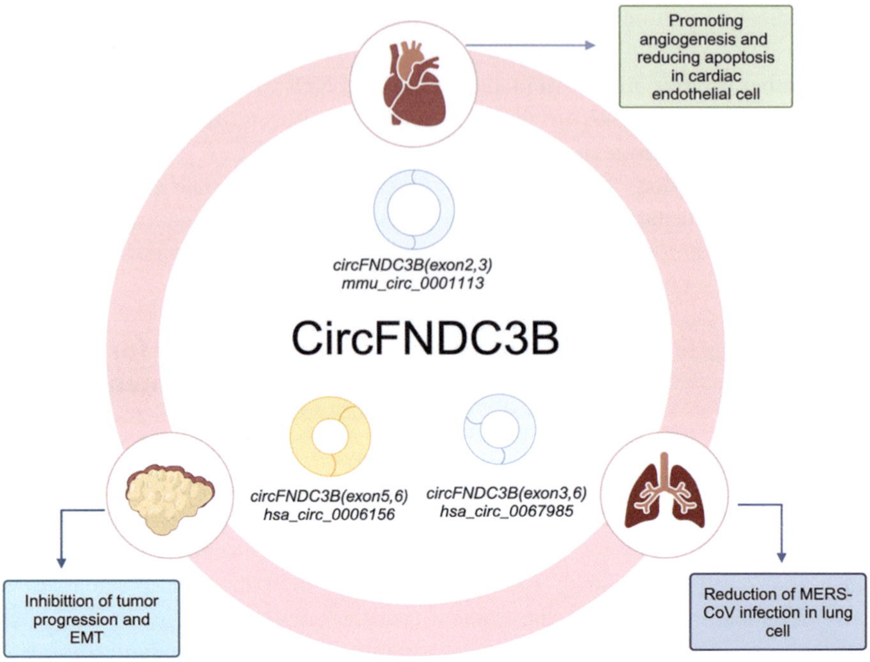

Fig. 2.2 The role of circFNDC3B in disease progression. mmu_circ_0001113, originating from exons 2 and 3 of the *FNDC3B* gene, which is labeled with FNDC3B (exons 2 and 3); hsa_circ_0006156, originating from exons 5 and 6 of the FNDC3B, which is labeled with FNDC3B (exons 5 and 6); hsa_circ_0067985, originating from exons 3 to 6 of FNDC3B, which is labeled with FNDC3B (exons 3 to 6). The features were created with BioRender.com

cue cardiac dysfunction mediated by doxorubicin [36]. Drawing from circBase, the annotated circRNAs from the INSR locus include a spectrum of types: single-exon circRNAs, multi-exon circRNAs, multi-exon circRNAs with varied BSJs, and intron-derived circRNAs.

1. Single-exon circRNAs: mmu_circ_0014773, which originates from exon 2 of the *INSR* gene (according to the ENSMUST00000091291 transcript) is proposed to be named circINSR(2) under the new guidelines (Fig. 2.3a).
2. Muti-exon circRNAs: mmu_circ_0014769, derived from exons 12 and 13 of INSR (according to the same transcript), is suggested to be named circINSR(12,13) (Fig. 2.3b).
3. Multi-exon circRNAs with diverse BSJs: mmu_circ_0014771 and mmu_circ_0001658, both derived from exons 3 to 5 of INSR (according to the same transcript), can be distinguished in their nomenclature to reflect their distinct exonic composition: circINSR(3) and circINSR(3,4,5) (Fig. 2.3c).
4. Intron-derived circRNAs: mmu_circ_0014774, which stems from intron 2 of *INSR* (according to the same transcript), can be named with the alternative prefix "ci" to indicate its origin from intron lariats, resulting in the name ciINSR(2) (Fig. 2.3d).

Although the majority of circRNAs can be categorized as single-exon, multi-exon, those with diverse BSJs, or intron-derived, there exists a subset with unique structural features. These circRNAs will have retained introns (RIs), alternative splicing at the 5′ or 3′ ends, and unannotated exons. The nomenclature guidelines address these special cases as follows:

1. CircRNAs that contain an RI are recommended to be labeled with the "RI" suffix to indicate the presence of a retained intron within the circRNA structure.

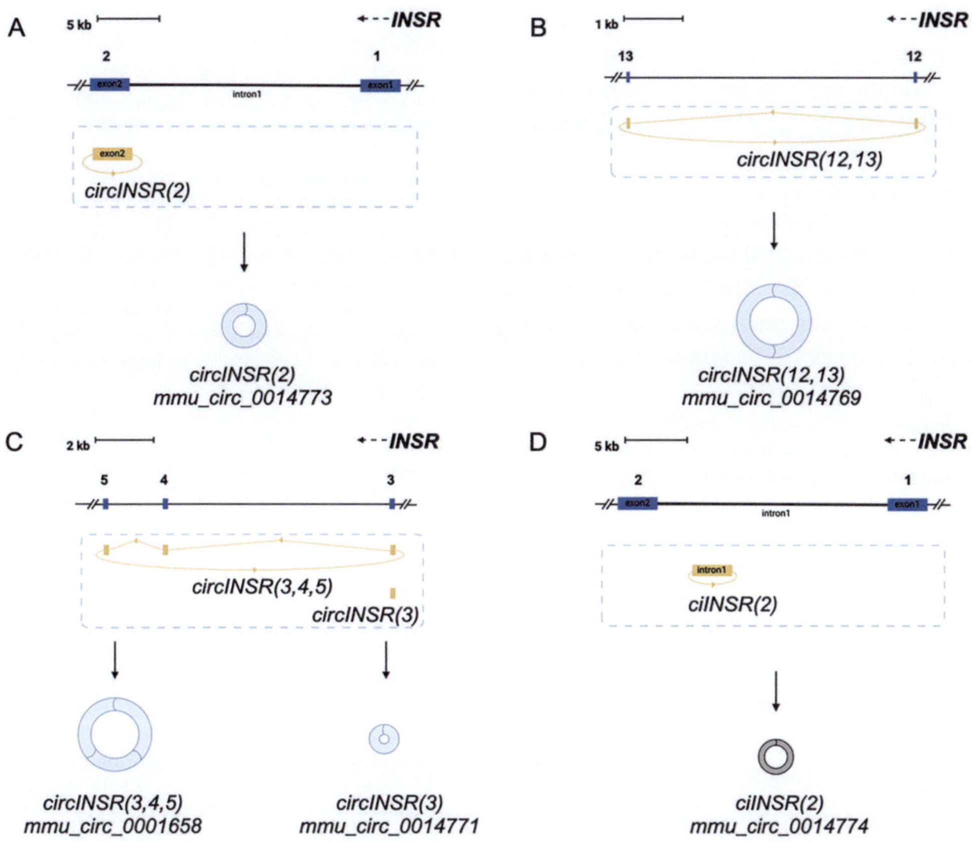

Fig. 2.3 An illustrative nomenclature system for circular RNAs originating from the *INSR* gene. (**a**) A schematic graph of circ-INSR derived from the single exon of the *INSR* gene. (**b**) A schematic graph of circ-INSR derived from the muti-exon of INSR. (**c**) A schematic graph of circ-INSR variants with diverse backsplicing junction (BSJ) sites. (**d**) A schematic graph of circ-INSR originating from the intronic region of INSR. The features were created with BioRender.com

2. For circRNAs featuring alternative splicing at either the 5′ or 3′ ends, the guidelines propose appending an "S" for the shorter version or an "L" for the longer version to the exon number involved in the alternative splicing.
3. When a circRNA includes unannotated exons, the guideline of nomenclature suggested using "NE" (new exon) in the circRNA name.

5 Prospect and Outlook

The proposed nomenclature system for circRNAs marks a pivotal advancement, streamlining communication within the research community with a unified and methodical approach. This system is crafted to embrace the intricacies and variations inherent in circRNA biology, offering clarity and precision in scientific discourse. Nevertheless, the evolution of the circRNA field may necessitate periodic revisions to the naming conventions, particularly for circRNAs with unique structures or features. For example, the new exon will be annotated when the reference gene will also be updated.

Furthermore, the classification of circRNAs originating from noncoding RNA sources, such as tRNAs, rRNAs, and mitochondrial RNAs, as well as those with the potential for coding [37–39], remains an open challenge that demands the development of a unique naming protocol. As we progress toward refining the nomenclature, it is still necessary to present a thorough set of details in the publications, including the precise genomic coordinates, the genome assembly referenced, the full circular RNA sequence, and any existing nomenclature or identifiers recognized by circRNA databases.

With the collective support of all authors, reviewers, and the editorial teams, the integration of standard criteria for circRNA naming will improve the precision and understanding of circRNAs within the circRNA community.

Acknowledgments The authors would like to acknowledge the use of the AI writing assistant developed by Moonshot AI that was used during the editing process for enhancing the clarity and precision of the English language..

Conflict of Interest D.C.L., B.C., and T.T. have filed and partly licensed patents for ncRNAs including one patent for Circ-INSR. T.T. is the founder and CSO/CMO of Cardior Pharmaceuticals GmbH, a wholly owned subsidiary of Novo Nordisk A/S (outside of this chapter).

Funding This work was supported through funding from the Shanghai Pujiang Program (No. 23PJ1412200 to D.C.L) and National Natural Science Foundation of China (82400423 to D.C.L).

References

1. Lu D, Thum T (2019) RNA-based diagnostic and therapeutic strategies for cardiovascular disease. Nat Rev Cardiol 16(11):661–674. https://doi.org/10.1038/s41569-019-0218-x
2. Nilsen TW, Graveley BR (2010) Expansion of the eukaryotic proteome by alternative splicing. Nature 463(7280):457–463. https://doi.org/10.1038/nature08909
3. Szabo L, Salzman J (2016) Detecting circular RNAs: bioinformatic and experimental challenges. Nat Rev Genet 17(11):679–692. https://doi.org/10.1038/nrg.2016.114
4. Zhang Y, Xue W, Li X, Zhang J, Chen S, Zhang JL, Yang L, Chen LL (2016) The biogenesis of nascent circular RNAs. Cell Rep 15(3):611–624. https://doi.org/10.1016/j.celrep.2016.03.058
5. Enuka Y, Lauriola M, Feldman ME, Sas-Chen A, Ulitsky I, Yarden Y (2016) Circular RNAs are long-lived and display only minimal early alterations in response to a growth factor. Nucleic Acids Res 44(3):1370–1383. https://doi.org/10.1093/nar/gkv1367
6. Kristensen LS, Andersen MS, Stagsted LVW, Ebbesen KK, Hansen TB, Kjems J (2019) The biogenesis, biology and characterization of circular RNAs. Nat Rev Genet 20(11):675–691. https://doi.org/10.1038/s41576-019-0158-7
7. Yang L, Wilusz JE, Chen LL (2022) Biogenesis and regulatory roles of circular RNAs. Annu Rev Cell Dev Biol 38:263–289. https://doi.org/10.1146/annurev-cellbio-120420-125117
8. Sanger HL, Klotz G, Riesner D, Gross HJ, Kleinschmidt AK (1976) Viroids are single-stranded covalently closed circular RNA molecules existing as highly base-paired rod-like structures. Proc Natl Acad Sci USA 73(11):3852–3856. https://doi.org/10.1073/pnas.73.11.3852
9. Capel B, Swain A, Nicolis S, Hacker A, Walter M, Koopman P, Goodfellow P, Lovell-Badge R (1993) Circular transcripts of the testis-determining gene Sry in adult mouse testis. Cell 73(5):1019–1030. https://doi.org/10.1016/0092-8674(93)90279-y
10. Nigro JM, Cho KR, Fearon ER, Kern SE, Ruppert JM, Oliner JD, Kinzler KW, Vogelstein B (1991) Scrambled exons. Cell 64(3):607–613. https://doi.org/10.1016/0092-8674(91)90244-s
11. Liu CX, Chen LL (2022) Circular RNAs: characterization, cellular roles, and applications. Cell 185(12):2016–2034. https://doi.org/10.1016/j.cell.2022.04.021
12. Vromman M, Anckaert J, Bortoluzzi S, Buratin A, Chen CY, Chu Q, Chuang TJ, Dehghannasiri R, Dieterich C, Dong X, Flicek P, Gaffo E, Gu W, He C, Hoffmann S, Izuogu O, Jackson MS, Jakobi T, Lai EC, Nuytens J, Salzman J, Santibanez-Koref M, Stadler P, Thas O, Vanden Eynde E, Verniers K, Wen G, Westholm J, Yang L, Ye CY, Yigit N, Yuan GH, Zhang J, Zhao F, Vandesompele J, Volders PJ (2023) Large-scale benchmarking of circRNA detection tools reveals large differences in sensitivity but not in precision. Nat Methods 20(8):1159–1169. https://doi.org/10.1038/s41592-023-01944-6
13. Glazar P, Papavasileiou P, Rajewsky N (2014) circBase: a database for circular RNAs. RNA 20(11):1666–1670. https://doi.org/10.1261/rna.043687.113
14. Dong R, Ma XK, Li GW, Yang L (2018) CIRCpedia v2: an updated database for comprehensive circular RNA annotation and expression comparison. Genomics Proteomics Bioinformatics 16(4):226–233. https://doi.org/10.1016/j.gpb.2018.08.001
15. Liu M, Wang Q, Shen J, Yang BB, Ding X (2019) Circbank: a comprehensive database for circRNA with standard nomenclature. RNA Biol 16(7):899–905. https://doi.org/10.1080/15476286.2019.1600395
16. Wu W, Ji P, Zhao F (2020) CircAtlas: an integrated resource of one million highly accurate circular RNAs from 1070 vertebrate transcriptomes. Genome Biol 21(1):101. https://doi.org/10.1186/s13059-020-02018-y
17. Memczak S, Jens M, Elefsinioti A, Torti F, Krueger J, Rybak A, Maier L, Mackowiak SD, Gregersen LH,

Munschauer M, Loewer A, Ziebold U, Landthaler M, Kocks C, le Noble F, Rajewsky N (2013) Circular RNAs are a large class of animal RNAs with regulatory potency. Nature 495(7441):333–338. https://doi.org/10.1038/nature11928
18. Maass PG, Glazar P, Memczak S, Dittmar G, Hollfinger I, Schreyer L, Sauer AV, Toka O, Aiuti A, Luft FC, Rajewsky N (2017) A map of human circular RNAs in clinically relevant tissues. J Mol Med (Berl) 95(11):1179–1189. https://doi.org/10.1007/s00109-017-1582-9
19. Rybak-Wolf A, Stottmeister C, Glazar P, Jens M, Pino N, Giusti S, Hanan M, Behm M, Bartok O, Ashwal-Fluss R, Herzog M, Schreyer L, Papavasileiou P, Ivanov A, Ohman M, Refojo D, Kadener S, Rajewsky N (2015) Circular RNAs in the mammalian brain are highly abundant, conserved, and dynamically expressed. Mol Cell 58(5):870–885. https://doi.org/10.1016/j.molcel.2015.03.027
20. Zhang XO, Dong R, Zhang Y, Zhang JL, Luo Z, Zhang J, Chen LL, Yang L (2016) Diverse alternative back-splicing and alternative splicing landscape of circular RNAs. Genome Res 26(9):1277–1287. https://doi.org/10.1101/gr.202895.115
21. Guarnerio J, Bezzi M, Jeong JC, Paffenholz SV, Berry K, Naldini MM, Lo-Coco F, Tay Y, Beck AH, Pandolfi PP (2016) Oncogenic role of fusion-circRNAs derived from cancer-associated chromosomal translocations. Cell 165(2):289–302. https://doi.org/10.1016/j.cell.2016.03.020
22. Wang K, Long B, Liu F, Wang JX, Liu CY, Zhao B, Zhou LY, Sun T, Wang M, Yu T, Gong Y, Liu J, Dong YH, Li N, Li PF (2016) A circular RNA protects the heart from pathological hypertrophy and heart failure by targeting miR-223. Eur Heart J 37(33):2602–2611. https://doi.org/10.1093/eurheartj/ehv713
23. Devaux Y, Creemers EE, Boon RA, Werfel S, Thum T, Engelhardt S, Dimmeler S, Squire I, Cardiolinc N (2017) Circular RNAs in heart failure. Eur J Heart Fail 19(6):701–709. https://doi.org/10.1002/ejhf.801
24. Wang K, Gan TY, Li N, Liu CY, Zhou LY, Gao JN, Chen C, Yan KW, Ponnusamy M, Zhang YH, Li PF (2017) Circular RNA mediates cardiomyocyte death via miRNA-dependent upregulation of MTP18 expression. Cell Death Differ 24(6):1111–1120. https://doi.org/10.1038/cdd.2017.61
25. Hansen TB, Wiklund ED, Bramsen JB, Villadsen SB, Statham AL, Clark SJ, Kjems J (2011) miRNA-dependent gene silencing involving Ago2-mediated cleavage of a circular antisense RNA. EMBO J 30(21):4414–4422. https://doi.org/10.1038/emboj.2011.359
26. Hansen TB, Jensen TI, Clausen BH, Bramsen JB, Finsen B, Damgaard CK, Kjems J (2013) Natural RNA circles function as efficient microRNA sponges. Nature 495(7441):384–388. https://doi.org/10.1038/nature11993
27. Yan B, Zhang W, Mao XW, Jiang LY (2018) Circular RNA ciRS-7 correlates with advance disease and poor prognosis, and its down-regulation inhibits cells proliferation while induces cells apoptosis in non-small cell lung cancer. Eur Rev Med Pharmacol Sci 22(24):8712–8721. https://doi.org/10.26355/eurrev_201812_16636
28. Yang L, Bin Z, Hui S, Rong L, You B, Wu P, Han X, Qian H, Xu W (2019) The role of CDR1as in proliferation and differentiation of human umbilical cord-derived mesenchymal stem cells. Stem Cells Int 2019:2316834. https://doi.org/10.1155/2019/2316834
29. Yang X, Li S, Wu Y, Ge F, Chen Y, Xiong Q (2020) The circular RNA CDR1as regulate cell proliferation via TMED2 and TMED10. BMC Cancer 20(1):312. https://doi.org/10.1186/s12885-020-06794-5
30. Garikipati VNS, Verma SK, Cheng Z, Liang D, Truongcao MM, Cimini M, Yue Y, Huang G, Wang C, Benedict C, Tang Y, Mallaredy V, Ibetti J, Grisanti L, Schumacher SM, Gao E, Rajan S, Wilusz JE, Goukassian D, Houser SR, Koch WJ, Kishore R (2019) Circular RNA CircFndc3b modulates cardiac repair after myocardial infarction via FUS/VEGF-A axis. Nat Commun 10(1):4317. https://doi.org/10.1038/s41467-019-11777-7
31. Zeng W, Zhu JF, Guo J, Huang GJ, Ai LS, Zeng Y, Liao WJ (2022) M(6)A-modified circFNDC3B inhibits colorectal cancer stemness and metastasis via RNF41-dependent ASB6 degradation. Cell Death Dis 13(11):1008. https://doi.org/10.1038/s41419-022-05451-y
32. Pan Z, Cai J, Lin J, Zhou H, Peng J, Liang J, Xia L, Yin Q, Zou B, Zheng J, Qiao L, Zhang L (2020) A novel protein encoded by circFNDC3B inhibits tumor progression and EMT through regulating snail in colon cancer. Mol Cancer 19(1):71. https://doi.org/10.1186/s12943-020-01179-5
33. Zhang X, Chu H, Wen L, Shuai H, Yang D, Wang Y, Hou Y, Zhu Z, Yuan S, Yin F, Chan JF, Yuen KY (2020) Competing endogenous RNA network profiling reveals novel host dependency factors required for MERS-CoV propagation. Emerg Microbes Infect 9(1):733–746. https://doi.org/10.1080/22221751.2020.1738277
34. Chen LL, Bindereif A, Bozzoni I, Chang HY, Matera AG, Gorospe M, Hansen TB, Kjems J, Ma XK, Pek JW, Rajewsky N, Salzman J, Wilusz JE, Yang L, Zhao F (2023) A guide to naming eukaryotic circular RNAs. Nat Cell Biol 25(1):1–5. https://doi.org/10.1038/s41556-022-01066-9
35. Morales J, Pujar S, Loveland JE, Astashyn A, Bennett R, Berry A, Cox E, Davidson C, Ermolaeva O, Farrell CM, Fatima R, Gil L, Goldfarb T, Gonzalez JM, Haddad D, Hardy M, Hunt T, Jackson J, Joardar VS, Kay M, Kodali VK, McGarvey KM, McMahon A, Mudge JM, Murphy DN, Murphy MR, Rajput B, Rangwala SH, Riddick LD, Thibaud-Nissen F, Threadgold G, Vatsan AR, Wallin C, Webb D, Flicek P, Birney E, Pruitt KD, Frankish A, Cunningham F, Murphy TD (2022) A joint NCBI and EMBL-EBI transcript set for clinical genomics and research. Nature 604(7905):310–315. https://doi.org/10.1038/s41586-022-04558-8

36. Lu D, Chatterjee S, Xiao K, Riedel I, Huang CK, Costa A, Cushman S, Neufeldt D, Rode L, Schmidt A, Juchem M, Leonardy J, Buchler G, Blume J, Gern OL, Kalinke U, Wen Tan WL, Foo R, Vink A, van Laake LW, van der Meer P, Bar C, Thum T (2022) A circular RNA derived from the insulin receptor locus protects against doxorubicin-induced cardiotoxicity. Eur Heart J 43(42):4496–4511. https://doi.org/10.1093/eurheartj/ehac337
37. Chen R, Yang T, Jin B, Xu W, Yan Y, Wood N, Lehmann HI, Wang S, Zhu X, Yuan W, Chen H, Liu Z, Li G, Bowen TS, Li J, Xiao J (2023) CircTmeff1 promotes muscle atrophy by interacting with TDP-43 and encoding a novel TMEFF1-339aa protein. Adv Sci (Weinh) 10(17):e2206732. https://doi.org/10.1002/advs.202206732
38. Wu Z, Sun H, Wang C, Liu W, Liu M, Zhu Y, Xu W, Jin H, Li J (2020) Mitochondrial genome-derived circRNA mc-COX2 functions as an oncogene in chronic lymphocytic leukemia. Mol Ther Nucleic Acids 20:801–811. https://doi.org/10.1016/j.omtn.2020.04.017
39. Zhu Y, Gui W, Lin X, Li H (2020) Knock-down of circular RNA H19 induces human adipose-derived stem cells adipogenic differentiation via a mechanism involving the polypyrimidine tract-binding protein 1. Exp Cell Res 387(2):111753. https://doi.org/10.1016/j.yexcr.2019.111753

Part II

Bioinformatics for Circular RNAs

Circular RNAs: Biogenesis and Functions

3

Circular RNA detection and bioinformatic analysis

Hüseyin Vural

Abstract

Circular RNAs (circRNAs) are noncoding RNAs (ncRNAs) that show conservation across evolutionary lines and display specific expression patterns in tissues. They perform distinctive cellular roles and are promising biomarkers in therapeutic contexts. An ever-growing area of circRNA types and biological functions is closely related to recent advancements in sequencing technologies and bioinformatics tools. Practical uses of circRNAs have emerged in tandem with technical advancements, including their use as biomarkers for human disease. Therefore, to improve the diagnosis and prognosis of diseases associated with circRNAs and to have greater knowledge of the biological processes involved in circRNAs, it is imperative to create algorithms for the precise and sensitive detection of circRNAs.

Keywords

CircRNAs · RNA sequencing · CircRNA detection · Bioinformatics

H. Vural (✉)
Department of Computer Engineering, Adiyaman University, Adiyaman, Turkey
e-mail: hvural@adiyaman.edu.tr

1 Introduction

In the 1970s, the distinctive structure of "single-stranded and covalently closed circular RNA molecules" was initially identified in plant viruses [1]. Over the subsequent four decades, various confirmations regarding circular RNAs (circRNAs) dispelled the misconception of them being aberrant RNA splicing remnants. Instead, the mechanism involving backsplicing junctions (BSJs) was established, wherein a downstream 5′ splice-donor site attacks an upstream 3′ splice-acceptor site, forming a 3′–5′ phosphodiester bond [2]. Recent years have witnessed a revolution in high-throughput RNA sequencing (RNA-seq) and subsequent biochemical analyses, uncovering the abundance and diversity of circRNAs [3]. CircRNAs are essential endogenous RNAs that fall into two categories: coding circRNAs and noncoding circRNAs (ncRNAs) [4]. While most circRNAs, known as EcircRNAs, comprise one or several exons and reside in the cytoplasm, others, such as intronic circRNAs and exonic–intronic circRNAs, are found in the nucleus [5].

CircRNAs lack polyadenylation (poly(A)) and capping, rendering them resistant to exonuclease and RNase R-mediated degradation [6]. Recognizing that the presence of circRNA is not incidental, an increasing number of researchers are directing their attention to its biological functions [7]. Particularly noteworthy is the seminal

J. Xiao (ed.), *Advances in Circular RNAs*, Advances in Experimental Medicine and Biology 1485,
https://doi.org/10.1007/978-981-96-9428-0_3

work of another research that proposed its role as a sponge for microRNAs in 2011 [8]. CircRNAs can engage with RNA-binding proteins (RBPs), serving as scaffolds in protein complex assembly or recruiting specific proteins to distinct cellular locations [9]. Furthermore, certain circRNAs can modulate gene expression and govern alternative splicing, and a subset of endogenous circRNAs can even participate in protein translation [10].

Additionally, researchers have unearthed the significant involvement of circRNAs in various health conditions, including diabetes mellitus, cardiovascular diseases, chronic inflammatory diseases, neurological disorders, and cancer [11]. Numerous circRNAs have been identified with abnormal expression in tumor tissues such as hepatocellular carcinoma (HCC) and gastric cancer [12]. Mounting evidence indicates that the aberrant expression of circRNA can significantly impact the initiation and progression of cancer [13]. Consequently, circRNAs hold immense potential as novel biomarkers and therapeutic targets [14]. The increasing number of studies on circRNAs is shown in Fig. 3.1 [15].

Developments in high-throughput sequencing and bioinformatics tools have opened up new avenues for circRNA comprehension. In this study, we summarized the algorithms developed for circRNA detection and subsequently discussed the challenges associated with circRNA detection using RNA sequencing and bioinformatics analysis.

2 The Formation Process of CircRNAs

Pre-messenger RNA (pre-mRNA) is mainly backspliced to create circRNA, which is a closed RNA transcript [16]. Furthermore, during intron self-splicing, circRNAs can come from a variety of sources, such as transfer RNAs, ribosomal RNAs, mitochondrial RNAs, and small nuclear RNAs [16–18]. CircRNAs derived from exons, introns, untranslated regions (UTRs), or antisense sequences have been identified in recent studies using advanced RNA-seq technologies. These circRNAs are categorized into four main categories: intronic circRNAs, exonic–intronic circRNAs, exonic circRNAs, and others [19]. Backsplicing of pre-mRNA is the main process that generates circRNAs [20]. It involves joining a splice acceptor upstream of the 3′ splice site with a splice donor downstream of the 5′ splice site to form a circular shape with a 3′–5′ phosphodiester bond at

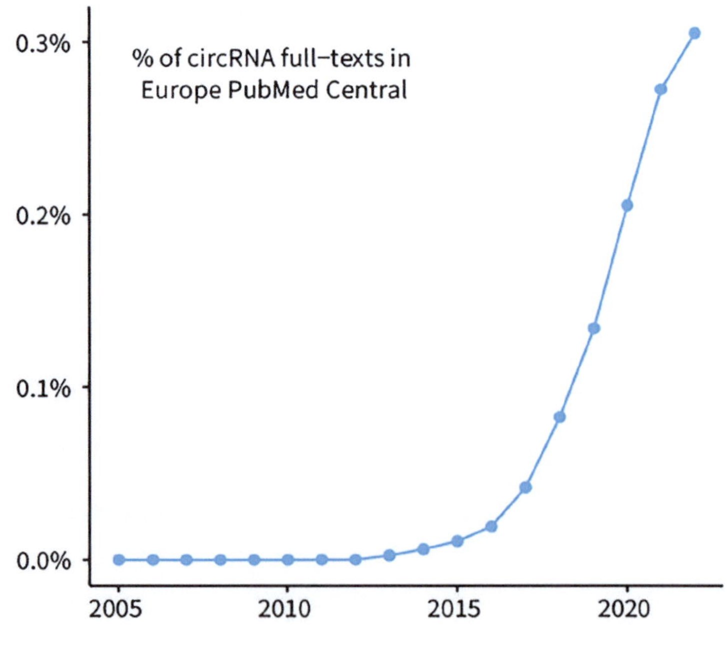

Fig. 3.1 The proportionate rise in European PubMed Central articles mentioning circRNAs.

the backsplicing junction (BSJ) [19]. Special intron sequences and RBPs may promote splicing and induce the synthesis of circRNA, therefore aiding in the generation of circRNA [21]. By encouraging nucleophilic attack and splicing, some inverted repeat sequences in introns, like Alu, can also increase the creation of circRNAs [22]. Nevertheless, little is known about the precise metabolic milieu and regulatory elements required for circRNA occurrence. Notably, RNA pairing rivalry across flanking introns can drive the production of numerous circRNAs from a single gene [23].

Recent studies have clarified a series of circRNA regulatory mechanisms, with an emphasis on epigenetic regulation, including translation, circRNA–miRNA sponges, circRNA:DNA hybrids, directing A-to-I editing, and circRNA–protein interactions [24]. CircRNAs in the cytoplasm function as miRNA sponges, preventing miRNA regulatory activities. As an example, ciRS-7 upregulates the expression of insulin receptor substrate 2 (IRS2), epidermal growth factor receptor (EGFR), and other genes by sponging up miR-7 [25]. Additionally, circRNAs interact with RBPs by acting as inhibitors or antagonists of those proteins [26]. They frequently interact with protein kinase R (PKR) and form duplexes, which inhibits innate immunity [27]. CircRNAs also perform A-to-I editing on mRNAs, modify RNA stability, and create R-loop structures [28]. These results clarify the several epigenetic regulatory functions of circRNAs.

3 Key Features of CircRNA Detection Algorithms

The primary challenge hindering circRNA development is the absence of effective detection strategies. Achieving precise and sensitive circRNA detection methods is crucial for understanding circRNA biological processes and advancing circRNA-related disease diagnosis and prognosis. Unlike mRNA and its counterpart, circRNA lacks distinct 5′ and 3′ free ends, posing a challenge for differentiation through standard RNA-seq methods. Additionally, the nonlinear conformation of circRNA results in poor migration properties, making it challenging to separate through highly cross-linked electrophoretic gels [29]. The combination of low expression levels and interference from linear isoforms in biological samples further complicates the identification and detection of circRNA [30].

Based on the origin of the genomic region that determines the final circRNA sequence, different kinds of circRNAs are recognized. Generally, three main types of circularization processes arise from RBPs and intron pairing-driven circularization. These categories include combinations of several exons or intronic sections in the circRNAs.

- All of the exons implicated in the circularization event make up exonic circRNAs or EcircRNAs.
- Exons processed after an alternative splicing event are included in alternative EcircRNAs.
- Exonic–intronic circRNAs keep the exons that make up the final circRNA's sequence in addition to an intronic section.

Lariat-driven circularization, which involves base pairing between splicing site motif sequences inside exonic and intronic sequences, usually results in the production of both single EcircRNAs or intronic circular RNAs. These characteristics are important considerations for circRNA prediction, and different approaches are being developed. These techniques are based on evaluating the structure of RNA-seq reads by mapping them against a reference genome or utilizing machine learning techniques that consider characteristics found in the sequence of a transcript, whether it codes for a protein or not [31].

In the swiftly developing realm of research, there is an ongoing requirement for precise detection and documentation of recently discovered circRNAs. Recently, the rapid advancement of bioinformatics algorithms and biomolecular technology has provided researchers with alternative analytical approaches to the detection of circRNAs. Below are the primary methods to address this problem.

4 CircRNA Detection Methods

Numerous circRNA detection methods have been created for circRNA research since 2012. They fall into three primary categories, which are as follows:

- RNA-seq
- Integrated
- Machine learning

Split-alignment-based (segmented-based) and pseudo-reference-based (candidate-based) are the two primary categories into which software often designed to study RNA-seq data can be placed. They differ in terms of the mapping of sequencing reads against a reference genome and/or gene annotation file [32]. While pseudo-reference techniques rely on gene annotation files to identify junction reads referred to as annotation, segmented-based tools find BSJs by aligning reads to the reference genome and are referred to as de novo tools. As an input, circRNA-enriched RNA-seq datasets are preferred by most identifying methods. Using paired-end (PE) read sequencing is a further tactic to enhance the detection of decoy reads, which are usually disregarded as alignment artifacts. This may be useful in removing circRNAs that are falsely positive. Nevertheless, single-end (SE) reads are typically included in a large portion of useful RNA-seq data, such as cross-linking immunoprecipitation sequencing (CLIP-Seq) data [33]. In our study, we included the most frequently used studies among these three categories, and these studies are shown in the Table 3.1.

4.1 RNA-seq-Based Methods

RNA-seq offers a thorough comprehension of the eukaryotic transcriptome and an in-depth investigation of the diverse RNA molecule types present within the cell. To improve the required depth of sequencing for circRNA analysis, RNA-seq of total RNA with ribosomal RNA depletion is frequently utilized. This is commonly followed by an RNase R step, which effectively eliminates linear RNAs. Even though this method works well, different findings can be obtained based on the quality of the material, the sequencing, and the circRNA expression levels [34, 35].

Due to the abundance of RNA-seq data, many algorithms have been developed for the detection of circRNAs based on RNA-seq. Computational approaches are useful in investigating the biology of these ubiquitous and diversified circRNAs because they simplify the analysis of high-throughput RNA-seq data and offer benefits for circRNA expression profiling [32]. Here, we outline some of the frequently employed RNA-seq-based methods for detecting circRNAs.

CIRI [36] is a technique that Gao et al. [37] introduced. It uses a novel chiastic clipping signal algorithm to detect circRNAs completely. Using a variety of filtering methods, CIRI can accurately and impartially identify circRNAs from transcriptomic data. Its main application is in the annotation and detection of circRNAs from RNA-seq data. In contrast to other approaches, CIRI uses a unique algorithm to identify paired cross-clip signals in Burrows–Wheeler Aligner Maximum Exact Match (BWA-MEM) sequence alignments/maps and combines them with systematic filtering to remove false positives.

Zhang et al. [21] created a circRNA detection tool that was the first to reveal the regulatory mechanism of complementary sequences in the synthesis of exon-derived circRNAs. This tool demonstrated that complementary sequences can be paired competitively to regulate varied cyclization. This provides a new theoretical understanding of the complexity and diversity of gene expression at the transcriptional and posttranscriptional levels. In the human embryonic stem cell line H9, almost 10,000 circRNAs were found using a specific nuclease to enrich circRNAs in conjunction with computational analysis software. This finding revealed that the corresponding sequence of intron RNA mediates exon cyclization. Selective production of either linear RNAs or circRNAs can be achieved through competitive pairing of complementary sequences between distinct areas.

CircSeq is a computational pipeline created by Guo et al. [38] to discover and quantify the

Table 3.1 Some circRNA detection tools

Name	Category	TT	AL	AC	PL	References
CIRI	PE, SE	De novo	BWA-MEM	BSJ	Perl	[32]
CIRCexplorer	PE	De novo, annotation	STAR- BWA	BSJ	Python	[37]
CircSeq	SE	De novo	Bowtie	BSJ	R	[21]
CircRNA Identifier 2 (CIRI2)	PE	De novo	BWA	BSJ	Perl	[38]
(DCC) detect circRNAs from chimeric reads	PE, SE	Annotation	STAR	BSJ	R, python	[41]
CircRNA_finder	PE, SE	De novo	STAR	BSJ	Perl	[42]
UROBORUS	PE, SE	Annotation	Bowtie	BSJ	Perl	[43]
CIRCexplorer3	PE	Annotation	TopHat-fusion	BSJ	Python	[46]
CIRIquant	PE	Annotation	BWA	BSJ	R	[49]
Find_circ	SE	De novo	Bowtie2	BSJ	Python	[51]
Find_circ2	PE	Annotation	BWA-Bowtie2	BSJ	R	[40]

PE paired end, *SE* single end, *BWA-MEM* Burrows–Wheeler Aligner Maximum Exact Match, *STAR-BWA* Spliced Transcripts Alignment to a Reference-BWA, *BSJ* backsplicing junction

relative abundance of circRNAs from RNA-seq databases. Its purpose is to find and describe mammalian circRNAs. CircSeq does not rely on existing gene annotation for circRNA detection, in contrast to other detection techniques. Its use in the ENCODE project to analyze RNA-seq data that was not poly(A)-selected demonstrated its capacity to categorize and thoroughly describe more than 7000 human circRNAs.

One of the most widely used methods for finding circRNAs is CircRNA Identifier 2 (CIRI2) [39]. CircRNA Identifier 2 (CIRI2) uses a two-step scanning procedure to identify potential BSJs and eliminate false positives resulting from tandem forward-splicing junctions (FSJs). This process is based on the alignment output of BWA-MEM [40]. First, it examines the reads' paired chiastic clipping (PCC) signal to determine whether the reads are compatible with BSJ templates, which are obtained from the readings' CIGAR (Compact Idiosyncratic Gapped Alignment Report) in the BWA alignment. To improve accuracy, it also takes into account the paired-end mapping signal, even if SE data could not have this information. Additionally, it incorporates canonical GT/AG splice sites from exon boundaries optionally. Next, to mitigate the effect of imbalanced sequences and lower the false discovery rate, CIRI2 employs an adapted maximum likelihood estimate method based on multiple seed matching to determine the segment location and distinguish BSJ reads from FSJ reads.

BWA-MEM is used by Find_circ2 [41] to achieve read alignment. Essentially, to identify putative circRNA splicing events, this tool takes 20-bp anchors from both ends of the unmapped reads that were retrieved from the first alignment and looks at their mapping positions. It involves extra measures to improve specificity, like examining the splice site's GU/AG signal and permitting no more than two mismatches for a BSJ supporting read. Interestingly, Find_circ2 does not make use of PE mapping data.

DCC [42] maps chimerically aligned reads using the STAR (Spliced Transcripts Alignment to a Reference) aligner and finds BSJ reads among them. It employs several filters, such as excluding candidates mapped to homologous or repetitive regions, verifying the GT/AT splicing signal at suspected BSJs, and aligning the inner-circle region of the mate read in PE sequencing.

The foundation of circRNA_finder [43] is STAR alignment. Chimeric junction reads undergo filtering based on parameters including matching GT/AG splice sites, thus permitting a maximum of three mismatches for BSJ supporting reads and enforcing a maximum distance of

100 kb between the acceptor and donor anchors after STAR alignment. It requires that the mate read of a BSJ supporting read stays inside the hypothesized circRNA region for PE RNA-seq data.

UROBORUS [44] extracts unmapped reads for additional processing by aligning TopHat [45] and Bowtie [46]. To create artificial PE reads, it cuts the unmapped reads to produce 20-bp sequences from both ends. To find BSJ reads aligned to the connecting area of two backspliced exons, these short PE reads are then realigned to the genome. Lastly, several filters are applied, such as minimum number of supporting reads, maximum gene distance, and chromosomal position.

For RNA-seq data, trimmed PE FASTQ files are required as well as genome reference and annotation files for input into CIRCexplorer3 [47]. It first uses the flexible aligner HISAT2 [48] to map reads to the genome reference, with a focus on finding reads that cross splice junctions in coding transcripts. Reads that are not mapped are then realigned to the same reference genome. It is possible to identify putative circRNA readings using genome reference annotation and the quick and easy-to-use aligner STAR [49]. Next, each BSJ event's precise start and finish positions are identified, and the BSJ is mapped back to the annotation file. Details about circRNA annotation, including circRNA ID, chromosome, start and finish coordinates, strand, read count within the circRNA, flanking introns, and other information important for further analysis, are included in the final output file.

A modified version of CIRI2, called CIRIquant [50], starts by aligning readings with HISAT2 [48]. De novo circRNA identification from unmapped reads is the first step, after which BWA is used to map the reads to the reference genome [53]. The Sequence Alignment/Map (SAM) files then go through two scanning cycles. In the first scan, circRNA and BSJ candidates are found using paired chiastic clipping (PCC) signals. These candidates are then further filtered using splicing junction signals (GT/AG) and PE mapping (PEM). After BSJ reads are grouped and circRNA candidates are registered, the Sequence Alignment/Map (SAM) file is scanned once more for further circRNA candidates. This time, a filter is added to remove false positives that are mistakenly assigned to homologous genes and repetitive sequences. Additionally, the output file that is produced contains details on the circRNA annotation, including circRNA ID, chromosomal, start and finish coordinates, strand, read count, flanking introns, and other pertinent data.

The algorithm for de novo circRNA prediction is found in Find_circ [41]. To get the best Phred Quality Score, input files must be reduced. First, Bowtie2 is used to align reads to the reference genome [52], producing SAM format files with unmapped reads. The process of splicing identification then entails removing 20-mers from both ends of the unmapped reads and aligning them with the reference genome to determine distinct anchor sites. The output file includes extra columns with data and statistics in addition to all identified splice junctions in BED format.

4.2 Integrated Methods

The issue of variations in expected circRNA numbers and excessive false positives brought on by the disparities in algorithmic approaches employed by various tools can be addressed by multi-tool approaches for circRNA detection. Certain detection tools fall under the multi-tool category since they combine the output of current stable tools with their integration. Integration of various circRNA detection tools has been shown to lower the false-positive rate [6, 53]. Certain circRNAs are sensitive to RNase R and can be severely reduced by treatment with RNase R. Consequently, it is problematic to identify these RNase R-sensitive circRNAs only using RNase R-treated library preparation-based techniques [54]. As a result, multi-tool approaches for circRNA detection can help achieve this goal by addressing the problem of variations in anticipated circRNA counts and high false positives brought on by variations in the algorithmic techniques employed by various tools.

Among these tools is a pipeline for circRNA discovery, quantification, and annotation called CirComPara [55], which combines the results from CIRCexplorer, CIRI, Find_circ, and Segemehl. With circ_battle, users can additionally enhance the sensitivity and specificity of circRNA detection by combining or comparing various circRNA prediction algorithms. In another study [53], users could enhance the sensitivity and specificity of circRNA identifications by combining or comparing various circRNA prediction methods [21, 36, 41, 44]. An example is RAISE (Realignment and Analysis of Internal Structure and Expression of circRNAs) [56], which can also quantify circRNA abundance and forecast circRNA structure. CircRNAwrap [57] is an even more extensive pipeline tool that can predict transcripts and identify and estimate abundance.

4.3 Machine Learning-Based Methods

Some circRNA detection tools use machine learning approaches to predict circRNA utilizing circRNA knowledge and attributes of known circRNA to build a classification model, even though the majority of circRNA detection tools require RNA-seq datasets as input. Model learning from real circRNAs is made possible by machine learning methods. More features can be employed as machine learning input to create more dependable classification models as circRNA knowledge grows, and most tools can accommodate more rules as filter steps. The common traits of machine learning models to detect circRNAs are Alu repeats, sequence motifs, and structure motifs [58, 59].

A study that uses a machine learning algorithm is PredcircRNA [59], which uses a multiple kernel learning technique to separate circRNAs from other long noncoding RNAs (lncRNAs). With the help of a stem cell dataset, WebCircRNA [60] can be trained to predict circRNAs that are unique to stem cells. DeepCirCode [61] employs a convolutional neural network with the flanking 50-nt sequence of the backsplicing site as input to predict human circRNAs. Additionally, this tool provides a mechanism for visualizing circRNA production aspects as sequence motifs. PredicircRNATool [62] analyzes structural and thermodynamic features in flanking regions to predict circRNAs. The first deep learning model for circRNA prediction is called DeepCirCode. The recently published StackCirRNAPred [63] uses a stacking method to distinguish circRNAs from other lncRNAs, while CirRNAPL [64] predicts circRNAs based on structural traits and sequence composition. These instruments can be used in conjunction with well-known split bases and detection techniques based on references that are found in the literature.

5 Challenges in RNA-seq and Bioinformatics Analysis

Over the past decade, there has been a significant increase in scientific interest in circRNAs, resulting in a more than 90% rise in related publications. This surge in interest is mirrored by the development of numerous tools for circRNA detection, alongside other bioinformatics resources, contributing to a deeper understanding of this emerging field in RNA biology research. Despite the availability of various algorithms and tools for circRNA analysis, only a few systematic evaluations of their performance have been conducted [34, 53, 65]. Five distinct tools [21, 37, 41, 43, 66] exhibited a moderate overlap of 16.8% in their predictions, showing considerable changes in the circRNA landscape based on the selected algorithm, according to research [67]. Notably, a single technique accounted for 40% of the anticipated circRNAs. RNase R treatment of the samples was performed to validate the true positive candidates among these tool-specific predictions. The results showed that, except for CIRCexplorer [21] and MapSplice [66], over half of the putative circRNAs were sensitive to RNase R treatment, indicating that the other half were false positives. Therefore, using different algorithms is recommended to reduce the possibility of missing potential circRNAs and finding false-positive artifacts.

Sensitivity and precision are evaluated as part of the bioinformatics instrument performance evaluation process. Sensitivity gauges how well the algorithm can link reads to circRNAs, which affects the total amount of circRNAs that are predicted. Contrarily, precision indicates algorithmic correctness and can be improved by filtering circRNA candidates using read count criteria. In a study, researchers conducted a benchmarking study [65] in which 11 circRNA detection techniques were assessed. The results showed that while Segemehl [68], Find_circ [41], and UROBORUS [44] performed less well, NCLScan [69], MapSplice [66], CIRCexplorer [21], detect circRNAs from chimeric reads (DCC) [42], and PTESFinder [70] showed lower false-positive rates. In contrast to other tools, NCLScan [58] and CIRCexplorer [21] showed better precision, while CIRI [37], Segemehl [68], PTESFinder [70], and CIRCexplorer [21] were typically more sensitive. Various techniques yielded various results when measuring sensitivity for PE data, where both reads span the same BSJ. These results showed significant diversity in the distances between splicing sites involved in backsplicing areas, ranging from several kilobases to hundreds of nucleotides.

Subsequently, in 2018, Hansen [53] evaluated how preprocessed reads affected algorithm performance; this study includes adaptor trimming and low-quality read removal. The benchmarking showed that while the median expression was relatively constant for the majority of algorithms, some algorithms benefited more from read trimming than others, leading to a slight increase in sensitivity as indicated by read count. Most methods showed a median expression of 14 to 20 reads; the least sensitive methods were detect circRNAs from chimeric reads (DCC) [42], circRNA_finder [43], and UROBORUS [44], with median readings per circRNA of 11, 9, and 5, respectively. The initial quality of the RNA-seq dataset affects sensitivity levels. Circall [71] is a newly developed tool that offers better true positive detection than CIRI2 [39], CIRCexplorer [21], MapSplice [66], and Fnd_circ [41] when it comes to detecting EcircRNAs from PE RNA-Seq data.

6 Conclusions

A growing class of noncoding RNAs known as circRNAs is being recognized for its role in biological functions. Even with the abundance of bioinformatics tools for circRNA detection and quantification, there are still difficulties in circRNA research because there is no standard nomenclature system and because different species, genes, tissues, and developmental stages exhibit different expression patterns. The main goal of ongoing research is to clarify the mechanisms of circRNA action, which include peptide coding, interactions, and sponge effects. A detailed examination of circRNAs is hindered by the limits of current bioinformatics techniques, which include the incapacity to accurately assess expression levels based on the entire circRNA sequence and to take into consideration mature isoform sequences that originate from the same BSJ. Direct RNA-seq may be able to address these issues. To advance circRNA research, more effective data integration and customized algorithms are required. Furthermore, developing a common nomenclature system is necessary to enable reliable comparisons of circRNA expression in many illnesses, tissues, and circumstances. It is crucial to note that numerous articles have investigated different aspects of computational techniques for the in silico functional study and detection of circRNAs. However, in contrast to these assessments, our analysis includes technical specifics and offers a more approachable summary of the top circRNA detection tools. The increasing role that circRNA is playing in both health and sickness highlights the need for more specialized circRNA tools to meet these new needs as well as possible future uses.

References

1. Sanger HL, Klotz G, Riesner D, Gross HJ, Kleinschmidt AK (1976) Viroids are single-stranded covalently closed circular RNA molecules existing as highly base-paired rod-like structures. Proc Nat I Acad Sci 73(11):3852–3856. https://doi.org/10.1073/pnas.73.11.3852

2. Patop IL, Wust S, Kadener S (2019) Past, present, and future of circRNAs. EMBO J 38(16):e100836. https://doi.org/10.15252/embj.2018100836
3. Westholm JO, Miura P, Olson S, Shenker S, Joseph B, Sanfilippo P, Celniker SE, Graveley BR, Lai EC (2014) Genome-wide analysis of drosophila circular RNAs reveals their structural and sequence properties and age-dependent neural accumulation. Cell Rep 9(5):1966–1980. https://doi.org/10.1016/j.celrep.2014.10.062
4. Lu YL, Lin C, Zhang J, Shen Z (2021) Translation role of circRNAs in cancers. J Clin Lab Anal 35(7):e23866. https://doi.org/10.1002/jcla.23866
5. Ivanov A, Memczak S, Wyler E, Torti F, Porath HT, Orejuela MR, Piechotta M, Levanon EY, Landthaler M, Dieterich C, Rajewsky N (2015) Analysis of intron sequences reveals hallmarks of circular RNA biogenesis in animals. Cell Rep 10(2):170–177. https://doi.org/10.1016/j.celrep.2014.12.019
6. Kristensen LS, Andersen MS, Stagsted LVW, Ebbesen KK, Hansen TB, Kjems J (2019) The biogenesis, biology and characterization of circular RNAs. Nat Rev Genet 20(11):675–691. https://doi.org/10.1038/s41576-019-0158-7
7. Meng S, Zhou H, Feng Z, Xu Z, Tang Y, Li P, Wu M (2017) CircRNA: functions and properties of a novel potential biomarker for cancer. Mol Cancer 16(1):94. https://doi.org/10.1186/s12943-017-0663-2
8. Salmena L, Poliseno L, Tay Y, Kats L, Pandolfi PP (2011) A ceRNA hypothesis: the Rosetta stone of a hidden RNA language. Cell 146(3):353–358. https://doi.org/10.1016/j.cell.2011.07.014
9. Lu M (2020) Circular RNA: functions, applications and prospects. ExRNA 2:1–7. https://doi.org/10.1186/s41544-019-0046-5
10. Salzman J (2016) Circular RNA expression: its potential regulation and function. Trends Genet 32(5):309–316. https://doi.org/10.1016/j.tig.2016.03.002
11. Wang X, Dong J, Li X, Cheng Z, Zhu Q (2021) CPSF4 regulates circRNA formation and microRNA mediated gene silencing in hepatocellular carcinoma. Oncogene 40(25):4338–4351. https://doi.org/10.1038/s41388-021-01867-6
12. Zhang C, Hu J, Yu Y (2020) CircRNA is a rising Star in researches of ocular diseases. Front Cell Dev Biol 8:850. https://doi.org/10.3389/fcell.2020.00850
13. Goodall GJ, Wickramasinghe VO (2021) RNA in cancer. Nat Rev Cancer 21(1):22–36. https://doi.org/10.1038/s41568-020-00306-0
14. Santer L, Bar C, Thum T (2019) Circular RNAs: a novel class of functional RNA molecules with a therapeutic perspective. Mol Ther 27(8):1350–1363. https://doi.org/10.1016/j.ymthe.2019.07.001
15. Vromman M, Anckaert J, Bortoluzzi S, Buratin A, Chen CY, Chu Q, Chuang TJ, Dehghannasiri R, Dieterich C, Dong X, Flicek P, Gaffo E, Gu W, He C, Hoffmann S, Izuogu O, Jackson MS, Jakobi T, Lai EC, Nuytens J, Salzman J, Koref MS, Stadler P, Thas O, Eynde EV, Verniers K, Wen G, Westholm J, Yang L, Ye CY, Yigit N, Yuan GH, Zhang J, Zhao F, Vandesompele J (2023) Large-scale benchmarking of circRNA detection tools reveals large differences in sensitivity but not in precision. Nat Methods 20:1159–1169. https://doi.org/10.1038/s41592-023-01944-6
16. Yang L, Wilusz JE, Chen LL (2022) Biogenesis and regulatory roles of circular RNAs. Annu Rev Cell Dev Biol 38(1):263–289. https://doi.org/10.1146/annurev-cellbio-120420-125117
17. Zhao Q, Liu J, Deng H, Ma R, Liao JY, Liang H, Hu J, Li J, Guo Z, Cai J, Xu X, Gao Z, Su S (2020) Targeting mitochondria-located circRNA SCAR alleviates NASH via reducing mROS output. Cell 183(1):76–93.e22. https://doi.org/10.1016/j.cell.2020.08.009
18. Tang X, Ren H, Guo M, Qian J, Yang Y, Gu C (2021) Review on circular RNAs and new insights into their roles in cancer. Comput Struct Biotechnol J 19:910–928. https://doi.org/10.1016/j.csbj.2021.01.018
19. Chen LL (2020) The expanding regulatory mechanisms and cellular functions of circular RNAs. Nat Rev Mol Cell Biol 21(8):475–490. https://doi.org/10.1038/s41580-020-0243-y
20. Chen LL, Bindereif A, Bozzoni I, Chang HY, Matera AG, Gorospe M, Hansen TB, Kjems J, Ma XK, Pek JW, Rajewsky N, Salzman J, Wilusz JE, Yang L, Zhao F (2023) A guide to naming eukaryotic circular RNAs. Nat Cell Biol 25(1):1–5. https://doi.org/10.1038/s41556-022-01066-9
21. Zhang XO, Wang HB, Zhang Y, Lu X, Chen LL, Yang L (2014) Complementary sequence-mediated exon circularization. Cell 159(1):134–147. https://doi.org/10.1016/j.cell.2014.09.001
22. Conn SJ, Pillman KA, Toubia J, Conn VM, Salmanidis M, Phillips CA, Roslan S, Schreiber AW, Gregory PA, Goodall GJ (2015) The RNA binding protein quaking regulates formation of circRNAs. Cell 160(6):1125–1134. https://doi.org/10.1016/j.cell.2015.02.014
23. Ashwal-Fluss R, Meyer M, Pamudurti NR, Ivanov A, Bartok O, Hanan M, Evantal N, Memczak S, Rajewsky N, Kadener S (2014) circRNA biogenesis competes with pre-mRNA splicing. Mol Cell 56(1):55–66. https://doi.org/10.1016/j.molcel.2014.08.019
24. Chen L, Wang Y, Lin J, Song Z, Wang Q, Zhao W, Yan W, Xiu X, Deng Y, Li X, Li Q, Wang X, Li J, Liu X, Liu K, Zhou J, Li K, Liu Y, Liao S, Deng Q, Xu C, Sun Q, Wu S, Zhang K, Guan MX, Zhou T, Sun F, Cai X, Huang C, Shan G (2022) Exportin 4 depletion leads to nuclear accumulation of a subset of circular RNAs. Nat Commun 13(1):5769. https://doi.org/10.1038/s41467-022-33356-z
25. Zhao J, Tao Y, Zhou Y, Qin N, Chen C, Tian D, Xu L (2015) MicroRNA-7: a promising new target in cancer therapy. Cancer Cell Int 15:103. https://doi.org/10.1186/s12935-015-0259-0
26. Li XX, Xiao L, Chung HK, Ma XX, Liu X, Song JL, Jin CZ, Rao JN, Gorospe M, Wang JY (2020) Interaction between HuR and circPABPN1 modulates

autophagy in the intestinal epithelium by altering ATG16L1 translation. Mol Cell Biol 40(6):e00492–e0e519. https://doi.org/10.1128/MCB.00492-19
27. Wang X, Chen T, Li C, Li W, Zhou X, Li Y, Luo D, Zhang N, Chen B, Wang L, Zhao W, Fu S, Yang Q (2022) CircRNA-CREIT inhibits stress granule assembly and overcomes doxorubicin resistance in TNBC by destabilizing PKR. J Hematol Oncol 15(1):122. https://doi.org/10.1186/s13045-022-01345-w
28. Yi Z, Qu L, Tang H, Liu Z, Liu Y, Tian F, Wang C, Zhang X, Feng Z, Yu Y, Yuan P, Yi Z, Zhao Y, Wei W (2022) Engineered circular ADAR-recruiting RNAs increase the efficiency and fidelity of RNA editing in vitro and in vivo. Nat Biotechnol 40(6):946–955. https://doi.org/10.1038/s41587-021-01180-3
29. Tabak HF, Van der Horst G, Smit J, Winter AJ, Mul Y, Groot Koerkamp MJ (1988) Discrimination between RNA circles, interlocked RNA circles and lariats using two-dimensional polyacrylamide gel electrophoresis. Nucleic Acids Res 16(14A):6597–6660. https://doi.org/10.1093/nar/16.14.6597
30. Szabo L, Salzman J (2016) Detecting circular RNAs: bioinformatic and experimental challenges. Nat Rev Genet 17(11):679–692. https://doi.org/10.1038/nrg.2016.114
31. Camilo R, Juan PS, Nicolás S, Vinicius MC (2023) Computational approaches for circRNAs prediction and in silico characterization. Brief Bioinform 24:(3):bbad154. https://doi.org/10.1093/bib/bbad154
32. Gao Y, Zhao F (2018) Computational strategies for exploring circular RNAs. Trends Genet 34(5):389–400. https://doi.org/10.1016/j.tig.2017.12.016
33. Zucko D, Boris-Lawrie K (2020) Circular RNAs are regulators of diverse animal transcriptomes: one health perspective. Front Genet 11:999. https://doi.org/10.3389/fgene.2020.00999
34. Holdt LM, Stahringer A, Sass K, Pichler G, Kulak NA, Wilfert W, Kohlmaier A, Herbst A, Northoff BN, Nicolaou A, Gabel G, Beutner F, Scholz M, Thiery J, Musunuru K, Krohn K, Mann M, Teupser D (2016) Circular non-coding RNA ANRIL modulates ribosomal RNA maturation and atherosclerosis in humans. Nat Commun 7:12429. https://doi.org/10.1038/ncomms12429
35. Vausort M, Salgado-Somoza A, Zhang L, Leszek P, Scholz M, Teren A, Burkhardt R, Thiery J, Wagner DR, Devaux Y (2016) Myocardial infarction-associated circular RNA predicting left ventricular dysfunction. J Amer College Cardiol 68(11):1247–1248. https://doi.org/10.1016/j.jacc.2016.06.040
36. Stoddard BL (2014) Homing endonucleases from mobile group I introns: discovery to genome engineering. Mob DNA 5(1):7. https://doi.org/10.1186/1759-8753-5-7
37. Gao Y, Wang J, Zhao F (2015) CIRI: an efficient and unbiased algorithm for de novo circular RNA identification. Genome Biol 16(1):4. https://doi.org/10.1186/s13059-014-0571-3
38. Guo JU, Agarwal V, Guo H, Bartel DP (2014) Expanded identification and characterization of mammalian circular RNAs. Genome Biol 15(7):409. https://doi.org/10.1186/s13059-014-0409-z
39. Gao Y, Zhang J, Zhao F (2018) Circular RNA identification based on multiple seed matching. Brief Bioinform 19(5):803–810. https://doi.org/10.1093/bib/bbx014
40. Li H (2013) Aligning sequence reads, clone sequences and assembly contigs with BWA-MEM. arXiv:1303.3997. https://doi.org/10.48550/arXiv.1303.3997
41. Memczak S, Jens M, Elefsinioti A, Torti F, Krueger J, Rybak A, Maier L, Mackowiak SD, Gregersen LH, Munschauer M, Loewer A, Ziebold U, Landthaler M, Kocks C, Noble F, Rajewsky N (2013) Circular RNAs are a large class of animal RNAs with regulatory potency. Nature 495(7441):333–338. https://doi.org/10.1038/nature11928
42. Cheng J, Metge F, Dieterich C (2016) Specific identification and quantification of circular RNAs from sequencing data. Bioinformatics 32(7):1094–1096. https://doi.org/10.1093/bioinformatics/btv656
43. Westholm JO, Miura P, Olson S, Shenker S, Joseph B, Sanfilippo P, Celniker SE, Graveley BR, Lai EC (2014) Genome-wide analysis of drosophila circular RNAs reveals their structural and sequence properties and age-dependent neural accumulation. Cell Rep 9(5):1966–1980. https://doi.org/10.1016/j.celrep.2014
44. Song X, Zhang N, Han P, Moon BS, Lai RK, Wang K, Lu W (2016) Circular RNA profile in gliomas revealed by identification tool UROBORUS. Nucleic Acids Res 44(9):e87. https://doi.org/10.1093/nar/gkw075
45. Trapnell C, Pachter L, Salzberg SL (2009) TopHat: discovering splice junctions with RNA-Seq. Bioinformatics 25(9):1105–1111. https://doi.org/10.1093/bioinformatics/btp120
46. Langmead B, Trapnell C, Pop M, Salzberg SL (2009) Ultrafast and memory-efficient alignment of short DNA sequences to the human genome. Genome Biol 10(3):R25. https://doi.org/10.1186/gb-2009-10-3-r25
47. Ma XK, Wang MR, Liu CX, Dong R, Carmichael GG, Chen LL, Yang L (2019) CIRCexplorer3: a CLEAR pipeline for direct comparison of circular and linear RNA expression. Genomics Proteomics Bioinformatics 17(5):511–521. https://doi.org/10.1016/j.gpb.2019.11.004
48. Kim D, Paggi JM, Park C, Bennett C, Salzberg SL (2019) Graph-based genome alignment and genotyping with HISAT2 and HISAT-genotype. Nat Biotechnol 37(8):907–915. https://doi.org/10.1038/s41587-019-0201-4
49. Dobin A, Davis CA, Schlesinger F, Drenkow J, Zaleski C, Jha S, Batut P, Chaisson M, Gingeras TR (2013) STAR: ultrafast universal RNA-seq aligner. Bioinformatics 29(1):15–21. https://doi.org/10.1093/bioinformatics/bts635

50. Zhang J, Chen S, Yang J, Zhao F (2020) Accurate quantification of circular RNAs identifies extensive circular isoform switching events. Nat Commun 11(1):90. https://doi.org/10.1038/s41467-019-13840-9
51. Li H, Durbin R (2009) Fast and accurate short read alignment with burrows-wheeler transform. Bioinformatics 25(14):1754–1760. https://doi.org/10.1093/bioinformatics/btp324
52. Langmead B, Salzberg SL (2012) Fast gapped-read alignment with bowtie 2. Nat Methods 9(4):357–359. https://doi.org/10.1038/nmeth.1923
53. Hansen TB (2018) Improved circRNA identification by combining prediction algorithms. Front Cell Dev Biol 6:20. https://doi.org/10.3389/fcell.2018.00020
54. You X, Conrad TO (2016) Acfs: accurate circRNA identification and quantification from RNA-Seq data. Sci Rep 6:38820. https://doi.org/10.1038/srep38820
55. Gaffo E, Bonizzato A, Kronnie GT, Bortoluzzi S (2017) CirComPara: a multi-method comparative bioinformatics pipeline to detect and study circRNAs from RNA-seq data. Non-Coding RNA 3(1):8. https://doi.org/10.3390/ncrna3010008
56. Li L, Zheng YC, Kayani MUR, Xu W, Wang GQ, Sun P, Ao N, Zhang LN, Gu ZQ, Wu LC, Zhao HT (2017) Comprehensive analysis of circRNA expression profiles in humans by RAISE. Int J Oncol 51(6):1625–1638. https://doi.org/10.3892/ijo.2017.4162
57. Li L, Bu D, Zhao Y (2019) CircRNAwrap – a flexible pipeline for circRNA identification, transcript prediction, and abundance estimation. FEBS Lett 593(11):1179–1189. https://doi.org/10.1002/1873-3468.13423
58. Jeck WR, Sorrentino JA, Wang K, Slevin MK, Burd CE, Liu J, Marzluff WF, Sharpless NE (2013) Circular RNAs are abundant, conserved, and associated with ALU repeats. RNA 19(2):141–157. https://doi.org/10.1261/rna.035667.112
59. Pan X, Xiong K (2015) PredcircRNA: computational classification of circular RNA from other long non-coding RNA using hybrid features. Mol BioSyst 11(8):2219–2226. https://doi.org/10.1039/c5mb00214a
60. Pan X, Xiong K, Anthon C, Hyttel P, Freude KK, Jensen LJ, Gorodkin J (2018) WebCircRNA: classifying the circular RNA potential of coding and noncoding RNA. Genes 9(11):536. https://doi.org/10.3390/genes9110536
61. Wang J, Wang L (2019) Deep learning of the back-splicing code for circular RNA formation. Bioinformatics 35(24):5235–5242. https://doi.org/10.1093/bioinformatics/btz382
62. Liu Z, Han J, Lv H, Liu J, Liu R (2016) Computational identification of circular RNAs based on conformational and thermodynamic properties in the flanking introns. Comput Biol Chem 61(1):221–225. https://doi.org/10.1016/j.compbiolchem.2016.02.003
63. Wang X, Liu Y, Li J, Wang G (2022) StackCirRNAPred: computational classification of long circRNA from other lncRNA based on stacking strategy. BMC Bioinf 23(1):563. https://doi.org/10.1186/s12859-022-05118-7
64. Niu M, Zhang J, Li Y, Wang C, Liu Z, Ding H, Zou Q, Ma Q (2020) CirRNAPL: a web server for the identification of circRNA based on extreme learning machine. Comput Struct Biotechnol J 18(1):834–842. https://doi.org/10.1016/j.csbj.2020.03.028
65. Zeng X, Lin W, Guo M, Zou Q (2017) A comprehensive overview and evaluation of circular RNA detection tools. PLoS Comput Biol 13(6):e1005420. https://doi.org/10.1371/journal.pcbi.1005420
66. Wang K, Singh D, Zeng Z, Coleman SJ, Huang Y, Savich GL, He X, Mieczkowski P, Grimm S, Perou CM, MacLeod JN, Chiang DY, Prins JF, Liu J (2010) MapSplice: accurate mapping of RNA-seq reads for splice junction discovery. Nucleic Acids Res 38(18):e178. https://doi.org/10.1093/nar/gkq622
67. Hansen TB, Venø MT, Damgaard CK, Kjems J (2016) Comparison of circular RNA prediction tools. Nucleic Acids Res 44(6):e58. https://doi.org/10.1093/nar/gkv1458
68. Hoffmann S, Otto C, Kurtz S, Sharma CM, Khaitovich P, Vogel J, Stadler PF, Hackermüller J (2009) Fast mapping of short sequences with mismatches, insertions and deletions using index structures. PLoS Comput Biol 5(9):e1000502. https://doi.org/10.1371/journal.pcbi.1000502
69. Chuang TJ, Wu CS, Chen CY, Hung LY, Chiang TW, Yang MY (2016) NCLscan: accurate identification of non-co-linear transcripts (fusion, trans-splicing and circular RNA) with a good balance between sensitivity and precision. Nucleic Acids Res 44(3):e29. https://doi.org/10.1093/nar/gkv1013
70. Izuogu OG, Alhasan AA, Alafghani HM, Santibanez-Koref M, Elliott DJ, Jackson MS (2016) PTESFinder: a computational method to identify post-transcriptional exon shuffling (PTES) events. BMC Bioinf 17:31. https://doi.org/10.1186/s12859-016-0881-4
71. Nguyen DT, Trac QT, Nguyen TH, Nguyen HN, Ohad N, Pawitan Y, Vu TN (2021) Circall: fast and accurate methodology for discovery of circular RNAs from paired-end RNA-sequencing data. BMC Bioinf 22:495. https://doi.org/10.1186/s12859-021-04418-8

Online Databases in Circular RNAs

4

Sachiko Kuwamoto-Imanishi and Hodaka Fujii

Abstract

Circular RNAs (circRNAs) have been attracting attention as biomarkers of various diseases, including cancer, because, as non-coding RNA (ncRNA)/protein sponges, they regulate gene expression and can be detected in various body fluids. Several comprehensive databases have been published to allow easy identification of target circRNAs.

In all, 43 circRNA-related databases have been established and are listed in Database Commons since the development of the first comprehensive database of circRNAs, Circ2Traits, in 2013. Some of these databases, including Circ2Traits, provide annotation, ncRNA/protein–circRNA interactions and disease associations, and information on protein synthesis potentials and conservation among species. Additionally, databases that include analyzed RNA sequencing (RNA-seq) data or manually curated literature reporting the expression of circRNAs in specific tissues, organs, and body fluids have been established, making the exploration of target circRNAs easier. Currently, in addition to animals, plant circRNAs are being analyzed, and they cover a surprisingly large number of species.

In this chapter, we summarize a selection of popular databases [circBase, CircInteractome, CircNet, CircR2Disease, the tissue-specific circular RNA database (TSCD), and the cancer-specific circRNA database (CSCD)] discussed in the previous edition and introduce some new circRNA databases listed in Database Commons. Furthermore, we discuss some reported challenges related to the nomenclature and updating of those databases.

S. Kuwamoto-Imanishi · H. Fujii (✉)
Department of Biochemistry and Genome Biology, Hirosaki University Graduate School of Medicine, Hirosaki, Aomori Prefecture, Japan
e-mail: hodaka@hirosaki-u.ac.jp

Keywords

circRNA · Databases · Interactome · Ontology · Conservation · Sequence annotation · Function annotation · Disease association · Bioinformatics

Abbreviations

BLAST	Basic local alignment search tool
BSJ	Backsplice junction
CLIP-seq	Cross-linking immunoprecipitation sequence
CSCD	Cancer-specific circRNA database
circRNA	Circular RNA

J. Xiao (ed.), *Advances in Circular RNAs*, Advances in Experimental Medicine and Biology 1485,
https://doi.org/10.1007/978-981-96-9428-0_4

ENCORI/starBase	Encyclopedia of RNA Interactomes/starBase
FPKM	Fragments per kilobase of transcript per million mapped reads
FPM	Fragments mapped to BSJs per million mapped fragments
GO	Gene Ontology
miRNA	MicroRNA
MNDR	Mammalian ncRNA-Disease Repository
IRES	Internal ribosome entry site
lncRNA	Long non-coding RNA
KEGG	Kyoto Encyclopedia of Genes and Genomes
MS	Mass spectrometry
m^6A	N^6-methyladenosine
ncRNA	Non-coding RNA
ORF	Open reading frame
RBP	RNA-binding protein
siRNA	Small interfering RNA
TSCD	Tissue-specific circular RNA database

1 Introduction

Circular RNAs (circRNAs) regulate gene expression and protein activity either directly or indirectly via microRNAs (miRNAs) [1–4] and RNA-binding proteins (RBPs) [5, 6]. Some circRNAs promote or suppress oncogenesis [7, 8] and the immune response [9] and can be detected in blood, urine, saliva, and exosomes [10–15]. Owing to these characteristics, circRNAs have been considered potential novel biomarkers or therapeutic agents for various diseases, including cancer and cardiovascular and autoimmune diseases [16–18].

As circRNAs have been identified in plants, fungi, and protists other than in humans and mice [19], circRNA databases cover numerous organisms. Moreover, complicated interactions with ncRNAs and proteins have been revealed, resulting in large amounts of circRNA data. Therefore, comprehensive databases that integrate information on circRNAs facilitate the exploration of circRNAs of interest. Since the development of Circ2Traits [20] (not available at present) in 2013, the first comprehensive circRNA database, the number of databases specializing in circRNAs has continued to increase, and some ncRNA databases have added entries of circRNAs as they have been renewed [21–23].

As of November 2023, 43 circRNA databases are listed in Database Commons, a "curated catalog of worldwide biological databases" (https://ngdc.cncb.ac.cn/databasecommons/) [24]. CircRNA databases are accurately maintained through either manual curation or the prediction of circRNAs from RNA-seq data using algorithms or both.

Some of these databases, such as Circ2Traits, provide information about ncRNA–circRNA interactions, protein–circRNA interactions, disease associations, protein synthesis potential, conservation across species, and subcellular localization [25–27]. Others store the results of expression analyses in specific tissues, organs, and body fluids [25, 28, 29]. The establishment of a new circRNA database that includes all this information in one place will further help the search for target circRNAs.

To efficiently search for circRNAs with large amounts of data, information is subdivided into several categories. In this chapter, we have classified annotations into six categories as follows, and each is discussed below.

Sequence/function annotation
Interactome
Expression
Disease associations
Conservation
Others

Some databases feature more than one annotation. For example, circAtlas provides all six of the above annotations [30]. In addition to the major databases described in previous editions, the 16 databases described in this chapter have a median z-index ≥6.7 (calculated by dividing total citations by database age) in Database Commons [24]. Plant-specific databases are discussed separately in Sect. 8.

We have also discussed some recently reported challenges and prospects (e.g., nomenclature and update frequency). All databases described here are listed in Table 4.1.

Table 4.1 Available online circRNA databases

Database name	Version	Organisms (in total)	CircRNA detection tools/methods	Main function	Download format	z-index	Latest update	Remarks
ENCORI/ starBase	–	23 species	Manually curated	Sect. 2	API	228.67	2023	circBase IDs are used
circBase	–	6 species	Manually curated	Sect. 2	BED FASTA	86.89	2017	Original IDs are used
LncRNADisease	3.0	4 species	Manually curated	Sect. 5	TSV XLSX	67.6	2023	circBase IDs and symbols are used
CIRCpedia	2.0	6 species	CIRCexploer2 MapSplice	Sect. 2 Sect. 4	CSV FASTA	66	2018	Original IDs are used
exoRBase	2.0	Human	ASJACIRI2	Sect. 2 Sect. 4	TXT	46.6	2021	circBase IDs and original IDs are used
Lnc2Cancer	3.0	Human	Manually curated	Sect. 5	CSV XLSX	42	2020	Symbols are used as circRNA IDs
CSCD	2.0	Human	CIRI2 find_circ circRNA_finder CIRCexplorer	Sect. 2 Sect. 5	TXT	37.6	2018	Loci were used as circRNA IDs
circAtlas	3.0	10 species	CIRI2 DCC CIRCexplorer2 MapSplice	Sect. 2—Sect. 7	BED	32.67	2023	Original IDs are used
circRNADb	–	Human	Manually curated	Sect. 2	FASTA GTF	28.86	2016	Filtered raw circRNA dataset obtained from related literature, supported by only one read. Algorithms used are undescribed
CircNet	2.0	Human	CIRI2 DCC CIRCexplorer2 MapSplice	Sect. 2 Sect. 4 Sect. 5	CSV SVG (boxplots that display expression)	27.43	2021	circBase IDs and loci are available to search
TSCD	–	Human Mouse	CIRI find_circ circRNAfinder	Sect. 2	TXT	27	2017	Loci were used as circRNA IDs
DeepBase	3.0	14 species	Downloaded from circBase find_circ	Sect. 2 Sect. 4	CSV	18.92	2021	Original IDs are used
RNALocate	2.0	104 species	Manually curated	Sect. 7	TXT XLSX FASTA RAR	16.5	2021	circBase IDs and exoRBase IDs are used

(continued)

Table 4.1 (continued)

Database name	Version	Organisms (in total)	CircRNA detection tools/methods	Main function	Download format	z-index	Latest update	Remarks
RNADisease	4.0	117 species	Manually curated	Sect. 5	XLSX CSV	16.1	2022	circBase IDs, symbols, and original IDs are used
PlantcircBase	–	16 species	Manually curated CIRI CIRCexplorer PcircRNA_finder circRNA_ finderKNIFE find_circ BLASTcirc	Sect. 2 Sect. 6 Sect. 8	FASTATXT	13.67	2022	Original IDs are used
EVLncRNAs	3.0	162 species	Manually curated	Sect. 3 Sect. 5	XLS	13.4	2023	Original IDs are used
CircR2Disease	–	Human	Manually curated	Sect. 5	XLSX	13.4	2018	circBase IDs are used
NPInter	5.0	More than ten species	Manually curated	Sect. 3 Sect. 5	TXT	13	2022	circBase IDs and original IDs are used
TransCirc	–	Human	Downloaded from circAtlas CIRI-AS CIRI-full	Sect. 2	TSV FASTA	11	2021	circAtlas IDs and original IDs are used
AtCircDB	–	*Arabidopsis*	CIRCexplorer2, CIRI2	Sect. 2 Sect. 8	XLSX	8.43	2017	Original IDs are used
NoncoRNA	–	Human	Manually curated	Sect. 5	TXT XLS	6.67	2019	circBase IDs are used
CircInteractome	–	Human	Downloaded from circBase	Sect. 2 Sect. 7	XLS	4.2	2020	circBase IDs are used
CircFunBase	–	15 species	Manually curated	Sect. 2 Sect. 5 Sect. 6	FASTA API		2019	circBase IDs are used
Circbank	–	Human	Downloaded from circBase	Sect. 2 Sect. 4 Sect. 6	FASTA BED		2019	circBase IDs and original IDs are used

CropCircDB	–	Rice Corn	CIRCexplorer2 CIRI2	Sect. 2 Sect. 8	CSV		2018	Original IDs are used

Column descriptions: "Database name" provides the name of the online circRNA database available. "Organisms" provides the organisms covered by the database. "circRNA detection tools/methods" provides the methods and algorithms used to identify circRNAs. "Main functions" describes the main functions provided by the databases. Each number corresponds to the section numbers. "Download format" provides the data format that the user can select. "z-index" provides the index related to the citation described in "Database Commons." "Latest update" refers to the year in which the database was last updated. "Remarks" provides useful information about the database.

CSCD cancer-specific circRNA Database, *TSCD* tissue-specific circular RNA database

2 Sequence/Function Annotation

In this Sect. 2 we describe databases that integrate information about circRNA sequences, including their loci, their original genes, their spliced lengths, the number of backsplicing junction (BSJ) reads, their binding sites for ncRNAs and proteins, and their functional elements, such as their internal ribosome entry sites (IRESs), open reading frames (ORFs), and N^6-methyladenosine (m^6A) modification. CircRNA functions, such as miRNA/protein sponge and protein coding, can be predicted by annotating these sequential elements. As such, some sequence annotation databases also offer circRNA function prediction. We describe circBase [31], CIRCpedia [28, 32], CircFunBase [33], CircNet [34], CircRNADb [35], Circbank [36], tissue-specific circular RNA database (TSCD) [37], ENCORI/starBase (Encyclopedia of RNA Interactomes/starBase) [38, 39], CircInteractome [40], and TransCirc [41].

The circBase database was established in 2014 and specializes in sequence annotation. It collects circRNAs curated manually based on nine large-scale studies and provides genomic coordinates, BSJ sequences, and functional elements for various cell lines or cells in six organisms. The evidence for their expression can be accessed, downloaded, and searched in a genomic context [31].

In addition to the general annotations described in circBase, some databases provide more characteristic information. In TSCD, human and mouse tissue-specific circRNAs curated by three algorithms (CIRI, circRNA_finder [42], and find_circ) can be explored. Furthermore, general information, circRNA types (exonic or intronic), and derived regions [mRNAs or long non-coding RNAs (lncRNAs)] are featured in this database. Additionally, lists of putative miRNA-binding sites, predicted by TargetScan [43], are available.

CircFunBase describes identification methods and detection tools. CircRNADb, manually curated from four large-scale studies, focuses on exonic circRNAs and provides the coding potential predicted by calculating the R-score and the existence of a pseudoknot associated with IRES prediction. The Circbank database also provides information on coding potential, which is calculated using the Coding-Potential Assessment Tool [44]. In TransCirc, which specializes in the coding potential of circRNAs, the coding potential is derived by integrating predictions [ORF and IRES sequences, translation initiation sites, m^6A sites, and mass spectrometry (MS) data].

CircInteractome provides a bioinformatics prediction of miRNA-binding sites using TargetScan and ranks the binding sites according to context+ scores [43]. In this site, circRNAs are curated based on circBase. RBP-binding sites are predicted based on cross-linking immunoprecipitation sequence (CLIP-seq) datasets. The authors of the database emphasize that the binding sites identified are predicted from CLIP-seq datasets and require experimental validation.

ENCORI/starBase is a comprehensive ncRNA database that includes circRNA. The miRNA-binding sites are predicted by four algorithms (miRanda-mirSVR [45], TargetScan, PITA [46], Pictar 2.0 [47], and RNA22 [48]), and RBP-binding sites are derived from published CLIP-seq datasets. CircBase IDs have been assigned to circRNAs on this page so that they can be easily linked to general annotation information.

CircNet specializes in human cancer circRNAs, which are bioinformatically curated using four algorithms (CIRI2 [49, 50], find_circ [3], CIRCexplor2 [51], and DCC [52]) to make the results more reliable (Table 4.1). Putative ORFs, binding miRNAs, and miRNA-/RBP-binding sites are provided. The circBase IDs are assigned to identified circRNAs. The miRNA-binding sites are predicted using PITA, miRanda [53], and TargetScan. Unfortunately, network rendering and miRNA-binding site information are not available online (miRNA-binding site information can be downloaded).

3 Interactome

CircRNAs can form large networks with ncRNAs (miRNAs and lncRNAs), mRNAs, DNAs, and proteins. Therefore, it is important to capture the network comprehensively. This section describes databases that render networks such as NPInter [26] and circAtlas [30,25].

NPInter collects RNA interactions (including circRNA–miRNA, circRNA–protein, and circRNA–mRNA) and disease and functional annotations identified experimentally by literature mining and in the data of high-throughput sequencing experiments. The curators have compiled information about interactions according to RNA type, description, related disease annotation, and references on the "Molecule Profile" page. The network is visualized as a node-link diagram. It is unique, in that an original ID given by the curators is assigned to each interaction.

circAtlas is a bioinformatically curated database constructed using 2,674 RNA-seq datasets and contains datasets of 33 normal tissues collected from six vertebrate species. The curation tool is the same algorithm used in CircNet (Table 4.1). In circAtlas, mRNA–circRNA, circRNA–miRNA, and circRNA–RBP interaction networks are integrated and can be visualized as a node-link diagram. Unlike many databases, this network is used to predict the potential functions of circRNAs based on GO (Gene Ontology) and KEGG (Kyoto Encyclopedia of Genes and Genomes) pathway analysis.

4 Expression

Since circRNAs have attracted attention as biomarkers, the identification of cell- and tissue-specific circRNAs and analysis of their expression in normal and diseased samples have begun. This section focuses on databases that analyze the expression levels of each circRNA.

CIRCpedia [28, 32], DeepBase [54, 55], CircNet, circAtlas, and exoRBase [29] provide information on circRNA expression levels and their analysis.

The DeepBase database was designed to facilitate the analysis of the expression, evolution, and function of various ncRNAs, including circRNAs, by mining over 46,000 RNA-seq datasets from tissue, cancer, and exosome samples. The targets of DeepBase are mainly lncRNAs and small RNAs, and the information related to circRNAs is limited to expression in human cell lines and exosomes. The expression levels of each circRNA are calculated from RNA-seq data as fragments per kilobase of transcript per million mapped reads (FPKM). The expression levels in different cell lines and exosomes are normalized by z-score or mean [56], and the results are shown as a heatmap.

The circRNAs in CIRCpedia are curated using CIRCexplorer2 and MapSplice [57] based on more than 180 RNA-seq datasets. Expression is presented as fragments mapped to BSJs per million mapped fragments (FPM). Cell lines and species in the list can be selected for analysis, and results are outputted as a plot or in heatmap format.

The circNet database is also distinctive in displaying the expression level of each miRNA that binds to a target circRNA as a boxplot (available for download as an .svg file). The circAtlas database contains the expression information of circRNAs, and users can search for not only the expression levels of target circRNAs in each tissue as a boxplot but also circRNAs with high expression levels in each tissue. circAtlas implements FPM as a quantitative value for circRNA expression levels.

The exoRBase database is a repository of RNAs that localize to extracellular vesicles obtained from human blood, urine, cerebrospinal fluid, and bile samples. CircRNAs were identified and annotated using the Assembling Splice Junctions Analysis program [58], CIRI2, and circBase, and a quantitative value for circRNA expression is shown as counts per million (CPM). Some features allow the expression levels in each sample to be displayed as a heat map or a boxplot.

5 Disease Associations

CircRNAs have been reported to be associated with cancers, autoimmune diseases, Alzheimer's disease [59], non-alcoholic fatty liver disease [60], and others. Moreover, owing to their stability [3] and presence in various body fluids, they have drawn attention as potential biomarkers [18].

However, experimentally validated circRNA–disease associations have not been completely integrated. Some circRNA–disease-associated databases are linked to disease ontology databases and are divided into multiple and specific diseases, such as the cancer database.

The databases Circ2Disease [61], circRNADisease [62], CircR2Disease [63], circAtlas, NPInter, LncRNADisease [23, 64], and RNADisease [65] accumulate associations with multiple diseases through manual curation based on experimentally confirmed reports. Circ2Disease can be used to search for circRNAs associated with 54 diseases and allows the rendering of the circRNA–miRNA–protein (gene)–disease network, and evidential references are included with the general annotation.

CircRNADisease features circRNAs associated with 330 diseases in 12 species and furnishes descriptions of the functions of the circRNAs in different diseases. Circ2Mut, a tool for exploring circRNA mutations, can also be used.

CircR2Disease [63] and EVLncRNAs [22] are similar databases to Circ2Disease and CircRNAD and are curated using literature published through March 31, 2018, and May 15, 2020, respectively. In CircR2Disease, the circBase ID is tied to the MalaCards ID [66] as the disease ID. It has a simple page structure that only provides disease names, expression patterns, experimental methods for validation, and references for target circRNAs. EVLncRNAs are mainly used for searching lncRNAs, and they have identified 188 circRNAs associated with various diseases. Diseases and molecules related to target circRNA can be comprehensively browsed. These associations are visualized as a node-link diagram.

The circAtlas database lists diseases associated with circRNAs by integrating three databases, namely, circad [67], CircR2Disease, and circRNADisease. NPInter lists diseases associated with target circRNAs. The disease names refer to those in the Mammalian ncRNA-Disease Repository (MNDR) [68].

LncRNADisease is curated using experimentally validated literature to search for circRNA–disease associations in humans, mice, rats, and rabbits. Each circRNA–disease pair is assigned a confidence score based on experimental evidence. All circRNA–disease associations are classified into general and causal associations, evaluating how circRNAs can directly contribute to the development/progression of corresponding diseases. Additionally, disease names were referred from Disease Ontology (DO; http://www.disease-ontology.org) [69] and Medical Subject Headings (MeSH; https://www.ncbi.nlm.nih.gov/mesh) [70]. The relevance is shown in a node-link diagram, and the center node is the disease or target circRNA.

Like LncRNADisease, RNADisease is a disease-associated database curated using experimentally validated literature. It covers 18 types of RNAs in 117 species and provides RNA disease prediction tools. CircRNA ID disease information, including drug resistance, is retrieved from DO, MeSH, and KEGG DISEASE. CircRNA IDs on this site are referred from circBase IDs and Circbank IDs.

In contrast to the comprehensive disease databases described above, cancer-specific circRNA database (CSCD) [71, 72], CircNet, Lnc2Cancer [73], and NoncoRNA [74] are cancer-associated circRNA databases.

CSCD is a malignant disease-specific database that is bioinformatically curated. It is unique in predicting 207 RBP-binding sites from the CLIP-seq dataset and integrating immunohistochemistry staining and mass spectrometry (MS) data from HPA [75] and dbDEPC [76] to analyze RBP expression in different cancers. Rendering is expanded to general annotation, and much of the information is organized visually.

Lnc2Cancer is a manually curated database based on experimentally supported literature. The authors categorize human circRNAs based on the three following elements:

- Regulatory mechanisms (miRNAs, transcription factors, variants, methylation, and enhancers)
- Biological functions (cell proliferation, apoptosis, autophagy, epithelial–mesenchymal transition, immunity, and coding ability)
- Clinical applications (metastasis, recurrence, circulation, drug resistance, and prognosis)

As the main target of this site is lncRNA, the provided single-cell web and RNA-seq web tools do not support circRNA.

NoncoRNA is an ncRNA database, including human circRNAs associated with drug sensitivity/resistance in various human cancers, curated from experimentally validated literature. The database contains 8,233 entries of the associations between 5,568 ncRNAs and 154 drugs in 134 cancers. The authors unified the information, including drug IDs from DrugBank [77] and circRNA IDs from circBase. Disease Ontology and DrugBank were used to unify cancer and drug annotations. Some circRNA disease association evidence collected from different databases can be inconsistent due to different disease association criteria or differences in disease population samples [78].

6 Conservation

Circular RNAs, or circRNAs, are present in diverse organisms, from archaea and plants to yeast [19]. Interestingly, orthologs of circRNAs found in the human brain have also been identified in the brains of mice and flies [79]. Despite this, the evolutionary processes of circRNAs remain unclear. To shed light on the evolutionary process of circRNAs, conservation annotation has been applied to a wide range of species, which has been crucial for understanding circRNA evolution.

For those interested in advanced conservation analysis, the circAtlas database is an excellent resource. To predict ortholog circRNAs, 50 bp fragments on both sides of the BSJs of circRNAs were extracted and aligned with the full-length sequences of the orthologous genes and circRNA BSJ sequences using BLAT [80]. In addition, the authors have developed a unique scheme called the multiple conservation score scheme to assess the conservation of circRNAs at the species, tissue, and individual (or sample) levels. The database allows users to select any of ten species and obtain a list of circRNAs conserved among those species.

7 Others

Databases with functions other than those described above are introduced in this section.

7.1 Design of Primers for PCR Amplification and Small Interfering RNA (siRNA) Against CircRNAs

CircInteractome provides the design tools for divergent primer sets for PCR amplification and small interfering RNA (siRNA) against circRNAs for loss-of-function experiments. Each divergent primer set was designed using Primer3 [81] and the National Center for Biotechnology Information (NCBI) primer design tool, and a region-spanning BSJ was used as the template. siRNAs with a length of 21 nt were designed to target a BSJ spanning a region of 5–16 nt on either side of the junction. Using Basic Local Alignment Search Tool (BLAST), users need to confirm whether the siRNAs are specific. Links to IDT Inc. are also provided so that the designed siRNAs can be ordered (a link to Dharmacon Inc. is also provided but is inaccessible).

7.2 Secondary Structure Prediction

Based on the assembled complete circRNA sequences, circAtlas predicts the secondary structure using the RNA-fold algorithm of the Vienna RNA 2.0 package [82]. The predicted secondary structure is visualized.

7.3 Localization

RNALocate [27] is a repository of more than 213,000 RNA subcellular localization entries at 171 locations in 104 species. CircBase and exoRBase IDs can be used to identify circRNAs, and localization data are integrated by linking the IDs to five other databases, namely, CSCD,

EVmiRNA [83], exoRBase, PomBase [84], and TAIR [85].

8 Plant CircRNA Databases

PlantcircBase [86, 87] features manually curated circRNAs from validated literature and those that have been identified bioinformatically. It covers 21 plant species, and some primer sets of experimentally validated circRNAs are provided. In addition, "Visualize" and "Network" tools allow the visualization of the position of the circRNA in the genome and the prediction of circRNA–miRNA and circRNA–mRNA networks. Although PlantcircNet [88] and GreencircRNA [89] have been reported to offer similar functions, we experienced difficulty accessing them despite changing browsers (Chrome, Microsoft Edge, and Safari) and making multiple attempts.

Other noteworthy databases include CropCircDB [90] and AtCircDB [91]. CropCircDB is the only database that includes circRNAs related to stress responses, such as drought, salinity, and cold, and covers rice and maize.

AtCircDB is a specialized database featuring *Arabidopsis* circRNAs based on bioinformatics analysis and RNA-seq dataset annotation, which allows users to search for tissue-specific circRNAs. Although other species-specific databases are available, many have been established for less than 5 years, resulting in lower z-indexes. Therefore, we have excluded them from this chapter.

9 Challenges and Prospects of Circular RNA Databases

The development of databases has brought about a great deal of progress in the field, but, at the same time, it has given rise to several problems that need to be resolved. One significant issue is the lack of a uniform nomenclature, which has hindered data integration [78, 92]. Each database uses different identification systems, such as circBase IDs, chromosomal coordinates, or their nomenclature, making the integration of data across the platforms difficult. Although Circbank has proposed a new ID system based on the host gene and circRNA position [36], there is still no uniform nomenclature. The absence of a standardized nomenclature based on full-length sequences is a significant concern.

Another challenge is that circRNAs in some databases are putative, meaning that they have not been experimentally verified [92]. Similarly, the identification of miRNA-binding and protein-coding regions also needs experimental verification. Although more manually curated databases have appeared recently, they do not cover all circRNAs. Furthermore, the lack of sequence analysis using divergent or convergent primers makes it challenging to accurately predict and identify circRNAs [92].

The expression and detection of circRNA can vary considerably depending on a number of factors, including sample types, sequencing methods, and circRNA detection tools. Thus, detecting circRNA in a particular tissue or cell type does not mean it is specifically expressed in that tissue/cell type, making it difficult to compare data across multiple datasets [92].

Universal methods for identifying circRNAs have not yet been established, and it is crucial to confirm their existence using multiple validation methods. Therefore, creating a library of circRNAs, the full-length sequences of which have been experimentally identified, and developing a pipeline that allows a more accurate prediction of their existence is required [92].

Another challenge is that some databases are not updated regularly [92]. Incorrect data and out-of-date tools that no longer work appropriately have been neglected. For instance, search results are not viewable when the interface is created in Adobe Flash, which is no longer supported. Furthermore, some databases, such as PlantcircNet, GreenCircRNA, and Circ2Traits, are constantly or frequently inaccessible. Sometimes, the browsers cause access problems; thus, in our opinion, it would be helpful in future updates of these database websites to indicate on the top pages which browsers work optimally with the database.

10 Conclusion

As the number of circRNA databases, manually curated based on reliable literature or experiments, has increased, the functions they provide have been expanded and classified into five categories. Although reliable databases have been developed, many circRNAs are still putative and need confirmation. Additionally, the challenges of nomenclature and updating remain. These problems should be resolved shortly.

Acknowledgments We thank the members of the Fujii lab for their help on our circRNA studies.

Competing Interests The authors declare no competing financial interests.

References

1. Barrett SP, Parker KR, Horn C, Mata M, Salzman J (2017) ciRS-7 exonic sequence is embedded in a long non-coding RNA locus. PLoS Genet 13(12):e1007114. https://doi.org/10.1371/journal.pgen.1007114
2. Hansen TB, Jensen TI, Clausen BH, Bramsen JB, Finsen B, Damgaard CK, Kjems J (2013) Natural RNA circles function as efficient microRNA sponges. Nature 495(7441):384–388. https://doi.org/10.1038/nature11993
3. Memczak S, Jens M, Elefsinioti A, Torti F, Krueger J, Rybak A, Maier L, Mackowiak SD, Gregersen LH, Munschauer M, Loewer A, Ziebold U, Landthaler M, Kocks C, le Noble F, Rajewsky N (2013) Circular RNAs are a large class of animal RNAs with regulatory potency. Nature 495(7441):333–338. https://doi.org/10.1038/nature11928
4. Piwecka M, Glažar P, Hernandez-Miranda LR, Memczak S, Wolf SA, Rybak-Wolf A, Filipchyk A, Klironomos F, Cerda Jara CA, Fenske P, Trimbuch T, Zywitza V, Plass M, Schreyer L, Ayoub S, Kocks C, Kühn R, Rosenmund C, Birchmeier C, Rajewsky N (2017) Loss of a mammalian circular RNA locus causes miRNA deregulation and affects brain function. Science (1979) 357(6357):eaam8526. https://doi.org/10.1126/science.aam8526
5. Ashwal-Fluss R, Meyer M, Pamudurti NR, Ivanov A, Bartok O, Hanan M, Evantal N, Memczak S, Rajewsky N, Kadener S (2014) circRNA biogenesis competes with pre-mRNA splicing. Mol Cell 56(1):55–66. https://doi.org/10.1016/j.molcel.2014.08.019
6. Abdelmohsen K, Panda AC, Munk R, Grammatikakis I, Dudekula DB, De S, Kim J, Noh JH, Kim KM, Martindale JL, Gorospe M (2017) Identification of HuR target circular RNAs uncovers suppression of PABPN1 translation by *CircPABPN1*. RNA Biol 14(3):361–369. https://doi.org/10.1080/15476286.2017.1279788
7. Zhang J, Liu H, Hou L, Wang G, Zhang R, Huang Y, Chen X, Zhu J (2017) Circular RNA_LARP4 inhibits cell proliferation and invasion of gastric cancer by sponging miR-424-5p and regulating LATS1 expression. Mol Cancer 16(1):1–16. https://doi.org/10.1186/s12943-017-0719-3
8. Guarnerio J, Zhang Y, Cheloni G, Panella R, Mae Katon J, Simpson M, Matsumoto A, Papa A, Loretelli C, Petri A, Kauppinen S, Garbutt C, Nielsen GP, Deshpande V, Castillo-Martin M, Cordon-Cardo C, Dimitrios S, Clohessy JG, Batish M, Pandolfi PP (2019) Intragenic antagonistic roles of protein and circRNA in tumorigenesis. Cell Res 29(8):628–640. https://doi.org/10.1038/s41422-019-0192-1
9. Chen YG, Kim MV, Chen X, Batista PJ, Aoyama S, Wilusz JE, Iwasaki A, Chang HY (2017) Sensing self and foreign circular RNAs by intron identity. Mol Cell 67(2):228–238.e5. https://doi.org/10.1016/j.molcel.2017.05.022
10. Memczak S, Papavasileiou P, Peters O, Rajewsky N (2015) Identification and characterization of circular RNAs as a new class of putative biomarkers in human blood. PLoS One 10(10):e0141214. https://doi.org/10.1371/JOURNAL.PONE.0141214
11. Vo JN, Cieslik M, Zhang Y, Shukla S, Xiao L, Zhang Y, Wu YM, Dhanasekaran SM, Engelke CG, Cao X, Robinson DR, Nesvizhskii AI, Chinnaiyan AM (2019) The landscape of circular RNA in cancer. Cell 176(4):869–881.e13. https://doi.org/10.1016/j.cell.2018.12.021
12. Preußer C, Hung LH, Schneider T, Schreiner S, Hardt M, Moebus A, Santoso S, Bindereif A (2018) Selective release of circRNAs in platelet-derived extracellular vesicles. J Extracell Vesicles 7(1):1424473. https://doi.org/10.1080/20013078.2018.1424473
13. Bahn JH, Zhang Q, Li F, Chan TM, Lin X, Kim Y, Wong DTW, Xiao X (2015) The landscape of MicroRNA, piwi-interacting RNA, and circular RNA in human saliva. Clin Chem 61(1):221–230. https://doi.org/10.1373/clinchem.2014.230433
14. Russo F, Di Bella S, Vannini F, Berti G, Scoyni F, Cook HV, Santos A, Nigita G, Bonnici V, Laganà A, Geraci F, Pulvirenti A, Giugno R, De Masi F, Belling K, Jensen LJ, Brunak S, Pellegrini M, Ferro A (2018) MiRandola 2017: a curated knowledge base of non-invasive biomarkers. Nucleic Acids Res 46(D1):D354–D359. https://doi.org/10.1093/nar/gkx854
15. Li Y, Zheng Q, Bao C, Li S, Guo W, Zhao J, Chen D, Gu J, He X, Huang S (2015) Circular RNA is enriched and stable in exosomes: a promising biomarker for cancer diagnosis. Cell Res 25(8):981–984. https://doi.org/10.1038/cr.2015.82
16. Liu CX, Li X, Nan F, Jiang S, Gao X, Guo SK, Xue W, Cui Y, Dong K, Ding H, Qu B, Zhou Z, Shen N, Yang L, Chen LL (2019) Structure and degradation

of circular RNAs regulate PKR activation in innate immunity. Cell 177(4):865–880.e21. https://doi.org/10.1016/j.cell.2019.03.046
17. Lu D, Thum T (2019) RNA-based diagnostic and therapeutic strategies for cardiovascular disease. Nat Rev Cardiol 16(11):661–674. https://doi.org/10.1038/s41569-019-0218-x
18. Liu C-X, Chen L-L (2022) Circular RNAs: characterization, cellular roles, and applications. Cell 185(12):2016–2034. https://doi.org/10.1016/j.cell.2022.04.021
19. Wang PL, Bao Y, Yee M-C, Barrett SP, Hogan GJ, Olsen MN, Dinneny JR, Brown PO, Salzman J (2014) Circular RNA is expressed across the eukaryotic tree of life. PLoS One 9(6):e90859. https://doi.org/10.1371/journal.pone.0090859
20. Ghosal S, Das S, Sen R, Basak P, Chakrabarti J (2013) Circ2Traits: a comprehensive database for circular RNA potentially associated with disease and traits. Front Genet 4:283. https://doi.org/10.3389/fgene.2013.00283
21. Gao Y, Wang P, Wang Y, Ma X, Zhi H, Zhou D, Li X, Fang Y, Shen W, Xu Y, Shang S, Wang L, Wang L, Ning S, Li X (2019) Lnc2Cancer v2.0: updated database of experimentally supported long non-coding RNAs in human cancers. Nucleic Acids Res 47(D1):D1028–D1033. https://doi.org/10.1093/nar/gky1096
22. Zhou B, Ji B, Liu K, Hu G, Wang F, Chen Q, Yu R, Huang P, Ren J, Guo C, Zhao H, Zhang H, Zhao D, Li Z, Zeng Q, Yu J, Bian Y, Cao Z, Xu S, Yang Y, Zhou Y, Wang J (2021) EVLncRNAs 2.0: an updated database of manually curated functional long non-coding RNAs validated by low-throughput experiments. Nucleic Acids Res 49(D1):D86–D91. https://doi.org/10.1093/nar/gkaa1076
23. Bao Z, Yang Z, Huang Z, Zhou Y, Cui Q, Dong D (2019) LncRNADisease 2.0: an updated database of long non-coding RNA-associated diseases. Nucleic Acids Res 47(D1):D1034-D1037. https://doi.org/10.1093/nar/gky905
24. Ma L, Zou D, Liu L, Shireen H, Abbasi AA, Bateman A, Xiao J, Zhao W, Bao Y, Zhang Z (2022) Database commons: a catalog of worldwide biological databases. Genomics Proteomics Bioinformatics 21:1054. https://doi.org/10.1016/j.gpb.2022.12.004
25. Wu W, Ji P, Zhao F (2020) CircAtlas: an integrated resource of one million highly accurate circular RNAs from 1070 vertebrate transcriptomes. Genome Biol 21(1):1–14. https://doi.org/10.1186/s13059-020-02018-y
26. Zheng Y, Luo H, Teng X, Hao X, Yan X, Tang Y, Zhang W, Wang Y, Zhang P, Li Y, Zhao Y, Chen R, He S (2023) NPInter v5.0: ncRNA interaction database in a new era. Nucleic Acids Res 51(1 D):D232–D239. https://doi.org/10.1093/nar/gkac1002
27. Cui T, Dou Y, Tan P, Ni Z, Liu T, Wang D, Huang Y, Cai K, Zhao X, Xu D, Lin H, Wang D (2022) RNALocate v2.0: an updated resource for RNA subcellular localization with increased coverage and annotation. Nucleic Acids Res 50(D1):D333–D339. https://doi.org/10.1093/nar/gkab825
28. Dong R, Ma XK, Li GW, Yang L (2018) CIRCpedia v2: an updated database for comprehensive circular RNA annotation and expression comparison. Genomics Proteomics Bioinformatics 16(4):226–233. https://doi.org/10.1016/j.gpb.2018.08.001
29. Lai H, Li Y, Zhang H, Hu J, Liao J, Su Y, Li Q, Chen B, Li C, Wang Z, Li Y, Wang J, Meng Z, Huang Z, Huang S (2022) exoRBase 2.0: an atlas of mRNA, lncRNA and circRNA in extracellular vesicles from human biofluids. Nucleic Acids Res 50(D1):D118–D128. https://doi.org/10.1093/nar/gkab1085
30. Wu W, Zhao F, Zhang J (2024) circAtlas 3.0: a gateway to 3 million curated vertebrate circular RNAs based on a standardized nomenclature scheme. Nucleic Acids Res 52(D1):D52–D60. https://doi.org/10.1093/nar/gkad770
31. Glažar P, Papavasileiou P, Rajewsky N (2014) CircBase: A database for circular RNAs. RNA 20(11):1666–1670. https://doi.org/10.1261/rna.043687.113
32. Zhang XO, Dong R, Zhang Y, Zhang JL, Luo Z, Zhang J, Chen LL, Yang L (2016) Diverse alternative back-splicing and alternative splicing landscape of circular RNAs. Genome Res 26(9):1277–1287. https://doi.org/10.1101/gr.202895.115
33. Meng X, Hu D, Zhang P, Chen Q, Chen M (2019) CircFunBase: a database for functional circular RNAs. Database 2019:baz003. https://doi.org/10.1093/database/baz003
34. Chen Y, Yao L, Tang Y, Jhong JH, Wan J, Chang J, Cui S, Luo Y, Cai X, Li W, Chen Q, Huang HY, Wang Z, Chen W, Chang TH, Wei F, Lee TY, Da HH (2022) CircNet 2.0: an updated database for exploring circular RNA regulatory networks in cancers. Nucleic Acids Res 50(D1):D93–D101. https://doi.org/10.1093/nar/gkab1036
35. Chen X, Han P, Zhou T, Guo X, Song X, Li Y (2016) circRNADb: a comprehensive database for human circular RNAs with protein-coding annotations. Sci Rep 6(1):34985. https://doi.org/10.1038/srep34985
36. Liu M, Wang Q, Shen J, Yang BB, Ding X (2019) Circbank: a comprehensive database for circRNA with standard nomenclature. RNA Biol 16(7):899–905. https://doi.org/10.1080/15476286.2019.1600395
37. Xia S, Feng J, Lei L, Hu J, Xia L, Wang J, Xiang Y, Liu L, Zhong S, Han L, He C (2017) Comprehensive characterization of tissue-specific circular RNAs in the human and mouse genomes. Brief Bioinform 18(6):984–992. https://doi.org/10.1093/bib/bbw081
38. Yang J-H, Li J-H, Shao P, Zhou H, Chen Y-Q, Qu L-H (2011) starBase: a database for exploring microRNA–mRNA interaction maps from Argonaute CLIP-Seq and Degradome-Seq data. Nucleic Acids Res 39(suppl_1):D202–D209. https://doi.org/10.1093/nar/gkq1056
39. Li J-H, Liu S, Zhou H, Qu L-H, Yang J-H (2014) starBase v2.0: decoding miRNA-ceRNA, miRNA-ncRNA and protein–RNA interaction networks

from large-scale CLIP-Seq data. Nucleic Acids Res 42(D1):D92–D97. https://doi.org/10.1093/nar/gkt1248
40. Dudekula DB, Panda AC, Grammatikakis I, De S, Abdelmohsen K, Gorospe M (2016) Circinteractome: a web tool for exploring circular RNAs and their interacting proteins and microRNAs. RNA Biol 13(1):34–42. https://doi.org/10.1080/15476286.2015.1128065
41. Huang W, Ling Y, Zhang S, Xia Q, Cao R, Fan X, Fang Z, Wang Z, Zhang G (2021) TransCirc: an interactive database for translatable circular RNAs based on multi-omics evidence. Nucleic Acids Res 49(D1):D236–D242. https://doi.org/10.1093/nar/gkaa823
42. Westholm JO, Miura P, Olson S, Shenker S, Joseph B, Sanfilippo P, Celniker SE, Graveley BR, Lai EC (2014) Genome-wide analysis of drosophila circular RNAs reveals their structural and sequence properties and age-dependent neural accumulation. Cell Rep 9(5):1966–1980. https://doi.org/10.1016/j.celrep.2014.10.062
43. Agarwal V, Bell GW, Nam J-W, Bartel DP (2015) Predicting effective microRNA target sites in mammalian mRNAs. elife 4(AUGUST2015):e05005. https://doi.org/10.7554/eLife.05005
44. Wang L, Park HJ, Dasari S, Wang S, Kocher J-P, Li W (2013) CPAT: coding-potential assessment tool using an alignment-free logistic regression model. Nucleic Acids Res 41(6):e74. https://doi.org/10.1093/nar/gkt006
45. Betel D, Koppal A, Agius P, Sander C, Leslie C (2010) Comprehensive modeling of microRNA targets predicts functional non-conserved and non-canonical sites. Genome Biol 11(8):R90. https://doi.org/10.1186/gb-2010-11-8-r90
46. Kertesz M, Iovino N, Unnerstall U, Gaul U, Segal E (2007) The role of site accessibility in microRNA target recognition. Nat Genet 39(10):1278–1284. https://doi.org/10.1038/ng2135
47. Krek A, Grün D, Poy MN, Wolf R, Rosenberg L, Epstein EJ, MacMenamin P, Da Piedade I, Gunsalus KC, Stoffel M, Rajewsky N (2005) Combinatorial microRNA target predictions. Nat Genet 37(5):495-500. https://doi.org/10.1038/ng1536
48. Miranda KC, Huynh T, Tay Y, Ang Y-S, Tam W-L, Thomson AM, Lim B, Rigoutsos I (2006) A pattern-based method for the identification of MicroRNA binding sites and their corresponding heteroduplexes. Cell 126(6):1203–1217. https://doi.org/10.1016/j.cell.2006.07.031
49. Gao Y, Zhang J, Zhao F (2018) Circular RNA identification based on multiple seed matching. Brief Bioinform 19(5):803–810. https://doi.org/10.1093/bib/bbx014
50. Gao Y, Wang J, Zhao F (2015) CIRI: an efficient and unbiased algorithm for de novo circular RNA identification. Genome Biol 16(1):4. https://doi.org/10.1186/s13059-014-0571-3
51. Dong R, Ma X-K, Chen L-L, Yang L (2019) Genome-wide annotation of circRNAs and their alternative Back-splicing/splicing with CIRCexplorer pipeline. Methods Mol Biol 1870:137–149. https://doi.org/10.1007/978-1-4939-8808-2_10
52. Cheng J, Metge F, Dieterich C (2016) Specific identification and quantification of circular RNAs from sequencing data. Bioinformatics 32(7):1094–1096. https://doi.org/10.1093/bioinformatics/btv656
53. Enright AJ, John B, Gaul U, Tuschl T, Sander C, Marks DS (2003) MicroRNA targets in drosophila. Genome Biol 5(1):R1. https://doi.org/10.1186/gb-2003-5-1-r1
54. Zheng LL, Li JH, Wu J, Sun WJ, Liu S, Wang ZL, Zhou H, Yang JH, Qu LH (2016) DeepBase v2.0: identification, expression, evolution and function of small RNAs, LncRNAs and circular RNAs from deep-sequencing data. Nucleic Acids Res 44(D1):D196–D202. https://doi.org/10.1093/nar/gkv1273
55. Xie F, Liu S, Wang J, Xuan J, Zhang X, Qu L, Zheng L, Yang J (2021) DeepBase v3.0: expression atlas and interactive analysis of ncRNAs from thousands of deep-sequencing data. Nucleic Acids Res 49(D1):D877–D883. https://doi.org/10.1093/nar/gkaa1039
56. Djebali S, Davis CA, Merkel A, Dobin A, Lassmann T, Mortazavi A, Tanzer A, Lagarde J, Lin W, Schlesinger F, Xue C, Marinov GK, Khatun J, Williams BA, Zaleski C, Rozowsky J, Röder M, Kokocinski F, Abdelhamid RF, Alioto T, Antoshechkin I, Baer MT, Bar NS, Batut P, Bell K, Bell I, Chakrabortty S, Chen X, Chrast J, Curado J, Derrien T, Drenkow J, Dumais E, Dumais J, Duttagupta R, Falconnet E, Fastuca M, Fejes-Toth K, Ferreira P, Foissac S, Fullwood MJ, Gao H, Gonzalez D, Gordon A, Gunawardena H, Howald C, Jha S, Johnson R, Kapranov P, King B, Kingswood C, Luo OJ, Park E, Persaud K, Preall JB, Ribeca P, Risk B, Robyr D, Sammeth M, Schaffer L, See L-H, Shahab A, Skancke J, Suzuki AM, Takahashi H, Tilgner H, Trout D, Walters N, Wang H, Wrobel J, Yu Y, Ruan X, Hayashizaki Y, Harrow J, Gerstein M, Hubbard T, Reymond A, Antonarakis SE, Hannon G, Giddings MC, Ruan Y, Wold B, Carninci P, Guigó R, Gingeras TR (2012) Landscape of transcription in human cells. Nature 489(7414):101–108. https://doi.org/10.1038/nature11233
57. Wang K, Singh D, Zeng Z, Coleman SJ, Huang Y, Savich GL, He X, Mieczkowski P, Grimm SA, Perou CM, MacLeod JN, Chiang DY, Prins JF, Liu J (2010) MapSplice: accurate mapping of RNA-seq reads for splice junction discovery. Nucleic Acids Res 38(18):e178. https://doi.org/10.1093/nar/gkq622
58. Zhao J, Li Q, Li Y, He X, Zheng Q, Huang S (2019) ASJA: a program for assembling splice junctions analysis. Comput Struct Biotechnol J 17:1143–1150. https://doi.org/10.1016/j.csbj.2019.08.001
59. Dube U, Del-Aguila JL, Li Z, Budde JP, Jiang S, Hsu S, Ibanez L, Fernandez MV, Farias F, Norton J, Gentsch J, Wang F, Dominantly Inherited Alzheimer

Network (DIAN), Salloway S, Masters CL, Lee J-H, Graff-Radford NR, Chhatwal JP, Bateman RJ, Morris JC, Karch CM, Harari O, Cruchaga C (2019) An atlas of cortical circular RNA expression in Alzheimer disease brains demonstrates clinical and pathological associations. Nat Neurosci 22(11):1903–1912. https://doi.org/10.1038/s41593-019-0501-5

60. Zhao Q, Liu J, Deng H, Ma R, Liao J-Y, Liang H, Hu J, Li J, Guo Z, Cai J, Xu X, Gao Z, Su S (2020) Targeting mitochondria-located circRNA SCAR alleviates NASH via reducing mROS output. Cell 183(1):76–93.e22. https://doi.org/10.1016/j.cell.2020.08.009
61. Yao D, Zhang L, Zheng M, Sun X, Lu Y, Liu P (2018) Circ2Disease: a manually curated database of experimentally validated circRNAs in human disease. Sci Rep 8(1):1–6. https://doi.org/10.1038/s41598-018-29360-3
62. Sun Z-Y, Yang C-L, Huang L-J, Mo Z-C, Zhang K-N, Fan W-H, Wang K-Y, Wu F, Wang J-G, Meng F-L, Zhao Z, Jiang T (2024) circRNADisease v2.0: an updated resource for high-quality experimentally supported circRNA-disease associations. Nucleic Acids Res 52(D1):D1193–D1200. https://doi.org/10.1093/nar/gkad949
63. Fan C, Lei X, Tie J, Zhang Y, Wu FX, Pan Y (2022) CircR2Disease v2.0: an updated web server for experimentally validated circRNA–disease associations and its application. Genomics Proteomics Bioinformatics 20(3):435–445. https://doi.org/10.1016/j.gpb.2021.10.002
64. Lin X, Lu Y, Zhang C, Cui Q, Tang Y-D, Ji X, Cui C (2024) LncRNADisease v3.0: an updated database of long non-coding RNA-associated diseases. Nucleic Acids Res 52(D1):D1365–D1369. https://doi.org/10.1093/nar/gkad828
65. Chen J, Lin J, Hu Y, Ye M, Yao L, Wu L, Zhang W, Wang M, Deng T, Guo F, Huang Y, Zhu B, Wang D (2023) RNADisease v4.0: an updated resource of RNA-associated diseases, providing RNA-disease analysis, enrichment and prediction. Nucleic Acids Res 51(D1):D1397–D1404. https://doi.org/10.1093/nar/gkac814
66. Rappaport N, Nativ N, Stelzer G, Twik M, Guan-Golan Y, Stein TI, Bahir I, Belinky F, Morrey CP, Safran M, Lancet D (2013) MalaCards: an integrated compendium for diseases and their annotation. Database (Oxford) 2013:bat018. https://doi.org/10.1093/database/bat018
67. Rophina M, Sharma D, Poojary M, Scaria V (2020) Circad: a comprehensive manually curated resource of circular RNA associated with diseases. Database (Oxford) 2020:baaa019. https://doi.org/10.1093/database/baaa019
68. Ning L, Cui T, Zheng B, Wang N, Luo J, Yang B, Du M, Cheng J, Dou Y, Wang D (2021) MNDR v3.0: mammal ncRNA-disease repository with increased coverage and annotation. Nucleic Acids Res 49(D1):D160–D164. https://doi.org/10.1093/nar/gkaa707
69. Schriml LM, Munro JB, Schor M, Olley D, McCracken C, Felix V, Baron JA, Jackson R, Bello SM, Bearer C, Lichenstein R, Bisordi K, Dialo NC, Giglio M, Greene C (2022) The human disease ontology 2022 update. Nucleic Acids Res 50(D1):D1255–D1261. https://doi.org/10.1093/nar/gkab1063
70. Bhattacharya S, Ha-Thuc V, Srinivasan P (2011) MeSH: a window into full text for document summarization. Bioinformatics 27(13):i120–i128. https://doi.org/10.1093/bioinformatics/btr223
71. Xia S, Feng J, Chen K, Ma Y, Gong J, Cai F, Jin Y, Gao Y, Xia L, Chang H, Wei L, Han L, He C (2018) CSCD: a database for cancer-specific circular RNAs. Nucleic Acids Res 46(D1):D925–D929. https://doi.org/10.1093/nar/gkx863
72. Feng J, Chen W, Dong X, Wang J, Mei X, Deng J, Yang S, Zhuo C, Huang X, Shao L, Zhang R, Guo J, Ma R, Liu J, Li F, Wu Y, Han L, He C (2022) CSCD2: an integrated interactional database of cancer-specific circular RNAs. Nucleic Acids Res 50(D1):D1179–D1183. https://doi.org/10.1093/nar/gkab830
73. Gao Y, Shang S, Guo S, Li X, Zhou H, Liu H, Sun Y, Wang J, Wang P, Zhi H, Li X, Ning S, Zhang Y (2021) Lnc2Cancer 3.0: an updated resource for experimentally supported lncRNA/circRNA cancer associations and web tools based on RNA-seq and scRNA-seq data. Nucleic Acids Res 49(D1):D1251–D1258. https://doi.org/10.1093/nar/gkaa1006
74. Li L, Wu P, Wang Z, Meng X, Zha C, Li Z, Qi T, Zhang Y, Han B, Li S, Jiang C, Zhao Z, Cai J (2020) NoncoRNA: a database of experimentally supported non-coding RNAs and drug targets in cancer. J Hematol Oncol 13(1):13–16. https://doi.org/10.1186/s13045-020-00849-7
75. Thul PJ, Åkesson L, Wiking M, Mahdessian D, Geladaki A, Ait Blal H, Alm T, Asplund A, Björk L, Breckels LM, Bäckström A, Danielsson F, Fagerberg L, Fall J, Gatto L, Gnann C, Hober S, Hjelmare M, Johansson F, Lee S, Lindskog C, Mulder J, Mulvey CM, Nilsson P, Oksvold P, Rockberg J, Schutten R, Schwenk JM, Sivertsson Å, Sjöstedt E, Skogs M, Stadler C, Sullivan DP, Tegel H, Winsnes C, Zhang C, Zwahlen M, Mardinoglu A, Pontén F, von Feilitzen K, Lilley KS, Uhlén M, Lundberg E (2017) A subcellular map of the human proteome. Science 356(6340):eaal3321. https://doi.org/10.1126/science.aal3321
76. He Y, Zhang M, Ju Y, Yu Z, Lv D, Sun H, Yuan W, He F, Zhang J, Li H, Li J, Wang-Sattler R, Li Y, Zhang G, Xie L (2012) dbDEPC 2.0: updated database of differentially expressed proteins in human cancers. Nucleic Acids Res 40(D1):D964–D971. https://doi.org/10.1093/nar/gkr936
77. Wishart DS, Feunang YD, Guo AC, Lo EJ, Marcu A, Grant JR, Sajed T, Johnson D, Li C, Sayeeda Z, Assempour N, Iynkkaran I, Liu Y, Maciejewski A, Gale N, Wilson A, Chin L, Cummings R, Le D, Pon A, Knox C, Wilson M (2018) DrugBank 5.0: a major update to the DrugBank database for 2018.

Nucleic Acids Res 46(D1):D1074–D1082. https://doi.org/10.1093/nar/gkx1037

78. Chen L, Wang C, Sun H, Wang J, Liang Y, Wang Y, Wong G (2021) The bioinformatics toolbox for circRNA discovery and analysis. Brief Bioinform 22(2):1706–1728. https://doi.org/10.1093/bib/bbaa001
79. Rybak-Wolf A, Stottmeister C, Glažar P, Jens M, Pino N, Giusti S, Hanan M, Behm M, Bartok O, Ashwal-Fluss R, Herzog M, Schreyer L, Papavasileiou P, Ivanov A, Öhman M, Refojo D, Kadener S, Rajewsky N (2014) Circular RNAs in the mammalian brain are highly abundant, conserved, and dynamically expressed. Mol Cell 58(5):870–885. https://doi.org/10.1016/j.molcel.2015.03.027
80. Kent WJ (2002) BLAT —the BLAST –like alignment tool. Genome Res 12(4):656–664. https://doi.org/10.1101/gr.229202
81. Untergasser A, Cutcutache I, Koressaar T, Ye J, Faircloth BC, Remm M, Rozen SG (2012) Primer3--new capabilities and interfaces. Nucleic Acids Res 40(15):e115. https://doi.org/10.1093/nar/gks596
82. Lorenz R, Bernhart SH, Höner zu Siederdissen C, Tafer H, Flamm C, Stadler PF, Hofacker IL (2011) ViennaRNA package 2.0. Algorithms Molecul Biol 6(1):26. https://doi.org/10.1186/1748-7188-6-26
83. Liu T, Zhang Q, Zhang J, Li C, Miao Y-R, Lei Q, Li Q, Guo A-Y (2019) EVmiRNA: a database of miRNA profiling in extracellular vesicles. Nucleic Acids Res 47(D1):D89–D93. https://doi.org/10.1093/nar/gky985
84. Lock A, Rutherford K, Harris MA, Wood V (2018) PomBase: the scientific resource for fission yeast. Methods Mol Biol 1757:49–68. https://doi.org/10.1007/978-1-4939-7737-6_4
85. Berardini TZ, Reiser L, Li D, Mezheritsky Y, Muller R, Strait E, Huala E (2015) The arabidopsis information resource: making and mining the "gold standard" annotated reference plant genome. Genesis 53(8):474–485. https://doi.org/10.1002/dvg.22877
86. Chu Q, Zhang X, Zhu X, Liu C, Mao L, Ye C, Zhu QH, Fan L (2017) PlantcircBase: a database for plant circular RNAs. Mol Plant 10(8):1126–1128. https://doi.org/10.1016/j.molp.2017.03.003
87. Xu X, Du T, Mao W, Li X, Ye CY, Zhu QH, Fan L, Chu Q (2022) PlantcircBase 7.0: full-length transcripts and conservation of plant circRNAs. Plant Commun 3(4):100343. https://doi.org/10.1016/j.xplc.2022.100343
88. Zhang P, Meng X, Chen H, Liu Y, Xue J, Zhou Y, Chen M (2017) PlantCircNet: a database for plant circRNA–miRNA–mRNA regulatory networks. Database (Oxford) 2017:bax089. https://doi.org/10.1093/database/bax089
89. Zhang J, Hao Z, Yin S, Li G (2020) GreenCircRNA: a database for plant circRNAs that act as miRNA decoys. Database 2020:baaa039. https://doi.org/10.1093/database/baaa039
90. Wang K, Wang C, Guo B, Song K, Shi C, Jiang X, Wang K, Tan Y, Wang L, Wang L, Li J, Li Y, Cai Y, Zhao H, Sun X (2019) CropCircDB: a comprehensive circular RNA resource for crops in response to abiotic stress. Database (Oxford) 2019(1):baz053. https://doi.org/10.1093/database/baz053
91. Ye J, Wang L, Li S, Zhang Q, Zhang Q, Tang W, Wang K, Song K, Sablok G, Sun X, Zhao H (2019) AtCircDB: a tissue-specific database for Arabidopsis circular RNAs. Brief Bioinform 20(1):58–65. https://doi.org/10.1093/bib/bbx089
92. Vromman M, Vandesompele J, Volders PJ (2021) Closing the circle: current state and perspectives of circular RNA databases. Brief Bioinform 22(1):288–297. https://doi.org/10.1093/bib/bbz175

Prediction of Circular RNA Secondary Structures and Their Targets

5

Stephan H. Bernhart, Jörg Fallmann, Ronny Lorenz, and Peter F. Stadler

Abstract

At the level of secondary structure, circular RNAs (circRNAs) can be understood in terms of base pairing, base-pair stacking, and entropic loop contribution in the same way as linear RNAs and intermolecular RNA–RNA interactions. The folding problem of circular RNAs can thus be solved by dynamic programming algorithms in essentially the same manner. In this chapter, we review the similarities and differences between circular and linear RNAs with a focus on the software tools provided by the `ViennaRNA` package. Comparative analysis of RNA structures can also be generalized to circular RNA molecules. However, the task of constructing pairwise and multiple alignments of circular sequences is more difficult than those of their linear counterpart, whence fewer and less convenient software solutions are available. This chapter has also touched upon recent developments such as applications of chemical probing to circular RNAs and prediction of secondary structures on the interaction of circular RNAs with other molecules.

Keywords

Secondary structure · Dynamic programming · Accessibility · Circular alignment · Consensus structures

S. H. Bernhart (✉) · J. Fallmann
Bioinformatics Group, Department of Computer Science, and Interdisciplinary Center for Bioinformatics, Leipzig University, Leipzig, Germany
e-mail: berni@bioinf.uni-leipzig.de; fall@bioinf.uni-leipzig.de

R. Lorenz
Bioinformatics Group, Department of Computer Science, and Interdisciplinary Center for Bioinformatics, Leipzig University, Leipzig, Germany

Institute for Theoretical Chemistry, University of Vienna, Wien, Austria
e-mail: ronny@bioinf.uni-leipzig.de

P. F. Stadler
Bioinformatics Group, Department of Computer Science, and Interdisciplinary Center for Bioinformatics, Leipzig University, Leipzig, Germany

Institute for Theoretical Chemistry, University of Vienna, Wien, Austria

Max Planck Institute for Mathematics in the Sciences, Leipzig, Germany

Center for Non-coding RNA in Technology and Health, University of Copenhagen, Frederiksberg C, Denmark

Facultad de Ciencias, Universidad National de Colombia, Bogotá, Colombia

Santa Fe Institute, Santa Fe, NM, USA
e-mail: studla@bioinf.uni-leipzig.de

J. Xiao (ed.), *Advances in Circular RNAs*, Advances in Experimental Medicine and Biology 1485,
https://doi.org/10.1007/978-981-96-9428-0_5

1 Introduction

Circular RNAs (circRNAs) have received rapidly increasing attention in recent years, and much effort has been expended in their identification with high-throughput methods. A recent benchmarking study of methods for identifying circRNAs [1] has highlighted their abundance and diversity and has suggested that they may still be underreported. While some RNA folding software can handle circular RNAs, this is usually limited to a few core functionalities. More generally, circular secondary structures have not received much systematic attention. This may come as a surprise, given that circRNAs have been implicated in functions such as miRNA sponging, in which structure clearly plays a key role.

Circular RNA molecules fall into several broad classes [2]:

(i) The most abundant class by far, often referred to simply as "circRNAs," is produced by backsplicing [3, 4] in eukaryotes. This class is highly abundant and diverse, with the current version 3.0 of the `circAltas` database reporting more than three million circRNAs from ten well-studied vertebrates alone [5]. A review of databases dedicated to circRNAs can be found in [6]. In line with their extensive diversity, circRNAs can function by different mechanisms [7]. Early functional studies identified some circRNAs as miRNA sponges or competing endogenous RNAs (ceRNAs) [8, 9], embedding them into extensive RNA-centric regulatory networks [10]. circRNAs may also function as part of RNA–protein complexes [11]. A rapidly increasing body of literature shows that many circRNAs are translated, expanding the diversity of the proteome; see [12] for a recent review. Some circRNAs, termed "CIciRNAs," retain introns upon circularization [13] and do so mostly in the nucleus [13]. Their diverse function may explain their widespread associations with human diseases, as reviewed, e.g., in [14], [15], and [16]. Virus-derived circRNAs, furthermore, may play important roles in infection [17].

(ii) Historically, viroids were the first circular RNAs described in the literature. They exhibit a rod-like secondary structure, do not code for proteins, and are typically plant pathogens; see [18] and [19] for reviews. Many plant viruses are accompanied by small, circular satellite RNAs that are similar to viroids [20]. A wider class of infectious RNA circles containing autocatalytic ribozymes are reviewed in [21]. These include retrozymes, which propagate via circular RNA intermediates up to 400 nt in length. An exceptional case is the only known retroviroid CarSV [22]. The circular genome of hepatitis delta virus (HDV) [23] is the best understood representative of Ribozyviria, which comprises a broad range of recently discovered HDV-like agents; see, e.g., Chang et al. [24] and Netter et al. [25]. A recent survey of metatranscriptomes has uncovered not only many new examples of these abovementioned viroid-like RNAs but also distinct groups of ribozyme-like satellite viruses and viroid-like RNAs with previously unknown combinations of ribozymes [26].

(iii) Circular excised introns were first observed for self-splicing group I [27] and group II introns [28]. A class of nonconventional introns, RNA secondary structures that bring their ends together, is processed into circular RNAs in euglenids [29]. A broad class of circular intronic RNAs (ciRNAs) or stable intronic sequence RNAs (sisRNAs) is produced from eukaryotic intron lariats that escape debranching and thus rapid degradation [30, 31]. In contrast to other circular RNAs, the cyclic structure of ciRNAs is closed by a 2′–5′ phosphodiester bond. They appear to frequently engage in feedback loops to control the expression of their parental genes [32], possibly due to their propensity for forming DNA–RNA hybrids (R-loops) that, in turn, makes them a target for RNase H1 cleavage [33].

(iv) Circular RNAs are produced by β-hydroxybutyrate (BHB) element-related enzymatic splicing mostly in archaea. A particular case is the identification of intermediates during ribosomal RNA processing by [34]. Several function ncRNAs are circularized by the same mechanism [35–37], presumably to increase their resilience against degradation and also to stabilize their (secondary) structures. A detailed reanalysis of transcriptome data from archaea suggested that there are also circularized RNAs that are not associated with recognizable BHB elements [38]. In archaeal species, circRNAs can indeed be formed by ligation catalyzed by a ligase belonging to the Rnl3 family [39].

A metatranscriptomics survey [26] also reported more than 200 circular RNAs with a length of about 4 kb that encode an RNA-dependent RNA polymerase and show similarities to ambiviruses. Similar circular RNAs were observed for mitovirus. Artificial circular RNA constructs are playing an increasing role in synthetic biology because of their resilience against exonucleases [40].

The influence of RNA structure on circRNA biogenesis, function, and decay has been reviewed in Busa and Leung [41]. Many endogenous circRNAs form imperfect RNA duplexes with a length of 16–26 nt that inhibit protein kinase R (PKR) that plays a key role in innate immunity. This mechanism appears to be important in autoimmune diseases such as systemic lupus erythematosus [42]. Secondary structures are also important for the translation of circular RNAs. Due to the absence of 5′ cap, translation requires internal ribosome entry sites (IRESs), which are usually highly structured RNA sequences [43].

2 Secondary Structures of Linear and Circular RNAs

The only conceptual difference between circular and linear nucleic acids is that the latter has two ends and thus a well-defined beginning and end, see Fig. 5.1. Since linear nucleic acids are oriented by the chemical differences of their 5′- and 3′-ends, the numbering of their nucleotides from 1 at the 5′-end to n at the 3′-end is uniquely defined. Circular nucleic acids are also oriented, but the starting point of their number is arbitrary. In both cases, secondary structures are defined by a set of crossing-free base pairs. This notion of crossing-free is also the same for linear and circular RNAs, as most easily seen by drawing the n nucleotides of the backbone as a regular n-gon. Secondary structures are then sets of nonintersecting chords (straight lines) connecting corners of the polygon. Whether or not positions 1 and n are connected by a covalent bond does not make a difference in this picture.

In the standard energy model, every "loop" (i.e., every facet in this planar drawing) provides an additive contribution that depends on the nucleotide sequence and the distribution of backbone edges and chords. In the energy model, the difference arises from the one loop that contains nucleotides 1 and n. This loop is called the *exterior loop*. In a linear nucleic acid, it is unconstrained and, thus, by definition, contributes 0 to the total folding energy. (We neglect the subtleties of the so-called dangling end contributions [44, 45] here.) In contrast, the "exterior loop" in a circular structure is a loop like any other and is thus scored accordingly. As a consequence, the folding energies of a linear RNA, obtained by cutting open the same circular RNA at different positions, will vary with the cutting position. Since the cutting can affect highly different loops, the energetic differences may be substantial (see Fig. 5.2). Combinatorial enumeration of linear and circular structures yields similar distribution of secondary structure elements. One of the most significant difference is a reduction of the asymptotic number of secondary structures by a factor n in response to the additional constraint introduced by circularization [46]. An

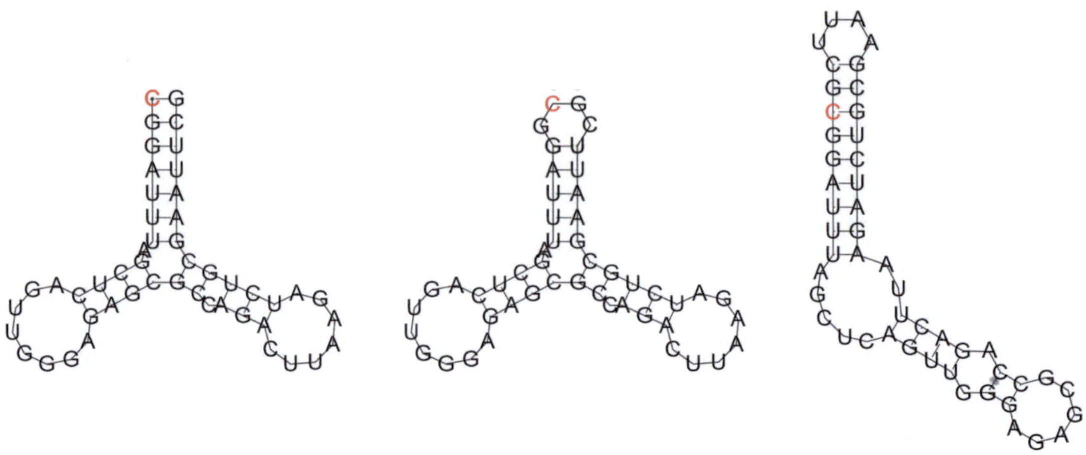

Fig. 5.1 Linear versus circular structure prediction. Left: Linear prediction of the minimum free energy secondary structure of an arbitrary RNA molecule, with the first base in red. Center: Circularization of the same structure by converting the exterior loop containing the first and last base into a hairpin loop due to the connection between the two. Right: Computed minimum free energy structure of the circular RNA molecule. The two minimum free energy structures (left and right) do not have any base pair in common

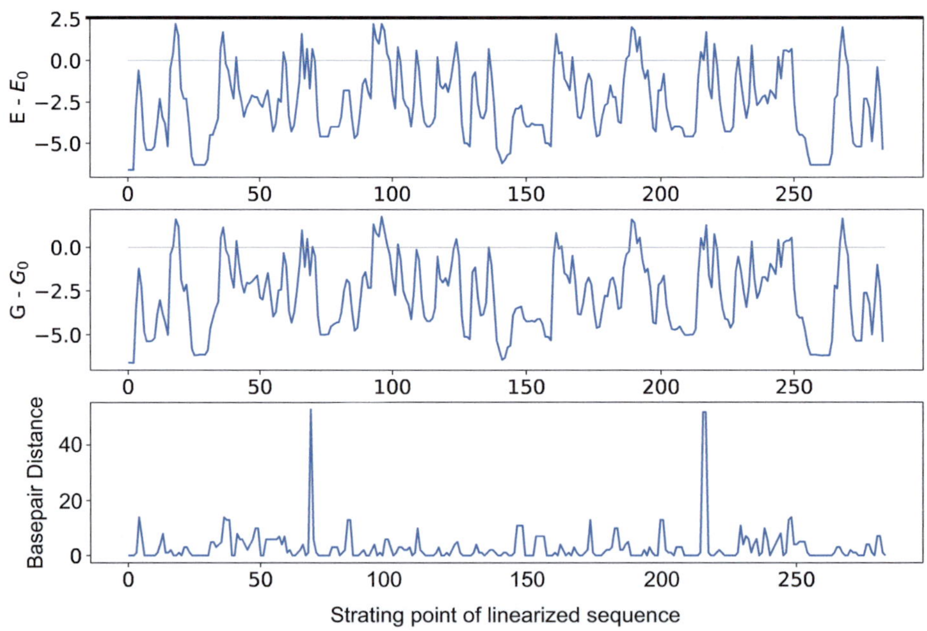

Fig. 5.2 The differences of RNA folding energies and partition function as well as base-pair distance of the circular RNA structure of citrus viroid IV virus and linear versions obtained by cutting at different positions along the sequence. Values are obtained by subtracting energy or partition function of the circular form from the respective measure of the linear molecule. Base-pair distance is calculated similarly with the function `bp distance` of the `ViennaRNA` Python API. (The citrus viroid IV virus genome is available at https://www.ncbi.nlm.nih.gov/datasets/genome/GCF_000852785.1)

important qualitative difference is the tendency of linear RNA structures to form approximate panhandles, i.e., base pairs between the 5′- and 3′-ends [47–49]. In circular RNAs, a similar effect is of course not possible.

There are several different approaches to computing RNA secondary structure from its sequence. The most commonly used, and, in our opinion, the most reliable is the thermodynamic model. It makes the—empirically very well-founded—assumption that the energy of an RNA secondary structure can be understood as the sum of local contributions from so-called individual "loops" within the structure. Each loop is a cycle in a graph-theoretical sense. Stabilizing contributions arise from the stacking of base pairs, while entropic effects of loops, including hairpin, interior, and multiloops, are primarily destabilizing. These loops consist of a closing base pair, unpaired stretches of sequences, and interior delimiting base pair(s) in the case of interior and multibranched loops. In addition, there are specific contributions of unpaired nucleotides at the ends of helices. A "standard energy model" of this type has been compiled from an extremely large number of precise calorimetric measurements [50] (Fig. 5.2).

The energy model is compatible with a set of dynamic programming recursions that elaborate on the circular matching problem. This approach is based on the simple observation that the absence of crossing base pairs in the secondary structure implies that every base pair (i, j) subdivides the structure into an interior part and an exterior part. No basepairs reach from the interior to the exterior, since base pairs do cross in secondary structures. The secondary structure can thus be decomposed according to the rules $S \to \bullet S$, $S \to (S)S$, or $S \to (S)$, where S is a self-contained secondary structure, (…) denotes a base pair, and • is an unpaired base. Energy minimization in this simple model, the amount to the recursion

$$E_{ij} = \min e_i + E_{i+1,j}, \min_{i<k<j-1} E_{i+1,k-1} + E_{k+1,j} + \varepsilon_{ik}, E_{i+1,j-1} + \varepsilon_{ij} \quad (5.1)$$

with base case $E_{ii} = e_i$. Here, E_{ij} is minimum energy of all secondary structures on the sequence interval from i to j, e_i is the energy contribution of the unpaired base i, and ε_{ik} is the energy contribution of the base pair (i, k) [51]. We note that at this simplified level of modeling with a position and base-pairwise energy model, there is no difference between circular and linear RNAs, i.e., the same minimum is obtained independent of the point at which a circular sequence is "cut open." The main algorithmic difference required to handle a realistic energy model derives from the need to differentiate different loop types. To this end, additional variables C_{ij} designate secondary structures enclosed by the base pair (i, j). These replace the simple contributions $E_{i+1,k-1} + \varepsilon_{ik}$ in Eq.(5.1). In addition, a specific decomposition is introduced for multibranched loops (see Fig. 5.3 for details). The latter enforces a linear dependence of the energy contribution on size and branching order to enable efficient computations [53].

The folding algorithm implemented in `mfold` [54, 55] treats linear RNAs as a special case of circular RNAs by design. In contrast, the `ViennaRNA` package [56, 57] optimizes memory consumption for linear RNAs because these are much more frequently used. Evaluation of the external loop then requires an additional set of recursions that are simply skipped for linear input sequences. In partition function calculations [58], the circular part involves a postprocessing step in the forward recursion and a preprocessing step in the backward recursion [52]. The details are summarized in Fig. 5.3. `ViennaRNA` web services [59] provide easy access to the folding algorithm for both linear and circular RNAs. A web service dedicated to circular RNAs and to making it easy to introduce constraints interactively is `cRNAsp12` [60]. Both `mfold` software and its non-free successor `UNAfold` [61] do not seem to be under active development. Hence, they lack most of the recent advances in RNA secondary structure prediction, such as incorporation of chemical RNA structure probing data. Unfortunately, `RNAstructure` [62], another frequently used, comprehensive, actively developed and maintained prediction tool set,

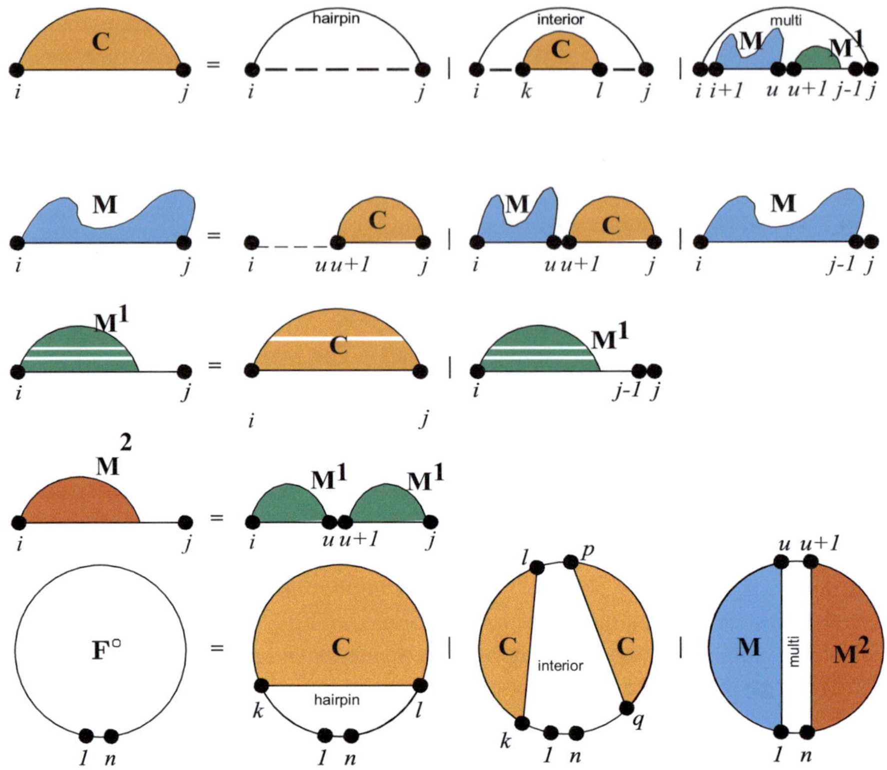

Fig. 5.3 RNA folding recursions for circular RNAs. Structures enclosed by base pair, C, are hairpins, interior loops, or multibranched loops. The latter are decomposed into a substructures with at least one component, M, and exactly one component, M^1. The exterior loop F° follows the same decomposition, also requiring information of substructures with exactly two components, M^2. (Adapted from Hofacker and Stadler [52])

does not implement the required steps for circular RNAs.

A specific complication in circular RNAs arises from potential rotational symmetries. If the circular sequence is the concatenation of two or more copies of the same string, then they can also form secondary structures with symmetries that reduce the number of physically distinct conformations. The relevant quantity is the length ℓ of the orbit determined by the symmetry. A symmetry amounts to a (entropic) destabilizing contribution to the free energy of $RT \ln \ell$ that is not accounted for in the standard energy model. Since the effect is fairly small and biologically relevant sequences in most cases do not exhibit exact symmetries, this effect is usually neglected in minimum free energy (MFE) calculations. A detailed analysis shows that the chance that the predicted MFE structure is symmetric decreases with sequence length [63]. For shorter sequences, it is possible to enumerate all structures within a narrow energy range above the ground state [64]. In the `ViennaRNA` package, this functionality is also available for circular sequences. This is usually sufficient to find the lowest nonsymmetric structure if the predicted MFE structure has a symmetry. The true ground state can then be determined by comparing the energetically most favorable structure for each type of symmetry after applying the appropriate symmetry correction. When precise thermodynamic parameters are required, it is advisable to compute the ensemble free energy for a Boltzmann distribution instead of using MFEs. The algorithms for computing partition functions, however, inherently undercount symmetric structures, thereby exactly compensating for symmetry effect. A detailed formal treatment can be found in [65].

Several extensions and modifications to the standard model of energy-directed RNA folding have been introduced in recent years. While mostly discussed in the literature for linear RNAs, they also concern circular RNAs. In the following paragraphs, we briefly touch upon the most important facets.

2.1 G-quadruplexes in Circular RNAs

G-quadruplexes may play a role in circRNA biogenesis and function since they are known to be involved in the regulation of both splicing and translation, as recently reviewed, e.g., in [66]. For linear RNA molecules, the `ViennaRNA` package already provides an option to include G-quadruplexes in the secondary structure prediction [67]. With the next release of the `ViennaRNA package`, this feature will be additionally extended to circular input sequences.

2.2 Structure Probing in Circular RNAs

External evidence for chemical or enzymatic probing of RNA secondary structures can be readily included in energy-directed folding algorithms in the form of pseudo-energies. These reward local structural features that conform the experimental data. In most cases, a profile of position specific reactivities r_i is obtained experimentally, which is converted into position-specific pseudo-energy contributions g_i to which base i is paired (or unpaired). To this end, empirical formulas can be used. In a more principled approach, g_i is related to the externally determined probability $p(r_i)$ that position i with a signal r_i has the property in question. Following Zarringhalam et al. [68],

$$g_i = -\text{RT}\ln\frac{p(r_i)}{1-p(r_i)} \tag{5.2}$$

The function $p(r)$ is estimated from empirical data by comparing measures probing signals to reference RNAs with a known structure; see, e.g., [69]. The `ViennaRNA` package provides different ways to convert the experimental signal into pseudo-energies and include it in the folding algorithm [70].

A systematic study of the secondary structure of circular RNAs using SHAPE-MaP has been reported in [71], showing that the secondary structures of circularized products indeed differ drastically from those attained by the same sequence embedded in a larger, linear mRNA. The study has used a recent version of the `ViennaRNA` package to compute the base-pairing probabilities using a single command:

```
RNAfold -p -d2 --circ
--shape=⟨reactivities⟩ --shapeMethod=D
< ⟨seq⟩
```

2.3 Chemical Modifications in Circular RNAs

N6-methyladenosine (m^6A) is a frequent chemical modification that affects RNA fate. It is widespread in circRNAs. On the other hand, circRNAs regulate proteins that regulate m^6A modifications [72]. Such chemical modifications affect the energy contributions of base pairs and loops in a nontrivial manner; see Tanzer et al. [73] and the references therein. For m^6A, stacking and dangling end parameters are tabulated in Kierzek et al. [74]. Such modified parameters can be passed to the `ViennaRNA` package and thus are also available for use with circular RNAs. Modification of the energy parameters in response to temperature and salt concentrations [75] can be incorporated in the same manner and pertain to circular RNAs without changes.

3 Consensus Structures of Circular RNAs

3.1 Alignments of Cyclic Sequences

Functional secondary structures are often conserved over long evolutionary time scales. In this situation, they can be detected in multiple sequence alignments. In circular RNAs, the alignment problem itself differs significantly from linear RNAs. A pairwise linear alignment of two sequences x and y can be thought of as a sequence of pairs of the form (x_i, x_j), $(-, y_j)$, and $(x_i, -)$ that contains all letters of both x and y exactly once and preserves their order. An in/del in the alignment is the maximal sequence of consecutive pairs of the form $(x_i, -)$ or of the form $(-, y_j)$. If x and y are circular sequences, then only the circular order needs to be preserved in the alignment. Furthermore, in/dels may wrap around the ends of the alignment. Denoting the *cyclic shift operator* by $\sigma(x) = (x_2, \ldots, x_{n-1}, x_n, x_1)$, the cyclic alignment problem thus consists of finding optimally scoring linear alignment of cyclically shifted sequences $\sigma^p(x)$ and $\sigma^q(y)$ for the optimal choices of p and q, respectively [76–79]. A more detailed discussion on the similarities and differences of cyclic versus linear alignment problems can be found in [80]. For the special case of linear gap cost functions, there is an $O(n^2 \ln n)$ algorithm that makes use of the fact that optimal alignment paths cannot cross [78]. This is not true for more realistic gap cost models, including the affine gap costs employed, e.g., in `clustalW` [81].

In [82], we devised a cubic time algorithm for the affine gap cost model that is based on Gotoh's quadratic time algorithm for alignments of linear sequences with affine gap costs [83]. It considers the alignments of the longer sequence x with all $O(n)$ cyclic shifts $\sigma^q(y)$ of the shorter sequence y and distinguishes the three possible ends of these linearized alignments: (i) (mis)match $x_n,(\sigma(y))_m$, (ii) deletion $(x_n, -)$, and (iii) insertion $(-,(\sigma^q(y)_m))$. In the first case, the score of the cyclic and linear alignments coincides. The insertion case (ii) does not need to be considered since a trailing insertion in y can be rotated to the beginning of the string, a configuration that is considered anyway since all $\sigma^q(y)$ values are considered. The correct scoring in the deletion case is obtained by starting the alignment in the deletion state, which can be achieved by initializing the dynamic programming matrices to exclude an initial match or insertion state. This yields a simple $O(nm \times \min(n, m))$ algorithm for the cyclic alignment problem with affine gap costs.

The computation effort can be considerably reduced by heuristics that reduce the number of cyclic shifts that are considered. Usually, one attempts to find the best cyclic shift. The simplest approach is probably used in `cyclops` [82]: the optimal local alignment between x and y is computed and the match in the center of this local alignment is then used to "anchor" the alignment of cyclic sequences. Better linear time heuristics are described in [84]. `CSA` [85] introduces a cyclic suffix array to determine the largest chain of non-repeated blocks that are present in all sequences to simultaneous estimate the best rotation of all cyclic sequences. A fast alternative is to use different distance measures based on block-wise q-gram comparisons. For this measure, the optimal rotation of y can be computed exactly in $O(bm + n)$ time with the help of a suffix array, depending on the number β of blocks [86]. MARS [87] elaborates on this idea to provide cyclic shifts simultaneously for multiple input sequences.

Progressive multiple alignments can be constructed by generalizing the pairwise cyclic alignment algorithms to profiles in exactly the same way as with linear sequences. An implementation using the sum of pair scoring for cyclic sequences is available in the `cyclope` package [82]. Alternatively, best shifts are computed for all sequences, e.g., with `CSA` or `MARS`, and the shifted sequences are then used as input for one of the many tools for the multiple alignment of linear sequences.

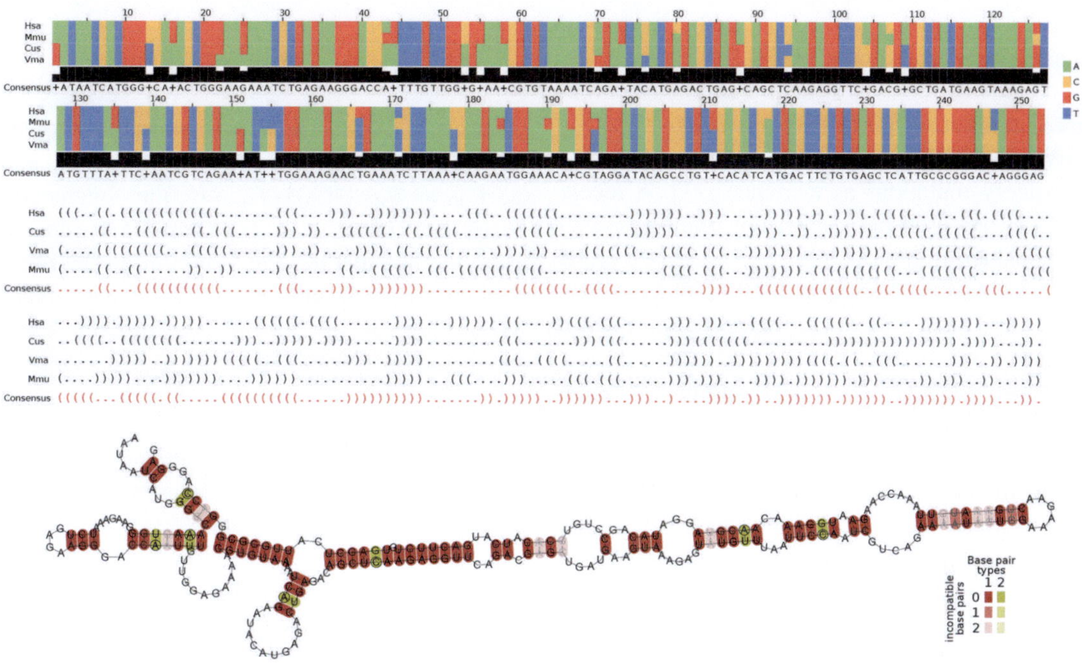

Fig. 5.4 Top: Alignment of the human circular RNA hsa circEZH2 001 from the riboCIRC database [88] (Hsa), its mouse homolog from the same database (Mmu), and homolog sequences from *Catharus ustulatus* (Cus) and *Vidua macroura* (Vma). The circular RNA is built from the second and third exon of the *EZH2* gene. Center: Circular structure predictions for single RNAs and circular consensus structures computed by `RNAalifold` show large differences. Bottom: `RNAalifold` predicted circular consensus structures with color-coded numbers of both compatible and incompatible types of base pairs

3.2 From Alignments to Consensus Secondary Structures

The RNA folding algorithms outlined in Sect. 3 can be applied not only to a input sequence but also to an sequence alignment: the indices in the recursions then refer to columns in an multiple sequence alignment instead of a single sequence. The algorithms then compute either a single consensus structure comprising the consensus base pairs of the aligned input sequences (using the algorithm for minimum free energy computations) or a matrix base-pairing probability matrix (using McCaskill's partition function approach) (Fig. 5.4).

The only additional ingredient is an appropriately adapted energy model. In the simplest model, the average of the energy contributions of the sequences in the rows of the alignment is used. `RNAalifold` [89, 90], in addition, uses small pseudo-energies to reward covariations and/or penalize nonstandard base pairs. The `ViennaRNA` package also implements the circular folding extensions for alignments, both for energy minimization and partition function computations. Consensus structure circular RNAs can therefore be obtained using `RNAalifold-C`.

3.3 Detection of Conserved Secondary Structure Elements in Circular RNAs

At present, there is no software tool dedicated to detecting conserved secondary structure elements from circular alignments. Several tools for linear structures, however, can be used for a suitable rotation of the alignment, can be adapted to work with circular RNAs, or can be applied at least as excellent approximations.

R-scape takes a structure-annotated alignment as input and performs statistical tests for the significance of conserved base pairs, using compensatory mutations as evidence [91]. A recent version has used helix-level covariation instead of considering base pairs [92]. In either case, it suffices to rotate the alignment such that the ends are within an unpaired region, most conveniently a hairpin.

RNAz [93] and its recent reimplementation [94] scores overlapping windows of the alignment for evidence of conserved secondary structures. In contrast to R-scape, it not only relies on sequence covariation but also emphasizes thermodynamic considerations and uses a trained decision model to compare sequence variation and secondary structure variation [93]. For circular RNAs, the choice of a window-size W smaller than the sequence length n and duplicating the first W alignment columns at the end of the alignment can be used. This ensures that windows spanning across the end of the alignment are scored. The windows-based approach is also an approximation for linear RNAs, with expected errors deriving from neglecting adjacent or competing structures. We, therefore, do not expect a significantly different performance for circular RNAs, provided W is not too close to n.

alifoldz [95] tests whether the energy of the consensus structure (computed by RNAalifold) is significantly more stable than the consensus structure of a suitable randomized alignment. It would not be difficult to replace the prediction of linear consensus structures by their circular counterparts in order to obtain a variant dedicated to circRNAs.

4 Interactions with Circular RNAs

In an equilibrium setting, standard thermodynamic theory is applied to the binding of RNA M to some partner P, which may be an RNA, a protein, or any other ligand. For the binding reaction $M + P \rightleftharpoons MP$, we have

$$\frac{[MP]}{[M][P]} = \exp\left(- \quad G_{MP} / RT\right)$$

where [.] denotes concentrations, $\Delta\ G_{MP}$ is the free energy of the binding reactions, and $\Delta\ G_{MP} := G_{MP} - G_M - G_P$. If the interaction site(s) are known, $\Delta\ G_{MP}$ can be computed using the fact that the free energy is a state function. Suppose M and P only bind each other when they are in specific conformations M^* and P^*, respectively. Then, we can decompose $G_{MP} = G_{M \to M^*} + G_{P \to P^*} + G_{M^*P^*}$, where $\Delta G_{M \to M^*}$ and $\Delta G_{P \to P^*}$ are the free energies bringing M and P into their binding conformations and $\Delta G_{M^*P^*}$ is the free energy of binding given that M and P are already in the correct conformation (see Fig. 5.5). The opening energy $\Delta G_{M \to M^*}$ is computed readily from restricted partition functions: $G_{M \to M^*} = -RT \ln\left(Z_{M^*} / Z_M\right)$, where Z_{M^*} is the partition function over the restricted ensemble of structures in which the binding site is unpaired (or satisfies another structural constraint). If P is also a nucleic acid, $\Delta\ G_{P \to P^*}$ can be computed analogously. The ViennaRNA package can incorporate such constraints into its partition function computations for both linear and circular nucleic acids.

The term "$\Delta\ G_{M^*P^*}$" describes the free energy of the intermolecular interaction of the two exposed binding sites. For two nucleic acids, the structural constraints amount to leaving the binding site(s) unpaired. The term "$\Delta\ G_{M^*P^*}$" can then be computed directly by considering the intermolecular base pairs, using, e.g., RNAhybrid [96] or a similar tool. The same computational approaches have been used successfully for RNA–DNA interactions [97]. If the interaction partner P is not a nucleic acid, we usually lack a concise and complete model for computing $\Delta\ G_{MP}$ and additional assumptions are necessary. In the case of riboswitches, for example, one can decompose the reaction into the formation of a correct binding pocket on the RNA side (with free energy $\Delta\ G_{M \to M^*}$) and the binding of the partner molecule P to the correctly formed pocked (with energy $\Delta\ G_{M^*P}$) [98, 99]. The latter

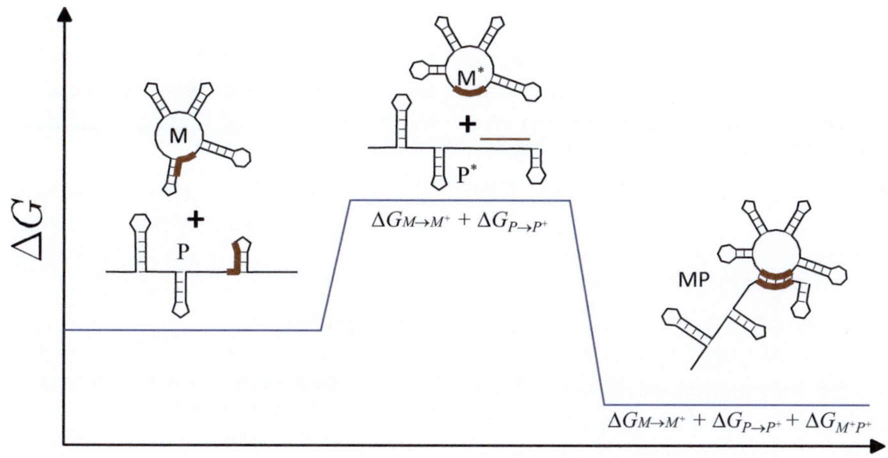

Fig. 5.5 Decomposition of binding free energies. In the first step, both interaction partners, here a circRNA M and a linear RNA P, refold to expose the binding site (highlighted region). The free energy differences are readily computed by means of partition functions restricted, keeping the binding sites unpaired. In the second step, the exposed binding sites pair with each other. The corresponding free energy change is computed by co-folding the two binding sites. The total binding energy is given by the sum of the free energy contributions

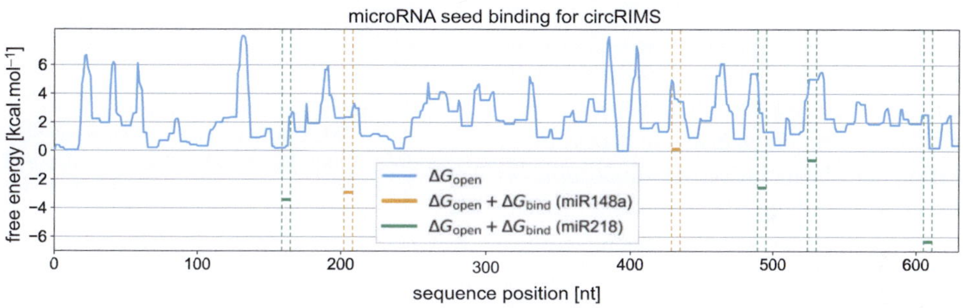

Fig. 5.6 CircRIMS was discovered due its association with gastric cancer metastasis and has been shown to act as a sponge for hsa-miR-148a-5p and hsa-miR-218-5p [100]. The opening energies (blue profile) computed using `RNAfold -p` constrained to keeping each interval of length 7 unpaired. For all potential binding sites of the mature microRNAs hsa-miR-148a-5p and hsa-miR-218-5p, the sum of opening energy and binding energy of the seed to the corresponding miRNA is shown, indicating four stable binding sites with significantly negative free energy of interaction. The two binding sites with energies close to 0 might be spurious

now implicitly subsumes any structural changes on the side of the partner molecule and must be inferred from experimental data. RNA-protein binding is amenable to analogous modeling.

It is also possible to compute the opening energies, usually referred to as accessibilities, for all sequence intervals $[i, j]$, with a total effort comparable to a single partition function computation [101–104]. This can be used for detecting binding sites between mRNAs and ncRNAs [101] and/or computing binding profiles of RNA-binding proteins [105]. At present, this functionality is not available for circular RNAs; it is subject to ongoing development, however, and will become available soon in the `ViennaRNA` package. Nevertheless, specific accessibilities can be computed by means of constrained partition functions (see Fig. 5.6 for an example).

It is also possible to combine thermodynamic interaction computations with considerations of sequence and structure conservation. To this end, copraRNA combines the significance measures of predicted interactions for several homologous sRNA-target pairs [106]. Currently, as handling circular RNAs is challenging linearized sequences are used as an approximation. A knowledge-based approach to interactions of circRNAs is provided by the `CircInteractome` web service [107], which collates information from public circRNA, miRNA, and RBP databases.

5 Discussion

With increasing information on functional circRNAs and the realization that many of them are relevant in a disease context, there is also renewed interest in the bioinformatics of circular nucleic acids. While the differences between linear and circular RNAs are moderate, they typically cannot be neglected. Many—but by no means all—functionalities of energy-directed folding software tools have also become available for circular input molecules. There are, however, several aspects for which the capability to handle circular input sequences has not (yet) been implemented. Most prominently, efficient accessibility computations and the inclusion of G-quadruplexes are subject to ongoing developments in the `ViennaRNA` package. The increasing importance of circular RNAs certainly serves as a motivation to revisit folding algorithms and to provide extensions of circular RNAs where they are not already available.

For some tasks beyond folding algorithms, however, circular RNAs cannot be handled with simply generalizations or extensions of linear RNAs. This is the case in particular for structural comparison. We have already seen above that circular alignments are different from the linear ones in several respects. We suspect that this will also be the case for structural alignments, both using simultaneous folding and alignment (i.e., variations on the Sankoff algorithm [108]), and considering alignments of trees or forests [109, 110]. To our knowledge, the corresponding circular problems have not been addressed in the literature so far.

Very little is known about pseudoknots in circular RNAs. A potentially conserved pseudoknot has been proposed as a suboptimal, likely functional structure in Pospiviroidae [111]. In general, pseudoknots may be rare in circular RNAs. In principle, circular RNAs could also form true knots. A survey of known structures, however, showed that knotted structures are virtually absent in linear RNAs [112], suggesting that they might also be virtually absent in the circular case.

References

1. Vromman M, Anckaert J, Bortoluzzi S, Buratin A, Chen CY, Chu Q, Chuang TJ, Dehghannasiri R, Dieterich C, Dong X, Flicek P, Gaffo E, Gu W, He C, Hoffmann S, Izuogu O, Jackson M, Jakobi T, Lai E, Nuytens J, Salzman J, Santibanez-Koref M, Stadler PF, Thas O, Vanden Eynde E, Verniers K, Wen G, Westholm J, Yang L, Ye CY, Yigit N, Yuan GH, Zhang J, Zhao F, Vandesompele J, Volders PJ (2023) Large-scale benchmarking of circRNA detection tools reveals large differences in sensitivity but not in precision. Nat Methods 20:1159–1169. https://doi.org/10.1038/s41592-023-01944-6
2. Lasda E, Parker R (2014) Circular RNAs: diversity of form and function. RNA 20:1829–1842. https://doi.org/10.1261/rna.047126.114
3. Nigro JM, Cho KR, Fearon ER, Kern SE, Ruppert JM, Oliner JD, Kinzler KW, Vogelstein B (1991) Scrambled exons. Cell 64(3):607–613. https://doi.org/10.1016/0092-8674(91)90244-s
4. Cocquerelle C, Daubersies P, Majérus MA, Kerckaert JP, Bailleul B (1992) Splicing with inverted order of exons occurs proximal to large introns. EMBO J 11(3):1095–1098. https://doi.org/10.1002/j.1460-2075.1992.tb05148.x
5. Wu W, Ji P, Zhao F (2020) CircAtlas: an integrated resource of one million highly accurate circular RNAs from 1070 vertebrate transcriptomes. Genome Biol 21(1):101. https://doi.org/10.1186/s13059-020-02018-y
6. Vromman M, Vandesompele J, Volders PJ (2020) Closing the circle: current state and perspectives of circular RNA databases. Brief Bioinform 22(1):288–297. https://doi.org/10.1093/bib/bbz175
7. Patop ILP, Wüst S, Kadener S (2019) Past, present, and future of circRNAs. EMBO J 38(16):e100836. https://doi.org/10.15252/embj.2018100836
8. Memczak S, Jens M, Elefsinioti A, Torti F, Krueger J, Rybak A, Maier L, Mackowiak SD, Gregersen LH, Munschauer M, Loewer A, Ziebold U, Landthaler M, Kocks C, le Noble F, Rajewsky N (2013) Circular RNAs are a large class of animal RNAs with regulatory potency. Nature 495:333–338. https://doi.org/10.1038/nature11928
9. Hansen TB, Jensen TI, Clausen BH, Bramsen JB, Finsen B, Damgaard CK, Kjems J (2013) Natural RNA circles function as efficient microRNA

sponges. Nature 495:384–388. https://doi.org/10.1038/nature11993

10. Kleaveland B, Shi CY, Stefano J, Bartel DP (2018) A network of noncoding regulatory RNAs acts in the mammalian brain. Cell 174(2):350–362. https://doi.org/10.1016/j.cell.2018.05.022
11. Huang A, Zheng H, Wu Z, Chen M, Huang Y (2020) Circular RNA-protein interactions: functions, mechanisms, and identification. Theranostics 10(8):3503–3517. https://doi.org/10.7150/thno.42174
12. Sinha T, Panigrahi C, Das D, Panda AC (2022) Circular RNA translation, a path to hidden proteome. Wiley Interdiscip Rev RNA 13(1):e1685. https://doi.org/10.1002/wrna.1685
13. Li Z, Huang C, Bao C, Chen L, Lin M, Wang X, Zhong G, Yu B, Hu W, Dai L, Zhu P, Chang Z, Wu Q, Zhao Y, Jia Y, Xu P, Liu H, Shan G (2015) Exon-intron circular RNAs regulate transcription in the nucleus. Nat Struct Mol Biol 22:256–264. https://doi.org/10.1038/nsmb.2959
14. Chen LL (2020) The expanding regulatory mechanisms and cellular functions of circular RNAs. Nat Rev Mol Cell Biol 21(8):475–490. https://doi.org/10.1038/s41580-020-0243-y
15. Verduci L, Tarcitano E, Strano S, Yarden Y, Blandino G (2021) CircRNAs: role in human diseases and potential use as biomarkers. Cell Death Dis 12:468. https://doi.org/10.1038/s41419-021-03743-3
16. Liu CX, Chen LL (2022) Circular RNAs: characterization, cellular roles, and applications. Cell 185(12):2016–2034. https://doi.org/10.1016/j.cell.2022.04.021
17. Zhang X, Liang Z, Wang C, Shen Z, Sun S, Gong C, Hu X (2022) Viral circular RNAs and their possible roles in virus-host interaction. Front Immunol 13:939768. https://doi.org/10.3389/fimmu.2022.939768
18. Venkataraman S, Badar U, Shoeb E, Hashim G, AbouHaidar M, Hefferon K (2021) An inside look into biological miniatures: molecular mechanisms of viroids. Int J Mol Sci 22(6):2795. https://doi.org/10.3390/ijms22062795
19. Wang Y (2021) Current view and perspectives in viroid replication. Curr Opin Virol 47:32–37. https://doi.org/10.1016/j.coviro.2020.12.004
20. Navarro B, Rubino L, Di Serio F (2017) Small circular satellite RNAs. In: Hadidi A, Flores R, Randles J, Palukaitis P (eds) Viroids and satellites. Elsevier, London, pp 659–669
21. de la Peña M, Ceprián R, Cervera A (2020) A singular and widespread group of mobile genetic elements: RNA circles with autocatalytic ribozymes. Cells 9(12):2555. https://doi.org/10.3390/cells9122555
22. Darós JA, Flores R (1995) Identification of a retroviroid-like element from plants. Proc Natl Acad Sci USA 92(15):6856–6860. https://doi.org/10.1073/pnas.92.15.6856
23. Kos A, Dijkema R, Arnberg AC, van der Meide PH, Schellekens H (1986) The hepatitis delta (δ) virus possesses a circular RNA. Nature 323(6088):558–560. https://doi.org/10.1038/323558a0
24. Chang WS, Pettersson JHO, Le Lay C, Shi M, Lo N, Wille M, Eden JS, Holmes EC (2019) Novel hepatitis D-like agents in vertebrates and invertebrates. Virus Evol 5(2):vez021. https://doi.org/10.1093/ve/vez021
25. Netter HJ, Barrios MH, Littlejohn M, Yuen LKW (2021) Hepatitis delta virus (HDV) and Delta-like agents: insights into their origin. Front Microbiol 12:652962. https://doi.org/10.3389/fmicb.2021.652962
26. Lee BD, Neri U, Roux S, Wolf YI, Camargo AP, Krupovic M, RNA Virus Discovery Consortium, Simonds P, Kyrpides N, Gophna U, Dolja VV, Koonin EV (2023) Mining metatranscriptomes reveals a vast world of viroid-like circular RNAs. Cell 186(3):646–661.e4. https://doi.org/10.1016/j.cell.2022.12.039
27. Grabowski PJ, Zaug AJ, Cech TR (1981) The intervening sequence of the ribosomal RNA precursor is converted to a circular RNA in isolated nuclei of Tetrahymena. Cell 23(2):467–476. https://doi.org/10.1016/0092-8674(81)90142-2
28. Mörl M, Schmelzer C (1990) Group II intron RNA-catalyzed recombination of RNA in vitro. Nucleic Acids Res 18(22):6545–6551. https://doi.org/10.1093/nar/18.22.6545
29. Gumińska N, Zakryś B, Milanowski R (2021) A new type of circular RNA derived from nonconventional introns in nuclear genes of euglenids. J Mol Biol 433(3):166758. https://doi.org/10.1016/j.jmb.2020.166758
30. Zhang Y, Zhang XO, Chen T, Xiang JF, Yin QF, Xing YH, Zhu S, Yang L, Chen LL (2013) Circular intronic long noncoding RNAs. Mol Cell 51(6):792–806. https://doi.org/10.1016/j.molcel.2013.08.017
31. Talhouarne GJS, Gall JG (2018) Lariat intronic RNAs in the cytoplasm of vertebrate cells. Proc Natl Acad Sci USA 115(34):E7970–E7977. https://doi.org/10.1073/pnas.1808816115
32. Pek JW (2018) Stable intronic sequence RNAs engage in feedback loops. Trends Genet 34(5):330–332. https://doi.org/10.1016/j.tig.2018.01.006
33. Li X, Zhang JL, Lei YN, Liu XQ, Xue W, Zhang Y, Nan F, Gao X, Zhang J, Wei J, Yang L, Chen LL (2021) Linking circular intronic RNA degradation and function in transcription by RNase H1. Sci China Life Sci 64(11):1795–1809. https://doi.org/10.1007/s11427-021-1993-6
34. Tang TH, Rozhdestvensky TS, d'Orval BC, Bortolin ML, Huber H, Charpentier B, Branlant C, Bachellerie JP, Brosius J, Hüttenhofer A (2002) RNomics in archaea reveals a further link between splicing of archaeal introns and rRNA processing. Nucleic Acids Res 30(4):921–930. https://doi.org/10.1093/nar/30.4.921

35. Lykke-Andersen J, Aagaard C, Semionenkov M, Garrett RA (1997) Archaeal introns: splicing, intercellular mobility and evolution. Trends Biochem Sci 22(9):326–331. https://doi.org/10.1016/s0968-0004(97)01113-4
36. Starostina NG, Marshburn S, Johnson LS, Eddy SR, Terns RM, Terns MP (2004) Circular box C/D RNAs in Pyrococcus furiosus. Proc Natl Acad Sci USA 101(39):14097–14101
37. Danan M, Schwartz S, Edelheit S, Sorek R (2012) Transcriptome-wide discovery of circular RNAs in archaea. Nucleic Acids Res 40(7):3131–3142. https://doi.org/10.1093/nar/gkr1009
38. Berkemer SJ, Höner zu Siederdissen C, Amman F, Wintsche A, Will S, Hofacker IL, Prohaska SJ, Stadler PF (2015) Processed small RNAs in archaea and BHB elements. Genomics Comp Biol 1(1):e18. https://doi.org/10.18547/gcb.2015.vol1.iss1.e18
39. Becker HF, L'Hermitte-Stead C, Myllykallio H (2019) Diversity of circular RNAs and RNA ligases in archaeal cells. Biochimie 164:37–44. https://doi.org/10.1016/j.biochi.2019.06.011
40. Liu L, Li W, Li J, Zhao D, Li S, Jiang G, Wang J, Chen X, Bi C, Zhang X (2023) Circular guide RNA for improved stability and CRISPR-Cas9 editing efficiency in vitro and in bacteria. ACS Synth Biol 12(2):350–359. https://doi.org/10.1021/acssynbio.2c00381
41. Busa VF, Leung AKL (2021) Thrown for a (stem) loop: how RNA structure impacts circular RNA regulation and function. Methods 196:56–67. https://doi.org/10.1016/j.ymeth.2021.02.019
42. Liu CX, Nan F, Jiang S, Gao X, Guo SK, Xue W, Cui Y, Dong K, Ding H, Qu B, Zhou Z, Shen N, Yang L, Chen LL (2019) Structure and degradation of circular RNAs regulate PKR activation in innate immunity. Cell 177(4):865–880. https://doi.org/10.1016/j.cell.2019.03.046
43. Chen CK, Cheng R, Demeter J, Chen J, Weingarten-Gabbay S, Jiang L, Snyder MP, Weissman JS, Segal E, Jackson PK, Chang HY (2021) Structured elements drive extensive circular RNA translation. Mol Cell 81(20):4300–4318. https://doi.org/10.1016/j.molcel.2021.07.042
44. Sugimoto N, Kierzek R, Turner DH (1987) Sequence dependence for the energetics of dangling ends and terminal base pairs in ribonucleic acid. Biochemistry 26(14):4554–4558. https://doi.org/10.1021/bi00388a058
45. Janssen S, Schudoma C, Steger G, Giegerich R (2011) Lost in folding space? Comparing four variants of the thermodynamic model for RNA secondary structure prediction. BMC Bioinformatics 12:429. https://doi.org/10.1186/1471-2105-12-429
46. Cuesta JA, Manrubia S (2017) Enumerating secondary structures and structural moieties for circular RNAs. J Theor Biol 419:375–382. https://doi.org/10.1016/j.jtbi.2017.02.024
47. Fang LT (2011) The end-to-end distance of RNA as a randomly self-paired polymer. J Theor Biol 280(1):101–107. https://doi.org/10.1016/j.jtbi.2011.04.010
48. Yoffe AM, Prinsen P, Gelbart WM, Ben-Shaul A (2011) The ends of a large RNA molecule are necessarily close. Nucl Acids Res 39(1):292–299. https://doi.org/10.1093/nar/gkq642
49. Clote P, Ponty Y, Steyaert JM (2012) Expected distance between terminal nucleotides of RNA secondary structures. J Math Biol 65(3):581–599. https://doi.org/10.1007/s00285-011-0467-8
50. Turner DH, Mathews DH (2010) NNDB: the nearest neighbor parameter database for predicting stability of nucleic acid secondary structure. Nucl Acids Res 38:D280–D282. https://doi.org/10.1093/nar/gkp892
51. Nussinov R, Piecznik G, Griggs JR, Kleitman DJ (1978) Algorithms for loop matching. SIAM J Appl Math 35(1):68–82. https://doi.org/10.1137/0135006
52. Hofacker IL, Stadler PF (2006) Memory efficient folding algorithms for circular RNA secondary structures. Bioinformatics 22(10):1172–1176. https://doi.org/10.1093/bioinformatics/btl023
53. Zuker M, Stiegler P (1981) Optimal computer folding of large RNA sequences using thermodynamics and auxiliary information. Nucleic Acids Res 9:133–148. https://doi.org/10.1093/nar/9.1.133
54. Zuker M (1989) The use of dynamic programming algorithms in RNA secondary structure prediction. In: Waterman MS (ed) Mathematical methods for DNA sequences. CRC Press, Boca Raton, pp 159–184
55. Zuker M (1989) On finding all suboptimal foldings of an RNA molecule. Science 244(4900):48–52. https://doi.org/10.1126/science.2468181
56. Hofacker IL, Fontana W, Stadler PF, Bonhoeffer LS, Tacker M, Schuster P (1994) Fast folding and comparison of RNA secondary structures. Monatsh Chem 125:167–188. https://doi.org/10.1007/BF00818163
57. Lorenz R, Bernhart SH, Höner zu Siederdissen C, Tafer H, Flamm C, Stadler PF, Hofacker IL (2011) ViennaRNA Package 2.0. Algorithms Mol Biol 6:26. https://doi.org/10.1186/1748-7188-6-26
58. McCaskill JS (1990) The equilibrium partition function and base pariring probabilities for RNA secondary structures. Biopolmers 29(6–7):1105–1119. https://doi.org/10.1002/bip.360290621
59. Gruber AR, Bernhart SH, Lorenz R (2015) The ViennaRNA web services. Methods Mol Biol 1269:307–326. https://doi.org/10.1007/978-1-4939-2291-8_19
60. Wang F, Lei WL, Li B, Xie L, Tong Y, Xu X (2023) cRNAsp12 web server for the prediction of circular RNA secondary structures and stabilities. Int J Mol Sci 24(4):3822. https://doi.org/10.3390/ijms24043822
61. Markham NR, Zuker M (2008) Unafold. In: Keith JM (ed) Bioinformatics: structure, function and applications. Humana Press, Totowa, pp 3–31. https://doi.org/10.1007/978-1-60327-429-6_1

62. Reuter JS, Mathews DH (2010) RNAstructure: software for RNA secondary structure prediction and analysis. BMC Bioinformatics 11:129. https://doi.org/10.1186/1471-2105-11-129
63. Hofacker IL, Reidys CM, Stadler PF (2012) Symmetric circular matchings and RNA folding. Discret Math 312(1):100–112. https://doi.org/10.1016/j.disc.2011.06.004
64. Wuchty S, Fontana W, Hofacker IL, Schuster P (1999) Complete suboptimal folding of RNA and the stability of secondary structures. Biopolymers 49(2):145–165. https://doi.org/10.1002/(SICI)1097-0282(199902)49:2<145::AID-BIP4>3.0.CO;2-G
65. Dirks RM, Bois JS, Schaeffer JM, Winfree E, Pierce NA (2007) Thermodynamic analysis of interacting nucleic acid strands. SIAM Rev 49(1):65–88. https://doi.org/10.1137/060651100
66. Sahayasheela VJ, Sugiyama H (2024) RNA G-quadruplex in functional regulation of noncoding RNA: challenges and emerging opportunities. Cell Chem Biol 31(1):53–70. https://doi.org/10.1016/j.chembiol.2023.08.010
67. Lorenz R, Bernhart SH, Qin J, Höner zu Siederdissen C, Tanzer A, Amman F, Hofacker IL, Stadler PF (2013) 2D meets 4G: G-quadruplexes in RNA secondary structure prediction. IEEE Trans Comput Biol Bioinform 10(4):832–844. https://doi.org/10.1109/TCBB.2013.7
68. Zarringhalam K, Meyer MM, Dotu I, Chuang JH, Clote P (2012) Integrating chemical footprinting data into RNA secondary structure prediction. PLoS One 7(10):e45160. https://doi.org/10.1371/journal.pone.0045160
69. Kolberg T, von Löhneysen S, Ozerova I, Wellner KW, Hartmann RK, Stadler PF, Mörl M (2023) Led-seq—ligation-enhanced double-end sequence-based structure analysis of RNA. Nucleic Acids Res 51(11):e63. https://doi.org/10.1093/nar/gkad312
70. Lorenz R, Luntzer D, Hofacker IL, Stadler PF, Wolfinger MT (2016) SHAPE directed RNA folding. Bioinformatics 32(1):145–147. https://doi.org/10.1093/bioinformatics/btv523
71. Guo SK, Nan F, Liu CX, Yang L, Chen LL (2021) Mapping circular RNA structures in living cells by SHAPE-MaP. Methods 196:47–55. https://doi.org/10.1016/j.ymeth.2021.01.011
72. Xu T, He B, Sun H, Nie J (2022) Novel insights into the interaction between N6-methyladenosine modification and circular RNA. Mol Ther Nucleic Acids 27:824–837. https://doi.org/10.1016/j.omtn.2022.01.007
73. Tanzer A, Hofacker IL, Lorenz R (2019) RNA modifications in structure prediction—status quo and future challenges. Methods 156:32–39. https://doi.org/10.1016/j.ymeth.2018.10.019
74. Kierzek E, Zhang X, Watson RM, Kennedy SD, Szabat M, Kierzek R, Mathews DH (2022) Secondary structure prediction for RNA sequences including N6-methyladenosine. Nat Commun 13:1271. https://doi.org/10.1038/s41467-022-28817-4
75. Yao HT, Lorenz R, Hofacker IL, Stadler PF (2023) Mono-valent salt corrections for RNA secondary structures in the ViennaRNA package. Algorithms Mol Biol 18:8. https://doi.org/10.1186/s13015-023-00236-0. bioRxiv doi: 10.1101/2023.04.07.536000
76. Bunke H, Bühler U (1993) Applications of approximate string matching to 2D shape recognition. Pattern Recogn 26(12):1797–1812. https://doi.org/10.1016/0031-3203(93)90177-X
77. Gregor J, Thomason MG (1993) Dynamic programming alignment of sequences representing cyclic patterns. IEEE Trans Pattern Anal Mach Intell 15(2):129–135. https://doi.org/10.1109/34.192484
78. Maes M (1990) On a cyclic string-to-string correction problem. Inform Process Lett 35(2):73–78. https://doi.org/10.1016/0020-0190(90)90109-B
79. Mollineda RA, Vidal E, Casacuberta F (2002) Cyclic sequence alignments: approximate versus optimal techniques. Int J Pattern Rec Artif Intel 16(3):291–299. https://doi.org/10.1142/S0218001402001678
80. Will S, Stadler PF (2014) A common framework for linear and cyclic multiple sequence alignment problems. In: Brown D, Morgenstern B (eds) WABI 2014, Lecture notes in computer science, vol 8701, pp 135–147. https://doi.org/10.1007/978-3-662-44753-6_11
81. Thompson JD, Higgs DG, Gibson TJ (1994) CLUSTALW: improving the sensitivity of progressive multiple sequence alignment through sequence weighting, position specific gap penalties, and weight matrix choice. Nucl Acids Res 22(22):4673–4680. https://doi.org/10.1093/nar/22.22.4673
82. Mosig A, Hofacker IL, Stadler PF (2006) Comparative analysis of cyclic sequences: viroids and other small circular RNAs. In: Giegerich R, Stoye J (eds) Proceedings GCB 2006, Lecture notes in informatics, vol P-83, pp 93–102
83. Gotoh O (1982) An improved algorithm for matching biological sequences. J Mol Biol 162(3):705–708. https://doi.org/10.1016/0022-2836(82)90398-9
84. Barton C, Iliopoulos CS, Pissis SP (2014) Fast algorithms for approximate circular string matching. Algorithms Mol Biol 9:9. https://doi.org/10.1186/1748-7188-9-9
85. Fernandes F, Pereira L, Freitas AT (2009) CSA: an efficient algorithm to improve circular DNA multiple alignment. BMC Bioinformatics 10:230. https://doi.org/10.1186/1471-2105-10-230
86. Grossi R, Iliopoulos CS, Mercas R, Pisanti N, Pissis SP, Retha A, Vayani F (2016) Circular sequence comparison: algorithms and applications. Algorithms Mol Biol 11:12. https://doi.org/10.1186/s13015-016-0076-6
87. Ayad LAK, Pissis SP (2017) MARS: improving multiple circular sequence alignment using refined sequences. BMC Genomics 18:86. https://doi.org/10.1186/s12864-016-3477-5
88. Li H, Xie M, Wang Y, Yang L, Xie Z, Wang H (2021) riboCIRC: a comprehensive database of translat-

able circrnas. Genome Biol 22(1):79. https://doi.org/10.1186/s13059-021-02300-7
89. Hofacker IL, Fekete M, Stadler PF (2002) Secondary structure prediction for aligned RNA sequences. J Mol Biol 319(5):1059–1066. https://doi.org/10.1016/S0022-2836(02)00308-X
90. Bernhart SH, Hofacker IL, Will S, Gruber AR, Stadler PF (2008) RNAalifold: improved consensus structure prediction for RNA alignments. BMC Bioinformatics 9:474. https://doi.org/10.1142/s0219720008003886
91. Rivas E, Clements J, Eddy SR (2017) A statistical test for conserved RNA structure shows lack of evidence for structure in lncRNAs. Nat Methods 14:45–48. https://doi.org/10.1038/nmeth.4066
92. Rivas E (2023) RNA covariation at helix-level resolution for the identification of evolutionarily conserved RNA structure. PLoS Comput Biol 19(7):e1011262. https://doi.org/10.1371/journal.pcbi.1011262
93. Washietl S, Hofacker IL, Stadler PF (2005) Fast and reliable prediction of noncoding RNAs. Proc Natl Acad Sci USA 102(7):2454–2459. https://doi.org/10.1073/pnas.0409169102
94. Klapproth C, Zöztsche S, Kühnl F, Fallmann J, Stadler PF, Findeiss S (2023) Tailored machine learning models for functional RNA detection in genome-wide screens. Nucleic Acids Res Genom Bioinf 5(3):lqad072. https://doi.org/10.1093/nargab/lqad072
95. Washietl S, Hofacker IL (2004) Consensus folding of aligned sequences as a new measure for the detection of functional RNAs by comparative genomics. J Mol Biol 342(1):19–39. https://doi.org/10.1016/j.jmb.2004.07.018
96. Krüger J, Rehmsmeier M (2006) RNAhybrid: microRNA target prediction easy, fast and flexible. Nucleic Acids Res 34:W451–W454. https://doi.org/10.1093/nar/gkl243
97. Lorenz R, Hofacker IL, Bernhart SH (2012) Folding RNA/DNA hybrid duplexes. Bioinformatics 28(19):2530–2531. https://doi.org/10.1093/bioinformatics/bts466
98. Wachsmuth M, Domin GD, Lorenz R, Serfling R, Findeiß S, Stadler PF, Mörl M (2015) Design criteria for synthetic riboswitches acting on transcription. RNA Biol 12(2):221–231. https://doi.org/10.1080/15476286.2015.1017235
99. Findeiß S, Wachsmuth M, Mörl M, Stadler PF (2014) Design of transcription regulating riboswitches. Methods Enzymol 550:1–22. https://doi.org/10.1016/bs.mie.2014.10.029
100. Lin J, Zhang Y, Zeng X, Xue C, Lin X (2020) CircRNA CircRIMS acts as a microRNA sponge to promote gastric cancer metastasis. ACS Omega 5(36):23237–23246. https://doi.org/10.1021/acsomega.0c02991
101. Mückstein U, Tafer H, Hackermüller J, Bernhart S, Stadler PF, Hofacker IL (2006) Thermodynamics of RNA-RNA binding. Bioinformatics 22(10):1177–1182. https://doi.org/10.1093/bioinformatics/btl024
102. Busch A, Richter A, Backofen R (2008) IntaRNA: efficient prediction of bacterial sRNA targets incorporating target site accessibility and seed regions. Bioinformatics 24:2849–2856. https://doi.org/10.1093/bioinformatics/btn544
103. Mückstein U, Tafer H, Bernhard SH, Hernandez-Rosales M, Vogel J, Stadler PF, Hofacker IL (2008) Translational control by RNA-RNA interaction: improved computation of RNA-RNA binding thermodynamics. In: Elloumi M, Küng J, Linial M, Murphy RF, Schneider K, Toma CT (eds) BioInformatics research and development—BIRD 2008, Communications in computer and information science, vol 13. Springer, Berlin, pp 114–127. https://doi.org/10.1007/978-3-540-70600-7_9
104. Bernhart SH, Mückstein U, Hofacker IL (2011) RNA accessibility in cubic time. Algorithms Mol Biol 6:3. https://doi.org/10.1186/1748-7188-6-3
105. Sutandy FXR, Ebersberger S, Huang L, Busch A, Bach M, Kang HS, Fallmann J, Maticzka D, Backofen R, Stadler PF, Zarnack K, Sattler M, Legewie S, König J (2018) In vitro iCLIP-based modeling uncovers how splicing factor U2AF2 relies on regulation by cofactors. Genome Res 28(5):699–713. https://doi.org/10.1101/gr.229757.117
106. Wright PR, Richter AS, Papenfort K, Mann M, Vogel J, Hess WR, Backofen R, Georg J (2013) Comparative genomics boosts target prediction for bacterial small RNAs. Proc Natl Acad Sci USA 110(37):E3487–E3496. https://doi.org/10.1073/pnas.1303248110
107. Dudekula DB, Panda AC, Grammatikakis I, De S, Abdelmohsen K, Gorospe M (2016) CircInteractome: a web tool for exploring circular RNAs and their interacting proteins and microRNAs. RNA Biol 13(1):34–42. https://doi.org/10.1080/15476286.2015.1128065
108. Sankoff D (1985) Simultaneous solution of the RNA folding, alignment and protosequence problems. SIAM J Appl Math 45(5):810–825. https://doi.org/10.1137/0145048
109. Jiang T, Wang L, Zhang K (1995) Alignment of trees—an alternative to tree edit. Theor Comput Sci 143(1):137–148. https://doi.org/10.1016/0304-3975(95)80029-9
110. Hoechsmann M, Voß B, Giegerich R (2004) Pure multiple RNA secondary structure alignments: a progressive profile approach. IEEE/ACM Trans Comp Biol Bioinform 1(1):53–62. https://doi.org/10.1109/TCBB.2004.11
111. Gast FU (2003) A new structural motif in the left terminal domain of large viroids identified by covariation analysis. Virus Genes 26(1):19–23. https://doi.org/10.1023/a:1022369803448
112. Micheletti C, Di Stefano M, Orland H (2015) Absence of knots in known RNA structures. Proc Natl Acad Sci USA 112(7):2052–2057. https://doi.org/10.1073/pnas.1418445112

Circular RNA Identification and Characterization with CircRNAFlow: A Bioinformatics Approach

6

Edward A. Salinas and Yvonne J. K. Edwards

Abstract

Circular RNAs (circRNAs) play various roles in gene regulation and are known to be associated with the development of disease. They can act as microRNA sponges, interact with RNA-binding proteins, and even contain open reading frames (ORFs) that are translated to generate proteins. Scientists hope to better understand these covalently closed ring-like structures as they are not well-characterized.

Here, we present CircRNAFlow, a free and open-source Nextflow-based pipeline for in silico circRNA identification and characterization. This next-generation sequencing (NGS) pipeline takes paired-end FASTQ files as input and produces tables and graphical images that report identified circRNAs and their corresponding characterizations. Reports include CircTest to compare circRNA expression to host gene expression and across groups. Characterization includes ClusterProfiler, CRAFT (CircRNA Function prediction Tool), and deepTarget, which explore the regulatory roles of identified circRNAs.

In this chapter, we discuss the bioinformatics analyses carried out by the incorporated third-party software of CircRNAFlow [namely, FastQC, Flexbar, Bowtie2, STAR (Spliced Transcripts Alignment to a Reference), DCC (detect circRNAs from chimeric reads), CircTest, CRAFT, ClusterProfiler, and deepTarget]. This chapter, along with a companion GitHub repository, provides a tutorial on CircRNAFlow, including installation, running the software, and interpretation of the outputs. This chapter also provides some troubleshooting tips.

Keywords

Circular RNA · CircRNAFlow · Nextflow · Workflow development · RNA-seq · MicroRNA sponge · RNA-binding protein locus · Template for translation

E. A. Salinas
Department of Biochemistry and Molecular Genetics, Heersink School of Medicine, University of Alabama at Birmingham, Birmingham, AL, USA

Y. J. K. Edwards (✉)
Department of Biochemistry and Molecular Genetics, Heersink School of Medicine, University of Alabama at Birmingham, Birmingham, AL, USA

Gregory Fleming James Cystic Fibrosis Research Center, Heersink School of Medicine, University of Alabama at Birmingham, Birmingham, AL, USA

Department of Cell, Development and Integrative Biology, Heersink School of Medicine, University of Alabama at Birmingham, Birmingham, AL, USA
e-mail: yedwards@uab.edu

J. Xiao (ed.), *Advances in Circular RNAs*, Advances in Experimental Medicine and Biology 1485,
https://doi.org/10.1007/978-981-96-9428-0_6

1 Introduction

Circular RNAs (circRNAs) are single-stranded, covalently closed RNA molecules, which are generated through different mechanisms and found in all domains of life [1, 2]. Due to their close association with the development of human disease [3–9], the underlying biological mechanisms of circRNA biogenesis and degradation have recently been intensively investigated in eukaryotes [1, 10]. CircRNAs are generated from precursor mRNAs by backsplicing of exons when a splice donor (from the 5′ splice site) is covalently joined to an upstream splice acceptor (from the 3′ splice site) [10, 11]. Backsplicing is a noncanonical splicing event, which results in a covalently closed loop known as a backsplicing junction (BSJ). In contrast to their linear counterparts, circRNAs are unusually stable as the lack of a terminal end prevents them from being broken down through conventional RNA degradation pathways [6].

Recent studies have suggested that circRNAs play a crucial role in regulating gene expression. Some circRNAs provide a locus to bind microRNAs (miRNAs) or RNA-binding proteins (RBPs) [3, 12]. While circRNAs are generally categorized as noncoding, some of their segments may contain open reading frames that are subsequently translated to generate proteins and peptides [3, 12, 13]. CircRNAs are expressed in a tissue- and development stage-specific manner [3]. Due to these biochemical and biomolecular characteristics, they have been implicated in major cellular processes and, hence, in the pathophysiology of a range of diseases. CircRNAs may emerge as novel biomarkers for diagnosis, providing improved understanding of disease mechanisms [3, 4, 9, 14]..

In light of the above and motivated to study circRNAs , we developed the software CircRNAFlow, a Nextflow-based bioinformatics pipeline for in silico identification and characterization of circRNAs from next-generation sequencing (NGS) transcriptome data [1, 10]. CircRNAFlow was initially inspired by a protocol described by Jakobi and Dieterich [15]. Using Nextflow, this pipeline integrates the individual steps of the circRNA detection workflow to provide a single seamless process. CircRNAFlow identifies circRNAs and predicts circRNA functions such as miRNA sponge potential, RBP locus potential, and protein translation potential (Table 6.1). CircRNAFlow is designed for human or mouse transcriptome studies and NGS transcriptome data with paired-end read constructs.

CircRNAFlow is written in Nextflow [29], a workflow language and management system useful for authoring and executing bioinformatics pipelines; it thus inherits the benefits of all Nextflow-based pipeline features including resuming workflows by checkpointing, portability by utilizing containers, and scalability. If a workflow crashes due to malformed input data or faulty hardware, Nextflow's `resume` flag can be used to pick up where the pipeline left off, prior to any errors. CircRNAFlow can utilize containers such as Docker images [30] in Docker Hub and `*.sif` singularity images in Zenodo [27, 28]. Containers are software packages containing programs and an environment in which those programs can run. This combination enables portability. Finally, CircRNAFlow provides configuration files that enable it to run on both standalone Linux computers and high-performance computing clusters with the slurm scheduler [31]. Slurm supports batch-job submission of software, thereby enabling scaling, which is managed by Nextflow and orchestrated by Nextflow-based pipelines.

CircRNAFlow is among the first published circRNA pipelines using a formal pipeline language like Nextflow [14]. CircRNA analysis is an emerging area of study in RNA biology, and CircRNAFlow is available as a free resource to the scientific community to help advance the field.

Section 6.2 of this chapter presents details of CircRNAFlow and the component bioinformatics tools. Section 6.3 includes a tutorial on how to install and run CircRNAFlow. Section 6.4 provides a description of CircRNAFlow output. Finally, Section 6.5 concludes with a summary of

Table 6.1 A list of CircRNAFlow software dependencies

Software	Version	Note(s)	References
FastQC	0.11.9	Basic QC of FASTQ files. GPLv3 license	[16]
Flexbar	3.5.0	Adapter removal. BSD-3 license	[17]
Bowtie2	2.3.0	Ribosomal RNA removal. GPLv3 license	[18]
STAR	2.7.9a	Alignment to reference genome. MIT license	[19]
DCC	0.5.0	CircRNA identification. GPLv3 license	[20]
CircTest	Commit 0f5a86c	CircRNA differential expression. https://github.com/dieterich-lab/CircTest	[20]
ClusterProfiler	4.4.4	Function overrepresentation analysis of the circRNA host genes. https://github.com/YuLab-SMU/clusterProfiler	[21]
CRAFT	Note[a]	Prediction of circRNA interactions with miRNA and RBP. Prediction of circRNA open reading frames. GPLv3 license	[13]
MiRanda[b]	3.3a	MiRNA binding site prediction in circRNAs. GPL license	[22]
PITA[b]	Note[c]	MiRNA binding site prediction in circRNAs. GPL license	[23]
ORF-finder[b]	0.4.3	Open reading frame prediction https://www.ncbi.nlm.nih.gov/orffinder/	[13]
BeRBP[b]	Note[c]	RBP binding loci prediction in circRNAs https://bioinfo.vanderbilt.edu/beRBP/download.html	[24]
deepTarget	Note[d]	MiRNA binding site prediction in circRNAs. GPLv3 license	[25, 26]

[a]CRAFT [13] is based on the docker image annadalmolin/craft:v1.0
[b]The software is an incorporated component of CRAFT
[c]Un-versioned
[d]deepTarget [25] was downloaded from Zenodo [27, 28] via Kipoi [26] (https://zenodo.org/records/1466129)

this chapter and a discussion of the next steps for CircRNAFlow development.

2 Bioinformatics Characterization of CircRNAs with CircRNAFlow

In this section, the component programs in the CircRNAFlow pipeline are described, focusing on their individual roles and respective positions within the larger workflow (Fig. 6.1 and Table 6.1).

CircRNAFlow was first published in 2023 as part of a search for circRNAs in the whole blood transcriptomes of cystic fibrosis (CF) patients [14]. While CircRNAFlow originally began in DSL1, an older dialect of Nextflow, the pipeline has been updated to utilize Nextflow DSL2. This chapter highlights the DSL2 version of the pipeline (Fig. 6.1 and Table 6.1). DSL2 provides improved readability compared to DSL1. Importantly, in contrast to DSL1, DSL2 is more recent and compatible with the current versions of Nextflow (see https://www.nextflow.io/docs/latest/dsl1.html#migrating-from-dsl-1).

The design of CircRNAFlow was inspired by the initial protocol that was published [15]. CircRNAFlow extends that protocol by incorporating Nextflow pipeline technology and including additional circRNA characterization steps. CircRNAFlow is designed to take whole transcriptome sequencing data and produce various reports including quality assessment, circRNA identification, circRNA characterization, and more.

CircRNAs can be identified through NGS RNA-seq data with detection of the backsplicing junction. CircRNAFlow incorporates software designed for this purpose, e.g., DCC [20]. Observation of differential expression relative to the host gene can assist in this regard [15]. Once circRNAs are identified in the NGS data, subsequent characterization steps such as miRNA binding analysis (Table 6.1) can contribute to understanding the role(s) circRNAs may play in gene regulation.

It should be noted that CircRNAFlow is a predictive software workflow and, as such, there will sometimes be errors in predictions. Though

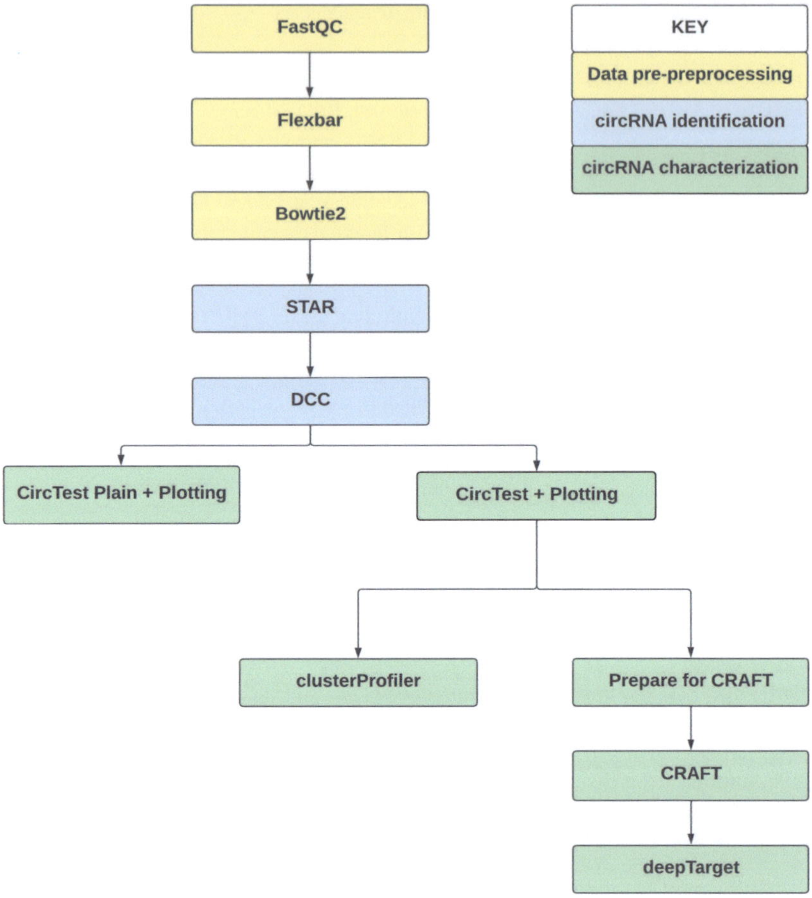

Fig. 6.1 A schematic flowchart highlighting the conceptual steps in CircRNAFlow. The data preprocessing steps include FastQC for quality control, Flexbar for adapter removal, and Bowtie2 to align against rRNA reference for rRNA removal. The data pre-preprocessing steps are shown as the yellow boxes. Next, STAR and DCC are run to align reads, find backsplicing junctions, and identify circRNAs. These circRNA identification steps are shown in the blue boxes. Finally, the characterization steps shown in the green boxes include CircTest, plotting, ClusterProfiler, CRAFT, and deepTarget. Table 6.1 provides more details.

results with circAtlas [32, 33] identifiers might be considered genuine, scientists should always consider using lab-based experimental techniques such as primers and PCR to experimentally validate circRNAs [1, 10, 34, 35].

As discussed below, CircRNAFlow begins with several quality control (QC) and filtering stages (Fig. 6.1 and Table 6.1). CircRNAFlow proceeds to circRNA identification and concludes with subsequent analyses and characterization steps (Figs. 6.1 and 6.2 and Table 6.1).

2.1 Initial QC Steps of CircRNAFlow

Basic QC CircRNAFlow begins with quality control and data quality assessment processes (Fig. 6.1). First, it incorporates FastQC [16] to perform basic quality checks of the NGS data. This allows the user to determine whether the quality and quantity of the data is sufficient for the intended analysis [1, 10].

Adapter Removal Next, CircRNAFlow applies Flexbar [17] for the removal of adapter sequences (Fig. 6.1 and Table 6.1). The input adapter file, in FastA format, is specified as a parameter to CircRNAFlow. Some of the settings used for Flexbar are inherited from the Jakobi and Dieterich (2018) protocol [15]; we set a minimum read length of 30, no uncalled bases, quality trimming at the tail, and a minimum quality threshold of 28 (Sanger quality).

rRNA Removal The last QC step is a call to Bowtie2 [18] to align the NGS data to a ribosomal RNA (rRNA) database (Fig. 6.1 and Table 6.1). Sequences that align to rRNA are then

```
quick_start_output
├── circtest
├── circtest_plain
├── circtest_plotting
├── circtest_plotting_plain
├── cluster_profiler
├── craft
│   ├── 15_49876278_49879459_ATP8B4_MINUS_M
│   │   └── functional_predictions
│   ├── 15_49876278_49879459_ATP8B4_MINUS_O
│   │   └── functional_predictions
│   ├── 15_49876278_49879459_ATP8B4_MINUS_R
│   │   └── functional_predictions
│   ├── 15_49876278_49885925_ATP8B4_MINUS_M
│   │   └── functional_predictions
│   ├── 15_49876278_49885925_ATP8B4_MINUS_O
│   │   └── functional_predictions
│   └── 15_49876278_49885925_ATP8B4_MINUS_R
│       └── functional_predictions
├── craft_prep
├── DCC
├── deeptarget
├── FastQC
├── flexbar_logs
├── rRNA_cleaned_star_input
│   └── adapter_removed_rRNA_filtered
├── star_output_first
├── star_output_pair
└── star_output_second

28 directories
```

Fig. 6.2 The output directory structures generated by CircRNAFlow. The directory structure of the `quick_start_output` directory as seen by the output of the Linux command `tree` invoked with `-d` configuration to focus on the directory structure

filtered out. Previous analyses have shown that circRNA identification on rRNA-depleted data produces a lower false-positive rate in identifying circRNA [36]. The main steps of the pipeline (Fig. 6.1) are applied to the reads that do not align to the reference rRNA.

2.2 Core CircRNA Identification Steps of CircRNAFlow

Alignment and CircRNA Identification DCC, which detects circRNAs from chimeric reads [20], is the core circRNA identification tool used by CircRNAFlow (Fig. 6.1). It analyzes RNA-seq NGS data to identify alignments across possible backsplicing junctions. Before DCC is invoked by CircRNAFlow, the STAR aligner [19] is used to map short read NGS data to the reference genome for subsequent analysis by DCC. STAR is invoked by CircRNAFlow with special parameters that promote and enable circRNA identification. STAR's use of the GTF file with `–sjdbGTFfile` helps identify splice junctions. DCC can make use of a repeat file in GTF format; this file indicates repetitive regions of the genome and can be used to reduce false-positive results.

2.3 CircRNA Characterization Steps of CircRNAFlow

Differential Expression Analysis The first characterization step after circRNA identification (Fig. 6.1) utilizes CircTest [20]. CircTest performs a statistical analysis, applying a beta-binomial test to determine whether the circRNA is differentially expressed compared to its host gene and across samples. The inputs to the test are the read counts for both circRNA and the linear gene. CircTest is run in two modes: one with default parameters and the other in which thresholds are allowed to vary (Tables 6.2 and 6.3). For each CircTest invocation, a custom plotting module is invoked to

Table 6.2 CircTest parameter names and explanations

Parameter	Plain value (default)	Short description
Filtercount	5	Minimum read count filtering
Filtersample	4	A minimum number of samples necessary to have at least filtercount number of reads in the circRNA counts
Percentage	100%	A minimum percentage of circRNAs accounting for the transcripts in a cohort/group
Alpha	0.05	adjusted p-value/FDR (false discovery rate) cutoff

The parameters used to prefilter data before being subjected to CircTest [20]. The values used in the `plain` analyses are the program's default values. A brief description of the parameters is given in the rightmost column

generate bar plots of BSJ read counts for differentially regulated circRNAs (Fig. 6.3). Additionally, the circAtlas database [32, 33] is incorporated into CircRNAFlow to facilitate identification of both previously identified and publicly annotated circRNAs; CircRNAFlow annotates any known circRNA in the given circAtlas database with the circAtlas identifier (Fig. 6.3).

Functional Overrepresentation Analyses The second characterization step of CircRNAFlow (Fig. 6.1) utilizes ClusterProfiler [21]. ClusterProfiler performs a statistically based gene ontology (GO) and Kyoto Encyclopedia of Genes and Genomes (KEGG) pathway overrepresentation analysis to identify common themes among differentially expressed circRNA host genes (Fig. 6.4). The functions enrichGO and enrichKEGG [21] for overrepresentation analyses are at the center of ClusterProfiler's use in CircRNAFlow. The host genes of circRNAs that are differentially expressed at various fold changes and adjusted P-value significance thresholds are used as inputs to these functions (Fig. 6.4).

Prediction of Molecular Interactions and Coding Potential Following CircTest, CRAFT is run (Fig. 6.1). CRAFT is a *CircRNA Function prediction Tool* [13]. Based on sequence analysis, CRAFT identifies the miRNAs and RBPs likely to bind to the circRNA (Figs. 6.5 and 6.6). CRAFT also identifies open reading frames present in the circRNAs. CRAFT can be run in three modes: `M` (miRNA), `O` (ORF), and `R` (RNA-binding protein).

When run in the `M` mode, CRAFT looks for miRNA binding loci using MiRanda 3.3a [22] and PITA [23]. MiRanda first performs a sequence complementarity search using dynamic-programming. Second, it completes a thermodynamic calculation to explore the energetics of miRNA/circRNA interactions. PITA also performs energy calculations.

(a) For `O`, ORF finder (https://www.ncbi.nlm.nih.gov/orffinder/) is run with a threshold of ≥30 nt on the positive strand.
(b) In the `R` mode, CRAFT searches for interactions with RNA-binding proteins using the tool called binding estimation for human RBPs (beRBP) [24]. Both protein-specific and generic models are calculated, after which, random forest models and position-weighted matrices (PWMs) are incorporated to make final estimates.

Only circRNAs that pass a default CircTest-adjusted *p*-value threshold of 1.0 and a top-n threshold of 30 are submitted to CRAFT. The adjusted p-value threshold is a statistical threshold, while the top-n threshold is strictly a way to limit the number of CRAFT invocations and is applied after sorting by the adjusted *p*-value.

Finally, CircRNAFlow employs a deep learning model called deepTarget [25, 26] (Fig. 6.1) to yield probability estimates of whether a given miRNA binds to a specific locus of a circRNA (Table 6.4). CircRNAFlow

Table 6.3 Example output from CircTest

Chr	Start	End	Gene	sig_p	Group_1 ratio_mean	Group_2 ratio_mean	HC1	HC2	HC3	CF1	CF2	CF3	*p*-val
15	49,876,278	49,879,459	ATP8B4	0.002	0.026	0.260	0	3	0	5	21	21	0.001
15	49,876,278	49,885,925	ATP8B4	0.537	0.102	0.137	0	2	2	1	4	6	0.537

Example output from CircTest for two circRNAs from the *ATP8B4* host gene using thresholds filtercount 1, filtersample 1, percentage 0.01, and alpha 1. sig_p is the adjusted p-value, while the raw p-value is under column p-val. The ratio means, compared against each other, are the values from the ratio a/(a + b) where a and b are read counts from the circRNA and the host gene, respectively. Columns HC1, HC2, and HC3 show the circRNA counts for the healthy controls (Fig. 6.3). CF1, CF2, and CF3 show the circRNA counts for the CF cases (Fig. 6.3). For both records, the start–end regions are both exonic and are on the negative strand (Fig. 6.3). Additionally, both have `OverallRegion` values of `CDS,three_prime_utr,gene,exon,transcript`

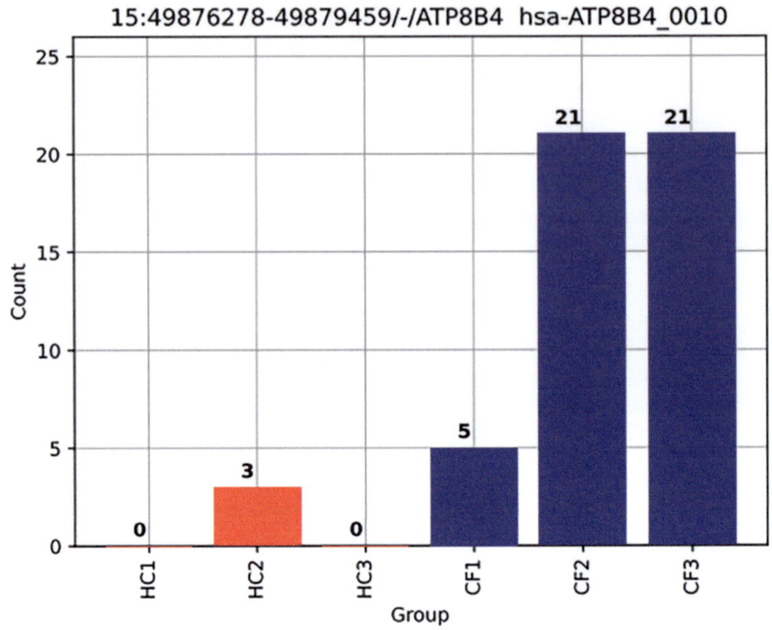

Fig. 6.3 Example plots generated by CircRNAFlow plotting modules [14]. Dysregulated circRNAs in CF samples compared with healthy controls. Plots of BSJ counts for upregulated circRNAs in CF compared to non-CF in circRNA hsa-ATP8B4_0010. The circRNA read counts (Table 6.3) in the healthy control group (three samples) versus the CF group (three samples) are shown in red and blue bars, respectively. The mean comparison was statistically significant (Table 6.3), and the blue bars are much taller.

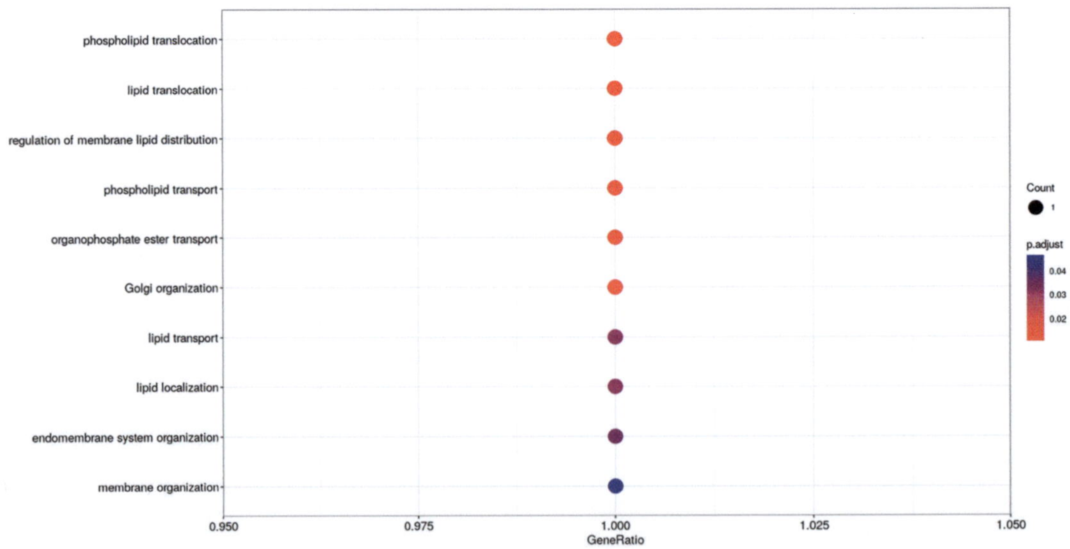

Fig. 6.4 Example ClusterProfiler output. The plot file named `cp_input.tsv.sig.thresh.0.05.BP.tiff` from the ClusterProfiler step of the pipeline is available in the uncompressed file `cp_outputs.zip` output directory named `cluster_profiler`. A gene ontology biological process overrepresentation plot for the host gene (*ATP8B4*) of the differentially expressed circRNA (hsa_ATP8B4_0010). The *y*-axis shows the overrepresented biological process gene ontologies. The *x*-axis shows the GeneRatio: the overlap between the genes of interest and the gene set divided by the number of all unique genes of interest [21]. In this example analysis, only one gene, *ATP8B4*, is input to the analysis, so the GeneRatio is always 1.0.

```
quick_start_output/craft$find . -iname "*.html"
./15_49876278_49879459_ATP8B4_MINUS_M/graphical_output/15:49876278-49879459/functional_predictions_15:49876278-49879459.html
./15_49876278_49879459_ATP8B4_MINUS_O/graphical_output/general/functional_predictions_all_circRNAs.html
./15_49876278_49879459_ATP8B4_MINUS_R/graphical_output/general/functional_predictions_all_circRNAs.html
./15_49876278_49885925_ATP8B4_MINUS_M/graphical_output/15:49876278-49885925/functional_predictions_15:49876278-49885925.html
./15_49876278_49885925_ATP8B4_MINUS_O/graphical_output/general/functional_predictions_all_circRNAs.html
./15_49876278_49885925_ATP8B4_MINUS_R/graphical_output/general/functional_predictions_all_circRNAs.html
```

Fig. 6.5 An example of the HTML CRAFT output. If the Linux command `find . -iname '*.html'` is run from the quick_start_output CRAFT directory, its output lists the six HTML output files generated by CRAFT.

CircRNA	Gene name	RBP	Number of RBP binding sites	CircRNA length	Density	Uniprot
All	All	All	All	All	All	All
15:49876278-49879459	ATP8B4	NOVA2	1	466	0.21	Q9UNW9
15:49876278-49879459	ATP8B4	ZFP36L1	1	466	0.21	A0A024R658,B3KNA8,Q07352
15:49876278-49879459	ATP8B4	ZFP36L2	1	466	0.21	P47974
15:49876278-49879459	ATP8B4	ZFP36	1	466	0.21	M0QY76,P26651

Showing 1 to 4 of 4 entries

Previous 1 Next

Copy CSV Excel PDF Print

Search:

Fig. 6.6 Example CRAFT output. A screenshot of a table from the RBP prediction by CRAFT in the file 15_49876278_49879459_ATP8B4_MINUS_R/graphical_output/general/functional_predictions_all_circRNAs.html. The table is in the RNA-binding protein (RBP) binding sites prediction section of the file and lists the four RBPs predicted to interact with the circRNA hsa-ATP8B4_0010 (Fig. 6.3). The first column provides the circRNA genome locus (Fig. 6.3). The second column identifies the host gene. The third column displays the RBP. The fourth column highlights the number of RBP interactions. The fifth column gives the circRNA length in nucleotides. The sixth column indicates density, which gives the fraction of the circRNA covered by the binding site. Finally, the seventh column offers Uniprot [37] protein accession identifiers with links to the Uniprot [37] for the RBP

uses mirbase [38] a reference of miRNAs with deepTarget. The deepTarget model was trained with an encoder–decoder architecture that eliminates the need for sequence alignments. Final estimates are made by way of a recurrent neural network (RNN) [25]. The versions of all CircRNAFlow's software dependences are listed in Table 6.1.

3 Practicum: A Short Walk-Through Exercise and Troubleshooting Notes

The online GitHub repository provides the most up-to-date information for installing and running CircRNAFlow. The referenced README, along with this chapter, offer guidelines and troubleshooting tips for initiation of execution and application to other datasets.

Table 6.4 Example output from deepTarget

summary_id	microRNA_id	microRNA_seed	microRNA_num_candidate_positions	has_any_trans_bsj	seed_position	trans_bsj_indicator	pred_0	pred_1
119	hsa-miR-181a-2-3p	CAGUGG	2	True	19	True	0.045	0.955
144	hsa-miR-212-3p	ACUGUU	2	True	7	True	0.661	0.339
193	hsa-miR-132-3p	ACUGUU	2	True	7	True	0.005	0.995
34	hsa-miR-23a-3p	AUGUGA	1	False	228	False	0.063	0.937
43	hsa-miR-26b-3p	GAACAG	1	False	241	False	0.001	0.991
50	hsa-miR-30a-5p	GUUUAC	2	False	46	False	0.362	0.632

Some example output from the file
`15_49876278_49879459_ATP8B4_MINUS.dt_results.norm.tsv`. For the first three rows, the seed position is near the start of the circRNA. As a result, the binding site spans the backsplicing junction as indicated by the `True` indicator in the `trans_bsj_indicator` column. For the first row, under the pred_1 column, hsa-miR-181a-2-3p has a 95.52% predicted probability to bind at the indicated seed. The `microRNA_num_candidate_positions` column indicates the number of sites where the miRNA might bind and the number of rows for it in the table

3.1 The Online Quick Start

The authors consider the GitHub repository for CircRNAFlow to be a living and evolving repository for the Nextflow code, the supporting scripts, and the documentation. The scripts, the workflow, and the documentation will continue to be updated after this chapter is published. The below link points to the quick start page in the repository, which provides a user guide to download CircRNAFlow and the associated test data and run the CircRNAFlow pipeline on the test data. The quick start includes the script `run_pipe_demo.sh`. The user may want to modify the `PROFILES` variable in the script. Guidance is offered on setting the `PROFILES` variable. The next section, about `PROFILES`, includes additional notes on using and selecting profiles.

The quick start guide for the DSL2 version of CircRNAFlow is available at: https://github.com/yjkedwards/circRNAFlow/tree/main/quick_start_2

The quick start guide also provides links to additional sample data, which can be downloaded from Zenodo [27, 28]. Section 6.4 of this chapter covers output from CircRNAFlow, based on a successful run with the downloaded test data, after following the notes and instructions.

The reader should first consult the README page using the above link, especially during the first run. Although the above-referenced quick start page offers some directions and guidance for running CircRNAFlow, additional troubleshooting notes are provided below to highlight potentially confusing or otherwise problematic steps.

3.1.1 Notes About the Provided Test Data

The test dataset is provided in three files. The first file contains the sample input data for the pipeline. This test dataset was extracted from a project comparing cystic fibrosis (CF) samples to healthy controls [14]. CircRNAFlow was applied in that project and a set of differentially expressed circRNAs was discovered. From that set, based on the intersection of statistically significant results from CircTest (with sufficiently low *p*-values), ClusterProfiler (non-empty tests of enrichment), and an invocation of CRAFT (yielding some beRBP [24] results), as well as membership in circAtlas [32, 33], two circRNAs were selected to become reference data: hsa-ATP8B4_0010 and hsa-ATP8B4_0053 from the *ATP8B4* gene (Table 6.3; Figs. 6.3 and 6.4). ATP8B4 is a component of the P4-ATPase flippase complex. Mammalian P4-ATPases localize to various subcellular organelles or the plasma membrane, where they translocate various lipids (Fig. 6.4). The test data from a CF study [14, 39] shows some of the top most enriched biological process gene ontologies include phospholipid translocation, lipid translocation, regulation of membrane lipid distribution, phospholipid transport, lipid transport, and lipid localization (Fig. 6.4). Lipid translocation is an important process for many cellular activities [40, 41]. Abnormal lipid levels have been observed in the tissues of CF patients [42, 43]. It is likely that ATP8B4 and/or the circRNA hsa-ATP8B4_0010 have some role in dysregulated lipids in CF [14] (Figs. 6.3 and 6.4).

The second data file available for download contains reference input data for deepTarget and CRAFT (formatted BLAST data [44] and hg38 reference data); the third data file contains singularity image files (`*.sif`), which can be used with singularity to run CircRNAFlow without downloading any images from Docker Hub.

Notes About Using Your Own Data When using your own data, you may need to edit the provided `demo.config` configuration file (or write your own) or the provided `run_pipe_demo.sh` script. For additional information on Nextflow configuration files and their format, please consult https://www.nextflow.io/docs/latest/config.html#configuration-file. The provided configuration file and bash script are available as a model for any new configuration file or script.

3.2 Troubleshooting: Initiating Execution of the Workflow

3.2.1 Profiles and Profile Selection (Container and Executor)

The provided Nextflow configuration file makes use of Nextflow configuration profile capabilities. Some key ideas of profiles are as follows:

(a) Each profile includes any number of settings within it.
(b) A single profile can be selected by simply specifying its name.
(c) Multiple profiles can be invoked by separating their names with commas when passing them to the Nextflow `-profile` command-line parameter.

See https://www.nextflow.io/docs/latest/config.html#config-profiles for additional information about Nextflow profiles.

The provided configuration file contains several profiles for the executor. The executor is the computer system that runs the processes of the workflow, for example, a high-performance computer cluster or a laptop. The configuration file provided also contains several profiles for the containers. The containers are software environments in which the processes run, for example, docker containers or singularity containers. The config file supports workflow-wide profile use, meaning that once a profile is selected, it applies to the entire workflow and not just some portion of the workflow. Accordingly, in the `run_pipe_demo.sh` script, users should specify exactly two profiles: one profile for an executor and the other for a container system. A comma should separate them in the `PROFILES` variable.

The provided config file supports two profiles for the executor: `slurm` and `local`.

`local` Executor PROFILE The local executor profile is used to execute pipeline steps on the same computer that runs Nextflow. If using this profile, consider setting the `queueSize` variable in the config file. For our 256-GB, 24-core server, we used 6 for the `queueSize`.

`slurm` Executor PROFILE The `slurm` executor is used to run pipeline steps via the `slurm` scheduler [31]. If using the `slurm` executor, users will need to edit the process-specific queue variables in the config file. The variable sets the partition to which the jobs are submitted.

The config file supports three container modes: `docker`, `singularity`, and `singularity_local_sifs`.

`docker` and `singularity` Container PROFILEs If using the docker container mode, no other changes are necessary. The docker software should simply be available and installed on the computer. Similarly, if singularity is used, then it must be installed on the computer system. Whether using the docker or singularity container profile, the images are pulled from Docker Hub. In the case of singularity, they are first converted to a singularity format after download.

singularity_local_sifs PROFILE The `singularity_local_sifs` profile is available and uses copies of `*.sif` singularity image files instead of downloading container data from Docker Hub. To use this mode, download the file in the link named `lumacaftor_small_test_data_JUST_SIFs_DSL2.tar.gz` and uncompress it via the following Linux/bash command:

```
tar -xvzf lumacaftor_small_test_data_
JUST_SIFs_DSL2.tar.gz
```

Paths to the `*.sif` files should be set correctly in the config file and will need to be adjusted if the decompression is made in any directory other than the `quick_start_2` directory. When the `singularity` profile is used, `singularity` converts the downloaded images from the Docker Hub to singularity format. During this conversion, instances have been observed where there is a transient error during the conversion. By providing the `*.sif` files and using this profile, the possibility of encountering those errors is eliminated. If those errors are observed, simply rerun the pipeline using the `-resume` flag (which is already present in the `run_pipe_demo.sh` script) to resume the pipeline where the error occurred.

3.3 Troubleshooting: When Using Other Data

Several steps involved in using other data are discussed below including using other FASTQ data as well as setting up other comparisons. In all cases, readers are encouraged to inspect the provided test data to observe data that are in the proper format. This proper format, in cases where there is confusion, may serve as an example upon which users can model their own input to yield a successful execution of CircRNAFlow.

3.3.1 Troubleshooting: FASTQ File Input Names

The FASTQ input file names should follow a naming scheme. As a trade-off to not using a sample sheet, names of input files are used to pair `R1` and `R2 fastq.gz` data together. File names should be named not unlike one of the files used in the test data, i.e., `SRR8383065_R1_f.fastq.gz`, and follow the three rules in series:

(a) Start alphanumeric (no spaces, no underscores, just letters and numbers), for example, SRR8383065.
(b) Second, have `_R1_` (or `_R2_`).
(c) Finally, end with `f.fastq.gz`.

For example, one of the R1 input data is named `SRR8383065_R1_f.fastq.gz`, where `SRR8383065` corresponds to the first part above, `_R1_` to the second, and, finally, `f.fastq.gz` to the last.

3.3.2 Troubleshooting: On Defining Comparisons with Your Own Data

This section discusses the two files used to set up the groups and comparisons and refers to an additional example file for comparisons.

As discussed in the section in README called On Defining Comparisons, the two files, `cohort_comp_conf.json` and `comp_list.txt`, are used to define comparisons. If using the provided sample data, then these files are ready and neither modification nor opening them is necessary. However, if using your own data, editing them will be necessary.

The text file lists comparisons to be made. The JSON file [45] lists additional information including associating sample names with FASTQ files. To understand how to use these files, first consult the sample download data (in the file `lumacaftor_small_test_data.circrnaflow.tar.gz`) referenced in README. Second, the JSON file is objectively the more complex of the two. The repository contains an additional example file for this in the file `cohort_comp_conf_BIG_example.json`.

Additional Example File cohort_comp_conf_BIG_example.json This JSON file defines five comparisons, namely, `main`, `cf`, `males`, `females`, and `wt`, where `main` compares CF samples to non-CF controls; `cf` compares CF males with CF females; `males` compare CF males with non-CF males; and, finally, `wt` compares non-CF males with non-CF females.

The Main Comparison in cohort_comp_conf_BIG_example.json In the `main` comparison, there are six CF samples and six non-CF samples. The value 6 for the `main` key in the `comparison_counts` section of the file indicates that the first six belongs to the CF group for the samples involved in the `main` comparison. The remaining 12 – 6 = 6, by process of elimination, leaves the remaining six samples for the non-CF control group. These 12 samples are listed in lines 34 to 46 of the file in the section called `cohort_comps`. The `cohort_comps` section of this file is a dictionary of comparisons, which assigns a list of samples (as values) involved in each comparison (the key). Next in the `file_samp_ren` section of the file, sample names and FASTQ file names get their associations. These are used to rename columns in data generated by DCC for viewing CircTest output and for plotting. It should be noted that this JSON file was defined before the naming scheme for `fastq.gz` files was defined, so there should be no concern that the file names in it violate the naming scheme. Finally, in the section named `comp_names`, the pair of cohorts in each comparison is named. The main part of this section imposes the labels of the CF and healthy controls in lines 74–76 of the JSON file. These labels are used in the plotting routines after CircTest.

Additional Example File comp_list_BIG_example.txt The JSON file `cohort_comp_conf_BIG_example.json` contains five comparisons. These are listed in the corresponding `comp_list` file `comp_list_BIG_example.txt`.

3.3.3 Troubleshooting: CircAtlas

Version 2 of the circAtlas database [32, 33], acquired from the circAtlas website in 2022, is used. Provided that the bed format does not change, any newer version of the file should be compatible with CircRNAFlow. Any different files should be formatted as a tab-separated file, which includes column headers. The headers should be `Chro`, `Start`, `End`, `Stand`, and `Name`. The fourth column is five characters long and does not contain an "`r`." Additionally, data in the Start and End fields should be integers (per the bed format, https://genome.ucsc.edu/FAQ/FAQformat.html). The `Stand` field contents should be either "+" or "−."

3.3.4 Troubleshooting: Setup for CRAFT

Compared to circAtlas, the CRAFT data setup is more complex. The setup of CRAFT for the pipeline was the most complex part of the pipeline's composition. Per the documentation in the CRAFT repository (https://github.com/annadalmolin/CRAFT), the setup requires several files or sets of files:

(a) `params.txt` file (whose contents provide parameters for the components of CRAFT, e.g., beRBP [24]),
(b) `list_backsplice.txt`, `backsplice_gene_name.txt`, and `path_files.txt` (these are handled by CircRNAFlow, and the user does not need to be concerned with making them).
(c) Reference data files (e.g. genome, fasta, gtf, formatted BLAST index files [44])

Since the three files listed in point b above are generated by CircRNAFlow, we first offer notes on the `params.txt` and then the naming, content, and generation of the reference data.

params.txt Passed to CircRNAFlow via `craft_params`, the `params.txt` configuration file has its content filled according to the instructions described in the CRAFT GitHub repository. We note that `hg38` on line 5 of the `params.txt` must match the prefix for the genome in the `.fa` file in `ref_data/craft_input` (`hg38.fa`) in the case of the test data. The pipeline parameter `craft_ref_path_file` path is relative to the craft data. Since, as shown with the example data, the file is located at `ref_data/craft_input/Homo_sapiens.GRCh38.dna.primary_assembly.fa`, accordingly, `craft_ref_path_file` is simply named as `Homo_sapiens.GRCh38.dna.primary_assembly.fa`.

Reference Data Files The naming and placement of the reference data is potentially the most confusing aspect of setting up data for CRAFT.

One might suppose that CRAFT analysis could be done with a single reference file. Though perhaps possible, to mimic the contents and follow the example in the CRAFT GitHub repository, we decided to use multiple, but differently named, copies of the same reference genome. That is why two `.fa` files are present under `ref_data/craft_input/`.

The reference data files must be in their own separate directory. In the provided sample data, that directory is `ref_data/craft_input`. The two reference genomes (`hg38.fa` and `Homo_sapiens.GRCh38.dna.primary_assembly.fa`) should also be there, and it should also contain a copy of BLAST-formatted indices [44] for use by beRBP. The indices can be created, for example, by commands like `makeblastdb -dbtype nucl -in mm10.fa -out mm10` followed by `makembindex -input mm10 -iformat blastdb` in the case of using mm10 genome reference data when using mouse models.

4 Exploring Analyses Results

After executing the pipeline using the provided configuration file, an output directory called `quick_start_output` (Fig. 6.2) is created, which is modifiable in the configuration file. It contains outputs generated from most of the steps of the pipeline. Some steps, for example, the Flexbar step, have no or partial output in the `quick_start_ouput` directory. The FASTQ files, having been processed by Flexbar, are considered intermediate/temporary files and are absent from the formal output, but their log files, comprising report logs, are part of the formal output. An example of the output directory structure, generated by the Linux `tree` command, is seen in Fig. 6.2.

Some outputs of the pipeline are in compressed zip files. If a zip file is found in the output directory, simply uncompress it by running the Linux command `unzip FILE` and replacing `FILE` with the actual file name.

4.1 Output from the Initial QC Steps of CircRNAFlow

FastQC Inside the FastQC subdirectory, the zip files of the FastQC output from each of the FASTQ files can be found. Un-compression of the files yields a file named `fastqc_data.txt` containing a short tab-separated-values report file. A more human reader-friendly html file `fastqc_report.html` is also available. Opening the html file with a Web browser enables human-readable viewing and perusal. The FastQC website https://www.bioinformatics.babraham.ac.uk/projects/fastqc/ in its Example Reports section has links to examples of good data and bad data. If adapter problems are observed in the output, supplying an adapter FASTA file for Flexbar on subsequent runs of the pipeline may affect downstream results.

Flexbar In the `flexbar_logs` directory, logs from running Flexbar [17] are available. The first section of any Flexbar log file lists the filtering criteria including passed command-line arguments. The adapter removal statistics section contains notes about adapter filtering and related statistics. The last section on filtering statistics contains count information about reads that are filtered or trimmed. Some percentage information is available for some statistics (such as remaining reads).

Bowtie2 (rRNA Filtering) The final QC step of CircRNAFlow is mapping the data utilizing Bowtie2 [18] against an rRNA database. For each FASTQ read pair, the directory `rRNA_cleaned_star_input` contains three files: first, the `bowtie2.err` file provides a short report of the alignment and any error and, second and third, the R1 and R2 read pair files. The `.err` file includes information such as the number of reads input and a few notes about concordance. The data that did not align with the rRNA database are also available in the output directory under a subdirectory named `adapter_removed_rRNA_filtered`. Though it is a temporary file, in the sense that it is subsequently input to STAR (Fig. 6.1), a decision was made to make it available here for convenience purposes as a break point so that the alignment/BAM data are available for inspection or for use elsewhere.

4.2 Output from CircRNA Identification Steps of CircRNAFlow

Alignment by STAR [19] and circRNA identification by DCC [20] are the core of the identification steps of CircRNAFlow (Fig. 6.1).

STAR For each sample, there are three separate STAR runs: first of the pair, second of the pair, and the pair. The corresponding outputs are in the three directories: `star_output_first`, `star_output_second`, and `star_output_pair`.

For each of these, three output files are available in each of the three directories, having the following suffixes:

(a) `Aligned.sortedByCoord.out.bam`
(b) `chimeric.out.junction`, and
(c) `SJ.out.tab`.

`Aligned.sortedByCoord.out.bam` contains the direct alignments from STAR; these data can be visualized, for example, with Integrative Genomics Viewer (IGV) [46] or similar software. The `*.chimeric.out.junction` files contain information on chimeric alignments. For more details on this file's format and contents, see section 5.4 of the STAR Manual (https://github.com/alexdobin/STAR/blob/master/doc/STARmanual.pdf). The `*.SJ.out.tab` files contain high-confidence collapsed splice junctions in tab-delimited format. For more details on this file's format, see section 4.4 of the STAR Manual.

DCC The directory named DCC contains the three principal outputs from DCC:

(a) `CircCoordinates`
(b) `CircRNACount`
(c) `LinearCount`

The `CircCoordinates` file is a tab-separated values file with one line for each identified circRNA. In addition to containing the typical chromosome, start, and stop information, this file contains a field for `JunctionType`, `Start-End Region`, and `OverallRegion`. The `JunctionType` is 0 if found by STAR, and any other value for a noncanonical junction: 1: GT/AG, 2: CT/AC, 3: GC/AG, 4: CT/GC, 5: AT/AC, 6: GT/AT (https://github.com/dieterich-lab/DCC). The `Start-End Region` field describes each side of the junction; `exon-exon` describes a junction between two exons, for example, while `intergenic-exon` describes one side being `intergenic` and the other `exonic`. The `OverallRegion` is text describing the region the circRNA covers.
The `CircRNACount` and `LinearCount` files give read counts for the circRNA and the host genes, respectively. Besides the typical chromosome, start, and stop fields, a column is present for each chimeric.out.junction file giving the corresponding count. These files are later used by CircTest for differential expression analysis. It is worth pointing out that these count files are the first outputs from the pipeline showing identified circRNAs.

4.3 Output from CircRNA Characterization Steps of CircRNAFlow

CircRNA are subjected to several characterization steps including CircTest [20], ClusterProfiler [21], custom plotting, CRAFT [13], and deepTarget [25, 26] (Fig. 6.1 and Table 6.1).

CircTest CircTest is run in two modes (Fig. 6.1), i.e., `regular` and `plain`, to compare read counts of circRNA to counts from linear genes across the groups. In the `regular` mode, various combinations of thresholds for parameters (Table 6.2) are used and, for each such combination, a unique output file comprising a table is generated. In contrast, the `plain` mode uses a single combination of default thresholds. For both cases, the output file name indicates the values of the parameters. The parameters and descriptions used in the `plain` mode are given in Table 6.2. These were gleaned from the files Circ.filter.R and Circ.test.R in the CircTest GitHub repository (https://github.com/dieterich-lab/CircTest).

In the case of the example test data using the thresholds filtercount 1, filtersample 1, percentage 0.01, and alpha 1, there is an adjusted p-value (column `sig_p` in Table 6.3) of 0.002 for the circRNA from the *ATP8B4* gene ending at position −459 (Table 6.3). Recall that the test compares read counts from the circRNA to the linear gene across groups using a beta-binomial test. The read counts for the circRNA are on the right in Table 6.3. Due to the small p-value, given that this and other records are present in the file, they are processed by ClusterProfiler [21]. The output from the `plain` runs is in a folder called `circtest_plain` and regular in a folder called `circtest`. Table 6.3 provides the read counts for the circRNA and for each replicate in the two groups (the CF cases and the healthy controls).

CircTest Plotting CircTest results are then subjected to a simple plotting script. A default, but changeable, p-value threshold of 0.1 is used; usage of the `plot_cutoff` parameter is the way to change it. Any record with a p-value greater than the threshold is not considered for plotting. The data plotted are the circRNA read count values from the corresponding sample columns in the CircTest output file. The titles include the locus, strand, and gene. If a circAtlas identifier is known, it is also included (Fig. 6.3).

The plotting step, in addition to plotting, performs a de-duplication step where output rows, which may be duplicated from CircTest, are grouped into a single record. Duplication may occur because of the combinations of CircTest parameters invoked on the data.
The `regular` CircTest results are also run through ClusterProfiler after two filtering preprocessing steps. In the first preprocessing step, fold changes are converted to $\log_2$. If CircTest returned results with multiple thresholds, only one representative is kept with an average p-value and an average $\log_2$-fold change value. The first preprocessing step is not tightly coupled with ClusterProfiler but run before the ClusterProfiler script. The second preprocessing step, however, is tightly coupled to ClusterProfiler being conducted in the same script that runs ClusterProfiler.

The second preprocessing step applies an adjusted p-value filter at each of the following five adjusted p-value thresholds: 0.001, 0.005, 0.010, 0.050, and 0.100. Once each of these filters is applied, records are grouped as having negative $\log_2$-fold change, positive $\log_2$-fold change, or any $\log_2$-fold change. In this way, 5x3 = 15 combinations of filter groupings are considered. For each such filter grouping, ClusterProfiler's enrichment analyses are run.

ClusterProfiler After the preprocessing for ClusterProfiler [21], ClusterProfiler is invoked for each of the three gene ontology categories with differential expression directional filters. The three gene ontology categories are BP (biological process), CC (cellular component), and MF (molecular function). In addition to GO overrepresentation, a KEGG overrepresentation analysis is also performed. For both GO and KEGG overrepresentations, if the data are not `null`, the results are plotted as a dot plot (Fig. 6.4). If either the word `up` or `down` is in the filename, then data in the corresponding circRNAs are filtered to be regulated either `up` or `down`, respectively. The absence of `up` or `down` in an output plot filename indicates that the circRNAs are either upregulated or downregulated, respectively. The output from ClusterProfiler is in a directory named `cluster_profiler` (Fig. 6.2).

Prepare for CRAFT As shown in the bottom right of 6.1, ClusterProfiler and prepare_for_craft/CRAFT/deepTarget run in separate threads of activity, enabled by Nextflow's DAG-centric workflow architecture. Prepare_for_craft runs, and pre-processes the data for CRAFT [13] and deepTarget [25, 26]. Any circRNAs output from CircTest, which is considered for plotting, is passed to the `prepare_for_craft` step. This step performs two major functions. First, it filters circRNAs by a CircTest-adjusted p-value threshold. Second, it applies a `top` threshold where the `top-n` circRNAs are kept. Default values are used for both thresholds to limit the number of CRAFT invocations. The default adjusted p-value threshold is 1.0; the default `top-n` threshold is

30. Once filtering is done, the `prepare_for_craft` step takes the filtered data and reformats them into strings and files in the format that CRAFT expects. More specifically, it creates a `backsplice_gene_name.txt` file and a `list_backsplice.txt` file for each circRNA. A pair of one of each such files is required by CRAFT. The output files from this step are in a directory named `craft_prep` (Fig. 6.2).

CRAFT CRAFT [13] is run in three modes per circRNA passing the filters. As discussed in the previous section, these are the `M` mode (searching for miRNA interactions), the `O` mode searching for open reading frames, and, finally, the `R` mode searching for RNA-binding protein interactions. For each circRNA, for each mode, a corresponding zip file is found in the `craft` directory of the `quick_start_output` directory (Fig. 6.2). In Linux, a bash command like `find *.zip|xargs -tI {} unzip {}` can be helpful to quickly `unzip` all the files one after another. Many files result from uncompressing, but the HTML files may be considered the primary output here. Each HTML file references other outputs. By running the Linux command `find . -iname '*.html,'` one can see the six HTML files produced from the two circRNAs, one per each of the `M`, `O`, and `R` modes. Figure 6.5 shows an example of the command's output and relative directory location (Fig. 6.2) of the CRAFT output files for the provided test data.

To see the output from the `R` analysis, open the file `15_49876278_49879459_ATP8B4_MINUS_R/graphical_output/general/functional_predictions_all_circRNAs.html` for the circRNA indicated by `Chr15:49876278-49879459`. Upon opening that file and navigating to the section titled "`RNA-binding protein (RBP) binding sites prediction`," the results can be viewed. The results (Fig. 6.6) highlight four RBPs predicted to interact with the hsa-ATP8B4_0010 circRNA (Fig. 6.3) from the *ATP8B4* gene.

A similar table follows this one, showing the RNA recognition element (RRE) start and end. Similar tables are available for miRNA binding sites and Circos-like plots [47] for open reading frames.

CRAFT's output includes the generation of a fasta file whose contents is a circRNA including the sequence. These are taken from CRAFT and Nextflow, via the Nextflow code `map{it.text}.unique()`, which takes each unique sequence and submits the sequences to deepTarget for further analysis.

deepTarget The deepTarget model [25, 26] can be invoked only when the circRNA contains a `seed` for a binding site. Seeds are miRNA-specific; if the circRNA contains no seed, then no analysis is performed. For each seed identified, a binding prediction is made.

For each circRNA, deepTarget is invoked once for each sequence found in the provided miRNA fasta file. The results from this analysis are found in the `deeptarget` directory in two `TSV` files:

(a) `*.dt_results.norm.tsv`: This file contains prediction results when a circRNA contains a seed. This file is more amenable for parsing by other software. Table 6.4 shows some example rows from this file.
(b) `*.dt_results.tsv`: For each miRNA in the database, this file contains a single record. Since some circRNAs can have multiple seeds, each row in this file can have results from multiple predictions. Some fields use JSON-like notation to show results and may not be readily parseable.

Both files contain the seed sequence and a field called `trans_bsj_indicator`. `trans_bsj_indicator` is Boolean in nature, that is `True` or `False`, and indicates whether binding spans the backsplicing junction or not (Table 6.4). The `summary_id` column is shared between both `TSV` files. It can be used as a `foreign-key` from the `.norm.tsv` file into the `dt_results.tsv` file where it is a primary key. Some keys can appear twice in the `.norm.tsv` file if a given miRNA has multiple seeds for the circRNA.

5 Future Directions and Conclusion

Section 6.1 of this chapter introduced CircRNAFlow and its design. Section 6.2 described the component programs in the CircRNAFlow pipeline and their respective positions in the pipeline. Section 6.3 presented a CircRNAFlow user guide including the online quick start tutorial plus troubleshooting tips for downloading and running the CircRNAFlow pipeline together with the provided test data. Section 6.4 provided a tour of some of the main outputs from CircRNAFlow. Section 6.5 provides a brief review, future directions, related works, and conclusions.

5.1 Brief Review, Future Directions, and Related Works

CircRNAFlow performs quality control checks and preprocesses FASTQ NGS paired-end data by adapter trimming and rRNA removal with FastQC, Flexbar, and Bowtie2. It identifies circRNA with STAR and DCC. CircRNAFlow characterizes circRNAs in the following ways: by computing differentially expressed circRNAs using CircTest; examining gene ontology and KEGG pathway overrepresentations of circRNA host genes using ClusterProfiler; predicting circRNA molecular interactions with RBPs or miRNAs utilizing CRAFT and deepTarget; and predicting protein translational potential in circRNAs utilizing CRAFT. These types of characterizations help explore the possible roles of identified circRNAs.

As a DSL2 Nextflow-based pipeline, CircRNAFlow enables circRNA analysis to be portable with containers, scalable with slurm execution environments, and resumable with its checkpointing capabilities. Since its initial publication in 2023, CircRNAFlow [14] has received several notable updates. First, its code has been upgraded to Nextflow DSL2 (see https://www.nextflow.io/docs/edge/dsl2.html). Second, CRAFT and deepTarget were incorporated. For CRAFT and deepTarget, their incorporations into CircRNAFlow were aided by publicly available docker images [26].

Currently, CircRNAFlow employs DCC for the critical task of circRNA identification. DCC is one of many tools for circRNA identification as discussed elsewhere [34]. Vromman et al. [34] also performed benchmarking analyses of several tools. To quote, "sensitivity and number of predicted circRNAs … are the most significant differentiators [among the software]" [34]. According to their analysis, using multiple detection tools can lower the false-positive rate. Considering these results and observations, a roadmap for improvements to CircRNAFlow is suggested. Exploring additional circRNA detection tools and how they might be incorporated into CircRNAFlow could be done next. These could be run in parallel to DCC. Subsequent steps might involve a filter (for example, a circRNA being called by multiple circRNA identification programs).

For a robust circRNA investigation, it is advisable to start with a good experimental analysis and design for your circRNA study by consulting best practice standards for circRNA research [1, 10]. Backsplicing junctions can be detected from whole transcriptomic data, including rRNA-depleted total RNA (with or without RNAase R treatment) [34]..

CircRNAFlow can be run as part of a suite of software for circRNA analysis. Other Nextflow-based pipelines such as the nf-core/circRNA (https://nf-co.re/circrna) are available to use along with CircRNAFlow. As discussed previously [34], use of more than one tool for circRNA identification to obtain a consensus prediction for circRNA identification is better than just one.

5.2 Conclusion

We presented CircRNAFlow, the background on its design, how to use it, CircRNAFlow output descriptions, and some troubleshooting tips. CircRNAFlow implementation in Nextflow enables its use at a scale on high-performance computing clusters as well as on individual laptops. The Nextflow checkpointing capability enables continuation of interrupted workflows. CircRNAFlow is a free and open-source software we developed, which can be used as part of a turnkey solution in circRNA analysis projects.

Acknowledgments The authors thank Dr Kim M. Keeling, PhD, Dr Barbara Tabak, PhD, Dr Simon M. Brocklehurst, PhD, and Mr Victor Macauley, MS, MBA, for consideration and providing helpful feedback on this chapter.

Conflict of Interest The authors declare no conflict of interest.

References

1. Nielsen AF, Bindereif A, Bozzoni I, Hanan M, Hansen TB, Irimia M, Kadener S, Kristensen LS, Legnini I, Morlando M, Jarlstad Olesen MT, Pasterkamp RJ, Preibisch S, Rajewsky N, Suenkel C, Kjems J (2022) Best practice standards for circular RNA research. Nat Methods 19(10):1208–1220. https://doi.org/10.1038/s41592-022-01487-2
2. Ebermann C, Schnarr T, Muller S (2020) Recent advances in understanding circular RNAs. F1000Res 9:655. https://doi.org/10.12688/f1000research.25060.1
3. Misir S, Wu N, Yang BB (2022) Specific expression and functions of circular RNAs. Cell Death Differ 29(3):481–491. https://doi.org/10.1038/s41418-022-00948-7
4. Verduci L, Tarcitano E, Strano S, Yarden Y, Blandino G (2021) CircRNAs: role in human diseases and potential use as biomarkers. Cell Death Dis 12(5):468. https://doi.org/10.1038/s41419-021-03743-3
5. Ruan H, Wang PC, Han L (2022) Characterization of circular RNAs with advanced sequencing technologies in human complex diseases. Wiley Interdiscip Rev RNA:e1759. https://doi.org/10.1002/wrna.1759
6. Chen Y, Li C, Tan C, Liu X (2016) Circular RNAs: a new frontier in the study of human diseases. J Med Genet 53(6):359–365. https://doi.org/10.1136/jmedgenet-2016-103758
7. Chen YM, Zheng YL, Su X, Wang XQ (2021) Crosstalk between MicroRNAs and circular RNAs in human diseases: a bibliographic study. Front Cell Dev Biol 9:754880. https://doi.org/10.3389/fcell.2021.754880
8. Li X, Yang L, Chen LL (2018) The biogenesis, functions, and challenges of circular RNAs. Mol Cell 71(3):428–442. https://doi.org/10.1016/j.molcel.2018.06.034
9. Zhang Z, Yang T, Xiao J (2018) Circular RNAs: promising biomarkers for human diseases. EBioMedicine 34:267–274. https://doi.org/10.1016/j.ebiom.2018.07.036
10. Dodbele S, Mutlu N, Wilusz JE (2021) Best practices to ensure robust investigation of circular RNAs: pitfalls and tips. EMBO Rep 22(3):e52072. https://doi.org/10.15252/embr.202052072
11. Lasda E, Parker R (2014) Circular RNAs: diversity of form and function. RNA 20(12):1829–1842. https://doi.org/10.1261/rna.047126.114
12. Liang Y, Liu N, Yang L, Tang J, Wang Y, Mei M (2021) A brief review of circRNA biogenesis, detection, and function. Curr Genomics 22(7):485–495. https://doi.org/10.2174/1389202922666210331130722
13. Dal Molin A, Gaffo E, Difilippo V, Buratin A, Tretti Parenzan C, Bresolin S, Bortoluzzi S (2022) CRAFT: a bioinformatics software for custom prediction of circular RNA functions. Brief Bioinform 23(2). https://doi.org/10.1093/bib/bbab601
14. Salinas EA, Macauley V, Keeling KM, Edwards YJK (2023) Discovery of dysregulated circular RNAs in whole blood transcriptomes from cystic fibrosis patients – implication of a role for cellular senescence in cystic fibrosis. J Cyst Fibros 22(4):683–693. https://doi.org/10.1016/j.jcf.2023.04.021
15. Jakobi T, Dieterich C (2018) Deep computational circular RNA analytics from RNA-seq data. Methods Mol Biol 1724:9–25. https://doi.org/10.1007/978-1-4939-7562-4_2
16. Andrews S (2019) FastQC: A quality control tool for high throughput sequence data, August 1. https://www.bioinformatics.babraham.ac.uk/projects/fastqc/
17. Roehr JT, Dieterich C, Reinert K (2017) Flexbar 3.0 – SIMD and multicore parallelization. Bioinformatics 33(18):2941–2942. https://doi.org/10.1093/bioinformatics/btx330
18. Langmead B, Salzberg SL (2012) Fast gapped-read alignment with Bowtie 2. Nat Methods 9(4):357–359. https://doi.org/10.1038/nmeth.1923
19. Dobin A, Davis CA, Schlesinger F, Drenkow J, Zaleski C, Jha S, Batut P, Chaisson M, Gingeras TR (2013) STAR: ultrafast universal RNA-seq aligner. Bioinformatics 29(1):15–21. https://doi.org/10.1093/bioinformatics/bts635
20. Cheng J, Metge F, Dieterich C (2016) Specific identification and quantification of circular RNAs from sequencing data. Bioinformatics 32(7):1094–1096. https://doi.org/10.1093/bioinformatics/btv656
21. Wu T, Hu E, Xu S, Chen M, Guo P, Dai Z, Feng T, Zhou L, Tang W, Zhan L, Fu X, Liu S, Bo X, Yu G (2021) clusterProfiler 4.0: a universal enrichment tool for interpreting omics data. Innovation (Camb) 2(3):100141. https://doi.org/10.1016/j.xinn.2021.100141
22. Enright AJ, John B, Gaul U, Tuschl T, Sander C, Marks DS (2003) MicroRNA targets in Drosophila. Genome Biol 5(1):R1. https://doi.org/10.1186/gb-2003-5-1-r1
23. Kertesz M, Iovino N, Unnerstall U, Gaul U, Segal E (2007) The role of site accessibility in microRNA target recognition. Nat Genet 39(10):1278–1284. https://doi.org/10.1038/ng2135

24. Yu H, Wang J, Sheng Q, Liu Q, Shyr Y (2019) beRBP: binding estimation for human RNA-binding proteins. Nucleic Acids Res 47(5):e26. https://doi.org/10.1093/nar/gky1294
25. Lee B, Baek J, Park S, Yoon S (2016) deepTarget: end-to-end Learning framework for microRNA target prediction using deep recurrent neural networks. In: Proceedings of the 7th ACM international conference on bioinformatics, computational biology, and health informatics, Seattle, WA, USA, 2 October 2016, pp 434–442
26. Avsec Z, Kreuzhuber R, Israeli J, Xu N, Cheng J, Shrikumar A, Banerjee A, Kim DS, Beier T, Urban L, Kundaje A, Stegle O, Gagneur J (2019) The Kipoi repository accelerates community exchange and reuse of predictive models for genomics. Nat Biotechnol 37(6):592–600. https://doi.org/10.1038/s41587-019-0140-0
27. European Organization for Nuclear Research (2013) Zenodo. CERN. https://www.zenodo.org/
28. Sicilia MA, García-Barriocana E, Sanchez-Alonso S (2017) Community curation in open dataset repositories: insights from Zenodo. Procedia Comput Sci 106:54–60
29. Di Tommaso P, Chatzou M, Floden EW, Barja PP, Palumbo E, Notredame C (2017) Nextflow enables reproducible computational workflows. Nat Biotechnol 35(4):316–319. https://doi.org/10.1038/nbt.3820
30. Merkel D (2014) Docker: lightweight Linux containers for consistent development and deployment. https://dl.acm.org/doi/10.5555/2600239.2600241.239
31. Yoo AB, Jette MA, Grondona M (2003) SLURM: simple Linux utility for resource management. In: Feitelson D, Rudolph L, Schwiegelshohn U (eds) Job scheduling strategies for parallel processing. Springer, Berlin/Heidelberg, pp 44–60. https://doi.org/10.1007/10968987_3
32. Wu W, Zhao F, Zhang J (2024) circAtlas 3.0: a gateway to 3 million curated vertebrate circular RNAs based on a standardized nomenclature scheme. Nucleic Acids Res 52(D1):D52–D60. https://doi.org/10.1093/nar/gkad770
33. Wu W, Ji P, Zhao F (2020) CircAtlas: an integrated resource of one million highly accurate circular RNAs from 1070 vertebrate transcriptomes. Genome Biol 21(1):101. https://doi.org/10.1186/s13059-020-02018-y
34. Vromman M, Anckaert J, Bortoluzzi S, Buratin A, Chen CY, Chu Q, Chuang TJ, Dehghannasiri R, Dieterich C, Dong X, Flicek P, Gaffo E, Gu W, He C, Hoffmann S, Izuogu O, Jackson MS, Jakobi T, Lai EC, Nuytens J, Salzman J, Santibanez-Koref M, Stadler P, Thas O, Vanden Eynde E, Verniers K, Wen G, Westholm J, Yang L, Ye CY, Yigit N, Yuan GH, Zhang J, Zhao F, Vandesompele J, Volders PJ (2023) Large-scale benchmarking of circRNA detection tools reveals large differences in sensitivity but not in precision. Nat Methods 20(8):1159–1169. https://doi.org/10.1038/s41592-023-01944-6
35. Jakobi T, Uvarovskii A, Dieterich C (2019) circtools-a one-stop software solution for circular RNA research. Bioinformatics 35(13):2326–2328. https://doi.org/10.1093/bioinformatics/bty948
36. Hansen TB, Veno MT, Damgaard CK, Kjems J (2016) Comparison of circular RNA prediction tools. Nucleic Acids Res 44(6):e58. https://doi.org/10.1093/nar/gkv1458
37. UniProt C (2023) UniProt: the universal protein knowledgebase in 2023. Nucleic Acids Res 51(D1):D523–D531. https://doi.org/10.1093/nar/gkac1052
38. Kozomara A, Birgaoanu M, Griffiths-Jones S (2019) miRBase: from microRNA sequences to function. Nucleic Acids Res 47(D1):D155–D162. https://doi.org/10.1093/nar/gky1141
39. Kopp BT, Fitch J, Jaramillo L, Shrestha CL, Robledo-Avila F, Zhang S, Palacios S, Woodley F, Hayes D Jr, Partida-Sanchez S, Ramilo O, White P, Mejias A (2020) Whole-blood transcriptomic responses to lumacaftor/ivacaftor therapy in cystic fibrosis. J Cyst Fibros 19(2):245–254. https://doi.org/10.1016/j.jcf.2019.08.021
40. Shin HW, Takatsu H (2022) Regulatory roles of N- and C-terminal cytoplasmic regions of P4-ATPases. Chem Pharm Bull (Tokyo) 70(8):524–532. https://doi.org/10.1248/cpb.c22-00042
41. Wang J, Molday LL, Hii T, Coleman JA, Wen T, Andersen JP, Molday RS (2018) Proteomic analysis and functional characterization of P4-ATPase phospholipid flippases from murine tissues. Sci Rep 8(1):10795. https://doi.org/10.1038/s41598-018-29108-z
42. Vandebrouck C, Ferreira T (2020) Glued in lipids: lipointoxication in cystic fibrosis. EBioMedicine 61:103038. https://doi.org/10.1016/j.ebiom.2020.103038
43. Cottrill KA, Farinha CM, McCarty NA (2020) The bidirectional relationship between CFTR and lipids. Commun Biol 3(1):179. https://doi.org/10.1038/s42003-020-0909-1
44. Camacho C, Coulouris G, Avagyan V, Ma N, Papadopoulos J, Bealer K, Madden TL (2009) BLAST+: architecture and applications. BMC Bioinformatics 10:421. https://doi.org/10.1186/1471-2105-10-421
45. Pezoa F, Reutter JL, Suarez F, Ugarte M, Vrgoc D (2016) Foundations of JSON schema WWW '16: Proceedings of the 25th International Conference on World Wide Web Pages 263 - 273 https://doi.org/10.1145/2872427.2883029
46. Robinson JT, Thorvaldsdottir H, Turner D, Mesirov JP (2023) igv.js: an embeddable JavaScript implementation of the integrative genomics viewer (IGV). Bioinformatics 39(1). https://doi.org/10.1093/bioinformatics/btac830
47. Krzywinski M, Schein J, Birol I, Connors J, Gascoyne R, Horsman D, Jones SJ, Marra MA (2009) Circos: an information aesthetic for comparative genomics. Genome Res 19(9):1639–1645. https://doi.org/10.1101/gr.092759.109

Part III

Biogenesis of Circular RNAs

Formation of Circular RNAs

7

Arvind Srinivasan and Marzena Wojciechowska

Abstract

Circular RNAs (circRNAs) constitute a class of covalently closed, single-stranded RNA molecules that form a continuous loop structure, setting them apart from linear RNA molecules. These circular entities arise through a distinct biogenetic process termed "backsplicing," which involves the covalent bonding of the 5′ site of an upstream exon with the 3′ site of the same or a downstream exon during pre-messenger RNA (pre-mRNA) splicing. Such a direct backsplicing model exists along with the exon- and lariat-skipping models of circRNA generation, giving rise to three major types of circRNAs, i.e., exonic (EcircRNAs), intronic (IcircRNAs), and exonic–intronic circRNAs (EIcircRNAs). A complex process of interaction between *cis*-acting elements and *trans*-acting factors governs circRNA synthesis. In particular, reverse complementary matches (RCMs) and RNA-binding proteins (RBPs) play critical roles in circRNA formation. Complementary sequences increase base pairing, which aids in the alignment of splice sites for backsplicing. RBPs can either accelerate or inhibit circRNA synthesis by binding to certain sequences or structures, thereby altering backsplicing efficiency. The careful balance of *cis*-acting components and *trans*-acting factors highlights the complex regulatory mechanisms driving circRNA synthesis, which contributes to the variety and functional relevance of these transcripts in cellular activities. In this chapter, we review the variables that influence circRNA biogenesis and highlight the origins of different types of these transcripts in humans.

Keywords

Circular RNA · Backsplicing · RNA splicing · *Cis*-regulatory elements · *Trans*-acting factors

1 Introduction

A circular RNAs (circRNAs) is a type of single-stranded RNA molecule that forms covalently closed loops. Their structure safeguards molecules against degradation mediated by exonucleases. Hence, in cells, they are more stable and have longer half-lives than linear RNAs. In 1976, the first circRNA molecules to be discovered were viroids [1]. A few years later, circular transcripts were also identified in the cytoplasm of eukaryotic cells using electron microscopy [2]. However, the first analysis of circRNAs was performed in the 1990s, and the example that has been exam-

A. Srinivasan · M. Wojciechowska (✉)
Department of Rare Diseases, Polish Academy of Sciences, Institute of Bioorganic Chemistry, Poznan, Poland
e-mail: mwojska@ibch.poznan.pl

J. Xiao (ed.), *Advances in Circular RNAs*, Advances in Experimental Medicine and Biology 1485,
https://doi.org/10.1007/978-981-96-9428-0_7

ined the most extensively since then is the one produced by the sex-determining region Y (*SRY*) gene in mice [3]. Even though single-stranded circRNAs have been observed in eukaryotic cells for decades, they were thought to be splicing errors and had been somewhat beneath the radar. However, thanks to large-scale high-throughput sequencing experiments conducted in this age of "big data," the significance of circRNAs was brought back to the forefront [4, 5]. As a result, these molecules have been identified on a large scale, first in humans and then in many other species, ranging from fruit flies, worms, diverse mammalian species, and in plants [4, 6–13].

CircRNAs can be derived from protein-coding or noncoding genomic loci and exhibit considerable variation in lengths [14]. Their production is thought to be conserved in all eukaryotes, is dependent on RNA polymerase II and a noncanonical splicing mechanism of pre-messenger RNAs (pre-mRNAs), and is often catalyzed by either the machinery of the spliceosome or by group I and group II ribozymes [15]. CircRNAs, in their most fundamental forms, are synthesized by backsplicing, which results in the covalent linkage of a downstream splice donor to an upstream splice acceptor, creating a head-to-tail splicing junction [4, 16]. A few different mechanisms have been postulated for the production of these transcripts (Fig. 7.1). One of them involves the circularization of a single exon, which is made possible by the presence of nearby repetitive sequences in flanking introns [3]. Alternatively, the removal of multiple consecutive exons from a spliced transcript or the selection of a less common splice donor or acceptor can contribute to the formation of circular RNA molecules [17, 18]. Modulating backsplicing events is known to include both *cis*-regulatory elements and *trans*-acting factors of RNA-binding proteins (RBPs).

There is a wide variety of circRNAs since they can be produced by either post-transcriptional or co-transcriptional mechanisms. The findings by Ashwal-Fluss and colleagues suggest that circRNAs are co-transcribed products since linear mRNA splicing and circRNA backsplicing may compete with one another [16]. However, in another in vitro experimental setup, it was demonstrated that pre-mRNA needs a stable 3′-end to promote circRNA formation, signifying that this process may occur after transcription is completed [19]. Three basic groups of circRNAs exist based on their structural components and mechanisms of circularization, i.e., exonic (EcircRNAs), intronic (IcircRNAs), and exonic–intronic (EIcircRNAs) [20, 21]. The structural variety of these transcripts is a key component of their cellular adaptability, which allows them to regulate protein activity, operate as microRNA or protein inhibitors (so-called sponges), and even be translated [22, 23]. Additionally, circRNAs have been linked to several diseases including cancer, neurological disorders, and cardiovascular diseases [20, 24, 25]. The expression patterns of these RNAs vary among tissues, cells, and developmental stages [26].

In this chapter, we present an overview of the biogenesis of different types of circRNAs and factors affecting their formation in higher eukaryotes.

2 Canonical and Alternative Splicing

Canonical splicing is the process of joining consecutive exons and removing introns, while alternative splicing (AS) results in the generation of multiple splice variants from one gene [27, 28]. From plants to humans, AS is a potent mechanism that enables transcriptome variability and can affect the expression levels of certain genes [29–31]. RNA splice variants that originate from AS can display a range of mRNA structures and their stabilities. It is believed that around 95% of human genes undergo AS, which highlights the significance of this process [29, 32, 33]. In general, alternative splicing may be broken down into five distinct categories [34]. The first type is exon skipping (cassette exon), which is the most prevalent kind of alternative splicing in mammalian cells. The second type of AS is known as mutually exclusive exons, which produce alternative isoforms by preserving only a single exon from a group of adjacent internal exons in the

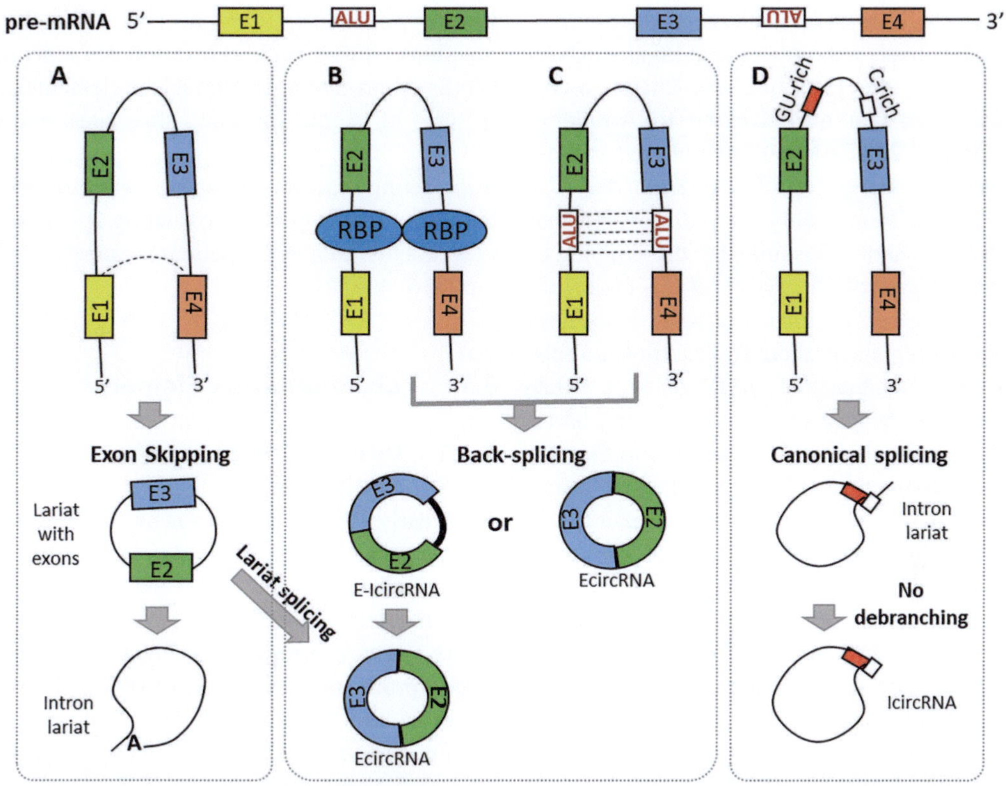

Fig. 7.1 Biogenesis of circRNAs. (**a**) Lariat-driven circularization or exon-skipping model. When a pre-mRNA is spliced, the upstream exon's 3′ hydroxyl interacts with the downstream exon's 5′ phosphate to establish a covalent connection, resulting in a lariat of exons and introns. The 5′ intron's 2′ hydroxyl interacts with the 3′ intron's 5′ phosphate, followed by a contact between the 3′ exon's 3′ hydroxyl and the 5′ phosphate of the 5′ exon, resulting in the formation of an EcircRNA. (**b**) RNA-binding protein (RBP)-driven circularization. RBPs enhance downstream and upstream intron contact, forming EcircRNA or EIciRNA. (**c**) *Base-pairing-driven circularization.* Inverse-repeating or complementary sequences like *Alu* elements couple downstream and upstream introns. Introns are deleted or maintained to generate EcircRNA or EIciRNA. (**d**) Biosynthesis of IcircRNAs via canonical splicing. The formation of intronic circles is primarily dependent on a 7-nt GU-rich motif and an 11-nt C-rich element in order to avoid debranching and exonucleolytic degradation.

final transcript. The third and fourth types are referred to as, respectively, an alternative 5′ splice site (alternative donor) and an alternative 3′ splice site (alternative acceptor). The fifth type is intron retention (IR), which refers to the process by which one or more introns are held in mRNA instead of being spliced out [35].

The skipping of exons during alternative splicing and the formation of circRNAs are two phenomena that are related to one another. It was hypothesized that certain circRNA-producing genes generated linear transcripts that skipped the alternative exons. Lariats containing skipped exons undergo additional splicing events that result in the circularization of the skipped exons (Fig. 7.1a). Zaphiropoulos et al. discovered a circularized transcript containing exons 2–4 from the rat cytochrome P450 *2C24* gene [36]. The same research group established the association between exon skipping and the generation of circRNAs during expression of the cytochrome P450 *2C18* gene in the human epidermis and the androgen-binding protein (*ABP*) gene in the rat testis [37], whereas Surono and coworkers identified over ten distinct circRNAs consisting of exons of the dystrophin transcript that were skipped due to alternative splicing [38]. Comprehensive RNA sequencing (RNA-seq) research in human cells have revealed that exons undergoing circularization are less likely to be detected in linear mRNAs from similar genes [39]. In 2016, Khan and colleagues found that mice lacking RNA-binding motif protein 20 (RBM20) have altered expression levels of several circRNAs from the *TTN* (titin) I-band. Remarkably, they discovered that circRNA production was linked to the skipping of the I-band region of titin. Exons of this region are part of the linear *TTN* transcript when RBM20 is absent, which prevents them from acting as a substrate for the synthesis of circRNAs [40]. Other studies of skipping events using plasmid-based circRNA expression vectors have found that the inhibiting components of the spliceosome might significantly raise the proportion of circular to linear RNAs [41–43]. Furthermore, as shown in *Saccharomyces pombe*, circRNAs could be synthesized via an exon-containing lariat intermediate created by exon skipping [44]. It should be emphasized, however, that not all skipped exons yield circRNAs.

The majority of circRNAs are created by the process of backsplicing, which involves a downstream 3′ splice-donor site attacking an upstream 5′ splice-acceptor site, resulting in the formation of a 3′–5′ phosphodiester link. Despite the lower efficiency of backsplicing compared to linear splicing, circRNAs can accumulate in cells of low proliferative capacity. For example, numerous circRNAs were found to be increased in neuronal and muscular tissues from mice and humans [5, 7, 45–48]. Their covalently closed ring shape likely protects these molecules from exonuclease-mediated destruction, resulting in their remarkable stability [5]. For most genomic loci producing circRNAs, there is likely a competition between canonical splicing and backsplicing events [4].

3 Regulation of the Biogenesis of Circular RNAs

CircRNAs are produced by a backsplicing process that is controlled by several *cis*-regulatory elements and *trans*-acting proteins. The *cis*-elements include, among others, intronic complementary sequences (ICSs) near the backsplicing junctions (BSJs), such as inverted *Alu* repeats of SINEs (short interspersed nuclear elements), the features of exonic–intronic splice sites, and the presence of cryptic splice sites [5], whereas the *trans*-acting factors comprise various RNA-binding proteins (RBPs), including spliceosome components and RNA helicases (cleavage factors) [16, 49, 50].

3.1 *Cis*-Regulatory Elements

3.1.1 Intronic Complementary Sequences

In the early 1990s, when it was first observed that the *Sry* gene generates predominantly circular transcripts, Capel and colleagues hypothesized that "inverted repeats" located in introns adjacent to the circle-forming exons might facilitate circRNA production by bringing splice sites into closer proximity via base pairing [3]. They developed this concept based on the presence of more than 15.5 kb of nearly completely complementary sequences flanking the 1.2-kb exon of the *Sry* locus. This model was validated in an in vitro experiment demonstrating that circularization of the *Sry* transcript requires as few as 400 nt of flanking inverted repeats [51]. Even though these lengthy repeats are present only in a small number of exons, they indicate that *cis* elements are among the regulatory factors regulating the biogenesis of circRNAs.

Large-scale sequencing studies of eukaryotic genomes have provided a great deal of detail on the presence of SINEs. In the human genome, *Alu* elements and mammalian-wide interspersed repeat (MIR) elements are the two most abundant families of SINEs [52]. Numerous inverted *Alu* repeats, each around 300 nucleotides in length, are found in introns that surround human exons [53]. Originally, it was postulated that the presence of several inverted *Alu* repeats aids circRNA biogenesis [5]. As a support for the assumption, Zhang and colleagues performed studies on POLR2A circRNAs by deleting or inserting fragments of sequences in introns flanking BSJs [43]. Their study also identified numerous circular transcripts to be created from a single gene by alternative circularization caused by differential pairing of inverted *Alu* repeats. In a similar mutagenesis experiment performed for circRNAs originating from *ZKSCAN1*, *HIPK3*, and *EPHB4*, Liang and Wilusz reported that short inverted repeats of about 30–40-nt long and miniature introns (<100 nt) were sufficient to facilitate the circularization of the intervening exons [19]. The concept of *Alu*-rich genes producing a diversity of exon-containing circRNAs was studied by Ottesen and coworkers, who discovered about 53 circRNAs from the survival of motor neuron (*SMN*) gene that contains more than 40 *Alu* elements occupying 39% of the transcribed region. In addition, the authors showed that the mouse *SMN* locus generates a much smaller pool of circRNAs compared to the human *SMN* locus as a result of a lower percentage of *Alu* elements [54]. A comparative analysis performed by Tang Y and coworkers found that a low fraction of human circRNAs was conserved and could be found in parallel mouse samples [55]. Respectively, about 85% and 60% of human and mouse circRNAs, respectively, exhibited species-specific expression patterns, indicating that the number of circRNA molecules would increase along with accumulated SINEs (*Alu* in humans) during species evolution [55].

Many proteins have been identified that can interfere with circRNA biogenesis (Fig. 7.1b). Single-stranded RNA-specific adenosine deaminase 1 (ADAR1) can stop circularization by interfering with inverted-repeat *Alu* elements. This protein edits adenosine to inosine, weakening intronic RNA pairs' stability and complementarity [8]. DHX9, an adenosine triphosphate (ATP)-dependent RNA helicase (A), works in tandem with ADAR1 to suppress circRNA synthesis [50, 54]. Kokot and colleagues have shown that ADAR2 decreases the expression of circAKAP13, thereby reducing the risk of cardiac failure [56].

3.1.2 Splice Site Dinucleotides and Splice Site Strength

Regardless of whether backsplicing differs from canonical splicing, the synthesis of the vast majority of circRNAs requires the presence of canonical splicing sites and the canonical spliceosome machinery [57–59]. Mutagenesis studies in circRNA expression vectors have indicated that both canonical splice sites and the spliceosomal machinery are required for the synthesis of circRNAs [16, 60]. They are formed as a result of the joining of the 3′ and 5′ ss of a portion of the linear transcript, typically with joining locations flanked by ag/gt splicing signals [61]. Analysis of the entire human genome showed that "ag" is enriched at the junctions in both 5′ ss and 5′ backsplice sites (bss), and "gt" is enriched at the junction in both 3′ ss and 3′bss [10, 60]. By mutating splice sites involved in backsplicing and canonical splicing, Gao and colleagues identified that this modification suppresses both circular and linear RNA expression from the same gene locus, suggesting the requirement for the same splice site signals in vivo [62]. Likewise, mutating the ag/gt dinucleotides at the BSJ was shown to diminish the generation of circRNAs in the minigene experiments [16]. In contrast, non-ag/gt splicing signals could also guide the formation of circular transcripts. For example, Liu and coworkers identified the presence of a noncanonical ac/ct splicing signal at the BSJ, where circRNAs originate from the internal exon and intron regions of transcripts in humans and mice [63]. They referred to these circRNAs as interior circRNAs (i-circRNAs). Similarly, noncanonical splice site signals were identified in other eukaryotes such as moths [64].

CircRNA biogenesis may also be influenced by splice site strength, with variations in the efficiency of splicing at flanking splice sites impacting the formation of circular RNA molecules. Zhang and colleagues identified novel exons forming circRNAs that had similar splice site strength compared to constitutive exons and cassette exons [65]. On the other hand, contradictory results were obtained by Di Liddo and coworkers, who showed that the strength of 5′ splice sites of the intron at donor backsplice sites was significantly higher than that of 5′ splice sites at flanking or randomly selected 5′ splice sites that did not engage in backsplicing [66]. Similar results were seen in heterogeneous nuclear ribonucleoprotein M (HNRNPM)-originating backspliced exons with stronger 5′ splice sites than the original flanking proximal exons, implying that exon recognition in circRNAs may be based on combinatorial control that takes splice site strength into account [67].

3.1.3 Intron Length and Exon Size

Based on experimental evidence, the production of circSry requires the presence of long introns and the complementary sequences flanking the region of circularization [51]. This observation was expanded by genome-wide RNA sequencing, which revealed that exons included in human circRNAs are surrounded by longer introns that are enriched in *Alu* elements or non-repeat inverted complementary sequences [43, 68]. The presence of long flanking introns and complementary sequences in the majority of backsplicing exons provides two crucial clues for mammalian circRNA synthesis. First, the circularization areas are identified using difficult-to-splice lengthy introns, and, second, the flanking reverse complementary sequences and interactions with proteins bring the two backsplicing sites together (Fig. 7.1b, c). The timing of circularization is determined by the length of flanking intronic repeats, and co-transcriptional backsplicing is probably promoted by repeats that are 300 nucleotides long or longer [69]. The association between intron length and circRNA expression might be due to a higher chance of complementary sequence motif acquisition in longer introns or it could be a feature of a complementary sequence motif-independent backsplicing process. The biogenesis of the so-called AUG circRNAs, a subclass of circRNAs driven by circularization of start-codon-containing exons and flanked by lengthy introns, has been demonstrated to be insensitive to DHX9 alterations [70].

When the 3′-tail and the 5′-head come from the same exon, a single-exon circRNA is formed, which is generally due to a co-transcriptional process [3, 17, 71, 72]. When the 5′-head comes from another upstream exon, an intermediate lariat with exons and intron(s) is generated [17, 22, 73]. In this situation, a circRNA with multiple exons can be formed when the intron(s) are removed posttranscriptionally by the typical linear splicing method [74].

3.1.4 Cryptic Splice Sites

Most exonic circRNAs are generated through the backsplicing of canonical donor and acceptor splice sites and contain one or more exons [17]. Moreover, only internal exons of a gene are subjected to backsplicing, with the first and last exons being excluded due to the lack of acceptor and donor splice sites, respectively [17, 43, 75]. Nevertheless, it was reported that in the *UGT1A* gene, the first 317-nt of its last exon was subjected to backsplicing and formed circUGT1A-33 through a cryptic exonic donor splice site. Similarly, a part of the last exon of *UGT8* was also backspliced into circUGT8–8 via a cryptic exonic donor splice site [18]. The cryptic donor splice sites for both last exons have the conserved splicing signal dinucleotide "gt" that conforms to the ag/gt splicing rule. Likewise, other studies have also shown the presence of cryptic donor/acceptor splice sites within exons that were frequently involved in backsplicing, leading to the assembly of exonic circRNAs that contain partial exon sequences [60], whereas noncanonical splicing sites inside introns could be used to construct intronic circRNAs. In such conditions, backsplicing occurs when an intronic cryptic splice-donor site pairs with a cryptic splice-acceptor site within the same or another intron. As a result, a circRNA molecule carrying the intronic region(s) is formed [18]. Backsplicing in exonic–intronic circRNAs entails using a cryptic splice-donor site in an exon or an intron and a cryptic splice-acceptor site in an intron or another exon. This causes the RNA molecule to circularize, resulting in exonic–intronic RNA (Table 7.1).

3.2 *Trans*-Acting Factors

The biogenesis of circRNAs is a complex process controlled by many *trans*-acting factors, including RNA-binding proteins (RBPs). Certain RBPs can promote or hinder circRNA formation (Table 7.2). Previous studies have showed that,

Table 7.1 *Cis*-acting elements in circRNAs biogenesis

Cis-acting element	Schematic representation	References
Reverse complementary motifs The presence of *ALU* elements within flanking introns facilitates circRNAs biogenesis	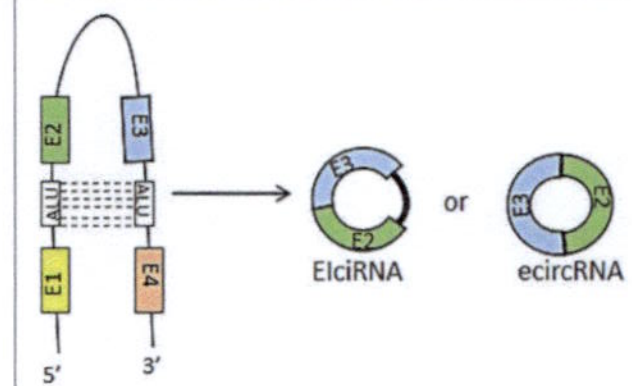	[3, 5, 19]
Intron length Long introns flanking cRNA exons may assist in their circularization	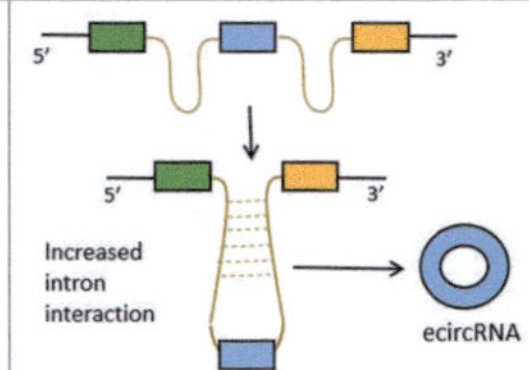	[69, 70, 112]
Exon length Large-sized exons (>300 bp) are more susceptible to forming single-exon circRNAs	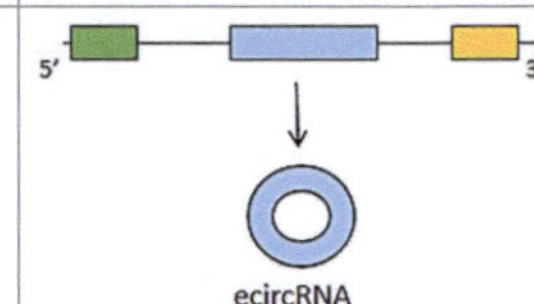	[3, 17, 71]
Splice site dinucleotide Mostly canonical splice signal ag/gt; in rare cases, noncanonical splice signals such as ac/ct and others	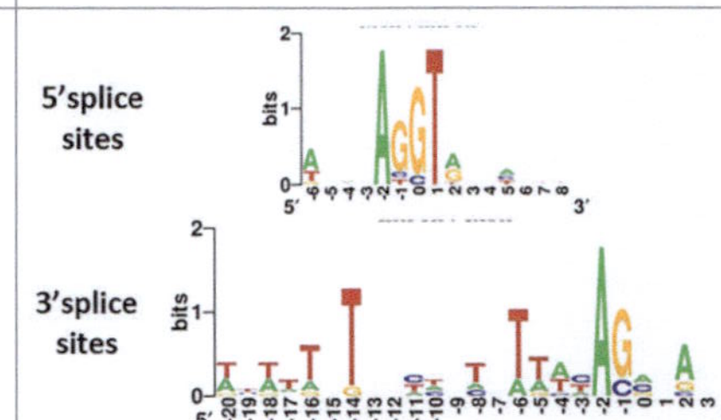	[41, 60, 63]

(continued)

Table 7.1 (contined)

Cis-acting element	Schematic representation	References
Backsplice site strength Backspliced exons with stronger 5′ splice sites than the original flanking proximal exons	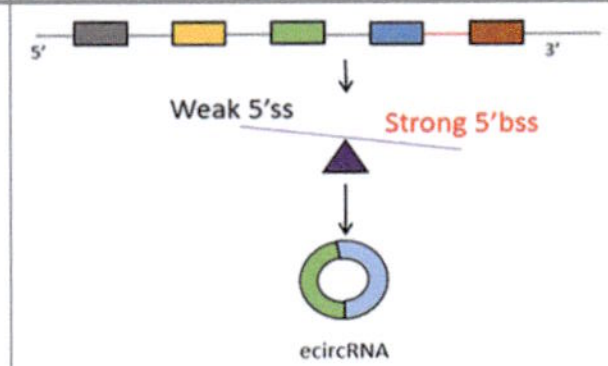	[66, 67]
Cryptic splice sites The presence of cryptic splice sites within introns and exons leads to the formation of intra-intronic circRNA (ciRNA), intra-exonic circRNA (ecircRNA), and exonic-intronic circRNA (EIciRNA)	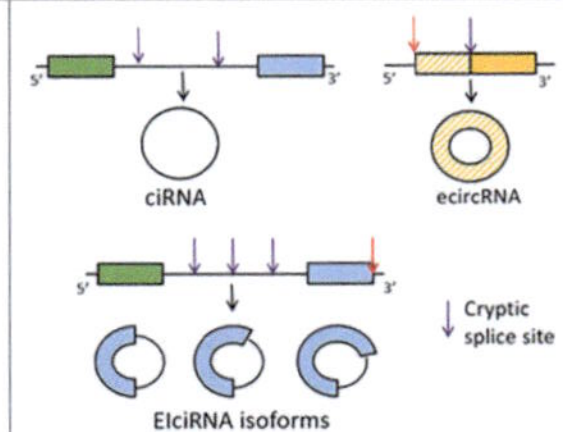	[18]

Table 7.2 *Trans*-acting proteins in circRNAs biogenesis, inhibition, and degradation

Protein	Target circRNA	Function[a]	Experimental model	References
EIF4A3	cSERPINE2	P	Human breast cancer tissue	[76]
	circ_0087429	I	Human epithelial–mesenchymal transition (EMT)	[77]
	circLMNB2	P	Lung adenocarcinoma	[78]
	circTOLLIP	P	Hepatocellular carcinoma	[79]
	circHomer1	P	Orbitofrontal cortex	[80]
	hsa_circ_0088088	P	MDA-MB-231 cells	[81]
	circSEPT9	P	Triple-negative breast cancer	[82]
QKI-5	hsa_circ_0098181	P	Human hepatocellular carcinoma	[83]
	circARFGEF2	P	Pancreatic ductal adenocarcinoma	[84]
ESRP1	circDOCK1	P	Human epithelial–mesenchymal transition (EMT)	[85]
	circ-TNPO3	P	Clear cell renal cell carcinoma	[86]
	circZNF652	P	Goblet cell metaplasia in allergic airway epithelial remodeling	[87]
	circUHRF1	P	Oral squamous cell carcinoma	[88]
	circCAMSAP1	P	Colorectal cancer	[89]
	circBIRC6	P	Human embryonic stem cells	[90]
QKI	Specific circular RNAs derived from titin (formin homology 2 domain containing 3) and striatin (calmodulin-binding protein 3)	P	Doxorubicin-mediated cardiotoxicity	[91]
	circRNAs: POLE2, SMARCA5, OXNAD1, SHPRH, SMAD2, ATXN2, DOCK1, GNB1	P	Epithelial–mesenchymal transition	[92]
DHX9	Circular RNAs containing inverted-repeat *Alu* elements	I	Human/embryonic stem cells	[50]
HNRNPL	circANKRD42	P	Idiopathic pulmonary fibrosis	[93]
	In all, 139 significantly upregulated and 93 downregulated circRNAs upon HNRNPL knockdown	P and I	Prostate cancer	[94]
NOVA2	circEfnb2	P	Mouse cortex	[95]
KAT5	circSMARCA5	P	Prostate cancer	[96]
FUS	circCNOT6L	P	Mouse and human spermatozoa	[97]
	circHIF1A	P	Triple-negative breast cancer	[98]
	In all, 136 circRNAs varied significantly with 111 were downregulated upon FUS knockdown	P and I	Amyotrophic lateral sclerosis	[16]
ADAR1	circNEIL3	I	Pancreatic ductal adenocarcinoma	[99]
	circARSP91	I	Hepatocellular carcinoma	[100]
E2F1	circSEPT9	P	Triple-negative breast cancer	[82]
SRSF10	circ-ATXN1	P	Glioma-associated endothelial cells	[101]
TNRC6A	circ0006916	P	16HBE and 16HBE-T cells	[102]
RBM20	circRNAs that originate from the I-band of the *Titin* gene	P	Dilated cardiomyopathy	[40]
RBM3	SCD (stearoyl-CoA desaturase)-circRNA 2	P	Hepatocellular carcinoma	[103]

(continued)

Table 7.2 (continued)

Protein	Target circRNA	Function[a]	Experimental model	References
NF90/NF110	circPOLR2A, circDHX34, circPDE3B	P	Viral infection model in HeLa cells	[104]
MBL/ MBNL1	circMbl	P	Drosophila S2 cells and murine homolog	[16]
c-Myb	circHIPK3	P	Colorectal cancer	[105]
hnRNP and SR proteins	Laccase2 circular RNA	P	Drosophila S2 and DL1 cell	[69]
RNA Pol II and U1 snRNP	circEIF3J, circPAIP2	P	HEK293 cells	[106]
RNA Pol II	ci-ankrd52, ci-sirt-7	P	Human cells (GM12878, HUVEC, HepG2, NHEK, H9, HeLa S3, and K562 cell lines)	[49]
RNase L	Global circRNAs	D	Systemic lupus erythematosus	[107]
Rrp44	Global circRNAs	D	Yeast	[108]
GW182	circRNAs: CaMKI, Dbp80, uex, laccase2, Haspin, ps, pan, dati, nipped-B, CG42663, PlexA, Ect4, mnb	D	Drosophila DL1 or S2 cells	[109]
PKR	circPOLR2A	D	Systemic lupus erythematosus	[107]
Ago2	CDR1as/ciRS-7	D	HEK293 cell lines	[110]
SFPQ	Distal-*Alu*-Long-Intron (DALI) circRNAs	P	HEK293T and HepG2 cell lines	[111]

[a]*P* promote circRNA biogenesis, *I* inhibit circRNA biogenesis, *D* degrade circRNA

among others, the muscleblind (MBL/MBNL1) protein is involved in circRNA biogenesis [16]. It was shown that the second exon of muscleblind can undergo circularization in flies to form circMbl. The MBL protein recognizes and binds to the sequences in the circMbl flanking introns, which, in turn, stimulates the synthesis of this circRNA. Another protein identified as a key regulator of circRNA biogenesis is splicing factor proline and glutamine-rich (SFPQ), particularly involved in interactions with long flanking introns that contain distal inverted *Alu* repeat elements [70]. circRNA biogenesis can also be influenced by various proteins, including spliceosome components and helicases, which play pivotal roles in modulating the splicing and circularization processes [49, 50, 73].

4 Types of CircRNAs

4.1 Exonic CircRNAs

An investigation of the mammalian transcriptome on a genome-wide scale using RNA-seq data found that EcircRNAs are the most frequently found when compared with the other types of circRNAs [5, 112, 113]. Additionally, they are evolutionarily conserved among different species [11, 112, 114]. The majority of exon-derived circRNAs bear multiple exons, while a minority originate from single exons [3, 17, 22, 72, 106]. EcircRNAs are also generated through the pairing of cryptic splice sites within an exon with the conventional splice sites that surround the same exon [18]. Several models were proposed to explain the potential origin of EcircRNAs. One of them is the exon-skipping model, which proposes the formation of a large lariat containing the skipped exon. This model is initiated by the circularization of exonic sequences through the backsplicing of a lariat intermediate formed during exon-skipping events in pre-mRNA splicing [5, 36–38]. Another model for EcircRNA biogenesis is driven by the re-splicing of mature mRNA. A second wave of canonical splicing can occur in certain mature mRNA molecules seen in cancer cells, like TSG101 and FHIT. Following this splicing event utilizing cryptic splice sites, the lariat solely consists of an exonic sequence [115]. The incidence of re-splicing, whether it yields exonic circRNAs or not, is unknown. EcircRNA biogenesis is also facilitated by intron pairing-driven circularization, a process in which complementary sequences within flanking introns bring together splice sites, leading to the formation of circular RNA molecules [5, 8, 19].

4.2 Intronic CircRNAs

During the splicing of pre-mRNA, noncoding introns are removed and exons are linked together [116]. The intron excision leads to the release of lariat molecules [117], which give rise to circRNAs generated due to the failure of intron lariat debranching. These circRNAs are characterized by the presence of the 2′–5′ bond that was generated by the intron lariat. The first evidence of intronic circular transcripts was published in 2013 [49]. The authors identified that intron lariats holding an 11-nt C-rich element near the 5′ splice site in addition to a 7-nt GU-rich region close to the branch point can produce IcircRNAs and withstand debranching (Fig. 7.1d). Several studies have shown that circRNAs produced from lariats are derived from introns that do not use an adenosine at the branch point but rather a "C" or, less frequently, a "G" or a "T" [49, 118–120]. The lariat debranching enzyme (DBR1) is incapable of hydrolyzing the 2′ − 5′ bond located at these residues [121]. Intronic circRNAs can also be generated using noncanonical splice sites inside introns [18]. When a cryptic donor splice site is located inside an intron and comes into contact with a cryptic acceptor splice site that is located within the same or another intron, backsplicing can take place. The outcome is the formation of a circRNA molecule that carries the intronic region [18]. Some intron-derived circRNAs comprise the whole intron sequence with a traditional 3′–5′ covalent link, and they were named "intron circles" [122]. Based on the studies by Taggart et al., a model was proposed for circularization to occur in one of two ways: (i) as a two-step debranching and ligation process or (ii) as a third nucleophilic attack on the 2′–5′ phosphate.

4.3 Exonic–Intronic CircRNAs

E-IcircRNAs are generated due to the improper removal of introns from freshly formed circRNAs, resulting in their retention in the final RNA product [106]. A recent study that used the long-read sequencing method to investigate the internal components of full-length circRNAs in human and mouse brains has identified that multi-exon circRNAs hold retained intron(s) [123]. When comparing all annotated introns with the circRNA-rich genomic areas, the authors discovered that shorter introns (<5000 bp) are more likely to be retained in circRNAs than longer introns. Backsplicing in E-IcircRNAs may also use cryptic splice-donor and splice-acceptor sites in exons or introns.

5 Challenges in the CircRNA Field

The biogenesis of circular RNAs encounters various challenges that span across all facets of molecular control. *Cis-* and *trans*-acting variables are critical in controlling the complex backsplicing process, in which a downstream splice-donor site meets an upstream splice-acceptor site to produce a closed-loop structure. It is a difficult task to identify the particular *cis*-acting sites within the RNA sequence and its surrounding regions that impact circRNA synthesis. Simultaneously, *trans*-acting RNA-binding proteins add to the regulatory complexity, and understanding their interactions with circRNA-forming exons remains difficult. The phenomenon of alternative splicing introduces diversity in circRNA synthesis and necessitates sophisticated knowledge of the regulatory signals regulating this process. The cell- and tissue-specific expression patterns of circRNAs provide difficulties, requiring an in-depth investigation of the variables driving this specificity. Moreover, the existence of multiple circRNA isoforms arising from alternative backsplicing processes adds complexity, necessitating a comprehensive investigation into their diverse functions across various cellular contexts. Exploring the evolutionary conservation of circRNA biogenesis across species is an area worth exploring, especially given the possible heterogeneity in regulatory networks. Furthermore, technical developments are critical for overcoming these hurdles, with circRNA analysis requiring enhanced high-throughput sequencing methods to improve accuracy and sensitivity, allowing precise quantification and discrimination from other RNA species.

Acknowledgments This work was supported by the Polish National Science Center (2019/33/B/NZ5/02473; 2020/39/B/NZ3/01811, and 2021/43/O/NZ1/01590).

Competing Financial Interests The authors declare no competing financial interests.

References

1. Sanger HL, Klotz G, Riesner D, Gross HJ, Kleinschmidt AK (1976) Viroids are single-stranded covalently closed circular RNA molecules existing as highly base-paired rod-like structures. Proc Natl Acad Sci USA 73(11):3852–3856. https://doi.org/10.1073/pnas.73.11.3852
2. Hsu MT, Coca-Prados M (1979) Electron microscopic evidence for the circular form of RNA in the cytoplasm of eukaryotic cells. Nature 280(5720):339–340. https://doi.org/10.1038/280339a0
3. Capel B, Swain A, Nicolis S, Hacker A, Walter M, Koopman P, Goodfellow P, Lovell-Badge R (1993) Circular transcripts of the testis-determining gene Sry in adult mouse testis. Cell 73(5):1019–1030. https://doi.org/10.1016/0092-8674(93)90279-y
4. Salzman J, Gawad C, Wang PL, Lacayo N, Brown PO (2012) Circular RNAs are the predominant transcript isoform from hundreds of human genes in diverse cell types. PLoS One 7(2):e30733. https://doi.org/10.1371/journal.pone.0030733
5. Jeck WR, Sorrentino JA, Wang K, Slevin MK, Burd CE, Liu J, Marzluff WF, Sharpless NE (2013) Circular RNAs are abundant, conserved, and associated with ALU repeats. RNA 19(2):141–157. https://doi.org/10.1261/rna.035667.112
6. Westholm JO, Miura P, Olson S, Shenker S, Joseph B, Sanfilippo P, Celniker SE, Graveley BR, Lai EC (2014) Genome-wide analysis of drosophila circular RNAs reveals their structural and sequence properties and age-dependent neural accumulation. Cell Rep 9(5):1966–1980. https://doi.org/10.1016/j.celrep.2014.10.062
7. Memczak S, Jens M, Elefsinioti A, Torti F, Krueger J, Rybak A, Maier L, Mackowiak SD, Gregersen LH, Munschauer M, Loewer A, Ziebold U, Landthaler M, Kocks C, le Noble F, Rajewsky N (2013) Circular RNAs are a large class of animal RNAs with regulatory potency. Nature 495(7441):333–338. https://doi.org/10.1038/nature11928
8. Ivanov A, Memczak S, Wyler E, Torti F, Porath HT, Orejuela MR, Piechotta M, Levanon EY, Landthaler M, Dieterich C, Rajewsky N (2015) Analysis of intron sequences reveals hallmarks of circular RNA biogenesis in animals. Cell Rep 10(2):170–177. https://doi.org/10.1016/j.celrep.2014.12.019
9. Liu Fan X, Zhang X, Wu X, Guo H, Hu Y, Tang F, Huang Y (2015) Single-cell RNA-seq transcriptome analysis of linear and circular RNAs in mouse preimplantation embryos. Genome Biol 16(1):148. https://doi.org/10.1186/s13059-015-070
10. Liang G, Yang Y, Niu G, Tang Z, Li K (2017) Genome-wide profiling of Sus scrofa circular RNAs across nine organs and three developmental stages. DNA Res 24(5):523–535. https://doi.org/10.1093/dnares/dsx022
11. Wang PL, Bao Y, Yee MC, Barrett SP, Hogan GJ, Olsen MN, Dinneny JR, Brown PO, Salzman J (2014) Circular RNA is expressed across the eukaryotic tree of life. PLoS One 9(6):e90859. https://doi.org/10.1371/journal.pone.0090859
12. Ye CY, Chen L, Liu C, Zhu QH, Fan L (2015) Widespread noncoding circular RNAs in plants. New Phytol 208(1):88–95. https://doi.org/10.1111/nph.13585
13. Lu T, Cui L, Zhou Y, Zhu C, Fan D, Gong H, Zhao Q, Zhou C, Zhao Y, Lu D, Luo J, Wang Y, Tian Q, Feng Q, Huang T, Han B (2015) Transcriptome-wide investigation of circular RNAs in rice. RNA 21(12):2076–2087. https://doi.org/10.1261/rna.052282.115
14. Szabo L, Salzman J (2016) Detecting circular RNAs: bioinformatic and experimental challenges. Nat Rev Genet 17(11):679–692. https://doi.org/10.1038/nrg.2016.114
15. Liu J, Liu T, Wang X, He A (2017) Circles reshaping the RNA world: from waste to treasure. Mol Cancer 16(1):58. https://doi.org/10.1186/s12943-017-0630-y
16. Ashwal-Fluss R, Meyer M, Pamudurti NR, Ivanov A, Bartok O, Hanan M, Evantal N, Memczak S, Rajewsky N, Kadener S (2014) circRNA biogenesis competes with pre-mRNA splicing. Mol Cell 56(1):55–66. https://doi.org/10.1016/j.molcel.2014.08.019
17. Ragan C, Goodall GJ, Shirokikh NE, Preiss T (2019) Insights into the biogenesis and potential functions of exonic circular RNA. Sci Rep 9(1):2048. https://doi.org/10.1038/s41598-018-37037-0
18. Hu DG, Mackenzie PI, Hulin JA, McKinnon RA, Meech R (2021) Circular RNAs of UDP-glycosyltransferase (*UGT*) genes expand the complexity and diversity of the UGT transcriptome. Mol Pharmacol 99(6):488–503. https://doi.org/10.1124/molpharm.120.000225
19. Liang D, Wilusz JE (2014) Short intronic repeat sequences facilitate circular RNA production. Genes

Dev 28(20):2233–2247. https://doi.org/10.1101/gad.251926.114
20. Qiu LP, Wu YH, Yu XF, Tang Q, Chen L, Chen KP (2018) The emerging role of circular RNAs in hepatocellular carcinoma. J Cancer 9(9):1548–1559. https://doi.org/10.7150/jca.24566
21. Yang L, Fu J, Zhou Y (2018) Circular RNAs and their emerging roles in immune regulation. Front Immunol 9:2977. https://doi.org/10.3389/fimmu.2018.02977
22. Huang A, Zheng H, Wu Z, Chen M, Huang Y (2020) Circular RNA-protein interactions: functions, mechanisms, and identification. Theranostics 10(8):3503–3517. https://doi.org/10.7150/thno.42174
23. Kristensen LS, Andersen MS, Stagsted LVW, Ebbesen KK, Hansen TB, Kjems J (2019) The biogenesis, biology and characterization of circular RNAs. Nat Rev Genet 20(11):675–691. https://doi.org/10.1038/s41576-019-0158-7
24. Li TR, Jia YJ, Wang Q, Shao XQ, Lv RJ (2017) Circular RNA: a new star in neurological diseases. Int J Neurosci 127(8):726–734. https://doi.org/10.1080/00207454.2016.1236382
25. Altesha MA, Ni T, Khan A, Liu K, Zheng X (2019) Circular RNA in cardiovascular disease. J Cell Physiol 234(5):5588–5600. https://doi.org/10.1002/jcp.27384
26. Patop IL, Wüst S, Kadener S (2019) Past, present, and future of circRNAs. EMBO J 38(16):e100836. https://doi.org/10.15252/embj.2018100836
27. Mercer TR, Gerhardt DJ, Dinger ME, Crawford J, Trapnell C, Jeddeloh JA, Mattick JS, Rinn JL (2011) Targeted RNA sequencing reveals the deep complexity of the human transcriptome. Nat Biotechnol 30(1):99–104. https://doi.org/10.1038/nbt.2024
28. Khan MR, Wellinger RJ, Laurent B (2021) Exploring the alternative splicing of long noncoding RNAs. Trends Genet 37(8):695–698. https://doi.org/10.1016/j.tig.2021.03.010
29. Castle JC, Zhang C, Shah JK, Kulkarni AV, Kalsotra A, Cooper TA, Johnson JM (2008) Expression of 24,426 human alternative splicing events and predicted cis regulation in 48 tissues and cell lines. Nat Genet 40(12):1416–1425. https://doi.org/10.1038/ng.264
30. Gueroussov S, Gonatopoulos-Pournatzis T, Irimia M, Raj B, Lin ZY, Gingras AC, Blencowe BJ (2015) An alternative splicing event amplifies evolutionary differences between vertebrates. Science 349(6250):868–873. https://doi.org/10.1126/science.aaa8381
31. Muhammad S, Xu X, Zhou W, Wu L (2022) Alternative splicing: an efficient regulatory approach towards plant developmental plasticity. Wiley Interdiscip Rev RNA 14(3):e1758. https://doi.org/10.1002/wrna.1758
32. Pan Q, Shai O, Lee LJ, Frey BJ, Blencowe BJ (2008) Deep surveying of alternative splicing complexity in the human transcriptome by high-throughput sequencing. Nat Genet 40(12):1413–1415. https://doi.org/10.1038/ng.259
33. Nilsen TW, Graveley BR (2010) Expansion of the eukaryotic proteome by alternative splicing. Nature 463(7280):457–463. https://doi.org/10.1038/nature08909
34. Wang Y, Liu J, Huang BO, Xu YM, Li J, Huang LF, Lin J, Zhang J, Min QH, Yang WM, Wang XZ (2015) Mechanism of alternative splicing and its regulation. Biomed Rep 3(2):152–158. https://doi.org/10.3892/br.2014.407
35. Kumari A, Sedehizadeh S, Brook JD, Kozlowski P, Wojciechowska M (2022) Differential fates of introns in gene expression due to global alternative splicing. Hum Genet 141(1):31–47. https://doi.org/10.1007/s00439-021-02409-6
36. Zaphiropoulos PG (1996) Circular RNAs from transcripts of the rat cytochrome P450 2C24 gene: correlation with exon skipping. Proc Natl Acad Sci USA 93(13):6536–6541. https://doi.org/10.1073/pnas.93.13.6536
37. Zaphiropoulos PG (1997) Exon skipping and circular RNA formation in transcripts of the human cytochrome P- 450 2C18 gene in epidermis and of the rat androgen binding protein gene in testis. Mol Cell Biol 17(6):2985–2993. https://doi.org/10.1128/MCB.17.6.2985
38. Surono A, Takeshima Y, Wibawa T, Ikezawa M, Nonaka I, Matsuo M (1999) Circular dystrophin RNAs consisting of exons that were skipped by alternative splicing. Hum Mol Genet 8(3):493–500. https://doi.org/10.1093/hmg/8.3.493
39. Kelly S, Greenman C, Cook PR, Papantonis A (2015) Exon skipping is correlated with exon circularization. J Mol Biol 427(15):2414–2417. https://doi.org/10.1016/j.jmb.2015.02.018
40. Khan MA, Reckman YJ, Aufiero S, van den Hoogenhof MM, van der Made I, Beqqali A, Koolbergen DR, Rasmussen TB, van der Velden J, Creemers EE, Pinto YM (2016) RBM20 regulates circular RNA production from the titin gene. Circ Res 119(9):996–1003. https://doi.org/10.1161/CIRCRESAHA.116.309568
41. Liang D, Tatomer DC, Luo Z, Wu H, Yang L, Chen LL, Cherry S, Wilusz JE (2017) The output of protein-coding genes shifts to circular RNAs when the pre-mRNA processing machinery is limiting. Mol Cell 68(5):940–954.e3. https://doi.org/10.1016/j.molcel.2017.10.034
42. Zhang Y, Xue W, Li X, Zhang J, Chen S, Zhang JL, Yang L, Chen LL (2016) The biogenesis of nascent circular RNAs. Cell Rep 15(3):611–624. https://doi.org/10.1016/j.celrep.2016.03.058
43. Zhang XO, Wang HB, Zhang Y, Lu X, Chen LL, Yang L (2014) Complementary sequence-mediated exon circularization. Cell 159(1):134–147. https://doi.org/10.1016/j.cell.2014.09.001

44. Barrett SP, Wang PL, Salzman J (2015) Circular RNA biogenesis can proceed through an exon-containing lariat precursor. elife 4:e07540. https://doi.org/10.7554/eLife.07540
45. Enuka Y, Lauriola M, Feldman ME, Sas-Chen A, Ulitsky I, Yarden Y (2016) Circular RNAs are long-lived and display only minimal early alterations in response to a growth factor. Nucleic Acids Res 44(3):1370–1383. https://doi.org/10.1093/nar/gkv1367
46. Hansen TB, Jensen TI, Clausen BH, Bramsen JB, Finsen B, Damgaard CK, Kjems J (2013) Natural RNA circles function as efficient microRNA sponges. Nature 495(7441):384–388. https://doi.org/10.1038/nature11993
47. Szabo L, Morey R, Palpant NJ, Wang PL, Afari N, Jiang C, Parast MM, Murry CE, Laurent LC, Salzman J (2015) Statistically based splicing detection reveals neural enrichment and tissue-specific induction of circular RNA during human fetal development. Genome Biol 16(1):126. https://doi.org/10.1186/s13059-015-0690-5
48. You X, Vlatkovic I, Babic A, Will T, Epstein I, Tushev G, Akbalik G, Wang M, Glock C, Quedenau C, Wang X, Hou J, Liu H, Sun W, Sambandan S, Chen T, Schuman EM, Chen W (2015) Neural circular RNAs are derived from synaptic genes and regulated by development and plasticity. Nat Neurosci 18(4):603–610. https://doi.org/10.1038/nn.3975
49. Zhang Y, Zhang XO, Chen T, Xiang JF, Yin QF, Xing YH, Zhu S, Yang L, Chen LL (2013) Circular intronic long noncoding RNAs. Mol Cell 51(6):792–806. https://doi.org/10.1016/j.molcel.2013.08.017
50. Aktaş T, Avşar Ilık İ, Maticzka D, Bhardwaj V, Pessoa Rodrigues C, Mittler G, Manke T, Backofen R, Akhtar A (2017) DHX9 suppresses RNA processing defects originating from the Alu invasion of the human genome. Nature 544(7648):115–119. https://doi.org/10.1038/nature21715
51. Dubin RA, Kazmi MA, Ostrer H (1995) Inverted repeats are necessary for circularization of the mouse testis Sry transcript. Gene 167(1–2):245–248. https://doi.org/10.1016/0378-1119(95)00639-7
52. Zhang XO, Pratt H, Weng Z (2021) Investigating the potential roles of SINEs in the human genome. Annu Rev Genomics Hum Genet 22:199–218. https://doi.org/10.1146/annurev-genom-111620-100736
53. Chen LL, DeCerbo JN, Carmichael GG (2008) Alu element-mediated gene silencing. EMBO J 27(12):1694–1705. https://doi.org/10.1038/emboj.2008.94
54. Ottesen EW, Luo D, Seo J, Singh NN, Singh RN (2019) Human survival motor neuron genes generate a vast repertoire of circular RNAs. Nucleic Acids Res 47(6):2884–2905. https://doi.org/10.1093/nar/gkz034
55. Tang Y, Bao J, Hu J, Liu L, Xu DY (2021) Circular RNA in cardiovascular disease: expression, mechanisms and clinical prospects. J Cell Mol Med 25(4):1817–1824. https://doi.org/10.1111/jcmm.16203
56. Kokot KE, Kneuer JM, John D, Rebs S, Möbius-Winkler MN, Erbe S, Müller M, Andritschke M, Gaul S, Sheikh BN, Haas J, Thiele H, Müller OJ, Hille S, Leuschner F, Dimmeler S, Streckfuss-Bömeke K, Meder B, Laufs U, Boeckel JN (2022) Reduction of A-to-I RNA editing in the failing human heart regulates formation of circular RNAs. Basic Res Cardiol 117(1):32. https://doi.org/10.1007/s00395-022-00940-9
57. Chen LL (2016) The biogenesis and emerging roles of circular RNAs. Nat Rev Mol Cell Biol 17(4):205–211. https://doi.org/10.1038/nrm.2015.32
58. Jeck WR, Sharpless NE (2014) Detecting and characterizing circular RNAs. Nat Biotechnol 32(5):453–461. https://doi.org/10.1038/nbt.2890
59. Yang L (2015) Splicing noncoding RNAs from the inside out. Wiley Interdiscip Rev RNA 6(6):651–660. https://doi.org/10.1002/wrna.1307
60. Starke S, Jost I, Rossbach O, Schneider T, Schreiner S, Hung LH, Bindereif A (2015) Exon circularization requires canonical splice signals. Cell Rep 10(1):103–111. https://doi.org/10.1016/j.celrep.2014.12.002
61. Zhang M, Wang T, Xiao G, Xie Y (2020) Large-scale profiling of RBP-circRNA interactions from public CLIP-Seq datasets. Genes (Basel) 11(1):54. https://doi.org/10.3390/genes11010054
62. Gao X, Ma XK, Li X, Li GW, Liu CX, Zhang J, Wang Y, Wei J, Chen J, Chen LL, Yang L (2022) Knockout of circRNAs by base editing back-splice sites of circularized exons. Genome Biol 23(1):16. https://doi.org/10.1186/s13059-021-02563-0
63. Liu X, Hu Z, Zhou J, Tian C, Tian G, He M, Gao L, Chen L, Li T, Peng H, Zhang W (2020) Interior circular RNA. RNA Biol 17(1):87–97. https://doi.org/10.1080/15476286.2019.1669391
64. Gao M, Liu Y, Wang Y, Zhang X, Dong S, Liu X (2019) Newly identified APN splice isoforms suggest novel splicing mechanisms may underlie circRNA circularization in moth. FEBS Open Bio 9(9):1521–1535. https://doi.org/10.1002/2211-5463.12689
65. Zhang XO, Dong R, Zhang Y, Zhang JL, Luo Z, Zhang J, Chen LL, Yang L (2016) Diverse alternative back-splicing and alternative splicing landscape of circular RNAs. Genome Res 26(9):1277–1287. https://doi.org/10.1101/gr.202895.115
66. Di Liddo A, de Oliveira Freitas Machado C, Fischer S, Ebersberger S, Heumüller AW, Weigand JE, Müller-McNicoll M, Zarnack K (2019) A combined computational pipeline to detect circular RNAs in human cancer cells under hypoxic stress. J Mol Cell Biol 11(10):829–844. https://doi.org/10.1093/jmcb/mjz094
67. Ho JS, Di Tullio F, Schwarz M, Low D, Incarnato D, Gay F, Tabaglio T, Zhang J, Wollmann H, Chen L, An O, Chan THM, Hall Hickman A, Zheng S, Roudko V, Chen S, Karz A, Ahmed M, He HH, Greenbaum BD, Oliviero S, Serresi M, Gargiulo

G, Mann KM, Hernando E, Mulholland D, Marazzi I, DKB W, Guccione E (2021) HNRNPM controls circRNA biogenesis and splicing fidelity to sustain cancer cell fitness. elife 10:e59654. https://doi.org/10.7554/eLife.59654
68. Tsirigos A, Rigoutsos I (2009) Alu and b1 repeats have been selectively retained in the upstream and intronic regions of genes of specific functional classes. PLoS Comput Biol 5(12):e1000610. https://doi.org/10.1371/journal.pcbi.1000610
69. Kramer MC, Liang D, Tatomer DC, Gold B, March ZM, Cherry S, Wilusz JE (2015) Combinatorial control of Drosophila circular RNA expression by intronic repeats, hnRNPs, and SR proteins. Genes Dev 29(20):2168–2182. https://doi.org/10.1101/gad.270421.115
70. Stagsted LV, Nielsen KM, Daugaard I, Hansen TB (2019) Noncoding AUG circRNAs constitute an abundant and conserved subclass of circles. Life Sci Alliance 2(3):e201900398. https://doi.org/10.26508/lsa.20190039
71. Santer L, Bär C, Thum T (2019) Circular RNAs: a novel class of functional RNA molecules with a therapeutic perspective. Mol Ther 27(8):1350–1363. https://doi.org/10.1016/j.ymthe.2019.07.001
72. Chu Q, Shen E, Ye C-Y, Fan L, Zhu Q-H (2018) Emerging roles of plant circular RNAs. J Plant Cell Dev 1:1–14. https://doi.org/10.14302/issn.2832-5311.jpcd-18-1955
73. Li J, Yang J, Zhou P, Le Y, Zhou C, Wang S, Xu D, Lin HK, Gong Z (2015) Circular RNAs in cancer: novel insights into origins, properties, functions and implications. Am J Cancer Res 5(2):472–480. https://doi.org/10.1186/s13048-019-0558-5
74. Errichelli L, Dini Modigliani S, Laneve P, Colantoni A, Legnini I, Capauto D, Rosa A, De Santis R, Scarfò R, Peruzzi G, Lu L, Caffarelli E, Shneider NA, Morlando M, Bozzoni I (2017) FUS affects circular RNA expression in murine embryonic stem cell-derived motor neurons. Nat Commun 8:14741. https://doi.org/10.1038/ncomms14741
75. Wilusz JE (2015) Repetitive elements regulate circular RNA biogenesis. Mob Genet Elements 5(3):1–7. https://doi.org/10.1080/2159256X.2015.1045682
76. Zhou B, Mo Z, Lai G, Chen X, Li R, Wu R, Zhu J, Zheng F (2023) Targeting tumor exosomal circular RNA cSERPINE2 suppresses breast cancer progression by modulating MALT1-NF-kB-IL-6 axis of tumor-associated macrophages. J Exp Clin Cancer Res 42(1):48. https://doi.org/10.1186/s13046-023-02620-5
77. Yang M, Hu H, Wu S, Ding J, Yin B, Huang B, Li F, Guo X, Han L (2022) EIF4A3-regulated circ_0087429 can reverse EMT and inhibit the progression of cervical cancer via miR-5003-3p-dependent upregulation of OGN expression. J Exp Clin Cancer Res 41(1):165. https://doi.org/10.1186/s13046-022-02368-4
78. Qiu M, Chen M, Lan Z, Liu B, Xie J, Li X (2022) Plasmacytoma variant translocation 1 stabilized by EIF4A3 promoted malignant biological behaviors of lung adenocarcinoma by generating circular RNA LMNB2. Bioengineered 13(4):10123–10140. https://doi.org/10.1080/21655979.2022.2063666
79. Liu Y, Song J, Zhang H, Liao Z, Liu F, Su C, Wang W, Han M, Zhang L, Zhu H, Zhang Z, Liang H, Zhang L, Zhang B, Chen X (2022) EIF4A3-induced circTOLLIP promotes the progression of hepatocellular carcinoma via the miR-516a-5p/PBX3/EMT pathway. J Exp Clin Cancer Res 41(1):164. https://doi.org/10.1186/s13046-022-02378-2
80. Hafez AK, Zimmerman AJ, Papageorgiou G, Chandrasekaran J, Amoah SK, Lin R, Lozano E, Pierotti C, Dell'Orco M, Hartley BJ, Alural B, Lalonde J, Esposito JM, Berretta S, Squassina A, Chillotti C, Voloudakis G, Shao Z, Fullard JF, Brennand KJ, Turecki G, Roussos P, Perlis RH, Haggarty SJ, Perrone-Bizzozero N, Brigman JL, Mellios N (2022) A bidirectional competitive interaction between circHomer1 and Homer1b within the orbitofrontal cortex regulates reversal learning. Cell Rep 38(3):110282. https://doi.org/10.1016/j.celrep.2021.110282
81. Liu Q, Dong H (2021) EIF4A3-mediated hsa_circ_0088088 promotes the carcinogenesis of breast cancer by sponging miR-135-5p. J Biochem Mol Toxicol 35(11):e22909. https://doi.org/10.1002/jbt.22909
82. Zheng X, Huang M, Xing L, Yang R, Wang X, Jiang R, Zhang L, Chen J (2020) The circRNA circSEPT9 mediated by E2F1 and EIF4A3 facilitates the carcinogenesis and development of triple-negative breast cancer. Mol Cancer 19(1):73. https://doi.org/10.1186/s12943-020-01183-9
83. Gao P, Yang Y, Li X, Zhao Q, Liu Y, Dong C, Zhang Y, Liu D (2023) Circular RNA hsa_circ_0098181 inhibits metastasis in hepatocellular carcinoma by activating the Hippo signaling pathway via interaction with eEF2. Ann Hepatol 28(5):101124. https://doi.org/10.1016/j.aohep.2023.101124
84. Kong Y, Luo Y, Zheng S, Yang J, Zhang D, Zhao Y, Zheng H, An M, Lin Y, Ai L, Diao X, Lin Q, Chen C, Chen R (2023) Mutant KRAS mediates circARFGEF2 biogenesis to promote lymphatic metastasis of pancreatic ductal adenocarcinoma. Cancer Res 83(18):3077–3094. https://doi.org/10.1158/0008-5472.CAN-22-3997
85. Liu D, Dredge BK, Bert AG, Pillman KA, Toubia J, Guo W, Dyakov BJA, Migault MM, Conn VM, Conn SJ, Gregory PA, Gingras AC, Patel D, Wu B, Goodall GJ (2023) ESRP1 controls biogenesis and function of a large abundant multiexon circRNA. Nucleic Acids Res 52:gkad1138. https://doi.org/10.1093/nar/gkad1138
86. Pan X, Huang B, Ma Q, Ren J, Liu Y, Wang C, Zhang D, Fu J, Ran L, Yu T, Li H, Wang X, Yang F, Liang C, Zhang Y, Wang S, Ren J, Li W, Wang Y, Xiao B (2022) Circular RNA circ-TNPO3 inhibits clear cell renal cell carcinoma metastasis by binding to IGF2BP2 and destabilizing SERPINH1

mRNA. Clin Transl Med 12(7):e994. https://doi.org/10.1002/ctm2.994

87. Wang X, Xu C, Cai Y, Zou X, Chao Y, Yan Z, Zou C, Wu X, Tang L (2023) CircZNF652 promotes the goblet cell metaplasia by targeting the miR-452-5p/JAK2 signaling pathway in allergic airway epithelia. J Allergy Clin Immunol 150(1):192–203. https://doi.org/10.1016/j.jaci.2021.10.041
88. Zhao W, Cui Y, Liu L, Qi X, Liu J, Ma S, Hu X, Zhang Z, Wang Y, Li H, Wang Z, Liu Z, Wu J (2019) Splicing factor derived circular RNA circUHRF1 accelerates oral squamous cell carcinoma tumorigenesis via feedback loop. Cell Death Differ 27(3):919–933. https://doi.org/10.1038/s41418-019-0423-5
89. Zhou C, Liu HS, Wang FW, Hu T, Liang ZX, Lan N, He XW, Zheng XB, Wu XJ, Xie D, Wu XR, Lan P (2020) circCAMSAP1 promotes tumor growth in colorectal cancer via the miR-328-5p/E2F1 axis. Mol Ther 28(3):914–928. https://doi.org/10.1016/j.ymthe.2019.12.008
90. Yu CY, Li TC, Wu YY, Yeh CH, Chiang W, Chuang CY, Kuo HC (2017) The circular RNA circBIRC6 participates in the molecular circuitry controlling human pluripotency. Nat Commun 8(1):1149. https://doi.org/10.1038/s41467-017-01216-w
91. Gupta SK, Garg A, Bär C, Chatterjee S, Foinquinos A, Milting H, Streckfuß-Bömeke K, Fiedler J, Thum T (2018) Quaking inhibits doxorubicin-mediated cardiotoxicity through regulation of cardiac circular RNA expression. Circ Res 122(2):246–254. https://doi.org/10.1161/CIRCRESAHA.117.311335
92. Conn SJ, Pillman KA, Toubia J, Conn VM, Salmanidis M, Phillips CA, Roslan S, Schreiber AW, Gregory PA, Goodall GJ (2015) The RNA binding protein quaking regulates formation of circRNAs. Cell 160(6):1125–1134. https://doi.org/10.1016/j.cell.2015.02.014
93. Xu P, Zhang J, Wang M, Liu B, Li R, Li H, Zhai N, Liu W, Lv C, Song X (2022) hnRNPL-activated circANKRD42 back-splicing and circANKRD42-mediated crosstalk of mechanical stiffness and biochemical signal in lung fibrosis. Mol Ther 30(6):2370–2387. https://doi.org/10.1016/j.ymthe.2022.01.045
94. Fei T, Chen Y, Xiao T, Li W, Cato L, Zhang P, Cotter MB, Bowden M, Lis RT, Zhao SG, Wu Q, Feng FY, Loda M, He HH, Liu XS, Brown M (2017) Genome-wide CRISPR screen identifies HNRNPL as a prostate cancer dependency regulating RNA splicing. Proc Natl Acad Sci USA 114(26):E5207–E5215. https://doi.org/10.1073/pnas.1617467114
95. Knupp D, Cooper DA, Saito Y, Darnell RB, Miura P (2021) NOVA2 regulates neural circRNA biogenesis. Nucleic Acids Res 49(12):6849–6862. https://doi.org/10.1093/nar/gkab523
96. Xie X, Sun FK, Huang X, Wang CH, Dai J, Zhao JP, Fang C, He W (2021) A circular RNA, circSMARCA5, inhibits prostate cancer proliferative, migrative, and invasive capabilities via the miR-181b-5p/miR-17-3p-TIMP3 axis. Aging (Albany NY) 13(15):19908–19919. https://doi.org/10.18632/aging.203408
97. Chioccarelli T, Falco G, Cappetta D, De Angelis A, Roberto L, Addeo M, Ragusa M, Barbagallo D, Berrino L, Purrello M, Ambrosino C, Cobellis G, Pierantoni R, Chianese R, Manfrevola F (2021) FUS driven circCNOT6L biogenesis in mouse and human spermatozoa supports zygote development. Cell Mol Life Sci 79(1):50. https://doi.org/10.1007/s00018-021-04054-8
98. Chen T, Wang X, Li C, Zhang H, Liu Y, Han D, Li Y, Li Z, Luo D, Zhang N, Zheng M, Chen B, Wang L, Zhao W, Yang Q (2021) CircHIF1A regulated by FUS accelerates triple-negative breast cancer progression by modulating NFIB expression and translocation. Oncogene 40(15):2756–2771. https://doi.org/10.1038/s41388-021-01739-z
99. Shen P, Yang T, Chen Q, Yuan H, Wu P, Cai B, Meng L, Huang X, Liu J, Zhang Y, Hu W, Miao Y, Lu Z, Jiang K (2021) CircNEIL3 regulatory loop promotes pancreatic ductal adenocarcinoma progression via miRNA sponging and A-to-I RNA-editing. Mol Cancer 20(1):51. https://doi.org/10.1186/s12943-021-01333-7
100. Shi L, Yan P, Liang Y, Sun Y, Shen J, Zhou S, Lin H, Liang X, Cai X (2017) Circular RNA expression is suppressed by androgen receptor (AR)-regulated adenosine deaminase that acts on RNA (ADAR1) in human hepatocellular carcinoma. Cell Death Dis 8(11):e3171. https://doi.org/10.1038/cddis.2017.556
101. Liu X, Shen S, Zhu L, Su R, Zheng J, Ruan X, Shao L, Wang D, Yang C, Liu Y (2020) SRSF10 inhibits biogenesis of circ-ATXN1 to regulate glioma angiogenesis via miR-526b-3p/MMP2 pathway. J Exp Clin Cancer Res 39(1):121. https://doi.org/10.1186/s13046-020-01625-8
102. Dai X, Zhang N, Cheng Y, Yang T, Chen Y, Liu Z, Wang Z, Yang C, Jiang Y (2018) RNA-binding protein trinucleotide repeat-containing 6A regulates the formation of circular RNA circ0006916, with important functions in lung cancer cells. Carcinogenesis 39(8):981–992. https://doi.org/10.1093/carcin/bgy061
103. Dong W, Dai ZH, Liu FC, Guo XG, Ge CM, Ding J, Liu H, Yang F (2019) The RNA-binding protein RBM3 promotes cell proliferation in hepatocellular carcinoma by regulating circular RNA SCD-circRNA 2 production. EBioMedicine 45:155–167. https://doi.org/10.1016/j.ebiom.2019.06.030
104. Li X, Liu CX, Xue W, Zhang Y, Jiang S, Yin QF, Wei J, Yao RW, Yang L, Chen LL (2017) Coordinated circRNA biogenesis and function with NF90/NF110 in viral infection. Mol Cell 67(2):214–227.e7. https://doi.org/10.1016/j.molcel.2017.05.023
105. Zeng K, Chen X, Xu M, Liu X, Hu X, Xu T, Sun H, Pan Y, He B, Wang S (2018) CircHIPK3 promotes colorectal cancer growth and metastasis by sponging miR-7. Cell Death Dis 9(4):417. https://doi.org/10.1038/s41419-018-0454-8

106. Li Z, Huang C, Bao C, Chen L, Lin M, Wang X, Zhong G, Yu B, Hu W, Dai L, Zhu P, Chang Z, Wu Q, Zhao Y, Jia Y, Xu P, Liu H, Shan G (2015) Exon-intron circular RNAs regulate transcription in the nucleus. Nat Struct Mol Biol 22(3):256–264. https://doi.org/10.1038/nsmb.2959
107. Liu CX, Li X, Nan F, Jiang S, Gao X, Guo SK, Xue W, Cui Y, Dong K, Ding H, Qu B, Zhou Z, Shen N, Yang L, Chen LL (2019) Structure and degradation of circular RNAs regulate PKR activation in innate immunity. Cell 177(4):865–880.e21. https://doi.org/10.1016/j.cell.2019.03.046
108. Schaeffer D, Tsanova B, Barbas A, Reis FP, Dastidar EG, Sanchez-Rotunno M, Arraiano CM, van Hoof A (2009) The exosome contains domains with specific endoribonuclease, exoribonuclease and cytoplasmic mRNA decay activities. Nat Struct Mol Biol 16(1):56–62. https://doi.org/10.1038/nsmb.1528
109. Jia R, Xiao MS, Li Z, Shan G, Huang C (2019) Defining an evolutionarily conserved role of GW182 in circular RNA degradation. Cell Discov 5:45. https://doi.org/10.1038/s41421-019-0113-y
110. Hansen TB, Wiklund ED, Bramsen JB, Villadsen SB, Statham AL, Clark SJ, Kjems J (2011) miRNA-dependent gene silencing involving Ago2-mediated cleavage of a circular antisense RNA. EMBO J 30(21):4414–4422. https://doi.org/10.1038/emboj.2011.359
111. Stagsted LVW, O'Leary ET, Ebbesen KK, Hansen TB (2021) The RNA-binding protein SFPQ preserves long-intron splicing and regulates circRNA biogenesis in mammals. elife 10:e63088. https://doi.org/10.7554/eLife.63088. PMID: 33476259; PMCID: PMC7819710
112. Chen I, Chen CY, Chuang TJ (2015) Biogenesis, identification, and function of exonic circular RNAs. Wiley Interdiscip Rev RNA 6(5):563–579. https://doi.org/10.1002/wrna.1294
113. Lyu D, Huang S (2017) The emerging role and clinical implication of human exonic circular RNA. RNA Biol 14(8):1000–1006. https://doi.org/10.1080/15476286.2016
114. Salzman J, Chen RE, Olsen MN, Wang PL, Brown PO (2013) Cell-type specific features of circular RNA expression. PLoS Genet 9(9):e1003777. https://doi.org/10.1371/annotation/f782282b-eefa-4c8d-985c-b1484e845855-w
115. Kameyama T, Suzuki H, Mayeda A (2012) Re-splicing of mature mRNA in cancer cells promotes activation of distant weak alternative splice sites. Nucleic Acids Res 40(16):7896–7906. https://doi.org/10.1093/nar/gks520
116. Keller W (1984) The RNA lariat: a new ring to the splicing of mRNA precursors. Cell 39:423–425. https://doi.org/10.1016/0092-8674(84)90449-5
117. Padgett RA, Konarska MM, Grabowski PJ, Hardy SF, Sharp PA (1984) Lariat RNA's as intermediates and products in the splicing of messenger RNA precursors. Science:898–903. https://doi.org/10.1126/science.6206566
118. Chan SN, Pek JW (2018) Stable intronic sequence RNAs (sisRNAs): an expanding universe. Trends Biochem Sci 44(3):258–272. https://doi.org/10.1016/j.tibs.2018.09.016
119. Osman I, Tay ML, Pek JW (2016) Stable intronic sequence RNAs (sisRNAs): a new layer of gene regulation. Cell Mol Life Sci 73(18):3507–3519. https://doi.org/10.1007/s00018-016-2256-4
120. Jin J, He X, Silva E (2020) Stable intronic sequence RNAs (sisRNAs) are selected regions in introns with distinct properties. BMC Genomics 21(1):287. https://doi.org/10.1186/s12864-020-6687-9
121. Jacquier A, Rosbash M (1986) RNA splicing and intron turnover are greatly diminished by a mutant yeast branch point. Proc Natl Acad Sci USA 83(16):5835–5839. https://doi.org/10.1073/pnas.83.16.5835
122. Taggart AJ, Lin CL, Shrestha B, Heintzelman C, Kim S, Fairbrother WG (2017) Large-scale analysis of branchpoint usage across species and cell lines. Genome Res 27(4):639–649. https://doi.org/10.1101/gr.202820.115
123. Rahimi K, Venø MT, Dupont DM, Kjems J (2021) Nanopore sequencing of brain-derived full-length circRNAs reveals circRNA-specific exon usage, intron retention and microexons. Nat Commun 12(1):4825. https://doi.org/10.1038/s41467-021-24975-z

RNA-Binding Proteins (RBPs) and Circular RNA Biogenesis

8

Shasha Shi and Rui Zhao

Abstract

Many protein-coding genes in eukaryotes have the capacity to produce circular RNAs (circRNAs). CircRNAs are stable and covalently closed products generated from backsplicing. Their biogenesis is regulated by multiple factors, with RNA-binding proteins (RBPs) playing a critical role. RBPs bind to flanking introns of the circularizing exons, bringing the intervening splice site into close proximity and thereby facilitating the process of backsplicing. In this chapter, we summarize the literature on RBP-mediated circularization in circRNA biogenesis, illustrating how RBPs influence RNA fate and gene regulation.

Keywords

CircRNA · Biogenesis · RNA-binding proteins · Backsplicing

S. Shi · R. Zhao (✉)
Department of Biochemistry and Molecular Genetics, University of Colorado Anschutz Medical Campus, Aurora, CO, USA
e-mail: rui.zhao@cuanschutz.edu

1 Introduction

Circular RNAs (circRNAs) are single-stranded, covalently closed-loop structures in which the 5′- and 3′-ends are joined together through backsplicing. Due to their unique circular form, it is seldom detected in mRNA sequencing data. Initially discovered in viroids in 1976 [1], only a limited number of genes were found to produce circRNAs over the next 30 years. Thus, circRNAs were considered "splicing noise." In 2013, Jeck et al. performed high-throughput sequencing on libraries generated from ribosome-depleted human fibroblasts and identified over 25,000 circRNAs [2]. Subsequent studies have demonstrated that circRNAs are found ubiquitously from viruses to mammals, exhibit high stability, and participate in diverse cellular processes, such as miRNA and protein sponging, as well as transcription and translation regulation [3].

The biogenesis of circRNAs involves a backsplicing event mediated by the spliceosome machinery and various protein factors, such as RNA-binding proteins (RBPs). RBPs have both positive and negative influences in this process. This chapter outlines the events involved in RBP-mediated circRNA generation and elucidates their crucial role in RNA circularization.

J. Xiao (ed.), *Advances in Circular RNAs*, Advances in Experimental Medicine and Biology 1485,
https://doi.org/10.1007/978-981-96-9428-0_8

2 Backsplicing Generates CircRNAs

The majority of circRNAs are generated from pre-messenger RNA (pre-mRNA) backsplicing. Unlike canonical splicing, where an upstream 5′ splice site (ss) is joined to a downstream 3′ ss across an intron, the 5′ splice site pairs with an upstream 3′ ss across an exon, forming circRNAs. The efficiency of backsplicing circularization from pre-mRNA is extremely low compared with that of canonical splicing in most endogenous human loci, as determined by analyzing the kinetics of nascent circRNA processing [4, 5]. However, the circular nature of circRNAs makes them resistant to most nucleases and renders them highly stable. An unbiased assessment of circRNA species in mammalian cells has demonstrated that circRNAs are abundant, stable, conserved, and functionally active [2, 6]. Remarkably, some circular RNAs are expressed at levels up to tenfold higher than their associated linear mRNAs [7]. CircRNAs generated by backsplicing have attracted widespread attention; however, the precise mechanism behind their generation through backsplicing has remained elusive for a long time. For example, it was known that the same splice sites and the spliceosome are involved in backsplicing, but it was unclear whether this is the same spliceosome (compositionally and structurally) required for canonical splicing and it was unknown what signal triggers a pre-mRNA to undergo backsplicing instead of canonical splicing.

A structural study on the *Saccharomyces cerevisiae* (yeast) spliceosomal E complex by our lab in 2019 unraveled the mystery of circRNA biogenesis via elucidating the role of exon definition in backsplicing [8]. The architecture of the yeast E complex assembled on intron-containing pre-mRNA revealed by our structure indicates that the same spliceosomal E and A complexes can be assembled on an exon (the exon definition complex) without the need of additional components or structural arrangements. Our model (Fig. 8.1), derived from the structure, suggests that when the exon (on which the exon definition complex is formed) is short, the tri-snRNP cannot join the spliceosome to form the pre-B complex efficiently due to steric hindrance. This may trigger the spliceosome to remodel from an exon-spanning to an intron-spanning complex, which then completes the canonical splicing. However, when the exon is long, the tri-snRNP will be able to join the spliceosome, undergo the rest of the splicing cycle, and carry out the two transesterification reactions to ligate the two ends of the exon, forming a circRNA. It is worth noting that the exon that the spliceosome initially assembles on does not need to be a single exon. Instead, it can contain multiple exons or reside in a lariat containing a skipped exon. In essence, backsplicing utilizes the same transesterification reaction and spliceosome machinery employed in canonical splicing, without requiring additional components or structure rearrangement, providing a simple model that unifies intron definition, exon definition, and backsplicing mechanisms.

3 RNA-Binding Proteins

RBPs are a diverse class of proteins with a multitude of binding substrates and functions in gene expression at different levels, including RNA production, maturation, localization, translation, and degradation of cellular RNAs. They are widely found in microorganism, plants, and animals. Based on the existence of established RNA-binding domains (RBDs), at least 1500 RBPs are identified in the human genome, accounting for about 7.5% of all protein-coding genes in humans [9].

RBPs interact with RNAs via their specific structured RBDs, such as RNA recognition motif (RRM), zinc finger (ZnF), K-homology domain (KH), DEAD box helicase domain, and dsRNA-binding motif (dsRBM) [10–14]. Many of those RBPs preferentially bind to a "bipartite motif" composed of one or two low-complexity base types spaced one or more bases apart, some of which serve as functional elements regulating the backsplicing process [15]. These interactions between RBPs and specific RNA motifs contribute significantly to the intricate regulation of RNA processing and function, including those of circRNAs.

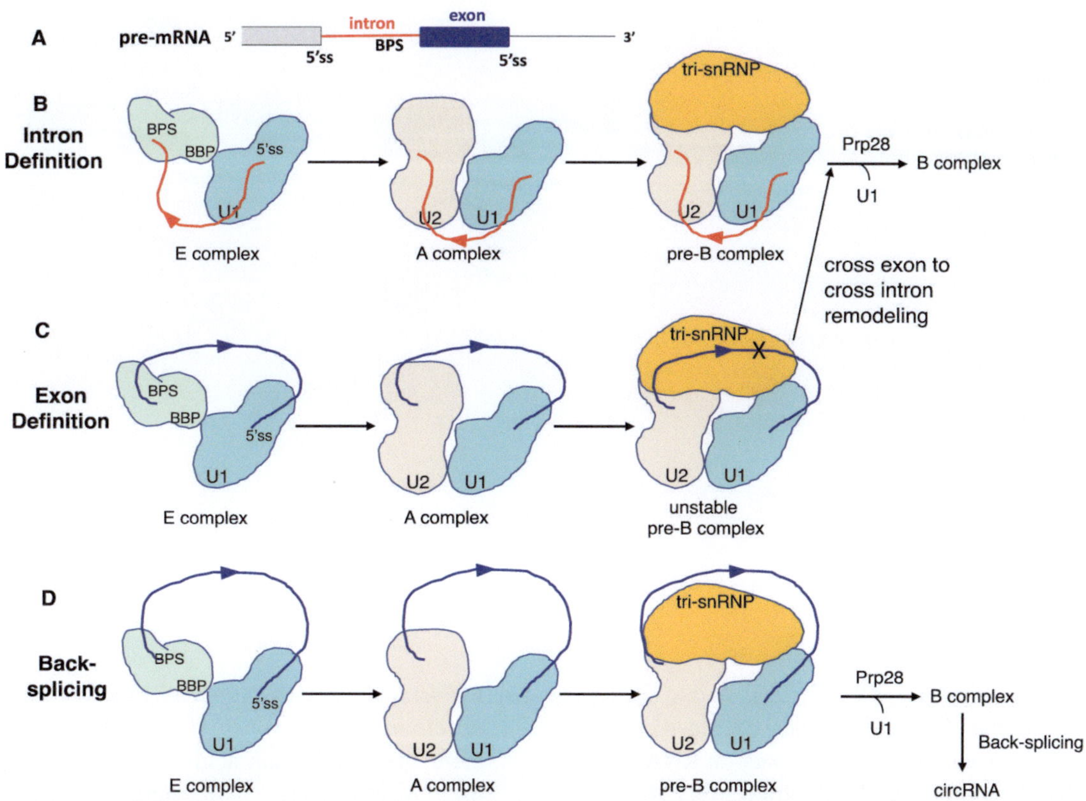

Fig. 8.1 A unified intron and exon definition and backsplicing model based on the yeast E complex structure. (**a**) A schematic of the pre-mRNA. (**b**) The steps from the E to B complex in a canonical assembly pathway across an intron (red). The arrow in pre-mRNA indicates the 5′-to-3′ direction. (**c**) The same spliceosomal E and A complexes as in panel (**b**) can assemble across an exon (blue) but cannot form the pre-B complex on short exons due to steric hindrance. This may trigger the spliceosome to remodel and become intron-spanning, which then carry out the canonical splicing cycle. (**d**) The tri-snRNP can join the spliceosome complex assemble on long exons to carry out backsplicing and form circRNAs

4 RBPs Promote Circular RNA Biogenesis

The formation of circRNAs by backsplicing is facilitated by a looping mechanism that brings backsplice sites into close proximity. This looping can be mediated by intronic complementary sequences in flanking introns, such as Alu elements in human cells [2, 16–21], or by protein factors binding to the flanking introns or the circularizing exons, such as RBPs [22, 23] (Fig. 8.2). RBP-mediated circRNA biogenesis has been well-summarized in a previous review in 2021, where a number of RBPs (MBNL1, QKI, hnRNPs and SR proteins, RBM20, DHX9, Sam68, NOVA2, SFPQ, NF90/NF110, hnRNPL, FUS, ESRP1, TNRC6A, and RBM3) and their corresponding circRNAs were discussed [3]. Since then, an increasing number of RBPs have been discovered to play essential roles in the production of circRNAs. Here, we focus on RBP-mediated circRNA biogenesis that was not discussed in the previous review [3] to provide a more up-to-date picture of the role of RBP in circRNA formation.

4.1 Eukaryotic Initiation Factor 4A-3 (EIF4A3)

RNA-binding protein EIF4A3 is a DEAD box family containing a Q motif (residues 38 to 60), a DEAD box motif (residues 187 to 190), a helicase adenosine triphosphate (ATP)-dependent

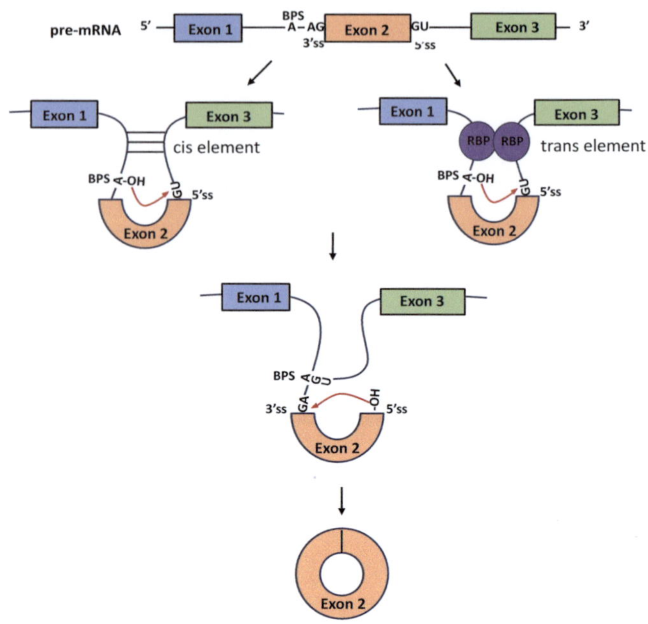

Fig. 8.2 CircRNA formation can be promoted by *cis*-elements such as intronic complementary sequences or *trans*-elements such as RBPs

domain (residues 69 to 293), and a helicase C-terminal domain (residues 250 to 411) [24]. EIF4A3 has a significant impact on RNA metabolism, including circRNA biogenesis [24–27].

(a) *EIF4A3 binds to both the upstream and downstream regions of circRNAs*: CircSEPT9, derived from exon 2 of the *SEPT9* gene via backsplicing, is highly expressed in triple-negative breast cancer (TNBC) [28]. EIF4A3 binds to two sites upstream of exon 2 (nt −165 to −136 and nt −126 to −97) and three sites downstream of exon 2 (nt 759–867, 888–925, and 963–986), as revealed by an RNA immunoprecipitation (RIP) assay [28]. Overexpression of EIF4A3 leads to an increased expression of circSEPT9, while its knockdown decreases circSEPT9 level. EIF4A3 binding to the flanking region of circSEPT9 promotes the expression of circSEPT9, thereby facilitating the carcinogenesis and development of TNBC [28]. Moreover, accumulated evidence highlights EIF4A3's role in promoting biogenesis of other circRNA by binding to the flanking regions of exons, including hsa_circ_0088088 [29], circ0008399 [30], circ-DOCK5 [31], circZNF609 [32], circIKBKB [33], circLMNB2 [34], circTOLLIP [35], circRNF13 [36], circCCNB1 [37], circWAC [38], and circSCAP [39].

(b) *EIF4A3 binds to backspliced junction sites or circularizing exons*: The EIF4A3 fragment spanning amino acids 161–212, residing within the N-terminal helicase ATP-binding domain, enhances circARHGAP29 (a 372-nt circRNA containing exons 21 and 22 of its parental gene) cyclization by interacting with three sequences downstream of the 3′ exon (nt 505 to 540, 580 to 647, and 906 to 969) and the backspliced junction site [40]. EIF4A3-induced circARHGAP29 is upregulated in prostate cancer cells and promotes docetaxel resistance. Inhibition of EIF4A3 results in a reduced expression of circARHGAP29 [40]. EIF4A3 that binds to circularizing exons are also involved in the production of other circRNAs, such as circCCAR1 and circSERPINE2 [41, 42].

(c) *EIF4A3 binds either upstream or downstream of circRNA*: The cancer-specific circRNA circPRKAR1B is derived from exons 5 to 9 within the PRKAR1B locus. EIF4A3 can bind to the downstream flanking sequence of exon 9 on PRKAR1B mRNA (nt 961–1000), as revealed by RIP assays [43]. Knockdown of EIF4A3 inhibits circPRKAR1B expression in the human osteosarcoma (HOS) cell. EIF4A3-induced expression of circPRKAR1B enhances HOS cell proliferation as well as migration and promotes epithelial–mesenchymal transition (EMT), subsequently promoting the processes of oncogenesis and metastasis [43]. circRAB3IP [44] and circASAP1 [45] are two additional circRNAs promoted by EIF4A3 binding to the downstream flanking region. EIF4A3 also binds to the upstream flanking sequence and facilitates the biogenesis of circ_0084615 [46], circ_GLIS2 [47], circFIP1L1 [48], circMMP9 [49], and circRHBDD1 [50].

4.2 Zinc Finger CCHC-Type and RNA-Binding Motif 1 (ZCRB1)

Alu elements are the most abundant ubiquitous sequence repeats in the human genome and have a significant role in the regulation of backsplicing [51, 52]. AluJb and AluSx are a pair of reverse complementary sequences of Alu elements located in the flanking intron within 1 kb of circHEATR5B [53]. ZCRB1 interacts with AluJb and AluSx and regulates the biogenesis of HEATR5B circRNA, which is derived from exons 19 to 33 [53]. circHEATR5B has been shown to encode a novel protein termed "HEATR5B-881aa" in glioblastoma multiforme (GBM) cells. Importantly, the expression levels of ZCRB1, circHEATR5B, and HEATR5B-881aa are positively correlated, which regulates aerobic glycolysis and proliferation in GBM cells [53].

4.3 Zinc Finger E-box Binding Homeobox 1 (ZEB1)

circNIPBL originated from exons 2 to 9 of the *NIPBL* gene, as verified by Sanger sequencing [54]. ZEB1 binds the flanking regions in intron 1 (nt 45,999–46,005) and intron 9 (nt 618–624) of NIPBL pre-mRNA, as revealed by RIP assays, suggesting ZEB1's involvement in promoting circNIPBL biogenesis via backsplicing [54]. Either upregulation or downregulation of ZEB1 in bladder cancer (BCa) results in increased or decreased expression of circNIPBL. Interestingly, circNIPBL can activate the Wnt/β-catenin signaling pathway by acting as a sponge of miR-16-2-3p. This action results in the upregulation of ZWB1 expression. This positive feedback loop of circNIPBL/miR-16-2-3p/Wnt5a/ZEB1 sustains the migration and invasion of BCa cells [54].

4.4 Splicing Factor Synergistic Lethal with U5 snRNA 7 (SLU7)

SLU7, an RNA-binding protein, serves as an integrative hub that influences gene expression at multiple levels, including epigenetic DNA remodeling, pre-mRNA splicing, modulation of promoter activity, and protein stability [55–57]. SLU7 has a zinc knuckle motif, four stretches of leucine repeats, and a lysine-rich domain. Its interaction with proximal AG sites of Alu elements regulates pre-mRNA splicing [58]. circCAPG is generated via backsplicing from exons 6 to 8 of CAPG mRNA, confirmed by Sanger sequencing [59]. The intronic regions flanking upstream exon 6 and downstream exon 8 contain three Alu elements: AluSp, AluSz, and AluJo. SLU7's binding to AluSp and AluSz is essential for the formation of circCAPG [59]. The expression of circCAPG induced by SLU7 encodes a polypeptide CAPG-171aa, which leads to the proliferation and metastasis of TNBC cells.

4.5 Sex-Lethal (SXL)

The RNA-binding protein SXL is encoded by the sex-determining gene *Sxl* in female *Drosophila* [60]. SXL comprises 354 amino acids, including two RRM domains spanning amino acids 112–294 [61]. Exogenous SXL, when expressed in male-derived S2 cells that lack endogenous SXL, greatly improves the backsplicing of circRNAs that promote sex differentiation, such as dmc*_579*, *_2185*, *_6411*, *_7247*, *_2797*, and *_3242* [62]. On the other hand, SXL RRM mutant (N122A, Y123A, and Q126A in RRM1 and R244A, V246A, and F248A in RRM2), which greatly reduces SXL's RNA-binding activity, cannot support the backsplicing event [62, 63].

Furthermore, 60 out of 129 female-specific differentially backspliced circRNAs have SXL-binding peaks near the backsplicing sites in female-derived Kc cells in photoactivatable ribonucleoside-enhanced cross-linking and immunoprecipitation (PAR-CLIP) experiments [62]. SXL binding to flanking exons and introns in pre-mRNAs has been shown to promote backsplicing, as confirmed by a splicing assay performed on mini-genes with targeted mutations in the SXL-binding sites [62]. These data demonstrate that SXL promotes the generation of sex-differentially backspliced circRNAs, a function that critically depends on its effective RNA-binding activity.

4.6 RNA-Binding Motif Single-Stranded Interacting Protein 3 (RBMS3)

RBMS3 is an RBP consisting of two RRM domains at its N-terminus [64]. As a tumor suppressor gene, it plays an essential role in cancer regulation. circHECTD1, a 1403-nt sequence, is transcribed from exons 23 to 26 of HECTD1 pre-mRNA, as verified by Sanger sequencing [65]. RBMS3 specifically binds both flanking introns of circHECTD1 and effectively promotes the generation of circHECTD1, while mutations in the RBMS3-binding region significantly decrease circHECTD1 formation [65]. RBMS3-induced circHECTD1 suppresses vasculogenic mimicry formation in glioblastoma multiforme.

4.7 EWS RNA-Binding Protein 1 (EWSR1)

Ewing Sarcoma RNA-binding Protein 1 (EWSR1) is an RNA-binding protein that contains an N-terminal serine–tyrosine–glycine–glutamine (SYGQ)-rich domain, a middle RRM domain, a C-terminal zinc finger domain, three RRG-rich regions (RRGs), and a proline–tyrosine (PY) region [66]. EWSR1 and fused in sarcoma (FUS), which belong to the TET family, are involved in neurodegenerative disorders [67–70]. FUS is a well-known RBP that binds to flanking introns and regulates the biogenesis of more than 200 circRNAs through backsplicing, as discussed in the previous review [3]. Similarly, EWSR1 has been found to be recruited to the upstream (nt −751 to −627) and downstream (nt 761–851) regions of circNEIL3, as tested by (RIP-qPCR) RNA Immunoprecipitation (RIP) with quantitative PCR (qPCR), facilitating the cyclization of circNEIL3, which is backspliced from exons 8 and 9 of the parental gene [71]. circNEIL3, upregulated by EWSR1 overexpression in glioma tissues, promotes glioma progression [71].

4.8 RNA-Binding Motif Protein 25 (RBM25)

RBM25 is an 843-residue protein consisting of an N-terminal RRM domain (residues 87 to 164), a central RE/RD-rich domain (ER rich domain, residues 285 to 664), and a C-terminal PWI domain (residues 734 to 843) [72]. RBM25 interacts with the exon definition splicing machinery primarily through its ER-rich domain to promote the inclusion of alternatively spliced exons [73]. Previous research has revealed that RBM25 associates with poly-G sequences in circAMOTL1L, derived from exons 2 and 3 of the *AMOTL1* gene, inducing its biogenies [74]. RBM25 also facilitates cCNTNAP3 circularization by directly binding to a "CGGGCA" sequence in the exonic splicing enhancer located in exon 9 [75]. Furthermore, a 2023 report unveiled that RBM25 binds to the flanking region of exons 2 and 5, consequently increasing the formation of cirKrt4 by backsplicing [76].

4.9 KH-Type Splicing Regulatory Protein (KHSRP, Also Called KSRP)

KHSRP is a 75-kDa RBP comprising four KH domains in the central region [77]. Due to its complex interactions with various RNA sequences and proteins, it regulates diverse functions in RNA metabolism. Recent reports have indicated that KHSRP knockdown in K562 cells results in a 15% decrease in the biogenesis of circRNAs with KHSRP-binding sites in both flanking introns but not for circRNAs lacking KHSRP-binding sites [78]. Thus, KHSRP promotes the biogenesis of a subset of circRNAs by binding to the KHSRP-binding sites within flanking introns in K562 cells [78].

4.10 Pre-mRNA Processing Factor 8 (PRP8)

circMaml2 is a novel circRNA molecule with a 1686-nt circular transcript generated from exon 2 of Maml2 pre-mRNA [79]. Recent research has suggested that mouse PRP8, a splicing factor and RNA-binding protein, is enriched in the upstream and downstream regions of exon 2, as revealed by RIP analysis [79]. Overexpression of PRP8 promotes the biogenesis of circMaml2, whereas knockdown of PRP8 decreases circMaml2 expression [79]. PRP8-induced circMaml2 facilitates intestinal mucosal repair through recruiting PTBP1 to upregulate Ebf1 expression and interacting with miR-683 to regulate Sec62 [79].

4.11 YTH N6-Methyladenosine RNA-Binding Protein C1 (YTHDC1)

YTHDC1 can regulate the backsplicing of circRNAs by acting as an N6-methyladenosine (m6A) reader. It is found to be enriched at the m^6A modifications on the exon from which circZNF609 is derived. Knockdown of YTHDC1 decreases the expression of circZNF609 in HeLa cells [80]. Furthermore, YTHDC1 binds to the region of circCDYL on its parental gene and promotes its circularization. YTHDC1 silencing leads to a decrease in circCDYL formation [81]. Taken together, N6-methyladenosine modification stimulates backsplicing processes via YTHDC1 to promote circRNA biogenesis.

4.12 HnRNPs

HnRNPs are a large family of RBPs with multiple family members sharing the same RBDs. RRM, quasi-RRM (qRRM), KH, and arginine–glycine–glycine (RGG) are four RBDs identified in hnRNP proteins [82]. Below are several hnRNP proteins that are involved in circRNA biogenesis.

(a) *AU-Binding Factor 1 (AFU1)*: AFU1, also known as hnRNP D, contains two RRM domains [83]. The AUF1 protein interacting with hsa_circ_0010467 (a 1430-nt-long circRNA generated from the backsplicing of exons 9 to 13 of the *EIF4G3* gene) was identified using an RIP assay and quantitative reverse transcription polymerase chain reaction (qRT-PCR) [84]. AFU1 is identified to bind to the upstream flanking region of exon 9 and regulates hsa_circ_0010467 biogenesis in ovarian cancer (OC) cells [84]. Increased biogenesis of hsa_circ_0010467 is found in OC tissues, which promotes platinum resistance in OC through the miR-637/LIF/STAT3 axis. Knockdown of AUF1 in OC cells inhibits the formation of spheroids. AFU1 was also found to regulate the generation of circMALAT1 in hepatocellular cancer stem cells by interacting with the flanking regions of circ-MALAT1, and silencing of AFU1 significantly reduced its biosynthesis [85].

(b) *Heterogeneous Nuclear Ribonucleoprotein F (hnRNPF)*: hnRNPF, an RBP with two qRRM domains, is found to bind to two highly reverse complementary Alu elements (Sc and Jb) located in the flanking introns of circPFKP, a PCa-related circRNA derived from exons 15 and 16 of the *PFKP* gene through backsplicing [86]. hnRNPF promotes the circularization of circPFKP, thereby promoting cell proliferation in prostate cancer.

(c) *Heterogeneous Nuclear Ribonucleoprotein U-Like 1 (hnRNPUL1)*: This is a member of the hnRNP family harboring an RGG box and it regulates DNA damage repair. hnRNPUL1 binds to circMAN1A2 in TE1 and EC9706 cells as revealed by an RNA pull-down and an RIP assay [87]. Knockdown of hnRNPUL1 remarkably decreases the expression of circMAN1A2 [87].

In addition to the RBPs outlined above, additional circRNAs mediated by RBPs discussed in the 2021 review have been identified [3]. For example, Quaking (QKI) regulates the biogenies of circSLC26A4 [88], circ-SHPRH [89], circPIKfyve [90], circARFGEF2 [91], circGlis3 [92], circPDIA4 [93], hsa_circ_0098181 [94], circFOXP1 [95], circRNA-SFMBT2 [96], circROCK1-E3/E4 [97], hsa_circ_0007919 [98], circVAMP3 [99], and multiple circRNAs in bisphenol A (BPA)-induced toxicity in GC-2 germ cells [100]. hnRNPL was found to promote the cyclization of circARHGAP35 [101], circCSPP1 [102], circLMO7 [103], circANKRD42 [104], circELP2 [105], and circFAM13B [106]. ESRP1 is involved in the circular RNA formation of circUHRF1 [107], circZNF652 [108], circ_0068162 [109], and circDOCK1(2–27) [110]. Knocking down NF90/NF110 or FUS significantly decrease the level of circPUS7, thereby suppressing cadmium-induced cell transformation, migration, and invasion [111]. DHX9 can bind to the flanking reverse complementary match sequences of target RNAs, inhibiting the pairing of these sequences and suppressing the biogenesis of circDCUN1D4 [112], cGGNBP2 [113], circDLC1 [114], Circ_0007379 [115], circBNC2 [116], circRNA-CREIT [117], circLMO1 [118], circSFMBT2 [119], circSMARCA5 [120], and circASH2 [121].

5 Concluding Remarks

CircRNAs were once considered by-products of mRNA production during mis-splicing. However, advancements in RNA sequencing (RNA-seq) have led to the identification of thousands of circRNAs. Many studies have demonstrated their significant roles in cell proliferation, differentiation, apoptosis, and cell death, revealing the biological significance of circRNAs. CircRNAs generated from backsplicing can be regulated through either *cis*-elements, such as complementary sequences in flanking introns, or *trans*-factors, such as RBPs. RBPs bind to specific motifs in flanking introns or circularizing exons, bringing spliced sites into proximity, thereby facilitating backsplicing. It is worth noting that RBPs can act synergistically with *cis*-elements to promote circRNA formation. For instance, the *cis*-element Alu is bound by ZCRB1, promoting the cyclization of circHEATR5B. In addition to promoting circRNA formation, some RBPs can suppress circRNA expression by disrupting the parring of the reverse complementary matches, such as DHX9. Additionally, ZEB1 binds to the promoter of ANKRD17, significantly enhancing the expression of circANKRD17 but not ANKRD17 [122], illustrating an unusual way of circRNA biogenesis regulation, although the exact mechanism is unclear.

Despite the significant progress made in the study of RBP-mediated circRNA biogenesis in recent years, a number of questions remain to be answered. Most current studies focus on an individual circRNA or RBP. It is unclear whether a single RBP binds to both flanking regions of a circRNA or two different RBPs bind to the flanking regions and come together either directly or through an intermediatory protein to facilitate circRNA biogenesis. It is possible that both scenarios exist. For example, EIF4A3 can bind to both flanking regions of an exon; however, it can also bind to only one side of a circRNA (either upstream or downstream). In addition, more than one RBP-binding site is often identified in the flanking regions of circularizing exons, but only some of these are used by an RBP to regulate circRNA biogenesis. How RBPs choose which site to use remains unclear. It is possible that a combination of RNA sequences, RNA secondary structural features, RBP–RNA binding affinity, and locations of RBPs in the flanking regions of the exon contribute to the final sequences used by a particular RBP.

Answering some of these questions may require the use of methods that have not been routinely used in the study of RBP and circRNA biogenesis so far. In most studies, the overexpression or underexpression of a circRNA is observed in certain diseases. A potential RBP is predicted based on potential binding sites and experimentally confirmed by pull-down, RIP, electrophoretic mobility shift assay (EMSA), and overexpression or knockdown of RBPs. While isolated structures of RBPs with their cognate RNA are often known, there are no structures of RBPs binding to flanking regions of circRNAs.

The combination of structural and other biophysical approaches (such as single molecule fluorescence to observe the binding of RBPs to both flanking regions of the exon) with frequently used biochemical and molecular biology approaches can help reveal molecular details and new mechanistic insights into the regulation of circRNAs by RBPs. Furthermore, a more systematic approach to evaluate all the circRNAs affected by a particular RBP or all the RBPs that regulate a certain circRNA in either diseased or healthy cells may provide a more complete picture of the regulation of circRNA biogenesis by RBPs. This will also help us achieve the ultimate goal of predicting how an RBP regulates circRNA biogenesis.

Acknowledgments This work was supported by NIH grants R35GM145289, R01GM126157, and R01GM130673 (to R.Z.).

References

1. Sanger HL, Klotz G, Riesner D, Gross HJ, Kleinschmidt AK (1976) Viroids are single-stranded covalently closed circular RNA molecules existing as highly base-paired rod-like structures. Proc Natl Acad Sci USA 73(11):3852–3856. https://doi.org/10.1073/pnas.73.11.3852
2. Jeck WR, Sorrentino JA, Wang K, Slevin MK, Burd CE, Liu J, Marzluff WF, Sharpless NE (2013) Circular RNAs are abundant, conserved, and associated with ALU repeats. RNA 19(2):141–157. https://doi.org/10.1261/rna.035667.112
3. Das A, Sinha T, Shyamal S, Panda AC (2021) Emerging role of circular RNA-protein interactions. Noncoding RNA 7(3). https://doi.org/10.3390/ncrna7030048
4. Zhang Y, Xue W, Li X, Zhang J, Chen S, Zhang JL, Yang L, Chen LL (2016) The biogenesis of nascent circular RNAs. Cell Rep 15(3):611–624. https://doi.org/10.1016/j.celrep.2016.03.058
5. Chen LL (2020) The expanding regulatory mechanisms and cellular functions of circular RNAs. Nat Rev Mol Cell Biol 21(8):475–490. https://doi.org/10.1038/s41580-020-0243-y
6. Cui J, Chen M, Zhang L, Huang S, Xiao F, Zou L (2022) Circular RNAs: biomarkers of cancer. Cancer Innov 1(3):197–206. https://doi.org/10.1002/cai2.28
7. Wilusz JE (2018) A 360 degrees view of circular RNAs: from biogenesis to functions. Wiley Interdiscip Rev RNA 9(4):e1478. https://doi.org/10.1002/wrna.1478
8. Li X, Liu S, Zhang L, Issaian A, Hill RC, Espinosa S, Shi S, Cui Y, Kappel K, Das R, Hansen KC, Zhou ZH, Zhao R (2019) A unified mechanism for intron and exon definition and back-splicing. Nature 573(7774):375–380. https://doi.org/10.1038/s41586-019-1523-6
9. Gerstberger S, Hafner M, Tuschl T (2014) A census of human RNA-binding proteins. Nat Rev Genet 15(12):829–845. https://doi.org/10.1038/nrg3813
10. Maris C, Dominguez C, Allain FH (2005) The RNA recognition motif, a plastic RNA-binding platform to regulate post-transcriptional gene expression. FEBS J 272(9):2118–2131. https://doi.org/10.1111/j.1742-4658.2005.04653.x
11. Gamsjaeger R, Liew CK, Loughlin FE, Crossley M, Mackay JP (2007) Sticky fingers: zinc-fingers as protein-recognition motifs. Trends Biochem Sci 32(2):63–70. https://doi.org/10.1016/j.tibs.2006.12.007
12. Valverde R, Edwards L, Regan L (2008) Structure and function of KH domains. FEBS J 275(11):2712–2726. https://doi.org/10.1111/j.1742-4658.2008.06411.x
13. Cordin O, Banroques J, Tanner NK, Linder P (2006) The DEAD-box protein family of RNA helicases. Gene 367:17–37. https://doi.org/10.1016/j.gene.2005.10.019
14. Fierro-Monti I, Mathews MB (2000) Proteins binding to duplexed RNA: one motif, multiple functions. Trends Biochem Sci 25(5):241–246. https://doi.org/10.1016/s0968-0004(00)01580-2
15. Dominguez D, Freese P, Alexis MS, Su A, Hochman M, Palden T, Bazile C, Lambert NJ, Van Nostrand EL, Pratt GA, Yeo GW, Graveley BR, Burge CB (2018) Sequence, structure, and context preferences of human RNA binding proteins. Mol Cell 70(5):854–867 e859. https://doi.org/10.1016/j.molcel.2018.05.001
16. Zhang XO, Wang HB, Zhang Y, Lu X, Chen LL, Yang L (2014) Complementary sequence-mediated exon circularization. Cell 159(1):134–147. https://doi.org/10.1016/j.cell.2014.09.001
17. Veno MT, Hansen TB, Veno ST, Clausen BH, Grebing M, Finsen B, Holm IE, Kjems J (2015) Spatio-temporal regulation of circular RNA expression during porcine embryonic brain development. Genome Biol 16:245. https://doi.org/10.1186/s13059-015-0801-3
18. Ivanov A, Memczak S, Wyler E, Torti F, Porath HT, Orejuela MR, Piechotta M, Levanon EY, Landthaler M, Dieterich C, Rajewsky N (2015) Analysis of intron sequences reveals hallmarks of circular RNA biogenesis in animals. Cell Rep 10(2):170–177. https://doi.org/10.1016/j.celrep.2014.12.019
19. Starke S, Jost I, Rossbach O, Schneider T, Schreiner S, Hung LH, Bindereif A (2015) Exon circularization requires canonical splice signals. Cell Rep 10(1):103–111. https://doi.org/10.1016/j.celrep.2014.12.002
20. Capel B, Swain A, Nicolis S, Hacker A, Walter M, Koopman P, Goodfellow P, Lovell-Badge R (1993) Circular transcripts of the testis-determining gene Sry in adult mouse testis. Cell 73(5):1019–1030. https://doi.org/10.1016/0092-8674(93)90279-y

21. Xiao MS, Ai YX, Wilusz JE (2020) Biogenesis and functions of circular RNAs come into focus. Trends Cell Biol 30(3):226–240. https://doi.org/10.1016/j.tcb.2019.12.004
22. Wang L, Xu GE, Spanos M, Li G, Lei Z, Sluijter JPG, Xiao J (2023) Circular RNAs in cardiovascular diseases: regulation and therapeutic applications. Research (Wash D C) 6:0038. https://doi.org/10.34133/research.0038
23. Chen LL, Yang L (2015) Regulation of circRNA biogenesis. RNA Biol 12(4):381–388. https://doi.org/10.1080/15476286.2015.1020271
24. Ye J, She X, Liu Z, He Z, Gao X, Lu L, Liang R, Lin Y (2021) Eukaryotic initiation factor 4A-3: a review of its physiological role and involvement in oncogenesis. Front Oncol 11:712045. https://doi.org/10.3389/fonc.2021.712045
25. Sakellariou D, Frankel LB (2021) EIF4A3: a gatekeeper of autophagy. Autophagy 17(12):4504–4505. https://doi.org/10.1080/15548627.2021.1985881
26. Budiman ME, Bubenik JL, Miniard AC, Middleton LM, Gerber CA, Cash A, Driscoll DM (2009) Eukaryotic initiation factor 4a3 is a selenium-regulated RNA-binding protein that selectively inhibits selenocysteine incorporation. Mol Cell 35(4):479–489. https://doi.org/10.1016/j.molcel.2009.06.026
27. Chan CC, Dostie J, Diem MD, Feng WQ, Mann M, Rappsilber J, Dreyfuss G (2004) eIF4A3 is a novel component of the exon junction complex. RNA 10(2):200–209. https://doi.org/10.1261/rna.5230104
28. Zheng X, Huang M, Xing L, Yang R, Wang X, Jiang R, Zhang L, Chen J (2020) The circRNA circSEPT9 mediated by E2F1 and EIF4A3 facilitates the carcinogenesis and development of triple-negative breast cancer. Mol Cancer 19(1):73. https://doi.org/10.1186/s12943-020-01183-9
29. Liu Q, Dong H (2021) EIF4A3-mediated hsa_circ_0088088 promotes the carcinogenesis of breast cancer by sponging miR-135-5p. J Biochem Mol Toxicol 35(11):e22909. https://doi.org/10.1002/jbt.22909
30. Wei W, Sun J, Zhang H, Xiao X, Huang C, Wang L, Zhong H, Jiang Y, Zhang X, Jiang G (2021) Circ0008399 interaction with WTAP promotes assembly and activity of the m(6)A methyltransferase complex and promotes cisplatin resistance in bladder cancer. Cancer Res 81(24):6142–6156. https://doi.org/10.1158/0008-5472.CAN-21-1518
31. Meng L, Zheng Y, Liu S, Ju Y, Ren S, Sang Y, Zhu Y, Gu L, Liu F, Zhao Y, Zhang X, Sang M (2021) ZEB1 represses biogenesis of circ-DOCK5 to facilitate metastasis in esophageal squamous cell carcinoma via a positive feedback loop with TGF-beta. Cancer Lett 519:117–129. https://doi.org/10.1016/j.canlet.2021.06.026
32. Guan C, Liu L, Zhao Y, Zhang X, Liu G, Wang H, Gao X, Zhong X, Jiang X (2021) YY1 and eIF4A3 are mediators of the cell proliferation, migration and invasion in cholangiocarcinoma promoted by circ-ZNF609 by targeting miR-432-5p to regulate LRRC1. Aging (Albany NY) 13(23):25195–25212. https://doi.org/10.18632/aging.203735
33. Xu Y, Zhang S, Liao X, Li M, Chen S, Li X, Wu X, Yang M, Tang M, Hu Y, Li Z, Yu R, Huang M, Song L, Li J (2021) Circular RNA circIKBKB promotes breast cancer bone metastasis through sustaining NF-kappaB/bone remodeling factors signaling. Mol Cancer 20(1):98. https://doi.org/10.1186/s12943-021-01394-8
34. Qiu M, Chen M, Lan Z, Liu B, Xie J, Li X (2022) Plasmacytoma variant translocation 1 stabilized by EIF4A3 promoted malignant biological behaviors of lung adenocarcinoma by generating circular RNA LMNB2. Bioengineered 13(4):10123–10140. https://doi.org/10.1080/21655979.2022.2063666
35. Liu Y, Song J, Zhang H, Liao Z, Liu F, Su C, Wang W, Han M, Zhang L, Zhu H, Zhang Z, Liang H, Zhang L, Zhang B, Chen X (2022) EIF4A3-induced circTOLLIP promotes the progression of hepatocellular carcinoma via the miR-516a-5p/PBX3/EMT pathway. J Exp Clin Cancer Res 41(1):164. https://doi.org/10.1186/s13046-022-02378-2
36. Zhao Q, Zhu Z, Xiao W, Zong G, Wang C, Jiang W, Li K, Shen J, Guo X, Cui J, Guo L, Wan R (2022) Hypoxia-induced circRNF13 promotes the progression and glycolysis of pancreatic cancer. Exp Mol Med 54(11):1940–1954. https://doi.org/10.1038/s12276-022-00877-y
37. Li X, Wang C, Chen G, Zou W, Deng Y, Zhou F (2022) EIF4A3-induced circCCNB1 (hsa_circ_0001495) promotes glioma progression by elevating CCND1 through interacting miR-516b-5p and HuR. Metab Brain Dis 37(3):819–833. https://doi.org/10.1007/s11011-021-00899-x
38. Huang WH, Yang Q, Zhang C (2022) eIF4A3-induced circWAC promotes breast cancer progression through mediating miR-599/E2F3 axis. Kaohsiung J Med Sci 38(4):321–335. https://doi.org/10.1002/kjm2.12496
39. Zhang Y, Qi W, Wu Y (2023) EIF4A3-induced circular RNA SCAP facilitates tumorigenesis and progression of non-small-cell lung cancer via miR-7/SMAD2 signaling. Environ Sci Pollut Res Int 30(24):65237–65249. https://doi.org/10.1007/s11356-023-26307-8
40. Jiang X, Guo S, Wang S, Zhang Y, Chen H, Wang Y, Liu R, Niu Y, Xu Y (2022) EIF4A3-induced circARHGAP29 promotes aerobic glycolysis in docetaxel-resistant prostate cancer through IGF2BP2/c-Myc/LDHA signaling. Cancer Res 82(5):831–845. https://doi.org/10.1158/0008-5472.CAN-21-2988
41. Hu Z, Chen G, Zhao Y, Gao H, Li L, Yin Y, Jiang J, Wang L, Mang Y, Gao Y, Zhang S, Ran J, Li L (2023) Exosome-derived circCCAR1 promotes CD8 + T-cell dysfunction and anti-PD1 resistance in hepatocellular carcinoma. Mol Cancer 22(1):55. https://doi.org/10.1186/s12943-023-01759-1
42. Zhou B, Mo Z, Lai G, Chen X, Li R, Wu R, Zhu J, Zheng F (2023) Targeting tumor exosomal circular RNA cSERPINE2 suppresses breast cancer progression by modulating MALT1-NF-κB-IL-6

axis of tumor-associated macrophages. J Exp Clin Cancer Res 42(1):48. https://doi.org/10.1186/s13046-023-02620-5

43. Feng ZH, Zheng L, Yao T, Tao SY, Wei XA, Zheng ZY, Zheng BJ, Zhang XY, Huang B, Liu JH, Chen YL, Shan Z, Yuan PT, Wang CG, Chen J, Shen SY, Zhao FD (2021) EIF4A3-induced circular RNA PRKAR1B promotes osteosarcoma progression by miR-361-3p-mediated induction of FZD4 expression. Cell Death Dis 12(11):1025. https://doi.org/10.1038/s41419-021-04339-7
44. Chen D, Wang Y, Yang F, Keranmu A, Zhao Q, Wu L, Han S, Xing N (2021) The circRAB3IP mediated by eIF4A3 and LEF1 contributes to enzalutamide resistance in prostate cancer by targeting miR-133a-3p/miR-133b/SGK1 pathway. Front Oncol 11:752573. https://doi.org/10.3389/fonc.2021.752573
45. Wei Y, Lu C, Zhou P, Zhao L, Lyu X, Yin J, Shi Z, You Y (2021) EIF4A3-induced circular RNA ASAP1 promotes tumorigenesis and temozolomide resistance of glioblastoma via NRAS/MEK1/ERK1-2 signaling. Neuro-Oncology 23(4):611–624. https://doi.org/10.1093/neuonc/noaa214
46. Jiang Z, Tai Q, Xie X, Hou Z, Liu W, Yu Z, Liang Z, Chen S (2021) EIF4A3-induced circ_0084615 contributes to the progression of colorectal cancer via miR-599/ONECUT2 pathway. J Exp Clin Cancer Res 40(1):227. https://doi.org/10.1186/s13046-021-02029-y
47. Xing W, Zhou PC, Zhang HY, Chen LM, Zhou YM, Cui XF, Liu ZG (2022) Circular RNA circ_GLIS2 suppresses hepatocellular carcinoma growth and metastasis. Liver Int 42(3):682–695. https://doi.org/10.1111/liv.15097
48. Zhou X, Yuan G, Wu Y, Yan S, Jiang Q, Tang S (2022) EIF4A3-induced circFIP1L1 represses miR-1253 and promotes radiosensitivity of nasopharyngeal carcinoma. Cell Mol Life Sci 79(7):357. https://doi.org/10.1007/s00018-022-04350-x
49. Wang R, Zhang S, Chen X, Li N, Li J, Jia R, Pan Y, Liang H (2020) Correction to: EIF4A3-induced circular RNA MMP9 (circMMP9) acts as a sponge of miR-124 and promotes glioblastoma multiforme cell tumorigenesis. Mol Cancer 19(1):153. https://doi.org/10.1186/s12943-020-01271-w
50. Cai J, Chen Z, Zhang Y, Wang J, Zhang Z, Wu J, Mao J, Zuo X (2022) CircRHBDD1 augments metabolic rewiring and restricts immunotherapy efficacy via m(6)A modification in hepatocellular carcinoma. Mol Ther Oncolytics 24:755–771. https://doi.org/10.1016/j.omto.2022.02.021
51. Daniel C, Behm M, Ohman M (2015) The role of Alu elements in the cis-regulation of RNA processing. Cell Mol Life Sci 72(21):4063–4076. https://doi.org/10.1007/s00018-015-1990-3
52. Häsler J, Strub K (2006) Alu elements as regulators of gene expression. Nucleic Acids Res 34(19):5491–5497. https://doi.org/10.1093/nar/gkl706
53. Song J, Zheng J, Liu XB, Dong WW, Yang CQ, Wang D, Ruan XL, Zhao YB, Liu LB, Wang P, Zhang MY, Liu YH (2022) A novel protein encoded by ZCRB1-induced circHEATR5B suppresses aerobic glycolysis of GBM through phosphorylation of JMJD5. J Exp Clin Cancer Res 41(1):ARTN 171. https://doi.org/10.1186/s13046-022-02374-6
54. Li Y, Kong Y, An M, Luo Y, Zheng H, Lin Y, Chen J, Yang J, Liu L, Luo B, Huang J, Lin T, Chen C (2023) ZEB1-mediated biogenesis of circNIPBL sustains the metastasis of bladder cancer via Wnt/beta-catenin pathway. J Exp Clin Cancer Res 42(1):191. https://doi.org/10.1186/s13046-023-02757-3
55. Zheng X, Chen J, Deng M, Ning K, Peng Y, Liu Z, Li X, Zhou Z, Tang H, Li Y, Kang T, Liu Z (2023) G3BP1 and SLU7 jointly promote immune evasion by downregulating MHC-I via PI3K/Akt activation in bladder cancer. Adv Sci (Weinh) 11(7):e2305922. https://doi.org/10.1002/advs.202305922
56. Elizalde M, Urtasun R, Azkona M, Latasa MU, Goñi S, García-Irigoyen O, Uriarte I, Segura V, Collantes M, Di Scala M, Lujambio A, Prieto J, Avila MA, Berasain C (2014) Splicing regulator SLU7 is essential for maintaining liver homeostasis. J Clin Invest 124(7):2909–2920. https://doi.org/10.1172/Jci74382
57. Garate-Rascon M, Recalde M, Rojo C, Fernandez-Barrena MG, Avila MA, Arechederra M, Berasain C (2022) SLU7: a new hub of gene expression regulation-from epigenetics to protein stability in health and disease. Int J Mol Sci 23(21). https://doi.org/10.3390/ijms232113411
58. Lev-Maor G, Sorek R, Shomron N, Ast G (2003) The birth of an alternatively spliced exon: 3′ splice-site selection in Alu exons. Science 300(5623):1288–1291. https://doi.org/10.1126/science.1082588
59. Song R, Guo P, Ren X, Zhou L, Li P, Rahman NA, Wolczynski S, Li X, Zhang Y, Liu M, Liu J, Li X (2023) A novel polypeptide CAPG-171aa encoded by circCAPG plays a critical role in triple-negative breast cancer. Mol Cancer 22(1):104. https://doi.org/10.1186/s12943-023-01806-x
60. Samuels M, Deshpande G, Schedl P (1998) Activities of the sex-lethal protein in RNA binding and protein:protein interactions. Nucleic Acids Res 26(11):2625–2637. https://doi.org/10.1093/nar/26.11.2625
61. Kanaar R, Lee AL, Rudner DZ, Wemmer DE, Rio DC (1995) Interaction of the sex-lethal Rna-binding domains with Rna. EMBO J 14(18):4530–4539. https://doi.org/10.1002/j.1460-2075.1995.tb00132.x
62. Fan YJ, Ding Z, Zhang Y, Su R, Yue JL, Liang AM, Huang QW, Meng YR, Li M, Xue Y, Xu YZ (2023) Sex-lethal regulates back-splicing and generation of the sex-differentially expressed circular RNAs. Nucleic Acids Res 51(10):5228–5241. https://doi.org/10.1093/nar/gkad280
63. Handa N, Nureki O, Kurimoto K, Kim I, Sakamoto H, Shimura Y, Muto Y, Yokoyama S (1999) Structural basis for recognition of the tra mRNA precursor by the sex-lethal protein. Nature 398(6728):579–585. https://doi.org/10.1038/19242

64. Jayasena CS, Bronner ME (2012) Rbms3 functions in craniofacial development by posttranscriptionally modulating TGF-beta signaling. J Cell Biol 199(3):453–466. https://doi.org/10.1083/jcb.201204138
65. Ruan X, Liu Y, Wang P, Liu L, Ma T, Xue Y, Dong W, Zhao Y, Tiange E, Lin H, Wang D, Yang C, Song J, Liu J, Deng M, An P, Lin Y, Yang J, Cui Z, Cao Y, Liu X (2023) RBMS3-induced circHECTD1 encoded a novel protein to suppress the vasculogenic mimicry formation in glioblastoma multiforme. Cell Death Dis 14(11):745. https://doi.org/10.1038/s41419-023-06269-y
66. Lee J, Nguyen PT, Shim HS, Hyeon SJ, Im H, Choi MH, Chung S, Kowall NW, Lee SB (1865) Ryu H (2019) EWSR1, a multifunctional protein, regulates cellular function and aging via genetic and epigenetic pathways. Biochim Biophys Acta Mol basis Dis 7:1938–1945. https://doi.org/10.1016/j.bbadis.2018.10.042
67. Svetoni F, Frisone P, Paronetto MP (2016) Role of FET proteins in neurodegenerative disorders. RNA Biol 13(11):1089–1102. https://doi.org/10.1080/15476286.2016.1211225
68. Merner ND, Girard SL, Catoire H, Bourassa CV, Belzi VV, Rivière JB, Hince P, Levert A, Dionne-Laporte A, Spiegelman D, Noreau A, Diab S, Szuto A, Fournier H, Raelson J, Belouchi M, Panisset M, Cossette P, Dupré N, Bernard G, Chouinard S, Dion PA, Rouleau GA (2012) Exome sequencing identifies mutations as a cause of essential tremor. Am J Hum Genet 91(2):313–319. https://doi.org/10.1016/j.ajhg.2012.07.002
69. Mackenzie IRA, Neumann M (2012) FET proteins in frontotemporal dementia and amyotrophic lateral sclerosis. Brain Res 1462:40–43. https://doi.org/10.1016/j.brainres.2011.12.010
70. Ticozzi N, Vance C, LeClerc AL, Keagle P, Glass JD, McKenna-Yasek D, Sapp PC, Silani V, Bosco DA, Shaw CE, Brown RH, Landers JE (2011) Mutational analysis reveals the homolog as a candidate gene for familial amyotrophic lateral sclerosis. Am J Med Genet B 156b(3):285–290. https://doi.org/10.1002/ajmg.b.31158
71. Pan Z, Zhao R, Li B, Qi Y, Qiu W, Guo Q, Zhang S, Zhao S, Xu H, Li M, Gao Z, Fan Y, Xu J, Wang H, Wang S, Qiu J, Wang Q, Guo X, Deng L, Zhang P, Xue H, Li G (2022) EWSR1-induced circNEIL3 promotes glioma progression and exosome-mediated macrophage immunosuppressive polarization via stabilizing IGF2BP3. Mol Cancer 21(1):16. https://doi.org/10.1186/s12943-021-01485-6
72. Gong D, Yang F, Li F, Qian D, Wu M, Shao Z, Wu M, Wu J, Shi Y (2013) Crystal structure and functional characterization of the human RBM25 PWI domain and its flanking basic region. Biochem J 450(1):85–94. https://doi.org/10.1042/BJ20121382
73. Carlson SM, Soulette CM, Yang Z, Elias JE, Brooks AN, Gozani O (2017) RBM25 is a global splicing factor promoting inclusion of alternatively spliced exons and is itself regulated by lysine monomethylation. J Biol Chem 292(32):13381–13390. https://doi.org/10.1074/jbc.M117.784371
74. Yang Z, Qu CB, Zhang Y, Zhang WF, Wang DD, Gao CC, Ma L, Chen JS, Liu KL, Zheng B, Zhang XH, Zhang ML, Wang XL, Wen JK, Li W (2019) Dysregulation of p53-RBM25-mediated circAMOTL1L biogenesis contributes to prostate cancer progression through the circAMOTL1L-miR-193a-5p-Pcdha pathway. Oncogene 38(14):2516–2532. https://doi.org/10.1038/s41388-018-0602-8
75. Wang H, Song X, Wang Y, Yin X, Liang Y, Zhang T, Xu L, Jiang F, Dong G (2020) CircCNTNAP3-TP53-positive feedback loop suppresses malignant progression of esophageal squamous cell carcinoma. Cell Death Dis 11(11):1010. https://doi.org/10.1038/s41419-020-03217-y
76. Ma C, Wang X, Zhang L, Zhu X, Bai J, He S, Mei J, Jiang J, Guan X, Zheng X, Qu L, Zhu D (2023) Super enhancer-associated circular RNA-CircKrt4 regulates hypoxic pulmonary artery endothelial cell dysfunction in mice. Arterioscler Thromb Vasc Biol 43(7):1179–1198. https://doi.org/10.1161/ATVBAHA.122.318842
77. Briata P, Chen CY, Giovarelli M, Pasero M, Trabucchi M, Ramos A, Gherzi R (2011) KSRP, many functions for a single protein. Front Biosci (Landmark Ed) 16(5):1787–1796. https://doi.org/10.2741/3821
78. Okholm TLH, Sathe S, Park SS, Kamstrup AB, Rasmussen AM, Shankar A, Chua ZM, Fristrup N, Nielsen MM, Vang S, Dyrskjot L, Aigner S, Damgaard CK, Yeo GW, Pedersen JS (2020) Transcriptome-wide profiles of circular RNA and RNA-binding protein interactions reveal effects on circular RNA biogenesis and cancer pathway expression. Genome Med 12(1):112. https://doi.org/10.1186/s13073-020-00812-8
79. Deng Y, Xu X, Meng F, Lou J, Liao Y, Li Q, Zhuang M, Sun Y (2022) PRP8-induced CircMaml2 facilitates the healing of the intestinal mucosa via recruiting PTBP1 and regulating Sec62. Cells 11(21). https://doi.org/10.3390/cells11213460
80. Di Timoteo G, Dattilo D, Centron-Broco A, Colantoni A, Guarnacci M, Rossi F, Incarnato D, Oliviero S, Fatica A, Morlando M, Bozzoni I (2020) Modulation of circRNA metabolism by m(6)A modification. Cell Rep 31(6):107641. https://doi.org/10.1016/j.celrep.2020.107641
81. Wei YP, Fu JB, Zhang HL, Ling Y, Tang XW, Liu SW, Yu M, Liu FY, Zhuang GK, Qian HH, Zhang KC, Yang PH, Yang XW, Yang Q, Ge SN, Zhang BH, Tan YX, Li L, Wang HY (2023) N6-methyladenosine modification promotes hepatocarcinogenesis through circ-CDYL-enriched and EpCAM-positive liver tumor-initiating exosomes. Iscience 26(10):ARTN 108022. https://doi.org/10.1016/j.isci.2023.108022
82. Geuens T, Bouhy D, Timmerman V (2016) The hnRNP family: insights into their role in health and disease. Hum Genet 135(8):851–867. https://doi.org/10.1007/s00439-016-1683-5
83. White EJ, Matsangos AE, Wilson GM (2017) AUF1 regulation of coding and noncoding RNA. Wiley

Interdiscip Rev RNA 8(2). https://doi.org/10.1002/wrna.1393
84. Wu Y, Xu M, Feng Z, Wu H, Wu J, Ha X, Wu Y, Chen S, Xu F, Wen H, Li S, Wu X (2023) AUF1-induced circular RNA hsa_circ_0010467 promotes platinum resistance of ovarian cancer through miR-637/LIF/STAT3 axis. Cell Mol Life Sci 80(9):256. https://doi.org/10.1007/s00018-023-04906-5
85. Chen L, Kong RJ, Wu C, Wang S, Liu ZX, Liu SP, Li SP, Chen T, Mao CB, Liu SR (2020) Circ-MALAT1 functions as both an mRNA translation brake and a microRNA sponge to promote self-renewal of hepatocellular cancer stem cells (vol 7, 1900949, 2019). Adv Sci 7(9):ARTN 2000890. https://doi.org/10.1002/advs.202000890
86. Wang S, Chao F, Zhang C, Han D, Xu G, Chen G (2022) Circular RNA circPFKP promotes cell proliferation by activating IMPDH2 in prostate cancer. Cancer Lett 524:109–120. https://doi.org/10.1016/j.canlet.2021.10.021
87. Li J, Sang M, Zheng Y, Meng L, Gu L, Li Z, Liu F, Wu Y, Li W, Shan B (2021) HNRNPUL1 inhibits cisplatin sensitivity of esophageal squamous cell carcinoma through regulating the formation of circ-MAN1A2. Exp Cell Res 409(1):112891. https://doi.org/10.1016/j.yexcr.2021.112891
88. Ji F, Du R, Chen T, Zhang M, Zhu Y, Luo X, Ding Y (2020) Circular RNA circSLC26A4 accelerates cervical cancer progression via miR-1287-5p/HOXA7 axis. Mol Ther Nucleic Acids 19:413–420. https://doi.org/10.1016/j.omtn.2019.11.032
89. Zhou M, Li L, Chen B, Pan S, Tu W, Hou Y, Chen P, Hernandez RR, Zhou X (2021) Circ-SHPRH suppresses cadmium-induced transformation of human bronchial epithelial cells by regulating QKI expression via miR-224-5p. Ecotoxicol Environ Saf 220:112378. https://doi.org/10.1016/j.ecoenv.2021.112378
90. Su H, Chu Q, Zheng WW, Chang RJ, Gao WY, Zhang L, Xu TJ (2021) Circular RNA circPIKfyve acts as a sponge of miR-21-3p to enhance antiviral immunity through regulation of MAVS in teleost fish. J Virol 95(8):ARTN e02296–20. https://doi.org/10.1128/JVI.02296-20
91. Kong Y, Luo Y, Zheng S, Yang J, Zhang D, Zhao Y, Zheng H, An M, Lin Y, Ai L, Diao X, Lin Q, Chen C, Chen R (2023) Mutant KRAS mediates circARFGEF2 biogenesis to promote lymphatic metastasis of pancreatic ductal adenocarcinoma. Cancer Res 83(18):3077–3094. https://doi.org/10.1158/0008-5472.CAN-22-3997
92. Liu Y, Yang Y, Xu C, Liu J, Chen J, Li G, Huang B, Pan Y, Zhang Y, Wei Q, Pandol SJ, Zhang F, Li L, Jin L (2023) Circular RNA circGlis3 protects against islet beta-cell dysfunction and apoptosis in obesity. Nat Commun 14(1):351. https://doi.org/10.1038/s41467-023-35998-z
93. Shen Y, Zhang N, Chai J, Wang T, Ma C, Han L, Yang M (2023) CircPDIA4 induces gastric cancer progression by promoting ERK1/2 activation and enhancing biogenesis of oncogenic circRNAs. Cancer Res 83(4):538–552. https://doi.org/10.1158/0008-5472.CAN-22-1923
94. Gao P, Yang Y, Li X, Zhao Q, Liu Y, Dong C, Zhang Y, Liu D (2023) Circular RNA hsa_circ_0098181 inhibits metastasis in hepatocellular carcinoma by activating the Hippo signaling pathway via interaction with eEF2. Ann Hepatol 28(5):101124. https://doi.org/10.1016/j.aohep.2023.101124
95. Yang J, He W, Gu L, Zhu L, Liang T, Liang X, Zhong Q, Zhang R, Nan A, Su L (2023) CircFOXP1 alleviates brain injury after acute ischemic stroke by regulating STAT3/apoptotic signaling. Transl Res 257:15–29. https://doi.org/10.1016/j.trsl.2023.01.007
96. Li Z, Li YM, Han DW, Wang XL, Li C, Chen T, Li WH, Liang YR, Luo D, Chen B, Wang LJ, Zhao WJ, Yang QF (2023) circRNA-SFMBT2 orchestrates ERα activation to drive tamoxifen resistance in breast cancer cells. Cell Death Dis 14(7):ARTN 482. https://doi.org/10.1038/s41419-023-06006-5
97. Liu Y, Qiu G, Luo Y, Li S, Xu Y, Zhang Y, Hu J, Li P, Pan H, Wang Y (2022) Circular RNA ROCK1, a novel circRNA, suppresses osteosarcoma proliferation and migration via altering the miR-532-5p/PTEN axis. Exp Mol Med 54(7):1024–1037. https://doi.org/10.1038/s12276-022-00806-z
98. Xu L, Ma X, Zhang X, Zhang C, Zhang Y, Gong S, Wu N, Zhang P, Feng X, Guo J, Zhao M, Ren Z, Zhang P (2023) hsa_circ_0007919 induces LIG1 transcription by binding to FOXA1/TET1 to enhance the DNA damage response and promote gemcitabine resistance in pancreatic ductal adenocarcinoma. Mol Cancer 22(1):195. https://doi.org/10.1186/s12943-023-01887-8
99. Min J, Li Y, Li X, Wang M, Li H, Bi Y, Xu P, Liu W, Ye X, Li J (2023) The circRNA circVAMP3 restricts influenza A virus replication by interfering with NP and NS1 proteins. PLoS Pathog 19(8):e1011577. https://doi.org/10.1371/journal.ppat.1011577
100. Li HM, Zhao YH, Shen QZ, Li HG (2022) Multiple circRNAs regulated by QKI5 conjointly sponge miR-214-3p to antagonize bisphenol A-induced spermatocyte toxicity. Acta Biochim Biophys Sin 54(8):1090–1099. https://doi.org/10.3724/abbs.2022101
101. Li Y, Chen B, Zhao J, Li Q, Chen S, Guo T, Li Y, Lai H, Chen Z, Meng Z, Guo W, He X, Huang S (2021) HNRNPL circularizes ARHGAP35 to produce an oncogenic protein. Adv Sci (Weinh) 8(13):2001701. https://doi.org/10.1002/advs.202001701
102. Lu J, Zhong C, Luo J, Shu F, Lv D, Liu Z, Tan X, Wang S, Wu K, Yang T, Zhong W, Wang B, Chen Y, Li Y, Jia Z, Zou Y, Zhong W, Mao X (2021) HnRNP-L-regulated circCSPP1/miR-520h/EGR1 axis modulates autophagy and promotes progression in prostate cancer. Mol Ther Nucleic Acids 26:927–944. https://doi.org/10.1016/j.omtn.2021.10.006
103. Cao JC, Zhang X, Xu PH, Wang HX, Wang S, Zhang L, Li Z, Xie L, Sun GL, Xia YW, Lv JL, Yang J, Xu ZK (2021) Circular RNA circLMO7 acts as a microRNA-30a-3p sponge to promote gastric cancer progression via the WNT2/β-catenin pathway.

J Exp Clin Cancer Res 40(1):ARTN 6. https://doi.org/10.1186/s13046-020-01791-9
104. Xu P, Zhang J, Wang M, Liu B, Li R, Li H, Zhai N, Liu W, Lv C, Song X (2022) hnRNPL-activated circANKRD42 back-splicing and circANKRD42-mediated crosstalk of mechanical stiffness and biochemical signal in lung fibrosis. Mol Ther 30(6):2370–2387. https://doi.org/10.1016/j.ymthe.2022.01.045
105. Zhang S, Tu D, Liu W, Li R, Jiang M, Yuan X, Luan J, Li H, Lv C, Song X (2023) circELP2 reverse-splicing biogenesis and function as a pro-fibrogenic factor by targeting mitochondrial quality control pathway. J Cell Mol Med 28(3):e18098. https://doi.org/10.1111/jcmm.18098
106. Lv J, Li K, Yu H, Han J, Zhuang J, Yu R, Cheng Y, Song Q, Bai K, Cao Q, Yang H, Yang X, Lu Q (2023) HNRNPL induced circFAM13B increased bladder cancer immunotherapy sensitivity via inhibiting glycolysis through IGF2BP1/PKM2 pathway. J Exp Clin Cancer Res 42(1):41. https://doi.org/10.1186/s13046-023-02614-3
107. Zhao W, Cui Y, Liu L, Qi X, Liu J, Ma S, Hu X, Zhang Z, Wang Y, Li H, Wang Z, Liu Z, Wu J (2020) Splicing factor derived circular RNA circUHRF1 accelerates oral squamous cell carcinoma tumorigenesis via feedback loop. Cell Death Differ 27(3):919–933. https://doi.org/10.1038/s41418-019-0423-5
108. Wang X, Xu C, Cai Y, Zou X, Chao Y, Yan Z, Zou C, Wu X, Tang L (2022) CircZNF652 promotes the goblet cell metaplasia by targeting the miR-452-5p/JAK2 signaling pathway in allergic airway epithelia. J Allergy Clin Immunol 150(1):192–203. https://doi.org/10.1016/j.jaci.2021.10.041
109. Chen S, Zong Y, Hou Z, Deng Z, Xia Z (2023) Splicing factor ESRP1 derived circ_0068162 promotes the progression of oral squamous cell carcinoma via the miR-186/JAG axis. Carcinogenesis 45(3):107–118. https://doi.org/10.1093/carcin/bgad082
110. Liu D, Dredge BK, Bert AG, Pillman KA, Toubia J, Guo W, Dyakov BJA, Migault MM, Conn VM, Conn SJ, Gregory PA, Gingras AC, Patel D, Wu B, Goodall GJ (2023) ESRP1 controls biogenesis and function of a large abundant multiexon circRNA. Nucleic Acids Res 52(3):1387–1403. https://doi.org/10.1093/nar/gkad1138
111. Pan S, Wang Q, Zhang Q, Zhou M, Li L, Zhou X (2021) A novel circular RNA, circPUS7 promotes cadmium-induced transformation of human bronchial epithelial cells by regulating Kirsten rat sarcoma viral oncogene homolog expression via sponging miR-770. Metallomics 13(7):mfab043. https://doi.org/10.1093/mtomcs/mfab043
112. Liang Y, Wang H, Chen B, Mao Q, Xia W, Zhang T, Song X, Zhang Z, Xu L, Dong G, Jiang F (2021) circDCUN1D4 suppresses tumor metastasis and glycolysis in lung adenocarcinoma by stabilizing TXNIP expression. Mol Ther Nucleic Acids 23:355–368. https://doi.org/10.1016/j.omtn.2020.11.012
113. Li H, Lan T, Liu H, Liu C, Dai J, Xu L, Cai Y, Hou G, Xie K, Liao M, Li J, Huang J, Yuan K, Wang G, Zeng Y, Wu H (2023) Erratum: IL-6-induced cGGNBP2 encodes a protein to promote cell growth and metastasis in intrahepatic cholangiocarcinoma. Hepatology 78(1):E19–E20. https://doi.org/10.1097/HEP.0000000000000040
114. Liu H, Lan T, Li H, Xu L, Chen X, Liao H, Chen X, Du J, Cai Y, Wang J, Li X, Huang J, Yuan K, Zeng Y (2021) Circular RNA circDLC1 inhibits MMP1-mediated liver cancer progression via interaction with HuR. Theranostics 11(3):1396–1411. https://doi.org/10.7150/thno.53227
115. Long F, Li L, Xie C, Ma M, Wu Z, Lu Z, Liu B, Yang M, Zhang F, Ning Z, Zhong C, Yu B, Liu S, Wan L, Tian B, Yang K, Guo Y, Chen M, Chou J, Li X, Hu G, Lin C, Zhang Y (2023) Intergenic CircRNA Circ_0007379 inhibits colorectal cancer progression by modulating miR-320a biogenesis in a KSRP-dependent manner. Int J Biol Sci 19(12):3781–3803. https://doi.org/10.7150/ijbs.85063
116. Wang P, Huang Z, Peng Y, Li H, Lin T, Zhao Y, Hu Z, Zhou Z, Zhou W, Liu Y, Hou FF (2022) Circular RNA circBNC2 inhibits epithelial cell G2-M arrest to prevent fibrotic maladaptive repair. Nat Commun 13(1):6502. https://doi.org/10.1038/s41467-022-34287-5
117. Wang X, Chen T, Li C, Li W, Zhou X, Li Y, Luo D, Zhang N, Chen B, Wang L, Zhao W, Fu S, Yang Q (2022) CircRNA-CREIT inhibits stress granule assembly and overcomes doxorubicin resistance in TNBC by destabilizing PKR. J Hematol Oncol 15(1):122. https://doi.org/10.1186/s13045-022-01345-w
118. Ou R, Lu S, Wang L, Wang Y, Lv M, Li T, Xu Y, Lu J, Ge RS (2022) Circular RNA circLMO1 suppresses cervical cancer growth and metastasis by triggering miR-4291/ACSL4-mediated ferroptosis. Front Oncol 12:858598. https://doi.org/10.3389/fonc.2022.858598
119. Liu H, Yan Y, Lin J, He C, Liao H, Li H, Zhou Z, Wang J, Mao K, Xiao Z (2022) Circular RNA circSFMBT2 downregulation by HBx promotes hepatocellular carcinoma metastasis via the miR-665/TIMP3 axis. Mol Ther Nucleic Acids 29:788–802. https://doi.org/10.1016/j.omtn.2022.08.008
120. Xue C, Wei J, Li M, Chen S, Zheng L, Zhan Y, Duan Y, Deng H, Xiong W, Li G, Li H, Zhou M (2022) The emerging roles and clinical potential of circSMARCA5 in cancer. Cells 11(19). https://doi.org/10.3390/cells11193074
121. Liu B, Shen H, He J, Jin B, Tian Y, Li W, Hou L, Zhao W, Nan J, Zhao J, Shen J, Yu H, Wang Y, Shan G, Shi L, Cai X (2023) Cytoskeleton remodeling mediated by circRNA-YBX1 phase separation suppresses the metastasis of liver cancer. Proc Natl Acad Sci USA 120(30):e2220296120. https://doi.org/10.1073/pnas.2220296120
122. Cai W, Zhang Q (2022) The transcription factor ZEB1 mediates the progression of epithelial ovarian cancer by promoting the transcription of CircANKRD17. J Biochem Mol Toxicol 36(8):e23086. https://doi.org/10.1002/jbt.23086

The Internal Ribosome Entry Site in Circular RNAs

9

Queenie Fernandes, Zeenath Safira K. M. Yoosuf, Sarra Mestiri, Laila Assami, Karama Makni-Maalej, Shahab Uddin, Maysaloun Merhi, and Said Dermime

Abstract

The internal ribosome entry site (IRES) within circular RNAs (circRNAs) has emerged as a dynamic and multifaceted regulatory element, challenging traditional notions of translation initiation in eukaryotic cells. This chapter reviews the current state of research and knowledge on IRES in circRNAs, exploring its molecular mechanisms, physiological roles, and therapeutic implications. As the field continues to evolve, research efforts are essential for a comprehensive understanding of the interplay between IRES and circRNAs. This work also highlights the future directions in the study of IRES within circRNAs, underscoring its significance in gene expression regulation and opening new avenues for therapeutic innovation.

Keywords

Circular RNA (circRNA) · Internal Ribosome Entry Site (IRES) · Cap-independent translation · Non-canonical translation; · RNA translation mechanisms · IRES trans-acting factors (ITAFs)

1 Introduction

The origin of circular RNAs (circRNAs) dates back to 1976, when they were discovered by Sanger's group as being covalently attached to

Q. Fernandes
Translational Cancer Research Facility, National Center for Cancer Care and Research, Hamad Medical Corporation, Doha, Qatar

College of Medicine, Qatar University, Doha, Qatar

Zeenath Safira K. M. Yoosuf · S. Dermime (✉)
Translational Cancer Research Facility, National Center for Cancer Care and Research, Hamad Medical Corporation, Doha, Qatar

College of Health and Life Sciences, Hamad Bin Khalifa University, Doha, Qatar
e-mail: sdermime@hamad.qa

S. Mestiri
Translational Cancer Research Facility, National Center for Cancer Care and Research, Hamad Medical Corporation, Doha, Qatar

Qatar Biomedical Research Institute (QBRI), Hamad Bin Khalifa University, Qatar Foundation, Doha, Qatar

L. Assami · K. Makni-Maalej · M. Merhi
Translational Cancer Research Facility, National Center for Cancer Care and Research, Hamad Medical Corporation, Doha, Qatar

S. Uddin
Translational Research Institute and Dermatology Institute, Academic Health System, Hamad Medical Corporation, Doha, Qatar

Laboratory Animal Research Center, Qatar University, Doha, Qatar

J. Xiao (ed.), *Advances in Circular RNAs*, Advances in Experimental Medicine and Biology 1485,
https://doi.org/10.1007/978-981-96-9428-0_9

single-stranded viroids [1]. Currently, the stable characteristics of circRNAs render them captivating contenders as functional entities in the body's circulating fluids [2]. circRNAs are classified as a subset of noncoding endogenous RNAs that modulate mammalian gene expression either at the transcriptional or posttranscriptional stage through their interaction with microRNAs (miRNAs) [3]. Yet, it still remains uncertain whether all circRNA molecules serve as miRNA sponges [4]. circRNAs predominantly originate from the circularization of exons or introns [5]. They have long been dismissed as uncommon isoforms arising from splicing anomalies or gene rearrangements [4]. Indeed, through backsplicing events, where an upstream splice acceptor is connected to a downstream splice donor, the 3′- and 5′-ends of a single-stranded RNA molecule are covalently joined [4]. Consequently, these molecules manifest as continuous loops.

Research delving into the translational potential of circRNAs has been ongoing since the 1970s [6]. However, until the 1990s, there was a belief that the eukaryotic translation machinery lacked the ability to translate circRNAs. In subsequent years, the rule was overturned with the identification of translation initiation facilitated by internal ribosome entry sites (IRESs) [3–5]. Additionally, the previously assumed incapacity of the eukaryotic ribosome to bind circRNAs was refuted in 1995 when researchers generated artificial circRNAs containing an IRES [6]. This group noted a noteworthy translation of circRNAs incorporating the IRES from the encephalomyocarditis virus (EMCV). This study conclusively highlighted two key findings that contradicted earlier hypotheses; first, the 40S ribosomal subunit was not obligatory to be recruited at the messenger ribonucleic acid (mRNA) 5′-end but can be recruited internally onto an IRES, and, second, translation of a circRNAs was indeed achievable [6]. Moreover, circRNAs are known as a product of backsplicing pre-messenger RNAs (pre-mRNAs) via spliceosomes, where the 5′- and 3′-ends are joined together, resulting in the loss of the 5′-cap and 3′-poly-adenosine tail [7]. Therefore, circRNAs rely on cap-independent or cap-dependent mechanisms to initiate translation through IRES or N^6-methyladenosine (m^6A), respectively [8]. Indeed, circRNAs undergo translation through the cap-independent mechanism facilitated by the methylation of the nitrogen at position 6, or, m^6A [9]. m^6A represents a reversible epitranscriptomic modification identified in numerous eukaryotic mRNAs [10]. In addition, IRESs are sequences that allow the recruitment of ribosomes to initiate the translation process [11]. The exact mechanism is largely unknown; however, it has been reported that eukaryotic translation initiation factor A3 (eIFA3) can drive translation initiation by binding ribosomes at the G-subunit of eIF3. This recruits the small subunit (40S) of the ribosome and initiates translation [12]. Another report stated that endogenous and synthetic IRES sequences can initiate translation in circRNAs [13]. Interestingly, this study determined two regulatory elements in IRES-dependent circRNA translation that include complementary active regions on the 18S ribosomal RNA (rRNA) and a stem-loop-structured RNA element at 40–60 nucleotides of the IRES sequence [13]. In addition, another study by Abe et al. reported rolling circle amplification in circRNAs with an infinite open reading frame (ORF) that did not contain elements for internal ribosome entry [14]. Similarly, another study reported the translation of a viral circRNA into a micropeptide through IRES-dependent translation [15]. It is also intriguing to note that infectious viruses like Middle East respiratory syndrome coronavirus (MERS-CoV), severe acute respiratory syndrome coronavirus (SARS-CoV), and severe acute respiratory syndrome coronavirus 2 (SARS-CoV-2) were also identified to possess two circRNAs that can be translated and contain IRESs [16].

Moreover, numerous investigations have showcased the ability of IRESs to facilitate the translation of particular categories of capped mRNAs when the cap-dependent initiation mechanism is obstructed, a scenario that is commonly prevalent during cellular stress [10]. Consequently, the significance of the IRES-dependent mechanism has become evident in the translational response to stress, and it is governed by specific proteins known as IRES *trans*-acting factors (ITAF) [17]. IRESs also play a role in the

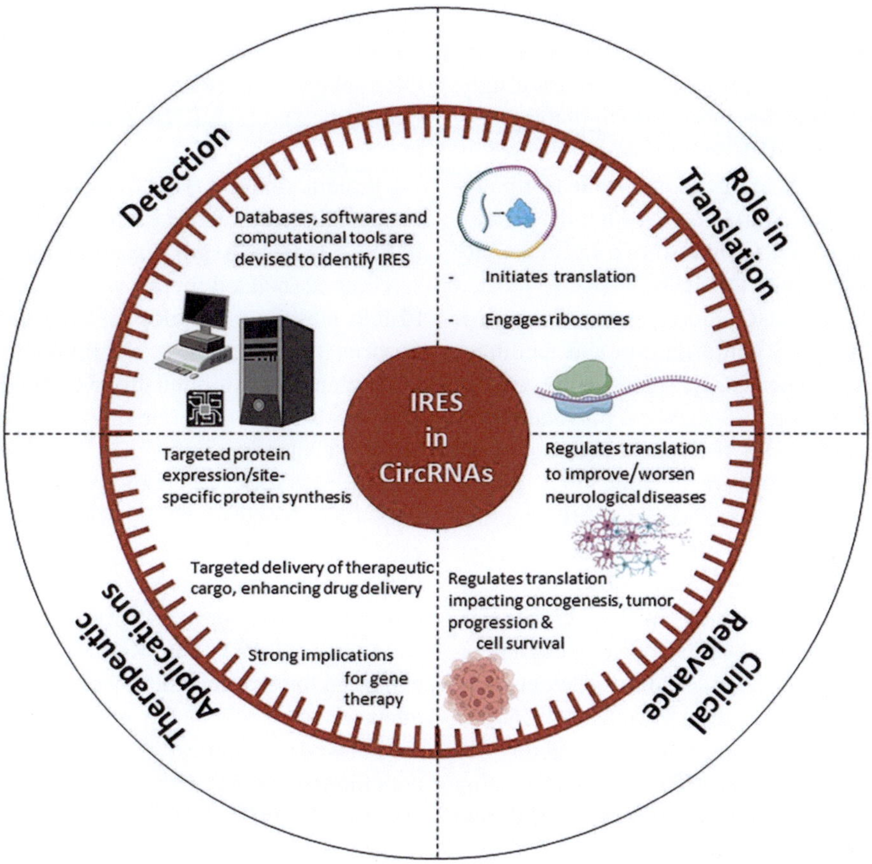

Fig. 9.1 A depiction of the multifaceted aspects of the IRESs in circRNAs.

heightened translation of these mRNAs in cancer cells, a phenomenon linked to aberrant rRNA modifications [18]. Thus, the translation of cellular circRNAs imparts comprehensive physiological significance to IRES-dependent translation. Highlights of the IRESs in the circRNAs discussed in this chapter are depicted in Fig. 9.1.

2 Functional Roles of IRESs in the Translation of circRNAs

The earliest evidence on the function of IRESs in circRNAs dates back to 1995, when it was reported that circRNA translation is m7G cap-independent and instead relies on IRESs [6, 19]. The experiment incorporated a synthetically engineered circRNA containing the ORF and IRES of EMCV, resulting in the production of an approximately 23-kDa protein [19]. Though the authors did not validate this function on natural circRNAs, numerous recent reports have extensively discussed IRES-mediated circRNA translation in viroids and in human and *Drosophila* cells [20]. IRESs facilitate circRNA translation irrespective of where they are positioned, either between or behind the ORF or within the untranslated region (UTR). Canonical eukaryotic mRNA translation depends on the interaction of cap-associated proteins, eIF4E, eIF4G, and eIF4A, which lead to the assembly of the 43S initiation complex for directing protein synthesis [21]. Conversely, in uncapped circRNA translation, IRES functions as an RNA scaffold and engages with initiation factors such as eIF4G or the eIF3 complex, which, in turn, anchors the 43S complex for protein translation [6]. Interestingly, emerging evidence has brought to light the potential regulation of IRES activity within viruses by ITAFs, notably heterogeneous nuclear ribonucleoproteins (hnRNPs) [17, 22]. An hnRNP demonstrates the ability to bind to two IRES activity sites present within the EMCV IRES, located upstream of the AUG codon, through which it stabilizes the IRES conformation and facilitates ribosome recruitment [23]. Additionally, certain RNA-binding proteins, such as polyadenylate-binding protein 1 (PABPC1) and heterogeneous nuclear ribonucleoprotein U (hnRNPU), play a role in accelerating the translation efficiency of IRES-like elements in circRNAs [24]. Collectively, ITAFs have been shown to govern IRES activity and influence circRNA biogenesis.

A group of researchers was able to engineer a single exon capable of undergoing efficient backsplicing, thereby generating a circRNA [25]. The resulting circRNA possessed the ability to undergo translation via an IRES, thus, producing a functional green fluorescent protein (GFP) that was instrumental in understanding circRNA function and biogenesis. This group observed that backsplicing is regulated by general splicing factors and specific *cis*-elements, showcasing regulatory mechanisms distinct from those observed in conventional splicing [25]. Moreover, it was found that multiple sites rich in m^6A mobilize protein translation from this circRNA reporter, especially under conditions of cellular stress [25]. Furthermore, in order to deepen comprehension of the IRES-mediated circRNA translation process, Meganck et al. investigated intron type and length, Alu element repeats, and exon length in circRNA formation and translation [26]. Introns guiding backsplicing commonly encompass inverted Alu elements or similar complementary sequences. The distance between these regions and splice donor/acceptor sites in circRNA backsplicing was explored using a previously reported system in which the homeodomain interacting protein kinase 3 (*HIPK3*) gene-derived introns induced circularization of a split GFP exon. This formed a circRNA uniting the GFP ORF upon transcription, and, subsequently, the translation initiation for the GFP ORF relied on an IRES within the circular green fluorescent protein RNA (circGFP RNA) [26]. Different versions of the homeodomain interacting protein kinase 3 (HIPK3) split GFP construct were engineered, with each incorporating distinct IRES elements (EMCV, poliovirus, Kaposi's sarcoma-associated herpesvirus (KSHV), viral FLICE inhibitory proteins (vFLIP), and hepatitis C virus (HCV)). These circRNAs with varied IRES elements exhibited differential GFP expression levels, with the poliovirus IRES-containing circRNA showing a fivefold higher expression compared to EMCV and KSHV IRES-containing ones [26]. Notably, HCV IRES-containing circRNA displayed no GFP expression. In addition, polyribosome fractions were isolated to evaluate translation efficiency. Accordingly, the circGFP RNA with EMCV and KSHV IRES elements exhibited lower translation, were largely untranslated, or were associated with monosomes and disomes. In contrast, the poliovirus IRES-containing circGFP RNA displayed higher levels associated with multiple ribosomes, which correlated with increased GFP expression [26]. These findings confirmed the lack of protein expression with the HCV IRES, where most circRNAs remained unbound, unlike the control linear GFP mRNA detected in polysome fractions. Furthermore, comparable RNA and protein levels among different cell lines indicated similar IRES-mediated translation efficiency of these circRNA vectors in vitro. This report also demonstrated that IRES impacted the circRNA splicing patterns and that the factors influencing circRNA translation are tissue-specific [26].

In line with this, Chen et al. devised a high-throughput reporter assay, employing an oligo-split-eGFP circRNA reporter construct to systematically evaluate, screen, and assess the IRES activity of RNA sequences facilitating circRNA translation [13]. As part of this analysis, crucial regulatory elements within the primary and secondary structures of circRNA IRES were identified; specifically, the complementarity to 18S rRNA and a 40–60 nucleotide secondary structure element (SuRE) on the IRES that are common features found among both exogenous and endogenous translating circRNAs [13]. Furthermore, by aligning these circRNA IRES sequences with human endogenous circRNA, potential protein-coding circRNAs and their encoded peptide sequences were determined. Validation through mass spectrometry (MS) peptidomic analysis has confirmed the circRNA IRES's ability to initiate translation in endogenous circRNAs, resulting in observable protein production. Furthermore, characterization was performed on circFGFR1p, a circRNA-encoded protein functioning as a negative regulator of fibroblast growth factor receptor 1 (FGFR1) via a dominant-negative mechanism, effectively suppressing cell growth during stress conditions [13]. The discovery of novel circular IRES sequences offers an opportunity to develop advanced circular RNA-based translation platforms for future use. In a recent publication, Fan et al. have developed a cell-based system to delineate circRNA translation prevalence and mechanisms [24]. Through extensive screening, 97 hexamers were identified that displayed AU-rich consensus motifs and significant enrichment in human circRNAs compared to linear RNAs. Multiple transacting factors are known to bind to these IRES-like elements and have been proven sufficient to initiate circRNA translation. Mass spectrometry analysis revealed numerous circRNA-coded peptides, highlighting widespread circRNA translation, though most of them exhibited low abundance owing to their rapid degradation [24]. Notably, approximately 50% of translatable endogenous circRNAs underwent rolling circle translation. Additionally, mutations in the identified IRES-like element within one circRNA led to decreased translation, emphasizing its role in circRNA translation initiation [24].

3 Identification and Characterization of IRESs in circRNAs

Comprehensive databases and computational tools are devised to identify IRESs within circRNA sequences. With an aim to detect endogenous IRES elements within selected circRNAs, many studies have utilized specialized databases like IRESite, IRESbase, and IRESfinder [27–29]. At present, several databases such as CircPrimer 2 [30], CircInteractome [31], CircRNADb [32], CircAtlas [33], and MStoCIRC [34] have cataloged circRNAs that potentially encode circproteins based on the presence of IRES sequences and open reading frames (ORFs). Recently, the TransCirc database has been established, combining seven pieces of evidence concerning circRNA translation; there are, namely, ribosome profiling, translation initiation site mapping, IRES presence, m6A modification, ORF existence, sequence composition score, and mass spectrometry data [35].

Several other studies have reiterated the significance of IRESs in circRNA translation. A study utilized the IRES Prediction System and CircRNADb to predict the protein-coding potential of sheep circRNAs [36]. The analysis identified 40 sheep circRNAs that harbored at least one IRES element and ORF. Soybean circRNA sequences were blasted against all IRES sequences on the IRESite platform to evaluate the protein coding potential of the circRNAs [37]. The analysis showed that at least one IRES element along with an ORF was contained within 165 soybean circRNAs, thus reflecting on their capacity to encode polypeptides or proteins [38]. Another study aimed to comprehend the function of circRNAs identified in early-stage diabetic cardiomyopathy mice [39]. IRESfinder software and Circatlas 2.0 were utilized to detect the presence of potential IRES and ORF elements within the circRNA sequences. Among the differentially expressed circRNAs, nine were identified to contain a potential IRES element, with mmu_circ_0001160 exhibiting relatively high scores for both IRES and ORF elements. This suggests that mmu_circ_0001160 may essentially encode a polypeptide, thus indicating a novel functional role that prompts further investigation [39]. In addition, bioinformatics analysis of 1766 differentially expressed circRNA in the abdominal adipose tissue of Gushi chickens demonstrated that 62.71% of circRNAs have IRES elements and that ten of them are capable of encoding proteins [40]. A similar examination to study the role of circRNAs in the skeletal muscle development of Gushi chickens reported that 1295 circRNAs contain IRES elements within their sequences, representing approximately 61.32% of the analyzed circRNAs. The ten primary circRNAs had high predicted scores for protein coding [41]. Moreover, IRESfinder software was used to investigate circRNA sequences and their translation potential in honeybee brain, following exposure to a sublethal quantity of dinotefuran. Out of 2050 circRNAs with a coding potential, 4 (circ_0002266, circ_0005080, circ_0010239, and circ_0005415) were found to encode proteins [42].

4 IRES-Mediated Translation in Various circRNAs

IRES-mediated translation in various circRNAs has emerged as a captivating area of research, shedding light on the intricate ways in which these noncoding RNAs can influence protein synthesis. Table 9.1 summarizes the roles of IRES-mediated translation in various circRNAs.

4.1 CircRNA Zinc Finger Protein 609 (circZNF609)

circZNF609 is an endogenous circRNA associated with polyribosomes that generates an approximately 30-kDa protein through splicing-dependent and IRES-mediated mechanisms [43]. The UTR element of circ-ZNF609, spanning from termination to initiation codons, facilitates IRES-dependent translation only when formed via splicing [43]. This suggests that factors loaded onto the transcript post-splicing may significantly impact ribosome recognition and translation initiation. The expressed protein regulates muscle cell proliferation and hence plays a role in myogenesis and Duchenne muscular dystrophy [43].

4.2 Muscleblind Circular RNA (circMbl)

Pamudurti et al. observed that circMbl isoforms share the translation start site with Mbl mRNA and undergo translation via the IRES activity present in the UTR sequence [45]. Mass spectrometry analysis in *Drosophila* head extracts confirmed the production of 10- and 37-kDa proteins from circMbl1 and circMbl3 isoforms, respectively, derived from the muscleblind gene [45]. Additionally, the study also determined IRES activity through a double-luciferase assay, revealing its regulation through the overexpression of the eukaryotic translation initiation factor 4E-binding protein 1 (4E-BP), which is commonly known to inhibit cap-dependent translation [45].

4.3 β-Catenin-Derived Circular RNA (circβ-Catenin)

circ β-catenin undergoes translation via IRES-mediated mechanisms, resulting in a newly identified 370 amino acid protein called circβ-catenin-370aa that initiates at the same AUG site as the linear β-catenin mRNA [52]. The expression of circβ-catenin-370aa contributes to the stimulation of liver cell growth and malignant characteristics, which is attributed to the stabilization of β-catenin, which is achieved by inhibiting GSK3β-induced β-catenin phosphorylation and subsequent degradation [52].

4.4 SNF2 Histone Linker PHD RING Helicase-Deriver Circular RNA (circ-SHPRH)

The circular variant of the SNF2 histone linker PHD RING helicase (*SHPRH*) gene encodes a novel protein referred to as SHPRH-146aa [46]. This is the first eukaryotic circRNA that comprises an ORF overlapping with a start–stop codon, which enables IRES-mediated translation. circ-SHPRH exhibited high expression in normal brain tissue but significantly lower levels in glioblastoma tissues [46].

Table 9.1 IRES-mediated translation in various circRNAs

CircRNA	IRES activity	Encoded protein	Role in pathology
circZNF609	IRES activity mediated by the UTR element spanning from termination to initiation codons	Poorly characterized transcription factor belonging to the zinc finger protein family	Upregulation post-heat-shock stress [43] Regulation of muscle cell proliferation and hence plays a role in myogenesis and Duchenne muscular dystrophy [44]
circMbl	IRES activity present in the UTR sequence	10- and 37-kDa proteins from circMbl1 and circMbl3 isoforms	Cellular stress [43]
circβ-catenin	IRES activity initiates at the same AUG site as the linear β-catenin mRNA	circβ-catenin-370aa	Promotes liver cancer cell growth through the activation of the Wnt pathway [45]
circSHPRH	IRES-mediated translation of ORF overlapping with a start–stop codon	SHPRH-146aa	Highly expressed in normal brain tissue [46] Significantly decreased levels in glioblastoma tissues [46] Tumor suppressor activity by acting as a decoy [46]
circPINTexon2	IRES-mediated translation of 87-aa sORF	PINT-87aa	Inhibition of glioblastoma cell proliferation [47] Suppression of transcriptional elongation of various oncogenes [47]
circ-AKT3	IRES activity mediated by overlapping start–stop codons	AKT3-174aa	Reduces radiation resistance and proliferation of glioblastoma cells [48] Decreased tumorigenicity of Glioblastoma (GBM) cells [48]
circFBXW7	IRES-mediated translation of the spanning junction ORF	FBXW7-185aa	Inhibits cancer cell proliferation and slows cell cycle progression [49]
circLINC-PINT	IRES-mediated translation of ORF	PINT87aa	Controls cell proliferation and tumorigenesis in cancer cells [47]
circFAM188B	IRES-mediated translation of ORF	circFAM188B-103aa	Promotes proliferation and suppresses differentiation of chicken skeletal muscle satellite cells [38]
circ-EIF6	IRES-mediated translation of 675-ntORF	EIF6-224aa	Promotes proliferation and metastasis of triple-negative breast cancer (TNBC) [38]
circEPS15	IRES-mediated translation of the spanning junction ORF	Unknown	Overexpressed in hepatocellular carcinoma and modulates tumor invasion and migration [50]
circ-HGF	IRES-mediated translation of the cross-junctional ORF	C-HGF-119aa	Upregulated in glioblastoma [51]

4.5 Circular RNA-Derived from Exon 2 of Long Intergenic Nonprotein-Coding RNA p53-Induced Transcript (LINC-PINT) (circPINTexon2)

It is indeed intriguing to note that the IRES within this circRNA is capable of facilitating cap-independent translation, resulting in the generation of a novel protein known as p53-induced transcript (PINT)-87aa [47]. Notably, this translation mechanism was found to be absent in the linear long intergenic nonprotein-coding RNA p53-induced transcript (LINC-PINT). Interestingly, PINT-87aa played a crucial role in inhibiting the proliferation of glioblastoma cells

[47]. Its mode of action involved an interaction with the polymerase-associated factor complex (PAF1c), leading to the suppression of transcriptional elongation of various oncogenes [47].

4.6 Circular RNA Originating from the AKT Serine/Threonine Kinase 3 Gene Locus (circ-AKT3)

It was observed that circular AKT3 utilizes IRES for translation, generating a novel 174 amino acid polypeptide known as AKT3-174aa [48]. Overexpression of AKT3-174aa in glioblastoma cells manifested reduced radiation resistance and cell proliferation. The proposed mechanism suggested that AKT3-174aa competitively interacts with phosphorylated 3-phosphoinositide-dependent kinase 1 (PDK1), consequently leading to reduced AKT-thr308 phosphorylation, thus negatively modulating the phosphoinositide 3-kinase (PI3K)/AKT regulatory pathway [48].

4.7 Parent Gene of F-box and WD Repeat Domain Containing 7-Derived Circular RNA (circ-FBXW7)

A novel 21-kDa F-box protein (FBXW7-185aa) is encoded by the spanning junction ORF in circ-FBXW7 aided through IRES-mediated translation [49]. The increased expression of FBXW7-185aa within cancer cells led to the inhibition of cell proliferation and slowed cell cycle progression. Conversely, suppressing FBXW7-185aa contributed to the expansion of malignant characteristics both in vitro and in vivo [49]. This was achieved by diminishing the stability of the MYC oncogene (c-Myc) via countering the actions of ubiquitin carboxyl-terminal hydrolase 28 (USP28) that otherwise stabilize c-Myc. Clinical samples from glioblastoma patients exhibited decreased levels of circ-FBXW7 and FBXW7-185aa compared to their respective tumor-adjacent tissues [49]. Higher expression of circ-FBXW7 correlated positively with improved overall survival rates in glioblastoma patients [49].

4.8 circFAM188B

The circFAM188B circRNA is known to contain an ORF and is commonly found in the cellular cytoplasm [38]. Interestingly, IRES assays identified an IRES sequence suggesting its potential in encoding a protein, which was then experimentally validated to be circFAM188B [38]. The protein was found to play a role in chicken myogenesis similar to its parent transcript, promoting proliferation and suppressing differentiation of chicken skeletal muscle satellite cells [38].

4.9 Circular RNA Eukaryotic Translation Initiation Factor 6 (circ-EIF6)

This circRNA was identified in triple-negative breast cancer (TNBC) and was found to be correlated with patients' characteristics and prognoses [53]. circ-EIF6 translates into a novel 224-aa peptide (EIF6-224aa) in an IRES-dependent manner and is expressed within TNBC cell lines and tissues. In order to assess its role in oncogenic functions, a circRNA called circ-EIF6-overexpression vector (circ-EIF6-OV) that lacks the IRES was constructed for comparison purposes [53]. Interestingly, only EIF6-224aa promoted proliferation and metastasis of TNBC cells both in vitro and in vivo [53].

4.10 Circular RNA Derived from Epidermal Growth Factor Receptor Pathway Substrate 15 (circEPS15)

Various circRNAs are known to be differentially expressed in hepatocellular carcinoma (HCC) samples and adjacent normal liver tissues. Quantitative reverse transcriptase polymerase chain reaction (qRT-PCR) analysis identified a downregulated circEPS15 expression in HCC tissues [50]. circEPS15 contains a spanning junction open reading frame that encodes a novel protein, validated by western blotting and bioinformatics analysis. Interestingly, low circEPS15 levels correlated with poorer overall survival in HCC patients [50].

4.11 Hepatocyte Growth Factor-Derived Circular RNA (circ-HGF)

circ-HGF is a circRNA that mediates the translation of a cross-junctional ORF, which encodes a 119 amino acid protein called C-HGF (hepatocyte growth factor variant), through an IRES-mediated pathway [51]. This protein is significantly upregulated in glioblastoma, as compared to normal brain tissue. The C-HGF protein secreted from the glioblastoma cells is capable of activating mesenchymal–epithelial transition factor (c-MET) signaling pathways upon interaction with the c-MET receptor. Interestingly, knockdown of C-HGF suppressed c-MET signaling and inhibited cell growth, motility, and invasiveness [51]. Furthermore, modulation of C-HGF expression in patient-derived xenograft cell models demonstrated a significant regulation of various cellular characteristics and markedly affected in vitro tumor growth [51].

5 Clinical Relevance and Applications of circRNA IRESs in Human Diseases

Intriguingly, dysregulation of IRES-dependent translation activity has been associated with several diseases. Indeed, IRESs play a crucial role in diverse cellular activities including cell proliferation, differentiation, migration, apoptosis, and stress response [54]. It has been reported that several proteins control IRES-dependent translation by promoting or inhibiting its activity [17]. ITAFs are the most well-characterized regulators [55]. Importantly, some circRNAs require the IRES-driven mechanism to be translated [24]. To date, several translated circRNAs have been revealed to play pivotal roles in different human pathologies. Here, we provide insights into the role of IRESs in the development and progression of human pathogenesis.

5.1 Neurodegenerative Diseases

Alzheimer's disease (AD) is the most common chronic neurodegenerative disease, leading to dementia in the elderly [56]. The accumulation of extracellular amyloid-β (Aβ) plaques in the brain is the major cause of AD [57]. CircRNAs may promote AD progression via the regulation of neuroinflammation, oxidative stress, autophagy, and the production and degradation of Aβ [58]. Importantly, the accumulation of circRNAs in the neural subcellular compartments during the aging process has been reported in some studies [59, 60]. Recently, researchers have discovered that several circRNAs can mediate AD by regulating the expression of Aβ [61]. Interestingly, Lo et al. identified 147 differentially expressed circRNAs in brain samples from AD patients with different severity levels [62]. The identified circRNAs may be considered as novel biomarkers of AD severity and potential therapeutic targets. Moreover, meta-analysis reports have demonstrated a significant association between the expression of a group of circRNA and AD diagnosis, clinical dementia severity, and neuropathological severity [63, 64]. Similarly, RNA sequencing of cortical tissue in AD patients has revealed that upregulation of circHOMER1 (an adult brain-enriched circRNA) and circCDR1-AS was significantly correlated with dementia severity [65]. Furthermore, Ravanidis et al. discovered six circRNAs differentially expressed in peripheral blood mononuclear cells (PBMCs) and suggested them as possible biomarkers for the diagnosis of Parkinson's disease (PD). The expression of these six circRNAs significantly decreased in AD patients compared to healthy controls [66]. On the other hand, Li et al. investigated the role of p53 IRES in the translation of p44, an androgen receptor coactivator and its implication in AD [67]. Interestingly, clinical reports showed that patients with late-onset AD expressed higher levels of p44 [67, 68].

5.2 Parkinson's Disease (PD)

PD is the second most prevalent neurodegenerative disease associated with the death of dopaminergic neurons in the substantia nigra that controls motor function [69]. Phosphatase and tensin homolog (PTEN)-induced putative kinase-1 (PINK1) is a serine/threonine kinase that plays crucial biological roles in protecting cells from stress-induced mitochondrial dysfunction while also promoting cellular survival [70]. It has been reported that PINK1 also plays a cytoprotective role in neurons [71]. PINK1 gene mutations have been associated with autosomal recessive PD [72]. Wild-type (WT) PINK1 deficiency was shown to impair cell bioenergetics and mitochondrial homeostasis, revealed by an abnormal mitochondrial morphology, lower adenosine triphosphate (ATP) production, and increased oxidative stress [73]. Moreover, PINK1 was found to play a protective role through the activation of the eukaryotic translation initiation factor 4E-binding protein 1 (4E-BP1) followed by the upregulation of IRES-dependent translation. However, PINK1 deficiency was found to induce an overexpression of 4E-BP1 that failed to upregulate IRES-dependent translational activity, leading to PINK1-associated PD [74]. Other studies have also shown that circRNAs such as circTLK1 and circSLC8A1 could be involved in neuronal injury [75, 76].

5.3 Cancer

Recently, IRES elements have attracted much attention for their pivotal roles in cancer. Contextually, several studies have demonstrated that IRES-mediated translation indeed impacts oncogenesis, cancer progression, and tumor cell survival [77]. Here, we attempt to highlight the role of IRES-mediated translation in cancer, the oncogenes and tumor suppressor genes regulated by IRES, and the conditions under which such a mechanism is initiated.

p53 is an important tumor suppressor gene that plays a vital role in cellular responses under conditions of cellular stress [78]. Following DNA damage or stress, the upregulation of p53 is known to induce a cell cycle arrest, DNA repair, and programmed cell death [78]. The dysregulation of p53 has been reported in several types of cancers and is particularly implicated in tumorigenesis [79]. Interestingly, the upregulation of p53 levels upon DNA damage has been shown to be directly controlled by IRES-mediated translation [80]. Other studies have demonstrated that p53 activity could be modulated by two IRES transacting factors: translational control protein 80 (TCP80) and the RNA helicase (RHA) [81, 82]. The knockdown or overexpression of these proteins was found to be directly associated with a change in p53 levels [80].

Bmi1, a member of the polycomb-repressive complex 1, is known to play a critical role in cell cycle regulation, cell senescence, chemoresistance, recurrence of cancer stem cells, and tumor progression [83]. It is upregulated in tumor cells and tissues and is often considered a hallmark of poor prognosis in various human cancers [84]. Recently, Wang et al. have shown that the translation of Bmi1 is controlled by two IRESs located in the 5′-UTR of Bmi1. Importantly, the mutation of these two IRESs partially inhibited the translation of Bmi1 in nasopharyngeal carcinoma [83]. On the other hand, hnRNPM is an IRES-interacting factor that controls the expression of genes involved in metabolic processes and cancer neoplasia [85]. Interestingly, the elevation of hnRNPM levels in colon cancer patient specimens was associated with poor prognosis. Genome-wide transcriptomics and translatomics analyses of the patients' samples revealed a unique set of hnRNPM-targeted genes involved in promoting tumorigenesis [85]. These findings demonstrated that IRESs may represent attractive targets for cancer therapy. Several strategies have been used to target cellular IRES translational activity and have shown promising results. Holmes et al. identified a novel compound 11 (C11) capable of inhibiting c-MYC IRES translation. The combination of C11 with PP242 (a mechanistic target of rapamycin kinase inhibitor

PP242) demonstrated a synergistic antitumor response in glioblastoma [86]. Another study showed that the sustained inhibition of IRES-mediated translation induced a sudden loss of viability in human triple-negative breast carcinoma and in glioblastoma cells [87].

6 Therapeutic Applications of IRESs in CircRNAs

The therapeutic applications of IRESs within circRNAs have emerged as a promising avenue for gene expression control and for targeted interventions. Indeed, IRES elements in circRNAs provide unique opportunities for therapeutic strategies, and several studies have shed light on their potential applications. One notable application lies in the ability to achieve targeted protein expression. By harnessing the IRES-dependent translation mechanism, circRNAs can be engineered to enable site-specific protein synthesis [88, 89]. This targeted expression holds potential for designing therapeutic circRNAs that produce specific proteins in desired tissues or under particular conditions. For instance, therapeutic circRNAs incorporating IRES elements may be designed to respond selectively to disease states, offering a tailored approach to gene modulation [90]. Furthermore, IRES-containing circRNAs have been explored as carriers for therapeutic cargo, such as small interfering RNAs (siRNAs) or microRNAs [91]. This approach allows for targeted delivery of therapeutic payloads, enhancing the precision and efficacy of drug delivery [91]. In the context of cancer therapy, aberrant translation mechanisms, including IRES-dependent translation, play a role in cancer progression [77]. Thus, therapeutic circRNAs targeting IRES elements may hold promise in selectively inhibiting or promoting the translation of specific genes associated with cancer development.

Evidently, the implications of the IRES in synthetically engineered circRNAs, particularly within the domain of gene therapy is indeed noteworthy. The utility of engineered exogenous circRNAs featuring an ORF for a specific gene, combined with an IRES to initiate cap-independent translation, presents an innovative prospect for forthcoming gene therapy applications [88, 89]. It has been observed that IRESs show distinct sensitivity across different cell types; however, the coxsackievirus B3 (CVB3) IRES demonstrated superior performance in all cell types that have been tested [89]. Meganck et al. underscored the potential of utilizing adeno-associated viral (AAV)-based vectors in circRNA expression as a tool to explore both circRNA functionality and tissue-specific regulation within animal models [26, 88]. A highly sensitive and quantitative split nanoluciferase (NanoLuc) reporter was instrumented from a template of a previously published backsplicing circRNA split GFP reporter to enhance the efficiency of analyzing IRES-mediated translation [92]. IRES-directed NanoLuc expression remained stable or increased under cell stress, in contrast to a capped control reporter that exhibited reduced translation. NanoLuc expression was detected from potential cellular IRESs and the Zika virus 5′-UTR region. These findings demonstrate that the constructed IRES reporter can verify, identify, and quantify the sequences' ability to mediate circRNA translation. Chen et al. established a systematic approach aimed at rapid assembly and evaluation of factors influencing protein production from synthetic circRNAs [90]. This group particularly optimized five elements: vector topology, 5′- and 3′-untranslated regions (UTRs), internal ribosome entry sites (IRESs), and synthetic aptamers, to enhance circRNA translation. The same group achieved a substantial increase in circRNA protein yields, demonstrating several 100-fold improvements over mRNA translation in vitro. Furthermore, this optimized approach showed enhanced and sustained protein production in vivo, thus proving its applicability across various transgenes [90]. In a recent report, a design strategy has been formulated by Kameda et al. to create synthetic circRNA devices capable of modulating circRNA translation based on the detection of specific target miRNAs and RNA-binding proteins [90]. These devices, termed "miRNA- and RNA-binding protein-responsive circRNA switches," were engineered by modifying UTRs and the coxsackievirus B3 (CVB3) IRES. The constructed synthetic circRNA circuits, which selectively activated translation by detecting and binding to endogenous miRNAs and protein-responsive circRNAs, exhibited superior persistence in expression levels compared to circuits based on linear mRNAs. They efficiently expressed reporter proteins without causing significant cell cytotoxicity or immunogenicity [93]. Another study evidenced an effective system for circRNA preparation using a self-splicing ribozyme derived from an optimized Tetrahymena thermophila group I intron [94]. Employing an optimized coxsackievirus B3 (CVB3) IRES sequence, the ribozyme-based circular mRNA expression system was demonstrated with GFP, thereby, successfully facilitating the translation of circularized mRNA [94]. The same Tetrahymena thermophila group I intron-mediated self-circularization construct was also designed to generate circRNAs through an autocatalytic end-to-end self-targeting and splicing (STS) reaction within a single RNA molecule to overcome the need for a *trans*-splicing reaction involving two separate RNAs [95]. Optimization strategies for circRNA generation, assessment of gene expression, stability, and evaluation of innate immune responses induced by the circRNAs in cells were studied. There was an efficient protein expression and stability with minimal induction of innate immunity upon the utilization of the circRNA with CVB3 IRES. Lately, the Programmable miRNA-Responsive Internal Ribosome Entry Site Translation Activation and Repression (PROMITAR) platform was initiated to regulate translation within circRNAs, introducing a novel strategy for precise translation control [96]. This platform allows for both upregulation and downregulation of translation via miRNA-based mechanisms within a single RNA construct. Through the PROMITAR platform, logic gates and cell-type classifier circRNAs were engineered, effectively identifying distinct mammalian cell types. The therapeutic potential of this platform was manifested by encoding cytotoxic proteins in the designed circRNAs for targeted cancer cell elimination. Table 9.2 summarizes the therapeutic applications of IRES in circRNA; yet, while these indeed showcase the potential of IRESs in circRNAs, continued research and exploration are essential for a comprehensive understanding of circRNA biology and translation regulation.

7 Conclusion and Future Prospects

In conclusion, the exploration of IRESs in circRNAs has unveiled a fascinating landscape of regulatory mechanisms and potential therapeutic applications. The versatility of IRES elements within circRNAs has been demonstrated in various studies, showcasing their ability to modulate translation initiation, enable targeted protein expression, and respond to cellular stress.

The field of noncanonical translatomics has garnered significant attention; however, there still exist critical unresolved issues. The knowledge surrounding the presence of IRES, and its associated RNA-binding proteins involved in circRNA translation, remains a challenge. Given that various RNA-binding proteins interact with circRNAs, the possibility of enhanced cap-independent translation through ITAFs is still unexplored [20, 97]. Moreover, the precise regulatory mechanisms governing natural circRNA translation also remain elusive. Furthermore, according to Chen et al., a stem-loop-structured RNA element in IRES called SuRE is implied to specifically facilitate the translation of circRNAs over linear RNAs [13]. However, detailed elucidation of the involvement of ITAFs and SuRE in regulating IRES activity in natural circRNA translation within cells is warranted. In addition, there are notable concerns regarding the interpretation of circRNA translation initiation mechanisms when utilizing expression vectors of a certain type. Recent studies have questioned the translation of circular mRNA templates, suggesting the likelihood of protein products arising from *trans*-splicing events [98]. These issues underscore the potential pitfalls of employing circRNA expression vectors for identifying the IRES in circRNAs. Moreover, the inadvertent activity of cryptic promoters within

Table 9.2 Therapeutic applications of IRESs in circRNAs

Mechanism	Advantages	Potential therapeutic application
Addition of nucleotides spacers separating the IRES and gene of interest from the splicing scar	Significantly increased protein production	Designing therapeutic circRNAs that enhance/inhibit the production of specific proteins [90]
IRES domain truncations	Ablated protein production	Designing therapeutic circRNAs that inhibit the production of specific proteins [90]
Synthetic IRES engineered with an eIF4G-binding aptamer	Enhanced circRNA translation	Translation of specific genes associated with cancer development [88] Diseases biomarkers [90]
Synthetic shuffled IRESs	Improved IRES translational activity Expanded repertoire of usable IRESs	
Constructed IRES identifier reporter (e.g., luciferase and GFP)	IRES element identification, verification, and quantification to mediate circRNA translation Identification of cell or tissue specificity of such protein/peptide translated from circRNAs	Disease biomarkers [91]
Synthetic circular RNA switches and circuits	Controlled protein expression in mammalian cells Enhanced translation and reduced immunogenicity Reduced cytotoxicity and enhanced cell viability	Potential for synthetic biology, circRNA-based therapies, and cellular engineering [93, 96]
IRES translation activation and repression (PROMITAR) platform	Construction of logic gates and cell-type classifier circRNAs Successful identification of desired mammalian cell types Expansion of the synthetic biology toolbox for RNA translational control	Targeted cancer cell killing by encoding a cytotoxic protein in engineered circRNAs [96] Potential approach for precision medicine [96]
CircRNA engineering by end-to-end self-targeting and splicing reaction	Rapid engineering RNA circularization system Efficient translation of gene of interest (GOI) in cells and negligible innate immune responses	Preparation of circular mRNAs to express therapeutic proteins [95] Effective choice for therapeutic use as well as vaccination against viral infection [95]
Insertion specific sequence of IRES in circRNA	Improved IRES translational activity Unique rolling circle translation Increased translation efficiency Reduced RNA-associated toxicity and innate immunogenicity	Production of therapeutic peptides/proteins as peptide/protein replacement therapy and vaccines [14]

an embedded sequence might lead to the production of transcripts harboring introns that can partake in *trans*-splicing reactions. This emphasizes the necessity for caution when using circRNA expression vectors for IRES discovery, thus advocating for experimental approaches in eliminating *trans*-splicing interference.

As we move forward, continued research will be pivotal to unraveling the full spectrum of possibilities offered by IRESs in circRNAs. Exploring the dynamic interplay between circular RNAs and IRES elements holds the key to unlocking new dimensions in gene expression control and developing tailored therapeutic interventions. The evolving landscape of circRNA research, enriched by the contributions of IRESs, promises exciting prospects for the future of molecular medicine and targeted therapies.

Author Contributions Conceptualization: SD, MM, and QF. Writing (original draft): QF, ZSKMY, SM, and LA. Data organization: KMM. Graphical representation: QF. Review and editing: MM and SD.

Conflict of Interest The authors declare that this work was conducted in the absence of any financial relationships that could be construed as a potential conflict of interest.

References

1. Liu L, Wang J, Khanabdali R, Kalionis B, Tai X, Xia S (2017) Circular RNAs: isolation, characterization and their potential role in diseases. RNA Biol 14(12):1715–1721. https://doi.org/10.1080/15476286.2017.1367886
2. Gao Y, Wang J, Zhao F (2015) CIRI: an efficient and unbiased algorithm for de novo circular RNA identification. Genome Biol 16(1). https://doi.org/10.1186/s13059-014-0571-3
3. Hansen TB, Kjems J, Damgaard CK (2013) Circular RNA and miR-7 in cancer. Cancer Res 73(18). https://doi.org/10.1158/0008-5472.CAN-13-1568
4. Barrett SP, Salzman J (2016) Circular RNAs: analysis, expression and potential functions. Development 143(11):1838–1847. https://doi.org/10.1242/dev.128074
5. Chen I, Chen C, Chuang T (2015) Biogenesis, identification, and function of exonic circular RNAs. WIREs RNA 6(5):563–579. https://doi.org/10.1002/wrna.1294
6. Prats AC, David F, Diallo LH, Roussel E, Tatin F, Garmy-Susini B, Lacazette E (2020) Circular RNA, the key for translation. Int J Mol Sci 21(22):8591. https://doi.org/10.3390/ijms21228591
7. Marques R, Lacerda R, Romão L (2022) Internal ribosome entry site (IRES)-mediated translation and its potential for novel mRNA-based therapy development. Biomedicines 10(8):1865. https://doi.org/10.3390/biomedicines10081865
8. Wen SY, Qadir J, Yang BB (2022) Circular RNA translation: novel protein isoforms and clinical significance. Trends Mol Med 28(5):405–420. https://doi.org/10.1016/j.molmed.2022.03.003
9. Diallo LH, Tatin F, David F, Godet AC, Zamora A, Prats AC, Garmy-Susini B, Lacazette E (2019) How are circRNAs translated by non-canonical initiation mechanisms? Biochimie 164:45–52. https://doi.org/10.1016/j.biochi.2019.06.015
10. Lacerda R, Menezes J, Romão L (2016) More than just scanning: the importance of cap-independent mRNA translation initiation for cellular stress response and cancer. Cell Mol Life Sci 74(9):1659–1680. https://doi.org/10.1007/s00018-016-2428-2
11. Lei M, Zheng G, Ning Q, Zheng J, Dong D (2020) Translation and functional roles of circular RNAs in human cancer. Mol Cancer 19(1). https://doi.org/10.1186/s12943-020-1135-7
12. Chang J, Shin MK, Park J, Hwang HJ, Locker N, Ahn J, Kim D, Baek D, Park Y, Lee Y, Boo SH, Kim HI, Kim YK (2023) An interaction between eIF4A3 and eIF3g drives the internal initiation of translation. Nucleic Acids Res 51(20):10950–10969. https://doi.org/10.1093/nar/gkad763
13. Chen CK, Cheng R, Demeter J, Chen J, Weingarten-Gabbay S, Jiang L, Snyder MP, Weissman JS, Segal E, Jackson PK, Chang HY (2021) Structured elements drive extensive circular RNA translation. Mol Cell 81(20):4300–4318.e13. https://doi.org/10.1016/j.molcel.2021.07.042
14. Abe N, Matsumoto K, Nishihara M, Nakano Y, Shibata A, Maruyama H, Shuto S, Matsuda A, Yoshida M, Ito Y, Abe H (2015) Rolling circle translation of circular RNA in living human cells. Sci Rep 5(1):16435. https://doi.org/10.1038/srep16435
15. Zhang Y, Zhu M, Zhang X, Dai K, Liang Z, Pan J, Zhang Z, Cao M, Xue R, Cao G, Hu X, Gong C (2022) Micropeptide vsp21 translated by Reovirus circular RNA 000048 attenuates viral replication. Int J Biol Macromol 209:1179–1187. https://doi.org/10.1016/j.ijbiomac.2022.04.136
16. Yang S, Zhou H, Liu M, Jaijyan D, Cruz-Cosme R, Ramasamy S, Subbian S, Liu D, Xu J, Niu X, Li Y, Xiao L, Tyagi S, Wang Q, Zhu H, Tang Q (2022) SARS-CoV-2, SARS-CoV, and MERS-CoV encode circular RNAs of spliceosome-independent origin. J Med Virol 94(7):3203–3222. https://doi.org/10.1002/jmv.27734
17. Godet AC, David F, Hantelys F, Tatin F, Lacazette E, Garmy-Susini B, Prats AC (2019) IRES trans-acting factors, key actors of the stress response. Int J Mol Sci 20(4):924. https://doi.org/10.3390/ijms20040924
18. Marcel V, Ghayad S, Belin S, Therizols G, Morel AP, Solano-Gonzàlez E, Vendrell J, Hacot S, Mertani H, Albaret M, Bourdon JC, Jordan L, Thompson A, Tafer Y, Cong R, Bouvet P, Saurin JC, Catez F, Prats AC, Diaz JJ (2013) p53 acts as a safeguard of translational control by regulating fibrillarin and rRNA methylation in cancer. Cancer Cell 24(3):318–330. https://doi.org/10.1016/j.ccr.2013.08.013
19. Chen CY, Sarnow P (1995) Initiation of protein synthesis by the eukaryotic translational apparatus on circular RNAs. Science 268(5209):415–417. https://doi.org/10.1126/science.7536344
20. Sinha T, Panigrahi C, Das D, Chandra Panda A (2021) Circular RNA translation, a path to hidden proteome. WIREs RNA 13(1). https://doi.org/10.1002/wrna.1685
21. Sonenberg N, Hinnebusch AG (2009) Regulation of translation initiation in eukaryotes: mechanisms and biological targets. Cell 136(4):731–745. https://doi.org/10.1016/j.cell.2009.01.042
22. Rossbach O, Hung LH, Khrameeva E, Schreiner S, König J, Curk T, Zupan B, Ule J, Gelfand MS, Bindereif A (2014) Crosslinking-immunoprecipitation (iCLIP) analysis reveals global regulatory roles of hnRNP L. RNA Biol 11(2):146–155. https://doi.org/10.4161/rna.27991
23. Romanelli M, Diani E, Lievens P (2013) New insights into functional roles of the polypyrimidine tract-

binding protein. Int J Mol Sci 14(11):22906–22932. https://doi.org/10.3390/ijms141122906
24. Fan X, Yang Y, Chen C, Wang Z (2022) Pervasive translation of circular RNAs driven by short IRES-like elements. Nat Commun 13(1). https://doi.org/10.1038/s41467-022-31327-y
25. Yang Y, Wang Z (2018) Constructing GFP-based reporter to study back splicing and translation of circular RNA. Methods Mol Biol (Clifton, NJ) 1724:107–118. https://doi.org/10.1007/978-1-4939-7562-4_9
26. Meganck RM, Borchardt EK, Castellanos Rivera RM, Scalabrino ML, Wilusz JE, Marzluff WF, Asokan A (2018) Tissue-dependent expression and translation of circular RNAs with recombinant AAV vectors in vivo. Mol Ther Nucleic Acids 13:89–98. https://doi.org/10.1016/j.omtn.2018.08.008
27. Zhao J, Li Y, Wang C, Zhang H, Zhang H, Jiang B, Guo X, Song X (2020) IRESbase: a comprehensive database of experimentally validated internal ribosome entry sites. Genomics Proteomics Bioinformatics 18(2):129–139. https://doi.org/10.1016/j.gpb.2020.03.001
28. Mokrejs M, Masek T, Vopálensky V, Hlubucek P, Delbos P, Pospísek M (2010) IRESite—a tool for the examination of viral and cellular internal ribosome entry sites. Nucleic Acids Res 38(Database issue):D131–D136. https://doi.org/10.1093/nar/gkp981
29. Zhao J, Wu J, Xu T, Yang Q, He J, Song X (2018) IRESfinder: identifying RNA internal ribosome entry site in eukaryotic cell using framed k-mer features. J Genet Genomics 45(7):403–406. https://doi.org/10.1016/j.jgg.2018.07.006
30. Zhong S, Feng J (2022) CircPrimer 2.0: a software for annotating circRNAs and predicting translation potential of circRNAs. BMC Bioinform 23(1):215. https://doi.org/10.1186/s12859-022-04705-y
31. Dudekula DB, Panda AC, Grammatikakis I, De S, Abdelmohsen K, Gorospe M (2016) CircInteractome: a web tool for exploring circular RNAs and their interacting proteins and microRNAs. RNA Biol 13(1):34–42. https://doi.org/10.1080/15476286.2015.1128065
32. Chen X, Han P, Zhou T, Guo X, Song X, Li Y (2016) circRNADb: a comprehensive database for human circular RNAs with protein-coding annotations. Sci Rep 6(1):34985. https://doi.org/10.1038/srep34985
33. Wu W, Ji P, Zhao F (2020) CircAtlas: an integrated resource of one million highly accurate circular RNAs from 1070 vertebrate transcriptomes. Genome Biol 21(1):101. https://doi.org/10.1186/s13059-020-02018-y
34. Cao Z, Li G (2022) MStoCIRC: a powerful tool for downstream analysis of MS/MS data to predict translatable circRNAs. Front Mol Biosci 9:791797. https://doi.org/10.3389/fmolb.2022.791797
35. Huang W, Ling Y, Zhang S, Xia Q, Cao R, Fan X, Fang Z, Wang Z, Zhang G (2021) TransCirc: an interactive database for translatable circular RNAs based on multi-omics evidence. Nucleic Acids Res 49(D1):D236–D242. https://doi.org/10.1093/nar/gkaa823
36. Li C, Li X, Ma Q, Zhang X, Cao Y, Yao Y, You S, Wang D, Quan R, Hou X, Liu Z, Zhan Q, Liu L, Zhang M, Yu S, Ni W, Hu S (2017) Genome-wide analysis of circular RNAs in prenatal and postnatal pituitary glands of sheep. Sci Rep 7(1):16143. https://doi.org/10.1038/s41598-017-16344-y
37. Chen L, Ding X, Zhang H, He T, Li Y, Wang T, Li X, Jin L, Song Q, Yang S, Gai J (2018) Comparative analysis of circular RNAs between soybean cytoplasmic male-sterile line NJCMS1A and its maintainer NJCMS1B by high-throughput sequencing. BMC Genomics 19(1):663. https://doi.org/10.1186/s12864-018-5054-6
38. Yin H, Shen X, Zhao J, Cao X, He H, Han S, Chen Y, Cui C, Wei Y, Wang Y, Li D, Zhu Q (2020) Circular RNA CircFAM188B encodes a protein that regulates proliferation and differentiation of chicken skeletal muscle satellite cells. Front Cell Dev Biol 8:522588. https://doi.org/10.3389/fcell.2020.522588
39. Dong S, Tu C, Ye X, Li L, Zhang M, Xue A, Chen S, Zhao Z, Cong B, Lin J, Shen Y (2020) Expression profiling of circular RNAs and their potential role in early-stage diabetic cardiomyopathy. Mol Med Rep 22(3):1958–1968. https://doi.org/10.3892/mmr.2020.11248
40. Jin W, Zhao Y, Zhai B, Li Y, Fan S, Yuan P, Sun G, Jiang R, Wang Y, Liu X, Tian Y, Kang X, Li G (2021) Characteristics and expression profiles of circRNAs during abdominal adipose tissue development in Chinese Gushi chickens. PLoS One 16(4):e0249288–e0249288. https://doi.org/10.1371/journal.pone.0249288
41. Yuan P, Zhao Y, Li H, Li S, Fan S, Zhai B, Li Y, Han R, Liu X, Tian Y, Kang X, Zhang Y, Li G (2022) CircRNAs related to breast muscle development and their interaction regulatory network in Gushi chicken. Genes (Basel) 13(11):1974. https://doi.org/10.3390/genes13111974
42. Huang M, Dong J, Guo H, Xiao M, Wang D (2022) Identification of circular RNAs and corresponding regulatory networks reveals potential roles in the brains of honey bee workers exposed to dinotefuran. Pestic Biochem Physiol 180:104994. https://doi.org/10.1016/j.pestbp.2021.104994
43. Legnini I, Di Timoteo G, Rossi F, Morlando M, Briganti F, Sthandier O, Fatica A, Santini T, Andronache A, Wade M, Laneve P, Rajewsky N, Bozzoni I (2017) Circ-ZNF609 is a circular RNA that can be translated and functions in myogenesis. Mol Cell 66(1):22–37.e9. https://doi.org/10.1016/j.molcel.2017.02.017
44. Yang Y, Fan X, Mao M, Song X, Wu P, Zhang Y, Jin Y, Yang Y, Chen LL, Wang Y, Wong CCL, Xiao X, Wang Z (2017) Extensive translation of circular RNAs driven by N 6 -methyladenosine. Cell Res 27(5):626–641. https://doi.org/10.1038/cr.2017.31

45. Pamudurti NR, Bartok O, Jens M, Ashwal-Fluss R, Stottmeister C, Ruhe L, Hanan M, Wyler E, Perez-Hernandez D, Ramberger E, Shenzis S, Samson M, Dittmar G, Landthaler M, Chekulaeva M, Rajewsky N, Kadener S (2017) Translation of CircRNAs. Mol Cell 66(1):9–21.e7. https://doi.org/10.1016/j.molcel.2017.02.021
46. Zhang M, Huang N, Yang X, Luo J, Yan S, Xiao F, Chen W, Gao X, Zhao K, Zhou H, Li Z, Ming L, Xie B, Zhang N (2018) A novel protein encoded by the circular form of the SHPRH gene suppresses glioma tumorigenesis. Oncogene 37(13):1805–1814. https://doi.org/10.1038/s41388-017-0019-9
47. Zhang M, Zhao K, Xu X, Yang Y, Yan S, Wei P, Liu H, Xu J, Xiao F, Zhou H, Yang X, Huang N, Liu J, He K, Xie K, Zhang G, Huang S, Zhang N (2018) A peptide encoded by circular form of LINC-PINT suppresses oncogenic transcriptional elongation in glioblastoma. Nat Commun 9(1):4475. https://doi.org/10.1038/s41467-018-06862-2
48. Xia X, Li X, Li F, Wu X, Zhang M, Zhou H, Huang N, Yang X, Xiao F, Liu D, Yang L, Zhang N (2019) A novel tumor suppressor protein encoded by circular AKT3 RNA inhibits glioblastoma tumorigenicity by competing with active phosphoinositide-dependent Kinase-1. Mol Cancer 18(1):131. https://doi.org/10.1186/s12943-019-1056-5
49. Yang Y, Gao X, Zhang M, Yan S, Sun C, Xiao F, Huang N, Yang X, Zhao K, Zhou H, Huang S, Xie B, Zhang N (2018) Novel role of FBXW7 circular RNA in repressing glioma tumorigenesis. J Natl Cancer Inst 110(3):304–315. https://doi.org/10.1093/jnci/djx166
50. Jiang B, Tian M, Li G, Sadula A, Xiu D, Yuan C, Bing Y (2022) circEPS15 overexpression in hepatocellular carcinoma modulates tumor invasion and migration. Front Genet 13:804848. https://doi.org/10.3389/fgene.2022.804848
51. Saunders JT, Kumar S, Benavides-Serrato A, Holmes B, Benavides KE, Bashir MT, Nishimura RN, Gera J (2023) Translation of circHGF RNA encodes an HGF protein variant promoting glioblastoma growth through stimulation of c-MET. J Neuro-Oncol 163(1):207–218. https://doi.org/10.1007/s11060-023-04331-5
52. Liang W-C, Wong C-W, Liang P-P, Shi M, Cao Y, Rao S-T, Tsui SK-W, Waye MM-Y, Zhang Q, Fu W-M, Zhang J-F (2019) Translation of the circular RNA circβ-catenin promotes liver cancer cell growth through activation of the Wnt pathway. Genome Biol 20(1):84. https://doi.org/10.1186/s13059-019-1685-4
53. Li Y, Wang Z, Su P, Liang Y, Li Z, Zhang H, Song X, Han D, Wang X, Liu Y, Yang J, Chen B, Wang L, Zhao W, Yang Q (2022) circ-EIF6 encodes EIF6-224aa to promote TNBC progression via stabilizing MYH9 and activating the Wnt/beta-catenin pathway. Mol Ther 30(1):415–430. https://doi.org/10.1016/j.ymthe.2021.08.026
54. Stoneley M, Willis AE (2004) Cellular internal ribosome entry segments: structures, trans-acting factors and regulation of gene expression. Oncogene 23(18):3200–3207. https://doi.org/10.1038/sj.onc.1207551
55. Lee KM, Chen CJ, Shih SR (2017) Regulation mechanisms of viral IRES-driven translation. Trends Microbiol 25(7):546–561. https://doi.org/10.1016/j.tim.2017.01.010
56. Cortes-Canteli M, Iadecola C (2020) Alzheimer's disease and vascular aging: JACC focus seminar. J Am Coll Cardiol 75(8):942–951. https://doi.org/10.1016/j.jacc.2019.10.062
57. Walsh DM, Klyubin I, Fadeeva JV, Cullen WK, Anwyl R, Wolfe MS, Rowan MJ, Selkoe DJ (2002) Naturally secreted oligomers of amyloid β protein potently inhibit hippocampal long-term potentiation in vivo. Nature 416(6880):535–539. https://doi.org/10.1038/416535a
58. Zhang Y, Zhao Y, Liu Y, Wang M, Yu W, Zhang L (2020) Exploring the regulatory roles of circular RNAs in Alzheimer's disease. Transl Neurodegener 9(1):35. https://doi.org/10.1186/s40035-020-00216-z
59. Gruner H, Cortés-López M, Cooper DA, Bauer M, Miura P (2016) CircRNA accumulation in the aging mouse brain. Sci Rep 6(1):38907. https://doi.org/10.1038/srep38907
60. Rybak-Wolf A, Stottmeister C, Glažar P, Jens M, Pino N, Giusti S, Hanan M, Behm M, Bartok O, Ashwal-Fluss R, Herzog M, Schreyer L, Papavasileiou P, Ivanov A, Öhman M, Refojo D, Kadener S, Rajewsky N (2014) Circular RNAs in the mammalian brain are highly abundant, conserved, and dynamically expressed. Mol Cell 58(5):870–885. https://doi.org/10.1016/j.molcel.2015.03.027
61. Urdánoz-Casado A, Sánchez-Ruiz de Gordoa J, Robles M, Roldan M, Macías Conde M, Acha B, Blanco-Luquin I, Mendioroz M (2023) circRNA from APP gene changes in Alzheimer's disease human brain. Int J Mol Sci 24(5):4308. https://doi.org/10.3390/ijms24054308
62. Lo Ij, Hill J, Vilhjálmsson BJ, Kjems J (2020) Linking the association between circRNAs and Alzheimer's disease progression by multi-tissue circular RNA characterization. RNA Biol 17(12):1789–1797. https://doi.org/10.1080/15476286.2020.1783487
63. Cochran KR, Veeraraghavan K, Kundu G, Mazan-Mamczarz K, Coletta C, Thambisetty M, Gorospe M, De S (2021) Systematic identification of circrnas in Alzheimer's disease. Genes (Basel) 12(8):1258. https://doi.org/10.3390/genes12081258
64. Puri S, Hu J, Sun Z, Lin M, Stein TD, Farrer LA, Wolozin B, Zhang X (2023) Identification of circRNAs linked to Alzheimer's disease and related dementias. Alzheimers Dement 19(8):3389–3405. https://doi.org/10.1002/alz.12960
65. Dube U, Del-Aguila JL, Li Z, Budde JP, Jiang S, Hsu S, Ibanez L, Fernandez MV, Farias F, Norton J, Gentsch J, Wang F, Allegri R, Amtashar F, Benzinger T, Berman S, Bodge C, Brandon S, Brooks W, Buck J, Buckles V, Chea S, Chrem P, Chui H, Cinco J, Clifford J, D'Mello M, Donahue T, Douglas J, Edigo N, Erekin-Taner N, Fagan A, Farlow M, Farrar A,

Feldman H, Flynn G, Fox N, Franklin E, Fujii H, Gant C, Gardener S, Ghetti B, Goate A, Goldman J, Gordon B, Gray J, Gurney J, Hassenstab J, Hirohara M, Holtzman D, Hornbeck R, DiBari SH, Ikeuchi T, Ikonomovic S, Jerome G, Jucker M, Kasuga K, Kawarabayashi T, Klunk W, Koeppe R, Kuder-Buletta E, Laske C, Levin J, Marcus D, Martins R, Mason NS, Maue-Dreyfus D, McDade E, Montoya L, Mori H, Nagamatsu A, Neimeyer K, Noble J, Norton J, Perrin R, Raichle M, Ringman J, Roh JH, Schofield P, Shimada H, Shiroto T, Shoji M, Sigurdson W, Sohrabi H, Sparks P, Suzuki K, Swisher L, Taddei K, Wang J, Wang P, Weiner M, Wolfsberger M, Xiong C, Xu X, Salloway S, Masters CL, Lee JH, Graff-Radford NR, Chhatwal JP, Bateman RJ, Morris JC, Karch CM, Harari O, Cruchaga C (2019) An atlas of cortical circular RNA expression in Alzheimer disease brains demonstrates clinical and pathological associations. Nat Neurosci 22(11):1903–1912. https://doi.org/10.1038/s41593-019-0501-5

66. Ravanidis S, Bougea A, Karampatsi D, Papagiannakis N, Maniati M, Stefanis L, Doxakis E (2021) Differentially expressed circular RNAs in peripheral blood mononuclear cells of patients with Parkinson's disease. Mov Disord 36(5):1170–1179. https://doi.org/10.1002/mds.28467
67. Li M, Pehar M, Liu Y, Bhattacharyya A, Zhang SC, O'Riordan KJ, Burger C, D'Adamio L, Puglielli L (2015) The amyloid precursor protein (APP) intracellular domain regulates translation of p44, a short isoform of p53, through an IRES-dependent mechanism. Neurobiol Aging 36(10):2725–2736. https://doi.org/10.1016/j.neurobiolaging.2015.06.021
68. Pehar M, Ko MH, Li M, Scrable H, Puglielli L (2014) P44, the "longevity-assurance" isoform of P53, regulates tau phosphorylation and is activated in an age-dependent fashion. Aging Cell 13(3):449–456. https://doi.org/10.1111/acel.12192
69. Dickson DW (2018) Neuropathology of Parkinson disease. Parkinsonism Relat Disord 46:S30–S33. https://doi.org/10.1016/j.parkreldis.2017.07.033
70. Pridgeon JW, Olzmann JA, Chin LS, Li L (2007) PINK1 protects against oxidative stress by phosphorylating mitochondrial chaperone TRAP1. PLoS Biol 5(7):e172. https://doi.org/10.1371/journal.pbio.0050172
71. Grigoruţă M, Dagda RK, Díaz-Sánchez ÁG, Martínez-Martínez A (2020) Psychological distress and lack of PINK1 promote bioenergetics alterations in peripheral blood mononuclear cells. Sci Rep 10(1):9820. https://doi.org/10.1038/s41598-020-66745-9
72. Song S, Jang S, Park J, Bang S, Choi S, Kwon KY, Zhuang X, Kim E, Chung J (2013) Characterization of PINK1 (PTEN-induced putative kinase 1) mutations associated with Parkinson disease in mammalian cells and drosophila. J Biol Chem 288(8):5660–5672. https://doi.org/10.1074/jbc.M112.430801
73. Heeman B, Van den Haute C, Aelvoet SA, Valsecchi F, Rodenburg RJ, Reumers V, Debyser Z, Callewaert G, Koopman WJH, Willems PHGM, Baekelandt V (2011) Depletion of PINK1 affects mitochondrial metabolism, calcium homeostasis and energy maintenance. J Cell Sci 124(7):1115–1125. https://doi.org/10.1242/jcs.078303
74. Lin W, Wadlington NL, Chen L, Zhuang X, Brorson JR, Kang UJ (2014) Loss of PINK1 attenuates HIF-1α induction by preventing 4E-BP1-dependent switch in protein translation under hypoxia. J Neurosci 34(8):3079–3089. https://doi.org/10.1523/JNEUROSCI.2286-13.2014
75. Chen W, Hou C, Wang Y, Hong L, Wang F, Zhang J (2022) Circular RNA circTLK1 regulates dopaminergic neuron injury during Parkinson's disease by targeting miR-26a-5p/DAPK1. Neurosci Lett 782:136638. https://doi.org/10.1016/j.neulet.2022.136638
76. Hanan M, Simchovitz A, Yayon N, Vaknine S, Cohen-Fultheim R, Karmon M, Madrer N, Rohrlich TM, Maman M, Bennett ER, Greenberg DS, Meshorer E, Levanon EY, Soreq H, Kadener S (2020) A Parkinson's disease CircRNAs resource reveals a link between circSLC8A1 and oxidative stress. EMBO Mol Med 12(11):e13551. https://doi.org/10.15252/emmm.20201355
77. Faye MD, Holcik M (2015) The role of IRES trans-acting factors in carcinogenesis. Biochim Biophys Acta Gene Regul Mech 1849(7):887–897. https://doi.org/10.1016/j.bbagrm.2014.09.012
78. Pflaum J, Schlosser S, Müller M (2014) P53 family and cellular stress responses in cancer. Front Oncol 4:285. https://doi.org/10.3389/fonc.2014.00285
79. Issaeva N (2019) P53 signaling in cancers. Cancers (Basel) 11(3):332. https://doi.org/10.3390/cancers11030332
80. Halaby M-J, Harris BRE, Miskimins WK, Cleary MP, Yang D-Q (2015) Deregulation of internal ribosome entry site-mediated p53 translation in cancer cells with defective p53 response to DNA damage. Mol Cell Biol 35(23):4006–4017. https://doi.org/10.1128/MCB.00365-15
81. Ray PS, Grover R, Das S (2006) Two internal ribosome entry sites mediate the translation of p53 isoforms. EMBO Rep 7(4):404–410. https://doi.org/10.1038/sj.embor.7400623
82. Candeias MM, Powell DJ, Roubalova E, Apcher S, Bourougaa K, Vojtesek B, Bruzzoni-Giovanelli H, Fåhraeus R (2006) Expression of p53 and p53/47 are controlled by alternative mechanisms of messenger RNA translation initiation. Oncogene 25(52):6936–6947. https://doi.org/10.1038/sj.onc.1209996
83. Wang H, Zhu Y, Hu L, Li Y, Liu G, Xia T, Xiong D, Luo Y, Liu B, An Y, Li M, Huang Y, Zhong Q, Zeng M (2020) Internal ribosome entry sites mediate cap-independent translation of Bmi1 in nasopharyngeal carcinoma. Front Oncol 10:1678. https://doi.org/10.3389/fonc.2020.01678
84. Silva J, García V, García JM, Peña C, Domínguez G, Díaz R, Lorenzo Y, Hurtado A, Sánchez A, Bonilla F (2007) Circulating Bmi-1 mRNA as a possible prognostic factor for advanced breast cancer patients.

Breast Cancer Res 9(4):R55. https://doi.org/10.1186/bcr1760
85. Chen TM, Lai MC, Li YH, Chan YL, Wu CH, Wang YM, Chien CW, Huang SY, Sun HS, Tsai SJ (2019) hnRNPM induces translation switch under hypoxia to promote colon cancer development. EBioMedicine 41:299–309. https://doi.org/10.1016/j.ebiom.2019.02.059
86. Holmes B, Lee J, Landon KA, Benavides-Serrato A, Bashir T, Jung ME, Lichtenstein A, Gera J (2016) Mechanistic target of rapamycin (mtor) inhibition synergizes with reduced internal ribosome entry site (ires)-mediated translation of cyclin d1 and c-myc mrnas to treat glioblastoma. J Biol Chem 291(27):14146–14159. https://doi.org/10.1074/jbc.M116.726927
87. Vaklavas C, Grizzle WE, Choi H, Meng Z, Zinn KR, Shrestha K, Blume SW (2016) IRES inhibition induces terminal differentiation and synchronized death in triple-negative breast cancer and glioblastoma cells. Tumour Biol 37(10):13247–13264. https://doi.org/10.1007/s13277-016-5161-4
88. Meganck RM, Liu J, Hale AE, Simon KE, Fanous MM, Vincent HA, Wilusz JE, Moorman NJ, Marzluff WF, Asokan A (2021) Engineering highly efficient backsplicing and translation of synthetic circRNAs. Mol Ther Nucleic Acids 23:821–834. https://doi.org/10.1016/j.omtn.2021.01.003
89. Wesselhoeft RA, Kowalski PS, Anderson DG (2018) Engineering circular RNA for potent and stable translation in eukaryotic cells. Nat Commun 9(1):2629. https://doi.org/10.1038/s41467-018-05096-6
90. Chen R, Wang SK, Belk JA, Amaya L, Li Z, Cardenas A, Abe BT, Chen C-K, Wender PA, Chang HY (2023) Engineering circular RNA for enhanced protein production. Nat Biotechnol 41(2):262–272. https://doi.org/10.1038/s41587-022-01393-0
91. Liu X, Zhang Y, Zhou S, Dain L, Mei L, Zhu G (2022) Circular RNA: an emerging frontier in RNA therapeutic targets, RNA therapeutics, and mRNA vaccines. J Control Release 348:84–94. https://doi.org/10.1016/j.jconrel.2022.05.043
92. Sehta P, Wilhelm A-M, Lin S-J, Urman MA, MacNeil HA, Fuchs G (2022) A split NanoLuc reporter quantitatively measures circular RNA IRES translation. Genes (Basel) 13(2):357. https://doi.org/10.3390/genes13020357
93. Kameda S, Ohno H, Saito H (2023) Synthetic circular RNA switches and circuits that control protein expression in mammalian cells. Nucleic Acids Res 51(4):e24. https://doi.org/10.1093/nar/gkac1252
94. Cui J, Zhang L, Zhang Z, Luo X, Liu Y, Li C, Huang W, Zou L, Yu X, Xiao F (2023) A precise and efficient circular RNA synthesis system based on a ribozyme derived from Tetrahymena thermophila. Nucleic Acids Res 51(14):e78. https://doi.org/10.1093/nar/gkad554
95. Lee KH, Kim S, Song J, Han SR, Kim JH, Lee S-W (2023) Efficient circular RNA engineering by end-to-end self-targeting and splicing reaction using Tetrahymena group I intron ribozyme. Mol Ther Nucleic Acids 33:587–598. https://doi.org/10.1016/j.omtn.2023.07.034
96. Ning H, Liu G, Li L, Liu Q, Huang H, Xie Z (2023) Rational design of microRNA-responsive switch for programmable translational control in mammalian cells. Nat Commun 14(1):7193. https://doi.org/10.1038/s41467-023-43065-w
97. Martínez-Salas E, Lozano G, Fernandez-Chamorro J, Francisco-Velilla R, Galan A, Diaz R (2013) RNA-binding proteins impacting on internal initiation of translation. Int J Mol Sci 14(11):21705–21726. https://doi.org/10.3390/ijms141121705
98. Ho-Xuan H, Glažar P, Latini C, Heizler K, Haase J, Hett R, Anders M, Weichmann F, Bruckmann A, Van den Berg D, Hüttelmaier S, Rajewsky N, Hackl C, Meister G (2020) Comprehensive analysis of translation from overexpressed circular RNAs reveals pervasive translation from linear transcripts. Nucleic Acids Res 48(18):10368–10382. https://doi.org/10.1093/nar/gkaa704

Part IV

Molecular Mechanisms and Gene Regulation of Circular RNAs

Circular RNAs Binding to RNA-Binding Proteins

10

Mandakini Singh, Kirthik Roshan, Srinivasan Muthuswamy, Ajay Kumar, and Santosh Kumar

Abstract

High-throughput RNA sequencing yields novel noncoding RNAs called circular RNAs (circRNAs). They are extensively distributed across the transcriptomes of humans and other metazoans. Research has demonstrated their critical functions in multiple cellular developmental processes and diseases, including cancer, cognitive disorders, and neurodegenerative disorders (NDDs). Aside from this, circRNAs carry out cellular functions by associating with RNA-binding proteins (RBPs) that are unique for each cell type. Given that they are a vast family of proteins, RBPs regulate gene expression through various processes like stabilization, localization, splicing, nuclear export, and RNA translation. The impact of circRNAs and the RBP complex on circRNA synthesis has become more apparent in recent years. As per new findings, circRNAs interact with RBPs and modify the way in which other RNAs and proteins operate. Numerous diseases are also linked to these relationships. Because of this, research regarding the circRNA and RBP relationship is crucial to comprehend the origins of a number of disorders. This chapter describes the newly discovered mechanisms underlying the interactions between circRNAs and RBPs as well as their functional significance in both illnesses and biological processes.

Keywords

CircRNA · RBP · MicroRNA sponging · Protein scaffold · Translatable circRNA

1 Introduction

Circular RNAs (circRNAs) are newly found, noncoding, endogenous RNAs and feature single-stranded, covalently closed loop-like structures. They are present in most of the species [1, 2]. Initially, they were described as viroids, which are plant pathogens. Later, they were discovered using electron microscopy for the first time from human HeLa cells [3, 4]. The majority of metazoans, including humans, have been shown to carry circRNAs, according to advances in bioinformatics, in terms of high-throughput RNA sequencing [5–8]. They have originated from the pre-mRNA

Authors Mandakini Singh and Kirthik Roshan are equally contributed to this chapter.

M. Singh · K. Roshan · S. Muthuswamy
S. Kumar (✉)
Department of Life Science, National Institute of Technology (NIT) Rourkela, Odisha, India
e-mail: kumarsantosh@nitrkl.ac.in

A. Kumar
Department of Zoology, Institute of Science, Banaras Hindu University, Varanasi, Uttar Pradesh, India

J. Xiao (ed.), *Advances in Circular RNAs*, Advances in Experimental Medicine and Biology 1485,
https://doi.org/10.1007/978-981-96-9428-0_10

by back splicing, in which 3′-end of the pre-mRNA is ligated to its own 5′-end to form covalently closed circular structure. They create covalently closed loop-like structures and lack polyadenylation at the 3′-end and 5′–3′ polarity, unlike mRNAs terminating with a 3′ polyadenylated tail and 5′ caps. This prevents RNase R-mediated degradation [8–10]. Circular RNAs were formerly considered to be the errors in splicing with "scrambled exon" patterns [2, 11]. However, according to multiple studies, they are associated with a variety of purposes, including acting as protein scaffolds. Nonetheless, circRNA has been found to have a variety of tasks, such as translating into polypeptides and sponging microRNAs (miRNAs) [12–16].

Because of their longer stability as compared to other RNAs, circRNAs are huge in number than their canonical equivalents in linear forms derived from the same genes. research regarding circRNA has demonstrated that they exhibit tissue and developmental stage specific expression and are extensively present across diverse species [17, 18]. According to Memczak et al. (2013), they are frequently identified as gene regulators at the transcriptional or posttranscriptional stage [19]. Primarily, RNA types include tRNA intronic circular RNAs (tricRNAs), exonic–intronic circRNAs (EIciRNAs), circular exonic circRNAs (ecircRNAs), and circular intronic RNAs (ciRNAs). ecircRNAs are cytoplasmic circRNAs that perform microRNA (miRNA) sponging through miRNA response elements (MREs). In contrast, nucleus-residing ciRNAs and EIciRNAs may modulate gene transcription and are plentiful inside the nucleus [17, 20–22]. A time- and tissue-specific conserved tRNA motif serves as the basis for the synthesis of tricRNA inside the nucleus, and these circRNAs are dispersed throughout the cytoplasm [23–27]. Furthermore, these RNAs have been found to be connected to a myriad of illnesses, such as viral infections, stress, osteoarthritis, diabetes, cancer, and cardiovascular disease [28, 29].

Several recent literatures have focused on the interconnection between circRNA and RNA-binding proteins to comprehend circRNA-regulatory mechanisms. Following the discovery of backsplicing events, the regulatory connections between RBPs and circRNAs are determined by cross-linking and immunoprecipitation sequencing (CLIP-seq) experimental data analysis [30]. By interconnecting with RBPs, these circRNAs create the ribonucleoprotein (RNP) complex. Research has found that the circRNA–RBP interaction is a crucial regulator of circRNA biogenesis [31–34]. These RBPs and circRNAs control how other proteins interact with downstream target mRNAs [28]. Therefore, more research into the circRNA–RBP interaction is required for deciphering the molecular pathways accounting for the circRNA functions. The etiology of various diseases may be greatly illuminated by studying the circRNA–RBP interaction since numerous illnesses are linked to circRNAs.

2 Functions of CircRNAs Through Interaction with RBPs

CircRNAs play physiological roles in controlling immunological responses, cell proliferation, and carcinogenesis in addition to their biological involvement in gene transcription and translation. They have the ability to regulate biological processes, including protein expression, as well as function as miRNA sponges and protein modulators [35]. CircRNAs function as protein recruiters, protein scaffolds, protein sponges or decoys, and miRNA sponges and also affect the target mRNA's fate [36]. For these aforementioned processes, the interaction of circRNAs with regulatory RBPs is crucial. In addition, some circRNAs carry an internal ribosome entry site (IRES) and undergo translation [37].

2.1 Translation of CircRNAs

Despite being noncoding RNAs, circRNAs feature entire exons, indicating that they may have a function similar to translation [38–41]. 39 have found 37 circRNAs with at least one unique ribosome footprinting (RFP) read by utilizing

RFP datasets to search the circular RNA open reading frame (ORF). This suggests that the translating ribosomes and circRNAs are related [39, 41]. The translation of circRNA does not depend on the common regulatory proteins of translation including eukaryotic translation initiation factor 4E-binding protein (4E-BP) because this process is usually followed by 5′-capping [38, 40]. The first study in 1995 implied that the IRES of the encephalomyocarditis virus is crucial for the translation of synthetic circRNAs [12, 42]. Furthermore, Yang et al. (2017) have reported that the prevalent RNA modification, viz., N6-methyladenosine (m6A), stimulates circRNA translation inside human cells, implying a function for the proteins generated from circRNA in response to external stress. The eukaryotic translation initiation factor 4 gamma 2 (eIF4G2) and the reader of m6A modification, YTH N6-methyladenosine RNA-binding protein F3 (YTHDF3) initiate the process, methyltransferase-like 3/14 (METTL3/14) promotes it, and demethylase fat mass and obesity-associated protein (FTO) inhibits it [42, 43]. Through m6A reader YTHDF3 depletion, researchers have observed a notable reduction of circRNA translation. It is proposed that YTHDF3 plays a role in attracting eIF4G2 to m6A-containing circRNAs and thereafter translating them (Fig. 10.1) [41, 43].

Several studies have described that circular RNAs are translated under different physiological conditions, including cellular stress in various cellular compartments, and act as regulatory protein products within the cell [39, 44, 45]. One example of the protein-coding circRNA in eukaryotic organisms is CircZNF609, which undergoes translation after a cap-independent splicing process. It is found to be connected to myogenesis in mice [44, 46, 47]. Its translation is initiated by employing zinc finger protein 609 (ZNF609) mRNA start codon and is terminated by the stop codon, which is generated upon circularization. CircZNF609 is found to be upregulated in the differentiated myoblast obtained from Duchenne muscular dystrophy patients [44, 48]. Pamudruti et al. have provided evidence that circMBL3 could translate fly head proteins obtained from mass spectrometric analysis of an MBL3 immunoprecipitate [39, 47].

Circular RNA SHPRH is generated from the SNF2 histone linker PHD RING helicase (*SHPRH*) gene [49]. Backsplicing of the *SHPRH* gene, at exons 26 to 29, results in this 400-nucleotide circRNA. The "UGAUGA" motif in this gene has both an overlapping start codon (AUG) and a stop codon (UGA). circSHPRH, in eukaryotic cells, is the first circular RNA with such features for protein translation. This circRNA codes for the SHPRH-146aa protein, which is 146 amino acids in length. It exhibits suppression of tumorigenic activity in glioma. Further studies have suggested that SHPRH-146aa prevents the proteasomal degradation of the SHPRH protein followed by ubiquitination [50].

Another circRNA, named circFBXW7, originated from the gene F-box and WD repeat domain-containing 7 (FBXW7). This gene comprises 620 nucleotides and is a tumor suppressor E3 ligase. Backsplicing of this gene, from the region exon 3 to exon 4, results in the synthesis of circFBXW7. This circular RNA codes for a peptide called FBXW7-185aa, which is 185 amino acids in length. It is anticipated to be involved in tumor suppressing in glioma [34]. It competitively binds to the deubiquitinating enzyme ubiquitin specific peptidase 28 (USP28), which has an antagonistic effect on the cellular Myc (c-Myc) stabilization induced by USP28. Another work has reported that FBXW7-185aa is able to regulate the growth and further development of breast carcinoma, which is of triple-negative type. This protein induces degradation of c-Myc and enhances the abundance of circFBXW7 in order to prevent cancer cell proliferation and further development [50, 51].

Backsplicing of the long intergenic noncoding RNA p53-induced transcript (LINC-PINT) leads to the emergence of circLINC-PINT. This circular RNA is derived from exon 2 of LINC-PINT upon circularization. It is abundantly present inside the nucleus and codes for peptide PINT87aa, comprising 87 amino acids [49]. This protein executes the role of a tumor suppressor during cell proliferation and tumorigenesis in

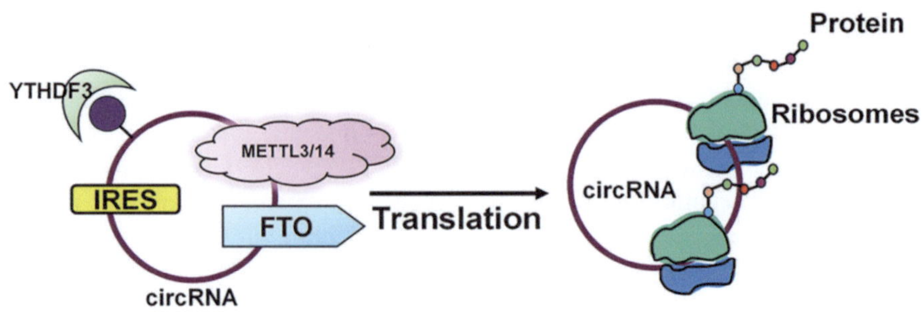

Fig. 10.1 m6A modification, along with IRES, is responsible for the translation of circRNAs

certain cancers. In order to prevent the transcription of multiple oncogenes, it can interact with the complex, viz., RNA polymerase II-associated factor 1 homolog (PAF1), and also assures the proper nuclear localization of the PAF1 complex.

Some studies have indicated a significant association between hepatocellular carcinoma (HCC) and the pathway of Wnt/β-catenin signaling through translation of circβ-catenin. This signaling pathway is initiated by degradation of β-catenin after its glycogen synthase kinase 3-beta (GSK3β)-mediated phosphorylation [52, 53]. circβ-catenin is an 1129-nucleotide circRNA molecule generated from the backsplicing of exons 2 to 7 of the catenin beta 1 (*CTNNB1*) gene. As per circBase, among 24 circRNAs originating from the *CTNNB1* gene, only circβ-catenin is found to be linked to HCC [54].

Zheng et al. have reported that backsplicing of gene, protein phosphatase 1 regulatory subunit 12A (PPP1R12A) gives rise to a new circRNA, circPPP1R12A [55]. This circRNA has an open reading frame (ORF) of 216 nucleotides and codes for circPPP1R12A-73aa, a 73 amino acid peptide. Just like circSHPRH, it also has overlapping start and stop codons for protein translation. circPPP1R12A is highly expressed in colon cancer showing tumor suppressor activity. Studies on the functional aspect of circPPP1R12A-73aa have described that it initiates the pathway of Hippo-YAP signaling and elevates the growth along with further development of colon cancer cells [50].

A circRNA, named circAKT3, emerges from the AKT serine/threonine kinase 3 (*AKT3*) gene. This RNA is transcribed from the *AKT3* gene region containing exons 3 to 7 and is composed of 524 nucleotides. It encodes the AKT3-174aa protein with a length of 174 amino acids. Analysis of circAKT3 expression has revealed that AKT3-17aa has the potential to exhibit tumor-suppressing activity. This protein, as a molecular decoy, restricts AKT3-thr308 phosphorylation by interacting with phosphorylated 3-phosphoinositide-dependent kinase 1 (PDK1). This phenomenon shows the negative regulation mediated by AKT3-174aa protein while strengthening the PI3K/AKT signaling [56].

2.2 Protein Sponging and Decoying

Acting as protein sponges, circRNAs control gene expression by including binding sites for one or more proteins. For instance, circular RNAs are stably linked to the RNA polymerase II (RNA pol II) enzyme and argonaute (AGO) proteins [19, 25, 57]. Because of their inherent stability, they can enable stable contact and the binding of numerous proteins. For instance, circular RNA Forkhead box O 3 (circFoxo3), forms the triad complex consisting of circFoxo3, p21, and CDK2 by interacting with cyclin-dependent kinase 2 (CDK2) and p21 (also known as cyclin-dependent kinase inhibitor 1 or CDK-interacting protein 1). As a result, it is possible to halt both the biologi-

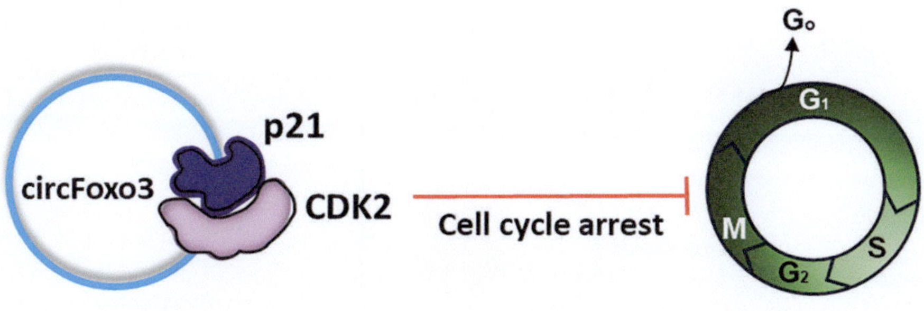

Fig. 10.2 Protein decoying by circRNAs to regulate RBP-dependent functions. The cell cycle progression is inhibited by the circFoxo3–p21–CDK2 ribonucleoprotein complex.

cal impacts of protein CDK2 and advancement of cell cycle (Fig. 10.2) [14]. Muscleblind (MBL) protein is sequestered by circMBL. This circRNA also prevents the interaction between its parent mRNA and MBL protein [58, 59].

CircRNAs, as protein sponges, can form ribonucleoprotein complexes with certain proteins by leveraging particular areas. RBPs have a broad effect on the life cycle of RNA transcripts, such as miRNAs and mRNAs, influencing their production as well as destruction. Specifically, circRNAs can interact with RBPs, which have RNA recognition and binding domains for their RNA targets [60, 61]. By interacting with translational activators, circRNAs can control how their corresponding mRNA is translated. circPABPN1, originating from pre-mRNA poly(A)-binding protein nuclear 1 (PABPN1), interactswith human antigen R (HuR). PABPN1 mRNA translation is suppressed by the substantial interaction of HuR with circPABPN1, which stops HuR from associating with mRNA PABPN1 (Fig. 10.3) [60, 62]. Another example of circRNA exhibiting the protein adsorption phenomenon is circRIF3J, which is an EIciRNA. It increases the parent gene transcription via connecting RNA pol II with the U1 small nuclear ribonucleoprotein (U1snRNP) complex. The relation of circRNAs with several RBPs, including Quaking (QKI), argonaute, IMP3 (also called IGF2BP3, insulin-like growth factor 2 mRNA-binding protein 3), and others, has become conspicuous by additional lines of evidence [31, 63–66].

H19, an imprinted circRNA, is a key element in lipid metabolism. The overexpressed H19 provides protection against elevated insulin sensitivity and obesity [67]. However, the interaction between polypyrimidine tract-binding protein 1(PTBP1) RBP and H19 alters lipid homeostasis in the liver [68, 69].

Studies have detected Cia-cGAS circular RNA from the nucleus of long-term hematopoietic stem cells (LT-HSCs). It antagonizes cyclic guanosine monophosphate-adenosine monophosphate (cGMP-AMP) synthase (cGAS)) activity [70]. The cGAS, as a double-stranded DNA cytosolic sensor, triggers stimulators of the interferon gene (*STING*) pathway in the bone marrow by synthesizing cGAMP. This results in type I interferon production associated with antiviral and anticancer activities. Under homeostatic conditions, Cia-cGAS downregulates cGAS and reduces type I interferon production, made possible by the sequestration of nuclear-localized cGAS from self-DNA by Cia-cGAS. As a result, LT-HSCs are protected from cGAS-mediated exhaustion [71]. Furthermore, studies have disclosed that a novel circular RNA XPO1, originating from the *XPO1* gene, promotes the development of lung adenocarcinoma (LUAD). It is also capable of enhancing CTNNB1 mRNA's stability by binding to IGF2BP1. CTNNB1 mRNA is a key regulator of tumorigenesis and malignant progression in lung cancer [72, 73].

Some human circRNAs are either processed as nuclear intron lariats or as the by-products of backsplicing (with retained introns). Their cellular localization speculates their potential functions. It has been observed that nuclear-localized circRNAs are associated with transcriptional

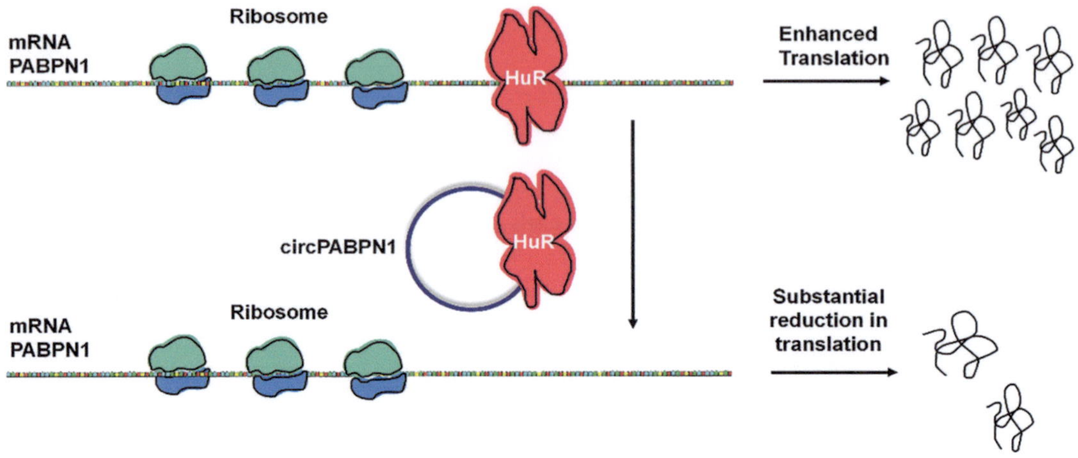

Fig. 10.3 CircRNA sponging of translational activators from mRNA reduces the efficiency of protein synthesis. By harboring HuR, circPABPN1 seems to suppress PABPN1 mRNA translation

regulation. For instance, EIciPAIP2 and EIciEIF3J are two exonic–intronic circular RNAs that improve their parental gene transcription by connecting with U1 snRNP. Exhaustion of these RNAs reduces the transcription levels of the polyadenylate-binding protein-interacting protein 2 (*PAIP2*) and eukaryotic translation initiation factor 3 subunit J (*EIF3J*) genes. This discovery reveals a unique function of nuclear-localized circRNAs: they modulate gene expression inside the nucleus [25, 74, 75].

In tumor protein p53 (p53) mutant cells, the circular RNA cyclin B1 (CCNB1) can interact with H2AX and suppress the process of tumorigenesis in mutant p53 cells. This study has shown that circ-CCNB1 can counteract the effects of p53 mutations in breast cancer as a result of the interaction between CCNB1 and CDK1 proteins [76]. AGO2 is a crucial element in the formation of the miRNA-induced silencing complex. CircAGO2, an intronic circular RNA derived from the *AGO2* gene, makes a connection with RBP HuR and facilitates HuR-mediated repression of the functions mediated by the AGO2–miRNA complex in gastric cancer (GC) [77].

Researchers have described differentially expressed 169 circRNAs, obtained from the sequencing of circRNAs present in pancreatic ductal adenocarcinoma (PDAC) cells. They have validated that Forkhead box K2 (FOXK2) circRNA is remarkably upregulated in both primary tumors and PDAC cells. CircFOXK2 promotes clonogenic ability, invasion, cell growth, and metastasis in PDAC. It also sponges miR-942 and promotes the expression of paired box 6 (PAX6), ankyrin 1 (ANK1), and glial cell-derived neurotrophic factor (GDNF). Further circRNA-pulldown and mass spectrometric analyses of circFOXK2 have yielded 94 interacting RBPs linked to myriads of molecular processes comprising mRNA splicing, cell adhesion, and structural molecular activity. Some researchers have illustrated that circFOXK2, by forming a complex with Y-box-binding protein 1 (YBX1) and heterogeneous nuclear ribonucleoprotein K (hnRNPK), elevates oncogenic protein expression, e.g., pyridoxal kinase (PDXK) and the NUF2 component of the NDC80 kinetochore complex in PDAC cells [78].

An interaction between miRNA effector argonaute (AGO) and circRNAs can degrade the function of AGO by binding to circRNAs such as Cerebellar Degeneration-Related protein 1 antisense (CDR1as) and sex determining region Y (SRY), as AGO is usually associated with inhibiting mRNA translation [19, 79]. The attachment of circPAIP2 with the RNA polymerase II complex alters the enzyme activity, which, in turn, affects the final transcriptional regulation [74, 80].

CircScrib570 is an antisense circRNA, which is associated with the suppression of mRNA translation and pre-mRNA splicing. It assists in cancer cell growth and development through upregulating the expression of E-cadherin and downregulating vimentin and N-cadherin expression [81]. Wang et al. have illustrated that circACTN4, transcribed from the α-Actinin-4 (*ACTN4*) gene, upregulates myelocytomatosis-protein arginine *N*-methyltransferase (MYC) mRNA transcription and enhances the expression of certain proteins associated with cell cycle, e.g., cyclin D2 (CCND2) and cyclin-dependent kinase 4 (CDK4). This leads to breast cancer development and its metastasis, especially after the activation of the proto-oncogene, MYC [82, 83].

Circ0005276 has been found to interact with the RBP fused in sarcoma (FUS) in prostate cancer (PCa). Interestingly, FUS and circ0005276 have already been investigated for their role in regulating the transcription of X-linked inhibitor of apoptosis protein (XIAP). The rescue assays were performed by researchers in order to denote the connection of XIAP with FUS or circ0005276-mediated progression of PCa. These studies have revealed that XIAP, FUS, and circ0005276 are interconnected in PCa progression. Circ0005276 and FUS upregulate XIAP expression, which leads to tumorigenesis [84].

Studies have identified circDLC1, which is derived from the exons 14 to 16 of the deleted in liver cancer 1 (*DLC1*) gene. It acts as a KIAA1429-regulated circRNA. KIAA1429 is an m6A methyltransferase. circDLC1 is closely associated with prognosis of liver cancer. In addition, this circRNA also intervenes in the interaction between mRNA matrix metallopeptidase 1 (MMP1) and HuR protein in HCC cells by directly interacting with RBP HuR. This leads to the progression of HCC pathogenesis [85].

circDNMT1 activates autophagy of breast cancer cells by promoting the translocation of ARE/poly(U)-binding/degradation factor 1 (AUF1) and p53 into the nucleus [86]. Downregulating p53 by circDNMT1 increases cellular autophagy, leading to cell growth, survival, and tumor growth. AUF1 has been reported to bind to the 3′-untranslated region (UTR) of DNA methyltransferase 1(DNMT1) mRNA and boost its degradation. Translocation of AUF1 into the nucleus, by circDNMT1, leads to increased stability and translation of DNMT1 mRNA. This ultimately promotes autophagy by inhibiting p53 transcription. CircMTO1 halts activation of Eg5 (also called kinesin family member 11 or KIF11) by sequestering tumor necrosis factor (TNF) receptor-associated factor 4 (TRAF4) in breast carcinoma. This leads to a decline in the viable population of cancer cells [87].

Another example is a circular antisense noncoding RNA in the INK4 locus (circANRIL). It interacts with peccadillo homolog 1 (PES1), a crucial assembly factor for the 60S-ribosomal subunit. This interaction suppresses the biogenesis of ribosomes and exonuclease-induced pre-rRNA processing in human macrophages and vascular smooth muscle cells [35, 88]. One of the research works has reported that the interaction of proteins with other molecules is intervened by circANRIL as it competitively binds to these proteins. CircANRIL originates from the antisense noncoding RNA in the INK4 Locus (*ANRIL*) gene localized in the region of chromosome 9p21. Exon 7 of this gene is covalently attached to its exon 5, and circANRIL is generated. This circular RNA binds to PES1 and influences the biogenesis process of ribosomes [88]. One of the functions of circANRIL is atheroprotection in coronary artery diseases. PES1, one of the constituents of the PeBoW complex, has been reported to have strong binding with circANRIL [89]. As a result, an interruption occurs in the construction of the PeBoW complex by its components. This leads to the failure of the functions assisted by this complex. Consequently, induction of nucleolar stress and p53 accumulation leads to an elevation in the apoptotic process and a decline in cell proliferation that restrains the development of atherosclerosis [90]. In this way, circANRIL contributes to atheroprotection by controlling the pathway associated with atherogenesis [88, 91].

Apoptosis is also regulated by circRNAs binding to RBPs. Studies have revealed that circFoxo3 stimulates the interplay between p53 and murine double minute 2 (MDM2) in order to execute p53 degradation. This circRNA further binds to the protein and represses the MDM2-mediated ubiquitination of Foxo3. This triggers the p53-upregulated modulator of apoptosis (*PUMA*) gene expression and influences apoptosis in breast carcinoma [92, 93]. Another research related to circRNA has suggested that circUBAP2 has been found to sponge miR-143 and upregulate the activity of B-cell lymphoma 2 (Bcl-2) to suppress osteosarcoma cell apoptosis [94].

Some studies have reported that microRNA miR-7 is capable of suppressing the advancement of cancer cells by inhibiting the phosphatidylinositol-4,5-bisphosphate 3-kinase catalytic subunit delta (PIK3CD) and cyclin E1 (CCNE1) activities in hepatocellular carcinoma (HCC) [95, 96]. Some workers have indicated that circ100284 is found in skin cancer induced by arsenic, promoting cancer cell cycle progression by binding to miR-217. This circRNA enhances the action of cyclin D1 and zeste homolog 2 (EZH2), leading to metastatic transformation [97, 98].

The ciRNAs, residing inside the nucleus, perform as the transcriptional regulators of the host genes through their interaction with the complex of RNA Pol II. The circular RNAs ci-sirt7 and ci-ankrd52 promote the transcription of sirtuin 7 (*SIRT7*) and ankyrin repeat domain 52 (*ANKRD52*) genes, respectively. Depleting ci-sirt7 and ci-ankrd52 decreases the transcriptional efficiency of their parental genes [25]. EicircRNAs interact with their host genes through their promoter region and stay in the nucleus. They interact with the complex of U1 snRNA–U1A–U1C and, subsequently, with RNA Pol II in order to elevate the transcriptional efficiency [99, 100].

2.3 Protein Scaffolds

Different types of protein-binding sites are found on circRNAs. As a result, promoting the interaction with various proteins, they can aid in protein complex stability and protein scaffolding [36]. Furthermore, scaffold building may trigger the protein's nuclear or cytoplasmic translocation through the nuclear pores [101]. For instance, it is found that circNDUFB2 enhances the connection between insulin-like growth factor 2 mRNA-binding proteins (IGF2BPs) and tripartite motif protein 25 (TRIM25), hence facilitating the ubiquitination of IGF2BPs. IGF2BPs are then degraded, resulting in the slow growth of tumors [102]. Additionally, it is reported that circPDE4B operates as a scaffold for resistance to the inhibitors cholinesterase 8A (RIC8A) and MID1 (also called TRIM18, tripartite motif protein 18) in human cartilage samples, facilitating their interaction and resulting in RIC8A's proteasomal degradation and prevention of cartilage deterioration (Fig. 10.4) [103]. Furthermore, in lung cancer, circDCUN1D4 forms a complex containing RBP HuR and encourages the cytoplasmic transport of HuR, preventing the metastasis of cancer cells [38, 40].

Circular RNAs interact with many proteins on a regular basis during their life cycle. By assembling the stress-related proteins hypoxia-inducible factor 1α (HIF1α) and focal adhesion kinase (FAK) with anti-senescence-related proteins like differentiation-1 (ID-1) and E2F tran-

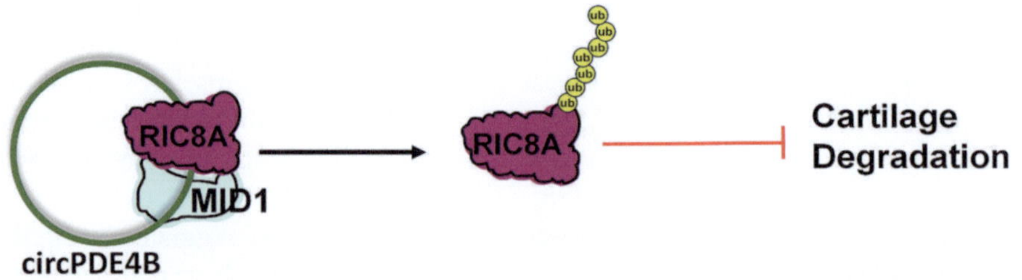

Fig. 10.4 CircRNAs support the stability of protein complexes by serving as scaffolds and helping them execute their functions. CircPDE4B protects RIC8A and MID1 from their inhibitors by performing scaffolding

scription factor 1 (E2F1) in the cytoplasm, circFoxo3 assists in cellular senescence of cardiac cells. This prevents HIF1α translocation into the nucleus and also FAK localization to the mitochondria of stressed cells [104]. One of the examples of protein scaffolding is the development of a triad complex by circACC1 with the regulatory subunits of the AMP-activated protein kinase β and γ (AMPKβ and AMPKγ, respectively). In this way, circACC1 further aids in metabolic adaptation in response to serum deprivation and promotes enzymatic catalysis of the metabolic holoenzyme AMPK [105]. circAMOTL1 is another illustration of this kind of control, and it is ubiquitously expressed in the cardiac tissue of human neonates. It interacts with phosphoinositide-dependent kinase 1 (PDK1) and the AKT1 kinase (also called PKB, protein kinase B) at the same time. It causes PDK1 to phosphorylate AKT1, which facilitates AKT1 transport into the nucleus of cardiomyocytes. This leads to the protection of cardiomyocytes against doxorubicin (Dox)-induced cardiomyopathy [16]. circNfix controls the development of cardiomyocytes in mice. It enhances the interaction between the RBP Y-box binding protein 1(YBX1) and an E3 ubiquitin protein ligase, neural precursor cell-expressed developmentally downregulated 4-like (NEDD4l). This leads to ubiquitination of YBX1 and repression of cardiomyocytes proliferation [106, 107].

According to research by William et al., human breast cancer patient samples have significant expression levels of the circRNA circSKA3. The function of this circRNA was then methodically assessed using gain- and loss-of-function methods. Immunoprecipitation of cell lysates has demonstrated that the tyrosine kinase substrate with 5 SH3 domains (Tks5) and integrin subunit beta 1 (ITGB1), along with other proteins, bind circSKA3. Fluorescence microscopy and cellular fractionation have also been used to verify and validate the co-localization of circSKA3–Tks5–ITGB1 and its location on the cell membrane. This work has shown that interaction among circSKA3, ITGB1, and Tks5 induces the development of invadopodium, which is linked to stromal migration, intravasation, and basement membrane breakdown, all of which are critical for the invasion and metastasis of cancer. ITGB1 is a cell surface receptor that mediates cell adhesion, dissociation, and migration. It has several known ligands. Tks5 participates in the src-dependent invadopodium formation process as an adapter protein. Tumor infiltration and invadopodium development can be induced only by Tks5 phosphorylation. The co-localizing Tks5 and puncta of actin are the indications of invasive porosity, which may be utilized to measure cancer cell invasiveness. Tks5 and ITGB1 do not directly interact with one another, but circSKA3 acts as a bridge for their interaction and facilitates the development of invadopodia. Further research is needed to determine whether circSKA3 also binds other compounds linked to invadopodia, as these organisms include additional molecules [108].

Evidence has indicated that circSMARCA5 may indirectly regulate serine and arginine-rich splicing factor 3 (SRSF3) mRNA expression by tethering SRSF1 within glioblastoma multiforme (GBM) cells. Splicing factors like SRSF1, SRSF3, and PTBP1 are oncogenes and are positively associated with GBM cell migration. SRSF1 forms a self-regulatory network with SRSF3 that itself regulates PTBP1 expression. This leads to the metastasis of GBM cells. However, circSMARCA5 inhibits cancer cell migration by binding to splicing factor serine and arginine-rich splicing factor 1 (SRSF1). In this way, this circRNA modulates the SRSF3 expression mediated by SRSF1 and subsequently inhibits the oncogenic activity of PTBP1 [109].

2.4 Recruiters of Proteins

Through the process of recruitment of proteins, circRNAs may also direct proteins to specific cellular sites [36]. CircMRPS35 has been shown to repress gastric cancer (GC) cell development in recent ex vivo and in vitro analyses. It provides a platform for facilitating the interaction between the KAT7 histone acetyltransferase and the promoter regions of genes like *FOXO3* and *FOXO1*. This causes the H4K5 acetylation at their pro-

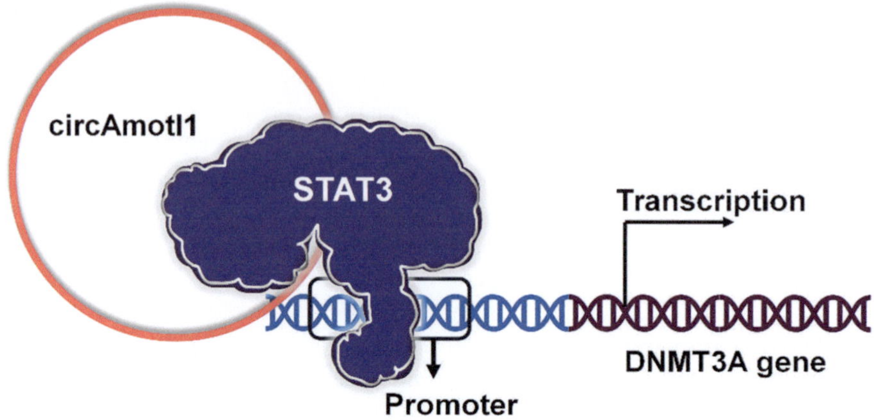

Fig. 10.5 Recruitment of proteins by circRNA, at the gene promoter, leads to the activation of gene transcription. CircAmotl1 recruits the STAT3 protein at the DNMT3A gene promoter to enhance its transcription

moter region, which will control histone modification [110]. CircCwc27 has been reported to have the ability to interact with RBP purine-rich element-binding protein A (Pur-α) directly and retain this protein in the cytoplasm. It inhibits the localization of the protein Pur-α onto the promoters of Alzheimer's disease-related genes, viz., neurotrophic tyrosine kinase receptor type 1, regulatory inhibitor subunit 1B, LIM homeobox 8, amyloid precursor protein, and protein phosphatase 1 [40, 111]. These findings were made from a non-oncological perspective.

Ten-eleven translocation methylcytosine dioxygenase 1 (TET1) is recruited by a circRNA FECR1 to the promoter region of the Friend leukemia virus integration 1 (*FLI1*) gene. This results in active transcription and CpG island demethylation, leading to breast cancer metastasis [112]. Furthermore, circAmotl1 coordinates the recruitment of signal transducer and activator of transcription 3 (STAT3) molecule to the DNA methyltransferase 3 alpha (*DNMT3A*) gene promoter and facilitates the transcription of the *DNMT3A* gene during the wound repair process (Fig. 10.5). Subsequently, circMYBL2 modulates the translation of FMS-like tyrosine kinase-3 (FLT3) mRNA by sequestering the RBP polypyrimidine tract-binding protein 1 (PTBP1), which aids in the mutation of the *FLT3* gene, viz., internal tandem duplication (ITD), in advancement of acute myeloid leukemia (AML) [113, 114].

To prevent matrix metalloproteinase-9 (MMP9) mRNA from being degraded, circ-MMP9 may physically interact with miR-149 and AUF1. This would enhance the expression of MMP9 and provide assistance for the rapid growth and metastatic development of squamous cell carcinoma (OSCC) cells. In animal models, the metastasis of oral squamous cell carcinoma (OSSC) cells is found to be reduced upon knockdown of circ-MMP9. This suggests the therapeutic aspect of circ-MMP9 to intervene in the dramatically increased cell proliferation and migration of OSCC cells. AUF1, being an mRNA destabilizer, is associated with targeting the "(A/U) U (A/U)" motif on the 3′-UTR of MMP9 mRNA in order to degrade this mRNA. However, the interaction between AUF and circ-MMP9 prevents this degradation, which, in turn, enhances the stability of MMP9 mRNA. It is also noticed that there is no contact between miR-149 on the 3′-UTR of MMP9 mRNA and the binding sites of AUF1. This indicates that each protein functions individually to degrade MMP9 mRNA. Other studies have also suggested that circ-DNMT1 can prevent DNMT1 mRNA decay by interconnecting with AUF1, resulting in breast cancer progression [115].

Circ-DONSON is expressed abundantly in the tissues and cell lines of gastric cancer (GC). The direct interaction between the SNF2L subunit of the NURF complex and circ-DONSON GC leads to its association with GC. Additionally, the enrichment of this NURF complex on the SOX4 promoter is reliant on circ-DONSON. Gene expression is regulated by the NURF complex, a crucial element for chromatin remodeling. Researchers have revealed that the reduction of NURF inhibits SOX4 transcription and also reduces the levels of SOX4 mRNA. Therefore, SOX4 signaling in GC is activated by the circ-DONSON/NURF axis [116].

In retinoblastoma (RB) cells and tissues, the circular RNA circ-0075804 has a loop shape, is expressed more often, and can positively control the E2F transcription factor 3 (E2F3) expression in RB. On the other hand, the knockdown of E2F3 has minimal effect on circ-0075804 expression. Furthermore, deficiency of circ-0075804 leads to the inhibition of RB cells and, ultimately, to cell death. HNRNPK is found to attach with circ-0075804 and favorably control the expression of E2F3 in RB cells. Circ-0075804 appears to be a viable target for a study on RB pathogenesis since it enhances the stability of E2F3 and elevates the growth of RB cells by combining HNRNPK [117].

Circ-CUX1 aids in the transactivation process of the Myc-associated zinc finger protein (MAZ) by interacting with the Ewing sarcoma breakpoint region 1/EWS RNA-binding protein 1 (EWSR1). This changes the transcription of the Cut homeobox gene 1 (*CUX1*) and other genes linked to the advancement of neuroblastoma (NB). The inhibition of circ-CUX1–EWSR1 interaction by an inhibitory peptide (EIP-22) and circ-CUX1 knockdown mediated by lentivirus inhibit the aggressiveness, tumorigenesis, and aerobic glycolysis of NB cells [118].

circFNDC3B downregulates E-cadherin expression and takes part in the cell growth and metastatic transformation of gastric cancer (GC). It also upregulates the expression of CD44. After circFNDC3B, CD44 has intricate regulatory systems that affect the growth of cancer. CD44, a cell surface transmembrane protein, promotes cell motility and reacts to the cellular milieu in addition to serving as a cell adhesion molecule. The study conducted by Hong et al. has reported that circFNDC3B, CD44 mRNA, and IGF2BP3 all interact, although the mechanism behind their interaction is unclear [119].

3 Conclusion

CircRNAs have a well-constructed configuration that makes them abundant in both plant and animal species. According to high-throughput sequencing, circRNAs influence transcriptional or posttranscriptional activities by interacting with RBPs. Additionally, they have a connection to a myriad of illnesses, such as autoimmune diseases, cancer, and other anomalies impacting the brain. They have been found to be novel biomarkers of cancer and other serious disorders in addition to controlling the process of carcinogenesis. Therefore, studies on circRNA and RBP interconnection will be useful in exploring the molecular mechanism behind the emergence of various diseases and also in developing efficient treatments against them.

Acknowledgments Mandakini Singh acknowledges the University Grants Commission, India, for providing research fellowship (F. No.16-9 (June 2019)/NET/CSIR). Santosh Kumar is thankful to the Indian Council of Medical Research (File no.: 52/23/2020-BIO/BMS) for providing research funding. Srinivasan Muthuswamy and Kirthik Roshan M acknowledge research support provided by the Indian Council of Medical Research (File No.5/13/48/2022/NCD-III), India.

Competing Financial Interest The authors disclose that they have no known financial conflicting interests.

References

1. Ma Z, Shuai Y, Gao X, Wen X, Ji J (2020) Circular RNAs in the tumour microenvironment. Mol Cancer 19(8):1–22. https://doi.org/10.1186/s12943-019-1113-0
2. Zhou WY, Cai ZR, Liu J, Wang DS, Ju HQ, Xu RH (2020) Circular RNA: metabolism, functions and interactions with proteins. Mol Cancer 19(172):1–19. https://doi.org/10.1186/s12943-020-01286-3
3. Hsu M-T, Coca-Prados M (1979) Electron microscopic evidence for the circular form of RNA in the cytoplasm of eukaryotic cells. Nature 280(5720):339–340. https://doi.org/10.1038/280339a0

4. Sanger HL, Klotz G, Riesner D, Gross HJ, Kleinschmidt AK, Pflanzenvirologie AL, Sanger IH, Riesner D (1976) Viroids are single-stranded covalently closed circular RNA molecules existing as highly base-paired rod-like structures (electron microscopy/endgroup analysis/ultracentrifugation/thermal denaturation). Proc Natl Acad Sci 73(11):3852–3856. https://doi.org/10.1073/pnas.73.11.3852
5. Fan X, Zhang X, Wu X, Guo H, Hu Y, Tang F, Huang Y (2015) Single-cell RNA-seq transcriptome analysis of linear and circular RNAs in mouse preimplantation embryos. Genome Biol 16(148):1–17. https://doi.org/10.1186/s13059-015-0706-1
6. Salzman J, Gawad C, Wang PL, Lacayo N, Brown PO (2012) Circular RNAs are the predominant transcript isoform from hundreds of human genes in diverse cell types. PLoS One 7(2):1–12. https://doi.org/10.1371/journal.pone.0030733
7. Westholm JO, Miura P, Olson S, Shenker S, Joseph B, Sanfilippo P, Celniker SE, Graveley BR, Lai EC (2014) Genome-wide analysis of drosophila circular RNAs reveals their structural and sequence properties and age-dependent neural accumulation. Cell Rep 9(5):1966–1980. https://doi.org/10.1016/j.celrep.2014.10.062
8. Zhao W, Cheng Y, Zhang C, You Q, Shen X, Guo W, Jiao Y (2017) Genome-wide identification and characterization of circular RNAs by high throughput sequencing in soybean. Sci Rep 7(5636):1–11. https://doi.org/10.1038/s41598-017-05922-9
9. Suzuki H, Tsukahara T (2014) A view of pre-mRNA splicing from RNase R resistant RNAs. Int J Mol Sci 15(6):9331–9342. https://doi.org/10.3390/ijms15069331
10. Chen LL, Yang L (2015) Regulation of circRNA biogenesis. RNA Biol 12(4):381–388. https://doi.org/10.1080/15476286.2015.1020271
11. Nigro JM, Cho KR, Fearon ER, Kern SE, Ruppert JM, Oliner JD, Kinzler KW, Vogelstein B (1991) Scrambled exons. Cell 64(3):607–613. https://doi.org/10.1016/0092-8674(91)90244-s
12. Chen C, Sarnow P (1995) Initiation of protein synthesis by the eukaryotic translational apparatus on circular RNAs. Science 268(5209):415–417. https://doi.org/10.1126/science.7536344
13. Denzler R, Agarwal V, Stefano J, Bartel DP, Stoffel M (2014) Assessing the ceRNA hypothesis with quantitative measurements of miRNA and target abundance. Mol Cell 54(5):766–776. https://doi.org/10.1016/j.molcel.2014.03.045
14. Du WW, Yang W, Liu E, Yang Z, Dhaliwal P, Yang BB (2016) Foxo3 circular RNA retards cell cycle progression via forming ternary complexes with p21 and CDK2. Nucleic Acids Res 44(6):2846–2858. https://doi.org/10.1093/nar/gkw027
15. Kristensen LS, Hansen TB, Venø MT, Kjems J (2018) Circular RNAs in cancer: opportunities and challenges in the field. Oncogene 37(5):555–565. https://doi.org/10.1038/onc.2017.361
16. Zeng Y, Du WW, Wu Y, Yang Z, Awan FM, Li X, Yang W, Zhang C, Yang Q, Yee A, Chen Y, Yang F, Sun H, Huang R, Yee AJ, Li RK, Wu Z, Backx PH, Yang BB (2017) A circular RNA binds to and activates AKT phosphorylation and nuclear localization reducing apoptosis and enhancing cardiac repair. Theranostics 7(16):3842–3855. https://doi.org/10.7150/thno.19764
17. Jeck WR, Sorrentino JA, Wang K, Slevin MK, Burd CE, Liu J, Marzluff WF, Sharpless NE (2013) Circular RNAs are abundant, conserved, and associated with ALU repeats. RNA 19(2):141–157. https://doi.org/10.1261/rna.035667.112
18. Salzman J, Chen RE, Olsen MN, Wang PL, Brown PO (2013) Cell-type specific features of circular RNA expression. PLoS Genet 9(9):1–15. https://doi.org/10.1371/journal.pgen.1003777
19. Memczak S, Jens M, Elefsinioti A, Torti F, Krueger J, Rybak A, Maier L, Mackowiak SD, Gregersen LH, Munschauer M, Loewer A, Ziebold U, Landthaler M, Kocks C, Le Noble F, Rajewsky N (2013) Circular RNAs are a large class of animal RNAs with regulatory potency. Nature 495(7441):333–338. https://doi.org/10.1038/nature11928
20. Lasda E, Parker R (2014) Circular RNAs: diversity of form and function. RNA 20(12):1829–1842. https://doi.org/10.1261/rna.047126.114
21. Qiu LP, Wu YH, Yu XF, Tang Q, Chen L, Chen KP (2018) The emerging role of circular RNAs in Hepatocellular carcinoma. J Cancer 9(9):1548–1559. https://doi.org/10.7150/jca.24566
22. Yu B, Shan G (2016) Functions of long noncoding RNAs in the nucleus. Nucleus 7(2):155–166. https://doi.org/10.1080/19491034.2016.1179408
23. Khan SU, Khan MU, Khan MI, Fadahunsi AA, Khan A, Gao S, Bilal M, Li F (2022) Role of circular RNAs in disease progression and diagnosis of cancers: an overview of recent advanced insights. Int J Biol Macromol 220:973–984. https://doi.org/10.1016/j.ijbiomac.2022.08.085
24. Schmidt CA, Giusto JD, Bao A, Hopper AK, Gregory Matera A (2019) Molecular determinants of metazoan tricRNA biogenesis. Nucleic Acids Res 47(12):6452–6465. https://doi.org/10.1093/nar/gkz311
25. Zhang Y, Zhang XO, Chen T, Xiang JF, Yin QF, Xing YH, Zhu S, Yang L, Chen LL (2013) Circular intronic long noncoding RNAs. Mol Cell 51(6):792–806. https://doi.org/10.1016/j.molcel.2013.08.017
26. Singh D, Kesharwani P, Alhakamy NA, Siddique HR (2022) Accentuating CircRNA-miRNA-transcription factors axis: a conundrum in cancer research. Front Pharmacol 12(784801):1–16. https://doi.org/10.3389/fphar.2021.784801
27. Lu Z, Filonov GS, Noto JJ, Schmidt CA, Hatkevich TL, Wen Y, Jaffrey SR, Gregory Matera A (2015) Metazoan tRNA introns generate stable circular RNAs in vivo. RNA 21(9):1554–1565. https://doi.org/10.1261/rna.052944.115

28. Das A, Sinha T, Shyamal S, Panda AC (2021) Emerging role of circular RNA–protein interactions. Noncoding RNA 7(48):1–20. https://doi.org/10.3390/ncrna7030048
29. Gu A, Jaijyan DK, Yang S, Zeng M, Pei S, Zhu H (2023) Functions of circular RNA in human diseases and illnesses. Noncoding RNA 9(38):1–28. https://doi.org/10.3390/ncrna9040038
30. Zhang K, Pan X, Yang Y, Shen H-B (2019) CRIP: predicting circRNA-RBP-binding sites using a codon-based encoding and hybrid deep neural networks. RNA 25(12):1604–1615. https://doi.org/10.1261/rna.070565.119
31. Conn SJ, Pillman KA, Toubia J, Conn VM, Salmanidis M, Phillips CA, Roslan S, Schreiber AW, Gregory PA, Goodall GJ (2015) The RNA binding protein quaking regulates formation of circRNAs. Cell 160(6):1125–1134. https://doi.org/10.1016/j.cell.2015.02.014
32. Errichelli L, Dini Modigliani S, Laneve P, Colantoni A, Legnini I, Capauto D, Rosa A, De Santis R, Scarfò R, Peruzzi G, Lu L, Caffarelli E, Shneider NA, Morlando M, Bozzoni I (2017) FUS affects circular RNA expression in murine embryonic stem cell-derived motor neurons. Nat Commun 8(14741):1–11. https://doi.org/10.1038/ncomms14741
33. Khan MAF, Reckman YJ, Aufiero S, Van Den Hoogenhof MMG, Van Der Made I, Beqqali A, Koolbergen DR, Rasmussen TB, Van Der Velden J, Creemers EE, Pinto YM (2016) RBM20 regulates circular RNA production from the titin gene. Circ Res 119(9):996–1003. https://doi.org/10.1161/CIRCRESAHA.116.309568
34. Yang Y, Gao X, Zhang M, Yan S, Sun C, Xiao F, Huang N, Yang X, Zhao K, Zhou H, Huang S, Xie B, Zhang N (2018) Novel role of FBXW7 circular RNA in repressing glioma tumorigenesis. J Natl Cancer Inst 110(3):304–315. https://doi.org/10.1093/jnci/djx166
35. Liu X, Zhang Y, Zhou S, Dain L, Mei L, Zhu G (2022) Circular RNA: an emerging frontier in RNA therapeutic targets, RNA therapeutics, and mRNA vaccines. J Control Release 348:84–94. https://doi.org/10.1016/j.jconrel.2022.05.043
36. Huang A, Zheng H, Wu Z, Chen M, Huang Y (2020) Circular RNA-protein interactions: functions, mechanisms, and identification. Theranostics 10(8):3506–3517. https://doi.org/10.7150/thno.42174
37. Karen A, Emily E (2017) More uses for genomic junk. Nature 543(7644):183–185. https://doi.org/10.1038/543183a
38. Liang Y, Wang H, Chen B, Mao Q, Xia W, Zhang T, Song X, Zhang Z, Xu L, Dong G, Jiang F (2021) circDCUN1D4 suppresses tumor metastasis and glycolysis in lung adenocarcinoma by stabilizing TXNIP expression. Mol Ther Nucleic Acids 23:355–368. https://doi.org/10.1016/j.omtn.2020.11.012
39. Pamudurti NR, Bartok O, Jens M, Ashwal-Fluss R, Stottmeister C, Ruhe L, Hanan M, Wyler E, Perez-Hernandez D, Ramberger E, Shenzis S, Samson M, Dittmar G, Landthaler M, Chekulaeva M, Rajewsky N, Kadener S (2017) Translation of CircRNAs. Mol Cell 66(1):9–21. https://doi.org/10.1016/j.molcel.2017.02.021
40. Dawoud A, Ihab Zakaria Z, Hisham Rashwan H, Braoudaki M, Youness RA (2023) Circular RNAs: new layer of complexity evading breast cancer heterogeneity. Noncoding RNA Res 8(1):60–74. https://doi.org/10.1016/j.ncrna.2022.09.011
41. Zang J, Lu D, Xu A (2018) The interaction of circRNAs and RNA binding proteins: an important part of circRNA maintenance and function. J Neurosci Res 98(1):87–97. https://doi.org/10.1002/jnr.24356
42. Lei K, Bai H, Wei Z, Xie C, Wang J, Li J, Chen Q (2018) The mechanism and function of circular RNAs in human diseases. Exp Cell Res 368(2):147–158. https://doi.org/10.1016/j.yexcr.2018.05.002
43. Yang Y, Fan X, Mao M, Song X, Wu P, Zhang Y, Jin Y, Yang Y, Chen LL, Wang Y, Wong CCL, Xiao X, Wang Z (2017) Extensive translation of circular RNAs driven by N 6 -methyladenosine. Cell Res 27(5):626–641. https://doi.org/10.1038/cr.2017.31
44. Legnini I, Di Timoteo G, Rossi F, Morlando M, Briganti F, Sthandier O, Fatica A, Santini T, Andronache A, Wade M, Laneve P, Rajewsky N, Bozzoni I (2017) Circ-ZNF609 is a circular RNA that can be translated and functions in myogenesis. Mol Cell 66(1):22–37. https://doi.org/10.1016/j.molcel.2017.02.017
45. Kristensen LS, Andersen MS, Stagsted LVW, Ebbesen KK, Hansen TB, Kjems J (2019) The biogenesis, biology and characterization of circular RNAs. Nat Rev Genet 20(11):675–691. https://doi.org/10.1038/s41576-019-0158-7
46. Zhang Z-h, Wang Y, Zhang Y, Zheng SF, Feng T, Tian X, Abudurexiti M, Da WZ, Zhu WK, Su JQ, Zhang HL, Shi GH, Wang ZL, Cao DL, Ye DW (2023) The function and mechanisms of action of circular RNAs in Urologic Cancer. Mol Cancer 22(61):1–24. https://doi.org/10.1186/s12943-023-01766-2
47. Wang X, Fang L (2018) Advances in circular RNAs and their roles in breast cancer. J Exp Clin Cancer Res 37(206):1–12. https://doi.org/10.1186/s13046-018-0870-8
48. van Zonneveld AJ, Kölling M, Bijkerk R, Lorenzen JM (2021) Circular RNAs in kidney disease and cancer. Nat Rev Nephrol 17(12):814–826. https://doi.org/10.1038/s41581-021-00465-9
49. Zhang M, Zhao K, Xu X, Yang Y, Yan S, Wei P, Liu H, Xu J, Xiao F, Zhou H, Yang X, Huang N, Liu J, He K, Xie K, Zhang G, Huang S, Zhang N (2018a) A peptide encoded by circular form of LINC-PINT suppresses oncogenic transcriptional elongation in glioblastoma. Nat Commun 9(1):1–17. https://doi.org/10.1038/s41467-018-06862-2
50. Lei M, Zheng G, Ning Q, Zheng J, Dong D (2020) Translation and functional roles of circular RNAs in human cancer. Mol Cancer 19(30):1–9. https://doi.org/10.1186/s12943-020-1135-7

51. Ye F, Gao G, Zou Y, Zheng S, Zhang L, Ou X, Xie X, Tang H (2019) circFBXW7 inhibits malignant progression by sponging miR-197-3p and encoding a 185-aa protein in triple-negative breast cancer. Mol Ther Nucleic Acids 18:88–98. https://doi.org/10.1016/j.omtn.2019.07.023
52. Liang WC, Wong CW, Liang PP, Shi M, Cao Y, Rao ST, Tsui SKW, Waye MMY, Zhang Q, Fu WM, Zhang JF (2019) Translation of the circular RNA circβ-catenin promotes liver cancer cell growth through activation of the Wnt pathway. Genome Biol 20(84):1–12. https://doi.org/10.1186/s13059-019-1685-4
53. Nejak-Bowen KN, Monga SPS (2011) Beta-catenin signaling, liver regeneration and hepatocellular cancer: sorting the good from the bad. Semin Cancer Biol 21(1):44–58. https://doi.org/10.1016/j.semcancer.2010.12.010
54. Glažar P, Papavasileiou P, Rajewsky N (2014) CircBase: a database for circular RNAs. RNA 20(11):1666–1670. https://doi.org/10.1261/rna.043687.113
55. Zheng X, Chen L, Zhou Y, Wang Q, Zheng Z, Xu B, Wu C, Zhou Q, Hu W, Wu C, Jiang J (2019) A novel protein encoded by a circular RNA circPPP1R12A promotes tumor pathogenesis and metastasis of colon cancer via Hippo-YAP signaling. Mol Cancer 18(47):1–13. https://doi.org/10.1186/s12943-019-1010-6
56. Huang X, Li Z, Zhang Q, Wang W, Li B, Wang L, Xu Z, Zeng A, Zhang X, Zhang X, He Z, Li Q, Sun G, Wang S, Li Q, Wang L, Zhang L, Xu H, Xu Z (2019) Circular RNA AKT3 upregulates PIK3R1 to enhance cisplatin resistance in gastric cancer via miR-198 suppression. Mol Cancer 18(71):1–20. https://doi.org/10.1186/s12943-019-0969-3
57. Jeck WR, Sharpless NE (2014) Detecting and characterizing circular RNAs. Nat Biotechnol 32(5):453–461. https://doi.org/10.1038/nbt.2890
58. Ashwal-Fluss R, Meyer M, Pamudurti NR, Ivanov A, Bartok O, Hanan M, Evantal N, Memczak S, Rajewsky N, Kadener S (2014) CircRNA biogenesis competes with pre-mRNA splicing. Mol Cell 56(1):55–66. https://doi.org/10.1016/j.molcel.2014.08.019
59. Zhang M, Xin Y (2018) Circular RNAs: a new frontier for cancer diagnosis and therapy. J Hematol Oncol 11(21):1–9. https://doi.org/10.1186/s13045-018-0569-5
60. Abdelmohsen K, Panda AC, Munk R, Grammatikakis I, Dudekula DB, De S, Kim J, Noh JH, Kim KM, Martindale JL, Gorospe M (2017) Identification of HuR target circular RNAs uncovers suppression of PABPN1 translation by CircPABPN1. RNA Biol 14(3):361–369. https://doi.org/10.1080/15476286.2017.1279788
61. Chen X, Yang T, Wang W, Xi W, Zhang T, Li Q, Yang A, Wang T (2019) Circular RNAs in immune responses and immune diseases. Theranostics 9(2):588–607. https://doi.org/10.7150/thno.29678
62. Wang M, Yu F, Li P (2018) Circular RNAs: characteristics, function and clinical significance in hepatocellular carcinoma. Cancers 10(258):1–22. https://doi.org/10.3390/cancers10080258
63. Li Y, Zheng Q, Bao C, Li S, Guo W, Zhao J, Chen D, Gu J, He X, Huang S (2015) Circular RNA is enriched and stable in exosomes: a promising biomarker for cancer diagnosis. Cell Res 25(8):981–984. https://doi.org/10.1038/cr.2015.82
64. Hansen TB, Jensen TI, Clausen BH, Bramsen JB, Finsen B, Damgaard CK, Kjems J (2013) Natural RNA circles function as efficient microRNA sponges. Nature 495(7441):384–388. https://doi.org/10.1038/nature11993
65. Cai H, Li Y, Niringiyumukiza JD, Su P, Xiang W (2019) Circular RNA involvement in aging: an emerging player with great potential. Mech Ageing Dev 178:16–24. https://doi.org/10.1016/j.mad.2018.11.002
66. Schneider T, Hung LH, Schreiner S, Starke S, Eckhof H, Rossbach O, Reich S, Medenbach J, Bindereif A (2016) CircRNA-protein complexes: IMP3 protein component defines subfamily of circRNPs. Sci Rep 6(1):1–11. https://doi.org/10.1038/srep31313
67. Schmidt E, Dhaouadi I, Gaziano I, Oliverio M, Klemm P, Awazawa M, Mitterer G, Fernandez-Rebollo E, Pradas-Juni M, Wagner W, Hammerschmidt P, Loureiro R, Kiefer C, Hansmeier NR, Khani S, Bergami M, Heine M, Ntini E, Frommolt P, Zentis P, Ørom UA, Heeren J, Blüher M, Bilban M, Kornfeld JW (2018) LincRNA H19 protects from dietary obesity by constraining expression of monoallelic genes in brown fat. Nat Commun 9(1):1–16. https://doi.org/10.1038/s41467-018-05933-8
68. Liu C, Yang Z, Wu J, Zhang L, Lee S, Shin DJ, Tran M, Wang L (2018) Long noncoding RNA H19 interacts with polypyrimidine tract-binding protein 1 to reprogram hepatic lipid homeostasis. Hepatology 67(5):1768–1783. https://doi.org/10.1002/hep.29654
69. Zhu Y, Gui W, Lin X, Li H (2020) Knock-down of circular RNA H19 induces human adipose-derived stem cells adipogenic differentiation via a mechanism involving the polypyrimidine tract-binding protein 1. Exp Cell Res 387(2):1–10. https://doi.org/10.1016/j.yexcr.2019.111753
70. Chen Q, Sun L, Chen ZJ (2016) Regulation and function of the cGAS-STING pathway of cytosolic DNA sensing. Nat Immunol 17(10):1142–1149. https://doi.org/10.1038/ni.3558
71. Xia P, Wang S, Ye B, Du Y, Li C, Xiong Z, Qu Y, Fan Z (2018) A circular RNA protects dormant hematopoietic stem cells from DNA sensor cGAS-mediated exhaustion. Immunity 48(4):688–701. https://doi.org/10.1016/j.immuni.2018.03.016
72. Ding X, Zhang S, Li X, Feng C, Huang Q, Wang S, Wang S, Xia W, Yang F, Yin R, Xu L, Qiu M, Li M, Wang J (2018) Profiling expression of coding genes, long noncoding RNA, and circular RNA in lung

adenocarcinoma by ribosomal RNA-depleted RNA sequencing. FEBS Open Bio 8(4):544–555. https://doi.org/10.1002/2211-5463.12397
73. Huang Q, Guo H, Wang S, Ma Y, Chen H, Li H, Li J, Li X, Yang F, Qiu M, Zhao S, Wang J (2020) A novel circular RNA, circXPO1, promotes lung adenocarcinoma progression by interacting with IGF2BP1. Cell Death Dis 11(1031):1–13. https://doi.org/10.1038/s41419-020-03237-8
74. Li Z, Huang C, Bao C, Chen L, Lin M, Wang X, Zhong G, Yu B, Hu W, Dai L, Zhu P, Chang Z, Wu Q, Zhao Y, Jia Y, Xu P, Liu H, Shan G (2015) Exon-intron circular RNAs regulate transcription in the nucleus. Nat Struct Mol Biol 22(3):256–264. https://doi.org/10.1038/nsmb.2959
75. Jin J, Sun H, Shi C, Yang H, Wu Y, Li W, Dong Y-h, Cai L, Meng X-m (2020) Circular RNA in renal diseases. J Cell Mol Med 24(12):6523–6533. https://doi.org/10.1111/jcmm.15295
76. Fang L, Du WW, Awan FM, Dong J, Yang BB (2019) The circular RNA circ-Ccnb1 dissociates Ccnb1/Cdk1 complex suppressing cell invasion and tumorigenesis. Cancer Lett 459:216–226. https://doi.org/10.1016/j.canlet.2019.05.036
77. Chen Y, Yang F, Fang E, Xiao W, Mei H, Li H, Li D, Song H, Wang J, Hong M, Wang X, Huang K, Zheng L, Tong Q (2019) Circular RNA circAGO2 drives cancer progression through facilitating HuR-repressed functions of AGO2-miRNA complexes. Cell Death Differ 26(7):1346–1364. https://doi.org/10.1038/s41418-018-0220-6
78. Wong CH, Lou UK, Li Y, Chan SL, Tong JHM, To KF, Chen Y (2020) CircFOXK2 promotes growth and metastasis of pancreatic ductal adenocarcinoma by complexing with RNA-binding proteins and sponging MiR-942. Cancer Res 80(11):2138–2149. https://doi.org/10.1158/0008-5472.CAN-19-3268
79. Hansen TB, Wiklund ED, Bramsen JB, Villadsen SB, Statham AL, Clark SJ, Kjems J (2011) MiRNA-dependent gene silencing involving Ago2-mediated cleavage of a circular antisense RNA. EMBO J 30(21):4414–4422. https://doi.org/10.1038/emboj.2011.359
80. Chen S, Zhao Y (2018) Circular RNAs: characteristics, function and role in human cancer. Histol Histopathol 33(9):887–893. https://doi.org/10.14670/HH-11-969
81. Ma J, Du WW, Zeng K, Wu N, Fang L, Lyu J, Yee AJ, Yang BB (2021) An antisense circular RNA circSCRIB enhances cancer progression by suppressing parental gene splicing and translation. Mol Ther 29(9):2754–2768. https://doi.org/10.1016/j.ymthe.2021.08.002
82. Wang X, Xing L, Yang R, Chen H, Wang M, Jiang R, Zhang L, Chen J (2021) The circACTN4 interacts with FUBP1 to promote tumorigenesis and progression of breast cancer by regulating the expression of proto-oncogene MYC. Mol Cancer 20(91):1–21. https://doi.org/10.1186/s12943-021-01383-x
83. Wang Z, Deng H, Jin Y, Luo M, Huang J, Wang J, Zhang K, Wang L, Zhou J (2023) Circular RNAs: biology and clinical significance of breast cancer. RNA Biol 20(1):859–874. https://doi.org/10.1080/15476286.2023.2272468
84. Feng Y, Yang Y, Zhao X, Fan Y, Zhou L, Rong J, Yu Y (2019) Circular RNA circ0005276 promotes the proliferation and migration of prostate cancer cells by interacting with FUS to transcriptionally activate XIAP. Cell Death Dis 10(792):1–14. https://doi.org/10.1038/s41419-019-2028-9
85. Liu H, Lan T, Li H, Xu L, Chen X, Liao H, Chen X, Du J, Cai Y, Wang J, Li X, Huang J, Yuan K, Zeng Y (2021) Circular RNA circDLC1 inhibits MMP1-mediated liver cancer progression via interaction with HuR. Theranostics 11(3):1396–1411. https://doi.org/10.7150/thno.53227
86. Du WW, Yang W, Li X, Awan FM, Yang Z, Fang L, Lyu J, Li F, Peng C, Krylov SN, Xie Y, Zhang Y, He C, Wu N, Zhang C, Sdiri M, Dong J, Ma J, Gao C, Hibberd S, Yang BB (2018) A circular RNA circ-DNMT1 enhances breast cancer progression by activating autophagy. Oncogene 37(44):5829–5842. https://doi.org/10.1038/s41388-018-0369-y
87. Liu Y, Dong Y, Zhao L, Su L, Luo J (2018) Circular RNA-MTO1 suppresses breast cancer cell viability and reverses monastrol resistance through regulating the TRAF4/Eg5 axis. Int J Oncol 53(4):1752–1762. https://doi.org/10.3892/ijo.2018.4485
88. Holdt LM, Stahringer A, Sass K, Pichler G, Kulak NA, Wilfert W, Kohlmaier A, Herbst A, Northoff BH, Nicolaou A, Gäbel G, Beutner F, Scholz M, Thiery J, Musunuru K, Krohn K, Mann M, Teupser D (2016) Circular non-coding RNA ANRIL modulates ribosomal RNA maturation and atherosclerosis in humans. Nat Commun 7(12429):1–14. https://doi.org/10.1038/ncomms12429
89. Grimm T, Hölzel M, Rohrmoser M, Harasim T, Malamoussi A, Gruber-Eber A, Kremmer E, Eick D (2006) Dominant-negative Pes1 mutants inhibit ribosomal RNA processing and cell proliferation via incorporation into the PeBoW-complex. Nucleic Acids Res 34(10):3030–3043. https://doi.org/10.1093/nar/gkl378
90. Guevara NV, Kim H-S, Antonova EI, Chan L (1999) The absence of p53 accelerates atherosclerosis by increasing cell proliferation in vivo. Nat Med 5(3):335–339. https://doi.org/10.1038/6585
91. Luo J, Liu H, Luan S, Li Z (2019) Guidance of circular RNAs to proteins' behavior as binding partners. Cell Mol Life Sci 76(21):4233–4243. https://doi.org/10.1007/s00018-019-03216-z
92. Yang W, Du WW, Li X, Yee AJ, Yang BB (2016) Foxo3 activity promoted by non-coding effects of circular RNA and Foxo3 pseudogene in the inhibition of tumor growth and angiogenesis. Oncogene 35(30):3919–3931. https://doi.org/10.1038/onc.2015.460
93. Du WW, Fang L, Yang W, Wu N, Awan FM, Yang Z, Yang BB (2017) Induction of tumor apoptosis

through a circular RNA enhancing Foxo3 activity. Cell Death Differ 24(2):357–370. https://doi.org/10.1038/cdd.2016.133
94. Zhang H, Wang G, Ding C, Liu P, Wang R, Ding W, Tong D, Wu D, Li C, Wei Q, Zhang X, Li D, Liu P, Cui H, Tang H, Ji F (2017) Increased circular RNA UBAP2 acts as a sponge of miR-143 to promote osteosarcoma progression. Oncotarget 8(37):61687–61697. https://doi.org/10.18632/oncotarget.18671
95. Fang Y, Xue JL, Shen Q, Chen J, Tian L (2012) MicroRNA-7 inhibits tumor growth and metastasis by targeting the phosphoinositide 3-kinase/Akt pathway in hepatocellular carcinoma. Hepatology 55(6):1852–1862. https://doi.org/10.1002/hep.25576
96. Herrero-Sánchez MC, Rodríguez-Serrano C, Almeida J, San Segundo L, Inogés S, Santos-Briz Á, García-Briñón J, Corchete LA, San Miguel JF, Del Cañizo C, Blanco B (2016) Targeting of PI3K/AKT/mTOR pathway to inhibit T cell activation and prevent graft-versus-host disease development. J Hematol Oncol 9(113):1–14. https://doi.org/10.1186/s13045-016-0343-5
97. Xue J, Liu Y, Luo F, Lu X, Xu H, Liu X, Lu L, Yang Q, Chen C, Fan W, Liu Q (2017) Circ100284, via miR-217 regulation of EZH2, is involved in the arsenite-accelerated cell cycle of human keratinocytes in carcinogenesis. Biochim Biophys Acta Mol basis Dis 1863(3):753–763. https://doi.org/10.1016/j.bbadis.2016.12.018
98. Zhang Z, Xie Q, He D, Ling Y, Li Y, Li J, Zhang H (2018) Circular RNA: new star, new hope in cancer. BMC Cancer 18(834):1–10. https://doi.org/10.1186/s12885-018-4689-7
99. Hsiao KY, Sun HS, Tsai SJ (2017) Circular RNA—new member of noncoding RNA with novel functions. Exp Biol Med 242(11):1136–1141. https://doi.org/10.1177/1535370217708978
100. Li M, Ding W, Sun T, Tariq MA, Xu T, Li P, Wang J (2018) Biogenesis of circular RNAs and their roles in cardiovascular development and pathology. FEBS J 285(2):220–232. https://doi.org/10.1111/febs.14191
101. Feng Z, Meng S, Zhou H, Xu Z, Tang Y, Li P, Liu C, Huang Y, Wu M (2019b) Functions and potential applications of circular RNAs in cancer stem cells. Front Oncol 9(500):1–10. https://doi.org/10.3389/fonc.2019.00500
102. Li B, Zhu L, Lu C, Wang C, Wang H, Jin H, Ma X, Cheng Z, Yu C, Wang S, Zuo Q, Zhou Y, Wang J, Yang C, Lv Y, Jiang L, Qin W (2021) circNDUFB2 inhibits non-small cell lung cancer progression via destabilizing IGF2BPs and activating anti-tumor immunity. Nat Commun 12(295):1–15. https://doi.org/10.1038/s41467-020-20527-z
103. Shen S, Yang Y, Shen P, Ma J, Fang B, Wang Q, Wang K, Shi P, Fan S, Fang X (2021) CircPDE4B prevents articular cartilage degeneration and promotes repair by acting as a scaffold for RIC8A and MID1. Ann Rheum Dis 80(9):1209–1219. https://doi.org/10.1136/annrheumdis-2021-219969
104. Du WW, Yang W, Chen Y, Wu ZK, Foster FS, Yang Z, Li X, Yang BB (2017) Foxo3 circular RNA promotes cardiac senescence by modulating multiple factors associated with stress and senescence responses. Eur Heart J 38(18):1402–1412. https://doi.org/10.1093/eurheartj/ehw001
105. Li Q, Wang Y, Wu S, Zhou Z, Ding X, Shi R, Thorne RF, Zhang XD, Hu W, Wu M (2019b) CircACC1 regulates assembly and activation of AMPK complex under metabolic stress. Cell Metab 30(1):157–173. https://doi.org/10.1016/j.cmet.2019.05.009
106. Huang S, Li X, Zheng H, Si X, Li B, Wei G, Li C, Chen Y, Chen Y, Liao W, Liao Y, Bin J (2019a) Loss of super-enhancer-regulated circRNA Nfix induces cardiac regeneration after myocardial infarction in adult mice. Circulation 139(25):2857–2876. https://doi.org/10.1161/CIRCULATIONAHA.118.038361
107. Chen LL (2020) The expanding regulatory mechanisms and cellular functions of circular RNAs. Nat Rev Mol Cell Biol 21(8):475–490. https://doi.org/10.1038/s41580-020-0243-y
108. Du WW, Yang W, Li X, Fang L, Wu N, Li F, Chen Y, He Q, Liu E, Yang Z, Awan FM, Liu M, Yang BB (2020) The circular RNA circSKA3 binds integrin β1 to induce invadopodium formation enhancing breast cancer invasion. Mol Ther. 28(5):1287–1298. https://doi.org/10.1016/j.ymthe.2020.03.002
109. Barbagallo D, Caponnetto A, Cirnigliaro M, Brex D, Barbagallo C, D'Angeli F, Morrone A, Caltabiano R, Barbagallo GM, Ragusa M, Di Pietro C, Hansen TB, Purrello M (2018) CircSMARCA5 inhibits migration of glioblastoma multiforme cells by regulating a molecular axis involving splicing factors SRSF1/SRSF3/PTB. Int J Mol Sci 19(2):1–13. https://doi.org/10.3390/ijms19020480
110. Jie M, Wu Y, Gao M, Li X, Liu C, Ouyang Q, Tang Q, Shan C, Lv Y, Zhang K, Dai Q, Chen Y, Zeng S, Li C, Wang L, He F, Hu C, Yang S (2020) CircMRPS35 suppresses gastric cancer progression via recruiting KAT7 to govern histone modification. Mol Cancer 19(56):1–16. https://doi.org/10.1186/s12943-020-01160-2
111. Song C, Zhang Y, Huang W, Shi J, Huang Q, Jiang M, Qiu Y, Wang T, Chen H, Wang H (2022) Circular RNA Cwc27 contributes to Alzheimer's disease pathogenesis by repressing Pur-α activity. Cell Death Differ 29(2):393–406. https://doi.org/10.1038/s41418-021-00865-1
112. Chen N, Zhao G, Yan X, Lv Z, Yin H, Zhang S, Song W, Li X, Li L, Du Z, Jia L, Zhou L, Li W, Hoffman AR, Hu JF, Cui J (2018) A novel FLI1 exonic circular RNA promotes metastasis in breast cancer by coordinately regulating TET1 and DNMT1. Genome Biol 19(218):1–14. https://doi.org/10.1186/s13059-018-1594-y
113. Sun Y-M, Wang W-T, Zeng Z-C, Chen T-Q, Han C, Pan Q, Huang W, Fang K, Sun L-Y, Zhou Y-F, Luo X-Q, Luo C, Du X, Chen Y-Q (2019) circ-

MYBL2, a circRNA from MYBL2, regulates FLT3 translation by recruiting PTBP1 to promote FLT3-ITD AML progression. Blood J Am Soc Hematol 134(18):1533–1546. https://doi.org/10.1182/blood.2019000802
114. Yang ZG, Awan FM, Du WW, Zeng Y, Lyu J, Wu D, Gupta S, Yang W, Yang BB (2017b) The circular RNA interacts with STAT3, increasing its nuclear translocation and wound repair by modulating Dnmt3a and miR-17 function. Mol Ther 25(9):2062–2074. https://doi.org/10.1016/j.ymthe.2017.05.022
115. Xia B, Hong T, He X, Hu X, Gao Y (2019) A circular RNA derived from MMP9 facilitates oral squamous cell carcinoma metastasis through regulation of MMP9 mRNA stability. Cell Transplant 28(12):1614–1623. https://doi.org/10.1177/0963689719875409
116. Ding L, Zhao Y, Dang S, Wang Y, Li X, Yu X, Li Z, Wei J, Liu M, Li G (2019) Circular RNA circ-DONSON facilitates gastric cancer growth and invasion via NURF complex dependent activation of transcription factor SOX4. Mol Cancer 18(45):1–11. https://doi.org/10.1186/s12943-019-1006-2
117. Zhao W, Wang S, Qin T, Wang W (2020) Circular RNA (circ-0075804) promotes the proliferation of retinoblastoma via combining heterogeneous nuclear ribonucleoprotein K (HNRNPK) to improve the stability of E2F transcription factor 3 E2F3. J Cell Biochem 121(7):3516–3525. https://doi.org/10.1002/jcb.29631
118. Li H, Yang F, Hu A, Wang X, Fang E, Chen Y, Li D, Song H, Wang J, Guo Y, Liu Y, Li H, Huang K, Zheng L, Tong Q (2019) Therapeutic targeting of circ- CUX 1 / EWSR 1/ MAZ axis inhibits glycolysis and neuroblastoma progression. EMBO Mol Med 11(12):1–17. https://doi.org/10.15252/emmm.201910835
119. Hong Y, Qin H, Li Y, Zhang Y, Zhuang X, Liu L, Lu K, Li L, Deng X, Liu F, Shi S, Liu G (2019) FNDC3B circular RNA promotes the migration and invasion of gastric cancer cells via the regulation of E-cadherin and CD44 expression. J Cell Physiol 234(11):19895–19910. https://doi.org/10.1002/jcp.28588

Circular RNAs Regulate Transcription and Translation

11

Shahidee Zainal Abidin

Abstract

Circular RNAs (circRNAs) are non-coding RNAs that are single-stranded and closed covalently. They are found in various organisms, from viruses to mammals. CircRNAs are involved in diverse biological processes, functioning as microRNA (miRNA) decoys and transcriptional modulators and interacting with RNA-binding proteins. Recent studies have shown that certain cytoplasmic circRNAs are capable of being translated into peptides. This discovery has shed light on the significance of circRNAs in cellular physiological processes. Moreover, there are several potential mechanisms for circRNA translation, including internal ribosome entry site (IRES)-driven translation and N6-methyladenosines (m6A)-mediated cap-independent translation. Thus, it indicates that circRNAs have a diverse role in regulating biological processes, particularly in various diseases. Furthermore, multiple studies have shown that circRNAs display unique expression patterns and play crucial roles in multiple diseases, suggesting their potential utility as diagnostic biomarkers and targets for therapeutic intervention.

Keywords

Circular RNA · Human antigen R · RNA-binding protein · N6-methyladenosines · Internal ribosome entry site

1 Introduction

Circular RNAs (circRNAs) were initially discovered as secondary products of mRNA splicing in 1976 [1]. Due to advanced technology and bioinformatics, an increasing number of circRNAs have been discovered and demonstrated to have functional roles in both physiological and pathological cellular processes [2]. Many studies have shown that circRNA can exert control over gene expression by functioning as competitive endogenous RNAs (ceRNAs) or protein sponges [3–6]. A small number of circRNAs have internal ribosome entry sites (IRESs) and open reading frames (ORFs) that enable them to translate unique peptides or proteins [7]. So far, numerous circRNAs have been discovered to have significant functions in encoding proteins across various.

Advanced technologies have made it possible to thoroughly study circRNAs for their identification and putative roles [8–10]. These technologies include ribosomal RNA depletion RNA sequencing, long-read sequencing, and enhanced circRNA mining bioinformatics algorithms. CircRNAs are an abundant and conserved class

S. Z. Abidin (✉)
Faculty of Science and Marine Environment,
Universiti Malaysia Terengganu,
Terengganu, Malaysia
e-mail: shahidee.zainal@umt.edu.my

J. Xiao (ed.), *Advances in Circular RNAs*, Advances in Experimental Medicine and Biology 1485,
https://doi.org/10.1007/978-981-96-9428-0_11

of RNAs, and they are extensively expressed in specific cell types and developmental stages [11–14]. CircRNAs have the ability to act as transcriptional regulators, exerting control over the expression of genes [15, 16]. In addition, they can function as microRNA (miRNA) sponges and regulate gene expression through the miRNA-mRNA regulatory pathway [3, 4]. Furthermore, multiple studies have shown the presence of peptides encoded by circRNAs. Thus, this provides a new insight into the cellular physiological processes of circRNAs.

Moreover, proteins serve as direct mediators in nearly all essential functions. The interactions between circRNA and proteins have significantly revitalised the study of circRNAs and enhanced our understanding of their biological importance. Nevertheless, there is a dearth of studies on the interaction between circRNA and proteins, which continues to captivate scientists due to its intriguing mechanisms. Hence, this article provides a comprehensive overview of the significant advancements in understanding the metabolism and roles of circRNAs.

2 Interaction with RNA-Binding Proteins

Thousands of circRNAs are expressed in human cells and exhibit the role of miRNA sponges. One of them is the circular transcript ciRS-7 (circular sponge for miR-7, also known as CDR1as), which is significantly expressed in human and animal brain tissue [4]. There are more than 70 binding sites for miR-7 in CiRS-7. Hence, ciRS-7 has the capability to obstruct the inhibitory effect of miR-7 on its mRNA targets while remaining entirely impervious to miRNA-induced destabilisation of these targets. Besides acting as a miRNA sponge, the circRNAs also serve as sponges for RNA-binding proteins (RBPs), subsequently affecting mRNA translation. For example, circ-CUX1 forms a bond with EWS RNA-binding protein 1 (EWSR1) to enhance its interaction with MAZ (MYC-associated zinc finger protein), glycolysis, and the advancement of neuroblastoma [17]. In addition, the binding of circ-CTNNB1 to DEAD-box polypeptide 3 (DDX3) enhances its direct interaction with transcription factor Yin Yang 1 (YY1) [18].

Another example of the RBPs is the Human antigen R (HuR), a member of the embryonic lethal abnormal visual protein (ELAV) family. As an RBP, HuR enhances the stability of target genes' mRNA by binding to poly-U elements or AU-rich elements (AREs) in the 3′-untranslated region (3'-UTR) [19]. HuR also plays a significant role in tumour formation, recurrence, invasion, and metastasis [20], and it has high levels of expression in breast, colorectal, gastric, and prostate cancer. Interestingly, HuR forms associations with numerous circRNAs and subsequently regulates the translation and stability of mRNAs.

One of the circRNAs, known as circAGO2 (hsa_circ_0135889), is markedly increased in several cancer tissues, such as gastric, colon, prostate, and neuroblastoma. The circAGO2 directly interacts with the HuR protein, leading to its accumulation on the 3'-UTR of target genes. This interaction reduces the binding of AGO2, inhibits the gene silence mediated by AGO2/miRNA, and promotes the development of tumours and aggressiveness [21]. CircAGO2 did not affect AGO2 translation or transcription in cancer cells. On the other hand, circAGO2 controls the activity of target genes by promoting the HuR-inhibited actions of AGO2/miRNA complexes. CircAGO2 regulates the translocation of HuR from the nucleus to the cytoplasm, resulting in an increased concentration of HuR on the 3'-UTR. This leads to a reduction in AGO2 binding and suppresses the activities of AGO2-miRNA complexes.

Furthermore, a circRNA located in the nucleus known as circ-HuR (hsa_circ_0049027) is found to be reduced in gastric cancer. The excessive expression of circ-HuR inhibits the proliferation and invasiveness of gastric cancer cells both in vitro and in vivo. Circ-HuR functions by impeding the accumulation of CNBP on the HuR promoter, leading to a decrease in HuR expression. In addition, the circ-HuR has the ability to block the transcriptional activity of CNBP. It acts as a natural inhibitor by preventing CNBP from binding to the HuR promoter, leading to the down-regulation of HuR and its target genes involved in cancer progression [22]. In a different study, the interaction between HuR and circPABPN1 inhibits the translation of the autophagy-related gene 16L1 (ATG16L1). The translation of this mRNA is stimulated by the binding of HuR to its 3'-UTR in the intestinal epithelium. Nevertheless, the circPABPN1 inhibited the interaction between HuR and Atg16l1, reducing the abundance of ATG16L1 [23].

Besides HuR, another RBP involved in mechanisms of action circRNA is ELAV1. For example, the circZNF609 regulates the stability and translation of CKAP5, UPF2, and SRRM1 mRNAs by recruiting the ELAVL1 RNA-binding protein [24]. ELAVL1 is a widely distributed protein interacting with RNA, and other cellular targets may vie for its binding. CircZNF609, possessing seven ELAVL1-binding sites, functions as a transporter for the delivery of this protein onto mRNA. For CKAP5 mRNA, which consists of one binding site for ELAVL1 in its 3'-UTR [25], this interaction is essential to contend with other cellular targets of ELAVL1. The precise stoichiometry between circZNF609 and other RNAs is not as crucial due to the dynamic nature of this process, where circZNF609 can undergo multiple cycles of ELAVL1 upload and download.

The impact of circZNF609 on the regulation of CKAP5 protein levels and its subsequent influence on the dynamic functions of the mitotic and proliferation is significant, particularly in relation to the increased expression of this RNA in various tumours [26–29]. These findings demonstrate that the circZNF609/CKAP5 axis, previously identified in the rhabdomyosarcoma cell line, is also found in various other cancer cell lines. Currently, circRNAs have primarily been associated with cancer due to their ability to sponge miRNAs [30]. Nevertheless, this characteristic limits their functionality mostly because of the unique miRNA profile that defines each specific type of cancer.

3 Interaction of Circular RNA Chromatin and N6-Methyladenosine

CircRNAs located in the nucleus regulate transcription, alternative splicing, and chromatin looping processes. The *CircSEP3* gene in Arabidopsis thaliana regulates the splicing process for the *SEPALLATA3* gene, often known as SEP3. SEP3 is a crucial MADS-box transcription factor that plays a vital role in forming floral homeotic phenotypes [31]. CircSEP3 is generated through the fusion of exon 6 of SEP3, resulting in a chimeric molecule consisting of both RNA and DNA sequences that are complementary to each other. As a consequence, transcription is temporarily halted, leading to the exclusion of exon 6. This results in the production of a modified SEP3 mRNA through alternative splicing.

On the other hand, suppression of circNOTCH1 increased the m6A modification on Nothch1 mRNA. circNOTCH1 was unable to alter NOTCH1 expression when METTL14 was depleted. METTL14, a member of the m6A 'writers' group, enhances the RNA m6A alterations that influence the functionality and stability of cellular mRNAs [32]. The G protein-coupled oestrogen receptor (GPER) can elevate the expression of circNOTCH1. This circNOTCH1 could then compete with NOTCH1 mRNA for binding to METTL14. Due to the absence of m6A alteration catalysed by METTL14 on the NOTCH1 mRNA, the stability of the NOTCH1 mRNA increased, making it more susceptible to protein translation. Furthermore, GPER has the ability to inhibit the phosphorylation of YAP1 and enhance the transcriptional regulation of YAP1-TEAD on QKI. This transacting RNA-binding factor plays a role in the production of circRNA. This facilitates the formation of circNOTCH1. Collectively, GPER has the ability to control the expression of NOTCH1 using the YAP1/QKI/circ-NOTCH1/m6A methylated NOTCH1 pathway [33].

4 Regulation of Transcription by Circular RNAs

Recent evidence indicates that circRNAs have a substantial role in transcriptional regulation by functioning as miRNA sponges or regulating the target genes. Overexpressing circ-Dnmt1 resulted in increased cellular proliferation and viability, as well as induced cellular autophagy in breast cancer cells. In addition to that, it also suppressed cellular senescence and facilitated tumour growth. These effects were modulated by directly binding to p53 and Auf1, subsequently facilitating their nuclear translocation. The nuclear translocation of p53 and Auf1 triggers cellular autophagy and leads to upregulation of Dnmt1 expression, suppressing p53 transcription [34]. Another study showed overexpression of circACTN4 (hsa_circ_0050900) in breast cancer cells. This overexpression was strongly associated with advanced clinical stage, metastasis, and unfavourable prognosis in patients with breast cancer. The high level of circACTN4 enhanced the proliferation and invasiveness of breast cancer cells both in vitro and in vivo. Conversely, reducing the levels of circACTN4 had the opposite effects. A previous study has indicated that circACTN4 has the ability to competitively interact with FUBP1 and hinder the interaction between FUBP1 and FIR, hence promoting MYC transcription [35].

Moreover, in endometrial cancer, the circular RNA hsa-circ-0000437 exhibits a notable decrease in expression. The hsa-circ-0000437 consists of a short open reading frame that encodes a functional peptide called CORO1C-47aa. The overexpression of CORO1C-47aa can effectively hinder the process of angiogenesis at its early stage by lowering the proliferation, migration, and differentiation of endothelial cells by competing with the transcription factor TACC3 to bind to ARNT and, therefore, suppressing the expression of VEGF. CORO1C-47aa directly interacted with ARNT via the PAS-B domain, inhibiting the binding between ARNT and TACC3, resulting in decreased VEGF production and decreased angiogenesis [36].

Another circRNA, circPDS5B, is highly expressed in Ischemic stroke (IS) patients and transient middle cerebral artery occlusion (tMCAO) mice [37]. The presence of CircPDS5B hindered the process of angiogenesis by recruiting hnRNPL to suppress the expression of VEGFA. The study revealed that circPDS5B can recruit hnRNPL to enhance the expression and stability of Runx1 and ZNF24 mRNA. Consequently, this leads to the deactivation of the pro-angiogenesis protein VEGFA. The depletion of hnRNPL elevated the pro-angiogenic effect of circPDS5B inhibition in tMCAO mice and OGD/R-challenged BMECs.

Additionally, the circRNA circTshz2-2, specific to neurons, influences the cell cycle of neurons and the spatial memory function in the brain. Neurons exhibited elevated levels of circTshz2-2 expression, and the suppression circTshz2-2 resulted in enhanced neurite length and increased neural complexity. This suggests that circTshz2-2 has a suppressive function in neural structure development [38]. The circTshz2-2 inhibits the transcription of Bdnf by interacting with the YY1 transcriptional repressive complex. Conversely, when circTshz2-2 dysregulation leads to an increase in BDNF expression by disrupting the development of the YY1 transcriptional repressor complex, a high level of BDNF leads to a halt in the G2/M phase of the cell cycle through the TrkB/Cyclin B2/CDK1 signalling pathway.

The disruption of circTshz2-2 results in dysregulation of the kinetochore, leading to the formation of polyploid neurons [38]. The dysregulation of circTshz2-2 impacts neuronal structure by causing neurite elongation by relocating the cytokinesis-related MTOC within the neurite. This indicates that circTshz2-2 regulates the G2/M cell cycle during neuronal development. In addition, circTshz2-2 enhances spatial memory capacity by regulating the expression of G2/M checkpoint protein in the brain. CircTshz2-2 reduction inhibited the expression of Cyclin B2 and CDK1, hence modulating spatial memory capacity in both normal and obese brains.

Interestingly, FOXO3 has been reported to promote the transcription of SPON1 and the generation of spon1-related circRNA, such as circSPON1. FOXO3 directly interacted with SPON1 and its promoter, leading to the production of SPON1 circRNA. This interaction also hindered the translocation of Smad3 into the nucleus by interacting with Smad3 and acting as a sponge for Smad7-targeted miRNAs, thus promoting the expression of Smad7. Consequently, this process suppressed the advancement of pulmonary fibrosis [39].

5 Regulation of Translation by Circular RNAs

The majority of circRNAs are unable to undergo protein translation due to the absence of a 5′ cap structure [40, 41]. Given that most circRNAs are located in the cytoplasm and originate from exons responsible for protein coding, it appears likely that circRNAs can undergo translation. Synthetic circRNAs can undergo translation in both in vitro and in vivo through mechanisms that do not rely on the presence of a 5′ cap structure for ribosomes to attach to [42, 43]. An example of this is the UTR section of circ-ZNF609, which spans from the termination codon to the initiation codon. This region can promote translation using the IRES mechanism [44]. In addition, m6A, the most common RNA base modification, has been found to promote the efficient initiation of protein translation from circRNAs in various circumstances [45].

Furthermore, the circMTHFD2L exhibited decreased expression in gastric cancer and produced a translated protein of 248 amino acids (CM-248aa) [46]. The circMTHFD2L structure was anticipated to be lacking the m6A modification site. Thus, the diminished expression of CM-248aa was attributed to the reduced expression of circMTHFD2L. Nevertheless, the translation of circRNA may be controlled by IRES and m6A modification. There are currently two mechanisms for understanding by which circRNA encodes proteins. One functions as the equivalent of parental proteins, such as MAPK1-109aa produced by circMAPK1, while AIXN1-295aa produced by circAXIN1, which have been observed in gastric cancer and competitively interact with upstream proteins alongside their parental protein [47, 48]. Another mechanism involves utilising a distinct binding function, exemplified by circ-DIDO1-encoded DIDO1-529aa, which directly interacts with poly (ADP-ribose) polymerase 1 and hinders its activity [49]. The current study shows that CM-248aa effectively suppressed the proliferation and metastasis of GC cells both in vivo and in vitro through its distinctive binding mechanism. The functioning of this mechanism relied on the nuclear localisation of CM-248aa, which was different from its parent protein. Mechanistically, CM-248aa acts as a competitive inhibitor of the SET-protein phosphatase 2A interaction by specifically targeting the acidic domain of the SET nuclear oncogene. This resulted in the stimulation of dephosphorylation of AKT, extracellular signal-regulated kinase, and P65 [46].

Due to the partial overlap in amino acid sequence between the translated circRNA and the protein produced by the parent gene, the circRNA can compete with enzymes to control the post-translational modification of the whole protein. Recently, it was discovered that circFBXW7 encodes the FBXW7-185aa protein, inhibiting the progression of breast cancer [50]. This study revealed that FBXW7-185aa engaged in competitive interaction with the deubiquitinating enzyme USP28, hence inhibiting the binding of USP28 to FBXW7. Consequently, this interaction increased the expression of FBXW7 [50]. The *circ-SPHRH* gene encodes SHPRH-146aa, which enhances the expression of the full-length SHPRH protein by protecting it against ubiquitination [51]. Another study showed that circ-β-catenin produces β-catenin-370aa, which competes with GSK3β and acts as a decoy, preventing GSK3β from binding to full-length β-catenin. This ultimately hinders GSK3β-induced destruction of β-catenin, resulting in its antagonisation [52]. Another study has demonstrated that circ-AKT3 encodes AKT3-174aa, which competes with AKT isoforms to bind to pPDK1 and decrease AKT-Thr308 phosphorylation [53]. This suggests that AKT3-174aa has a negative regulatory function in moderating the intensity of the phosphatidylinositol 3-kinase (PI3K)/AKT signalling pathway.

Furthermore, ciRS-7 is not involved in regulating *APP* and *BACE1* gene expression. Instead, it reduces the protein levels of APP and BACE1 by promoting their degradation via the proteasome and lysosome. Consequently, overexpression of ciRS-7 minimises the generation of β-amyloid, indicating a potential neuroprotective role of ciRS-7. These findings also suggest that ciRS-7 modulates APP and BACE1 levels in a nuclear factor-κB (NF-κB)-dependent manner. The ciRS-7 inhibits the translation of NF-κB and induces its cytoplasmic localisation, thus derepressing the expression of UCHL1, which promotes APP and BACE1 degradation. Additionally, the APP can reduce the level of ciRS-7, revealing a mutual regulation of ciRS-7 and APP [54].

6 Summary

Despite their role as crucial mediators of various essential cellular processes, circRNAs were formerly classified as non-coding due to the arbitrary site restriction. Although in their early stages, recent studies have demonstrated that circular RNAs play crucial roles in human malignancies. The translation of circRNAs will reveal a hitherto unknown human proteome, allowing us to gain a deeper understanding of the significance of circRNAs in human cancer. This area has been inadequately investigated thus far. There is evidence that certain circRNAs carry out their biological activity through both the production of peptides and through RNA-based regulatory mechanisms. Moreover, the interactions between circRNAs and proteins result in proteins' modifications, influencing several biological processes. This has been a motivation to develop innovative clinical approaches that manage circRNAs by targeting their interactions with proteins.

References

1. Sanger HL, Klotz G, Riesner D, Gross HJ, Kleinschmidt AK (1976) Viroids are single-stranded covalently closed circular RNA molecules existing as highly base-paired rod-like structures. Proc Natl Acad Sci USA 73(11):3852–3856. https://doi.org/10.1073/pnas.73.11.3852
2. Du WW, Fang L, Yang W, Wu N, Awan FM, Yang Z, Yang BB (2017) Induction of tumor apoptosis through a circular RNA enhancing Foxo3 activity. Cell Death Differ 24(2):357–370. https://doi.org/10.1038/cdd.2016.133
3. Memczak S, Jens M, Elefsinioti A, Torti F, Krueger J, Rybak A, Maier L, Mackowiak SD, Gregersen LH, Munschauer M, Loewer A, Ziebold U, Landthaler M, Knocks C, le Noble F, Rajewsky N (2013) Circular RNAs are a large class of animal RNAs with regulatory potency. Nature 495:333–338
4. Hansen TB, Jensen TI, Clausen BH, Bramsen JB, Finsen B, Damgaard CK, Kjems J (2013) Natural RNA circles function as efficient microRNA sponges. Nature 495:384–388
5. Zheng Q, Bao C, Guo W, Li S, Chen B, Luo Y, Lyu D, Li Y, Shi G, Liang L, Gu J, He X, Huang S (2016) Circular RNA profiling reveals an abundant circHIPK3 that regulates cell growth by sponging multiple miRNAs. Nat Commun 7:11215. https://doi.org/10.1038/ncomms11215
6. Kristensen LS, Hansen TB, Venø MT, Kjems J (2018) Circular RNAs in cancer: opportunities and challenges in the field. Oncogene 37(5):555–565. https://doi.org/10.1038/onc.2017.361
7. Lei M, Zheng G, Ning Q, Zheng J, Dong D (2020) Translation and functional roles of circular RNAs in human cancer. Mol Cancer 19(1):30. https://doi.org/10.1186/s12943-020-1135-7
8. Salzman J, Gawad C, Wang PL, Lacayo N, Brown PO (2012) Circular RNAs are the predominant transcript isoform from hundreds of human genes in diverse cell types. PLoS One 7(2):e30733. https://doi.org/10.1371/journal.pone.0030733
9. Derks KW, Misovic B, van den Hout MC, Kockx CE, Gomez CP, Brouwer RW, Vrieling H, Hoeijmakers JH, van IWF, Pothof J (2015) Deciphering the RNA landscape by RNAome sequencing. RNA Biol 12(1):30–42. https://doi.org/10.1080/15476286.2015.1017202
10. Szabo L, Salzman J (2016) Detecting circular RNAs: bioinformatic and experimental challenges. Nat Rev Genet 17(11):679–692. https://doi.org/10.1038/nrg.2016.114
11. Salzman J, Chen RE, Olsen MN, Wang PL, Brown PO (2013) Cell-type specific features of circular RNA expression. PLoS Genet 9(9):e1003777. https://doi.org/10.1371/journal.pgen.1003777
12. Venø MT, Hansen TB, Venø ST, Clausen BH, Grebing M, Finsen B, Holm IE, Kjems J (2015) Spatio-temporal regulation of circular RNA expression during porcine

embryonic brain development. Genome Biol 16:245. https://doi.org/10.1186/s13059-015-0801-3

13. Yang Q, Wu J, Zhao J, Xu T, Zhao Z, Song X, Han P (2018) Circular RNA expression profiles during the differentiation of mouse neural stem cells. BMC Syst Biol 12(Suppl 8):128. https://doi.org/10.1186/s12918-018-0651-1
14. Nicolet BP, Engels S, Aglialoro F, van den Akker E, von Lindern M, Wolkers MC (2018) Circular RNA expression in human hematopoietic cells is widespread and cell-type specific. Nucleic Acids Res 46(16):8168–8180. https://doi.org/10.1093/nar/gky721
15. Jeck WR, Sorrentino JA, Wang K, Slevin MK, Burd CE, Liu J, Marzluff WF, Sharpless NE (2013) Circular RNAs are abundant, conserved, and associated with ALU repeats. RNA 19(2):141–157. https://doi.org/10.1261/rna.035667.112
16. Ashwal-Fluss R, Meyer M, Pamudurti NR, Ivanov A, Bartok O, Hanan M, Evantal N, Memczak S, Rajewsky N, Kadener S (2014) circRNA biogenesis competes with pre-mRNA splicing. Mol Cell 56(1):55–66. https://doi.org/10.1016/j.molcel.2014.08.019
17. Li H, Yang F, Hu A, Wang X, Fang E, Chen Y, Li D, Song H, Wang J, Guo Y, Liu Y, Li H, Huang K, Zheng L, Tong Q (2019) Therapeutic targeting of circ-CUX1/EWSR1/MAZ axis inhibits glycolysis and neuroblastoma progression. EMBO Mol Med 11(12):e10835. https://doi.org/10.15252/emmm.201910835
18. Yang F, Fang E, Mei H, Chen Y, Li H, Li D, Song H, Wang J, Hong M, Xiao W, Huang K, Zheng L, Tong Q (2019a) Cis-acting circ-CTNNB1 promotes β-catenin signaling and cancer progression via DDX3-mediated transactivation of YY1. Cancer Res 79(3):557–571. https://doi.org/10.1158/0008-5472.CAN-18-1559
19. Doller A, El-Sayed A, Huwiler A, Müller R, Radeke HH, Pfeilschifter J, Eberhardt W (2008) Posttranslational modification of the AU-rich element binding protein HuR by protein kinase Cdelta elicits angiotensin II-induced stabilization and nuclear export of cyclooxygenase 2 mRNA. Mol Cell Biol 28(8):2608–2625. https://doi.org/10.1128/MCB.01530-07
20. Vázquez-Chantada M, Fernández-Ramos D, Embade N, Martínez-Lopez N, Varela-Rey M, Woodhoo A, Luka Z, Wagner C, Anglim PP, Finnell RH, Caballería J, Lairf-Offringa IA, Gorospe M, Lu SC, Mato JM, Martínez-Chantar ML (2010) HuR/methyl-HuR and AUF1 regulate the MAT expressed during liver proliferation, differentiation, and carcinogenesis. Gastroenterology 138(5):1943–1953. https://doi.org/10.1053/j.gastro.2010.01.032
21. Chen Y, Yang F, Fang E, Xiao W, Mei H, Li H, Li D, Song H, Wang J, Hong M, Wang X, Huang K, Zheng L, Tong Q (2019) Circular RNA circAGO2 drives cancer progression through facilitating HuR-repressed functions of AGO2-miRNA complexes. Cell Death Differ 26(7):1346–1364. https://doi.org/10.1038/s41418-018-0220-6
22. Yang F, Hu A, Li D, Wang J, Guo Y, Liu Y, Li H, Chen Y, Wang X, Huang K, Zheng L, Tong Q (2019b) Circ-HuR suppresses HuR expression and gastric cancer progression by inhibiting CNBP transactivation. Mol Cancer 18(1):158. https://doi.org/10.1186/s12943-019-1094-z
23. Li X-X, Xiao L, Chung HK, Ma X-X, Liu X, Song J-L, Jin C-Z, Rao JN, Gotospe M, Wanga JY (2020) Interaction between HuR and circPABPN1 modulates autophagy in the intestinal epithelium by altering ATG16L1 translation. Mol Cell Biol 40(6):e00492–e00419. https://doi.org/10.1128/MCB.00492-19
24. Rossi F, Beltran M, Damizia M, Grelloni C, Colantoni A, Setti A, Di Timoteo G, Dattilo D, Centrón-Broco A, Nicoletti C, Fanciulli M, Lavia P, Bozzoni I (2022) Circular RNA ZNF609/CKAP5 mRNA interaction regulates microtubule dynamics and tumorigenicity. Mol Cell 82(1):75–89. https://doi.org/10.1016/j.molcel.2021.11.032
25. Lebedeva S, Jens M, Theil K, Schwanhäusser B, Selbach M, Landthaler M, Rajewsky N (2011) Transcriptome-wide analysis of regulatory interactions of the RNA-binding protein HuR. Mol Cell 43(3):340–352. https://doi.org/10.1016/j.molcel.2011.06.008
26. Jin C, Zhao W, Zhang Z, Liu W (2019) Silencing circular RNA circZNF609 restrains growth, migration and invasion by up-regulating microRNA-186-5p in prostate cancer. Artif Cells Nanomed Biotechnol 47(1):3350–3358. https://doi.org/10.1080/21691401.2019.1648281
27. Zhu L, Liu Y, Yang Y, Mao X-M, Yin Z-D (2019) CircRNA ZNF609 promotes growth and metastasis of nasopharyngeal carcinoma by competing with microRNA-150-5p. Eur Rev Med Pharmacol Sci 23(7):2817–2826
28. He Y, Huang H, Jin L, Zhang F, Zeng M, Wei L, Tang S, Chen D, Wang W (2020) CircZNF609 enhances hepatocellular carcinoma cell proliferation, metastasis, and stemness by activating the Hedgehog pathway through the regulation of miR-15a-5p/15b-5p and GLI2 expressions. Cell Death Dis 11(5):358. https://doi.org/10.1038/s41419-020-2441-0
29. Liu Z, Liu F, Wang F, Yang X, Guo W (2021) CircZNF609 promotes cell proliferation, migration, invasion, and glycolysis in nasopharyngeal carcinoma through regulating HRAS via miR-338-3p. Mol Cell Biochem 476(1):175–186. https://doi.org/10.1007/s11010-020-03894-5
30. Kristensen LS, Andersen MS, Stagsted LVW, Ebbesen KK, Hansen TB, Kjems J (2019) The biogenesis, biology and characterization of circular RNAs. Nat Rev Genet 20(11):675–691. https://doi.org/10.1038/s41576-019-0158-7
31. Conn VM, Hugouvieux V, Nayak A, Conos SA, Capovilla G, Cildir G, Jourdain A, Tergaonkar V, Schmid M, Zubieta C, Conn SJ (2017) A circRNA from SEPALLATA3 regulates splicing of its cognate mRNA through R-loop formation. Nat Plants 3:17053. https://doi.org/10.1038/nplants.2017.53

32. Yang X, Zhang S, He C, Xue P, Zhang L, He Z, Zang L, Feng B, Sun J, Zheng M (2020) METTL14 suppresses proliferation and metastasis of colorectal cancer by down-regulating oncogenic long noncoding RNA XIST. Mol Cancer 19(1):46. https://doi.org/10.1186/s12943-020-1146-4
33. Shen Y, Li C, Zhou L, Huang JA (2021) G protein-coupled oestrogen receptor promotes cell growth of non-small cell lung cancer cells via YAP1/QKI/circNOTCH1/m6A methylated NOTCH1 signalling. J Cell Mol Med 25(1):284–296. https://doi.org/10.1111/jcmm.15997
34. Du WW, Yang W, Li X, Awan FM, Yang Z, Fang L, Lyu J, Li F, Peng C, Xie Y, Zhang Y, He C, Wu N, Zhang C, Sdiri M, Dong J, Ma J, Gao C, Hibberd S, Yang BB (2018) A circular RNA circ-DNMT1 enhances breast cancer progression by activating autophagy. Oncogene 37(44):5829–5842. https://doi.org/10.1038/s41388-018-0369-y
35. Wang X, Xing L, Yang R, Chen H, Wang M, Jiang R, Zhang L, Chen J (2021) The circACTN4 interacts with FUBP1 to promote tumorigenesis and progression of breast cancer by regulating the expression of proto-oncogene MYC. Mol Cancer 20(1):91. https://doi.org/10.1186/s12943-021-01383-x
36. Li F, Cai Y, Deng S, Yang L, Liu N, Chang X, Jing L, Zhou Y, Li H (2021) A peptide CORO1C-47aa encoded by the circular noncoding RNA circ-0000437 functions as a negative regulator in endometrium tumor angiogenesis. J Biol Chem 297(5):101182. https://doi.org/10.1016/j.jbc.2021.101182
37. Jiang Z, Jiang Y (2023) Circular RNA CircPDS5B impairs angiogenesis following ischemic stroke through its interaction with hnRNPL to inactivate VEGF-A. Neurobiol Dis 181:106080. https://doi.org/10.1016/j.nbd.2023.106080
38. Yoon G, Lim Y-H, Jo D, Ryu J, Song J, Kim Y-K (2021) Obesity-linked circular RNA circTshz2-2 regulates the neuronal cell cycle and spatial memory in the brain. Mol Psychiatry 26(11):6350–6364. https://doi.org/10.1038/s41380-021-01303-x
39. Li H, Li J, Hu Y, Zhang R, Gu X, Wei Y, Zhang S, Chen X, Wei L, Li X, Gu S, Jin J, Huang H, Zhou H, Yang C (2023) FOXO3 regulates Smad3 and Smad7 through SPON1 circular RNA to inhibit idiopathic pulmonary fibrosis. Int J Biol Sci 19(10):3042–3056. https://doi.org/10.7150/ijbs.80140
40. Wang L, Park HJ, Dasari S, Wang S, Kocher JP, Li W (2013) CPAT: coding-potential assessment tool using an alignment-free logistic regression model. Nucleic Acids Res 41(6):e74. https://doi.org/10.1093/nar/gkt006
41. Sinha T, Panigrahi C, Das D, Chandra Panda A (2022) Circular RNA translation, a path to hidden proteome. Wiley Interdiscip Rev RNA 13(1):e1685. https://doi.org/10.1002/wrna.1685
42. Usman N, Cedergren R (1992) Exploiting the chemical synthesis of RNA. Trends Biochem Sci 17(9):334–339. https://doi.org/10.1016/0968-0004(92)90306-t
43. Petkovic S, Müller S (2018) Synthesis and engineering of circular RNAs. Methods Mol Biol 1724:167–180. https://doi.org/10.1007/978-1-4939-7562-4_14
44. Legnini I, Di Timoteo G, Rossi F, Morlando M, Briganti F, Sthandier O, Fatica A, Santini T, Andronache A, Wade M, Laneve P, Rajewsky N, Bozzoni I (2017) Circ-ZNF609 is a circular RNA that can be translated and functions in myogenesis. Mol Cell 66(1):22–27. https://doi.org/10.1016/j.molcel.2017.02.017
45. Wu J, Guo X, Wen Y, Huang S, Yuan X, Tang L, Sun H (2023) N6-methyladenosine modification opens a new chapter in circular RNA biology. Front Cell Dev Biol 9:709299. https://doi.org/10.3389/fcell.2021.709299
46. Liu H, Fang D, Zhang C, Zhao Z, Liu Y, Zhao S, Zhang N, Xu J (2023) Circular MTHFD2L RNA-encoded CM-248aa inhibits gastric cancer progression by targeting the SET-PP2A interaction. Mol Ther 31(6):1739–1755. https://doi.org/10.1016/j.ymthe.2023.04.013
47. Jiang T, Xia Y, Lv J, Li B, Li Y, Wang S, Xuan Z, Xie L, Qiu S, He Z, Wang L, Xu Z (2021) A novel protein encoded by circMAPK1 inhibits progression of gastric cancer by suppressing activation of MAPK signaling. Mol Cancer 20(1):66. https://doi.org/10.1186/s12943-021-01358-y
48. Peng Y, Xu Y, Zhang X, Deng S, Yuan Y, Luo X, Hossain MT, Zhu X, Du K, Hu F, Chen Y, Chang S, Feng X, Fan X, Ashktorab H, Smoot D, Meltzer SJ, Hou G, Wei Y, Li S, Qin Y, Jin Z (2021) A novel protein AXIN1-295aa encoded by circAXIN1 activates the Wnt/beta-catenin signaling pathway to promote gastric cancer progression. Mol Cancer 20:158. https://doi.org/10.1186/s12943-021-01457-w
49. Zhang Y, Jiang J, Zhang J, Shen H, Wang M, Guo Z, Zang X, Shi H, Gao J, Cai H, Fang X, Qian H, Xu W, Zhang X (2021) CircDIDO1 inhibits gastric cancer progression by encoding a novel DIDO1-529aa protein and regulating PRDX2 protein stability. Mol Cancer 20(1):101. https://doi.org/10.1186/s12943-021-01390-y
50. Ye F, Gao G, Zou Y, Zheng S, Zhang L, Ou X, Xie X, Tang H (2019) circFBXW7 inhibits malignant progression by sponging miR-197-3p and encoding a 185-aa protein in triple-negative breast cancer. Mol Ther Nucleic Acids 18:88–98. https://doi.org/10.1016/j.omtn.2019.07.023
51. Zhang M, Huang N, Yang X, Luo J, Yan S, Xiao F, Chen W, Gao X, Zhao K, Zhou H (2018) A novel protein encoded by the circular form of the SHPRH gene suppresses glioma tumorigenesis. Oncogene 37(13):1805–1814. https://doi.org/10.1038/s41388-017-0019-9
52. Liang W-C, Wong C-W, Liang P-P, Shi M, Cao Y, Rao S-T, Wing Tsui SK, Waye MM-Y, Zhang Q (2019) Translation of the circular RNA circβ-catenin promotes liver cancer cell growth through activation of the Wnt pathway. Genome Biol 20(1):84. https://doi.org/10.1186/s13059-019-1685-4

53. Xia X, Li X, Li F, Wu X, Zhang M, Zhou H, Huang N, Yang X, Xiao F, Liu D (2019) A novel tumor suppressor protein encoded by circular AKT3 RNA inhibits glioblastoma tumorigenicity by competing with active phosphoinositide-dependent kinase-1. Mol Cancer 18(1):131. https://doi.org/10.1186/s12943-019-1056-5

54. Shi Z, Chen T, Yao Q, Zheng L, Zhang Z, Wang J, Hu Z, Cui H, Han Y, Han X, Zhang K, Hong W (2017) The circular RNA ciRS-7 promotes APP and BACE1 degradation in an NF-κB-dependent manner. FEBS J 284(7):1096–1109. https://doi.org/10.1111/febs.14045

Part V

Circular RNAs as Potential Disease Biomarkers

Circular RNA in Exosomes and Its Clinical Significance

12

Mohammad Kamalabadi Farahani, Sepehr Zamani, and Seyede Salehe Attari

Abstract

Circular RNAs (circRNAs) are highly conserved novel class of endogenous noncoding RNAs. CircRNAs have a stable covalently closed ring structures with no 5′-end cap or 3′-end poly (A) tail. These circular structures are formed by reverse splicing, mainly by means of a noose structure or intron complementary pairing. In clinical applications of cells, exosomes, as a tiny discoid vesicles with a diameter of 40–100 nm that are secreted by cells under physiological and pathological conditions, are considered in recent years. Exosomes play an important role in cell–cell communication by carrying DNA, microRNAs, mRNAs, proteins, and circRNAs. In this chapter, we summarize the biological functions of exosomal circRNAs and further reveal the clinical significance of exosomal circRNAs and their potential roles in different diseases, providing a scientific basis for the diagnosis, treatment, and prognosis of a wide variety of diseases.

Keywords

Circular RNAs · Exosomes · Disease biomarker · Diagnosis · Prognosis

1 Introduction

Disease diagnosis is a key step in clinical practice. The detection of soluble biomarkers from biofluids became a critical method in the early diagnosis of diseases. Recently, exosomes, one of the major subtypes of extracellular vesicles, have received a lot of attention in clinical research. After contact and fusion of multiple vesicles with a cell membrane, they are nanosized round or oval capsules that release from cells [1]. After their first identification by Wolf in 1967 [2], today, it has been demonstrated that exosomes are carriers of nucleic acids, proteins, lipids, and biological molecules, which play a crucial role in intercellular communication and regulation. Exosomes received increased attention due to their wide distribution in body fluids and their ability to reflect physiological and pathological conditions. Most cell types can release exosomes of different sizes, compositions, and functions.

M. K. Farahani (✉)
Department of Tissue Engineering, School of Medicine, Shahroud University of Medical Sciences, Shahroud, Iran
e-mail: kamalabadi@shmu.ac.ir

S. Zamani · S. S. Attari
Department of Tissue Engineering, School of Medicine, Shahroud University of Medical Sciences, Shahroud, Iran

Student Research Committee, School of Medicine, Shahroud University of Medical Sciences, Shahroud, Iran

J. Xiao (ed.), *Advances in Circular RNAs*, Advances in Experimental Medicine and Biology 1485,
https://doi.org/10.1007/978-981-96-9428-0_12

More importantly, exosomes are widely present in almost all biofluids, including cell supernatant, blood, plasma, saliva, urine, serum, and breast milk. Exosomal contents, including RNAs, DNAs, proteins, and lipids, can participate in physiological processes such as intercellular communication and material transport [3].

Noncoding RNAs (ncRNAs) are a class of endogenous RNAs that do not code for proteins, including microRNAs (miRNAs), long noncoding RNAs (lncRNAs), and circular RNAs (circRNAs), that are involved in posttranscriptional regulation. Initially, circRNAs have been considered to be byproducts of aberrant splicing or intermediate products of intron lariats that have escaped degradation [4]. Rapid development of molecular detection techniques indicated that circRNA is a type of ncRNA molecule that is composed of hundreds or even thousands of nucleotides and can be directly connected to the 5′ and 3′ ends of linear RNA. CircRNA can also be generated by reverse splicing. The downstream 5′ splice site (splice donor) is connected to the upstream 3′ splice site (splice acceptor) found in prokaryotes, eukaryotes, and viruses [5]. According to their sequences, circRNAs are classified into three categories: intronic circRNAs, exonic circRNAs, and exon–intron circRNAs [6]. Recently, a new type of circRNA, called tRNA intronic circular RNAs, has been identified [7]. CircRNAs have a number of biological functions and play an essential role in the occurrence, development, invasion, diagnosis, or prognosis of tumors, and contribute to the development of cardiovascular disease as well, neurological, orthopedic, and other disorders [6].

There is particularly strong evidence of exosomal circRNAs regulating gene expression and function in recipient cells. Exosomal circRNAs can affect normal physiological metabolic activities and participate in the development of various diseases, including tumor growth, neurodegenerative disease, and metabolic syndrome. Exosomal circRNAs are a promising source of diagnostic biomarkers for human diseases [8]. In this chapter, we focus on recent studies exploring exosomal circRNAs as biomarkers for human diseases.

2 Circular RNAs

Circular RNAs (circRNAs) are a new class of mammalian ncRNAs with distinctive structures which differ from the others. They are composed of single-stranded RNA transcripts with a characteristic covalently closed structure, which is formed by a direct covalent coupling between their 5′ and 3′ ends. The closure of the loop with unreachable ends prevents circRNAs from being degraded by enzyme RNase R, and so their biochemical differentiation is comparable to that of different types of RNA [9]. In comparison to the function RNAs and ncRNAs, they also show certain additional properties. These include lack of N7-methylated guanosine capping structure at 5′ end and polyadenylated tail at the 3′ end [10]. CircRNA is primarily generated from precursor mRNA through a "back-splicing" process. Circularization begins with the formation of stem-loop structures, and then back-splicing continues by joining a splice donor to a splice acceptor downstream [11, 12]. Intronic repeats facilitate such reactions by providing for the possibility of joining splice sites in proximity through complementary sequences and couplings [11].

In 1979, a microscopic identification of the circular form of an RNA taken from cytoplasm of HeLa carcinoma cells led to the initial detection of circRNAs [13]. Although Sanger et al. are thought to have reported the first circular structures in viroids 3 years ago [14], today, thousands of genetic loci for circRNAs have been found in mammals through the use of modern and precise techniques to detect RNA [15]. In human cell, thousands of circRNAs could be found [16]; in addition, the number of recorded circRNAs continues to grow, and researchers need to develop databases that allow them to store and use these data [17]. Besides high expression, circRNAs show tissue and developmental stage specificity, which indicates a number of opportunities for them [18].

Based on their sources, circRNAs are classified into four subtypes: exonic circRNAs transcribed from exons (ecircRNA), intronic circRNA (ciRNA), exonic–intronic circRNA (EIciRNA) circularized with both exons and introns simultaneously, and intergenic circRNAs [18]. The main myths used to identify and recode circRNAs are computational approaches using high-throughput RNA sequencing (RNA-seq), quantitative reverse transcription-polymerase chain reaction (qRT-PCR), and conventional applications [19]. However, other types of circRNAs have been identified, which are made by various mechanisms instead of originating in the back sequencing method. These include viral circRNAs derived from host enzymes, circRNA intermediates formed when rRNAs are processed, some ncRNAs with circular conformations such as RNase P, and also circRNAs derived from exons and introns after splicing [20]. CircRNAs are primarily found in lower volumes and rarely accumulate compared to other ncRNA classes [15]; however, some of these have been shown to be significantly larger than their counterparts due to a variety of characteristics such as sequence preservation and chemical stability which suggest an important role for them [21].

Enormous studies have shown that circRNAs play an important role in the physiology. As decoys, they are known to function as regulators of gene expression by modulating the binding of miRNAs to target mRNAs, thereby influencing mRNA translation [22]. Instead of activating miRNAs, circRNAs exert their functions through interactions with proteins, regulation of gene transcription, mRNA translation, and epigenetic control mechanisms [23]. CircRNAs were also found to be involved in important processes such as embryonic development, metabolism, and cellular stress, rather than regulatory roles in cell cycle and proliferation [24]. Therefore, dysregulation of circRNAs is associated with several human diseases such as various cancers, neurodegenerative diseases, diabetes, and cardiovascular diseases [24]. The role of circRNAs in the development and progression of cancer has attracted considerable attention over recent years, as further transcripts with functions that affect tumor behavior have been explored along with potential applications for diagnosis and prognosis. Therefore, the general public and field scientists can be benefited from a review of roles and potential.

Through linear splicing, ciRNAs and EIciRNAs compete to regulate the transcription of parental genes. To regulate the transcription of parent genes, ciRNAs can act as regulators of RNA polymerase II [25]. EIciRNAs can interact with U1 small nuclear RNA (snRNP) to form the EIciRNA-U1 snRNP complex, which then interacts with the transcriptional complex of RNA Pol II to affect parental gene expression [26, 27]. Furthermore, noncoding intron transcripts such as ciankrd52 are positive regulators of RNA Pol II transcriptions and have cis-regulatory effects on their parental coding genes which are associated with the mechanism to extend RNA Pol II [28]. Subsequently, a recent study showed that circ-DAB1 upregulates the expression of the recombinant signal-binding protein of the immunoglobulin kappa J region (RBPJ), resulting in the transfer of RBPJ to the promoter of the DAB adapter protein 1 (DAB1) lead, it was shown to increase binding and thereby activation, transcription of the parental gene DAB1 [29]. CircRNAs are also capable of acting as sponges of proteins, combining with RNA-binding proteins, and forming RNA protein complexes that affect protein expression [30]. For example, the RNA splicing factor MBL is an RBP that binds to the second exon of its parent gene and promotes its cyclization to form circMBL in drosophila [26]. At the same time, multiple sites on circMBL bind to MBL protein and decrease its effective concentration [26]. CircRNAs may also interact with specific target proteins and participate in cell proliferation, differentiation, and apoptosis [31]. For example, circ-Foxo3, cyclin-dependent kinase inhibitor 1 (p21), and cyclin-dependent kinase 2 (CDK2) form the ternary complex circ-Foxo3-p21-CDK2, which inhibits the function of CDK2 and blocks the cell cycle. Moreover, it promotes the proliferation of cells by directly silencing endogenous circoFox3 [32]. Another report indicates that circ-ANRIL may bind to PES1 protein, thereby preventing preRNA binding and rRNA maturation mediated by exonuclease. Therefore, circANRIL inhibits ribosome biogenesis, leading to activation of p53, a subsequent increase in apoptosis, and a decrease in proliferative activity [33]. The other effects of circRNAs binding to proteins need to be studied in greater detail, due to their many functions. Initially, circRNAs were thought to be uncoding RNA molecules, but in recent years, a series of studies have shown that they are involved in protein translation [34–36], which disproved a long-believed concept. Initially, circRNAs were thought to be uncoding RNA molecules, but a number of studies in recent years have shown that they are involved in protein translation [37]. For example, circ-PINT can be translated via its IRES into the PINT87aa polypeptide, which contains 87 amino acids. PINT87aa binds to the polymerase-associated factor complex gene and inhibits the development of malignant glioma [38]. The IRES domain of circRNAs was subsequently predicted to be a binding site for many RBPs, including HUR and PTB, which modulate the translation of IRES element-driven proteins [39]. As research progresses, it has been demonstrated that circRNAs with no IRES components can be converted into a number of functional proteins. For example, N6-methyladenosine (m6A) not only affects mRNA translation under heat shock stress conditions [40] but also regulates the translation of circRNAs. The presumed mechanism is m6A modification by the METTL3/METTL14-WTAP protein complex and the regulation of m6A demodification by FTO, and the site containing the m6A modification ultimately initiates protein translation by recruiting YTHDF3 and thus eIF4G2 [41], which fills a gap in the study of the chemical modifications of circRNAs. In addition, it has been established that circRNAs can be used for protein translation in combination with overlap of codons, which is a very rare form of translation [42]. In circAKT3, a new functional protein, 3-174aa, is generated by overlap between codons. This protein affects the phosphorylation of AKT2/3 molecules by binding to p-PDK1 and then negatively regulates the PI3K/AKT signaling pathway, preventing brain tumors from developing [43].

3 Exosomes

Exosomes are single membrane vesicles with the same topology as the cell, a diameter of 30–200 nm, and an otherwise normal morphology. Even for exosomes secreted from a single-cell line, the exosome size varies significantly, by any measure. Nevertheless, there are inherent biases and estimates of exosomes size in each technique used for exposing [44–46]. Exosomes, which are approximately five to ten times smaller than the wavelength of light, cannot be detected by conventional light microscopy. On the other hand, if exosomes are labeled with fluorescent probes specific for exosome bilayer, exosome proteins, nucleic acids, or carbohydrates, conventional fluorescence microscopy can detect exosomes [47]. However, as exosomes are much smaller than the wavelength of light and consist primarily of water, it is still not clear whether they generate a sufficient scattering signature to enable flow techniques to detect them reliably. The use of a fluorescence trigger technique, which may also indicate exosome size in certain iterations, could be one possible solution [48]. However, the current requirement of 50 or more fluorophores per exosome to trigger a signal is limiting this approach. In addition, significant empirical bias in favor of larger exosomes and higher abundance antigens is likely to result from this approach for exosome analysis. Unbiased cryoelectron microscopy experiments have provided important insights into the range of exosomes and other EVs circulating in human biofluids, although accurate exosome measurements remain an inherently elusive goal [45]. Almost, all have an ordinary spheroid morphology. Of these, approximately half were within the expected range of 30–200 nm diameter of exosomes, with the remainder mostly larger microvesicles. In addition, trace amounts of more strange membranous structures have been seen, some with several membranes, a few having elongated tubule like morphologies, and others that are not so uncommon. Technically, these unusual structures may be caused by directed biological mechanisms, physical forces leading to fragmentation, and mechanical resealing; they consistently account for at least a portion of the EVs detected in biological fluids or tissue culture supernatants[49]. Other particles of lipoproteins are among the contaminating elements in a typical exosome preparation [50], exomeres [a class of small secreted particles consisting of dense protein assemblies coated with a single phospholipid shell [51]], RNA–protein particles, and protein aggregates. Exosomes have been described as having a density of ~1.1–1.2 g/mL [52]. The protein:lipid ratio, which differs considerably from vesicle to vesicle, is an influence on the density of these cells. In addition, it has been established that expression of only one exosomal cargo protein can lead to increased exosome density as well as significant changes in exosomes size and shape [53]. As a result of hydroxyapatite crystallization by osteogenic exosomes, exosome density may also be influenced by exosome metabolic pathways, for example [54]. These considerations show that the exosome size, shape, and density are largely dependent on the particular proteins, lipids, enzymes, or minerals present in each single exosome; however, these factors do not define characteristics of an exosome. The properties of the exosomes have a direct bearing on the way in which they are cleaned. As previously mentioned, exosomes and microvesicles are traditionally defined based on differential centrifugation, where microvesicles are pelleted with one or two 10,000× g spins for 30 min, whereas exosomes are pelleted with 10,000× g spins for 30 min. The supernatant of the g spin is pelleted by centrifugation at 70,000–100,000× g for 90–120 min. However, there are more than just exosomes in these exosome pellet fragments. This pellet may contain significant amounts of chylomicrons and other lipoprotein particles, especially if the subject has recently been fed, as long as the starting material is biofluid such as plasma. The 100,000× g pellet may also contain exomeres, large protein aggregates, ribonucleoprotein (RNP) complexes [e.g., Ago2-miRNAs (microRNAs)], and small microsomes created by mechanical shearing of cell membranes. In addition, the physics of centrifugation and the geometry of centrifuge tubes are important [55].

Exosomes are composed of a wide variety of transmembrane proteins, lipid anchoring membrane proteins, peripheral and adjacent membrane proteins, and soluble exosome proteins (Fig. 12.1). Some of the more important and notable proteins, which have been reliably identified in exosome membranes on their surface or lumen, are described in this part [56].

This part describes some of the most important and significant proteins that have been reliably identified in exosome membranes on their surface or lumen. Among them, exosomal protein components are mainly divided into two categories, one is the public components, which participate in the process of vesicle formation and secretion, that is, exosomes are ubiquitous, including membrane transport and fusion-related proteins (like Rab, GTPases), heat shock proteins (like HSP70, HSP90), four-transmembrane protein superfamily (like CD63,CD81), ESCRT complex–related proteins (like Tsg101, Alix), integrins, etc.; the other one is a specific component, which is closely related to its progenitor cells, that is, cell-specific, such as CD45 and MHC-II derived from antigen-presenting cells. Its applications are becoming more and more widespread as a result of the in-depth exosome study.

4 Circular RNA in Exosomes

By regulating a wide range of biological processes between cells, exosomes have become a critical mediator of intercellular communication, both locally and systemically. CircRNA is a novel member of endogenous noncoding RNAs with widespread distribution and a wide range of cellular functions. For their enrichment and stability in exosomes, circular RNAs have recently been identified. CircRNAs are abundant and stably expressed in the exosomes [57], and they have been shown in several studies to be capable of packaging and functioning within exosomes. For example, circ_0088300 has been shown to be driven by KHDRBS3 and packaged into exosomes [58], and hnRNPA2B1 has been reported to package small RNA molecules, including circNEIL3 and circCCAR1, into exosomes [59]. However, there is still a lack of clarity as to the mechanism whereby special circRNAs are selectively packaged for exosomes. Exosomal circRNAs have been shown to be commonly used as microRNA sponge and stabilizer [59] and protein sponges or decoys [60]. They may also be used to regulate gene transcription of parents or serve as templates for translation [61]. The precise roles

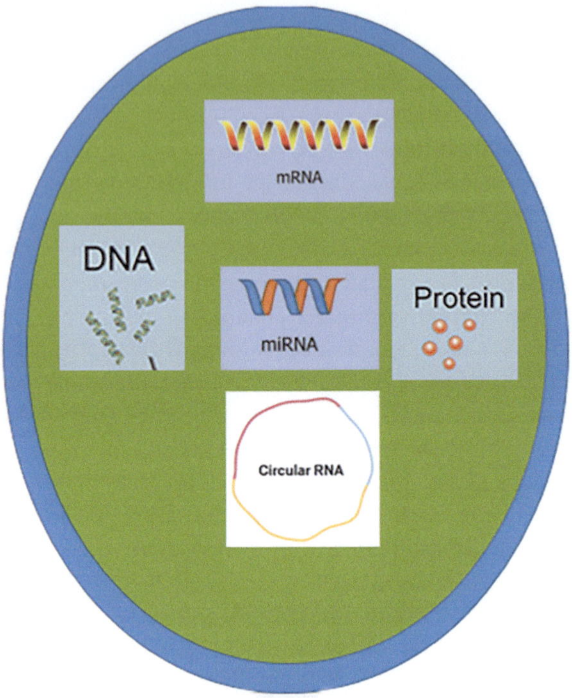

Fig. 12.1 Exosomes are tiny discoid vesicles with that are secreted by cells. Exosomes play an important role in cell–cell communication by carrying DNA, microRNAs, mRNAs, proteins, and circRNAs

and mechanisms of gene regulation remain largely elusive, and exosomal circRNAs have potential applications as disease biomarkers and novel therapeutic targets. In the following, we will talk more about the role and clinical importance of the exosomal circRNAs.

5 Clinical Significance of Circular RNA Embedded in Exosomes

5.1 Clinical Significance of Exosomal CircRNAs in Cancer

Cancer is one of the most difficult diseases in human society. Therefore, it is important for us to understand its pathogenesis and improve the cure rate. In cancers, tumor cell–derived exosomal circRNAs can act on target cells or organs through the transport of exosomes, and then participate in the regulation of tumor development and metastasis. Since exosomes exist in various body fluids and circRNAs in exosomes exhibit high stability, exosomal circRNAs have the potential as biomarkers for early and minimally invasive cancer diagnosis and prognosis judgment. Exosomal circRNAs derived from donor cells can modulate crosstalk with recipient cells locally or remotely to enhance cancer development and propagation, and play crucial roles in the tumor microenvironment (TME), leading to significant enhancement of tumor immunity, metabolism, angiogenesis, drug resistance, epithelial–mesenchymal transition (EMT), invasion, and metastasis.

Increasing evidence indicates that exosome-derived circRNAs act as signaling molecules to regulate cancer growth, angiogenesis, invasion, metastasis, and sensitivity to chemotherapy. Moreover, circulating exosomal circRNAs can reflect the progression and malignant characteristics of cancer, implying their great potential as promising, noninvasive biomarkers for cancer diagnosis and prognosis [6, 62]. Moreover, circRNAs are more abundant in exosomes than in the cell from which the exosomes originated. Colon, lung, stomach, breast, and cervical cancer cells secrete circRNA-containing exosomes. In exosomes present in the serum of patients with colorectal cancer, certain circRNAs were enriched compared with those in exosomes from healthy controls [63]. Therefore, exosomal circRNAs are implicated in the pathogenesis and/or progression of cancer.

5.1.1 Exosomal CircRNAs in Colorectal Cancer

Colorectal cancer (CRC) is one of the most malignant cancer types, characterized by elevated mortality rate and treatment resistance. Despite the progress achieved in the explanation of the molecular basis of the disease as well as introducing potential biomarkers in the clinical practice, further investigation is essential to identify innovative molecules that contribute to colorectal carcinogenesis. Increasing evidence indicates that exosomal circRNAs act as signaling molecules to regulate cancer proliferation, metastasis, and sensitivity to radio- and chemotherapy. Based on recent research, a number of exosomal circRNAs were implicated in CRC. Some of these circRNAs were shown to play role as biomarkers and may provide a novel pathway for cancer diagnosis. Some other exo-circRNAs were indicated to promote proliferation, metastasis, radioresistance, drug resistance, and stemness properties and therefore might have potential for monitoring disease progression or recurrence [64].

5.1.2 Exosomal Circular RNAs in Hepatocellular Carcinoma

Hepatocellular carcinoma (HCC) exacts a heavy disease burden and is currently the second most common cause of cancer-related deaths worldwide. HCC usually lacks obvious symptoms in the early stage, and most HCC patients are diagnosed at advanced stages with poor prognosis. Accumulating studies have demonstrated the promoter or suppressor roles of exosomal circRNAs in the carcinogenesis, progression, and metastasis of HCC. According to recent research, whether as diagnostic and prognostic markers or therapeutic monitoring markers of HCC, exosome-derived circRNAs have a good prospect, and exosomal circRNA-based therapies may be introduced for the precise treatment of HCC in the future [65].

5.1.3 Exosomal Circular RNAs in Gastrointestinal Malignancies

Despite the high prevalence of gastrointestinal tumors, early diagnosis and treatment of these tumors are limited by the lack of effective and specific biomarkers and therapeutic targets. Exosomal circRNAs play important roles in regulating cell proliferation, metastasis, angiogenesis, metabolism, and the immune microenvironment of gastrointestinal tumors and exhibit great potential as tumor biomarkers and antitumor targets or tools. However, no ideal tumor biomarkers of screening early GI cancers are currently available. Nowadays, several circRNAs and exosomal circRNAs have been used as potential biomarkers to screen GI cancers including pancreatic cancer (hsa_circ_0001649, circ_0007534, circ_0030235, circRHOT1, circZMYM2, circ-LDLRAD3, chr14:101402109-101464448C, chr4:52729603-52780244C, circ-IARS, and circ-PDE8A), HCC (circSETD3, circADAMTS13, hsa_circ_0007874, hsa_circ_104135, circFBLIM1, cSMARCA5, circRNA-100338, and circPTGR1), colorectal cancer (hsa_circ_0001178, hsa_circ_0000826, hsa_circ_0004771, circDDX17, circITGA7, and circHIPK3), and gastric cancer (hsa_circ_0074362, circNRIP1, circAKT3, circ-DONSON, circPSMC3, circ-KIAA1244, circPVRL3, circPVT1, hsa_circ_0000096, ciRS-133, hsa_circ_0001017, and hsa_circ_0061276). Accordingly, we emphasized that circRNAs and exosomal circRNAs have the potential high clinical diagnostic values for GI malignancies. Exosomal circRNAs are involved in the pathogenesis of cancer, especially metastasis, which is mainly ascribed to the frequently abnormal expression levels within neoplasms. More knowledge on the regulation and network interactions of exosomal circRNAs will help identify superior treatment strategies for the metastasis of gastrointestinal tumors [66].

5.1.4 Exosomal CircRNAs in Breast Cancer

Breast cancer is among the most common malignant cancers diagnosed in women worldwide. Despite the development of comprehensive treatments, breast cancer still has high mortality rates. Recent studies have unmasked critical roles for circRNAs in breast cancer as regulators of tumor initiation, progression, and metastasis. By interfering with several biological pathways, exosomal circRNAs can have either a proliferative or suppressive role in cancer. The involvement of exosomal circRNAs in breast cancer has been studied with consideration to tumor development and progression as well as its effects on therapeutic resistance. However, its exact mechanism is still unclear, and there have not been available clinical implications of exo-circRNAs in breast cancer [67].

5.2 Clinical Significance of Exosomal CircRNAs in Other Disease

Clinical applications of exosomal circular RNAs are more than diagnosis. Exosomes are small vesicles derived from cells used as cell-to-cell communication goods in numerous diseases including tumorigenesis, neurological diseases, cardiovascular diseases, and other diseases. Genome-wide RNA sequence analyses observed that circRNAs were stably improved in exosomes in association to parental cells. Little attention has been dedicated to exosomal circRNAs (exo-circRNAs). However, research has demonstrated that exo-circRNAs may have important regulatory functions because of their stability in cells and within exosomes. If well understood, the precise roles and mechanisms of exo-circRNAs might surge the impending clinical applications of these molecules as markers in the identification, prediction, and treatment of various diseases.

Not only do exosomal circRNAs play a vital role in tumors, but they also play an important role in diseases other than tumors, such as nervous system diseases and cardiovascular diseases. Apart from tumors, exosomal circRNAs play an important role in other diseases, and many of their functions have been specially reported in some studies. The potential role of exosome-derived circRNA in various diseases has been shown, and its use as a diagnostic marker and therapeutic point has been suggested for various diseases in addition to cancer, and studies in the field of biogenesis and biological functions of exosomes and circRNAs have taken place.

Human platelets are particularly rich in circRNA compared to other hematopoietic cell types, and since platelets are essential for central physiological processes such as hemostasis, wound healing, inflammation, and cancer metastasis, these findings could greatly expand the potential of circRNA as a prognostic and diagnostic biomarker [68]. About the roles of exosomes in CNS, where differentially expressed circRNAs might be related to the growth and repair of neurons, the development of the nervous system, and the transmission of nerve signals. The most highly correlated pathways identified were involved primarily with glutamatergic synapse and the cyclic guanosine monophosphate-protein kinase G signaling pathway [69].

Application of exosomal circular RNAs in cardiovascular diseases has been confirmed in many works. As natural nanoparticles, exosomes regulate a wide range of biological processes via modulation of its components, including circular RNAs (circRNAs). CircRNAs are a novel class of closed-loop single-stranded RNAs with a wide distribution and play diverse biological roles. Due to its stability in exosomes, exosomal circRNAs serve as biomarkers and pathogenic regulators and exert therapeutic potentials in some cardiovascular diseases, including atherosclerosis, acute coronary syndrome, ischemia/reperfusion injury, heart failure, and peripheral artery disease. CircHIPK3 in hypoxic exosomes (HPC-exos) plays a role in cardiac microvascular endothelial cells (CMVECs) under oxidative conditions through miR-29a-mediated IGF-1 expression, leading to a decrease in oxidative stress-induced CMVECs dysfunction. These data suggest that the exosomal circRNA in CMs is a potential target to control CMVECs dysfunction under oxidative conditions [70]. To explore the role of smooth muscle SIRT1 in endothelial angiogenesis after ischemia and the underlying mechanisms, a femoral artery ligation model using VSMC specific human SIRT1 transgenic (−Tg) and knockout (KO) mice was generated. −Tg VSMCs inhibited endothelial angiogenic activity induced by hypoxia via the exosomes containingcZFP609. The cZFP609 as a circRNA was delivered into ECs and detained HIF1α in the cytoplasm via its interaction with HIF1α, thereby inhibiting VEGFA expression and endothelial angiogenic functions [71].

Potential clinical value of circular RNAs as peripheral biomarkers for the diagnosis and treatment of major depressive disorder have been emphasized in various works. Circular RNAs (circRNAs) are expressed abundantly in the brain and are implicated in the pathophysiology of neuropsychiatric disease. However, the potential clinical value of circRNAs in major depressive disorder (MDD) remains unclear. About exosomal RNA, serum exosome-derived miR-139-5p is a potential biomarker for major depressive disorder [72].

In a recent study to further explore the pathogenesis of idiopathic membranous nephropathy (IMN), gene sequencing was used to analyze the differentially expressed circRNAs in exosomes of patients with IMN, which may lay the foundation for the research of circRNAs as a new class of exosome-based IMN diagnosis biomarkers. Differential expression study of circular RNAs in exosomes from serum and urine in patients with idiopathic membranous nephropathy revealed the significant differential and specific expression of circRNAs in the exosomes from patients with IMN. For example, MUC3A, which originated from chr7:100550808|100551062, could be considered a potential diagnostic biomarker of IMN. Furthermore, these figures may be used as a reference or supplement in the research of the pathogenesis of IMN [73].

In a study, Liu et al. showed that diabetes-related stress can upregulate cPWWP2A expression in pericytes. cPWWP2A can be transported from pericytes to ECs via exosomes. Dysregulated cPWWP2A leads to abnormal pericyte–endothelial cell (EC) interaction in vitro and retinal vascular dysfunction in vivo, suggesting that cPWWP2A is an important regulator in the pathogenesis of diabetic retinal vasculature. This study, by investigating diabetic microvascular complications, showed that exosomes can carry circular RNA-cPWWP2A, which is an endogenous miR-579 sponge, and has an indirect effect on retinal vessels in diabetic patients through different levels of expression of this circular RNA [74].

In a study, Liu et al. investigated the role of circRNA caspase recruitment domain family member 6 (circ-CARD6) in posterior capsular opacities (PCO) that develop after cataract surgery. They reported that exosome-delivered circ-CARD6 was highly expressed in PCO tissues and SRA01/04 cells treated with TGF-β2. Deletion of Circ-CARD6 also suppressed proliferation, metastasis, EMT process, and MAPK pathway, which was reversed by anti-miR-31 in TGF-β2-treated SRA01/04 cells. Meanwhile, circ-CARD6 repressed miR-31 that directly targeted FGF7 in TGF-β2-treated SRA01/04 cells. FGF7 overexpression alleviated miR-31 overexpression-induced suppression of proliferation, metastasis, EMT process, and MAPK pathway. In addition, circ-CARD6 regulated FGF7 expression with miR-31 sponge. They concluded that circ-CARD6 promotes PCO development through miR-31/FGF7 axis [75].

The differential expression of exosomal circular RNA (circRNA) in the serum of patients with proliferative diabetic retinopathy (PDR) as one of the main microvascular complications of diabetes, and preliminary investigation of the potential proangiogenic mechanism of specific exosomal circRNAs was ananlysed in a comprehensive study. Xinsheng Li et al in this work investigated the potential function of endothelial cell–derived circFndc3b induced by high glucose in human retinal vascular endothelial cells (HRVECs). The results showed the aberrant profile of exosomal circRNAs in PDR serum, and circFndc3b derived from high-glucose-induced endothelial cells may play an important role in PDR angiogenesis [76].

Increasing research indicates that circRNAs are now considered key regulators of cellular function.

Wang et al. studied the dysfunction of cardiac microvascular endothelial cells (CMVECs) under oxidative conditions. The results showed that CircHIPK3 in hypoxic exosomes released from hypoxic cardiomyocytes, through miR-29a-mediated IGF-1 expression, leads to reduction of CMVECs dysfunction caused by oxidative stress [70].

Yan et al. investigated exosomes derived from human umbilical cord mesenchymal stem cells (UMSC-Exo) in ischemic muscle damage and acknowledged that exosomes derived from human umbilical cord mesenchymal stem cells (UMSC-Exo) can prevent pyroptosis and repair ischemic muscle injury by releasing circular RNA through a novel exosome/circHIPK3/FOXO3a pathway [77].

About cZFP609, an exosomal circRNA with biological function affecting endothelial angiogenic activity, Dou et al stated that the angiogenic activity of endothelial cells after hypoxia occurs through cZFP609 exosome through the cZFP609/HIF1α/VEGFA pathway leads to the reprogramming of endothelial function. As a result, cZFP609 reduces hypoxia-induced VEGFA expression in endothelial cells and inhibits endothelial angiogenic functions, which can be suggested as a valuable indicator for evaluating the prognosis and clinical outcomes of ischemic diseases [71].

Cao et al., in a paper, determined the effects of bone marrow–derived mesenchymal stromal cell (BMSC)-derived exosomes modified with circ-Rtn4 (Rtn4-Exos) on TNF-α-induced cytotoxicity and apoptosis. The findings show that circRNA-derived exosomes (circ-Rtn4) attenuate TNF-α-induced cytotoxicity and apoptosis by a circ-Rtn4/miR-146a/TNF-α mechanism, suggesting that Rtn4-Exos may act as a new candidate for the treatment of osteoporosis [78].

Wang et al. by examining the expression of an exosomal circRNA, called circRNA-0077930, involved in promoting cellular senescence in vascular smooth muscle cells (VSMCs), stated that circRNA-0077930 with the mechanism of circRNA-0077930/miR-622/Kras leads to the occurrence of aging in vascular smooth muscles, which can be considered as an important factor for preventing diabetic vascular disease and creating drug targets [79].

He et al. reported that exosomal circular RNA can be used as a biomarker platform for early detection of immune-mediated demyelinating disease. By studying exosomal hsa_circ_0087862 and hsa_circ_0012077 in the CSF of patients with immune-mediated demyelinating diseases, they stated that GRIN2B is suggested as an important target gene of circRNAs in immune-mediated demyelinating diseases because it is regulated by circRNAs and promotes disease [80].

Ma and colleagues, in order to further investigate the pathogenesis of idiopathic membranous nephropathy (IMN), analyzed the differentially expressed circRNAs in exosomes of patients with IMN and claimed that the differential expression of exosomal circRNAs in the serum and urine of patients with IMN can be used as a new class of exosome-based IMN diagnostic biomarkers [73].

In a study, Liu et al. showed that diabetes-related stress can upregulate cPWWP2A expression in pericytes. The cPWWP2A can be transported from pericytes to ECs via exosomes. Dysregulated cPWWP2A leads to abnormal pericyte–endothelial cell (EC) interaction in vitro and retinal vascular dysfunction in vivo, suggesting that cPWWP2A is an important regulator in the pathogenesis of diabetic retinal vasculature. This study, by investigating diabetic microvascular complications, showed that exosomes can carry circular RNA-cPWWP2A, which is an endogenous miR-579 sponge, and has an indirect effect on retinal vessels in diabetic patients through different levels of expression of this circular RNA [74].

CircRNAs are a rising star in eye disease research. Recent studies identified the critical role of circRNAs in the pathogenesis of various eye diseases and highlighted circRNAs as promising biomarkers for diagnosis and assessment of progression and prognosis [81].

In a study, Liu et al. investigated the role of circRNA caspase recruitment domain family member 6 (circ-CARD6) in posterior capsular opacities (PCO) that develop after cataract surgery. They reported that exosome-delivered circ-CARD6 was highly expressed in PCO tissues and SRA01/04 cells treated with TGF-β2. Deletion of Circ-CARD6 also suppressed proliferation, metastasis, EMT process, and MAPK pathway, which was reversed by anti-miR-31 in TGF-β2-treated SRA01/04 cells. Meanwhile, circ-CARD6 repressed miR-31 that directly targeted FGF7 in TGF-β2-treated SRA01/04 cells. FGF7 overexpression alleviated miR-31 overexpression-induced suppression of proliferation, metastasis, EMT process, and MAPK pathway. In addition, circ-CARD6 regulated FGF7 expression with miR-31 sponge. They concluded that circ-CARD6 promotes PCO development through miR-31/FGF7 axis [75].

About the potential function of endothelial cell–derived circFndc3b induced by high glucose in human retinal vascular endothelial cells (HRVECs), investigations showed the aberrant profile of exosomal circRNAs in PDR serum, and circFndc3b derived from high-glucose-induced endothelial cells may play an important role in PDR angiogenesis [76].

Lin Ye et al., in a study to investigate molecular mechanisms in diabetic retinopathy (DR), explored the effects of exosomal circRNA derived from hypoxia-pretreated pericytes on high-glucose-induced endotheliocyte injuries. They constructed coculture models of hypoxia pericytes and high-glucose-induced endotheliocytes, then used the models to investigate the association between pericyte survival and the angiogenesis activity of endotheliocytes, and then recorded the changes that occurred in the endothelial circular RNA contained in exosomes released from hypoxia-treated pericytes. Next, they explored the effects of exosomal circRNA derived from hypoxia-pretreated pericytes (circEhmt1) on high-glucose-induced endotheliocyte injuries. Their results revealed a novel mechanism in which exosomal circRNAs serve as mediators of crosstalk between hypoxia-activated pericytes and high-glucose-induced endotheliocytes to produce retinal microvascular dysfunction [82].

Thousands of circRNAs are aberrantly expressed in diabetic retinopathy (DR) patients, animal models, or cell models. A number of circRNAs, such as cPWWP2A, circDNMT3B, circHIPK3, and circ_0005015, carried by exosomes, were shown to play an important role in DR by regulating angiogenesis, proliferation, apoptosis, and inflammatory response of various cells in the retina. Exosomes carrying these circRNAs play a role in the process of diabetic retinopathy by regulating neoangiogenesis, inflammatory response, and apoptosis in retinal cells and are used as biomarkers and therapeutic targets for DR [83, 84].

In a paper, Wang et al. investigated the role and mechanism of circRNA RNA polymerase III subunit A (POLR3A) in posterior capsular opacity (PCO) and reported that exosomes were involved in the transport of circ-POLR3A in SRA01/04 cells, and this Circ-POLR3A accelerates TGF-β2-induced cell survival, migration, and invasion of lens epithelial cells via the miR-31/TXNIP signaling cascade [85].

The use of exosomal circular RNAs in cardiovascular diseases has been confirmed in many works. Due to the stability of circRNAs in exosomes, exosomal circRNAs act as biomarkers and regulators of pathogenicity. The research conducted in the field of biogenesis and function of exosomal circRNAs, their role and therapeutic potentials in some cardiovascular diseases, including atherosclerosis (AS), acute coronary syndrome (ACS), ischemia/reperfusion injury (I/R), failure heart (HF), and peripheral artery disease (PAD) [86, 87].

Studies show that the expression of circRNAs in healthy and pathological cardiovascular tissues is different [88], and circRNAs are involved in the development of cardiovascular diseases (CVDs) by binding miRNAs or RNA-binding proteins, transforming into proteins and regulating transcription [89, 90]. It will be a confirmation of their potential as diagnostic and prognostic molecular markers. Among these exosomal circRNAs, as diagnostic biomarkers in serum or plasma for cardiovascular diseases, circ_0006896 [91] was detected in the serum of patients with atherosclerotic plaque, and circ_0005540 [92] and circNPHP4 [93] were detected in monocytes of patients with coronary artery disease.

6 Conclusions and Future Perspectives

Exosomes and circRNAs are both hot research topics. In this chapter, several important functions of circular RNA are summarized in detail. For example, circRNAs can act as miRNA sponges to affect gene expression, interact with Pol II to regulate the transcription of parental genes, act as a protein sponge to affect protein expression, and participate in protein translation through multiple pathways. Moreover, experiments have shown that circRNAs can enter exosomes. Although its regulatory mechanism is not completely clear, exosomal circRNAs play an important role in pathological processes such as tumors, nerves, cardiovascular diseases, and physiological processes such as inflammation and cell apoptosis. Exosomes are rich in lipid components such as cholesterol, neuramide, and sphingolipids, which are not easily degraded by proteases and ribonucleases, so circRNAs are stable in exosomes and can perform their functions. In addition, exosomes are found widely in various body fluids, such as blood, tears, saliva, sputum, semen, urine, and breast milk [94, 95]. Therefore, liquid biopsies can be performed; it is convenient and simple for clinicians to take specimens, and it is not traumatic for patients. Therefore, the study of exosomal circRNAs will provide a biological basis for the identification of new markers for early diagnosis, targeted therapy, and prognostic evaluation of various diseases in the future.

Even though significant breakthroughs have been made in the study of exosomes and circRNAs, there are still many unanswered questions. First, what conditions affect the interactions of circRNAs with other molecules, and what other types of modifications other than m6A in circRNAs are involved in protein translation? Second, are there differences between exosomes in cells and those in plasma, and do these differences affect the expression of circRNAs? Third, what is the specific mechanism by which circRNAs enter exosomes, and how are they cleared or involved in cell-to-cell communications? Fourth, blood, urine, sputum, and other liquids also contain many other impurities, including protein complexes and nucleic acid lysates, which interfere with the extraction of exosomes. Therefore, there are still many difficulties and challenges in the clinical application of exosomal circRNAs, and thus, continuous in-depth research and exploration are needed.

References

1. Raposo G, Stoorvogel W (2013) Extracellular vesicles: exosomes, microvesicles, and friends. J Cell Biol 200(4):373–383. https://doi.org/10.1083/jcb.201211138
2. Wolf P (1967) The nature and significance of platelet products in human plasma. Br J Haematol 13(3):269–288. https://doi.org/10.1111/j.1365-2141.1967.tb08741.x
3. Wang J, Yue B-L, Huang Y-Z, Lan X-Y, Liu W-J, Chen H (2022) Exosomal RNAs: novel potential biomarkers for diseases—a review. Int J Mol Sci 23(5):2461. https://doi.org/10.3390/ijms23052461
4. Zhu L, Li N, Sun L, Zheng D, Shao G (2021) Non-coding RNAs: the key detectors and regulators in cardiovascular disease. Genomics 113(1 Pt 2):1233–1246. https://doi.org/10.1016/j.ygeno.2020.10.024
5. Lasda E, Parker R (2014) Circular RNAs: diversity of form and function. RNA 20(12):1829–1842. https://doi.org/10.1261/rna.047126.114
6. Han Z, Chen H, Guo Z, Shen J, Luo W, Xie F, Wan Y, Wang S, Li J, He J (2022) Circular RNAs and their role in exosomes. Front Oncol 12:848341. https://doi.org/10.3389/fonc.2022.848341
7. Schmidt CA, Giusto JD, Bao A, Hopper AK, Matera AG (2019) Molecular determinants of metazoan tricRNA biogenesis. Nucleic Acids Res 47(12):6452–6465. https://doi.org/10.1093/nar/gkz311
8. Hosseini K, Ranjbar M, Pirpour Tazehkand A, Asgharian P, Montazersaheb S, Tarhriz V, Ghasemnejad T (2022) Evaluation of exosomal non-coding RNAs in cancer using high-throughput sequencing. J Transl Med 20(1):30. https://doi.org/10.1186/s12967-022-03231-y
9. Rahmati Y, Asemani Y, Aghamiri S, Ezzatifar F, Najafi S (2021) CiRS-7/CDR1as; an oncogenic circular RNA as a potential cancer biomarker. Pathol Res Pract 227:153639. https://doi.org/10.1016/j.prp.2021.153639
10. Taheri M, Najafi S, Basiri A, Hussen BM, Baniahmad A, Jamali E, Ghafouri-Fard S (2021) The role and

clinical potentials of circular RNAs in prostate cancer. Front Oncol 11:781414. https://doi.org/10.3389/fonc.2021.781414
11. Wilusz JE (2018) A 360 view of circular RNAs: from biogenesis to functions. Wiley Interdiscip Rev RNA 9(4):e1478. https://doi.org/10.1002/wrna.1478
12. Dubin RA, Kazmi MA, Ostrer H (1995) Inverted repeats are necessary for circularization of the mouse testis Sry transcript. Gene 167(1–2):245–248. https://doi.org/10.1016/0378-1119(95)00639-7
13. Hsu M-T, Coca-Prados M (1979) Electron microscopic evidence for the circular form of RNA in the cytoplasm of eukaryotic cells. Nature 280(5720):339–340. https://doi.org/10.1038/280339a0
14. Sanger HL, Klotz G, Riesner D, Gross HJ, Kleinschmidt AK (1976) Viroids are single-stranded covalently closed circular RNA molecules existing as highly base-paired rod-like structures. Proc Natl Acad Sci USA 73(11):3852–3856. https://doi.org/10.1073/pnas.73.11.3852
15. Zhang Y, Xue W, Li X, Zhang J, Chen S, Zhang J-L, Yang L, Chen L-L (2016) The biogenesis of nascent circular RNAs. Cell Rep 15(3):611–624. https://doi.org/10.1016/j.celrep.2016.03.058. Epub 2016 Apr 7
16. Liu Z, Ran Y, Tao C, Li S, Chen J, Yang E (2019) Detection of circular RNA expression and related quantitative trait loci in the human dorsolateral prefrontal cortex. Genome Biol 20(1):99. https://doi.org/10.1186/s13059-019-1701-8
17. Sayad A, Najafi S, Kashi AH, Hosseini SJ, Akrami SM, Taheri M, Ghafouri-Fard S (2022) Circular RNAs in renal cell carcinoma: functions in tumorigenesis and diagnostic and prognostic potentials. Pathol Res Pract 229:153720. https://doi.org/10.1016/j.prp.2021.153720
18. Su M, Xiao Y, Ma J, Tang Y, Tian B, Zhang Y, Li X, Wu Z, Yang D, Zhou Y (2019) Circular RNAs in cancer: emerging functions in hallmarks, stemness, resistance and roles as potential biomarkers. Mol Cancer 18(1):90. https://doi.org/10.1186/s12943-019-1002-6
19. Iparraguirre L, Prada-Luengo I, Regenberg B, Otaegui D (2019) To be or not to be: circular RNAs or mRNAs from circular DNAs? Front Genet 10:940. https://doi.org/10.3389/fgene.2019.00940
20. Chen L-L (2016) The biogenesis and emerging roles of circular RNAs. Nat Rev Mol Cell Biol 17(4):205–211. https://doi.org/10.1038/nrm.2015.32
21. Jeck WR, Sorrentino JA, Wang K, Slevin MK, Burd CE, Liu J, Marzluff WF, Sharpless NE (2013) Circular RNAs are abundant, conserved, and associated with ALU repeats. RNA 19(2):141–157. https://doi.org/10.1261/rna.035667.112
22. Bhat AA, Younes SN, Raza SS, Zarif L, Nisar S, Ahmed I, Mir R, Kumar S, Sharawat SK, Hashem S (2020) Role of non-coding RNA networks in leukemia progression, metastasis and drug resistance. Mol Cancer 19(1):57. https://doi.org/10.1186/s12943-020-01175-9
23. Zhao X, Cai Y, Xu J (2019) Circular RNAs: biogenesis, mechanism, and function in human cancers. Int J Mol Sci 20(16):3926. https://doi.org/10.3390/ijms20163926
24. Benmoussa A, Provost P (2019) Milk microRNAs in health and disease. Compr Rev Food Sci Food Saf 18(3):703–722. https://doi.org/10.1111/1541-4337.12424
25. Bose R, Ain R (2018) Regulation of transcription by circular RNAs. In: Circular RNAs: biogenesis and functions, pp 81–94. https://doi.org/10.1007/978-981-13-1426-1_7
26. Ashwal-Fluss R, Meyer M, Pamudurti NR, Ivanov A, Bartok O, Hanan M, Evantal N, Memczak S, Rajewsky N, Kadener S (2014) circRNA biogenesis competes with pre-mRNA splicing. Mol Cell 56(1):55–66. https://doi.org/10.1016/j.molcel.2014.08.019
27. Li Z, Huang C, Bao C, Chen L, Lin M, Wang X, Zhong G, Yu B, Hu W, Dai L (2015) Exon-intron circular RNAs regulate transcription in the nucleus. Nat Struct Mol Biol 22(3):256–264. https://doi.org/10.1038/nsmb.2959
28. Zhang Y, Zhang X-O, Chen T, Xiang J-F, Yin Q-F, Xing Y-H, Zhu S, Yang L, Chen L-L (2013) Circular intronic long noncoding RNAs. Mol Cell 51(6):792–806. https://doi.org/10.1016/j.molcel.2013.08.017
29. Okholm T (2020) Transcriptome-wide profiles of circular RNA and RNA-binding protein 680 interactions reveal effects on circular RNA biogenesis and cancer pathway expression. Genome Med 12(1):112. https://doi.org/10.1186/s13073-020-00812-8
30. Okholm T, Sathe S, Park S, Kamstrup A, Rasmussen A, Shankar A, Chua Z, Fristrup N, Nielsen M, Vang S (2020) Transcriptome-wide profiles of circular RNA and RNA-binding protein interactions reveal effects on circular RNA biogenesis and cancer pathway expression. Genome Med 12:112. https://doi.org/10.1186/s13073-020-00812-8
31. Zhang L, Wang Y, Zhang Y, Zhao Y, Li P (2021) Pathogenic mechanisms and the potential clinical value of circFoxo3 in cancers. Mol Ther Nucleic Acids 23:908–917. https://doi.org/10.1016/j.omtn.2021.01.010
32. Du WW, Yang W, Liu E, Yang Z, Dhaliwal P, Yang BB (2016) Foxo3 circular RNA retards cell cycle progression via forming ternary complexes with p21 and CDK2. Nucleic Acids Res 44(6):2846–2858. https://doi.org/10.1093/nar/gkw027
33. Holdt LM, Stahringer A, Sass K, Pichler G, Kulak NA, Wilfert W, Kohlmaier A, Herbst A, Northoff BH, Nicolaou A (2016) Circular non-coding RNA ANRIL modulates ribosomal RNA maturation and atherosclerosis in humans. Nat Commun 7:12429. https://doi.org/10.1038/ncomms12429
34. Zheng X, Chen L, Zhou Y, Wang Q, Zheng Z, Xu B, Wu C, Zhou Q, Hu W, Wu C (2019) A novel protein encoded by a circular RNA circPPP1R12A promotes tumor pathogenesis and metastasis of colon cancer via Hippo-YAP signaling. Mol Cancer 18(1):47. https://doi.org/10.1186/s12943-019-1010-6

35. Legnini I, Di Timoteo G, Rossi F, Morlando M, Briganti F, Sthandier O, Fatica A, Santini T, Andronache A, Wade M (2017) Circ-ZNF609 is a circular RNA that can be translated and functions in myogenesis. Mol Cell 66(1):22–37.e9. https://doi.org/10.1016/j.molcel.2017.02.017
36. Yang Y, Gao X, Zhang M, Yan S, Sun C, Xiao F, Huang N, Yang X, Zhao K, Zhou H, Huang S, Xie B, Zhang N (2017) Novel role of FBXW7 circular RNA in repressing glioma tumorigenesis. JNCI J Natl Cancer Inst 110(3):304–315. https://doi.org/10.1093/jnci/djx166
37. Chen C-y, Sarnow P (1995) Initiation of protein synthesis by the eukaryotic translational apparatus on circular RNAs. Science 268(5209):415–417. https://doi.org/10.1126/science.7536344
38. Zhang M, Zhao K, Xu X, Yang Y, Yan S, Wei P, Liu H, Xu J, Xiao F, Zhou H, Yang X, Huang N, Liu J, He K, Xie K, Zhang G, Huang S, Zhang N (2018) A peptide encoded by circular form of LINC-PINT suppresses oncogenic transcriptional elongation in glioblastoma. Nat Commun 9(1):4475. https://doi.org/10.1038/s41467-018-06862-2
39. Martínez-Salas E, Lozano G, Fernandez-Chamorro J, Francisco-Velilla R, Galan A, Diaz R (2013) RNA-binding proteins impacting on internal initiation of translation. Int J Mol Sci 14(11):21705–21726. https://doi.org/10.3390/ijms141121705
40. Zhou J, Wan J, Gao X, Zhang X, Jaffrey SR, Qian S-B (2015) Dynamic m6A mRNA methylation directs translational control of heat shock response. Nature 526(7574):591–594. https://doi.org/10.1038/nature15377
41. Yang Y, Fan X, Mao M, Song X, Wu P, Zhang Y, Jin Y, Yang Y, Chen L-L, Wang Y, Wong CCL, Xiao X, Wang Z (2017) Extensive translation of circular RNAs driven by N6-methyladenosine. Cell Res 27(5):626–641. https://doi.org/10.1038/cr.2017.31
42. Zhang M, Huang N, Yang X, Luo J, Yan S, Xiao F, Chen W, Gao X, Zhao K, Zhou H, Li Z, Ming L, Xie B, Zhang N (2018) A novel protein encoded by the circular form of the SHPRH gene suppresses glioma tumorigenesis. Oncogene 37(13):1805–1814. https://doi.org/10.1038/s41388-017-0019-9
43. Xia X, Li X, Li F, Wu X, Zhang M, Zhou H, Huang N, Yang X, Xiao F, Liu D (2019) A novel tumor suppressor protein encoded by circular AKT3 RNA inhibits glioblastoma tumorigenicity by competing with active phosphoinositide-dependent Kinase-1. Mol Cancer 18(1):131. https://doi.org/10.1186/s12943-019-1056-5
44. Chernyshev VS, Rachamadugu R, Tseng YH, Belnap DM, Jia Y, Branch KJ, Butterfield AE, Pease LF, Bernard PS, Skliar M (2015) Size and shape characterization of hydrated and desiccated exosomes. Anal Bioanal Chem 407(12):3285–3301. https://doi.org/10.1007/s00216-015-8535-3
45. Arraud N, Linares R, Tan S, Gounou C, Pasquet JM, Mornet S, Brisson AR (2014) Extracellular vesicles from blood plasma: determination of their morphology, size, phenotype and concentration. J Thromb Haemost 12(5):614–627. https://doi.org/10.1111/jth.12554
46. Sharma S, Das K, Woo J, Gimzewski JK (2014) Nanofilaments on glioblastoma exosomes revealed by peak force microscopy. J R Soc Interface 11(92):20131150. https://doi.org/10.1098/rsif.2013.1150
47. Booth AM, Fang Y, Fallon JK, Yang J-M, Hildreth JE, Gould SJ (2006) Exosomes and HIV Gag bud from endosome-like domains of the T cell plasma membrane. J Cell Biol 172(6):923–935. https://doi.org/10.1083/jcb.200508014
48. Nolan JP, Duggan E (2018) Analysis of individual extracellular vesicles by flow cytometry. Flow cytometry protocols. Methods Mol Biol 1678:79–92. https://doi.org/10.1007/978-1-4939-7346-0_5
49. Claude A (1943) The constitution of protoplasm. Science 97(2525):451–456. https://doi.org/10.1126/science.97.2525.451
50. Sódar BW, Kittel Á, Pálóczi K, Vukman KV, Osteikoetxea X, Szabó-Taylor K, Németh A, Sperlágh B, Baranyai T, Giricz Z (2016) Low-density lipoprotein mimics blood plasma-derived exosomes and microvesicles during isolation and detection. Sci Rep 18(6):24316. https://doi.org/10.1038/srep24316
51. Zhang H, Freitas D, Kim HS, Fabijanic K, Li Z, Chen H, Mark MT, Molina H, Martin AB, Bojmar L (2018) Identification of distinct nanoparticles and subsets of extracellular vesicles by asymmetric flow field-flow fractionation. Nat Cell Biol 20(3):332–343. https://doi.org/10.1038/s41556-018-0040-4
52. Raposo G, Nijman HW, Stoorvogel W, Liejendekker R, Harding CV, Melief CJ, Geuze HJ (1996) B lymphocytes secrete antigen-presenting vesicles. J Exp Med 183(3):1161–1172. https://doi.org/10.1084/jem.183.3.1161
53. Fang Y, Wu N, Gan X, Yan W, Morrell JC, Gould SJ (2007) Higher-order oligomerization targets plasma membrane proteins and HIV gag to exosomes. PLoS Biol 5(6):e158. https://doi.org/10.1371/journal.pbio.0050158
54. Bakhshian Nik A, Hutcheson JD, Aikawa E (2017) Extracellular vesicles as mediators of cardiovascular calcification. Front Cardiovasc Med 4:78. https://doi.org/10.3389/fcvm.2017.00078
55. Pegtel DM, Gould SJ (2019) Exosomes. Annu Rev Biochem 88:487–514. https://doi.org/10.1146/annurev-biochem-013118-111902
56. Kalluri R, LeBleu VS (2020) The biology, function, and biomedical applications of exosomes. Science 367(6478):eaau6977. https://doi.org/10.1126/science.aau6977
57. Wang Y, Liu J, Ma J, Sun T, Zhou Q, Wang W, Wang G, Wu P, Wang H, Jiang L, Yuan W, Sun Z, Ming L (2019) Exosomal circRNAs: biogenesis, effect and application in human diseases. Mol Cancer 18(1):116. https://doi.org/10.1186/s12943-019-1041-z
58. Shi H, Huang S, Qin M, Xue X, Guo X, Jiang L, Hong H, Fang J, Gao L (2021) Exosomal circ_0088300

derived from cancer-associated fibroblasts acts as a miR-1305 sponge and promotes gastric carcinoma cell tumorigenesis. Front Cell Dev Biol 9:676319. https://doi.org/10.3389/fcell.2021.676319. eCollection 2021
59. Zhang F, Jiang J, Qian H, Yan Y, Xu W (2023) Exosomal circRNA: emerging insights into cancer progression and clinical application potential. J Hematol Oncol 16(1):1–25. https://doi.org/10.1186/s13045-023-01452-21
60. Tong Y, Liu X, Xia D, Peng E, Yang X, Liu H, Ye T, Wang X, He Y, Xu H, Ye Z, Chen Z, Tang K (2021) Biological roles and clinical significance of exosome-derived noncoding RNAs in bladder cancer. Front Oncol 11:704703. https://doi.org/10.3389/fonc.2021.704703
61. Mahmoudi E, Kiltschewskij D, Fitzsimmons C, Cairns MJ (2020) Depolarization-associated CircRNA regulate neural gene expression and in some cases may function as templates for translation. Cells 9(1):25. https://doi.org/10.3390/cells9010025
62. Fanale D, Taverna S, Russo A, Bazan V (2018) Circular RNA in exosomes. In: Circular RNAs: biogenesis and functions, pp 109–117. https://doi.org/10.1007/978-981-13-1426-1_9
63. Hou J, Jiang W, Zhu L, Zhong S, Zhang H, Li J, Zhou S, Yang S, He Y, Wang D (2018) Circular RNAs and exosomes in cancer: a mysterious connection. Clin Transl Oncol 20(9):1109–1116. https://doi.org/10.1007/s12094-018-1839-y
64. Vakhshiteh F, Hassani S, Momenifar N, Pakdaman F (2021) Exosomal circRNAs: new players in colorectal cancer. Cancer Cell Int 21(1):483. https://doi.org/10.1186/s12935-021-02112-6
65. Zhao Y, Yao J (2022) The biological functions and clinical values of exosomal circular RNAs in hepatocellular carcinoma. Front Oncol 1:885214. https://doi.org/10.3389/fonc.2022.885214
66. Wang Y, Li Z, Xu S, Guo J (2020) Novel potential tumor biomarkers: circular RNAs and exosomal circular RNAs in gastrointestinal malignancies. J Clin Lab Anal 34(7):e23359. https://doi.org/10.1002/jcla.23359
67. Hussen BM, Mohamadtahr S, Abdullah SR, Hidayat HJ, Rasul MF, Hama Faraj GS, Ghafouri-Fard S, Taheri M, Khayamzadeh M, Jamali E (2023) Exosomal circular RNAs: new player in breast cancer progression and therapeutic targets. Front Genet 14:1126944. https://doi.org/10.3389/fgene.2023.1126944
68. D'Ambrosi S, Giannoukakos S, Antunes-Ferreira M, Pedraz-Valdunciel C, Bracht JW, Potie N, Gimenez-Capitan A, Hackenberg M, Fernandez Hilario A, Molina-Vila MA (2023) Combinatorial blood platelets-derived circRNA and mRNA signature for early-stage lung cancer detection. Int J Mol Sci 24(5):4881. https://doi.org/10.3390/ijms24054881
69. Xu Y, Kong S, Qin S, Shen X, Ju S (2020) Exosomal circRNAs: sorting mechanisms, roles and clinical applications in tumors. Front Cell Dev Biol 8:581558. https://doi.org/10.3389/fcell.2020.581558
70. Wang Y, Zhao R, Liu W, Wang Z, Rong J, Long X, Liu Z, Ge J, Shi B (2019) Exosomal circHIPK3 released from hypoxia-pretreated cardiomyocytes regulates oxidative damage in cardiac microvascular endothelial cells via the miR-29a/IGF-1 pathway. Oxidative Med Cell Longev 2019:7954657. https://doi.org/10.1155/2019/7954657
71. Dou YQ, Kong P, Li CL, Sun HX, Li WW, Yu Y, Nie L, Zhao LL, Miao SB, Li XK, Dong C, Zhang JW, Liu Y, Huo XX, Chi K, Gao X, Zhang N, Weng L, Yang H, Zhang F, Han M (2020) Smooth muscle SIRT1 reprograms endothelial cells to suppress angiogenesis after ischemia. Theranostics 10(3):1197–1212. https://doi.org/10.7150/thno.39320
72. Liang J-Q, Liao H-R, Xu C-X, Li X-L, Wei Z-X, Xie G-J, Cheng Y (2020) Serum exosome-derived miR-139-5p as a potential biomarker for major depressive disorder. Neuropsychiatr Dis Treat 16:2689–2693. https://doi.org/10.2147/NDT.S277392
73. Ma H, Xu Y, Zhang R, Guo B, Zhang S, Zhang X (2019) Differential expression study of circular RNAs in exosomes from serum and urine in patients with idiopathic membranous nephropathy. Arch Med Sci 15(3):738–753. https://doi.org/10.5114/aoms.2019.84690
74. Liu C, Ge H-M, Liu B-H, Dong R, Shan K, Chen X, Yao M-D, Li X-M, Yao J, Zhou R-M (2019) Targeting pericyte–endothelial cell crosstalk by circular RNA-cPWWP2A inhibition aggravates diabetes-induced microvascular dysfunction. PNAS 116(15):7455–7464. https://doi.org/10.1073/pnas.1814874116
75. Liu Y, Chen T, Zheng G (2021) Exosome-transmitted circ-CARD6 facilitates posterior capsule opacification development by miR-31/FGF7 axis. Exp Eye Res 207:108572. https://doi.org/10.1016/j.exer.2021.108572
76. Li X, Wang J, Qian H, Wu Y, Zhang Z, Hu Z, Xie P (2021) Serum exosomal circular RNA expression profile and regulative role in proliferative diabetic retinopathy. Front Genet 12:719312. https://doi.org/10.3389/fgene.2021.719312
77. Yan B, Zhang Y, Liang C, Liu B, Ding F, Wang Y, Zhu B, Zhao R, Yu X-Y, Li Y (2020) Stem cell-derived exosomes prevent pyroptosis and repair ischemic muscle injury through a novel exosome/circHIPK3/FOXO3a pathway. Theranostics 10(15):6728. https://doi.org/10.7150/thno.42259
78. Cao G, Meng X, Han X, Li J (2020) Exosomes derived from circRNA Rtn4-modified BMSCs attenuate TNF-α-induced cytotoxicity and apoptosis in murine MC3T3-E1 cells by sponging miR-146a. Biosci Rep 40(5):BSR20193436. https://doi.org/10.1042/BSR20193436
79. Wang S, Zhan J, Lin X, Wang Y, Wang Y, Liu Y (2020) CircRNA-0077930 from hyperglycaemia-stimulated vascular endothelial cell exosomes regulates senescence in vascular smooth muscle cells. Cell Biochem Funct 38(8):1056–1068. https://doi.org/10.1002/cbf.3543

80. He J, Ren M, Li H, Yang L, Wang X, Yang Q (2019) Exosomal circular RNA as a biomarker platform for the early diagnosis of immune-mediated demyelinating disease. Front Genet 10:860. https://doi.org/10.3389/fgene.2019.00860
81. Zhang C, Hu J, Yu Y (2020) CircRNA is a rising star in researches of ocular diseases. Front Cell Dev Biol 8:850. https://doi.org/10.3389/fcell.2020.00850
82. Ye L, Guo H, Wang Y, Peng Y, Zhang Y, Li S, Yang M, Wang L (2021) Exosomal circEhmt1 released from hypoxia-pretreated pericytes regulates high glucose-induced microvascular dysfunction via the NFIA/NLRP3 pathway. Oxidative Med Cell Longev 2021:8833098. https://doi.org/10.1155/2021/8833098
83. Zhang Z, Liang X, Zhou J, Meng M, Gao Y, Yi G, Fu M (2021) Exosomes in the pathogenesis and treatment of ocular diseases. Exp Eye Res 209:108626. https://doi.org/10.1016/j.exer.2021.108626
84. He M, Zhou R, Liu S, Cheng W, Wang W (2021) Circular RNAs: potential star molecules involved in diabetic retinopathy. Curr Eye Res 46(3):277–283. https://doi.org/10.1080/02713683.2020.1812086
85. Wang H, Zheng G, Sun M, Chi Y (2022) Circ-POLR3A accelerates TGF-β2-induced promotion in cell viability, migration, and invasion of lens epithelial cells via miR-31/TXNIP signaling cascade. J Biochem Mol Toxicol 36(9):e23144. https://doi.org/10.1002/jbt.23144
86. Sahoo S, Adamiak M, Mathiyalagan P, Kenneweg F, Kafert-Kasting S, Thum T (2021) Therapeutic and diagnostic translation of extracellular vesicles in cardiovascular diseases: roadmap to the clinic. Circulation 143(14):1426–1449. https://doi.org/10.1161/CIRCULATIONAHA.120.049254
87. Cui M, Han Y, Yang J, Li G, Yang C (2022) A narrative review of the research status of exosomes in cardiovascular disease. Ann Palliat Med 11(1):363–377. https://doi.org/10.21037/apm-21-3364
88. Aufiero S, Reckman YJ, Pinto YM, Creemers EE (2019) Circular RNAs open a new chapter in cardiovascular biology. Nat Rev Cardiol 16(8):503–514. https://doi.org/10.1038/s41569-019-0185-2
89. Rai AK, Lee B, Hebbard C, Uchida S, Garikipati VNS (2021) Decoding the complexity of circular RNAs in cardiovascular disease. Pharmacol Res 171:105766. https://doi.org/10.1016/j.phrs.2021.105766
90. Kishore R, Garikipati VNS, Gonzalez C (2020) Role of circular RNAs in cardiovascular disease. J Cardiovasc Pharmacol 76(2):128–137. https://doi.org/10.1097/FJC.0000000000000841
91. Wen Y, Chun Y, Lian ZQ, Yong ZW, Lan YM, Huan L, Xi CY, Juan LS, Qing ZW, Jia C, Ji ZH (2021) circRNA-0006896-miR1264-DNMT1 axis plays an important role in carotid plaque destabilization by regulating the behavior of endothelial cells in atherosclerosis. Mol Med Rep 23(5):1–13. https://doi.org/10.3892/mmr.2021.11950
92. Wu W-p, Pan Y-h, Cai M-y, Cen J-m, Chen C, Zheng L, Liu X, Xiong X-d (2020) Plasma-derived exosomal circular RNA hsa_circ_0005540 as a novel diagnostic biomarker for coronary artery disease. Dis Markers 2020:3178642. https://doi.org/10.1155/2020/3178642
93. Xiong F, Mao R, Zhang L, Zhao R, Tan K, Liu C, Xu J, Du G, Zhang T (2021) CircNPHP4 in monocyte-derived small extracellular vesicles controls heterogeneous adhesion in coronary heart atherosclerotic disease. Cell Death Dis 12(10):948. https://doi.org/10.1038/s41419-021-04253-y
94. Akers JC, Gonda D, Kim R, Carter BS, Chen CC (2013) Biogenesis of extracellular vesicles (EV): exosomes, microvesicles, retrovirus-like vesicles, and apoptotic bodies. J Neuro-Oncol 113(1):1–11. https://doi.org/10.1007/s11060-013-1084-8
95. Cappello F, Logozzi M, Campanella C, Bavisotto CC, Marcilla A, Properzi F, Fais S (2017) Exosome levels in human body fluids: a tumor marker by themselves? Eur J Pharm Sci 96:93–98. https://doi.org/10.1016/j.ejps.2016.09.010

Circular RNAs in Blood and Its Clinical Relevance

13

Farinaz Jafari Ghods

Abstract

Circular RNAs (circRNAs) have garnered considerable attention in recent years as potential biomarkers for liquid biopsies due to their unique characteristics offering unique advantages over linear RNAs. The significance of circRNAs, highlighting their stability, abundant expression, noninvasiveness, and cell/tissue specificity, makes circRNAs appealing candidates for liquid biopsies, facilitating real-time disease monitoring and guiding treatment decisions. Advancements in high-throughput RNA sequencing and experimental technologies have facilitated extensive identification and characterization of blood circRNAs (cell-free circRNAs and circRNAs in blood cells), revealing their roles in various human diseases. This chapter delves into the diverse roles of blood circRNAs across a spectrum of human diseases, including cancer, cardiovascular diseases, autoimmune disorders, neurological conditions, and pregnancy-related complications and shedding light on their cellular functions and regulatory mechanisms. It worth mentioning that individual circRNAs may sometimes lack sufficient sensitivity and specificity, necessitating the use of combinations for enhanced diagnostic value. Overall, it can be concluded that blood circRNAs hold immense promise as diagnostic and prognostic biomarkers in various human diseases. However, continued research efforts are needed to elucidate their full potential, optimize detection methods, and develop standardized diagnostic assays. By leveraging the unique properties of circRNAs, liquid biopsy approaches can revolutionize disease management, improve patient care, and ultimately enhance clinical outcomes.

Keywords

Circular RNA · Blood · Noninvasive sampling · Liquid biopsy

1 Introduction

Transcriptomic biomarkers refer to indexes of distinct RNA subtypes with tissue or cell-specific manner that can be objectively quantified through pattern-based RNA expression analysis and used to compare normal and pathogenic state. Due to abundant and dynamic expression, versatile biological functions, high stability, and specificity [1–4], as well as their availability in various peripheral blood components such as mononuclear cells, serum, plasma, platelets, erythrocytes, and extracellular vesicles

F. J. Ghods (✉)
Koç University Hospital, Koç University Research Center Translational Medicine (KUTTAM), Istanbul, Turkey

J. Xiao (ed.), *Advances in Circular RNAs*, Advances in Experimental Medicine and Biology 1485,
https://doi.org/10.1007/978-981-96-9428-0_13

[5–8], in recent years, blood-based noncoding small RNAs such as microRNAs (miRNAs) and circular RNAs (circRNAs) have attracted great attention to find out potential diagnostic, prognostic, predictive, or pharmacodynamics biomarkers for human diseases in liquid biopsy samples [9–11]. These small RNAs influence key processes within immune cells and tumor-associated immune responses, thereby modulating either protumor or tumor-suppressive activities [12]. For example, a group of studies have revealed that *miR-21* expression level increases in diverse cancer types such as gastric cancer (GC) and breast cancer (BC), which dysregulate proliferation, enhance invasion, and decrease apoptosis capability [12–14]. Based on recent findings, CCL28, NR3C2, and SNYPO2 have been predicted as miR-21 targets, highlighting their potential utility in gastric cancer monitoring and therapeutic developments [12]. Similarly, in another study, upregulation level of *miR-21* in drug resistance breast cancer cells (MCF7) was compared to normal breast cancer cells (MCF10), and when both cell types were transfected with *circ-21*, reduction in proliferation, migration, and colony formation in MCF7 cells and intensification of the antitumor activity of doxorubicin were shown [15]. Another important point in this era is that circRNAs follow the spatiotemporal expression pattern which is provided by the precise balance between circRNAs' generation/turnover rate controlled at both the transcriptional and posttranscriptional levels [16–18]. So, any disruption in this harmony may cause aberrant circRNAs expression in corresponding cells or tissues and disarrange normal health state [19]. Although the functional characterization of circRNAs has not been fully understood and is still at its early stages, this approach in turn can provide us significant clues to unrevealing the underlying mechanisms of myriads of human diseases alongside with related biological pathways.

2 Liquid Biopsy as Wealth Sources of Biomarker Molecules

Liquid biopsy is considered as an innovative technique for rapid, convenient, and nontissue specimen sampling, such as blood, urine, and other body fluids like saliva, pleural fluid, peritoneal fluid, and cerebral spinal fluid, that offers a wealth of analytes in the form of circular DNA and RNA molecules, exosomes, and circulating tumor cells. However, currently used techniques as the standard and undisputed methods such as tissue biopsies cannot be replaced by liquid biopsies for detection of prognostic and diagnostic biomarkers and for monitoring of any specific minor alterations associated with the diseases' pathological progression; still, liquid biopsy can provide myriad of data for caregivers to determine individualized treatment suite for each patient particularly due to its capability to indicate tumor heterogeneity in the metastatic setting and estimate the risk for metastatic relapse or metastatic progression, longitudinal monitoring of disease progression, and related resistance mechanisms [16, 20].

3 Circular RNAs and Their Stability Upon Sample Processing Delay

Since circRNAs are generally accepted to be more long lasting compared to their linear counterparts, they can be considered as attractive biomarker source for which standardization of sample processing in order to reduce technical noises and improve the data reproducibility is much more rudimental. According to variety of studies conducted on whole blood samples for transcriptional profiling studies, prolonged blood sample storage, delayed blood processing, and sample handling have negatively affected the multiomics studies' results as well as chromatin accessibility profiling data [21, 22]. In a recent study, isolated peripheral blood mononuclear cells were assessed for possible alteration in differentially expressed circRNAs' profile upon blood processing delay at six different incubation time points, and it was found that after incubation at 4 °C lasting to 48 h, the changes that have been created were negligible compared to extensive alterations in mRNAs and lncRNAs expression levels caused by the incubation time over 12 h. Although it is widely acknowledged that, due to their specific structural characteristics, circRNAs are stable in serum samples incubated at room temperature for up to 24 h, the degradation of circRNAs in kidney cancer tissues and cell cultures has been shown [23]. Hence, it can be concluded that stable structure of circRNAs is important where fast degradation of mRNA molecules may affect their eligibility for introducing as biomarkers.

4 Characterization of Blood Circular RNAs

Blood circRNAs can be categorized into blood cell-free circRNAs, and these circRNAs originate from and reside in blood cells. Cell-free circRNAs are those which are found in plasma, serum, and extracellular vesicles (EVs) that are secreted by blood cells or tissue cells [16]. It is worth mentioning that EVs include exosomes and microvesicles, and these cargo structures circulate in blood stream and migrate and interact with/or internalized to target cells [24]. The existence of circRNAs in exosomes was reported in 2015 for the first time, and since then, this kind of circRNA has been referred to as exo-circRNA [8]. CircRNAs that are carried by EVs are actually specifically and selectively released by origin cells, and some studies conducted in this field have shown that the circRNA to linear RNA counterpart ratio was significantly higher and is enriched in exosomes compared to origin cell [8, 25]. In confirmation of this, RNA-Seq analysis as well as RT-PCR measurements revealed that in spite of exosome-depleted serum samples, tumor-specific circRNAs are enriched in serum-derived exosomes [8, 26]. For instance, EVs isolated from serums of patients with endometrial cancer as well as matched controls and analyzed using RNA-seq technique. Due to obtained results, 209 upregulated circRNAs and 66 downregulated circRNAs were distinguished in cancer group compared to control group [26].

Apart from cell-free circRNAs, the presence of rare subtype of them known as fusion circRNAs (f-circRNAs) also has been proven by studies. This type of circRNAs is generated through chromosomal rearrangements which may lead to back splicing of chimeric mRNA transcripts derived from chromosomal translocations that have mainly been described in hematological cancers [27]. On the one hand, it was shown that f-circRNAs play a vital role in processes such as cellular transformation, cell viability, and therapy-induced resistance, and on the other hand, since gene translocations mostly happen in cancerous cells and due to their restricted amount found in blood, these molecules can provide an absolute superiority for indicating excellent specificity for a given pathology [27]. Just as cell-free miRNAs, cell-free circRNAs have corresponding tissue origin, and by binding to myriads of RNA-binding proteins as well as protein complexes, they are capable for playing role biological processes such as inflammation reactions, metastasis, hemostasis, and therapeutic outcome [28]. In a study conducted by Zhang et al., it was found that in hepatocellular carcinoma cancer (HCC) patients, natural killer cells' functions are arrested by plasma exosome-derived *circUHRF1* that degrades inhibitory *miR-449c-5p* and upregulates the expression level of TIM-3 that finally leads to resistance toward anti-PD1 immunotherapy [29]. In a study conducted in 2017, comparing plasma samples from patients with gastric cancer and plasma samples from health controls indicated that differentially expressed *hsa_circ_0000190* can be candidate as a novel noninvasive biomarker for the diagnosis of gastric cancer since it was downregulated in plasma samples from patients with gastric cancer, and intriguingly, its expression level was found to be significantly correlated with parameters such as tumor diameter, lymphatic metastasis, etc. [30]. Similarly, overexpression of *circFARSA* in non-small cell lung cancer (NSCLC) patients' plasma has been observed compared with control, and it has been suggested that *circFARSA* may play role as sponge for *miR-330-5p* and *miR-326*, leading to relieve their inhibitory effects on oncogene fatty acid synthase [31]. In another study, serum circRNA profiling of 231 lung cancer patients was compared with those of 41 healthy individuals. According to the results, increased expression level of *has_circ_0000190* was shown to be correlated to weak response to immunotherapy, and this molecule was introduced as a predictive biomarker that provides information on these patients' treatment outcome [32]. Furthermore, serum exosomal *circFLI1* has been suggested to be a prognostic biomarker in order to monitor the chemotherapy response in small cell lung cancer patients (SCLC) since its lower expression level has been found to be associated with longer disease remission onsets [33].

Just as other small low molecular weight RNA fractions, circRNAs also existed in anucleated and nucleated mature blood cells [6, 7, 34]. This circRNA category plays vital role in hematopoietic progenitors' differentiation regulation, and in other words, these circRNAs follow a cell-specific pattern and their expression level change due to maturation status. Since these circRNAs indicate the status of immune response, their expression levels are closely associated with antiviral and antibacterial responses [35–37]. Performing a series of cluster analysis, it has been demonstrated that circRNA-chr19 as an endogenous RNA (ceRNA) targets miR-30b-3p-regulated CLDN18 expression and can serve as promising targets for EBOV therapy [38]. For the case of the Hepatitis C Virus (HCV), in which life cycle is dependent on cellular *microRNA-122*, it is observed that artificial circRNA sponges sequester microRNA-122 and inhibit viral protein production in an HCV cell culture system [39]. Libraries derived from peripheral blood mononuclear cells of early HIV-infected individuals and matched controls were subjected to RNA sequencing and GO analysis, revealing 67 circRNAs that may regulate HIV-1 replication by modulating CCNK, CDKN1A, and IL-15 expression [40]. Aberrant expression of some circRNAs is detected in autoimmune cells. In SLE patients, circRNAs contribute to enhanced T-cell apoptosis, regulate the epigenetic alterations in DNA, and blockage of miRNAs' effect on corresponding genes [41–43]. By conducting a comparison between the children with SLE and healthy children, Li et al. showed that 74 circRNAs were upregulated, and 50 circRNAs were downregulated in children with SLE among which the *hsa_circ_0057762* level was positively associated with the SLEDAI-2 K score. According to obtained results, they concluded that *hsa_circ_0057762* and *hsa_circ_0003090* expression levels have roles in the pathogenesis of SLE and can be used to differentiate the patients with SLE from the healthy individuals [44]. In another study, it is found that PBMC-derived endogenous *circPOLR2A* upregulation has the ability to inhibit double-stranded RNA (dsRNA)-activated protein kinase (PKR) by forming 16–26 bp imperfect RNA duplexes in SLE patients. This phenomenon is completely opposite to what happens in normal individuals in whom circRNAs are being degraded via PKR-dependent RNase L activation following getting stimulated by viral infections in early cellular innate immune responses. This finding actually emphasizes on relatedness of circRNAs to innate immunity regulation mechanisms [45].

5 Other Human Disease–Associated Circulating Blood Circular RNAs

Except for abovementioned diseases, a number of studies have revealed that circRNAs are associated with the physiology and pathology of wide range of human diseases.

CircRNA microarray is performed on circRNAs isolated from peripheral blood mononuclear cells (PBMCs) of patients with Alzheimer's disease and healthy controls. After bioinformatics analysis, it has been shown that there is a strong association between differentially expressed circRNAs with inflammation, metabolism, and immune responses that play pivotal role in Alzheimer's disease development [46].

Using Parkinson's disease (PD) blood RNA-seq datasets related to two of the largest publicly available studies, Parkinson's Disease Biomarkers

Program (PDBP), and the Parkinson's Progression Markers Initiative (PPMI), 192 differentially expressed circRNAs in PD participants compared to healthy controls were identified following cross-sectional and longitudinal analysis. Among these, the level of five circRNAs does not change over time but due to the presence of the disease. Additionally, targeted miRNA analysis revealed that a subset of circRNAs is linked to pathways such as the dopaminergic synapse, long-term depression, Hippo signaling, and ubiquitin-mediated proteolysis; all of which have previously been associated with Parkinson's disease (PD) or are suspected to be disrupted in PD, has been revealed [47].

Circular RNA zinc finger protein 609 (*circ-ZNF609*) is an RNA that can be translated into a protein [48]. RNA sequencing analysis of samples from 25 patients with T-cell acute lymphoblastic leukemia (T-ALL), compared to healthy controls, revealed a marked overexpression of circ-ZNF609 specifically in immature T-ALL subtypes. Bioinformatics analysis indicates that in immature T-ALL, circ-ZNF609 binds to miR-181a-5p, which in turn targets O-fucosylpeptide 3-beta-N-acetylglucosaminyltransferase (LFNG), potentially enhancing leukemogenic activity via the Notch1 signaling pathway in T-ALL. So, using proper knocking down or knockout strategies could serve as potential therapeutic policies for treatments of T-ALL [49]. Extensive investigations claim that ferroptosis, that is defined as a form of iron and lipid peroxidation-dependent regulated cell death, is associated with development of some diseases [50]. A research on ALL disease showed that *miR-494-3p* binds to neuroepithelial cell transforming 1 (NET1) and induces the ferroptosis, while *circ_0000745* acts as a sponge for *miR-494-3p* leading to upregulation of NET1, inhibiting ferroptosis and promoting of cell cycle progression and glycolysis in ALL cells [51]. Moreover, Long et al., by assessing differentially expressed circRNAs obtained by microarray analysis on peripheral blood (PB) samples of AML patients, divulged that *circZBTB46* upregulates the expression of stearoyl-CoA desaturase 1 (SCD) most probably by acting as a miRNA sponge and inhibits the ferroptotic cell death in acute myeloid leukemia (AML) cells [52].

Analysis of circRNA profiles from coronary artery disease (CAD) patients and healthy controls revealed significant upregulation of *hsa-circ-0001879* and *hsa-circ-0004104* in CAD. Notably, increased expression of *hsa-circ-0004104* in THP-1-derived macrophages led to dysregulation of genes associated with atherosclerosis, reinforcing its potential role in CAD pathogenesis [53]. An atheroprotection role has been described for *CircANRIL,* which impairs exonuclease-mediated pre-rRNA processing along with ribosome biogenesis in vascular smooth muscle cells and macrophages and by binding to one of the essential 60S-preribosomal assembly factors called as pescadillo homologue 1 (PES1), and by this way, in contrary to its linear counterpart, its overexpression in cardiac patients' whole blood indicates a proatherogenic role [54]. A case–control study carried out on patients with essential hypertension (EH) has found a positive correlation between serum creatinine level and *has-circ-0037911* and concluded that higher expression *has-circ-0037911,* by changing the serum creatinine level, can influence the EH development [55]. A year later, another study conducted by microarray profiling of circular RNAs in EH patients and healthy control group reported that hyperexpression of *has-circ-0126991* can be correlated with the risk for EH incidence [56]. Despite the aforementioned studies in which upregulation of distinct circRNAs was found to be correlated with hypertension occurrence, another study performed on plasma samples of hypertensive patients and healthy controls showed that the expression level of a circRNA has been named *has-circ-0005870* and is being downregulated significantly in hypertensive patients [57].

Apart from hypertension, differential expression of circRNAs has also been observed across various cardiovascular conditions, including coronary heart disease (CHD), cardiomyopathies (CM), heart failure (HF), hypertrophic cardiomyopathy (HCM), and pediatric cases of CHD [58].

A systemic autoimmune disease known as rheumatoid arthritis (RA) is characterized by chronic synovial inflammation leading to various degrees of bone erosion and cartilage ablation. Microarray technique was used to assess the PBMC-derived circRNA expression profile on 10 RA patients and 10 matched healthy controls, and results demonstrated that 255 circRNAs were significantly upregulated, whereas 329 were downregulated in the RA cases and suggested that the possible potential role of these dysregulated circRNAs may have existed for occurrence and development of RA [59]. In an attempt to investigate the expression of *ciRS-7* in patients with RA, which has been suggested to act as a *miR-7* sponge, the expression level of *ciRS-7*, *miR-7*, and mTOR in PBMC samples of RA patients and healthy participants was measured. mTOR has been considered as one of the *miR-7* target genes. Significantly increased level of *ciRS-7* was shown that was parallel to mTOR expression level. mTOR level of expression had fallen dramatically following the upregulation of *miR-7*. It was concluded that enhanced *ciRS-7* may inhibit the miR-7 function and to a great extent alleviate the inhibitory action of *miR-7* on mTOR [60]. In this era, a recent study made inquiries about new biomarkers for the diagnosis of RA and screened differentially expressed circRNAs in PBMCs of RA patients, compared with those of systemic lupus erythematosus (SLE) patients as well as healthy participants. Significant downregulation of *has-circ-101328* in RA cases compared with SLE and healthy group was proved, and furthermore, it was revealed that significant correlation existed between *has-circ-101328* and positive C-reactive protein (CRP) level [61].

Interestingly in a study, blood circRNAs are used to delve into their association with advancing age, human aging phenotypes, and lifespan or cellular senescence. Changes in circRNA expression in relation to age were followed for over a 5-year period, and obtained data were combined with parental longevity score (PLS) and hand grip strength. Then, expression levels of 15 circRNAs were assessed in early passage and late passage primary human dermal fibroblasts, cardiomyocytes, astrocytes, and vascular endothelial cells. According to the results, associations between *circDEF6*, *circEP300*, *circFOXO3*, and *circFNDC3B* with parental longevity or hand grip strength were seen. Additionally, *circFOXO3* and *circEP300* also showed differential expression in one or more human senescent cell types. All suggested that circRNA may represent either biomarkers or future therapeutic targets for age-related disease [62]. In forensic investigations, when there is no eyewitness available, age estimation of biological evidence such as blood sampled from crime scenes provides valuable investigative leads. In a research, circRNAs have been used for establishing forensic age estimation models by providing additional novel age-related biomarkers [63]. In another study, Dluzen et al. compared human serum circRNAs' changes that may associate with normal aging in young (30–32 years) and old (80–85 years) female cases and identified that *has-circ-1305* and *has-circ-722* were upregulated in the elderly compared with young [64].

Major depressive disorder (MDD), characterized by influence on individuals' emotions, feelings, and behaviors, sometimes leads to suicide. Series of factors such as genetics, neuroendocrinology, epigenetic regulation, and social-psychological factors are believed to play role in its pathogenesis and show their effects in form of abnormalities in neural-endocrine-immune regulatory network, resulting in irregular neurotransmitters, hormones, and cytokines' secretion, all of which can lead to the occurrence of depression. Expression levels of circRNAs in PBMCs of patients with depression were compared with healthy controls, and differential expression of *has-circRNA-002143*, *has-circRNA-103636*, *has-circRNA-100679*, and *has-circRNA-104953* was detected. Among this circRNAs, only hsa-*circRNA-103636* was found to be significantly upregulated following antidepressant therapy. This observation suggests its potential utility as a diagnostic biomarker in individuals with depression [65]. In individuals with MDD, plasma level of *circDYM* is reduced. This circRNA typically acts as a sponge for *microRNA-9*, facilitating the ubiquitination of HSP90 and dampening microglial activation is revealed [66]. Applying high-throughput sequencing technology, differentially expressed circRNAs in female MDD patients were compared with healthy female controls, and five circRNAs, *has-circ-0020959*, *has-circ-0005959*, *has-circ-0033064*, *has-circ-0006862*, and *has-circ-0027732*, were found to be mostly related to developing the major depressive disorder in females [67]. By whole transcriptome sequencing on whole blood samples of MDD patients and comparing them with those of healthy individuals, their bioinformatics analysis confirmed that preliminarily proved and upregulated *has-circ-0079651*, *has-circ-0137187*, *has-circ-0006010*, and *has-circ-0113010* may act through the ceRNA mechanism in neuroplasticity of MDD [68]. Among the existence of a mass of dysregulated circRNAs in PBMCs between female MDD patients and healthy controls, a study by Bu et al. explored that *has-circ-0126218* was upregulated in female MDD patients [69]. And very recently, the expression level of *circRNA protein tyrosine kinase 2* (*circPTK2*) was measured by recruiting PBMC samples of MDD patients and healthy subjects. Due to results, decreased expression level of c*ircPTK2* was shown in MDD patients. Identification of the potential binding sequence of *miR-182-5p*, whose elevated levels have been previously observed in MDD patients [73], suggests that *circPTK2* may function as a sponge for miR-182-5p; thereby contributing to the beneficial effects of *circPTK2* [71].

Pregnancy-specific circRNAs have also been detected and compared between women with complicated pregnancy to those with an uncomplicated pregnancy in wide range of studies assessing maternal blood as well as exosomes isolated from umbilical cord blood. These studies specifically focused on individuals diagnosed with preeclampsia (PE) -defined as new-onset hypertension occurring after the 20th week of gestation- and gestational diabetes mellitus (GDM), both of which contribute to maternal and perinatal morbidity and mortality and are associated with damage to the kidneys, heart, brain, or liver. Matched healthy controls were also included for comparison.

In order to determine differently expressed circRNA as possible predictors for GDM development, elevated levels of *circ-0008285*, *circ-0026497*, *circ-0039480*, *circ-PNPT1*, and *circVEGFC* were tracked in maternal plasma and whole blood of women with GDM [72–75]. It is worth noting that high expression of *circVEGFC* has been found to be correlated with higher incidence rates of fetal malformation and hypertension. In a study by Huang et al., facilitation of the proliferation, migration, and angiogenesis of human umbilical vein endothelial cells (HG-HUVECs) through the *miR-1200/MEOX2* axis can be achieved by elimination of exosomal *circ-0074673*, and the existence of increased levels of *circ-0074673* exosomes isolated from umbilical cord blood of GDM cases is proof of this [76].

Comparing PE and healthy pregnant women, differentially expressed circRNAs obtained from whole blood samples were evaluated to clarify the possible involvement of circRNAs in the pathogenesis of PE. In either early-onset or late-onset PE patients, *has-circ-0007717*, *has-circ-0006460*, and *has-circ-0093055* were elevated significantly and related circRNA–miRNA–mRNA network analysis asserted that these circRNAs may participate in immune response-related pathways and cellular senescence [77].

The expression level of *hsa-circ-0049356* has been compared between seminal plasma and blood samples from patients with nonobstructive azoospermia (NOA)-the most severe form of male infertility-and samples from healthy controls and it has been explored that *hsa-circ-0049356* is overexpressed in NOA patients' whole blood samples in comparison to its significant downregulation in their seminal plasma samples versus to healthy controls and the different expression pattern in whole blood sample and seminal plasma samples got attributed to the existence of a blood-testis barrier and target prediction analysis revealed that a mRNA with regulatory role for guanyl-nucleotide exchange factor activity have vital role in germ cells' cytoskeleton rearrangement during spermatogenesis [78].

Several studies have attempted to identify new indicators of T2DM at early stages, and in this era, the expression profiles of circRNAs in T2DM and prediabetes patients remain to be explored, and their discriminating value, specificity, and sensitivity will be of great value.

In 2017, a research was performed on normal cases, prediabetes patients, and T2DM patients. To determine whether a subject was healthy or had prediabetes or T2DM, oral glucose tolerance test (OGTT) was used. RNA microarray analysis was performed on venous blood samples collected analysis was conducted and due to results *has-circ-0054633* presented a greatest diagnostic capability for pre-diabetes and T2DM and no differential expression of *has-circ-0054633* was detected attributable to the different gender and age groups. Authors hypothesized that *has-circ-0054633* may participate in the pathogenesis of T2DM by affecting the cellular metabolism, cell cycle, and mitotic cell cycle arrest. So, insufficient insulin secretion in these patients can be caused by impaired cell cycle progress, leading to decreased β cell proliferation [79]. It was found that of the five most abundant circRNAs in human islets, *CircCAMSAP1* is also associated with T2D status in human peripheral blood samples [80]. RNA sequencing of peripheral blood from T2DM patients experiencing thirst and fatigue revealed 21 differentially expressed circRNAs. Among these, downregulated circRNAs; *novel-circ-0016196, novel-circ-0016198, and novelcirc- 0005686*; were found to bind *miR-877-3p*, promoting cell apoptosis and suppressing insulin secretion. Conversely, upregulated *hsa_circ_0002590 and novel_circ_000372* were shown to inhibit insulin secretion by interacting with *miR-149-5p* [81].

Group of circRNAs with altered expression pattern was also detected in patients with type 2 diabetes mellitus with proliferative diabetic retinopathy (T2DR) compared with those without DR and healthy controls and suggested that these circRNAs may have role in pathogenesis of DR and can be considered as therapeutic targets in future studies [82]. Constructing the expression profiles of circRNAs by RNA sequencing in the peripheral blood samples obtained from type 2 diabetes mellitus patients with thirst and fatigue (D-T2DM) and healthy controls revealed that downregulated *novel_circ_0016196*, *novel_circ_0016198*, and *novel_circ_0005686* bind to *miR-877-3p,* leading to attenuation in mitochondrial apoptosis, while *hsa_circ_0002590* and *novel_circ_000372* with increased levels in D-T2DM cases inhibit insulin secretion by binding to *miR-149-5p* [81]. The expression characteristics of circular RNAs has been investigated in coronary artery disease patients, T2DM patients as well as coronary artery disease combined with T2DM patients and has been compared with those of healthy controls' peripheral blood samples obtained. Due to results, crude odds ratios and adjusted odds ratios of *hsa-circRNA11783-2* revealed significant alterations in either the coronary artery disease group or T2DM group. Although still further studies should be carried out, still it is known that *hsa-circRNA11783-2* has sponge role for *miR-608* and *miR-3907*. So the study on the relationships between these two miRNAs and T2DM is recommended [83].

The role of small noncoding RNAs with the cytokines' fluctuations in diseases such as diabetes has been the focus of a group of studies. In this regard, the relationship between cytokines and diabetes should not be ignored. Among them, interleukin 6 (IL-6) and tumor necrosis factor-α (TNF-α), two cytokines related to T2DM, can be mentioned. In an attempt for investigating the clinical value of *has_circ_CCNB1* and *has_circ_0009024* in T2DM patients, their expression level in their serum samples up- and downregulated, respectively, and these alterations were due to their positive and negative correlation with IL-6, TNF-α, and glucose level in studied cases [84]. In another comprehensive study, RNA sequencing and qPCR were used to reveal the expression profiles of circRNAs in peripheral leucocytes of patients with T2DM and followed by RegRNA and mirPath platforms analysis in order to analyze possible interactions with microRNAs. Significant upregulation of *circANKRD36* was seen in T2DM cases compared to control group. This increase was correlated with high glucose level as well as IL-6 level. This expression difference was also detected by comparing the expression level of circANKRD36 host gene in T2DM and healthy controls' cases.

Bioinformatics analysis and target analysis suggested three miRNAs, *hsa-miR-3614-3p*, *hsa-miR-498*, and *hsa-miR-501-5p,* as possible targets for *circANKRD36*. And all together, the authors concluded that *circANKRD36* perhaps is involved in T2DM and inflammation-associated pathways through interaction with abovementioned miRNAs [85].

For the case of T1DM, it is explored that *has-circ-0060450*, *circPPM1F*, and *has-circ-0002202* are peripheral blood mononuclear cells (PBMCs)-derived circRNAs that are upregulated in T1DM patients, and *has-circ-0060450* functions as sponge for *miR-199a-5p* to inhibit the JAK-STAT signaling pathway stimulated by type I IFN (IFN-I), leading to inhibition of macrophage-mediated inflammation [86]. In the following study, modulator role of circPPM1F for M1 macrophage activation and pancreatic islet inflammation was elucidated [87]. And finally, according to the computational and bioinformatics approaches' results, it was suggested that it may mediate IFN-I-induced macrophage inflammation [88]. DNAJ homolog subfamily C member 3 (DNAJC3)-originated *has-circ-0002473* is overexpressed in T1DM, and in contrast to DNAJC3, it can eliminate the translocation of the glucose-regulated protein 78-kDa (GRP78) to the plasma membrane to arrest the proapoptotic feedback loop in β-cells and preserve them [89].

As it is mentioned before, CircRNAs have been implicated in various diseases, and several specific circRNAs have been studied in the context of disease pathogenesis. Here, some of these diseases and related complications were reviewed, and ongoing research continues to uncover more circRNAs associated with diseases.

6 Conclusion

Circular RNAs (circRNAs) are gaining significant attention as potential biomarkers for liquid biopsies due to their unique properties such as their (a) stability, due to which they can be reliably detected in blood samples; (b) abundant expression, which makes them easily detectable and useful for diagnosis purposes; (c) noninvasiveness: by being detectable in a simple blood draw, this characteristic can be mentioned in bold since when clinical conditions have worsened, performing tissue biopsy becomes challenging or the accessibility of target tissue is limited; (d) disease- and tissue-specific nature: since accumulating evidence has demonstrated that circRNAs can be associated with the onset and progression of variety of diseases and their complications; (e) real-time monitoring, due to the possibility for performing liquid biopsies more frequently than tissue biopsies, that provide the opportunity for real-time monitoring of diseases and treatment efficacy; and (f) FDA approvals: currently, some circRNA-based liquid biopsy tests have received FDA approval and are used for guiding treatment, indicating circRNAs' growing acceptance to be used in clinical practice. It is worth mentioning that they significantly impact patient care and treatment strategies, and hence, liquid biopsy may reduce patents' stress. On the other hand, diagnostic abilities of circRNAs have been demonstrated by receiver operating characteristic ROC analysis performed in variety of studies. Circular RNAs (circRNAs) can be detected in various fractions of blood, including plasma and serum as well as cellular fraction including white blood cells (such as lymphocytes and monocytes) and red blood cells, and their enrichment and stability within exosomes make them intriguing candidates for disease diagnosis and therapeutic interventions.

Thanks to advancements in high-throughput RNA sequencing and experimental technologies, extensive identification and characterization of circRNAs have been done and provided valuable information about their cell- or tissue-specific expression patterns, epigenetic regulation, translation potential, localization, and metabolism, and their dysregulation has been linked to various human diseases, but sometimes, a circRNAs alone does not show enough sensitivity and specificity, and a combination of two or more circRNAs may need to enhance their diagnostic value. In summary, profiling circRNAs across different tissues and cell types requires comprehensive studies, and large-scale datasets and bioinformatics algorithms and detection methods, along with molecular techniques, need to be continually evolving to accelerate circRNA understanding.

Acknowledgments The author thanks Prof. Dr. Ahad Jafari Ghods, MD, FACP (Iran University of Medical Sciences) and Sana Maher, PhD candidate (Koc University, KUTTAM), for critical reading and comments on the manuscript.

Competing Financial Interests The author declares no competing financial interests.

References

1. Enuka Y, Lauriola M, Feldman ME, Sas-Chen A, Ulitsky I, Yarden Y (2016) Circular RNAs are long-lived and display only minimal early alterations in response to a growth factor. Nucleic Acids Res 44(3):1370–1383. https://doi.org/10.1093/nar/gkv1367
2. Jeck WR, Sorrentino JA, Wang K, Slevin MK, Burd CE, Liu J, Marzluff WF, Sharpless NE (2013) Circular RNAs are abundant, conserved, and associated with ALU repeats. RNA 19(2):141–157. https://doi.org/10.1261/rna.035667.112
3. Zhou T, Xie X, Li M, Shi J, Zhou JJ, Knox KS, Wang T, Chen Q, Gu W (2018) Rat Body Map transcriptomes reveal unique circular RNA features across tissue types and developmental stages. RNA 24(11):1443–1456. https://doi.org/10.1261/rna.067132.118
4. Zhang Z, Yang T, Xiao J (2018) Circular RNAs: promising biomarkers for human diseases. EBioMedicine 34:267–274. https://doi.org/10.1016/j.ebiom.2018.07.036
5. Memczak S, Papavasileiou P, Peters O, Rajewsky N (2015) Identification and characterization of circular RNAs as a new class of putative biomarkers in human blood. PLoS One 10(10):e0141214. https://doi.org/10.1371/journal.pone.0141214
6. Qian Z, Liu H, Li M, Shi J, Li N, Zhang Y, Zhang X, Lv J, Xie X, Bai Y et al (2018) Potential diagnostic power of blood circular RNA expression in active pulmonary tuberculosis. EBioMedicine 27:18–26. https://doi.org/10.1016/j.ebiom.2017.12.007
7. Alhasan AA, Izuogu OG, Al-Balool HH, Steyn JS, Evans A, Colzani M, Ghevaert C, Mountford JC, Marenah L, Elliott DJ et al (2016) Circular RNA enrichment in platelets is a signature of transcriptome degradation. Blood 127(9):e1–e11. https://doi.org/10.1182/blood-2015-06-649434
8. Li Y, Zheng Q, Bao C, Li S, Guo W, Zhao J, Chen D, Gu J, He X, Huang S (2015) Circular RNA is enriched and stable in exosomes: a promising biomarker for cancer diagnosis. Cell Res 25(8):981–984. https://doi.org/10.1038/cr.2015.82
9. Zemmour H, Planer D, Magenheim J, Moss J, Neiman D, Gilon D, Korach A, Glaser B, Shemer R, Landesberg G et al (2018) Noninvasive detection of human cardiomyocyte death using methylation patterns of circulating DNA. Nat Commun 9:1443. https://doi.org/10.1038/s41467-018-03961-y
10. Sole C, Arnaiz E, Manterola L, Otaegui D, Lawrie CH (2019) The circulating transcriptome as a source of cancer liquid biopsy biomarkers. Semin Cancer Biol 58:100–108. https://doi.org/10.1016/j.semcancer.2019.01.003
11. Pardini B, Sabo A, Birolo G, Calin G (2019) Noncoding RNAs in extracellular fluids as cancer biomarkers: the new frontier of liquid biopsies. Cancer 11(8):1170. https://doi.org/10.3390/cancers11081170
12. Ghafouri Kalajahi H, Yari A, Amini M, Catal T, Youshanlui M, Pourbagherian O, Sezer Ç, Ali MA (2023) Therapeutic effect of microRNA-21 on differentially expressed hub genes in gastric cancer based on systems biology. Sci Rep 13:21906. https://doi.org/10.1038/s41598-023-49225-8
13. Pfefer SR, Yang CH, Pfefer LM (2015) The role of miR-21 in cancer. Drug Dev Res 76(6):270–277. https://doi.org/10.1002/ddr.21257
14. Zhang Z, Li Z, Gao C, Ping C, Jianjun C, Wenzhong L, Shudong X, Hong L (2008) miR-21 plays a pivotal role in gastric cancer pathogenesis and progression. Lab Investig 88:1358–1366. https://doi.org/10.1038/labinvest.2008.94
15. Rama AR, Lara P, Mesas C, Quiñonero F, Vélez C, Melguizo C, Prados J (2022) Circular sponge against miR-21 enhances the antitumor activity of doxorubicin against breast cancer cells. Int J Mol Sci 23(23):14803. https://doi.org/10.3390/ijms232314803
16. Wen G, Zhou T, Gu W (2021) The potential of using blood circular RNA as liquid biopsy biomarker for human diseases. Protein Cell 12(12):911–946. https://doi.org/10.1007/s13238-020-00799-3
17. Li X, Yang L, Chen L-L (2018) The biogenesis, functions, and challenges of circular RNAs. Mol Cell 71(3):428–442. https://doi.org/10.1016/j.molcel.2018.06.034
18. Chen LL (2020) The expanding regulatory mechanisms and cellular functions of circular RNAs. Nature reviews. Mol Cell Biol 21(8):475–490. https://doi.org/10.1038/s41580-020-0243-y
19. Chen LL (2016) The biogenesis and emerging roles of circular RNAs. Nature reviews. Mol Cell Biol 17(4):205–211. https://doi.org/10.1038/nrm.2015.32
20. Russano M, Napolitano A, Ribelli G, Iuliani M, Simonetti S, Citarella F, Pantano F, Dell'Aquila E, Anesi C, Silvestris N, Argentiero A, Solimando AG, Vincenzi B, Tonini G, Santini D (2020) Liquid biopsy and tumor heterogeneity in metastatic solid tumors: the potentiality of blood samples. J Exp Clin Cancer Res 39(1):95. https://doi.org/10.1186/s13046-020-01601-2

21. Dvinge H, Ries RE, Ilagan JO, Stirewalt DL, Meshinchi S, Bradley RK (2014) Sample processing obscures cancer-specific alterations in leukemic transcriptomes. Proc Natl Acad Sci USA 111(47):16802–16807. https://doi.org/10.1073/pnas.1413374111
22. Massoni-Badosa R, Iacono G, Moutinho C, Kulis M, Palau N, Marchese D, Rodríguez-Ubreva J, Ballestar E, Rodriguez-Esteban G, Marsal S, Aymerich M, Colomer D, Campo E, Julià A, Martín-Subero JI, Heyn H (2020) Sampling time-dependent artifacts in single-cell genomics studies. Genome Biol 21(1):112. https://doi.org/10.1186/s13059-020-02032-0
23. Savage AK, Gutschow MV, Chiang T, Henderson K, Green R, Chaudhari M, Swanson E, Heubeck AT, Kondz N, Burley KC, Genge PC, Lord C, Smith T, Thomson Z, Beaubien A, Johnson E, Goldy J, Bolouri H, Buckner JH, Meijer P et al (2021) Multimodal analysis for human ex vivo studies shows extensive molecular changes from delays in blood processing. iScience 24(5):102404. https://doi.org/10.1016/j.isci.2021.102404
24. Revenfeld AL, Bæk R, Nielsen MH, Stensballe A, Varming K, Jørgensen M (2014) Diagnostic and prognostic potential of extracellular vesicles in peripheral blood. Clin Ther 36(6):830–846. https://doi.org/10.1016/j.clinthera.2014.05.008
25. Li Y, Zhao J, Yu S, Wang Z, He X, Su Y, Guo T, Sheng H, Chen J, Zheng Q, Li Y, Guo W, Cai X, Shi G, Wu J, Wang L, Wang P, He X, Huang S (2019) Extracellular vesicles long RNA sequencing reveals abundant mRNA, circRNA, and lncRNA in human blood as potential biomarkers for cancer diagnosis. Clin Chem 65(6):798–808. https://doi.org/10.1373/clinchem.2018.301291
26. Xu H, Gong Z, Shen Y, Fang Y, Zhong S (2018) Circular RNA expression in extracellular vesicles isolated from serum of patients with endometrial cancer. Epigenomics 10(2):187–197. https://doi.org/10.2217/epi-2017-0109
27. Guarnerio J, Bezzi M, Jeong JC, Paffenholz SV, Berry K, Naldini MM, Lo-Coco F, Tay Y, Beck AH, Pandolfi PP (2016) Oncogenic role of fusion-circRNAs derived from cancer-associated chromosomal translocations. Cell 166(4):1055–1056. https://doi.org/10.1016/j.cell.2016.07.035
28. Xu Y, Leng K, Yao Y, Kang P, Liao G, Han Y, Shi G, Ji D, Huang P, Zheng W, Li Z, Li J, Huang L, Yu L, Zhou Y, Jiang X, Wang H, Li C, Su Z, Tai S et al (2021) A circular RNA, cholangiocarcinoma-associated circular RNA 1, contributes to cholangiocarcinoma progression, induces angiogenesis, and disrupts vascular endothelial barriers. Hepatology (Baltimore, Md) 73(4):1419–1435. https://doi.org/10.1002/hep.31493
29. Zhang PF, Gao C, Huang XY, Lu JC, Guo XJ, Shi GM, Cai JB, Ke AW (2020) Cancer cell-derived exosomal circUHRF1 induces natural killer cell exhaustion and may cause resistance to anti-PD1 therapy in hepatocellular carcinoma. Mol Cancer 19(1):110. https://doi.org/10.1186/s12943-020-01222-5
30. Chen S, Li T, Zhao Q, Xiao B, Guo J (2017) Using circular RNA hsa_circ_0000190 as a new biomarker in the diagnosis of gastric cancer. Clin Chim Acta 466:167–171. https://doi.org/10.1016/j.cca.2017.01.025
31. Hang D, Zhou J, Qin N, Zhou W, Ma H, Jin G, Hu Z, Dai J, Shen H (2018) A novel plasma circular RNA circFARSA is a potential biomarker for non-small cell lung cancer. Cancer Med 7(6):2783–2791. https://doi.org/10.1002/cam4.1514
32. Luo YH, Yang YP, Chien CS, Yarmishyn AA, Ishola AA, Chien Y, Chen YM, Huang TW, Lee KY, Huang WC, Tsai PH, Lin TW, Chiou SH, Liu CY, Chang CC, Chen MT, Wang ML (2020) Plasma level of circular RNA hsa_circ_0000190 correlates with tumor progression and poor treatment response in advanced lung cancers. Cancer 12(7):1740. https://doi.org/10.3390/cancers12071740
33. Li L, Li W, Chen N, Zhao H, Xu G, Zhao Y, Pan X, Zhang X, Zhou L, Yu D, Li A, Hu JF, Cui J (2019) FLI1 exonic circular RNAs as a novel oncogenic driver to promote tumor metastasis in small cell lung cancer. Clin Cancer Res 25(4):1302–1317. https://doi.org/10.1158/1078-0432.CCR-18-1447
34. Gaffo E, Boldrin E, Dal Molin A, Bresolin S, Bonizzato A, Trentin L, Frasson C, Debatin KM, Meyer L, Kronnie G, Bortoluzzi S (2019) Circular RNA differential expression in blood cell populations and exploration of circRNA deregulation in pediatric acute lymphoblastic leukemia. Sci Rep 9:14670. https://doi.org/10.1038/s41598-019-50864-z
35. Nicolet BP, Engels S, Aglialoro F, van den Akker E, von Lindern M, Wolkers MC (2018) Circular RNA expression in human hematopoietic cells is widespread and cell-type specific. Nucleic Acids Res 46(16):8168–8180. https://doi.org/10.1093/nar/gky721
36. Awan FM, Yang BB, Naz A, Hanif A, Ikram A, Obaid A, Malik A, Janjua HA, Ali A, Sharif S (2021) The emerging role and significance of circular RNAs in viral infections and antiviral immune responses: possible implication as theranostic agents. RNA Biol 18(1):1–15. https://doi.org/10.1080/15476286.2020.1790198
37. Ng WL, Marinov GK, Liau ES, Lam YL, Lim YY, Ea CK (2016) Inducible RasGEF1B circular RNA is a positive regulator of ICAM-1 in the TLR4/LPS pathway. RNA Biol 13(9):861–871. https://doi.org/10.1080/15476286.2016.1207036
38. Wang ZY, Guo ZD, Li JM, Zhao ZZ, Fu YY, Zhang CM, Zhang Y, Liu LN, Qian J, Liu LN (2017) Genome-wide search for competing endogenous RNAs responsible for the effects induced by Ebola virus replication and transcription using a trVLP system. Front Cell Infect Microbiol 7:479. https://doi.org/10.3389/fcimb.2017.00479
39. Jost I, Shalamova LA, Gerresheim GK, Niepmann M, Bindereif A, Rossbach O (2018) Functional sequestration of microRNA-122 from Hepatitis C Virus by

circular RNA sponges. RNA Biol 15(8):1032–1039. https://doi.org/10.1080/15476286.2018.1435248
40. Zhang Y, Zhang H, An M, Zhao B, Ding H, Zhang Z, He Y, Shang H, Han X (2018) Crosstalk in competing endogenous RNA networks reveals new circular RNAs involved in the pathogenesis of early HIV infection. J Transl Med 16(1):332. https://doi.org/10.1186/s12967-018-1706-1
41. Li LJ, Zhu ZW, Zhao W, Tao SS, Li BZ, Xu SZ, Wang JB, Zhang MY, Wu J, Leng RX, Fan YG, Pan HF, Ye DQ (2018) Circular RNA expression profile and potential function of hsa_circ_0045272 in systemic lupus erythematosus. Immunology 155(1):137–149. https://doi.org/10.1111/imm.12940
42. Zhang C, Huang J, Chen Y, Shi W (2018) Low expression and clinical value of hsa_circ_0049224 and has_circ_0049220 in systemic lupus erythematous patients. Med Sci Monit 24:1930–1935. https://doi.org/10.12659/msm.906507
43. Zhang C, Wang X, Chen Y, Wu Z, Zhang C, Shi W (2018) The down-regulation of hsa_circ_0012919, the sponge for miR-125a-3p, contributes to DNA methylation of CD11a and CD70 in CD4+ T cells of systemic lupus erythematous. Clin Sci (London, England: 1979) 132(21):2285–2298. https://doi.org/10.1042/CS20180403
44. Li S, Zhang J, Tan X, Deng J, Li Y, Piao Y, Li C, Yang W, Mo W, Sun J, Sun F, Han T, Wang J, Kuang W, Li C (2019) Microarray expression profile of circular RNAs and mRNAs in children with systemic lupus erythematosus. Clin Rheumatol 38(5):1339–1350. https://doi.org/10.1007/s10067-018-4392-8
45. Liu CX, Li X, Nan F, Jiang S, Gao X, Guo SK, Xue W, Cui Y, Dong K, Ding H, Qu B, Zhou Z, Shen N, Yang L, Chen LL (2019) Structure and degradation of circular RNAs regulate PKR activation in innate immunity. Cell 177(4):865–880.e21. https://doi.org/10.1016/j.cell.2019.03.046
46. Li Y, Lv Z, Zhang J, Ma Q, Li Q, Song L, Gong L, Zhu Y, Li X, Hao Y, Yang Y (2020) Profiling of differentially expressed circular RNAs in peripheral blood mononuclear cells from Alzheimer's disease patients. Metab Brain Dis 35(1):201–213. https://doi.org/10.1007/s11011-019-00497-y
47. Beric A, Sun Y, Sanchez S, Martin C, Powell T, Adrian Pardo J, Sanford J, Botia JA, Cruchaga C, Ibanez L (2024) Circulating blood circular RNA in Parkinson's disease: a systematic study. medRxiv: the preprint server for health sciences, 2024012224301623. https://doi.org/10.1101/2024.01.22.24301623
48. Wang S, Wu J, Wang Z, Gong Z, Liu Y, Wang Z (2022) Emerging roles of circ-ZNF609 in multiple human diseases. Front Genet 13:837343. https://doi.org/10.3389/fgene.2022.837343
49. Buratin A, Paganin M, Gaffo E, Dal Molin A, Roels J, Germano G, Siddi MT, Serafin V, De Decker M, Gachet S, Durinck K, Speleman F, Taghon T, Te Kronnie G, Van Vlierberghe P, Bortoluzzi S (2020) Large-scale circular RNA deregulation in T-ALL: unlocking unique ectopic expression of molecular subtypes. Blood Adv 4(23):5902–5914. https://doi.org/10.1182/bloodadvances.2020002337
50. Ma TL, Chen JX, Zhu P, Zhang CB, Zhou Y, Duan JX (2022) Focus on ferroptosis regulation: exploring novel mechanisms and applications of ferroptosis regulator. Life Sci 307:120868. https://doi.org/10.1016/j.lfs.2022.120868
51. Yang X, Li Y, Zhang Y, Liu J (2022) Circ_0000745 promotes acute lymphoblastic leukemia progression through mediating miR-494-3p/NET1 axis. Hematology (Amsterdam, Netherlands) 27(1):11–22. https://doi.org/10.1080/16078454.2021.2008590
52. Long F, Lin Z, Long Q, Lu Z, Zhu K, Zhao M, Yang M (2023) CircZBTB46 protects acute myeloid leukemia cells from ferroptotic cell death by upregulating SCD. Cancer 15(2):459. https://doi.org/10.3390/cancers15020459
53. Wang L, Shen C, Wang Y, Zou T, Zhu H, Lu X, Li L, Yang B, Chen J, Chen S, Lu X, Gu D (2019) Identification of circular RNA Hsa_circ_0001879 and Hsa_circ_0004104 as novel biomarkers for coronary artery disease. Atherosclerosis 286:88–96. https://doi.org/10.1016/j.atherosclerosis.2019.05.006
54. Holdt LM, Stahringer A, Sass K, Pichler G, Kulak NA, Wilfert W, Kohlmaier A, Herbst A, Northoff BH, Nicolaou A, Gäbel G, Beutner F, Scholz M, Thiery J, Musunuru K, Krohn K, Mann M, Teupser D (2016) Circular non-coding RNA ANRIL modulates ribosomal RNA maturation and atherosclerosis in humans. Nat Commun 7:12429. https://doi.org/10.1038/ncomms12429
55. Bao X, Zheng S, Mao S, Gu T, Liu S, Sun J, Zhang L (2018) A potential risk factor of essential hypertension in case-control study: circular RNA hsa_circ_0037911. Biochem Biophys Res Commun 498(4):789–794. https://doi.org/10.1016/j.bbrc.2018.03.059
56. Liu L, Gu T, Bao X, Zheng S, Zha J, Zhang L (2019) Microarray profiling of circular RNA identifies hsa_circ_0126991 as a potential risk factor for essential hypertension. Cytogenet Genome Res 157(4):203–212. https://doi.org/10.1159/000500063
57. Wu N, Jin L, Cai J (2017) Profiling and bioinformatics analyses reveal differential circular RNA expression in hypertensive patients. Clin Exp Hypertens (New York, NY: 1993) 39(5):454–459. https://doi.org/10.1080/10641963.2016.1273944
58. Singh DD, Kim Y, Choi SA, Han I, Yadav DK (2023) Clinical significance of MicroRNAs, long non-coding RNAs, and CircRNAs in cardiovascular diseases. Cells 12(12):1629. https://doi.org/10.3390/cells12121629
59. Zheng F, Yu X, Huang J, Dai Y (2017) Circular RNA expression profiles of peripheral blood mononuclear cells in rheumatoid arthritis patients, based on microarray chip technology. Mol Med Rep 16(6):8029–8036. https://doi.org/10.3892/mmr.2017.7638
60. Tang X, Wang J, Xia X, Tian J, Rui K, Xu H, Wang S (2019) Elevated expression of ciRS-7 in peripheral blood mononuclear cells from rheumatoid

arthritis patients. Diagn Pathol 14(1):11. https://doi.org/10.1186/s13000-019-0783-7
61. Lu H, Yang Y, Kuang D, Liu P, Yang J (2022) Expression profile of circRNA in peripheral blood mononuclear cells of patients with rheumatoid arthritis. BMC Med Genet 15(1):77. https://doi.org/10.1186/s12920-022-01225-9
62. Haque S, Ames RM, Moore K, Pilling LC, Peters LL, Bandinelli S, Ferrucci L, Harries LW (2020) circRNAs expressed in human peripheral blood are associated with human aging phenotypes, cellular senescence and mouse lifespan. GeroScience 42(1):183–199. https://doi.org/10.1007/s11357-019-00120-z
63. Wang J, Wang C, Wei Y, Zhao Y, Wang C, Lu C, Feng J, Li S, Cong B (2022) Circular RNA as a potential biomarker for forensic age prediction. Front Genet 13:825443. https://doi.org/10.3389/fgene.2022.825443
64. Dluzen DF, Noren Hooten N, De S, Wood WH 3rd, Zhang Y, Becker KG, Zonderman AB, Tanaka T, Ferrucci L, Evans MK (2018) Extracellular RNA profiles with human age. Aging Cell 17(4):e12785. https://doi.org/10.1111/acel.12785
65. Cui X, Niu W, Kong L, He M, Jiang K, Chen S, Zhong A, Li W, Lu J, Zhang L (2016) hsa_circRNA_103636: potential novel diagnostic and therapeutic biomarker in major depressive disorder. Biomark Med 10(9):943–952. https://doi.org/10.2217/bmm-2016-0130
66. Song R, Bai Y, Li X, Zhu J, Zhang H, Shi Y, Li K, Wang B, Zhang H, Yang Y, Zhan Z (2020) Plasma circular RNA DYM related to major depressive disorder and rapid antidepressant effect treated by visual cortical repetitive transcranial magnetic stimulation. J Affect Disord 274:486–493. https://doi.org/10.1016/j.jad.2020.05.109
67. Zhang W, Yu Y, BT, Qiu X, Yang X, Qiao Z, Zhao M, Yang Y (2020) Bioinformatics analysis of peripheral blood circRNAs in female patients with major depressive disorder. Chin J Behav Med Brain Sci 12:725–729
68. Zhang D, Ji Y, Chen X, Chen R, Wei Y, Peng Q, Lin J, Yin J, Li H, Cui L, Lin Z, Cai Y (2022) Peripheral blood circular RNAs as a biomarker for major depressive disorder and prediction of possible pathways. Front Neurosci 16:844422. https://doi.org/10.3389/fnins.2022.844422
69. Bu T, Qiao Z, Wang W, Yang X, Zhou J, Chen L, Yang J, Xu J, Ji Y, Wang Y, Zhang W, Yang Y, Qiu X, Yu Y (2021) Diagnostic biomarker Hsa_circ_0126218 and functioning prediction in peripheral blood monocular cells of female patients with major depressive disorder. Front Cell Dev Biol 9:651803. https://doi.org/10.3389/fcell.2021.651803
70. Li YJ, Xu M, Gao ZH, Wang YQ, Yue Z, Zhang YX, Li XX, Zhang C, Xie SY, Wang PY (2013) Alterations of serum levels of BDNF-related miRNAs in patients with depression. PLoS One 8(5):e63648. https://doi.org/10.1371/journal.pone.0063648
71. Wang K, Yang Y, Wang Y, Jiang Z, Fang S (2024) CircPTK2 may be associated with depressive-like behaviors by influencing miR-182-5p. Behav Brain Res 462:114870. https://doi.org/10.1016/j.bbr.2024.114870
72. Chen H, Zhang S, Wu Y, Li Z, Wang D, Cai S, Wang Z (2021) The role of circular RNA circ_0008285 in gestational diabetes mellitus by regulating the biological functions of trophoblasts. Biol Res 54(1):14. https://doi.org/10.1186/s40659-021-00337-3
73. Jiang B, Zhang J, Sun X, Yang C, Cheng G, Xu M, Li S, Wang L (2022) Circulating exosomal hsa_circRNA_0039480 is highly expressed in gestational diabetes mellitus and may be served as a biomarker for early diagnosis of GDM. J Transl Med 20(1):5. https://doi.org/10.1186/s12967-021-03195-5
74. Zhang L, Zeng M, Tang F, Chen J, Cao D, Tang ZN (2021) Circ-PNPT1 contributes to gestational diabetes mellitus (GDM) by regulating the function of trophoblast cells through miR-889-3p/PAK1 axis. Diabetol Metab Syndr 13(1):58. https://doi.org/10.1186/s13098-021-00678-9
75. She W, Li T, Liu Y, Liu X (2021) CircRNA circVEGFC is highly expressed in gestational diabetes mellitus (GDM) and it is correlated with multiple adverse events. Diabetes Metab Syndr Obes 14:4409–4414. https://doi.org/10.2147/DMSO.S334728
76. Huang Y, Liang B, Chen X (2021) Exosomal circular RNA circ_0074673 regulates the proliferation, migration, and angiogenesis of human umbilical vein endothelial cells via the microRNA-1200/MEOX2 axis. Bioengineered 12(1):6782–6792. https://doi.org/10.1080/21655979.2021.1967077
77. Liu M, Luo X, Gao L, Shi M, Zhou R, Wang T (2022) Profiles of circular RNAs in human blood and their potential roles in preeclampsia. Arch Med Sci. https://doi.org/10.5114/aoms/145848
78. Derakhshan Z, Bahmanpour S, Alaee S, Fallahi J, Tabei SMB (2023) The role of circular RNAs in male infertility and reproductive cancers: a narrative review. Iran J Med Sci 48(6):527–541. https://doi.org/10.30476/IJMS.2022.95302.2661
79. Zhao Z, Li X, Jian D, Hao P, Rao L, Li M (2017) Hsa_circ_0054633 in peripheral blood can be used as a diagnostic biomarker of pre-diabetes and type 2 diabetes mellitus. Acta Diabetol 54:237–245. https://doi.org/10.1007/s00592-016-0943-0
80. Haque S, Ames RM, Moore K, Lee BP, Jeffery N, Harries LW (2020) Islet-expressed circular RNAs are associated with type 2 diabetes status in human primary islets and in peripheral blood. BMC Med Genet 13(1):64. https://doi.org/10.1186/s12920-020-0713-2
81. Lv B, Bao X, Li P, Lian J, Wu Y, An T, Zhang J, Yang X, Wang T, Zhu J, Hu Y, Jiang G, Gao S (2020) Transcriptome sequencing analysis of peripheral blood of type 2 diabetes mellitus patients with thirst and fatigue. Front Endocrinol 11:558344. https://doi.org/10.3389/fendo.2020.558344
82. Gu Y, Ke G, Wang L, Zhou E, Zhu K, Wei Y (2017) Altered expression profile of circular RNAs in the

serum of patients with diabetic retinopathy revealed by microarray. Ophthalmic Res 58(3):176–184. https://doi.org/10.1159/000479156
83. Li X, Zhao Z, Jian D, Li W, Tang H, Li M (2017) Hsa-circRNA11783-2 in peripheral blood is correlated with coronary artery disease and type 2 diabetes mellitus. Diab Vasc Dis Res 14(6):510–515. https://doi.org/10.1177/1479164117722714
84. Chen X, Yin J, Zhang F, Xiao T, Zhao M (2020) has_circ_CCNB1 and has_circ_0009024 function as potential biomarkers for the diagnosis of type 2 diabetes mellitus. J Clin Lab Anal 34(10):e23439. https://doi.org/10.1002/jcla.23439
85. Fang Y, Wang X, Li W, Han J, Jin J, Su F, Zhang J, Huang W, Xiao F, Pan Q, Zou L (2018) Screening of circular RNAs and validation of circANKRD36 associated with inflammation in patients with type 2 diabetes mellitus. Int J Mol Med 42(4):1865–1874. https://doi.org/10.3892/ijmm.2018.3783
86. Yang L, Han X, Zhang C, Sun C, Huang S, Xiao W, Gao Y, Liang Q, Luo F, Lu W, Fu J, Zhou Y (2020) Hsa_circ_0060450 negatively regulates type I interferon-induced inflammation by serving as miR-199a-5p sponge in type 1 diabetes mellitus. Front Immunol 11:576903. https://doi.org/10.3389/fimmu.2020.576903
87. Zhang C, Han X, Yang L, Fu J, Sun C, Huang S, Xiao W, Gao Y, Liang Q, Wang X, Luo F, Lu W, Zhou Y (2020) Circular RNA circPPM1F modulates M1 macrophage activation and pancreatic islet inflammation in type 1 diabetes mellitus. Theranostics 10(24):10908–10924. https://doi.org/10.7150/thno.48264
88. Yi X, Cheng X (2021) Understanding competitive endogenous RNA network mechanism in type 1 diabetes mellitus using computational and bioinformatics approaches. Diabetes Metab Syndr Obes 14:3865–3945. https://doi.org/10.2147/DMSO.S315488
89. Vig S, Buitinga M, Rondas D, Crèvecoeur I, van Zandvoort M, Waelkens E, Eizirik DL, Gysemans C, Baatsen P, Mathieu C, Overbergh L (2019) Cytokine-induced translocation of GRP78 to the plasma membrane triggers a pro-apoptotic feedback loop in pancreatic beta cells. Cell Death Dis 10(4):309. https://doi.org/10.1038/s41419-019-1518-0

Circular RNAs as Biomarkers in the Diagnosis of Human Diseases

14

Selçuk Özdemir

Abstract

Circular RNAs (circRNAs) are a class of noncoding RNAs (ncRNAs), which has the formation of reverse splicing of mRNAs. Interest in their regulatory function has recently been increasing. The most important features are its stability between species, its high abundance, and its evolutionary conservation. As a cellular function, circular RNAs are involved in transcriptional modulation and splicing from miRNA and protein sponges. In addition, they are differentially expressed in pathological conditions occurring in diseases, creating their potential as biomarkers. Resistance to RNases makes circular RNAs a less invasive biomarker. Studies on human tumors and other diseases have confirmed the regulatory role and expression profiles of circular RNAs in disease progenesis. This article discusses in detail the properties, functions, and mechanisms of action of circRNAs in human diseases. We also discuss the possibility of using circRNAs as potential therapeutic targets and biomarkers for human diseases.

S. Özdemir (✉)
Faculty of Veterinary Medicine, Department of Genetics, Atatürk University, Erzurum, Turkey
e-mail: selcuk.ozdemir@atauni.edu.tr

Keywords

CircRNA · NcRNA · Biomarker · Human diseases

1 Biomarker Potential of Circular RNAs

Recent advancements in molecular biology techniques have allowed researchers to delve into the coding and noncoding transcriptome. Circular RNAs (circRNAs), a novel class of noncoding RNAs, have garnered significant research interest. Unlike long noncoding RNAs (lncRNAs), which have a 5′ and 3′ structure, mRNAs, and microRNAs (miRNAs), circRNAs are characterized by a closed-loop structure formed through back-splicing. Their high stability, availability, and conservation in mammalian cells have made circRNAs a focal point of recent research. They regulate gene expression by acting as miRNA sponges, interacting with RNA-binding proteins (RBPs), and modulating transcription. Tissue-specific regulation of circRNA expression further enhances their potential as disease biomarkers [1–5].

The functions of circRNAs are crucial for further research. Recent evidence highlights several roles for circRNAs: (1) sponging miRNAs, (2) regulating gene transcription, (3) participating in translation, and (4) binding to proteins. These

J. Xiao (ed.), *Advances in Circular RNAs*, Advances in Experimental Medicine and Biology 1485,
https://doi.org/10.1007/978-981-96-9428-0_14

biological functions are promising for clinical applications [1].

Early diagnosis is vital, especially for high-mortality diseases like cancer and those that significantly impact quality of life, such as neurodegenerative diseases [2, 3]. Identifying new and reliable biomarkers is essential for early detection and effective treatment. Prioritizing less invasive biomarkers is crucial, and they must be easily accessible for clinical use [4, 5]. Biomarkers should be relevant to disease development, reflecting key disease signatures. High sensitivity and specificity are also critical [6, 7]. Stable biomarkers in body fluids are particularly valuable. The role of the biomarker candidate in disease pathogenesis enhances its reliability, with noncoding RNAs that regulate disease genes being prominent candidates. Among these, circRNAs are considered the most promising [7–9].

CircRNAs' potential as biomarkers is due to their stability in body fluids, high sensitivity, high specificity, and reproducibility. Consequently, there has been a surge in circular RNA-based biomarker studies in recent years.

2 Cancer-Associated CircRNA Biomarker Candidates

Cancer is a leading cause of death globally, with imaging and clinical testing being the primary diagnostic approaches. Among these, blood-based tests are preferred due to their noninvasive nature. Common blood test biomarkers include carcinoembryonic antigen (CEA), cancer antigen 19-9 (CA 19-9), and alpha-fetoprotein (AFP), though they have low specificity [10, 11]. Identifying effective biomarkers with high specificity and sensitivity for cancer is crucial. Recent studies highlight the potential of circular RNAs (circRNAs) as promising biomarkers for cancer [12].

CircRNAs, which play a regulatory role in cancer progression, are emerging as significant markers. Their ability to regulate gene expression enhances their candidacy as biomarkers. Significant circRNA-based biomarkers have been identified for various cancers, including gastric, lung, liver, colon, breast, and pancreatic cancers [13–15]. Studies have also linked circRNAs to thyroid cancer [16–18]. Developing stable and highly sensitive circRNA-based biomarkers for early diagnosis of these lethal cancers is promising for the future [16–18].

Lung cancer, one of the deadliest cancers, has general markers like epidermal growth factor receptor (EGFR), Kirsten rat sarcoma viral oncogene homolog (KRAS), and ROS proto-oncogene 1 (ROS1). CircRNAs offer more stability and specificity compared to traditional biomarkers. For instance, hsa_circ_0086414 is upregulated in lung cancer patients' plasma and is associated with EGFR mutation, suggesting it may activate EGFR mutations by binding to downstream miRNAs. Additionally, hsa_circ_0102537 might promote tumor signaling pathways such as PI3K-Akt, affecting lung cancer cell invasion and migration. Overexpression of hsa_circ_0001715 is associated with lung adenocarcinoma metastasis, demonstrating its diagnostic sensitivity [19–21].

For liver cancer, early biomarkers include AFP, AFP-L3, and PIVKA-II. AFP is effective but not reliable for all patients. Combining classic markers with plasma circRNAs, like hsa_circ_0001445, enhances diagnostic accuracy for hepatocellular carcinoma (HCC). This circRNA also acts as a miRNA sponge, inhibiting HCC cell growth and metastasis [22–41].

Colorectal cancer blood-based biomarkers include CEA, CA19-9, and EGFR, though their specificity is low. New circRNA candidates like hsa_circ_0001900, hsa_circ_0001178, and hsa_circ_0005927 show promise for early diagnosis [42–48].

Pancreatic cancer, highly aggressive with a short life expectancy, requires improved early diagnosis methods. The combination of plasma circ-LDLRAD3 and CEA is a promising biomarker candidate [49, 50].

For thyroid cancer, early detection is critical. CircRNAs hsa_circ_0124055 and hsa_circ_0101622 show potential as early detection markers [51].

Breast cancer, one of the most common cancers in women, benefits from early diagnosis. Among various circRNAs, hsa_circ_0001785 shows the best biomarker potential [52–54].

Advances in RNA sequencing accelerate circRNA research. Circ_0003945, a notable circRNA, regulates tumor cell proliferation, migration, invasion, apoptosis, autophagy, angiogenesis, and drug resistance, indicating a poorer prognosis in many cancers. It is also significant in nonmalignant conditions and immune-related diseases, such as autoimmune diseases, cancer, and infections [55–60].

CircRNAs play critical roles in immune responses, particularly in myeloid and lymphoid cells. They are involved in macrophage activation, with circCdr1as influencing pro- and antiinflammatory states. In dendritic cells (DCs), circRNAs affect immune functions and may play a role in autoimmune diseases. For example, circSnx5 modulates DC-driven immunity and tolerance, and melatonin affects BMDCs in colitis models by influencing noncoding RNAs, including circRNA-0520 and circRNA-2243 [61–70].

3 CircRNA-Based Biomarker for Neurodegenerative Diseases

Detecting circRNAs in the central nervous system is a crucial step, particularly for neurodegenerative diseases where therapeutic options are limited and biomarkers are scarce, complicating early diagnosis and treatment. Thus, identifying biomarkers for neurodegenerative diseases quickly is essential. Recently, candidate circRNA-based biomarkers have been identified for various neurodegenerative diseases, including Alzheimer's disease, Parkinson's disease, prion disease, glioblastoma, and ALS. A meta-analysis identified circRNAs associated with acute systemic stroke and elucidated their roles in molecular mechanisms [71–75].

Alzheimer's disease, an irreversible neurodegenerative condition leading to cognitive impairment, involves two key protein aggregates: beta-amyloid and tau in neurofibrillary tangles. Despite extensive research, there is still no effective treatment for this disease. Early diagnosis is critical, with current diagnostic approaches including amyloid and tau PET scans and CSF biomarkers [76–79].

Studies on circRNA profiles related to Alzheimer's disease have been ongoing since 2016, aiming to develop circRNA-based biomarkers. Numerous circRNAs have been confirmed to be expressed in the cortex, hippocampus, CSF, and plasma. Research has suggested that circPCCA and circAXL are associated with Mini-Mental State Examination (MMSE) scores and amyloid beta levels. Another significant circRNA biomarker candidate for Alzheimer's disease is circPSEN1. In plasma circRNA profile studies of Alzheimer's patients, many candidate circRNAs have been identified, including hsa_circ_0077001, hsa_circ_0022417, hsa_circ_0014356, hsa_circ_0014353, hsa_circ_0074533, and hsa_circ_0089894 [80–89].

4 Potential Circular RNAs in Diabetes

Diabetes encompasses a group of metabolic disorders characterized by hyperglycemia or elevated blood sugar levels. This condition can result in cardiovascular diseases and various complications such as retinopathy, nephropathy, and neuropathy, potentially causing irreversible damage. Recent research has identified several circRNAs in human pancreatic islets, including circRNA Cdr1as. CircRNA circHIPK3 was found to be upregulated in the retinas and retinal endothelial cells of diabetic patients. The precise relationship between circRNAs and diabetes is not yet fully understood, and further detailed studies are needed to identify candidate circRNAs associated with diabetes. One study found that circular RNA-ZNF532 regulates diabetes-induced retinal pericyte degeneration and vascular dysfunction. Additionally, hsa_circRNA_405498 and hsa_circRNA_100033 have been identified as promising differential diagnostic biomarkers for type 1 diabetes, with increased expression levels in patients with the condition. ROC curve analyses indicated acceptable diagnostic performance (AUC = 0.915, AUC = 0.993). HTS profiling studies revealed 441 differentially expressed circRNAs (366 upregulated, 75 downregulated) and 683 mRNAs (354 upregulated, 329 downregulated). Another study found that circVEGFC was highly expressed in patients with gestational diabetes mellitus (GDM) [90–96].

In recent years, researchers have suggested that noncoding RNAs play a role in regulating lipid and glucose metabolism. hsa_circRNA_102682 was found to be lowly expressed in the serum of pregnant women with GDM and was positively correlated with 1-h blood glucose levels. Additionally, hsa_circRNA_0054633 is involved in regulating blood glucose and glycosylation indicators, showing a significant correlation with hsa_circRNA_102682. This supports the hypothesis that hsa_circRNA_102682 is involved in glucose metabolism regulation. Furthermore, hsa_circRNA_102682 was significantly correlated with triglycerides (TG), apolipoprotein B (APOB), and apolipoprotein A1 (APOA1), suggesting its involvement in lipid metabolism during pregnancy. Lipid metabolic disorders, which can lead to various metabolic syndromes including diabetes, have become a pressing health issue. Numerous studies have shown that circRNAs are enriched in biological processes such as triglyceride catabolism and lipid storage, playing a role in regulating metabolic syndrome. Therefore, hsa_circRNA_102682 may participate in the regulation of both glucose and lipid metabolism, contributing to GDM, although its specific role requires further investigation [97–103].

A recent study by Liu et al. identified a novel circRNA associated with diabetic retinopathy (DR), cPWWP2A, which may act as a competing endogenous RNA interacting with miR-579 to promote DR-induced retinal vascular dysfunction by upregulating Angiopoietin 1, Occludin, and SIRT1. These findings suggest that CircRNA cPWWP2A could be a target for DR therapy. CircRNAs were found to be aberrantly expressed in the peripheral blood mononuclear cells (PBMCs) of patients with type 1 diabetes, with hsa_circRNA_100632 potentially serving as a diagnostic marker for the condition. Another study indicated that circRNA hsa_circ_0071336 is associated with type 2 diabetes through the miR-93-5p/GLUT4 axis [104–107].

5 CircRNA Biomarker Candidates in Liver Diseases

The liver is a vital organ responsible for digestion and detoxification. Changes in diet and lifestyle have led to an increase in liver diseases, such as nonalcoholic fatty liver disease (NAFLD), nonalcoholic steatohepatitis (NASH), liver cirrhosis, and hepatocellular carcinoma (HCC). Numerous studies have examined the role of circRNAs in liver function and disease, but their impact on molecular mechanisms remains unexplored. Research on circRNAs in liver diseases is complicated by the variety of liver cell types (e.g., hepatocytes, hepatic stellate cells, Kupffer cells, and immune cells). Additionally, the species-specific and tissue-specific nature of circRNAs limits their clinical applicability. Advanced bioinformatics approaches are needed to clarify the role of circRNAs in liver diseases and to develop new diagnostic and therapeutic strategies. However, significant challenges remain, requiring extensive experimental data and research efforts. In one study, circRNAs associated with alveolar echinococcus were identified and evaluated for their potential as biomarkers. Detecting hepatic alveolar echinococcosis (HAE) in patients is challenging due to its invasive and slow-growing nature, making early diagnosis crucial. CircRNAs, a type of noncoding RNA, have shown promise as biomarkers for various diseases. In this study, high-throughput sequencing was used on samples from nine HAE patients and nine controls to identify HAE-related circRNAs, followed by bioinformatic analysis to identify differentially expressed circRNAs and conduct pathway analysis. Validation of these circRNAs was performed using RT-PCR. The results showed 59 differentially expressed circRNAs, with 31 upregulated and 28 downregulated in HAE patients. The top five upregulated and downregulated circRNAs were validated by RT-qPCR, confirming their expression profiles. These exosomal circRNAs have potential as biomarkers for detecting HAE and understanding its pathogenesis [108–112].

The study identified exosomal circRNAs from serum in HAE patients compared to healthy controls, validated by RT-PCR. Significantly upregulated circRNAs included hsa_circ_2204, hsa_circ_2290, hsa_circ_1854, hsa_circ_2436, and hsa_circ_2249, while significantly downregulated circRNAs included hsa_circ_832, hsa_circ_160, hsa_circ_1332, hsa_circ_524, and hsa_circ_3988. Potential miRNA targets of these circRNAs were predicted, including the hsa-let-7a miRNA family [113].

Liver fibrosis involves excessive connective tissue proliferation and extracellular matrix deposition due to chronic liver diseases, such as chronic viral hepatitis, metabolic disorders, alcoholic liver disease (ALD), nonalcoholic steatohepatitis (NASH), and autoimmune imbalances. Liver fibrosis, a reversible wound-healing process, can progress to cirrhosis, liver failure, or hepatocellular carcinoma (HCC) if persistent. CircRNA-007371 has been linked to liver fibrosis, and circMTO1 (hsa_circ_0007874), derived from the MTO1 gene, is downregulated in HCC tissues, inhibiting HCC progression. Research by Zheng et al. on serum circRNAs as markers for liver fibrosis involved 360 treatment-naive CHB patients and 360 healthy controls [111, 113–120].

CircRNAs are gaining attention as potential disease markers due to their stable structures, evolutionary conservation, and presence in various body fluids like blood, saliva, serum, and exosomes. Previous studies have shown that circRNAs can promote cancer cell growth and metastasis, with circRNAs ciRS-7 and hsa_circ_0000069 identified as potential biomarkers for colorectal cancer. Differentially expressed miRNAs in HAE-infected liver tissues compared to normal tissues included miR-1237-3p, miR-33b-3p, and miR-483-3p (upregulated), and miR-4306 (downregulated). Although circRNAs' roles in cancer and other diseases have been widely studied, their involvement in HAE has not been thoroughly investigated. This study detected 2203 circRNAs, with 31 significantly upregulated and 28 down-regulated, indicating the potential of serum exosomal circRNAs as biomarkers for HAE due to their high sensitivity [121–128].

Bioinformatics analysis identified functions and mechanisms of circRNAs in HAE, linking them to cancer progression, proteoglycans in cancer, and focal adhesion, which is crucial for cell motility. Despite being studied for over 60 years, the cAMP signal still reveals new details, requiring further research to understand its role in individual cell types and its interactions with other pathways. The circRNAs detected in this study were associated with the TNF alpha pathway, highlighting their importance in inflammation. Three families of MAPKs (ERK, JNK/SAPK, and p38 MAPK) regulate cell cycle and proliferation-related proteins. The mechanistic target of rapamycin (mTOR) regulates essential cell processes, with dysregulation linked to cancer, diabetes, and aging. This study suggests that HAE-related circRNAs may play roles in autophagy and apoptosis [129–133].

6 Candidates for Circular RNA Associated with Toxicity

Pesticides are biological agents utilized to attract, seduce, destroy, or mitigate pests. They are now prevalent globally. Human activities lead to pesticides entering water bodies through runoff, seepage, and erosion. Additionally, pesticide residues can contaminate surface waters, soil, plants, residential areas, and animals via evaporation and wind erosion. Pesticides pose toxicity risks to both target and nontarget organisms. Among the most commonly used pesticides in agriculture, CYP and CPF are notable for their toxicity. Numerous studies have explored the toxicity of these chemicals. However, no research has yet investigated the relationship between acute toxicity caused by CYP and CPF in the brain and circRNAs. This study identifies and characterizes circRNAs associated with acute CYP and CPF toxicity in zebrafish brains using the RNA-seq method, and determines the molecular mechanisms involving these circRNAs [134–139].

Mitogen-activated protein kinases (MAPKs) are serine–threonine kinases that facilitate intracellular signaling and play a role in various cellular activities. The MAP kinase signaling pathway, which leads to gene expression changes, comprises three main steps: ERK1/2, JNK, and p38 MAP kinase. Several genes, including MKK1, MKK2, MKK3, MKK4, MKK6, and MKK7, regulate and activate the MAP kinase pathway. Interactions among different MAP kinase cascades activate various gene groups. The MAPK signaling pathway is crucial for synthesizing proteins needed for inflammation and is implicated in neurodegenerative diseases such as Alzheimer's, Parkinson's, and amyotrophic lateral sclerosis. Inhibitors of the p38 MAPK and JNK signaling pathways are considered potential neuroprotective drugs. Some agricultural pesticides are known to cause diseases including neurodegenerative, cardiovascular, respiratory, gastrointestinal, diabetic, dyslipidemia, and kidney diseases. Pesticide-induced oxidative stress can activate the MAPK signaling pathway in tissues. This study found that circRNAs from brains exposed to CYP and CPF significantly regulate the MAPK signaling pathway. This indicates that the acute toxicity of CYP and CPF activates the MAPK signaling pathway in zebrafish brains. Overactivation or misregulation of this pathway, which is critical for biological roles in the brain, can lead to serious pathological consequences. Therefore, the MAPK signaling pathway is a likely target for the brain toxicity caused by CYP and CPF. CircRNAs regulating this pathway could serve as candidate biomarkers for CYP and CPF toxicity [55, 140–147].

The toxic effects of pesticides are often mediated by apoptosis and redox (reduction/oxidation) signaling. Pesticides induce apoptosis through various signaling pathways, including those mediated by mitochondria and DNA damage (intrinsic pathways) and death receptors (extrinsic pathways). The extrinsic pathway is activated by the binding of death receptors, leading to receptor oligomerization, recruitment of adapter proteins, and activation of the caspase cascade, resulting in apoptotic cell death. The p53 pathway includes hundreds of genes and their products responding to various stress signals. CYP toxicity activates the apoptosis pathway by increasing p53 gene mRNA transcript levels, and CPF toxicity has a similar effect, enhancing apoptosis levels. This study shows that circRNAs associated with acute CYP and CPF exposure target the p53 signaling pathway and genes regulating apoptosis, indicating that these pesticides activate the apoptosis pathway in the brain [148–159].

Endocytosis is the process by which substances unable to pass through cell membranes are taken into cells, requiring energy. Monomers can pass through cell membranes, but polymer substances, such as starch, glycogen, and proteins, cannot. Endocytosis, comprising phagocytosis and pinocytosis, is essential for polymer uptake into cells. Endocytosis starts with the aggregation of endocytic proteins in the plasma membrane, forming a core structure for vesicle coating. This process determines where endocytic vesicles form and likely regulates the frequency of endocytic events. Endocytic events may occur randomly on the plasma membrane or concentrate in specific cell regions, such as synapses in neurons or budding daughter cells in yeast. If endocytosis is inactive, critical polymer substances for signal transduction, nutrient uptake, membrane recycling, cell growth, proliferation, and differentiation cannot enter the cell, leading to cell death. Endosomal trafficking is crucial in polarized cells like neurons, separating incoming from outgoing signals by forming and maintaining different membrane domains. Exposure to certain chemicals causes brain anomalies, including neuronal-glia heterotopias, cerebellar dysplasia, corpus callosum agenesis, hydrocephalus, and microcephaly. Previous studies have shown that CYP and CPF cause significant brain damage. This study reveals that acute CYP and CPF toxicity targets the endocytosis pathway in the brain, with circRNAs from CYP and CPF exposure targeting genes regulating this mechanism. These circRNAs could be biomarkers for endocytosis disruptions caused by CYP and CPF. Despite the importance of endocytosis in neuronal cells, there is no prior research on the effects of acute CYP and CPF exposure on this process in neurons. This study is the first to demonstrate that CYP and CPF toxicity targets endocytosis metabolism in the brain. However, chronic alcohol intake has been shown to disrupt endocytosis in brain cells [160–167].

7 CircRNAs as Potential Biomarkers in Asthma and Chronic Respiratory Diseases

Chronic respiratory diseases (CRDs) encompass a variety of lung conditions that impact the respiratory system, especially the airways and lungs. These diseases are marked by diminished lung function and, at times, permanent damage to respiratory tissues. In 2019, CRDs were the third leading cause of death worldwide, imposing a considerable social and economic burden, and presenting a major healthcare challenge [168,169, 170]. The most prevalent CRDs include asthma and chronic obstructive pulmonary disease (COPD).

Asthma presents as a complex and diverse airway condition, impacting approximately 260 million individuals worldwide [171, 172]. Its primary clinical features include airflow obstruction, breathing difficulties, coughing, chest tightness, and wheezing [173]. Although asthma can emerge at any stage of life, it is most prevalent in children. While the precise cause of asthma remains elusive, it is thought to result from a blend of genetic predispositions (notably in those with a family history of asthma) and environmental triggers (such as allergies, air pollution, cold air, and exercise). These factors contribute to pathological alterations in the airways, involving inflammation, excessive mucus production, airway restructuring, heightened airway responsiveness, and obstruction [174]. Airway remodeling entails substantial structural changes in the airways, including enlargement of airway smooth muscle cells (ASMCs), modifications in airway epithelial cells (AECs), subepithelial fibrosis, transformation of goblet cells, and increased expression of angiogenic factors [175]. While existing medications can effectively alleviate symptoms, a definitive cure for asthma remains elusive. Consequently, there is an urgent necessity to develop a therapy that can effectively address the underlying causes of the condition.

Chronic obstructive pulmonary disease (COPD) encompasses various lung ailments such as pulmonary emphysema and chronic bronchitis, leading to challenges in breathing and obstruction of airflow [176]. COPD significantly contributes to global mortality and morbidity, imposing substantial economic and social burdens worldwide [177]. While COPD can affect people across all age groups, it is more prevalent among older individuals [178]. Although smokers face a heightened risk of developing COPD, non-smokers can also be susceptible due to exposure to environmental pollutants, workplace dust and chemicals, respiratory infections, and genetic predispositions [179, 180]. Despite extensive endeavors to find effective treatments for COPD, the condition remains without a cure, and current therapeutic options primarily focus on symptom management, slowing disease progression, and improving quality of life [181]. Hence, there is an urgent need for the development of efficacious treatments for this debilitating ailment.

Circular RNAs (circRNAs) are emerging as promising biomarkers in asthma and COPD. For instance, analysis of circ_0002594 expression in CD4+ T cells of 83 asthmatic patients not using inhaled corticosteroids (ICS) showed a notably high area under the curve (AUC) value of 0.727 [182]. Similarly, an AUC value of 0.767 was obtained when assessing the diagnostic efficacy of circ_0002594 in 48 asthmatic patients undergoing ICS treatment [182]. Additionally, circ_0002594 expression exhibited positive correlations with exhaled nitric oxide levels, family history of the disease, positive skin prick test (SPT) results, and Th2 cytokine expression levels [182]. Furthermore, elevated expression of circSORT1 and circSERPINB1 was detected in induced sputum from 68 asthmatic patients compared to 20 healthy individuals, demonstrating heightened specificity and sensitivity in predicting asthma [183]. CircSORT1 expression correlated with various clinical parameters of asthma, including FeNO, EOS%, IL-17A, IFN-γ, and PC20, while CircSERPINB1 expression correlated with IL-6, IL-17A, IFN-γ, FEV1%, and FVC%. Moreover, significantly increased expression of circ_0005519 was observed in PBMCs and CD4+ T cells of 30 asthmatics compared to 24 healthy individuals, with this elevated expression associated with exhaled nitric oxide levels

[183]. In COPD, serum levels of circ_0001859 were found reduced in 38 COPD patients and 24 patients with acute exacerbation of COPD (AE-COPD) compared to 28 healthy controls, exhibiting higher specificity and sensitivity in predicting both conditions and correlating with FEV1% predicted. Furthermore, circ_004092 expression was increased in the blood of smokers with COPD (n = 22) or without COPD (n = 22) compared to nonsmokers (n = 22). These findings collectively suggest the potential of circRNAs as diagnostic and prognostic biomarkers for asthma and COPD, although further validation in larger sample cohorts is warranted.

Circular RNAs (circRNAs) present a hopeful avenue in asthma and COPD research, holding promise as both therapeutic targets and biomarkers. Despite notable advancements, significant hurdles persist. The limited number of identified circRNAs associated with asthma and COPD underscores the necessity for a more comprehensive circRNA profile in these diseases. These molecules play vital roles in diverse airway and immune cell functions, including airway epithelial cells (AECs), 16HBE, airway smooth muscle cells (ASMCs), CD4+ T cells, and macrophages in asthma, and 16HBE and AECs in COPD. Nevertheless, research on circRNA roles in ASMC dysfunction in COPD, a pivotal aspect of airway remodeling, remains unexplored. Further investigations are imperative to enrich our understanding of COPD. Additionally, while asthma and COPD studies have predominantly focused on circRNAs' ability to sponge microRNAs, other regulatory mechanisms like protein sponges and scaffolds have been overlooked. Furthermore, although most circRNA studies have been conducted in vitro, their in vivo roles remain largely uncharted. Only a handful, such as circZNF652 in asthma, and circFOXO3, circFOXOP1, and circBbs9 in COPD, have been examined in animal models, with no evaluation of their impact on lung function changes in asthma and COPD. Investigating the clinical utility of circRNAs as diagnostic tools, prognostic markers, and predictive biomarkers necessitates studies in extensive patient cohorts. Finally, circRNAs hold promise for targeted therapeutic interventions, requiring the identification of circRNAs with therapeutic potential and the development of strategies to modulate their expression or activity, potentially paving the way for innovative asthma and COPD treatments.

References

1. Zhong Y, Du Y, Yang X, Mo Y, Fan C, Xiong F, Ren D, Ye X, Li C, Wang Y, Wei F, Guo C, Wu X, Li X, Li Y, Li G, Zeng Z, Xiong W (2018) Circular RNAs function as ceRNAs to regulate and control human cancer progression. Mol Cancer 17(1):79. https://doi.org/10.1186/s12943-018-0827-8
2. Enuka Y, Lauriola M, Feldman ME, Sas-Chen A, Ulitsky I, Yarden Y (2016) Circular RNAs are long-lived and display only minimal early alterations in response to a growth factor. Nucleic Acids 44(3):1370–1383. https://doi.org/10.1093/nar/gkv1367
3. Conn SJ, Pillman KA, Gregory PA, Goodall Correspondence GJ (2015) The RNA binding protein quaking regulates formation of circRNAs in brief the RNA binding protein Quaking (QKI) promotes circRNA biogenesis during epithelial to mesenchymal transition, strongly arguing for their functions. Cell 160(6):1125–1134. https://doi.org/10.1016/j.cell.2015.02.014
4. Özdemir S, Arslan H (2022) circRNA-based biomarker candidates for acute cypermethrin and chlorpyrifos toxication in the brain of Zebrafish (Danio rerio). Chemosphere 298:134330. https://doi.org/10.1016/j.chemosphere.2022.134330
5. Mankoff DA, Pryma DA, Clark AS (2014) Molecular imaging biomarkers for oncology clinical trials. J Nucl Med 55(4):525–528. https://doi.org/10.2967/jnumed.113.126128
6. Ng WL, Mohd Mohidin TB, Shukla K (2018) Functional role of circular RNAs in cancer development and progression. RNA Biol 15(8):995–1005. https://doi.org/10.1080/15476286.2018.1486659
7. Meng H, Ruan J, Yan Z, Chen Y, Liu J, Li X, Meng F (2022) New progress in early diagnosis of atherosclerosis. Int J Mol Sci 23(16):8939. https://doi.org/10.3390/ijms23168939
8. Stenton SL, Prokisch H (2020) Genetics of mitochondrial diseases: identifying mutations to help diagnosis. EBioMedicine 56:102784. https://doi.org/10.1016/j.ebiom.2020.102784
9. Logroscino G, Urso D, Savica R (2022) Descriptive epidemiology of neurodegenerative diseases: what are the critical questions? Neuroepidemiology 56(5):309–318. https://doi.org/10.1159/000525639
10. Porsteinsson AP, Isaacson RS, Knox S, Sabbagh MN, Rubino I (2021) Diagnosis of early Alzheimer's disease: clinical practice in 2021. J Prev Alzheimers

Dis 8(3):371–386. https://doi.org/10.14283/jpad.2021.23

11. Li JJH, Wang W, Wang XQ, He Y, Wang SS, Yan YX (2019) A novel strategy of identifying circRNA biomarkers in cardiovascular disease by meta-analysis. J Cell Physiol 234(12):21601–21612. https://doi.org/10.1002/jcp.28817
12. Yang SQ, Peng L, De LL, Chen YZ, Liu MZ, Zhang C, Chen JW, Luo DY (2023) Identification of circRNA-miRNA-mRNA network as biomarkers for interstitial cystitis/bladder pain syndrome. Aging 15(21):12155–12170. https://doi.org/10.18632/aging.205170
13. Xu C, Jun E, Okugawa Y, Toiyama Y, Borazanci E, Bolton J, Taketomi A, Kim SC, Shang D, Von Hoff D, Zhang G, Goel A (2024) A circulating panel of circRNA biomarkers for the noninvasive and early detection of pancreatic ductal adenocarcinoma. Gastroenterology 166(1):178–190.e16. https://doi.org/10.1053/j.gastro.2023.09.050
14. Zhou WY, Cai ZR, Liu J, Wang DS, Ju HQ, Xu RH (2020) Circular RNA: metabolism, functions and interactions with proteins. Mol Cancer 19(1):172. https://doi.org/10.1186/s12943-020-01286-3
15. Yin Y, Long J, He Q, Li Y, Liao Y, He P, Zhu W (2019) Emerging roles of circRNA in formation and progression of cancer. J Cancer 10(21):5015–5021. https://doi.org/10.7150/jca.30828
16. Gao Y, Wang J, Zhou Y, Sheng S, Qian SY, Huo X (2018) Evaluation of serum CEA, CA19-9, CA72-4, CA125 and ferritin as diagnostic markers and factors of clinical parameters for colorectal cancer. Sci Rep 8(1):2732. https://doi.org/10.1038/s41598-018-21048-y
17. Dolscheid-Pommerich RC, Manekeller S, Walgenbach-Brünagel G, Kalff JC, Hartmann G, Wagner BS, Holdenrieder S (2017) Clinical performance of CEA, CA19-9, CA15-3, CA125 and AFP in gastrointestinal cancer using LOCITM-based assays. Anticancer Res 37(1):353–359. https://doi.org/10.21873/anticanres.11329
18. He M, Pan Y, You C, Gao H (2024) CircRNAs in cancer therapy tolerance. Clin Chim Acta 15(558):119684. https://doi.org/10.1016/j.cca.2024.119684
19. Pisignano G, Michael DC, Visal TH, Pirlog R, Ladomery M, Calin GA (2023) Going circular: history, present, and future of circRNAs in cancer. Oncogene 42(38):2783–2800. https://doi.org/10.1038/s41388-023-02780-w
20. Zhang Y, Zhang X, Xu Y, Fang S, Ji Y, Lu L, Xu W, Qian H, Liang ZF (2022) Circular RNA and its roles in the occurrence, development, diagnosis of cancer. Front Oncol 12:845703. https://doi.org/10.3389/fonc.2022.845703
21. Alkhathami AG, Sahib AS, Al Fayi MS, Fadhil AA, Jawad MA, Shafik SA, Sultan SJ, Almulla AF, Shen M (2023) Glycolysis in human cancers: emphasis circRNA/glycolysis axis and nanoparticles in glycolysis regulation in cancer therapy. Environ Res 234:116007. https://doi.org/10.1016/j.envres.2023.116007
22. Wang S, Dong Y, Gong A, Kong H, Gao J, Hao X, Liu Y, Wang Z, Fan Y, Liu C, Xu W (2021) Exosomal circRNAs as novel cancer biomarkers: challenges and opportunities. Int J Biol Sci 17(2):562–573. https://doi.org/10.7150/ijbs.48782
23. Bolha L, Ravnik-Glavač M, Glavač D (2017) Circular RNAs: biogenesis, function, and a role as possible cancer biomarkers. Int J Genomics 2017:6218353. https://doi.org/10.1155/2017/6218353
24. Hu F, Peng Y, Fan X, Zhang X, Jin Z (2023) Circular RNAs: implications of signaling pathways and bioinformatics in human cancer. Cancer Biol Med 20(2):104–128. https://doi.org/10.20892/j.issn.2095-3941.2022.0466
25. Zhou Q, Ju L-L, Ji X, Cao Y-L, Shao J-G, Chen L (2021) Plasma circRNAs as biomarkers in cancer. Cancer Manag Res 13:7325–7337. https://doi.org/10.2147/CMAR.S330228
26. Su M, Xiao Y, Ma J, Tang Y, Tian B, Zhang Y, Li X, Wu Z, Yang D, Zhou Y, Wang H, Liao Q, Wang W (2019) Circular RNAs in cancer: emerging functions in hallmarks, stemness, resistance and roles as potential biomarkers. Mol Cancer 18(1):90. https://doi.org/10.1186/s12943-019-1002-6
27. Zhu G, Chang X, Kang Y, Zhao X, Tang X, Ma C, Fu S (2021) CircRNA: a novel potential strategy to treat thyroid cancer (review). Int J Mol Med 48(5):201. https://doi.org/10.3892/ijmm.2021.5034
28. Zhang W, He Y, Zhang Y (2023) CircRNA in ocular neovascular diseases: Fundamental mechanism and clinical potential. Pharmacol Res 197:106946. https://doi.org/10.1016/j.phrs.2023.106946
29. Lei M, Zheng G, Ning Q, Zheng J, Dong D (2020) Translation and functional roles of circular RNAs in human cancer. Mol Cancer 19(1):30. https://doi.org/10.1186/s12943-020-1135-7
30. Li H (2023) circRNA: a promising all-around star in the future. Epigenomics 15(12):677–685. https://doi.org/10.2217/epi-2023-0141
31. Ge L, Sun Y, Shi Y, Liu G, Teng F, Geng Z, Chen X, Xu H, Xu J, Jia X (2022) Plasma circRNA microarray profiling identifies novel circRNA biomarkers for the diagnosis of ovarian cancer. J Ovarian Res 15(1):58. https://doi.org/10.1186/s13048-022-00988-0
32. Villalobos P, Wistuba II (2017) Lung cancer biomarkers. Hematol Oncol Clin North Am 31(1):13–29. https://doi.org/10.1016/j.hoc.2016.08.006
33. Chiou YH, Liou SH, Wong RH, Chen CY, Lee H (2015) Nickel may contribute to EGFR mutation and synergistically promotes tumor invasion in EGFR-mutated lung cancer via nickel-induced microRNA-21 expression. Toxicol Lett 237(1):46–54. https://doi.org/10.1016/j.toxlet.2015.05.019
34. Chou YT, Lin HH, Lien YC, Wang YH, Hong CF, Kao YR, Lin SC, Chang YC, Lin SY, Chen SJ, Chen HC, Der YS, Wu CW (2010) EGFR promotes lung

tumorigenesis by activating miR-7 through a Ras/ERK/Myc pathway that targets the Ets2 transcriptional repressor ERF. Cancer Res 70(21):8822–8831. https://doi.org/10.1158/0008-5472.CAN-10-0638
35. Lin S, Xiong W, Liu H, Pei L, Yi H, Guan Y (2020) Profiling and integrated analysis of differentially expressed circular RNAs in plasma exosomes as novel biomarkers for advanced-stage lung adenocarcinoma. Onco Targets Ther 13:12965–12977. https://doi.org/10.2147/OTT.S279710
36. Smith RA, Cokkinides V, Brawley OW (2009) Cancer screening in the United States, 2009: a review of current American Cancer Society guidelines and issues in cancer screening. CA Cancer J Clin 59(1):27–41. https://doi.org/10.3322/caac.20008
37. Park SJ, Jang JY, Jeong SW, Cho YK, Lee SH, Kim SG, Cha SW, Kim YS, Cho YD, Kim HS, Kim BS, Park S, Bang HI (2017) Usefulness of AFP, AFP-L3, and PIVKA-II, and their combinations in diagnosing hepatocellular carcinoma. Medicine 96(11):e5811. https://doi.org/10.1097/MD.0000000000005811
38. Lu GJ, Cui J, Qian Q, Hou ZB, Xie HY, Hu W, Hao KK, Xia N, Zhang Y (2020) Overexpression of hsa_circ_0001715 is a potential diagnostic and prognostic biomarker in lung adenocarcinoma. Onco Targets Ther 13:10775–10783. https://doi.org/10.2147/OTT.S274932
39. Qi F, Zhou A, Yan L, Yuan X, Wang D, Chang R, Zhang Y, Shi F, Han X, Hou J, Wei L, Zhang X (2020) The diagnostic value of PIVKA-II, AFP, AFP-L3, CEA, and their combinations in primary and metastatic hepatocellular carcinoma. J Clin Lab Anal 34(5):e23158. https://doi.org/10.1002/jcla.23158
40. Zhang X, Zhou H, Jing W, Luo P, Qiu S, Liu X, Zhu M, Liang C, Yu M, Tu J (2018) The circular RNA hsa_circ_0001445 regulates the proliferation and migration of hepatocellular carcinoma and may serve as a diagnostic biomarker. Dis Markers 2018:3073467. https://doi.org/10.1155/2018/3073467
41. Yu J, Xu QG, Wang ZG, Yang Y, Zhang L, Ma JZ, Sun SH, Yang F, Zhou WP (2018) Circular RNA cSMARCA5 inhibits growth and metastasis in hepatocellular carcinoma. J Hepatol 68(6):1214–1227. https://doi.org/10.1016/j.jhep.2018.01.012
42. He B, Zhang HQ, Xiong SP, Lu S, Wan YY, Song RF (2015) Changing patterns of serum CEA and CA199 for evaluating the response to first-line chemotherapy in patients with advanced gastric adenocarcinoma. Asian Pac J Cancer Prev 16(8):3111–3116. https://doi.org/10.7314/apjcp.2015.16.8.3111
43. Zhao S, Mei Y, Wang J, Zhang K, Ma R (2016) Different levels of CEA, CA153 and CA125 in milk and benign and malignant nipple discharge. PLoS One 11(6):e0157639. https://doi.org/10.1371/journal.pone.0157639
44. Chen Y, Gao SG, Chen JM, Wang GP, Wang ZF, Zhou B, Jin CH, Yang YT, Feng XS (2015) Serum CA242, CA199, CA125, CEA, and TSGF are biomarkers for the efficacy and prognosis of cryoablation in pancreatic cancer patients. Cell Biochem Biophys 71(3):1287–1291. https://doi.org/10.1007/s12013-014-0345-2
45. Campos-Da-paz M, Dórea JG, Galdino AS, Lacava ZGM, Almeida Santos M de FM (2018) Carcinoembryonic antigen (CEA) and hepatic metastasis in colorectal cancer: update on biomarker for clinical and biotechnological approaches. Recent Pat Biotechnol 2(4):269–279. https://doi.org/10.2174/187220831266618073110424
46. Lech G, Słotwiński R, Słodkowski M, Krasnodębski IW (2016) Colorectal cancer tumour markers and biomarkers: recent therapeutic advances. World J Gastroenterol 22(5):1745–1755. https://doi.org/10.3748/wjg.v22.i5.1745
47. Ye DX, Wang SS, Huang Y, Chi P (2019) A 3-circular RNA signature as a noninvasive biomarker for diagnosis of colorectal cancer. Cancer Cell Int 19:276. https://doi.org/10.1186/s12935-019-0995-7
48. Lin J, Cai D, Li W, Yu T, Mao H, Jiang S, Xiao B (2019) Plasma circular RNA panel acts as a novel diagnostic biomarker for colorectal cancer. Clin Biochem 74:60–68. https://doi.org/10.1016/j.clinbiochem.2019.10.012
49. Shen X, Chen Y, Li J, Huang H, Liu C, Zhou N (2021) Identification of Circ_001569 as a potential biomarker in the diagnosis and prognosis of pancreatic cancer. Technol Cancer Res Treat 20:1533033820983302. https://doi.org/10.1177/1533033820983302
50. Yang F, Liu DY, Guo JT, Ge N, Zhu P, Liu X, Wang S, Wang GX, Sun SY (2017) Circular RNA circ-LDLRAD3 as a biomarker in diagnosis of pancreatic cancer. World J Gastroenterol 23(47):8345–8354. https://doi.org/10.3748/wjg.v23.i47.8345
51. Sun JW, Qiu S, Yang JY, Chen X, Li HX (2020) Hsa_circ_0124055 and hsa_circ_0101622 regulate proliferation and apoptosis in thyroid cancer and serve as prognostic and diagnostic indicators. Eur Rev Med Pharmacol Sci 24(8):4348–4360. https://doi.org/10.26355/eurrev_202004_21016
52. Li X, Ma F, Wu L, Zhang X, Tian J, Li J, Cao J, Ma Y, Zhang L, Wang L (2020) Identification of Hsa_circ_0104824 as a potential biomarkers for breast cancer. Technol Cancer Res Treat 19:1533033820960745. https://doi.org/10.1177/1533033820960745
53. Hu Y, Song Q, Zhao J, Ruan J, He F, Yang X, Yu X (2020) Identification of plasma hsa_circ_0008673 expression as a potential biomarker and tumor regulator of breast cancer. J Clin Lab Anal 34(9):e23393. https://doi.org/10.1002/jcla.23393
54. Yin WB, Yan MG, Fang X, Guo JJ, Xiong W, Zhang RP (2018) Circulating circular RNA hsa_circ_0001785 acts as a diagnostic biomarker for breast cancer detection. Clin Chim Acta 487:363–368. https://doi.org/10.1016/j.cca.2017.10.011

55. Dhillon A, Hagan S, Rath O (2007) MAP kinase signalling pathways in cancer. Oncogene 26(22):3279–3290. https://doi.org/10.1038/sj.onc.1210421
56. Zhu J, Thompson CB (2019) Metabolic regulation of cell growth and proliferation. Nat Rev Mol Cell Biol 20(7):436–450. https://doi.org/10.1038/s41580-019-0123-5
57. Li X, Azhati B, Wang W, Rexiati M, Xing C, Wang Y (2021) Circular RNA UBAP2 promotes the proliferation of prostate cancer cells via the miR 1244/MAP 3K2 axis. Oncol Lett 21(6):486. https://doi.org/10.3892/ol.2021.12747
58. Ashwal-Fluss R, Meyer M, Pamudurti NR, Ivanov A, Bartok O, Hanan M, Evantal N, Memczak S, Rajewsky N, Kadener S (2014) circRNA biogenesis competes with pre-mRNA splicing. Mol Cell 56(1):55–66. https://doi.org/10.1016/j.molcel.2014.08.019
59. Bachmayr-Heyda A, Reiner AT, Auer K, Sukhbaatar N, Aust S, Bachleitner-Hofmann T, Mesteri I, Grunt TW, Zeillinger R, Pils D (2015) Correlation of circular RNA abundance with proliferation–exemplified with colorectal and ovarian cancer, idiopathic lung fibrosis and normal human tissues. Sci Rep 5:8057. https://doi.org/10.1038/srep08057
60. Zhang X, Ma L, Wan L, Wang H, Wang Z (2024) Circ_0003945: an emerging biomarker and therapeutic target for human diseases. Front Oncol 14:1275009. https://doi.org/10.3389/fonc.2024.1275009
61. Xie R, Zhang Y, Zhang J, Li J, Zhou X (2020) The role of circular RNAs in immune-related diseases. Front Immunol 11:545. https://doi.org/10.3389/fimmu.2020.00545
62. Gu J, Su C, Huang F, Zhao Y, Li J (2022) Past, present and future: the relationship between circular RNA and immunity. Front Immunol 13:894707. https://doi.org/10.3389/fimmu.2022.894707
63. Yan L, Chen GY (2020) Circular RNAs in immune response and viral infection. Trends Biochem Sci 45(12):1022–1034. https://doi.org/10.1016/j.tibs.2020.08.006
64. Liu Z, Ran Y, Tao C, Li S, Chen J, Yang E (2019) Detection of circular RNA expression and related quantitative trait loci in the human dorsolateral prefrontal cortex. Genome Biol 20(1):99. https://doi.org/10.1186/s13059-019-1701-8
65. Zurawska A, Mycko M, Selmaj K (2019) Circular RNAs as a novel layer of regulatory mechanism in multiple sclerosis. Neuroimmunology 334:576971. https://doi.org/10.1016/j.jneuroim.2019.576971
66. Xu M, Xie F, Tang X, Wang T, Wang S (2020) Insights into the role of circular RNA in macrophage activation and fibrosis disease. Pharmacol Res 156:104777. https://doi.org/10.1016/j.phrs.2020.104777
67. Gonzalez C, Cimini M, Cheng Z, Benedict C, Wang C (2022) Role of circular RNA cdr1as in modulation of macrophage phenotype. Life Sci 309:121003. https://doi.org/10.1016/j.lfs.2022.121003
68. Yang Q, Li F, He A, Yang BB (2021) Circular RNAs: expression, localization, and therapeutic potentials. Mol Ther 29(5):1683–1702. https://doi.org/10.1016/j.ymthe.2021.01.018
69. Chen Q, Mang G, Wu J, Sun P, Li T, Zhang H, Wang N, Tong Z, Wang W, Zheng Y, Tian J, Zhang M, Yu B (2020) Circular RNA circSnx5 controls immunogenicity of dendritic cells through the miR-544/SOCS1 axis and PU. 1 activity regulation. Mol Ther 28(11):2503–2518. https://doi.org/10.1016/j.ymthe.2020.07.001
70. Feng S, Xu Z, Zhang Z, Mo Y, Deng Y, Li L, Fei S (2022) RNA-Seq approach to investigate the effects of melatonin on bone marrow-derived dendritic cells from dextran sodium sulfate-induced colitis mice. Toxicology 481:153354. https://doi.org/10.1016/j.tox.2022.153354
71. Cheng Q, Wang J, Li M, Fang J, Ding H, Meng J, Zhang J, Fang X, Liu H, Ma C, Chen C, Zhang W (2022) CircSV2b participates in oxidative stress regulation through miR-5107-5p-Foxk1-Akt1 axis in Parkinson's disease. Redox Biol 56:102430. https://doi.org/10.1016/j.redox.2022.102430
72. Wu DP, Zhao YD, Yan QQ, Liu LL, Wei YS, Huang JL (2023) Circular RNAs: emerging players in brain aging and neurodegenerative diseases. J Pathol 259(1):1–9. https://doi.org/10.1002/path.6021
73. Xiao F, He Z, Wang S, Li J, Fan X, Yan T, Yang M, Yang D (2024) Regulatory mechanism of circular RNAs in neurodegenerative diseases. CNS Neurosci Ther 30(4):e14499. https://doi.org/10.1111/cns.14499
74. Moreno-García L, López-Royo T, Calvo AC, Toivonen JM, De La Torre M, Moreno-Martínez L, Molina N, Aparicio P, Zaragoza P, Manzano R, Osta R (2020) Competing endogenous RNA networks as biomarkers in neurodegenerative diseases. Int J Mol Sci 21(24):9582. https://doi.org/10.3390/ijms21249582
75. Zhang X, Wan M, Min X, Chu G, Luo Y, Han Z, Li W, Xu R, Luo J, Li W, Yang Y, Ma Y, Jiao L, Wang T (2023) Circular RNA as biomarkers for acute ischemic stroke: a systematic review and meta-analysis. CNS Neurosci Ther 29(8):2086–2100. https://doi.org/10.1111/cns.14220
76. Alzheimer's Association (2022) Alzheimer's disease facts and figures. Alzheimers Dement 18(4):700–789. https://doi.org/10.1002/alz.12638
77. Revised criteria for diagnosis and staging of Alzheimer's | AAIC. https://aaic.alz.org/diagnostic-criteria.asp. Accessed 10 May 2024
78. Dhillon S (2021) Aducanumab: first approval. Drugs 81(12):1437–1443. https://doi.org/10.1007/s40265-021-01569-z
79. Jack CR Jr, Bennett DA, Blennow K, Carrillo MC, Dunn B, Haeberlein SB, Holtzman DM, Jagust W, Jessen F, Karlawish J, Liu E, Molinuevo JL, Montine T, Phelps C, Rankin KP, Rowe CC, Scheltens

P, Siemers E, Snyder HM, Sperling R (2018) NIA-AA research framework: toward a biological definition of Alzheimer's disease. Alzheimers Dement 14(4):535–562. https://doi.org/10.1016/j.jalz.2018.02.018
80. Dube U, Del-Aguila JL, Li Z, Budde JP, Jiang S, Hsu S, Ibanez L, Fernandez MV, Farias F, Norton J, Gentsch J, Wang F, Dominantly Inherited Alzheimer Network (DIAN), Salloway S, Masters CL, Lee JH, Graff-Radford NR, Chhatwal JP, Bateman RJ, Morris JC, Karch CM, Harari O, Cruchaga C (2019) An atlas of cortical circular RNA expression in Alzheimer disease brains demonstrates clinical and pathological associations. Nat Neurosci 22(11):1903–1912. https://doi.org/10.1038/s41593-019-0501-5
81. Puri S, Hu J, Sun Z, Lin M, Stein TD, Farrer LA, Wolozin B, Zhang X (2023) Identification of circRNAs linked to Alzheimer's disease and related dementias. Alzheimers Dement 19(8):3389–3405. https://doi.org/10.1002/alz.12960
82. Li Y, Fan H, Sun J, Ni M, Zhang L, Chen C, Hong X, Fang F, Zhang W, Ma P (2020) Circular RNA expression profile of Alzheimer's disease and its clinical significance as biomarkers for the disease risk and progression. Int J Biochem Cell Biol 123:105747. https://doi.org/10.1016/j.biocel.2020.105747
83. Ren Z, Chu C, Pang Y, Cai H, Jia L (2022) A circular RNA blood panel that differentiates Alzheimer's disease from other dementia types. Biomark Res 10(1):63. https://doi.org/10.1186/s40364-022-00405-0
84. Nam K IL, Yoon G, Kim YK, Song J (2020) Transcriptome analysis of pineal glands in the mouse model of Alzheimer's disease. Front Mol Neurosci 12:318. https://doi.org/10.3389/fnmol.2019.00318
85. Chen HH, Eteleeb A, Wang C, Fernandez MV, Budde JP, Bergmann K, Norton J, Wang F, Ebl C, Morris JC, Perrin RJ, Bateman RJ, McDade E, Xiong C, Goate A, Farlow M, Chhatwal J, Schofield PR, Chui H, Harari O, Cruchaga C, Ibanez L (2022) Circular RNA detection identifies circPSEN1 alterations in brain specific to autosomal dominant Alzheimer's disease. Acta Neuropathol Commun 10(1):29. https://doi.org/10.1186/s40478-022-01328-5
86. Piscopo P, Manzini V, Rivabene R, Crestini A, Le Pera L, Pizzi E, Veroni C, Talarico G, Peconi M, Castellano AE, D'Alessio C, Bruno G, Corbo M, Vanacore N, Lacorte E (2022) A plasma circular RNA profile differentiates subjects with Alzheimer's disease and mild cognitive impairment from healthy controls. Int J Mol Sci 23(21):13232. https://doi.org/10.3390/ijms232113232
87. Meng S, Wang B, Li W (2022) CircAXL knockdown alleviates Aβ1-42-induced neurotoxicity in Alzheimer's disease via repressing PDE4A by releasing miR-1306-5p. Neurochem Res 47(6):1707–1720. https://doi.org/10.1007/s11064-022-03563-7
88. Liu L, Chen X, Chen YH, Zhang K (2020) Identification of circular RNA hsa_Circ_0003391 in peripheral blood is potentially associated with Alzheimer's disease. Front Aging Neurosci 12:601965. https://doi.org/10.3389/fnagi.2020.601965
89. Urdánoz-Casado A, Sánchez-Ruiz de Gordoa J, Robles M, Roldan M, Macías Conde M, Acha B, Blanco-Luquin I, Mendioroz M (2023) circRNA from APP gene changes in Alzheimer's disease human brain. Int J Mol Sci 24(5):4308. https://doi.org/10.3390/ijms24054308
90. Fan W, Pang H, Xie Z, Huang G, Zhou Z (2022) Circular RNAs in diabetes mellitus and its complications. Front Endocrinol (Lausanne) 13:885650. https://doi.org/10.3389/fendo.2022.885650
91. Zaiou M (2020) circRNAs signature as potential diagnostic and prognostic biomarker for diabetes mellitus and related cardiovascular complications. Cells 9(3):659. https://doi.org/10.3390/cells9030659
92. Fang X, Miao R, Wei J, Wu H, Tian J (2022) Advances in multi-omics study of biomarkers of glycolipid metabolism disorder. Comput Struct Biotechnol J 20:5935–5951. https://doi.org/10.1016/j.csbj.2022.10.030
93. Jiang Q, Liu C, Li CP, Xu SS, Di YM, Ge HM, Sun YN, Li XM, Zhang SJ, Shan K, Liu BH, Yao J, Zhao C, Yan B (2020) Circular RNA-ZNF532 regulates diabetes-induced retinal pericyte degeneration and vascular dysfunction. J Clin Invest 130(7):3833–3847. https://doi.org/10.1172/JCI123353
94. Zhang Z, Luo S, Xiao Z, Yin W, Shi X, Chen H, Xie Z, Liu Z, Li X, Zhou Z (2023) Hsa_circRNA_405498 and hsa_circRNA_100033 serve as potential biomarkers for differential diagnosis of type 1 diabetes. J Clin Endocrinol Metab 109(6):1464–1473. https://doi.org/10.1210/clinem/dgad761
95. Zhao W, Meng X, Liang J (2022) Analysis of circRNA-mRNA expression profiles and functional enrichment in diabetes mellitus based on high throughput sequencing. Int Wound J 19(5):1253–1262. https://doi.org/10.1111/iwj.13838
96. She W, Li T, Liu Y, Liu X (2021) CircRNA circVEGFC is highly expressed in gestational diabetes mellitus (GDM) and it is correlated with multiple adverse events. Diabetes Metab Syndr Obes 14:4409–4414. https://doi.org/10.2147/DMSO.S334728
97. Filardi T, Catanzaro G, Mardente S, Zicari A, Santangelo C, Lenzi A, Morano S, Ferretti E (2020) Non-coding RNA: role in gestational diabetes pathophysiology and complications. Int J Mol Sci 21(11):4020. https://doi.org/10.3390/ijms21114020
98. Zhang Y, Wu H, Wang F, Ye M, Zhu H, Bu S (2018) Long non-coding RNA MALAT1 expression in patients with gestational diabetes mellitus. Int J Gynaecol Obstet 140(2):164–169. https://doi.org/10.1002/ijgo.12384
99. Wu H, Wu S, Zhu Y, Ye M, Shen J, Liu Y, Zhang Y, Bu S (2019) Hsa-circRNA-0054633 is highly expressed in gestational diabetes mellitus and closely related

to glycosylation index. Clin Epigenetics 11(1):22. https://doi.org/10.1186/s13148-019-0610-8
100. Iqbal J, Al Qarni A, Hawwari A, Alghanem AF, Ahmed G (2018) Metabolic syndrome, dyslipidemia and regulation of lipoprotein metabolism. Curr Diabetes Rev 14(5):427–433. https://doi.org/10.2174/1573399813666170705161039
101. Li A, Huang W, Zhang X, Xie L, Miao X (2018) Identification and characterization of circRNAs of two pig breeds as a new biomarker in metabolism-related diseases. Cell Physiol Biochem 47(6):2458–2470. https://doi.org/10.1159/000491619
102. Guo J, Zhou Y, Cheng Y, Fang W, Hu G, Wei J, Lin Y, Man Y, Guo L, Sun M, Cui Q, Li J (2018) Metformin-induced changes of the coding transcriptome and non-coding RNAs in the livers of non-alcoholic fatty liver disease mice. Cell Physiol Biochem 45(4):1487–1505. https://doi.org/10.1159/000487575
103. Zhu Y, Gui W, Lin X, Li H (2020) Knock-down of circular RNA H19 induces human adipose-derived stem cells adipogenic differentiation via a mechanism involving the polypyrimidine tract-binding protein 1. Exp Cell Res 387(2):111753. https://doi.org/10.1016/j.yexcr.2019.111753
104. Liu C, Ge HM, Liu BH, Dong R, Shan K, Chen X, Yao MD, Li XM, Yao J, Zhou RM, Zhang SJ, Jiang Q, Zhao C, Yan B (2019) Targeting pericyte–endothelial cell crosstalk by circular RNA-cPWWP2A inhibition aggravates diabetes-induced microvascular dysfunction. Proc Natl Acad Sci USA 116(15):7455–7464. https://doi.org/10.1073/pnas.1814874116
105. Yan Q, He X, Kuang G, Ou C (2020) CircRNA cPWWP2A: an emerging player in diabetes mellitus. J Cell Commun Signal 14(3):351–353. https://doi.org/10.1007/s12079-020-00570-7
106. Yin W, Luo S, Qiu J, Xiao Z, Zhang Z, Xie Z, Li X, Zhou Z (2023) Identification of hsa_circRNA_100632 as a novel molecular biomarker for fulminant type 1 diabetes. Front Immunol 14:1075970. https://doi.org/10.3389/fimmu.2023.1075970
107. Yan Y, Dong J, Li Y, Lu Y, Yang K, Wang T, Zhang X, Xiao H (2022) CircRNA hsa_circ_0071336 is associated with type 2 diabetes through targeting the miR-93-5p/GLUT4 axis. FASEB J 36(5):e22324. https://doi.org/10.1096/fj.202200149RR
108. Du A, Li S, Zhou Y, Disoma C, Liao Y, Zhang Y, Chen Z, Yang Q, Liu P, Liu S, Dong Z, Razzaq A, Tao S, Chen X, Liu Y, Xu L, Zhang Q, Li S, Peng J, Xia Z (2022) M6A-mediated upregulation of circMDK promotes tumorigenesis and acts as a nanotherapeutic target in hepatocellular carcinoma. Mol Cancer 21(1):109. https://doi.org/10.1186/s12943-022-01575-z
109. Zeng Q, Liu CH, Wu D, Jiang W, Zhang N, Tang H (2023) LncRNA and circRNA in patients with non-alcoholic fatty liver disease: a systematic review. Biomol Ther 13(3):560. https://doi.org/10.3390/biom13030560
110. Wang P, Zhang Y, Deng L, Qu Z, Guo P, Liu L, Yu Z, Wang P, Liu N (2022) The function and regulation network mechanism of circRNA in liver diseases. Cancer Cell Int 22(1):141. https://doi.org/10.1186/s12935-022-02559-1
111. Chien Y, Tsai PH, Lai YH, Lu KH, Liu CY, Lin HF, Huang CS, Wu WW, Wang CY (2020) CircularRNA as novel biomarkers in liver diseases. J Chin Med Assoc 83(1):15–17. https://doi.org/10.1097/JCMA.0000000000000230
112. Özdemir S, Aksungur N, Altundaş N, Kara S, Korkut E, Özkaraca M, Sefa Mendil A, Öztürk G (2022) Genome-wide profiling of the expression of serum derived exosomal circRNAs in patients with hepatic alveolar echinococcosis. Gene 814:146161. https://doi.org/10.1016/j.gene.2021.146161
113. Campana L, Iredale J (2017) Regression of liver fibrosis. Semin Liver Dis 37(1):1–10. https://doi.org/10.1055/s-0036-1597816
114. Wang W, Dong R, Guo Y, He J, Shao C, Yi P, Yu F, Gu D, Zheng J (2019) CircMTO1 inhibits liver fibrosis via regulation of miR-17-5p and Smad7. J Cell Mol Med 23(8):5486–5496. https://doi.org/10.1111/jcmm.14432
115. Ohtani N, Kawada N (2019) Role of the gut–liver axis in liver inflammation, fibrosis, and cancer: a special focus on the gut microbiota relationship. Hepatol Commun 3(4):456–470. https://doi.org/10.1002/hep4.1331
116. Jiménez-Sousa E, Gómez-Moreno A, Pineda-Tenor D, Sánchez-Ruano J, Artaza-Varasa T, Martin-Vicente M, Fernández-Rodríguez A, Martínez I, Resino S (2019) Impact of DARC rs12075 variants on liver fibrosis progression in patients with chronic hepatitis C: a retrospective study. Biomol Ther 9(4):143. https://doi.org/10.3390/biom9040143
117. Brenner DA (2013) Reversibility of liver fibrosis. Gastroenterol Hepatol (N Y) 9(11):737–739
118. Cobb BR, Valsamakis A (2016) Chronic Hepatitis B, C, and D. Microbiol Spectr 4(4). https://doi.org/10.1128/microbiolspec.DMIH2-0025-2015
119. Zou GL, Zuo S, Lu S, Hu RH, Lu YY, Yang J, Deng KS, Wu YT, Mu M, Zhu JJ, Zeng JZ, Zhang BF, Wu X, Zhao XK, Li HY (2019) Bone morphogenetic protein-7 represses hepatic stellate cell activation and liver fibrosis via regulation of TGF-β/Smad signaling pathway. World J Gastroenterol 25(30):4222–4234. https://doi.org/10.3748/wjg.v25.i30.4222
120. Han D, Li J, Wang H, Su X, Hou J, Gu Y, Qian C, Lin Y, Liu X, Huang M, Li N, Zhou W, Yu Y, Cao X (2017) Circular RNA circMTO1 acts as the sponge of microRNA-9 to suppress hepatocellular carcinoma progression. Hepatology 66(4):1151–1164. https://doi.org/10.1002/hep.29270
121. Han B, Chao J, Yao H (2018) Circular RNA and its mechanisms in disease: from the bench to the

clinic. Pharmacol Ther 187:31–44. https://doi.org/10.1016/j.pharmthera.2018.01.010
122. Jeck WR, Sorrentino JA, Wang K, Slevin MK, Burd CE, Liu J, Marzluff WF, Sharpless NE (2013) Circular RNAs are abundant, conserved, and associated with ALU repeats. RNA 19(2):141–157. https://doi.org/10.1261/rna.035667.112
123. Memczak S, Jens M, Elefsinioti A, Torti F, Krueger J, Rybak A, Maier L, Mackowiak SD, Gregersen LH, Munschauer M, Loewer A, Ziebold U, Landthaler M, Kocks C, le Noble F, Rajewsky N (2013) Circular RNAs are a large class of animal RNAs with regulatory potency. Nature 495(7441):333–338. https://doi.org/10.1038/nature11928
124. Zheng LL, Li JH, Wu J, Sun WJ, Liu S, Wang ZL, Zhou H, Yang JH, Qu LH (2016) deepBase v2.0: identification, expression, evolution and function of small RNAs, LncRNAs and circular RNAs from deep-sequencing data. Nucleic Acids Res 44:D1–D202. https://doi.org/10.1093/nar/gkv1273
125. Li J, Yang J, Zhou P, Le Y, Zhou C, Wang S, Xu D, Lin HK, Gong Z (2015) Circular RNAs in cancer: novel insights into origins, properties, functions and implications. Am J Cancer Res 5(2):472–480
126. Memczak S, Papavasileiou P, Peters O, Rajewsky N (2015) Identification and characterization of circular RNAs as a new class of putative biomarkers in human blood. PLoS One 10(10). https://doi.org/10.1371/journal.pone.0141214
127. Dong Y, He D, Peng Z, Peng W, Shi W, Wang J, Li B, Zhang C, Duan C (2017) Circular RNAs in cancer: an emerging key player. J Hematol Oncol 10(1):2. https://doi.org/10.1186/s13045-016-0370-2
128. Ren B, Wang H, Ren L, Yangdan C, Zhou Y, Fan H, Lv Y (2019) Screening for microRNA-based diagnostic markers in hepatic alveolar echinococcosis. Medicine 98:37. https://doi.org/10.1097/MD.0000000000017156
129. Mitra SK, Hanson DA, Schlaepfer DD (2005) Focal adhesion kinase: in command and control of cell motility. Nat Rev Mol Cell Biol 6(1):56–68. https://doi.org/10.1038/nrm1549
130. Raker VK, Becker C, Steinbrink K (2016) The cAMP pathway as therapeutic target in autoimmune and inflammatory diseases. Front Immunol 7:123. https://doi.org/10.3389/fimmu.2016.00123
131. Aggarwal BB (2003) Signalling pathways of the TNF superfamily: a double-edged sword. Nat Rev Immunol 3(9):745–756. https://doi.org/10.1038/nri1184
132. Zhang W, Liu HT (2002) MAPK signal pathways in the regulation of cell proliferation in mammalian cells. Cell Res 12(1):9–18. https://doi.org/10.1038/sj.cr.7290105
133. Saxton RA, Sabatini DM (2017) mTOR Signaling in growth, metabolism, and disease. Cell 168(6):960–976. https://doi.org/10.1016/j.cell.2017.02.004
134. Nriagu J (2019) Encyclopedia of environmental health. Newnes
135. Pamies D, Martínez CE, Sogorb MA, Vilanova E (2011) Mechanism-based models in reproductive and developmental toxicology. In: Reproductive and developmental toxicology. Elsevier, pp 135–146. https://doi.org/10.1016/B978-0-12-382032-7.10011-6
136. Khan Z, Law FCP (2005) Adverse effects of pesticides and related chemicals on enzyme and hormone systems of fish, amphibians and reptiles: a review. Proc Pak Acad Sci 42:315–323
137. Dubus IG, Hollis JM, Brown CD (2000) Pesticides in rainfall in Europe. Environ Pollut 110(2):331–344. https://doi.org/10.1016/S0269-7491(99)00295-X
138. Özdemir S, Altun S, Arslan H (2018) Imidacloprid exposure cause the histopathological changes, activation of TNF-α, iNOS, 8-OHdG biomarkers, and alteration of caspase 3, iNOS, CYP1A, MT1 gene expression levels in common carp (*Cyprinus carpio* L.). Toxicol Rep 5:125–133. https://doi.org/10.1016/j.toxrep.2017.12.019
139. Altun S, Özdemir S, Arslan H (2017) Histopathological effects, responses of oxidative stress, inflammation, apoptosis biomarkers and alteration of gene expressions related to apoptosis, oxidative stress, and reproductive system in chlorpyrifos-exposed common carp (*Cyprinus carpio* L.). Environ Pollut 230:432–443. https://doi.org/10.1016/j.envpol.2017.06.085
140. McCubrey JA, LaHair MM, Franklin RA (2006) Reactive oxygen species-induced activation of the MAP kinase signaling pathways. Antioxid Redox Signal 8(9–10):1775–1789. https://doi.org/10.1089/ars.2006.8.1775
141. Torii S, Yamamoto T, Tsuchiya Y, Nishida E (2006) ERK MAP kinase in G1 cell cycle progression and cancer. Cancer Sci 97(8):697–702. https://doi.org/10.1111/j.1349-7006.2006.00244.x
142. Karin M, Gallagher E (2009) TNFR signaling: ubiquitin-conjugated TRAFfic signals control stop-and-go for MAPK signaling complexes. Immunol Rev 228(1):225–240. https://doi.org/10.1111/j.1600-065X.2008.00755.x
143. Pearson G, Robinson F, Beers Gibson T, Xu B, Karandikar M, Berman K, Cobb MH (2001) Mitogen-activated protein (MAP) kinase pathways: regulation and physiological functions. Endocr Rev 22(2):153–183. https://doi.org/10.1210/edrv.22.2.0428
144. Kim EK, Choi EJ (2010) Pathological roles of MAPK signaling pathways in human diseases. Biochim Biophys Acta (BBA) Mol Basis Dis 1802(4):396–405. https://doi.org/10.1016/j.bbadis.2009.12.009
145. Kaminska B, Gozdz A, Zawadzka M, Ellert-Miklaszewska A, Lipko M (2009) MAPK signal transduction underlying brain inflammation and gliosis as therapeutic target. Anat Rec (Hoboken) 292(12):1902–1913. https://doi.org/10.1002/ar.21047
146. Vanova N, Pejchal J, Herman D, Dlabkova A, Jun D (2018) Oxidative stress in organophosphate poi-

soning: role of standard antidotal therapy. J Appl Toxicol 38(8):1058–1070. https://doi.org/10.1002/jat.3605
147. Guignet M, Dhakal K, Flannery BM, Hobson BA, Zolkowska D, Dhir A, Bruun DA, Li S, Wahab A, Harvey DJ, Silverman JL, Rogawski MA, Lein PJ (2020) Persistent behavior deficits, neuroinflammation, and oxidative stress in a rat model of acute organophosphate intoxication. Neurobiol Dis 133:104431. https://doi.org/10.1016/j.nbd.2019.03.019
148. Park JH, Lee JE, Shin IC, Koh HC (2013) Autophagy regulates chlorpyrifos-induced apoptosis in SH-SY5Y cells. Toxicol Appl Pharmacol 268(1):55–67. https://doi.org/10.1016/j.taap.2013.01.013
149. Küçükler S, Çomaklı S, Özdemir S, Çağlayan C, Kandemir FM (2021) Hesperidin protects against the chlorpyrifos-induced chronic hepato-renal toxicity in rats associated with oxidative stress, inflammation, apoptosis, autophagy, and up-regulation of PARP-1/VEGF. Environ Toxicol 36(8):1600–1617. https://doi.org/10.1002/tox.23156
150. Ashkenazi A, Dixit VM (1998) Death receptors: signaling and modulation. Science 281(5381):1305–1308. https://doi.org/10.1126/science.281.5381.1305
151. Shi X, Gu A, Ji G, Li Y, Di J, Jin J, Hu F, Long Y, Xia Y, Lu C, Song L, Wang S, Wang X (2011) Developmental toxicity of cypermethrin in embryo-larval stages of zebrafish. Chemosphere 85(6):1010–1016. https://doi.org/10.1016/j.chemosphere.2011.07.024
152. Zeljezic D, Vinkovic B, Kasuba V, Kopjar N, Milic M, Mladinic M (2017) The effect of insecticides chlorpyrifos, α-cypermethrin and imidacloprid on primary DNA damage, TP 53 and c-Myc structural integrity by comet-FISH assay. Chemosphere 182:332–338. https://doi.org/10.1016/j.chemosphere.2017.05.010
153. Martínez M-A, Lopez-Torres B, Rodríguez J-L, Martínez M, Maximiliano J-E, Martínez-Larrañaga M-R, Anadón A, Ares I (2020) Toxicologic evidence of developmental neurotoxicity of Type II pyrethroids cyfluthrin and alpha-cypermethrin in SH-SY5Y cells. Food Chem Toxicol 137:111173. https://doi.org/10.1016/j.fct.2020.111173
154. Zhao H, Wang Y, Guo M, Mu M, Yu H, Xing M (2020) Grass carps co-exposed to environmentally relevant concentrations of cypermethrin and sulfamethoxazole bear immunodeficiency and are vulnerable to subsequent Aeromonas hydrophila infection. Environ Pollut 266:115156. https://doi.org/10.1016/j.envpol.2020.115156
155. Lee JE, Lim MS, Park JH, Park CH, Koh HC (2014) Nuclear NF-κB contributes to chlorpyrifos-induced apoptosis through p53 signaling in human neural precursor cells. Neurotoxicology 42:58–70. https://doi.org/10.1016/j.neuro.2014.04.001
156. Park JH, Ko J, Hwang J, Koh HC (2015) Dynamin-related protein 1 mediates mitochondria-dependent apoptosis in chlorpyrifos-treated SH-SY5Y cells. Neurotoxicology 51:145–157. https://doi.org/10.1016/j.neuro.2015.10.008
157. Jiao W, Han Q, Xu Y, Jiang H, Xing H, Teng X (2019) Impaired immune function and structural integrity in the gills of common carp (*Cyprinus carpio* L.) caused by chlorpyrifos exposure: through oxidative stress and apoptosis. Fish Shellfish Immunol 86:239–245. https://doi.org/10.1016/j.fsi.2018.08.060
158. Yuan S, Han Y, Ma M, Rao K, Wang Z, Yang R, Liu Y, Zhou X (2019) Aryl-phosphorus-containing flame retardants induce oxidative stress, the p53-dependent DNA damage response and mitochondrial impairment in A549 cells. Environ Pollut 250:58–67. https://doi.org/10.1016/j.envpol.2019.03.109
159. Shen W, Lou B, Xu C, Yang G, Yu R, Wang X, Li X, Wang Q, Wang Y (2020) Lethal toxicity and gene expression changes in embryonic zebrafish upon exposure to individual and mixture of malathion, chlorpyrifos and lambda-cyhalothrin. Chemosphere 239:124802. https://doi.org/10.1016/j.chemosphere.2019.124802
160. Henne WM, Boucrot E, Meinecke M, Evergren E, Vallis Y, Mittal R, McMahon HT (2010) FCHo proteins are nucleators of clathrin-mediated endocytosis. Science 328(5983):1281–1284. https://doi.org/10.1126/science.1188462
161. Nunez D, Antonescu C, Mettlen M, Liu A, Schmid SL, Loerke D, Danuser G (2011) Hotspots organize clathrin-mediated endocytosis by efficient recruitment and retention of nucleating resources. Traffic 12(12):1868–1878. https://doi.org/10.1111/j.1600-0854.2011.01273.x
162. Ehrlich M, Boll W, van Oijen A, Hariharan R, Chandran K, Nibert ML, Kirchhausen T (2004) Endocytosis by random initiation and stabilization of clathrin-coated pits. Cell 118(5):591–605. https://doi.org/10.1016/j.cell.2004.08.017
163. Merrifield CJ, Perrais D, Zenisek D (2005) Coupling between clathrin-coated-pit invagination, cortactin recruitment, and membrane scission observed in live cells. Cell 121(4):593–606. https://doi.org/10.1016/j.cell.2005.03.015
164. Kaksonen M, Roux A (2018) Mechanisms of clathrin-mediated endocytosis. Nat Rev Mol Cell Biol 19(5):313–326. https://doi.org/10.1038/nrm.2017.132
165. Schmidt MR, Haucke V (2007) Recycling endosomes in neuronal membrane traffic. Biol Cell 99(6):333–342. https://doi.org/10.1042/BC20070007
166. Streissguth AP, Barr HM, Olson HC, Sampson PD, Bookstein FL, Burgess DM (1994) Drinking during pregnancy decreases word attack and arithmetic scores on standardized tests: adolescent data from a population-based prospective study. Alcohol Clin Exp Res 18(2):248–254. https://doi.org/10.1111/j.1530-0277.1994.tb00009.x
167. Goodlett CR, Horn KH, Zhou FC (2005) Alcohol teratogenesis: mechanisms of damage and strategies

for intervention. Exp Biol Med 230(6):394–406. https://doi.org/10.1177/15353702-0323006-07
168. Marín MP, Esteban-Pretel G, Ponsoda X, Romero AM, Ballestín R, López C, Megías L, Timoneda J, Molowny A, Canales JJ, Renau-Piqueras J (2010) Endocytosis in cultured neurons is altered by chronic alcohol exposure. Toxicol Sci 115(1):202–213. https://doi.org/10.1093/toxsci/kfq040
169. Syamlal G, Bhattacharya A, Dodd KE (2020) Medical expenditures attributed to asthma and chronic obstructive pulmonary disease among workers—United States, 2011–2015. MMWR Morb Mortal Wkly Rep 69(26):809–814. https://doi.org/10.15585/mmwr.mm6926a1
170. Vos T, Lim SS, Abbafati C, Abbas KM, Abbasi M, Abbasifard M, Abbasi-Kangevari M, Abbastabar H, Abd-Allah F, Abdelalim A, Abdollahi M, Abdollahpour I, Abolhassani H, Aboyans V, Abrams EM, Abreu LG, Abrigo MR, Abu-Raddad LJ, Abushouk AI et al (2020) Global burden of 369 diseases and injuries in 204 countries and territories, 1990–2019: a systematic analysis for the Global Burden of Disease Study 2019. Lancet 396(10258):1204–1222. https://doi.org/10.1016/S0140-6736(20)30925-9
171. Ali MK, Schimmel K, Zhao L, Chen C-K, Dua K, Nicolls MR, Spiekerkoetter E (2022) The role of circular RNAs in pulmonary hypertension. Eur Respir J 60(6):2200012. https://doi.org/10.1183/13993003.00012-2022
172. Liu X, Ali MK, Dua K, Xu R (2022) The role of zinc in the pathogenesis of lung disease. Nutrients 14(10):2115. https://doi.org/10.3390/nu141021
173. Habib N, Pasha MA, Tang DD (2022) Current understanding of asthma pathogenesis and biomarkers. Cells 11(17):2764. https://doi.org/10.3390/cells11172764
174. Gysens F, Mestdagh P, de Bony de Lavergne E, Maes T (2022) Unlocking the secrets of long non-coding RNAs in asthma. Thorax 77(5):514–522. https://doi.org/10.1136/thoraxjnl-2021-218359
175. Curran DR, Cohn L (2010) Advances in mucous cell metaplasia. Am J Respir Cell Mol Biol 42(3):268–275. https://doi.org/10.1165/rcmb.2009-0151TR
176. Guo P, Li R, Piao TH, Wang CL, Wu XL, Cai HY (2022) Pathological mechanism and targeted drugs of COPD. Int J Chron Obstruct Pulmon Dis 17:1565–1575. https://doi.org/10.2147/COPD.S366126
177. Safiri S, Carson-Chahhoud K, Noori M, Nejadghaderi SA, Sullman MJM, Ahmadian Heris J, Ansarin K, Mansournia MA, Collins GS, Kolahi A-A, Kaufman JS (2022) Burden of chronic obstructive pulmonary disease and its attributable risk factors in 204 countries and territories, 1990-2019: results from the Global Burden of Disease Study 2019. BMJ 378:e069679. https://doi.org/10.1136/bmj-2021-069679
178. Matera MG, Hanania NA, Maniscalco M, Cazzola M (2023) Pharmacotherapies in older adults with COPD: challenges and opportunities. Drugs Aging 40(7):605–619. https://doi.org/10.1007/s40266-023-01038-0
179. He Y, Qian DC, Diao JA, Cho MH, Silverman EK, Gusev A, Manrai AK, Martin AR, Patel CJ (2023) Prediction and stratification of longitudinal risk for chronic obstructive pulmonary disease across smoking behaviors. Nat Commun 14(1):8297. https://doi.org/10.1038/s41467-023-44047-8
180. Feizi H, Alizadeh M, Nejadghaderi SA, Noori M, Sullman MJM, Ahmadian Heris J, Kolahi A-A, Collins GS, Safiri S (2022) The burden of chronic obstructive pulmonary disease and its attributable risk factors in the Middle East and North Africa region, 1990–2019. Respir Res 23(1):319. https://doi.org/10.1186/s12931-022-02242-z
181. Bollmeier SG, Hartmann AP (2020) Management of chronic obstructive pulmonary disease: a review focusing on exacerbations. Am J Health Syst Pharm 77(4):259–268. https://doi.org/10.1093/ajhp/zxz306
182. Huang Z, Cao Y, Zhou M, Qi X, Fu B, Mou Y, Wu G, Xie J, Zhao J, Xiong W (2019) Hsa_circ_0005519 increases IL-13/IL-6 by regulating hsa-let-7a-5p in CD4+ T cells to affect asthma. Clin Exp Allergy 49(8):1116–1127. https://doi.org/10.1111/cea.13445
183. Xu C, Du L, Guo Y, Liang Y (2023) Differential circular RNA expression profiles in induced sputum of patients with asthma. Int Arch Allergy Immunol 184(5):460–470. https://doi.org/10.1159/000528129

Circular RNAs as Novel Biomarkers and Application in Cardiovascular Diseases

15

Diego Franco and Fernando Bonet

Abstract

Cardiovascular diseases are the leading cause of death worldwide in the human population. Currently, both noninvasive and invasive conventional diagnostic methods are used in cardiovascular diseases. Noninvasive techniques such as electrocardiogram are widely accepted; yet, it is hindered by poor sensitivity and specificity. On the other hand, invasive examinations such as coronary arteriography and intravascular ultrasound are not resolute enough during the early stages of cardiovascular pathological changes. Consequently, there is an urgent need to identify noninvasive biomarkers that are highly sensitive, specific, and effective for the early-stage diagnosis of cardiovascular diseases. Noncoding RNAs have recently emerged as a novel layer of gene regulation, and their expression is widely impaired in pathological conditions, including cardiovascular diseases. In this context, circRNAs are highly stable molecules that are easy detectable in liquid biopsies, thus providing as good candidate for clinical diagnosis. Within this section, we provide a state-of-the art review of the current identification of circRNAs as biomarkers in distinct cardiovascular pathological conditions.

Keywords

Cardiogenesis · Congenital heart diseases · Cardiac pathology · Myocardial infarction

Cardiovascular diseases are the leading cause of death worldwide in the human population according to the World Health Organization, and both congenital and adult (acquired) cardiovascular pathologies contribute to it. Congenital cardiovascular diseases can be severely life-threatening such as tetralogy of Fallot or transposition of the great arteries, while others such as atrial septal defects or bicuspid aortic valve are compatible with life but predispose to additional cardiovascular diseases in the adulthood. On the other hand, adult cardiovascular diseases encompass a large array of cardiac abnormalities, including

D. Franco (✉)
Cardiovascular Research Group, Department of Experimental Biology, University of Jaen, Jaen, Spain

Medina Foundation, Technology Park of Health Sciences, Granada, Spain
e-mail: dfranco@ujaen.es

F. Bonet
Department of Medicine, School of Medicine, University of Cádiz (UCA), Cádiz, Spain

Research Unit, Biomedical Research and Innovation Institute of Cadiz (INiBICA), Puerta del Mar University Hospital, Cádiz, Spain

J. Xiao (ed.), *Advances in Circular RNAs*, Advances in Experimental Medicine and Biology 1485,
https://doi.org/10.1007/978-981-96-9428-0_15

therein as electrical, structural, and ischemic diseases.

Conventional diagnostic methods for cardiovascular disease are categorized as either noninvasive or invasive. Although noninvasive techniques, such as electrocardiogram (ECG) and Holter monitoring, are widely accepted, they are hindered by poor sensitivity and specificity. On the other hand, invasive examinations like coronary arteriography and intravascular ultrasound are not advantageous during the early stages of pathological changes. Consequently, there is an urgent need to identify noninvasive biomarkers that are highly sensitive, specific, and effective for the early-stage diagnosis of cardiovascular diseases. Within this section, we provide a state-of-the art review of the current identification of circRNAs as biomarkers in distinct cardiovascular pathological conditions, particularly in congenital heart diseases, atrial fibrillation, dilated and hypertrophic cardiomyopathy, coronary artery disease, and myocardial infarction, as detailed in the following sections. Additionally, their plausible applications will be highlighted.

1 CircRNAs in Cardiac Development and Congenital Heart Diseases

Circular RNAs (circRNAs) are endogenous noncoding RNAs involved in a variety of physiological and pathological processes, including heart diseases. Our current understanding of the molecular mechanisms driven by circRNAs during cardiac development is still scarce, although some studies revealed their essential role regulating their beating and contractile properties of developing cardiomyocytes as recently reported for circ-RCCD [1], circ_105039 [2], and circ-MGAT1 [3]. In recent years, understanding of the regulatory role of circRNAs has started to emerge also in distinct congenital heart diseases such as distinct cardiac septal defects and complex congenital malformations such as tetralogy of Fallot.

Liu et al. [4] reported differentially expressed circRNAs from cardiac samples of aborted fetuses with ventricular septal defects as compared to control samples by transcriptomic expression assays. Validation analyses identified five differentially expressed circRNAs (hsa_circRNA_002086, hsa_circRNA_007878, hsa_circRNA_100709, hsa_circRNA_101965, and hsa_circRNA_402565) that might be potentially served as VSD biomarkers [4]. However, no additional insights into their plausible mode of action are provided. More recently, Wu et al. [5] performed a transcriptional profile of circulating plasma circRNAs in patients with atrial and ventricular septal defects and identified 10 circRNAs significantly upregulated and 147 downregulated. Validation of three most downregulated circRNAs (hsa_circRNA_004183, hsa_circRNA_079265, and hsa_circRNA_105039) was carried out in a larger cohort of patients with atrial or ventricular septal defects and their suitability as biomarkers assessed, providing a high sensitivity. Furthermore, a bioinformatic-driven gene regulatory network, including circRNAs, microRNAs, and mRNAs, was constructed, including therein 43 circRNAs, 9 mRNAs, and 29 miRNAs, supporting the notion that key cardiogenic transcription factors such as Mef2c and Tbx3 might be potentially modulated [5].

Tetralogy of Fallot constitutes the most frequent type of cyanotic congenital heart disease, with a relative poor prognosis without corrective surgery. Yu et al. [6] reported the circRNA expression profiles of TOF patients by microarray analyses and identified 276 differentially expressed circRNAs, including 214 upregulated and 62 downregulated. Bioinformatic analyses constructed a gene regulatory network involving 19 circRNAs, 9 miRNAs, and 34 mRNAs. These authors reported the correlative expression of the hsa_circ_0007798/miR-199b-5p/HIF1A signaling axis, which might offer as a candidate diagnostic biomarkers for TOF [6]. Additional gene regulatory networks were reported by Kan et al. (2022) [7] by bioinformatically analyzing mRNA, microRNA, and circRNA expression profiles' datasets for the Gene Expression Omnibus (GEO) database. The hsa_circ_000601/hsa-miR-148a/BCL2L11 axis was identified as a key signaling pathway contributing to the pathogenesis of TOF as well as potential candidate diagnostic biomarkers [7]. Noteworthy, none of these two studies provide significant molecular mechanisms insights.

2 CircRNAs in Atrial Fibrillation

Atrial fibrillation (AF) is an abnormal heart rhythm characterized by rapid and irregular beating of the atria. AF represents the most frequent arrhythmogenic disease in human population with an estimated prevalence of 1–2% that raises up to 8–10% in the elderly. AF can be presented as a single self-terminating event, termed paroxysmal AF, or can be persistent or permanent if not self-terminating, thus requiring pharmacological and/or surgical intervention.

In recent years, the contribution of circRNAs to AF has been extensively examined. Several studies reported the circRNA transcriptomic analyses in both atrial appendages [8, 9] and peripheral blood monocytes [10] of AF patients. Zhang et al. [8] reported 20 upregulated and 3 downregulated circRNAs that are differentially expressed in both left and right atrial appendages. Among them, hsa_circ_0003965 has a significantly negative correlation with its parental gene TMEM245, and hsa_circ_0000075 and hsa_circ_0082096 might participate in the AF pathogenesis via Tgf-β signaling pathway regulation as revealed by a competing endogenous RNA network analyses [8]. On the other hand, Zhang et al. [9] reported 147 differentially expressed circRNAs (102 upregulated and 45 downregulated) in atrial appendages of AF patients as compared to sinus rhythm group, some of which were indeed distinctly expressed between left and right appendages [9]. Analyses of differentially expressed circRNAs in peripheral blood monocytes also identified 65 upregulated and 55 downregulated circRNAs [10]. While in all these studies, gene ontology analyses provided an entry site to discern the plausible mechanisms of circRNAs in AF pathogenesis, unfortunately, scarce experimental information is available about their plausible molecular mechanisms.

In addition to these circRNA transcriptomic analyses, gene regulatory networks involving circRNA, microRNA, and mRNAs have also been reported in AF, in several cases taking advantage of previously published transcriptomics analyses publicly available in different databases [11–17], while in other cases, new transcriptomic analyses of peripheral blood mononuclear cells [18] or heart tissues are provided. Additionally, circRNA transcriptomic and gene regulatory networks analyses have also been done in experimental rapid atrial pacing canine model [19]. Each of these studies reported distinct gene regulatory networks but with limited molecular mechanistic insights. For example, Liu et al. [13] speculated that hsa-circRNA-100053-hsa-miR-455-5p-TRPV1 and hsa-circRNA-005843-hsa-miR-188-5p-SPON1 interactions might be involved in AF pathogenesis, but no additional mechanistic insights are provided [13]. Importantly, Liu et al. [15] reported that sa_circ_0000672 expression was higher in peripheral blood monocytes of persistent AF patients with LA low voltage than those with LA normal voltage and furthermore confirmed by dual-luciferase assays that exerted biological functions as a sponge of miR-516a-5p to regulate expression of its target gene TRAF6, supporting experimental evidence of such gene regulatory network in AF pathophysiology [15].

Analyses of the circRNA gene regulatory networks have also been investigated in postoperative AF [20–22] as well as in distinct subtypes of AF [23–26]. In this context, hsa_circ_0006314 and hsa_circ_0055387 were found as potential biomarker of postoperative AF, supporting their role in the synthesis of extracellular matrix and inflammatory response as main processes involved in development and progression of AF [21]. Similarly, hsa_circRNA_001654 and hsa_circRNA_005899, as well as hsa-miR-19b-3p and hsa-miR-30a-5p regulating KLF10, were also significantly related to postoperative AF [20]. Importantly, distinct circRNA regulatory networks are operative between distinct phases of AF progression as reported by Costa et al. [23] and Sun et al. [25], and discrete circRNAs (hsa_circRNA002085, hsa_circRNA001321) can serve as biomarkers to distinguished nonvalvular persistent AF [26] or as biomarker of AF persistency such as hsa_circ_0004104 by promoting cardiac fibrosis via targeting MAPK and TGF-beta pathways [24]. As in the case of previous studies of circRNA gene regulatory networks in AF, limited mechanistic insights are also currently available in the transition between distinct AF subtypes.

Atrial fibrillation is frequently associated with both structural and electrical remodeling, supporting the notion that AF begets AF. Therefore, understanding the cellular and molecular substrates facilitating structural remodeling such as fibrosis or electrical remodeling is essential to halt AF progression. In this context, several studies have provided evidences of the functional role of distinct circRNAs in both AF-associated structural and electrical remodeling. In this context, Zhang et al. [27] reported that circCAMTA1 knockdown decreased atrial fibrosis by downregulating TGFBR1 expression throughout the modulation of miR-214-3p [27], while Wu et al. [28] demonstrated in a mouse experimental model of AF that mmu_circ_0005019 inhibited fibroblast cell proliferation and migration, while overexpression of mmu_circ_0005019 in cardiomyocytes promoted Kcnd1, Scn5a, and Kcnn3 expression. Importantly, mmu_circ_0005019 exerted its biological functions by acting as a miR-499-5p sponge, thus regulating the expression of its target gene Kcnn3 [28]. Additionally, circRNA_0263 and circRNA_1507, on the one hand side, and Cfa-circ002203 have also been associated to atrial fibrosis experimental rat and porcine models [29, 30], and their mechanisms, beyond competing with several plausible microRNAs, remain elusive.

Atrial fibrillation has been recently associated to be highly correlated to epicardial adipose tissue deposition [31–33]. Zheng et al. [34] reported thus an integrative network of circRNA, miRNA, and mRNA of epicardial adipose tissue in patients with atrial fibrillation, supporting their plausible role in AF onset. Additional evidence of the role of circRNAs in AF associated to other cardiovascular diseases has been also reported, such as those linked to valvular heart diseases [35], rheumatic heart disease [36] as well as new predictors of AF recurrence after surgical ablation in patients with valvular heart diseases [37].

3 CircRNAs in Dilated Cardiomyopathy (DCM)

Dilated cardiomyopathy (DCM) is a nonischemic heart muscle disease characterized by left ventricular enlargement and systolic dysfunction [38]. DCM is an important cause of sudden cardiac death and the most common cause of heart failure (HF) and heart transplantation [39]. DCM encompasses a broad group of diseases, acquired or genetic, which share a similar phenotype. However, evidence suggest that distinct DCM etiologies may produce different structural and/or functional phenotypes and may respond differently to therapy [40]. In up to 50% of cases, the exact cause remains unknown. Hence, discriminate different etiologies of DCM are crucial to improve patient care. Therefore, there is an urgency in finding novel specific diagnostic methods. Despite being promising tools to distinguish between DCM etiologies, due to its stability [41], the role of circRNAs as potential biomarkers for DCM remains poorly explored.

Sun and colleagues [42] performed microarrays to assess the differential expression of circRNAs in peripheral blood samples from pediatric DCM. qRT-PCR validation confirmed downregulation of has_circ_0067735 and has_circ_0069972, and upregulation of has_circ_0070186, suggesting a potential role as diagnostic biomarkers. Moreover, a functional analysis of these differentially expressed circRNAs indicated association with hypertrophy, remodeling, fibrosis, and autoimmunity [42]. A more recent report assessed the value of circRNA as biomarker to distinguish between etiologies. For this purpose, differentially circRNA expression was analyzed by qRT-PCR in plasma samples from idiopathic DCM, ischemic DCM, and in familial DCM patients, including LMNA and BCL2-associated athanogene 3 (BAG3) pathogenic variants. qRT-PCR analyses showed that, whereas hsa_circ_0003258, hsa_circ_0051238, and hsa_circ_0051239 were upregulated in LMNA-related DCM, hsa_circ_0089762 was specifically upregulated in the ischemic DCM cohort. ROC curve analysis confirmed the discriminative power of this circRNAs. Hsa_circ_0089762 reached the highest AUC value (0.92), and its expression was negatively associated with diastolic blood pressure and LVEF [43].

In summary, the exploration of circRNAs as potential diagnostic biomarkers for diverse DCM etiologies presents a promising avenue. The identified circRNAs exhibit discriminatory power, offering insights into the intricate landscape of DCM subtypes and paving the way for tailored diagnostic approaches and therapeutic interventions.

4 CircRNAs in Hypertrophic Cardiomyopathy (HCM)

Hypertrophic cardiomyopathy (HCM) is one of the most common inherited heart conditions with an estimated prevalence of 1 in 500 people [44]. HCM is characterized by increase in ventricular wall thickness. HCM is a common cause of sudden cardiac death mainly in young adults. The reason for death is often attributed to HF, thromboembolism, and, in some cases, disruptions in the heart's electrical signaling. Identifying HCM is challenging due to the lack of a consistent link between genetic makeup and observed symptoms within affected families. Hence, it is crucial to identify novel potential biomarkers for the diagnosis of HCM.

A first study investigated the expression pattern of several circRNAs in serum from 64 HCM patients and 53 healthy controls. Serum expression levels of circDNAJC6, circTMEM56, and circMBOAT2 circRNAs were found significantly downregulated in HCM patients. Importantly, these circRNAs demonstrated robust discriminatory capabilities for distinguishing between healthy individuals and those with different forms of HCM (nonobstructive (HNCM) or obstructive (HOCM). Notably, circTMEM56 and circDNAJC6 emerged as potential indicators of disease severity, particularly in HOCM patients, offering valuable insights for personalized therapeutic strategies. Thus, circRNAs emerged as novel biomarkers for HCM [45]. A second study adopts a systems biology approach, exploring the circular RNA-associated competing endogenous RNA (ceRNA) network in HCM. By analyzing plasma circRNA and mRNA expression profiles, the study identified that hsa_circ_0043762, hsa_circ_0036248, and hsa_circ_0071269 were upregulated in HCM. These circRNAs, integrated into a ceRNA network with miRNAs and mRNAs, revealed associations with crucial cardiac functions, including calcium-release channel activity and muscle filament sliding. This systematic analysis provides a global perspective on the complexity of circRNA interactions in HCM, offering potential regulatory candidates in the pathogenesis of this heritable cardiomyopathy [46].

Collectively, these studies contribute significantly to the emerging field of circRNA research in HCM. The identification of specific circRNAs as potential biomarkers not only holds promise for improved diagnostic precision but also opens avenues for understanding the molecular mechanisms and severity indicators in the intricate landscape of hypertrophic cardiomyopathy.

5 CircRNAs in Coronary Artery Disease (CAD)

Coronary artery disease (CAD) is one of the major cardiovascular diseases in humans. CAD is major cause of death in both the developed and developing countries, and several factors including genetic, environmental, and lifestyle pose as risk factors for the development of this cardiovascular disease. Several studies have documented circRNAs transcriptomic analyses in CAD patients. In a study aimed at identifying specific biomarkers for CAD using circRNA microarrays, five upregulated circRNAs were identified in CAD patients, with hsa_circ_0124644 exhibiting the highest fold change. Subsequent validation in an independent cohort highlighted its diagnostic potential with an area under the curve (AUC) of 0.769. Combining hsa_circ_0124644 with hsa_circ_0098964 further enhanced diagnostic efficacy, emphasizing their potential as a valuable biomarker for CAD [47]. In a single patient who died from heart attack, Pan et al. [48] documented 66 DEG circRNAs (54 upregulated, 12 downregulated) from epicardial coronary artery samples, suggesting that they might contribute to the pathology underlying coronary artery atherosclerosis [48]. A similar case was reported by Hou et al. (2020) [49] corresponding to a single patient with sudden death of myocardial infarction, highlighting the plausible use of hsa_circ_0016868, hsa_circ_0001364, and hsa_circ_0006731 as CAD biomarkers [49]. On the other hand, Ward et al. (2022) [50] reported 160 DE circRNAs as circulating cell-free RNAs from patients with stable CAD, some of which subsequently developed HF. No differences in the plasma RNA profile between stable CAD patients who developed HF compared with those who did not was observed [50]. Additionally, DE mRNAs and lncRNAs were also identified. A circRNA transcriptomic microarray analyses in CAD patients also identified 624 circRNAs and 171 circRNAs significantly upregulated and downregulated, respectively. Multivariant regression analyses demonstrated that the combination of two of these circRNAs, i.e., hsa_circ_0001879 and hsa_circ_0004104, together with CAD risk factors, had the better performance to discriminate CAD patients from healthy controls [51]. Wu et al. [52] analyzed the circRNA expression levels in exosomes CAD patients by RNA sequencing and identified that 355 circRNAs were differentially expressed (164 were upregulated, and 191 were downregulated). Among all of them, these authors reported that plasma exosomal hsa_circ_0005540 can be used as a promising diagnostic biomarker of CAD. Tong et al. [53] analyzed microarray data of circRNAs derived from exosomes of patients with CAD retrieved from exoRBase. Subsequent qRT-PCR validations and ROC analysis suggested circRNA0001785, circRNA0000973, circRNA0001741, and circRNA0003922 as potential biomarkers for the diagnosis of CAD [53]. Hsa_circ_0001946 was validated as significantly upregulated in peripheral blood mononuclear cells of CAD patients and revealed potential as an independent indicator of increased CAD risk. Further analysis, including Spearman correlations and ROC curves, underscored its diagnostic accuracy [54]. A multivariate logistic regression model highlighted circNPHP4 overexpression as a promising predictor for CAD risk. The inclusion of circNPHP4 in the CAD nomogram demonstrated beneficial risk prediction capabilities [55]. In all these studies, the authors suggested that these DE circRNAs may serve as diagnostic or therapeutic targets in CAD, yet validation analyses were not performed, and no mechanistic insights as how such circRNAs can contribute to CAD is reported. Finally, Fu et al. [56] investigated m6A-modified circRNAs in human coronary artery smooth muscle cells of CAD patients. These authors identified up- and downregulated m6A peaks that were corresponding to 463 and 243 up- and downregulated circRNAs, respectively. One of them, hsa_circZBTB46, was strongly correlated with the incidence of CAD, using a logistic regression model, supporting the notion that it can be used a CAD biomarker [56].

In addition to these transcriptomic analyses of differential expression of circRNAs in CAD patients, increasing number of studies are reported gene regulatory networks integrating both coding and noncoding RNAs [48, 57–61]. He et al. [57] reported RNA sequencing from coronary segments and peripheral blood mononuclear cells of patients with coronary atherosclerotic disease and identified 34 differentially expressed circRNAs. Four of them (hsa_circRPRD1A, hsa_circHERPUD2, hsa_circLMBR1, and hsa_circDHTKD1) were subsequently selected for further investigation, and circRNA–miRNA–mRNA network was composed, including 20 microRNAs and 66 mRNAs [57]. Ji et al. [58] performed RNA sequence analysis of circRNAs of peripheral blood mononuclear cells corresponding to CAD patients. A total of 13,160 downregulated and 12,905 upregulated circRNAs were identified in CAD. Bioinformatic tools were used to predict the microRNA (miRNA)-targeting interactions and to construct a triple network of differentially expressed gene–circRNA–miRNA–mRNA. Most of these studies reported bioinformatically inferred gene regulatory networks without experimental validation and without mechanistic insights into how such networks are indeed involved in CAD pathogenesis. Only few examples of novel signaling pathways have come to light, such as that reported by Pan et al. [48] demonstrating that circular RNAs promote TRPM3 expression by inhibiting hsa-miR-130a-3p or that reported by Wang et al. [51] demonstrating that the overexpression of hsa_circ_0004104 resulted in the dysregulation of atherosclerosis-related genes in THP-1-derived macrophages and the study by Dai et al. [62] demonstrating that circLDLR can regulate vascular smooth muscle cell proliferation and apoptosis by modulating miR-26-5p/KDM6A signaling.

Besides these transcriptomics and gene regulatory network analyses, identification of single circRNA as potential biomarkers for CAD has been reported from different sources such as plasma (hsa_circ_0001445 [63]), peripheral blood leukocytes and/or monocytes (circRNA ZNF609 [64]; BTBD7_hsa_circ_0000563 [65]) or coronary artery segments (BTBD7_hsa_circ_0000563 [66]). In addition, a single-nucleotide polymorphisms flanking circFOXO3 have also been identified as biomarker of CAD [67].

Understanding of the differential expression of circRNAs in CAD, and thus their use as biomarkers, has also extended to concomitant cardiovascular diseases, such as type 2 diabetes mellitus [68], hypertension [69], and Kawasaki disease [70]. Li et al. [68] reported that hsa-circRNA11783–2 correlated with CAD and type 2 diabetes mellitus, Razeghian-Jahromi et al. [69] identified a correlation between the expression ratio between circular and linear ANRIL in hypertensive patients with and without CAD, while Kim et al. [70] identified circWHAMMP1 and circZNF124 as plausible biomarkers of Kawasaki disease.

6 CircRNAs in Myocardial Infarction

Myocardial infarction (MI) is characterized by cardiomyocyte death resulting from decreased blood flow in coronary arteries [71]. It is usually predisposed by atherosclerosis and leads to the occlusion of coronaries. Globally, the incidence of MI and its mortality rate are increasing making it a major public health issue. The incidence of acute myocardial infarction (AMI) continues to rise annually, posing a substantial threat to both human health and quality of life [72]. Timely diagnosis of MI is crucial for initiating life-saving interventions promptly [73]. Despite biomarkers like creatine kinase isoenzyme (CKMB) and troponin I (TnI) are considered the gold standard for MI, the early diagnosis remains challenging. Therefore, there is a need to identify new biomarkers.

In the last decade, circRNAs have emerged as a promising biomarker for MI diagnosis. In 2016, Vausort et al. [74] observed lower expression levels of the circRNA myocardial infarction–associated circular RNA (MICRA) in blood samples from MI patients as compared to healthy controls. In addition, analyses based on logistic regression demonstrated that MICRA is a strong predictor of LV dysfunction as patients with low levels of MICRA were at high risk of LV dysfunction. Multivariate analysis confirmed the predictive power of MICRA. This association was validated in two independent patient cohorts [75]. Similarly, the circulating CDR1as was found enriched in serum from acute MI (AMI) patients and well correlated with AMI. In addition, CDR1as levels were remarkably upregulated in the myocardium of AMI mice. Receiver–operator characteristic (ROC) analysis of circulating CDR1as identified it as a potential biomarker of AMI [76]. Lin et al. [77] employed chip technology to identify 266 differentially expressed circRNAs in the peripheral blood of AMI patients. This study suggested that the variously expressed circRNAs may be novel potential biomarkers of the genesis and progression of MI. More recently, circRNA microarrays profiling in blood samples from AMI and mild coronary stenosis patients identified 154 differentially expressed circRNA between cohorts. RT-qPCR validation and ROC curve analysis indicated that circRNA_104761 could not only be an effective biomarker for AMI diagnosis, but also differentiate normal coronary artery, mild coronary artery stenosis, and AMI [78]. Cheng et al. [79] observed that the expression levels of circPostn were significantly enriched in the plasma from patients with MI relative to the healthy controls. However, its discriminatory diagnostic power was not evaluated. Zhao et al. [80] utilized a circRNA array to assess the expression levels of circRNAs in coronary blood samples obtained from individuals with MI patients and normal subjects. Differential expression analyses revealed significant higher levels of hsa_circRNA_001654 and hsa_circRNA_091761 in the blood of MI patients. This observation positions these circRNAs as potential biomarkers for the early diagnosis of MI. On the other hand, the circulating circular RNA cZNF292 combined with clinical information was reported to discriminate effectively between AMI and non-AMI patients. This was further confirmed in a validation cohort, identifying therefore cZNF292 as a potential biomarker for AMI [80]. Finally, in forensic autopsy cases, circSMARCC1 levels were found significantly elevated, whereas circLRBA levels were decreased in the groups with MI and early-MI compared to the normal control group. ROC curve analysis demonstrated that both circSMARCC1 and circLRBA possess a high discriminative power to distinguish between AMI and normal control cases. Moreover, a combination of these two circRNAs further enhanced the diagnostic efficacy of AMI. Consequently, circSMARCC1 and circLRBA emerge as potential biomarkers for the postmortem diagnosis of AMI [81].

7 Conclusions and Perspectives

CVDs remain a leading global cause of mortality, necessitating improved diagnostic tools for early intervention. The limitations of existing methods, both noninvasive and invasive, highlight the urgent need for innovative approaches. Due to the high stability, abundance, and spatiotemporal-specific expression in blood, cirRNAs have emerged as potential candidates for fulfilling this role.

The exploration of circRNAs in distinct cardiovascular pathologies, including congenital heart diseases, atrial fibrillation, DCM, HCM, CAD, and MI, demonstrates their diverse applications and their potential as disease-specific biomarkers.

However, a notable challenge lies in the lack of consistency among identified biomarkers across independent studies. Specifically, the circRNA markers proposed by different research groups, particularly in the context of CAD, exhibit variations. This inconsistency arises from differences in the methods of sample preparation as well as from the utilization of diverse sample types for analysis, such as whole blood, serum, or plasma, which may impact significantly in the composition of detected circRNAs. Moreover, variations in RNA extraction methods across studies contribute to differing results [82]. The absence of consensus on methodologies employed for circRNA quantification constitutes a major limiting factor. Each of the three main methods used to assess circRNA expression levels—qRT-PCR, microarray hybridization, and next-generation sequencing (NGS)—has its inherent limitations, thereby impacting result reproducibility [83]. Therefore, there is a necessity to standardize the process of identifying and validating blood circRNA biomarkers, including sample collection and preservation, circRNA extraction, library construction, and computational analysis [84, 85]. Adopting such a standardized procedure in future studies is imperative to assess the technical robustness and reproducibility of the suggested biomarkers both within and across different laboratories.

On the other hand, the circRNA biomarkers identified in these studies only represent preliminary signatures for CVD diseases. These results were obtained from cohorts exhibiting well-defined phenotypes, and the sample sizes are relatively small. Hence, a more extensive collection of clinical samples is necessary to validate their sensitivity and specificity within a larger cohort, particularly in distinguishing patients with similar clinical phenotypes. Additionally, their efficacy in disease diagnosis or prediction of disease outcomes requires evaluation in a prospective cohort within a clinical practice setting. Further investigations are imperative to examine and validate the utility of these biomarkers in clinical practice, including their potential to guide clinical decisions and enhance outcomes. While there are existing challenges and unresolved issues that need attention, the promising potential of integrating blood circRNA biomarkers into clinical practices presents novel and exciting opportunities for the field of liquid biopsy.

References

1. Liu Y, Gao J, Xu M, Zhou Q, Zhang Z, Ye J, Li R (2022) Circular RNA circ-RCCD promotes cardiomyocyte differentiation in mouse embryo development via recruiting YY1 to the promoter of MyD88. J Cell Mol Med 26(13):3616–3627. https://doi.org/10.1111/jcmm.17336
2. Yu B, Li M, Han SP, Yu Z, Zhu J (2021) Circular RNA hsa_circ_105039 promotes cardiomyocyte differentiation by sponging miR-17 to regulate cyclinD2 expression. Mol Med Rep 24(6):861. https://doi.org/10.3892/mmr.2021.12501
3. Chen P, Zhou C, Li B, Yang C (2020) Circular RNA MGAT1 regulates cell proliferation and apoptosis in hypoxia-induced cardiomyocytes through miR-34a/YAP1 axis. Int J Clin Exp Pathol 13(10):2474–2486
4. Liu H, Hu Y, Zhuang B, Yin J, Chen X-H, Wang J, Li M-M, Xu J, Wang X-Y, Yu Z-B, Han S-P (2018) Differential expression of CircRNAs in embryonic heart tissue associated with ventricular septal defect. Int J Med Sci 15(7):703–712. https://doi.org/10.7150/ijms.21660
5. Wu J, Li J, Liu H, Yin J, Zhang M, Yu Z, Miao H (2019) Circulating plasma circular RNAs as novel diagnostic biomarkers for congenital heart disease in children. J Clin Lab Anal 33(9):e22998. https://doi.org/10.1002/jcla.22998
6. Yu H, Wang X, Cao H (2021) Construction and investigation of a circRNA-associated ceRNA regulatory network in tetralogy of Fallot. BMC Cardiovasc Disord 21(1):437. https://doi.org/10.1186/s12872-021-02217-w
7. Kan Z, Yan W, Wang N, Fang Y, Gao H, Song Y (2022) Identification of circRNA-miRNA-mRNA regulatory network and crucial signaling pathway axis involved in tetralogy of Fallot. Front Genet 13:917454. https://doi.org/10.3389/fgene.2022.917454
8. Zhang P-P, Sun J, Li W (2020) Genome-wide profiling reveals atrial fibrillation-related circular RNAs in atrial appendages. Gene 728:144286. https://doi.org/10.1016/j.gene.2019.144286
9. Zhang Y, Shen H, Wang P, Min J, Yu Y, Wang Q, Wang S, Xi W, Nguyen QM, Xiao J, Wang Z (2020) Identification and characterization of circular RNAs in atrial appendage of patients with atrial fibrillation. Exp Cell Res 389(2):111821. https://doi.org/10.1016/j.yexcr.2020.111821
10. Ruan Z-B, Wang F, Bao T-T, Yu Q-P, Chen G-C, Zhu L (2020) Genome-wide analysis of circular RNA expression profiles in patients with atrial fibrillation. Int J Clin Exp Pathol 13(8):1933–1950
11. Chao X, Dai W, Li S, Jiang C, Jiang Z, Zhong G (2021) Identification of circRNA-miRNA-mRNA regulatory network and autophagy interaction network in atrial fibrillation based on bioinformatics analysis. Int J Gen Med 14:8527–8540. https://doi.org/10.2147/IJGM.S333752
12. Hu X, Chen L, Wu S, Xu K, Jiang W, Qin M, Zhang Y, Liu X (2019) Integrative analysis reveals key circular RNA in atrial fibrillation. Front Genet 10:108. https://doi.org/10.3389/fgene.2019.00108
13. Liu T, Zhang G, Wang Y, Rao M, Zhang Y, Guo A, Wang M (2020) Identification of circular RNA-MicroRNA-messenger RNA regulatory network in atrial fibrillation by integrated analysis. Biomed Res Int 2020:8037273. https://doi.org/10.1155/2020/8037273
14. Liu X, Zeng Y, Liu Z, Li W, Wang L, Wu M (2022) Bioinformatics analysis of the circRNA-miRNA-mRNA network for atrial fibrillation. Medicine 101(34):e30221. https://doi.org/10.1097/MD.0000000000030221
15. Liu X, Wu M, He Y, Gui C, Wen W, Jiang Z, Zhong G (2023) Construction and integrated analysis of the ceRNA network hsa_circ_0000672/miR-516a-5p/TRAF6 and its potential function in atrial fibrillation. Sci Rep 13(1):7701. https://doi.org/10.1038/s41598-023-34851-z
16. Wei F, Zhang X, Kuang X, Gao X, Wang J, Fan J (2022) Integrated analysis of circRNA-miRNA-mRNA-mediated network and its potential function in atrial fibrillation. Front Cardiovasc Med 9:883205. https://doi.org/10.3389/fcvm.2022.883205
17. Zhai Z, Qin T, Liu F, Han L, Zhou H, Li Q, Xia Z, Li J (2020) Identification of atrial fibrillation-related circular RNAs and constructing the integrative regulatory network of circular RNAs, microRNAs and mRNAs by bioinformatics analysis. Cell Mol Biol (Noisy-Le-Grand, France) 66(7):161–168
18. Wen J, Ruan Z-B, Wang F, Chen G-C, Zhu J-G, Ren Y, Zhu L (2023) Construction of atrial fibrillation-related circRNA/lncRNA-miRNA-mRNA regulatory network and analysis of potential biomarkers. J Clin Lab Anal 37(2):e24833. https://doi.org/10.1002/jcla.24833

19. Shangguan W, Liang X, Shi W, Liu T, Wang M, Li G (2018) Identification and characterization of circular RNAs in rapid atrial pacing dog atrial tissue. Biochem Biophys Res Commun 506(1):1–6. https://doi.org/10.1016/j.bbrc.2018.05.082
20. Chen X, Tang H, Lu K, Niu Z, Sheng W, Hwang HY, Pang PYK, Phillips JD, Khoynezhad A, Qu X, Li B, Han W (2023) Gene modules and genes associated with postoperative atrial fibrillation: weighted gene co-expression network analysis and circRNA-miRNA-mRNA regulatory network analysis. J Thorac Dis 15(9):4949–4960. https://doi.org/10.21037/jtd-23-1179
21. Chen Y, Ouyang T, Yin Y, Fang C, Tang C-E, Luo J, Luo F (2022) Analysis of infiltrated immune cells in left atriums from patients with atrial fibrillation and identification of circRNA biomarkers for postoperative atrial fibrillation. Front Genet 13:1003366. https://doi.org/10.3389/fgene.2022.1003366
22. Zhou Y, Wu Q, Ni G, Hong Y, Xiao S, Liu C, Yu Z (2022) Immune-associated pivotal biomarkers identification and competing endogenous RNA network construction in post-operative atrial fibrillation by comprehensive bioinformatics and machine learning strategies. Front Immunol 13:974935. https://doi.org/10.3389/fimmu.2022.974935
23. Costa MC, Cortez-Dias N, Gabriel A, de Sousa J, Fiúza M, Gallego J, Nobre Â, Pinto FJ, Enguita FJ (2019) circRNA-miRNA cross-talk in the transition from paroxysmal to permanent atrial fibrillation. Int J Cardiol 290:134–137. https://doi.org/10.1016/j.ijcard.2019.04.072
24. Gao Y, Liu Y, Fu Y, Wang Q, Liu Z, Hu R, Yang X, Chen M (2021) The potential regulatory role of hsa_circ_0004104 in the persistency of atrial fibrillation by promoting cardiac fibrosis via TGF-β pathway. BMC Cardiovasc Disord 21(1):25. https://doi.org/10.1186/s12872-021-01847-4
25. Sun H, Zhang J, Shao Y (2021) Integrative analysis reveals essential mRNA, long non-coding RNA (lncRNA), and circular RNA (circRNA) in paroxysmal and persistent atrial fibrillation patients. Anatol J Cardiol 25(6):414–428. https://doi.org/10.14744/AnatolJCardiol.2020.57295
26. Zhang Y, Ke X, Liu J, Ma X, Liu Y, Liang D, Wang L, Guo C, Luo Y (2019) Characterization of circRNA-associated ceRNA networks in patients with nonvalvular persistent atrial fibrillation. Mol Med Rep 19(1):638–650. https://doi.org/10.3892/mmr.2018.9695
27. Zhang L, Lou Q, Zhang W, Yang W, Li L, Zhao H, Kong Y, Li W (2023) CircCAMTA1 facilitates atrial fibrosis by regulating the miR-214-3p/TGFBR1 axis in atrial fibrillation. J Mol Histol 54(1):55–65. https://doi.org/10.1007/s10735-022-10110-9
28. Wu N, Li C, Xu B, Xiang Y, Jia X, Yuan Z, Wu L, Zhong L, Li Y (2021) Circular RNA mmu_circ_0005019 inhibits fibrosis of cardiac fibroblasts and reverses electrical remodeling of cardiomyocytes. BMC Cardiovasc Disord 21(1):308. https://doi.org/10.1186/s12872-021-02128-w
29. Liang X, Wang W, Liu Y, Liu Y, Wang M, Shangguan W, Miao S, Gu T, Liu E, Li G (2023) CircRNA_0263 and circRNA_1507 are dysregulated in a rat model of atrial fibrosis induced by chronic intermittent hypoxia. Am J Transl Res 15(1):63–81
30. Shangguan W, Gu T, Cheng R, Liu X, Liu Y, Miao S, Wang W, Song F, Wang H, Liu T, Liang X (2023) Cfa-circ002203 was upregulated in rapidly paced atria of dogs and involved in the mechanisms of atrial fibrosis. Front Cardiovasc Med 10:1110707. https://doi.org/10.3389/fcvm.2023.1110707
31. Conte M, Petraglia L, Cabaro S, Valerio V, Poggio P, Pilato E, Attena E, Russo V, Ferro A, Formisano P, Leosco D, Parisi V (2022) Epicardial adipose tissue and cardiac arrhythmias: focus on atrial fibrillation. Front Cardiovasc Med 9:932262. https://doi.org/10.3389/fcvm.2022.932262
32. Goudis CA, Vasileiadis IE, Liu T (2018) Epicardial adipose tissue and atrial fibrillation: pathophysiological mechanisms, clinical implications, and potential therapies. Curr Med Res Opin 34(11):1933–1943. https://doi.org/10.1080/03007995.2018.1462786
33. Takahashi N, Abe I, Kira S, Ishii Y (2023) Role of epicardial adipose tissue in human atrial fibrillation. J Arrhyth 39(2):93–110. https://doi.org/10.1002/joa3.12825
34. Zheng H, Peng Y, Wang P, Su P, Zhao L (2022) The integrative network of circRNA, miRNA and mRNA of epicardial adipose tissue in patients with atrial fibrillation. Am J Transl Res 14(9):6550–6562
35. Zhu X, Tang X, Chong H, Cao H, Fan F, Pan J, Wang D, Zhou Q (2020) Expression profiles of circular RNA in human atrial fibrillation with valvular heart diseases. Front Cardiovasc Med 7:597932. https://doi.org/10.3389/fcvm.2020.597932
36. Hu M, Wei X, Li M, Tao L, Wei L, Zhang M, Cheng H, Yuan Y (2019) Circular RNA expression profiles of persistent atrial fibrillation in patients with rheumatic heart disease. Anatol J Cardiol 21(1):2–10. https://doi.org/10.14744/AnatolJCardiol.2018.35902
37. Zhu X, Wang Y, Mo R, Chong H, Cao C, Fan F, Zhou Q, Wang D (2021) Left atrial appendage circular RNAs are new predictors of atrial fibrillation recurrence after surgical ablation in valvular atrial fibrillation patients. Heart Surg Forum 24(6):E968–E976. https://doi.org/10.1532/hsf.4125
38. Fatkin D (2011) Guidelines for the diagnosis and management of familial dilated cardiomyopathy. Heart Lung Circ 20(11):691–693. https://doi.org/10.1016/j.hlc.2011.07.008
39. Merlo M, Gentile P, Artico J, Cannatà A, Paldino A, De Angelis G, Barbati G, Alonge M, Gigli M, Pinamonti B, Ramani F, Zecchin M, Pirozzi F, Stolfo D, Sinagra G (2019) Arrhythmic risk stratification in patients with dilated cardiomyopathy and intermediate left ventricular dysfunction. J Cardiovasc Med (Hagerstown, Md) 20(5):343–350. https://doi.org/10.2459/JCM.0000000000000792

40. Orphanou N, Papatheodorou E, Anastasakis A (2022) Dilated cardiomyopathy in the era of precision medicine: latest concepts and developments. Heart Fail Rev 27(4):1173–1191. https://doi.org/10.1007/s10741-021-10139-0
41. Memczak S, Papavasileiou P, Peters O, Rajewsky N (2015) Identification and characterization of circular RNAs as a new class of putative biomarkers in human blood. PLoS One 10(10):e0141214. https://doi.org/10.1371/journal.pone.0141214
42. Sun W, Han B, Cai D, Wang J, Jiang D, Jia H (2020) Differential expression profiles and functional prediction of circular RNAs in pediatric dilated cardiomyopathy. Front Mol Biosci 7:600170. https://doi.org/10.3389/fmolb.2020.600170
43. Costa MC, Calderon-Dominguez M, Mangas A, Campuzano O, Sarquella-Brugada G, Ramos M, Quezada-Feijoo M, Pinilla JMG, Robles-Mezcua A, Del Aguila Pacheco-Cruz G, Belmonte T, Enguita FJ, Toro R (2021) Circulating circRNA as biomarkers for dilated cardiomyopathy etiology. J Mol Med (Berlin, Germany) 99(12):1711–1725. https://doi.org/10.1007/s00109-021-02119-6
44. Semsarian C, Ingles J, Maron MS, Maron BJ (2015) New perspectives on the prevalence of hypertrophic cardiomyopathy. JACC 65(12):1249–1254. https://doi.org/10.1016/j.jacc.2015.01.019
45. Sonnenschein K, Wilczek AL, de Gonzalo-Calvo D, Pfanne A, Derda AA, Zwadlo C, Bavendiek U, Bauersachs J, Fiedler J, Thum T (2019) Serum circular RNAs act as blood-based biomarkers for hypertrophic obstructive cardiomyopathy. Sci Rep 9(1):20350. https://doi.org/10.1038/s41598-019-56617-2
46. Guo Q, Wang J, Sun R, He Z, Chen Q, Liu W, Wu M, Bao J, Liu Z, Wang J, Zhang Y (2020) Comprehensive construction of a circular RNA-associated competing endogenous RNA network identified novel circular RNAs in hypertrophic cardiomyopathy by integrated analysis. Front Genet 11:764. https://doi.org/10.3389/fgene.2020.00764
47. Zhao Z, Li X, Gao C, Jian D, Hao P, Rao L, Li M (2017) Peripheral blood circular RNA hsa_circ_0124644 can be used as a diagnostic biomarker of coronary artery disease. Sci Rep 7:39918. https://doi.org/10.1038/srep39918
48. Pan R-Y, Liu P, Zhou H-T, Sun W-X, Song J, Shu J, Cui G-J, Yang Z-J, Jia EZ (2017) Circular RNAs promote TRPM3 expression by inhibiting hsa-miR-130a-3p in coronary artery disease patients. Oncotarget 8(36):60280–60290. https://doi.org/10.18632/oncotarget.19941
49. Hou C, Gu L, Guo Y, Zhou Y, Hua L, Chen J, He S, Zhang S, Jia Q, Zhao C, Zhang J, Xu G, Jia E (2020) Association between circular RNA expression content and severity of coronary atherosclerosis in human coronary artery. J Clin Lab Anal 34(12):e23552. https://doi.org/10.1002/jcla.23552
50. Ward Z, Schmeier S, Pearson J, Cameron VA, Frampton CM, Troughton RW, Doughty RN, Richards AM, Pilbrow AP (2022) Identifying candidate circulating RNA markers for coronary artery disease by deep RNA-sequencing in human plasma. Cells 11(20):3191. https://doi.org/10.3390/cells11203191
51. Wang L, Shen C, Wang Y, Zou T, Zhu H, Lu X, Li L, Yang B, Chen J, Chen S, Lu X, Gu D (2019) Identification of circular RNA Hsa_circ_0001879 and Hsa_circ_0004104 as novel biomarkers for coronary artery disease. Atherosclerosis 286:88–96. https://doi.org/10.1016/j.atherosclerosis.2019.05.006
52. Wu W-P, Pan Y-H, Cai M-Y, Cen J-M, Chen C, Zheng L, Liu X, Xiong XD (2020) Plasma-derived exosomal circular RNA hsa_circ_0005540 as a novel diagnostic biomarker for coronary artery disease. Dis Markers 2020:3178642. https://doi.org/10.1155/2020/3178642
53. Tong X, Zhao X, Dang X, Kou Y, Kou J (2023) circRNA, a novel diagnostic biomarker for coronary heart disease. Front Cardiovasc Med 10:1070616. https://doi.org/10.3389/fcvm.2023.1070616
54. Huang S, Zeng Z, Sun Y, Cai Y, Xu X, Li H, Wu S (2021) Association study of hsa_circ_0001946, hsa-miR-7-5p and PARP1 in coronary atherosclerotic heart disease. Int J Cardiol 328:1–7. https://doi.org/10.1016/j.ijcard.2020.12.026
55. Xiong F, Mao R, Zhang L, Zhao R, Tan K, Liu C, Xu J, Du G, Zhang T (2021) CircNPHP4 in monocyte-derived small extracellular vesicles controls heterogeneous adhesion in coronary heart atherosclerotic disease. Cell Death Dis 12(10):948. https://doi.org/10.1038/s41419-021-04253-y
56. Fu Y, He S, Li C, Gan X, Wang Y, Zhou Y, Jiang R, Zhang Q, Pan Y, Zhou H, Chen X, Jia E (2023) Detailed profiling of m6A modified circRNAs and synergistic effects of circRNA and environmental risk factors for coronary artery disease. Eur J Pharmacol 951:175761. https://doi.org/10.1016/j.ejphar.2023.175761
57. He S, Fu Y, Li C, Gan X, Wang Y, Zhou H, Jiang R, Zhang Q, Jia Q, Chen X, Jia E-Z (2023) Interaction between the expression of hsa_circRPRD1A and hsa_circHERPUD2 and classical coronary risk factors promotes the development of coronary artery disease. BMC Med Genom 16(1):131. https://doi.org/10.1186/s12920-023-01540-9
58. Ji W-F, Chen J-X, He S, Zhou Y-Q, Hua L, Hou C, Zhang S, Gan X-K, Wang Y-J, Zhou H-X, Li Q, Jia EZ (2021) Characteristics of circular RNAs expression of peripheral blood mononuclear cells in humans with coronary artery disease. Physiol Genomics 53(8):349–357. https://doi.org/10.1152/physiolgenomics.00020.2021
59. Lin F, Zhao G, Chen Z, Wang X, Lv F, Zhang Y, Yang X, Liang W, Cai R, Li J, Li M, Zhang G (2019) circRNA-miRNA association for coronary heart disease. Mol Med Rep 19(4):2527–2536. https://doi.org/10.3892/mmr.2019.9905
60. Miao L, Yin R-X, Zhang Q-H, Liao P-J, Wang Y, Nie R-J, Li H (2019) A novel circRNA-miRNA-mRNA network identifies circ-YOD1 as a biomarker for coronary artery disease. Sci Rep 9(1):18314. https://doi.org/10.1038/s41598-019-54603-2

61. Zhang Z, Qian H, Wang L, Tao Z, Cheng K, Wang K, Xie Y, Zhang L (2022) Construction of a circRNA-miRNA-mRNA regulatory network for coronary artery disease by bioinformatics analysis. Cardiol Res Pract 2022:4017082. https://doi.org/10.1155/2022/4017082
62. Dai H, Zhao N, Zheng Y (2022) CircLDLR modulates the proliferation and apoptosis of vascular smooth muscle cells in coronary artery disease through miR-26-5p/KDM6A Axis. J Cardiovasc Pharmacol 80(1):132–139. https://doi.org/10.1097/FJC.0000000000001275
63. Vilades D, Martínez-Camblor P, Ferrero-Gregori A, Bär C, Lu D, Xiao K, Vea À, Nasarre L, Sanchez Vega J, Leta R, Carreras F, Thum T, Llorente-Cortés V, de Gonzalo-Calvo D (2020) Plasma circular RNA hsa_circ_0001445 and coronary artery disease: performance as a biomarker. FASEB J 34(3):4403–4414. https://doi.org/10.1096/fj.201902507R
64. Liang B, Li M, Deng Q, Wang C, Rong J, He S, Xiang Y, Zheng F (2020) CircRNA ZNF609 in peripheral blood leukocytes acts as a protective factor and a potential biomarker for coronary artery disease. Ann Transl Med 8(12):741. https://doi.org/10.21037/atm-19-4728
65. Zhou H, Gan X, He S, Wang Y, Zhang S, Chen J, Zhou Y, Hou C, Hua L, Zhang Q, Jia E (2022) Identification of circular RNA BTBD7_hsa_circ_0000563 as a novel biomarker for coronary artery disease and the functional discovery of BTBD7_hsa_circ_0000563 based on peripheral blood mononuclear cells: a case control study. Clin Proteomics 19(1):37. https://doi.org/10.1186/s12014-022-09374-w
66. Chen J-X, Hua L, Zhao C-H, Jia Q-W, Zhang J, Yuan J-X, Zhang Y-J, Jin J-L, Gu M-F, Mao Z-Y, Sun H-J, Wang L-S, Ma W-Z, Jia E-Z (2020) Quantitative proteomics reveals the regulatory networks of circular RNA BTBD7_hsa_circ_0000563 in human coronary artery. J Clin Lab Anal 34(11):e23495. https://doi.org/10.1002/jcla.23495
67. Zhou Y-L, Wu W-P, Cheng J, Liang L-L, Cen J-M, Chen C, Liu X, Xiong XD (2020) CircFOXO3 rs12196996, a polymorphism at the gene flanking intron, is associated with circFOXO3 levels and the risk of coronary artery disease. Aging 12(13):13076–13089. https://doi.org/10.18632/aging.103398
68. Li X, Zhao Z, Jian D, Li W, Tang H, Li M (2017) Hsa-circRNA11783-2 in peripheral blood is correlated with coronary artery disease and type 2 diabetes mellitus. Diab Vasc Dis Res 14(6):510–515. https://doi.org/10.1177/1479164117722714
69. Razeghian-Jahromi I, Zibaeenezhad MJ, Karimi Akhormeh A, Dara M (2022) Expression ratio of circular to linear ANRIL in hypertensive patients with coronary artery disease. Sci Rep 12(1):1802. https://doi.org/10.1038/s41598-022-05731-9
70. Kim Y-K (2019) Analysis of circular RNAs in the coronary arteries of patients with Kawasaki disease. J Lip Atheros 8(1):50–57. https://doi.org/10.12997/jla.2019.8.1.50
71. Anderson JL, Morrow DA (2017) Acute myocardial infarction. N Eng J Med 376(21):2053–2064. https://doi.org/10.1056/NEJMra1606915
72. Salari N, Morddarvanjoghi F, Abdolmaleki A, Rasoulpoor S, Khaleghi AA, Hezarkhani LA, Shohaimi S, Mohammadi M (2023) The global prevalence of myocardial infarction: a systematic review and meta-analysis. BMC Cardiovasc Disord 23(1):206. https://doi.org/10.1186/s12872-023-03231-w
73. Jeong JH, Seo YH, Ahn JY, Kim KH, Seo JY, Chun KY, Lim YS, Park PW (2020) Performance of copeptin for early diagnosis of acute myocardial infarction in an emergency department setting. Ann Lab Med 40(1):7–14. https://doi.org/10.3343/alm.2020.40.1.7
74. Vausort M, Salgado-Somoza A, Zhang L, Leszek P, Scholz M, Teren A, Burkhardt R, Thiery J, Wagner DR, Devaux Y (2016) Myocardial infarction-associated circular RNA predicting left ventricular dysfunction. J Am Coll Cardiol 68(11):1247–1248. https://doi.org/10.1016/j.jacc.2016.06.040
75. Salgado-Somoza A, Zhang L, Vausort M, Devaux Y (2017) The circular RNA MICRA for risk stratification after myocardial infarction. Int J Cardiol Heart Vasc 17:33–36. https://doi.org/10.1016/j.ijcha.2017.11.001
76. Zhang Y, Sun L, Xuan L, Pan Z, Li K, Liu S, Huang Y, Zhao X, Huang L, Wang Z, Hou Y, Li J, Tian Y, Yu J, Han H, Liu Y, Gao F, Zhang Y, Wang S, Yang B (2016) Reciprocal changes of circulating long non-coding RNAs ZFAS1 and CDR1AS predict acute myocardial infarction. Sci Rep 6:22384. https://doi.org/10.1038/srep22384
77. Lin F, Yang Y, Guo Q, Xie M, Sun S, Wang X, Li D, Zhang G, Li M, Wang J, Zhao G (2020) Analysis of the molecular mechanism of acute coronary syndrome based on circRNA-miRNA network regulation. Evid Based Complement Alternat Med 2020:1584052. https://doi.org/10.1155/2020/1584052
78. Yang W, Sun L, Cao X, Li L, Zhang X, Li J, Zhao H, Zhan C, Zang Y, Li T, Zhang L, Liu G, Li W (2021) Detection of circRNA biomarker for acute myocardial infarction based on system biological analysis of RNA expression. Front Genet 12:686116. https://doi.org/10.3389/fgene.2021.686116
79. Cheng N, Wang M-Y, Wu Y-B, Cui H-M, Wei S-X, Liu B, Wang R (2020) Circular RNA POSTN promotes myocardial infarction-induced myocardial injury and cardiac remodeling by regulating miR-96-5p/BNIP3 Axis. Front Cell Dev Biol 8:618574. https://doi.org/10.3389/fcell.2020.618574
80. Zhao C, Liu J, Ge W, Li Z, Lv M, Feng Y, Liu X, Liu B, Zhang Y (2020) Identification of regulatory circRNAs involved in the pathogenesis of acute myocardial infarction. Front Genet 11:626492. https://doi.org/10.3389/fgene.2020.626492
81. Gao T, Peng X, Yang D, Hou C, Chen Z (2023) CircSMARCC1 and CircLRBA are potential biomarkers in forensic postmortem diagnosis of acute myocardial infarction. Legal Med (Tokyo, Japan) 60:102184. https://doi.org/10.1016/j.legalmed.2022.102184

82. Ortega-Pinazo J, Pacheco-Rodríguez MJ, Serrano-Castro PJ, Martínez B, Pinto-Medel MJ, Gómez-Zumaquero JM, Lago-Sampedro A, García-Díaz B, Estivill-Torrús G, Emilio Ferro Gallego P (2023) Comparing RNA extraction methods to face the variations in RNA quality using two human biological matrices. Mol Biol Rep 50(11):9263–9271. https://doi.org/10.1007/s11033-023-08761-2
83. Kristensen LS, Andersen MS, Stagsted LVW, Ebbesen KK, Hansen TB, Kjems J (2019) The biogenesis, biology and characterization of circular RNAs. Nat Rev Genet 20(11):675–691. https://doi.org/10.1038/s41576-019-0158-7
84. Anfossi S, Babayan A, Pantel K, Calin GA (2018) Clinical utility of circulating non-coding RNAs—an update. Nat Rev Clin Oncol 15(9):541–563. https://doi.org/10.1038/s41571-018-0035-x
85. Byron SA, Van Keuren-Jensen KR, Engelthaler DM, Carpten JD, Craig DW (2016) Translating RNA sequencing into clinical diagnostics: opportunities and challenges. Nat Rev Genet 17(5):257–271. https://doi.org/10.1038/nrg.2016.10

CircRNA Synthesis and Applications

16

Adam Greasley, Shuailong Li, KeXiang Liu, and Xiufen Zheng

Abstract

Circular RNA (circRNA) is becoming an increasingly popular field of study and topic of interest. With that, there is a renewed and heightened interest in generating circular RNA for both research and clinical use. CircRNA generation began in the 1990s using in vitro chemical ligation and has since expanded out to include in vitro enzymatic synthesis along with in vivo vector-based systems. Each of these methods has distinct advantages, limitations, and different function 1ality depending on application. Throughout this chapter, we explore and summarize different methods for generating pure circRNA constructs that can be readily utilized for different applications. We highlight the steps and aspects to be considered when selecting the method to synthesize circRNA based on yield, purity, and application. Overall, this chapter aims to provide the reader with an understanding of the current methods and applications of synthesizing circRNA both in vivo and in vitro for use in research and clinical settings.

Keywords

In vitro transcription · Enzymatic ligation · Permuted intron–exon · circRNA vector

A. Greasley
Department of Pathology and Laboratory Medicine, Western University, London, ON, Canada

Matthew Mailing Centre for Translational Studies, London Health Sciences Centre, London, ON, Canada

S. Li
Department of Pathology and Laboratory Medicine, Western University, London, ON, Canada

Department of Cardiovascular Surgery, The Second Hospital of Jilin University, Changchun, China

K. Liu
Department of Cardiovascular Surgery, The Second Hospital of Jilin University, Changchun, China

X. Zheng (✉)
Department of Pathology and Laboratory Medicine, Western University, London, ON, Canada

Matthew Mailing Centre for Translational Studies, London Health Sciences Centre, London, ON, Canada

Department of Surgery, Western University, London, ON, Canada

Department of Oncology, Western University, London, ON, Canada

Department of Microbiology &Immunology, Western University, London, ON, Canada

London Health Sciences Centre Research Institute, London, ON, Canada
e-mail: xzheng26@uwo.ca

J. Xiao (ed.), *Advances in Circular RNAs*, Advances in Experimental Medicine and Biology 1485,
https://doi.org/10.1007/978-981-96-9428-0_16

The generation of circular RNA (circRNA) is not a new or novel concept. Protocols dating back into the 1990s highlight the generation of structurally circular RNA for use as probes or tools in research application. Since 2010, there has been an increased interest in circRNA as a functional RNA molecule and a regulator of biological processes. With this comes a renewed interest in generating large, pure, and functional amounts of circRNA to be used as a therapeutic and clinical tool to treat disease. Currently, the focus to generate circRNA relies on three main methods: (1) in vitro chemical or enzymatical reaction, (2) vector-based expression systems, (3) self-splicing ribozymes. Throughout this chapter, we will cover methods, challenges, and the advantages and disadvantages of each to generate circRNA ranging both in vitro and in vivo.

1 Chemical and Enzymatic Circular RNA Generation

The first step in generating circRNA chemically or enzymatically is to generate a suitable precursor linear RNA to be used for circularization. Both methods generate a single-stranded linear RNA to be used for circularization, where the main differences are in product size, functional ends, and preprocessing required prior to circularization.

1.1 Chemical Synthesis of Precursor RNA

Chemical synthesis of RNA with an automated synthesizer allows the direct assembly of custom precursor RNA chains from phosphoramidite-building blocks to be used for circularization immediately (Fig. 16.1a) [1]. The use of an automated synthesizer allows for high accuracy of the target chain along with incorporation of any custom nucleotides desired. Custom nucleotides are often used to enhance stability, fluorescently tag the product, or mutate specific sites to enhance or diminish effects. In addition to high accuracy and flexibility, chemical synthesis also provides an easier platform to scale up yield for large-scale preparations. However, chemical synthesis of RNA fragments is severely limited by size of the strands that can be assembled. Typically, chemical synthesis is ideal for oligomers up to approximately ~50-nucleotides (nt) long. After this, yields diminish greatly, with ~100 nt being the upper acceptable limit. Protocols to join RNA fragments together in tandem are available. However, this merely yields RNA fragments up to 200 nt in length because joining more than two fragments in the proper orientation would severely limit yield. Many circRNA have lengths beyond 200 nt; therefore, making chemical synthesis is a relatively limited option for generation of precursor linear RNA.

1.2 T7 Polymerase Synthesis of Precursor RNA

In vitro transcription using bacteriophage RNA polymerases is a common method for producing RNA strands. RNA polymerases derived from T7, SP6, and T3 can be used, where T7 is the most common. T7 RNA polymerase utilizes a dsDNA template by binding to a promoter region at the 5′ end, using nucleotide triphosphates (NTPs) as building blocks and terminating when the enzyme reaches the 3′ end (Fig. 16.1b). The promoter region of the T7 prefers multiple consecutive guanosine residues at the transcription start site (TSS), which usually limits the first few nucleotides of the transcription to “GGG” [2]. This “GGG” inclusion of the T7 promoter severely limits the sequence specificity of T7 transcribed RNA for most circRNA, unless the sequence is rearranged to a portion that natively contains “GGG.” It is still possible to use only one guanosine residue at the +1 relative to the TSS to maintain improved sequence specificity, but this may severely impact yield.

Unlike chemical synthesis, T7-synthesized RNA can theoretically be any length. The major limitation of length is enzyme slipping off the dsDNA template resulting in short abrupt

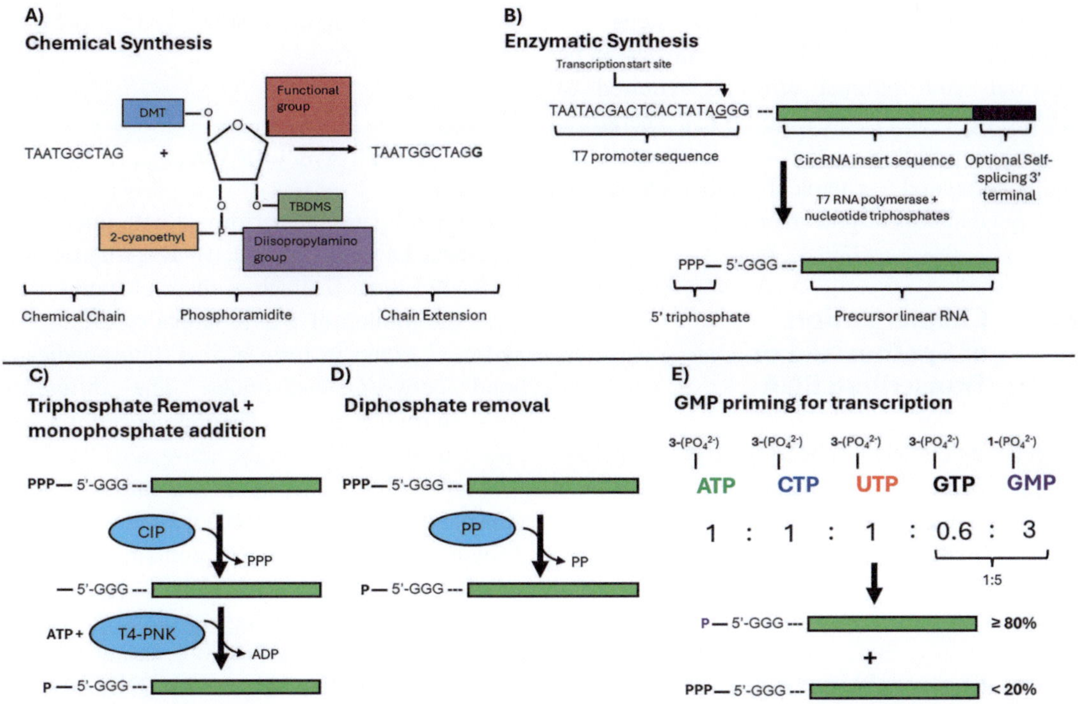

Fig. 16.1 Overview of precursor RNA synthesis using (**a**) chemical or (**b**) enzymatic techniques as well as post-synthesis processing for 5′ triphosphate modifications of enzymatically generated RNA, (**c**) calf alkaline phosphatase (CIP) treatment for triphosphate removal followed by T4 polynucleotide kinase (T4-PNK) treatment to add a monophosphate, (**d**) removal of a diphosphate using *E. coli* pyrophosphatase (PP), (**e**) GMP priming utilizing a 5:1 ratio of GMP:GTP during T7 transcription to generate predominantly monophosphate 5′ ends

transcripts [3]. In addition to enzyme slippage, T7 RNA polymerase is prone to adding —one to two nucleotides at the end which are not encoded in the DNA template [3, 4]. Although short abortive transcripts can be easily separated from the desired length using size exclusion such as polyacrylamide gel electrophoresis [5, 6] or high-performance liquid chromatography [7], separating —one to two nucleotide differences is more laborious and challenging. Two strategies exist to combat additional nucleotides at the 3′ end: (1) to include the hepatitis delta virus ribozyme sequence on the 3′ end of the dsDNA template which allows for precise RNA self-cleavage [8] and (2) addition of 2′-O-methyl at the 3′ end of the PCR template [4]. Both of these allow accurate 3′ termination of transcription.

In addition to 3′ considerations with in vitro-transcribed RNA, transcribed RNA requires 5′ modifications or additional steps to prepare the precursor strand for circularization. In vitro transcription utilizes NTPs resulting in a 5′ guanosine triphosphate (GTP), where joining of the 5′–3′ ends for circularization requires a guanosine monophosphate (GMP) to form proper RNA structure. There are a few strategies to prepare the precursor for circularization. Firstly, the triphosphates can be removed with a phosphatase followed by phosphorylation using ATP and T4 polynucleotide kinase, to introduce a monophosphate 5′ end (Fig. 16.1c). This two-step process can be further simplified using RNA pyrophosphatase from *Escherichia coli* to cleave the beta and gamma phosphates (Fig. 16.1d) [9, 10]. Lastly, GMP can be included in the in vitrotranscription mixture at a higher molar ratio than GTP [11]. This will leave products with predominantly GMP 5′ ends (Fig. 16.1e). Although the latter sacrifices some yield, it simplifies the process by removing additional steps prior to circularization.

Although more postprocessing steps are required, a yield of milligrams can easily be achieved, and minimal effort is required to achieve high purity and scale up [3]. Therefore, in vitro transcription is considered a more viable option compared to chemical synthesis to generate large amounts of precursor RNA.

2 Circularization of Synthesized or Transcribed RNA

To generate circRNA following precursor preparation, chemical or enzymatic methods are both options for the joining of the 5′ monophosphate to the 3′ hydroxyl group to create a continuous RNA structure.

2.1 Chemical Ligation

Chemical ligation protocols for joining RNA have been established since 1988. Chemical ligation is performed using cyanogen bromide (BrCN) or 1-ethyl-3-(3′dimethylaminopropyl) carbodiimide which act as electrophiles to catalyze phosphodiester bond formation (Fig. 16.2a) [12]. However, the use of chemically catalyzed nucleophilic attacks is not selective as to which hydroxyl group is attacked during ligation and can suffer from improper backbone formation. Instead of the standard 5′–3′ phosphodiester bond, a 5′–2′ bond can also be formed as a byproduct. In addition to the improper linkage, chemical ligation does not favor intermolecular (between strands) or intramolecular (within a strand) reactions. Since circularization of RNA to bring the loose ends into proximity to catalyze intramolecular reactions is less favorable, this would lead to low efficiency of circularization. Therefore, this method would be required to be coupled with conditions which favor RNA circularization prior to ligation (discussed later) [8, 13–15]. Together, the low efficiency, safety concerns due to the toxicity of the catalysts, and improper linkage have led researchers to heavily favor enzymatic ligation using T4 bacteriophage–derived enzymes.

2.2 Enzymatic Ligation

Enzymatic ligation is performed with three main enzymes: T4 DNA ligase I, T4 RNA ligase, and T4 RNA Ligase II. Each of these enzymes catalyzes the joining of a 5′ monophosphate to a 3′ hydroxyl group to form a 5′–3′ phosphodiester bond between nucleotides. The difference between each enzyme is the preference for a specific type of template in the active site.

T4 DNA ligase catalyzes the formation of a phosphodiester bond between a 5′ phosphate and 3′-hydroxyl group in a DNA containing double-stranded template (Fig. 16.2b) [16]. The enzyme requires complete homology at the active site to form this bond, and at least one strand must be DNA. To utilize this enzyme for RNA circularization and ligation, a ~10–20-nt DNA oligomer "splint" which is complementary to each end of the RNA strand is used [13, 17]. The splint accomplishes two tasks: (1) it allows for the ends of the RNA molecule to come into proximity, and (2) it creates a double-stranded template for the T4 DNA ligase to act on.

In contrast, T4 RNA ligase 1 catalyzes the formation of a phosphodiester bond between a 5′ phosphate and 3′-hydroxyl group of a single-stranded RNA or DNA template (Fig. 16.2b) [18]. Although no splint is required to catalyze intramolecular bonds, circularization efficiency is low due to favored intermolecular bonds and chain extension. Therefore, a splint is commonly used to improve circularization efficiency. However, since T4 RNA ligase 1 requires a single-stranded template, the splint must have a ~2–6-nt mismatch at the ligation site to allow for free ends and enzyme access. An alternative to using a splint is the addition of a complement-reverse complete motif: an inverted sequence repeat found at each end of the strand, that allows for strand folding and orientation of the free ends in proximity prior to ligation [10, 13, 14]. This was previously used to efficiently ligate circRNA

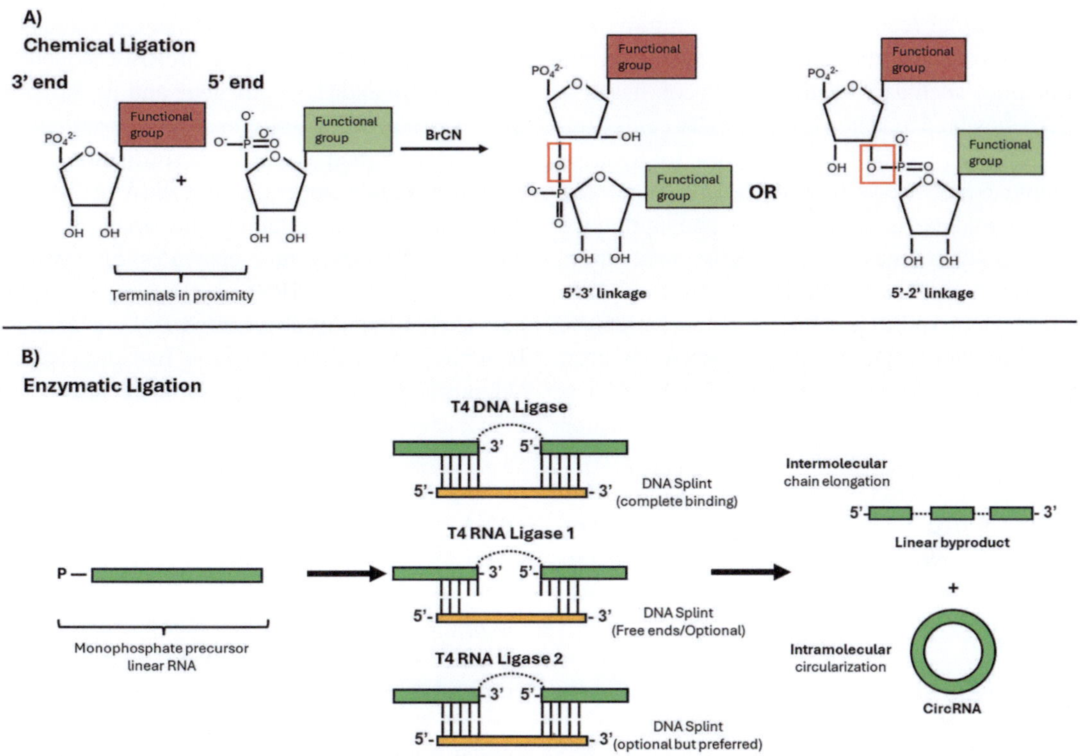

Fig. 16.2 Methods for ligating precursor RNA to generate circRNA. (**a**) Chemical ligation utilizes BrCN to catalyze the formation of a 5′–3′ phosphodiester bond and its unintended byproduct of a 5′–2′ linkage. (**b**) Enzymatic ligation with different ligases and the requirements for the presence and design of the DNA splint to generate circRNA

up to 4 kb in length [10]. However, if sequence specificity of the circRNA is crucial, a splint is preferred as inverted repeats may introduce addition base pairs at the splice site.

T4 RNA ligase 2 has also been utilized to generate circRNA. T4 RNA ligase 2 is unique as it catalyzes 5′ phosphate and 3′ hydroxyl bond formation of a double or single-stranded RNA template but has a preference for double-stranded templates (Fig. 16.2b) [13]. Splints have been used to assist in the formation of circRNA from T4 RNA ligase 2, but interestingly, in a more recent study, absence of a splint did not change the circularization efficiency [17]. A recent strategy uses mutations at the 3′ end to ensure that the 3′ terminus falls into a double-stranded region in proximity to the 5′ beginning through manipulation RNA secondary structure formation of the precursor. This allows the generation of circRNA with high yields (>80%) without the additional steps of splint addition or degradation [19]. However, this technique has not been tested with circRNA of significant size, and the translatability to a generalized circRNA synthesis protocol is yet to be seen.

2.3 Methods to Increase Circularization Efficiency

Regardless of ligation method, a major limitation remains the efficiency of circularization and isolation of purified circRNA. Methods and parameters have been suggested to increase the circularization rate and promote intramolecular ligation [8, 13–15, 17]. The most common is the utilization of a DNA splint as mentioned above. Lower reaction temperatures (~16 °C) have been reported to increase circularization efficiency in long overnight incubations and reduce intermolecular reactions by favoring intramolecular bonds. The addition of long-chained carbon structures, such as polyethylene glycol, has also been suggested to increase the circular conformation of the precursor strands prior to ligation, leading to increased efficiency. Ideally, all parameters would be used in tandem and optimized individually to maximize circRNA generation. Regardless, typical circularization reactions containing DNA splints and have efficiencies <50%; therefore, there is DNA and linear RNA that must be removed. The removal of the DNA splint is accomplished by treating the ligation reaction with DNase I. Linear RNA can be selectively degraded with RNase R, a ribonuclease that requires a free 3′ terminal for degradation, leaving circRNA intact. Following this, intact circRNA can be purified and validated to confirm generation of pure circRNA.

3 In Vivo CircRNA Generation

Although generating pure circRNA in a test tube to be used for cell treatment and in vivo application appears relatively simple, the lack of yield, consistency, and complications has led researchers to favor direct in vivo generation as the primary method for pathological and animal studies. To accomplish this, researchers have used the considerable progress made to understand circRNA biogenesis and engineered vectors to produce circRNA of interest. A variety of these vectors exist, which utilize different mechanisms of back-splicing such as inverted repeats or RNA-binding protein motifs, and are readily available for research use.

3.1 Vectors Utilizing Endogenous Sequences

Vectors for circRNA overexpression have been available for the last decade. The earliest of these is from the *ZKSCAN1* and *Laccase2* genes [20, 21]. These vectors often rely on Alu repetitive elements that exist in the intronic flanking regions to generate circRNA using the cell machinery. The premise behind these vectors is to utilize a gene with naturally high circRNA expression and circularization efficiency. The circRNA sequence, usually from exons, of this gene and its flanking introns are then cloned into a vector, and small deletions of 5′ and 3′ end are performed until the resulting vector generates circRNA with high efficiency while maintaining the smallest possible size. Together, this generates an intronic repetitive sequence (IRS)-circRNA vector capable of circularizing an inserted RNA sequence. However, these vectors exhibit two drawbacks: (1) preference for size of circRNA insert, (2) inconsistent back-splicing efficiency. *ZKSCAN1* vector was reduced to ~50-nt flanking regions leading to an inability to circularize larger circRNA, such as the *Hipk3* insert which is 1099-nt long [20]. To circumvent this, the *Laccase2* IRS-circRNA which contains longer IRS regions can circularize large inserts, but cannot reliably circularize sequences <300 nt [21]. Both IRS-circRNA vectors appear to rely on a relationship between the size of the insert and IRS, which is crucial for back-splicing efficiency. As a result, these IRS-circRNA vectors do not share robustness for the generation of any circRNA of interest.

More recently, steps were taken to specifically engineer a vector capable of robust and high circRNA overexpression utilizing intronic repeats from actin. Li et al. found that by using 100-nt flanking regions from the actin gene, it is possible to generate a robust vector capable of generating circRNA ranging from 138 to 1345 nt in length with high efficiency and low amounts of linear precursor RNA remaining [22]. This novel vector is a significant advantage for studying the effects of circRNA in pathology and allows for better understanding of their roles in disease without off-target effects.

3.2 Vectors Utilizing Engineered Sequences

In addition to using endogenous sequences, researchers have used elements of back-splicing to enhance vector-based systems. The RNA-binding protein quaking (QKI) has been shown to be able to pull intronic regions into proximity to allow for back-splicing to occur [23]. The QKI binding motif has successfully been used to circularize previously linear-only transcripts as well as enhance circularization of viral circRNA [24]. In theory, the QKI motif (AGTAACN$_{1\text{-}20}$TAAG) [23] can be applied to most vector-based systems to enhance efficiency. However, this has not been tested, and optimization regarding input location and compatibility with each system would be required.

Commercially available circRNA-generating vectors are also available as a source of in vivo circRNA production. Although the sequence of these is often proprietary, it is likely they use engineered inverted repeats with or without RNA-binding protein motifs to allow for an RNA secondary structure to bring the inserted sequence into proximity for splicing. From a research perspective, these constructs are convenient as companies often guarantee a level of expression, and vectors can be prepackaged into viral delivery systems for ease of use.

In general, all vector-based systems (Fig. 16.3) are effective at generating circRNA in their target cells, but challenges regarding their efficiency and off-target effects due to uncircularized precursor remain prevalent.

4 Self-Splicing Ribozymes for In Vitro and In Vivo Generation

The last class of molecules for circRNA synthesis we will discuss is ribozymes. Ribozymes are RNA sequences that can perform catalytic activity like proteins. They are typically classified into two categories: group I and group II introns, which can be found naturally throughout RNA in bacteria and yeast. Traditionally, these introns would self-splice themselves out resulting in linear exon–exon transcripts by using magnesium and a free guanosine residue. In 1992, it was discovered that moving the 5′ intron to the tail of the linear precursor and the 3′ intron to the 5′ lead, you could create a circular construct of interest. Thus, the permutated intron–exon (PIE) method was born (Fig. 16.4). Both group I and II introns have been successfully used to generate circRNA. The tRNA^{Leu} gene from *Anabaena* and the thymidylate synthase (*td*) gene from T4 bacteriophages have become staple sources as group I intron expression systems [25, 26], while the yeast mitochondrial gene al5γ is commonly used as a group II intronic template [27].

One downfall of group I introns is the need to maintain original exon boundaries, meaning that a portion of the original sequence must be maintained at the junction for splicing to occur (Fig. 16.4) [25]. However, this was overcome previously by selectively rearranging the precursor circRNA sequence to share similarity with the native intronic boundaries to allow for the same circularization efficiency [28]. The authors have also generated a program that can automatically score and predict successful circularization based on the linear precursor to simplify this work based on their findings.

Group II introns have the advantage of complete replacement at the exon site with the sequence of interest because the native exon boundaries are not required for circularization

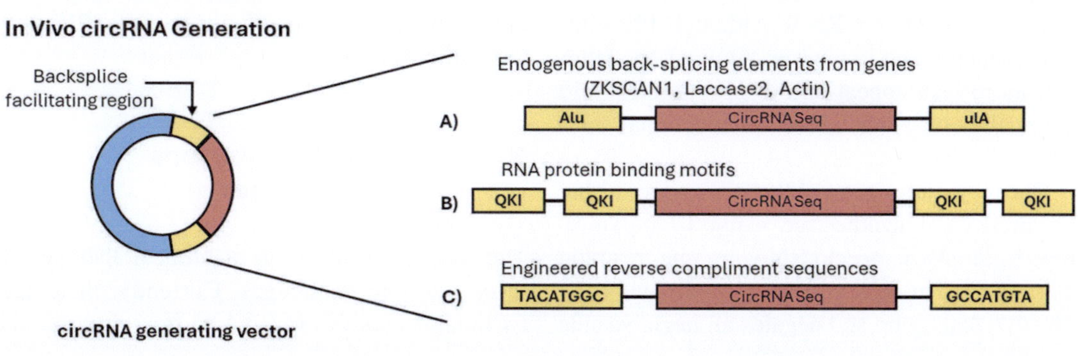

Fig. 16.3 Vector-based circRNA generating systems utilizing endogenous or engineered sequences. (**a**) Alu repeats, (**b**) RNA-binding protein, (**c**) engineered reverse repeats

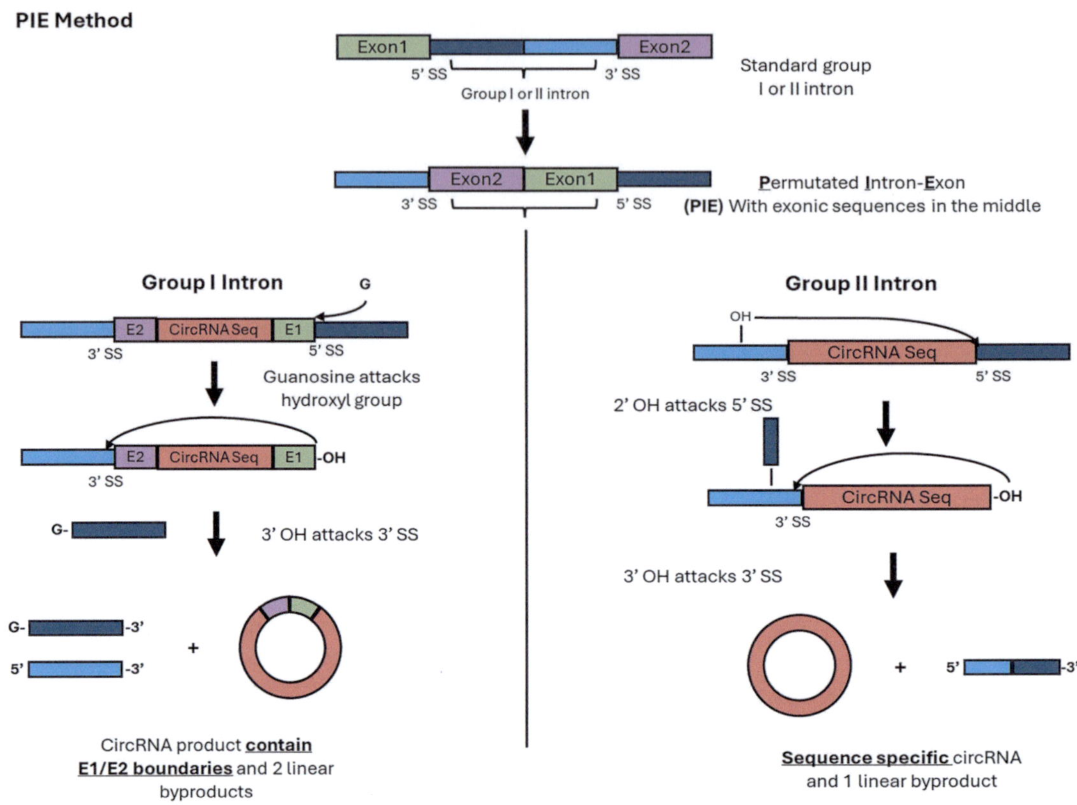

Fig. 16.4 Permutated intron–exon (PIE)-based circRNA generation. Group I intron-derived circRNA, which contains exon1/2 boundaries needed for splicing, compared to group II intron usage, which can have complete circRNA sequence specificity

(Fig. 16.4). More recently, Chen et al. reported a novel permutated group II intron and a framework that can efficiently produce pure circRNA in vitro which can easily be purified for downstream applications [29]. This framework allows for any researcher to make modifications to specific regions of the intronic boundaries in the linear precursor based on the sequence of their insert to generate circRNA. Although PIE constructs can be used in vitro and in vivo, the latter appears to lack appeal due to low efficiency in mammalian cells when used for functional studies [7, 22, 30].

One major advantage of the PIE system is that the circRNA is synthesized using T7 polymerases from DNA → circRNA in one reaction (Fig. 16.5). This allows for a simple and cost-effective scaling up and negates all negative side effects of the T7 polymerase reaction that affects chemical and enzymatic ligation. Since the circRNA precursor sequence is in the center of the transcript, the additional guanine residues at the 5′ end and addition residues at the 3′ end are negated by the targeted self-splicing. In addition, the lack of multiple processing steps prevents loss of yield due to handling, and instead, yield is only influenced by the efficiency of the PIE system utilized.

5 Clinical Applications and Challenges

The use of circRNA as a clinical therapeutic faces multiple challenges. Currently, there are two major uses of circRNA to treat disease: (1)

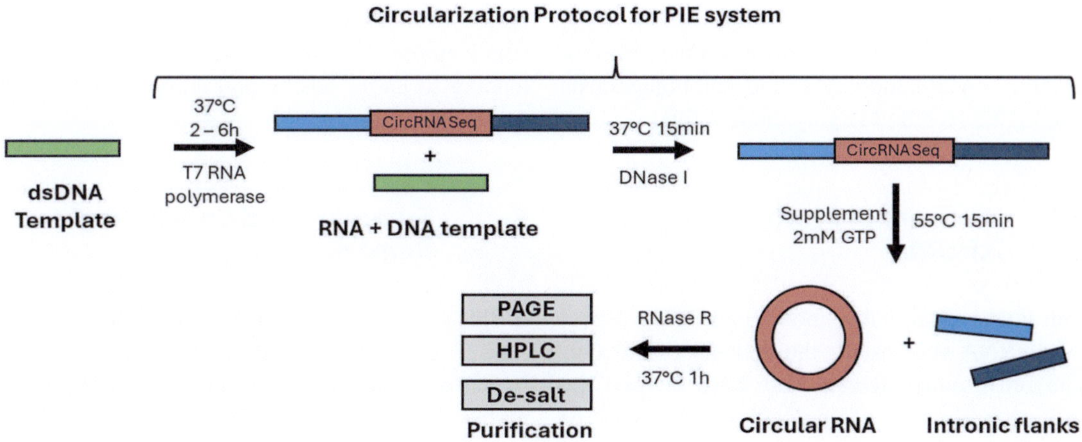

Fig. 16.5 Standard workflow of the PIE system. Highlighting simple reaction flow that allows generation and purification in one reaction tube

the use of circRNA as a therapeutic itself in its native role as miRNA and protein regulators and (2) the use of circRNA for stable protein expression [7, 29, 31], both requiring large amounts of pure and selective circRNA. To use circRNA as a protein expression system, a peptide sequence containing an internal ribosome entry site (IRES) is used in place of a circRNA sequence. This allows the ribosome to enter at the IRES and lead to concurrent rounds of translation as opposed to linear RNA which generates one peptide for each bound ribosome [11]. In addition to more efficient translation, circRNA poses lower immunogenicity than linear RNA [7]; however, it does not abolish it, and delivery of the circRNA is still required.

Major progress has been made to generate and utilize circRNA in a clinical setting since early attempts. However, there are still many challenges to overcome.

5.1 Purity of Synthesized CircRNA

The purity of the synthesized circRNA is an important aspect for use in clinical setting. Chemical and enzymatic ligations share low yields (<50%), resulting in large amounts of linear precursor that must be removed through digestion and purification. The PIE method results in higher efficiency in vitro and therefore requires less clean up. However, regardless of efficiency, there are linear fragments remaining from the precursor that must be removed even if a 100% circularization rate was achieved. Therefore, regardless of the methodology, a purification step is required prior to circRNA use, albeit the in vitro PIE method appears to be the least demanding with the highest yield returns.

5.2 Scalability

For circRNA synthesis to be a viable therapeutic strategy, the methodology of generation must be scalable to make synthesis cost-effective. Enzymatic synthesis using T7 RNA polymerase is a simple and cost-effective strategy to produce milligrams of RNA and is scalable. In vitro transcription can also be used to generate template for all ligation models, including the PIE method, assuming steps are taken to ensure accuracy of transcribed RNA.

It is also theoretically possible to scale up an in vivo-synthesized circRNA using bacteria or yeast systems, although this has not been tested. In this case, an IRS or PIE system would be utilized in a vector for synthesis and auto-splicing within a carrier cell to make large amounts of circRNA, similar to the production of insulin protein. However, the purification of the target circRNA would require specific probes, leading to reduced yield, increased time, and higher costs making this method largely unattractive.

5.3 Delivery of Synthesized circRNA

The final and arguably most challenging aspect of circRNA synthesis is the delivery of the synthesized product. Researchers have favored the in vivo generation of circRNA using a viral or transfection delivery model for animal and cell studies respectively. Viral therapy has been used effectively to target cancer using oncolytic viruses which could be used to deliver circRNA into target cells [32]. However, this is limited to circRNA which can affect cancer cells directly or the use of circRNA as a protein expression system in tumor cells. For circRNA to be effective as a robust therapeutic tool, there needs to be the ability to deliver circRNA synthesized from different methods to target many cell types. Currently, lipid coated nanoparticles are a robust and effective mechanism to deliver circRNA to different cell types [33]. Lipid nanoparticles have been utilized effectively in a streamlined workflow to deliver circRNA to target cells following in vitro PIE circRNA synthesis as mentioned above [29]. This represents potential for the utilization of these techniques in tandem. However, a major limitation of lipid nanoparticles is the small amount of circRNA that they can carry, leading to larger dosing and consequently more expensive treatment.

6 Summary and Concluding Remarks

In this chapter, we summarized different methods of generating circRNA both in vivo and in vitro from linear precursor generation to mature circRNA (Table 16.1). We highlighted advantages and disadvantages of each method along with clinical challenges that circRNA synthesis faces. Although strategies exist to improve synthesis, no solution to overcome all challenges is evident. It is clear that the synthesis of circRNA has made tremendous progress in the last decade, but no single technique is yet to be proven the most effective. It is likely that an approach utilizing a combination of techniques and synthesis steps will be required to generate clinical grade circRNA with high yield, efficiency, and specificity. Between this and other clinical challenges, more research is needed to test a circRNA synthesis and treatment workflow to determine the feasibility of circRNA as a therapeutic tool. Furthermore, if the reader wishes to continue with this topic, many in-depth and excellent reviews on circRNA synthesis can be found [8, 13–15, 17].

Table 16.1 Summarization of circRNA synthesis methods

Method	Advantages	Disadvantages
Chemical synthesis	High purity and customizability	Size limitations and costly with low yield to increase chain length
Enzymatic synthesis	Scalable and simple procedure	Lower accuracy, additional steps for purification and isolation of transcripts of interest
Chemical ligation	Simple and robust for all templates	Low efficiency and improper structural byproducts. Cytotoxic concerns
Enzymatic ligation	Better specificity for circRNA depending on conditions	Reasonable efficiency requires addition of DNA splints. Purification is more laborious
In vivo generation	Large amounts of circRNA generation from one backbone	Difficult to make robust circRNA vector. Gene therapy with vectors is challenging to deliver into cells. Limited control on dosage of circRNA produced
PIE method	In vitro and in vivo use. Scalable and high efficiency and purity	Limited in vivo use due to immunogenicity and off target effects

References

1. Ryczek M, Pluta M, Błaszczyk L, Kiliszek A (2022) Overview of methods for large-scale RNA synthesis. Appl Sci 12(3):1543. https://doi.org/10.3390/APP12031543/S1
2. Rong M, Biao H, Mcallister WT, Durbin RK (1998) Promoter specificity determinants of T7 RNA polymerase. Proc Natl Acad Sci USA 95(2):515–519. https://doi.org/10.1073/pnas.95.2.515
3. Milligan JF, Groebe DR, Witherell GW, Uhlenbeck OC (1987) Oligoribonucleotide synthesis using T7 RNA polymerase and synthetic DNA templates. Nucleic Acids Res 15(21):8783–8798. https://doi.org/10.1093/nar/15.21.8783
4. Kao C, Zheng M, Rüdisser S (1999) A simple and efficient method to reduce nontemplated nucleotide addition at the 3′ terminus of RNAs transcribed by T7 RNA polymerase. RNA 5(9):1268–1272. https://doi.org/10.1017/S1355838299991033
5. Breuer J, Rossbach O (2020) Production and purification of artificial circular RNA sponges for application in molecular biology and medicine. Methods Protoc 3(2):42. https://doi.org/10.3390/mps3020042
6. Petrov A, Wu T, Puglisi EV, Puglisi JD (2013) RNA purification by preparative polyacrylamide gel electrophoresis. Methods Enzymol 530:315–330. https://doi.org/10.1016/B978-0-12-420037-1.00017-8
7. Wesselhoeft RA, Kowalski PS, Parker-Hale FC, Huang Y, Bisaria N, Anderson DG (2019) RNA circularization diminishes immunogenicity and can extend translation duration in vivo. Mol Cell 74(3):508–520.e4. https://doi.org/10.1016/j.molcel.2019.02.015
8. Obi P, Chen YG (2021) The design and synthesis of circular RNAs. Methods 196:85–103. https://doi.org/10.1016/J.YMETH.2021.02.020
9. Deana A, Celesnik H, Belasco JG (2008) The bacterial enzyme RppH triggers messenger RNA degradation by 5′ pyrophosphate removal. Nature 451(7176):355–358. https://doi.org/10.1038/nature06475
10. Carmona EM (2019) Circular RNA: design criteria for optimal therapeutical utility. Doctoral dissertation, Harvard University, Graduate School of Arts & Sciences
11. Chen C, Sarnow P (1995) Initiation of protein synthesis by the eukaryotic translational apparatus on circular RNAs. Science 268(5209):415–417. https://doi.org/10.1126/SCIENCE.7536344
12. Dolinnaya NG, Sokolova NI, Ashirbekova DT, Shabarova ZA (1991) The use of BrCN for assembling modified DNA duplexes and DNA-RNA hybrids; comparison with water-soluble carbodiimide. Nucleic Acids Res 19(11):3067–3072. https://doi.org/10.1093/NAR/19.11.3067
13. Petkovic S, Müller S (2015) RNA circularization strategies in vivo and in vitro. Nucleic Acids Res 43(4):2454–2465. https://doi.org/10.1093/NAR/GKV045
14. Müller S, Appel B (2017) In vitro circularization of RNA. RNA Biol 14(8):1018–1027. https://doi.org/10.1080/15476286.2016.1239009
15. Chen X, Lu Y (2021) Circular RNA: biosynthesis in vitro. Front Bioeng Biotechnol 9:787881. https://doi.org/10.3389/FBIOE.2021.787881
16. Lohman GJS, Tabor S, Nichols NM (2011) DNA ligases. Curr Protoc Mol Biol 94(1):3.14.1–3.14.7. https://doi.org/10.1002/0471142727.MB0314S94
17. Abe N, Kodama A, Abe H (2018) Preparation of circular RNA in vitro. Methods Mol Biol 1724:181–192. https://doi.org/10.1007/978-1-4939-7562-4_15
18. Uhlenbeck OC, Gumport RI (1982) 2 T4 RNA ligase. Enzyme 15:31–58. https://doi.org/10.1016/S1874-6047(08)60274-7
19. Chen H, Cheng K, Liu X, An R, Komiyama M, Liang X (2020) Preferential production of RNA rings by T4 RNA ligase 2 without any splint through rational design of precursor strand. Nucleic Acids Res 48(9):e54–e54. https://doi.org/10.1093/NAR/GKAA181
20. Liang D, Wilusz JE (2014) Short intronic repeat sequences facilitate circular RNA production. Genes Dev 28(20):2233–2247. https://doi.org/10.1101/gad.251926.114
21. Kramer MC, Liang D, Tatomer DC, Gold B, March ZM, Cherry S, Wilusz JE (2015) Combinatorial control of Drosophila circular RNA expression by intronic repeats, hnRNPs, and SR proteins. Genes Dev 29(20):2168–2182. https://doi.org/10.1101/gad.270421.115
22. Li F, Lyu J, Yang Y, Yang Q, Santos C, Yang BB (2023) An improved model for circular RNA overexpression: using the actin intron reveals high circularization efficiency. Adv Genet 4(3):2200019. https://doi.org/10.1002/GGN2.202200019
23. Conn SJ, Pillman KA, Toubia J, Conn VM, Salmanidis M, Phillips CA, Roslan S, Schreiber AW, Gregory PA, Goodall GJ (2015) The RNA binding protein quaking regulates formation of circRNAs. Cell 160(6):1125–1134. https://doi.org/10.1016/j.cell.2015.02.014
24. Zhao J, Lee EE, Kim J, Yang R, Chamseddin B, Ni C, Gusho E, Xie Y, Chiang CM, Buszczak M, Zhan X, Laimins L, Wang RC (2019) Transforming activity of an oncoprotein-encoding circular RNA from human papillomavirus. Nat Commun 10(1):2300. https://doi.org/10.1038/s41467-019-10246-5
25. Puttaraju M, Been M (1992) Group I permuted intron-exon (PIE) sequences self-splice to produce circular exons. Nucleic Acids Res 20(20):5357–5364. https://doi.org/10.1093/NAR/20.20.5357
26. Ford E, Ares M (1994) Synthesis of circular RNA in bacteria and yeast using RNA cyclase ribozymes derived from a group I intron of phage T4. Proc Natl Acad Sci 91(8):3117–3121. https://doi.org/10.1073/PNAS.91.8.3117
27. Mikheeva S, Hakim-Zargar M, Carlson D, Jarrell K (1997) Use of an engineered ribozyme to produce a circular human exon. Nucleic Acids Res

25(24):5085–5094. https://doi.org/10.1093/NAR/25.24.5085

28. Rausch JW, Heinz WF, Payea MJ, Sherpa C, Gorospe M, Le Grice SFJ (2021) Characterizing and circumventing sequence restrictions for synthesis of circular RNA in vitro. Nucleic Acids Res 49(6):e35–e35. https://doi.org/10.1093/NAR/GKAA1256
29. Chen C, Wei H, Zhang K, Li Z, Wei T, Tang C, Yang Y, Wang Z (2022) A flexible, efficient, and scalable platform to produce circular RNAs as new therapeutics. bioRxiv 1–34. https://doi.org/10.1101/2022.05.31.494115
30. Ho-Xuan H, GlaÅar P, Latini C, Heizler K, Haase J, Hett R, Anders M, Weichmann F, Bruckmann A, Van Den Berg D, Hüttelmaier S, Rajewsky N, Hackl C, Meister G (2020) Comprehensive analysis of translation from overexpressed circular RNAs reveals pervasive translation from linear transcripts. Nucleic Acids Res 48(18):10368–10382. https://doi.org/10.1093/NAR/GKAA704
31. Xie J, Ye F, Deng X, Tang Y, Liang J-Y, Huang X, Sun Y, Tang H, Lei J, Zheng S, Zou Y (2023) Circular RNA: a promising new star of vaccine. J Transl Intern Med 11(4):372–381. https://doi.org/10.2478/JTIM-2023-0122
32. Lundstrom K (2023) Viral vectors in gene therapy: where do we stand in 2023? Viruses 15(3):698. https://doi.org/10.3390/V15030698
33. Hamilton AG, Swingle KL, Mitchell MJ (2023) Biotechnology: overcoming biological barriers to nucleic acid delivery using lipid nanoparticles. PLoS Biol 21(4):e3002105. https://doi.org/10.1371/JOURNAL.PBIO.3002105

Part VI

Circular RNAs and Human Diseases

Circular RNAs in Heart Diseases 17

Tingting Yang, Michail Spanos, Wensi Wan, and Junjie Xiao

Abstract

Circular RNAs (circRNAs) are a class of RNA molecules generated through back-splicing of precursor mRNAs, characterized by a covalent 3′,5′-phosphodiester bond at the junction site. They are broadly categorized into exonic circRNAs, exonic–intronic circRNAs, and intronic circRNAs. CircRNAs perform diverse biological functions in various diseases, including acting as miRNA sponges, modulating protein synthesis via interactions with RNA-binding proteins (RBPs), directly influencing gene expression, and, in some cases, undergoing translation into proteins. In the cardiovascular system, a plethora of circRNAs has been identified, contributing to pivotal processes involved in cardiovascular development and disease regulation. This chapter will summarize the latest research advancements on the role of circRNAs in cardiovascular diseases, such as pathological cardiac hypertrophy, myocardial infarction, myocardial ischemia-reperfusion injury, dilated cardiomyopathy, and cardiac fibrosis. Additionally, we will explore the potential clinical applications of circRNAs as diagnostic and prognostic biomarkers in cardiovascular diseases.

Keywords

Circular RNAs · Cardiovascular diseases · Biomarkers

Authors "Tingting Yang" and "Michail Spanos" have equally contributed to this chapter.

T. Yang · W. Wan · J. Xiao (✉)
Cardiac Regeneration and Ageing Lab, Institute of Geriatrics (Shanghai University), Affiliated Nantong Hospital of Shanghai University (The Sixth People's Hospital of Nantong) and School of Life Science, Shanghai University, Nantong, China

Institute of Cardiovascular Sciences, Shanghai Engineering Research Center of Organ Repair, Joint International Research Laboratory of Biomaterials and Biotechnology in Organ Repair (Ministry of Education), School of Life Science, Shanghai University, Shanghai, China
e-mail: junjiexiao@shu.edu.cn

M. Spanos
Cardiovascular Division of the Massachusetts General Hospital and Harvard Medical School, Boston, MA, USA

Albert Einstein College of Medicine, Department of Internal Medicine, NCB, Bronx, NY, USA

J. Xiao (ed.), *Advances in Circular RNAs*, Advances in Experimental Medicine and Biology 1485,
https://doi.org/10.1007/978-981-96-9428-0_17

1 Introduction

Cardiovascular diseases (CVDs) encompass a spectrum of disorders affecting the circulatory system, including atherosclerosis (AS), cardiac hypertrophy, myocardial infarction (MI), and heart failure (HF). The prevailing hypothesis suggests that the etiology of CVDs is closely linked to a specific array of risk factors, such as hyperglycemia, dyslipidemia, hypertension, obesity, and inflammation [1]. While lifestyle modifications are recommended as a preventive strategy to mitigate risk, pharmacological interventions employing conventional medications such as renin-angiotensin blockers, lipid-regulating agents, β-adrenergic receptor antagonists, and antithrombotic agents have exhibited limited efficacy in halting the progression of AS. Moreover, procedures like percutaneous coronary intervention (PCI) and coronary artery bypass grafting (CABG) are considered highly invasive [2]. Despite substantial advancements in disease prevention, diagnostic biomarkers, and medical therapeutics, CVDs remain the leading cause of morbidity and mortality globally. Hence, there exists a critical need for a comprehensive understanding of the molecular pathogenesis of CVDs, alongside the identification of prospective biomarkers and exploration of novel therapeutic targets. Ongoing basic science research is focused on identifying therapeutic targets for these diseases, highlighting the urgent need for further investigation.

Circular RNAs (circRNAs) have emerged as promising biomarkers and therapeutic targets for CVDs. Originally identified by Sanger et al. in 1976, circRNAs are a distinct subtype of noncoding RNAs, predominantly localized within the cytoplasm, exhibiting remarkable stability within the human body [3–6]. These molecules are ubiquitously expressed across diverse organisms and demonstrate widespread expression and evolutionary conservation among eukaryotes, with particularly pronounced conservation in humans and mice [7, 8]. CircRNAs play pivotal regulatory roles in modulating gene expression at both the transcriptional and posttranscriptional levels, interacting with microRNAs (miRNAs), messenger RNAs (mRNAs), and RNA-binding proteins (RBPs). Recent research has emphasized the notable influence of circRNAs on the development of numerous cardiovascular conditions, including AS, coronary artery disease, MI, myocardial hypertrophy, HF, dilated cardiomyopathy, and arrhythmias [9–12]. This chapter offers an in-depth exploration of the current understanding of circRNAs in the cardiovascular system, covering their biological origins, functional properties, and key roles in CVDs. It also investigates the molecular mechanisms that underlie their involvement in the development and progression of CVDs while examining the potential clinical applications of circRNAs as biomarkers and therapeutic targets, along with the associated risks and challenges.

2 Biogenesis of Circular RNAs

Initially regarded as byproducts of abnormal splicing with minimal functional relevance, circRNAs have since emerged as ubiquitous and abundantly expressed molecules across various eukaryotic organisms, particularly in mammals, where they orchestrate numerous biological processes [5, 13–15]. The biogenesis of circRNA deviates from the conventional splicing process of linear RNA, involving the formation of 3′, 5′-phosphodiester bonds at the splice junctions, which results from the reverse splicing process of precursor mRNA (pre-mRNA) [16–20]. CircRNA can be categorized into three distinct types: exonic circular RNA (exonic circRNA or EcircRNA), exon–intron circular RNA (exon–intron circRNA or EIciRNA), and circular intronic RNA (ciRNA). EcircRNAs are generated during cotranscriptional splicing of linear mRNA, while EIciRNAs retain intronic sequences within their circular structure, and ciRNAs are derived from intronic fragments generated during pre-mRNA splicing [21, 22].

Two primary mechanisms have been proposed for the formation of circRNAs: exon-skipping-driven circularization and intron-pairing-driven circularization (Fig. 17.1). Exon-skipping-driven circularization involves the canonical splicing and exon skipping of mRNA precursors, yielding linear RNAs with multiple exons that contain

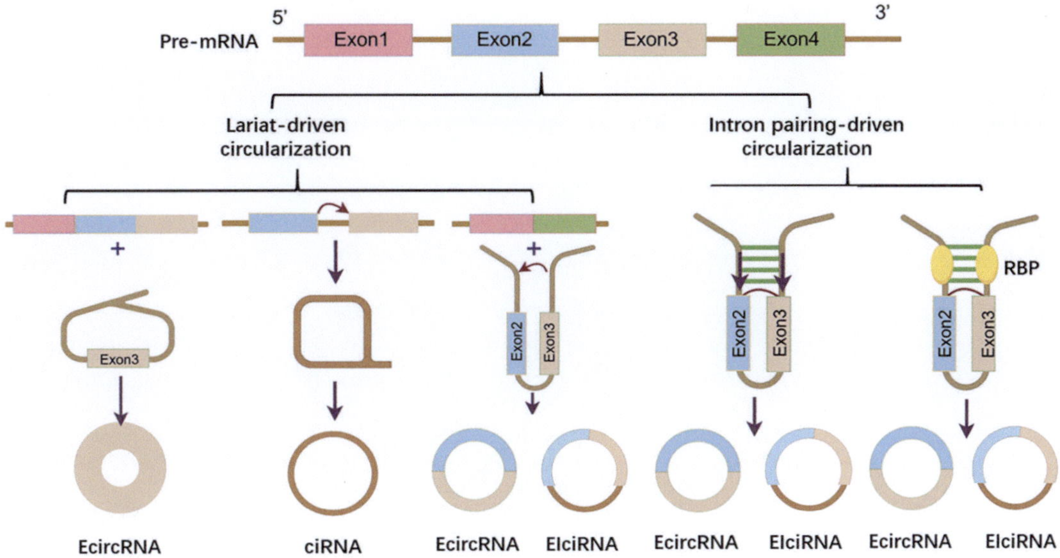

Fig. 17.1 Circular RNA formation mechanism (By Figdraw, ID: RIRIYcb8cc)

intronic lariats. Reverse splicing, in which the downstream 3′ splice acceptor of the intronic lariat pairs with the upstream 5′ splice donor, excises the intron and generates a circular RNA [23]. Intron-pairing-driven circularization operates through two mechanisms. First, complementary sequences within the introns' flanking exons facilitate the pairing of splice acceptors and donors into close proximity, facilitating exon circularization following intron excision [18]. Short interspersed nuclear elements (SINE), such as Alu elements, are known to promote circRNA circularization and are frequently used as circularization elements in exogenous circRNA expression systems [24]. Second, RBPs facilitate circularization by binding to intronic sequences flanking exons, which harbor RBP-binding sites [25, 26]. RBP-induced dimerization brings splice sites closer together, thereby promoting circularization [27]. Both exon-skipping and intron-pairing mechanisms can produce EcircRNAs and EIciRNAs. The formation of ciRNA differs from that of EcircRNA. Typically, intronic lariats generated during canonical splicing are rapidly degraded. However, when the 5′ splice site contains a 7-nt GU-rich sequence and the 3′ end harbors an 11-nt C-rich sequence, the intronic lariat is protected from degradation, leading to ciRNA formation [28]. Additionally, tRNA and viral RNAs can also undergo circularization to generate circRNA [29]. Extensive research has emphasized the involvement of circRNAs in a wide array of human diseases, including cancer, diabetes, neurological disorders, and cardiovascular conditions [25, 30–35].

3 Circular RNAs: Functional Mechanisms

CircRNAs, characterized by their abundance and evolutionary conservation, are implicated in several potential functions [36, 37]. Demonstrating a spectrum of autonomous mechanisms, they act as miRNA sponges, regulate protein synthesis through interactions with RBPs, directly influence gene expression, and, in some cases, undergo translation into proteins [38–41] (Fig. 17.2).

3.1 Circular RNAs Act as miRNA Sponges

One of the most extensively studied functions of circRNA is their role as miRNA sponges. miRNAs are a critical category of noncoding small RNAs that can bind to the 3′-untranslated region

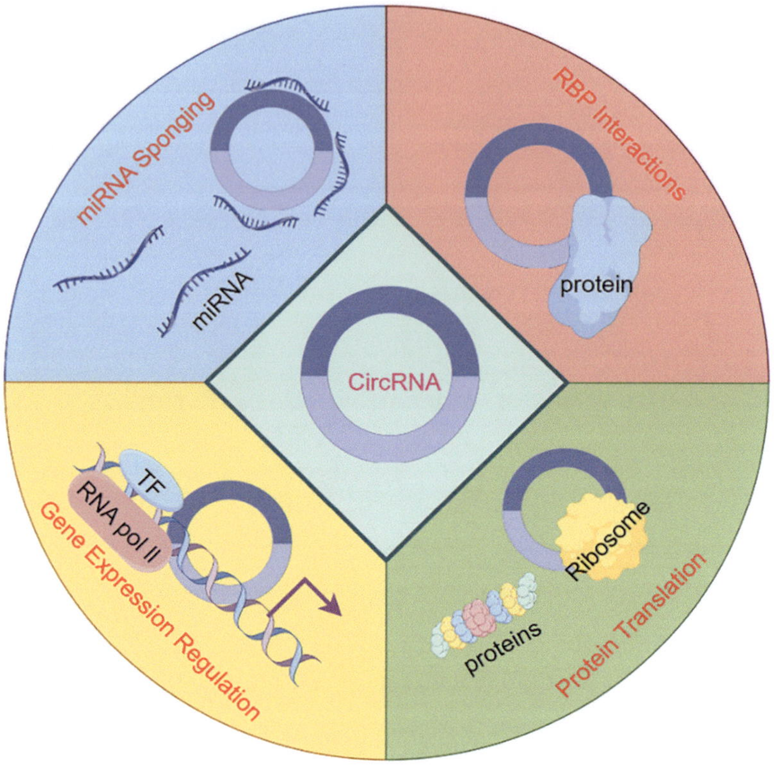

Fig. 17.2 Circular RNA function mechanism (By Figdraw, ID: AWYSIc47e4)

(3′UTR) of mRNAs, regulating gene expression, through protein degradation and translation inhibition [42]. Studies have shown that circRNA can bind to miRNAs and competitively inhibit their interaction with target mRNAs, thus functioning as miRNA sponges and indirectly regulating gene expression [43, 44]. By sequestering miRNAs, circRNAs deplete the pool of available miRNAs, leading to increased expression of miRNA-targeted mRNA. This increase allows the mRNAs to undergo ribosome-mediated translation, facilitating protein synthesis [45].

3.2 Circular RNAs and Regulation of Protein Synthesis Through RNA-Binding Protein Interactions

CircRNAs interact with RBPs via distinct binding motifs. Although circRNAs possess relatively fewer binding sites for proteins, their unique tertiary structures confer a strong and flexible binding capacity [39]. For instance, circAMOTL1 specifically interacts with the signal transducer and activator of transcription 3 (STAT3), promoting STAT3 expression and facilitating its nuclear translocation [46]. Additionally, circMbl, derived from the same pre-mRNA as the muscleblind-like splicing regulator 1 (MBL/MBNL1) protein, sequesters the MBL protein, influencing various physiological processes [21]. CircFoxo3 can bind with multiple RBPs, such as hypoxia-inducible factor and focal adhesion kinase, affecting cellular stress and aging [47]. Furthermore, circFoxo3 inhibits cell cycle progression by interacting with cyclin-dependent kinase 2 (CDK2) and P21, causing cell cycle arrest in the G1/S phase [48, 49]. Thus, circRNAs have the capacity to modulate the functions of specific proteins through their interactions with RBPs.

3.3 Circular RNAs and Direct Regulation of Gene Expression

In eukaryotic organisms, selective splicing mechanisms are essential for generating diverse proteins. CircRNAs are produced through a back-splicing process involving typical splicing machinery, which competes with pre-mRNA splicing and affects the expression of transcribed genes [21]. CircRNAs play a pivotal role in regulating gene expression during development and disease progression. This regulatory capacity highlights the importance of circRNAs in refining gene expression networks, thereby influencing various cellular functions. For instance, circ-EIF3J and circ-PAIP2 form complexes with U1 small nuclear ribonucleoprotein (snRNP) and subsequently interact with RNA polymerase II to facilitate parental gene transcription [50]. Additionally, certain ciRNAs exhibit feedback regulation on their parental genes. For example, ci-ankrd52 aggregates at its transcriptional locus, enhancing RNA polymerase II activity and facilitating positive transcriptional regulation [51]. This underscores the cis-regulatory role of ciRNAs in modulating the expression of their parental coding genes.

3.4 Circular RNAs Undergo Translation into Proteins

Although circRNAs belong to the category of noncoding RNAs, they lack the conventional 5′ cap or 3′ poly(A) tail required for translation regulation, as well as easily recognizable open reading frames (ORFs) [52]. Consequently, circRNAs were initially characterized as noncoding transcripts. However, emerging evidence challenges this view, suggesting that circRNAs have the potential to encode proteins. Specific sequences within circRNAs can facilitate the recruitment of ribosomes, allowing for the initiation of translation. For instance, the insertion of synthetic internal ribosome entry sites (IRES) upstream of the initiating codon enables translation to occur [53]. The methylation of adenosine nucleotides, specifically in the form of N6-methyladenosine (m6A), has been identified as a key mechanism that directly facilitates circRNA translation by binding to eukaryotic initiation factor 3 (eIF3) [52]. An example is circZNF609, which regulates the proliferation of myogenic cells. CircZNF609 contains an ORF at the initiation codon, which is identical to that of its linear transcript counterpart. Upon circularization, translation terminates at a stop codon. CircZNF609 associates with polysomes and undergoes protein translation in a splicing-dependent and cap-independent manner, illustrating the protein-encoding capacity of circRNAs in eukaryotes [54].

4 Circular RNAs in Heart Diseases

Several abundantly expressed circRNAs, including those derived from critical cardiac genes such as titin (TTN), ryanodine receptor 2 (RYR2), and dystrophin (DMD), underscore the significant regulatory role of circRNAs in maintaining normal cardiac function [55]. Moreover, the differential expression patterns of circRNAs have been closely associated with CVDs. For instance, Jakobi et al. observed that in adult mouse hearts, the majority of circRNAs were linked to host genes involved in cardiovascular pathology, such as Ppp2r3a, headcase homolog 1 (hesd1), and RYR2 [56]. Similarly, Maass et al. conducted a comprehensive analysis of circRNA profiles from the vena cava and right atrium of patients exhibiting diverse cardiac malformations, revealing distinct alterations in specific circRNA levels corresponding to different cardiovascular conditions [9]. Of particular interest was the identification of a circRNA subtype originating from the RYR2 gene, which was implicated in the pathogenesis of atrial fibrillation. Furthermore, a separate study identified distinct expression patterns of 6234 circRNAs in a fetal cohort diagnosed with congenital ventricular septal defect (VSD) compared to control fetuses [57]. Aberrant circRNA expression profiles have also been impli-

Table 17.1 Circular RNAs in cardiovascular diseases

Diseases	CircRNAs	Regulation	References
Pathological cardiac hypertrophy	HRCR	Down	[58]
	circSLC8A1	Up	[59]
	circRNA_000203	Up	[60]
Myocardial infarction	MICRA	Down	[61, 62]
	circ-cZNF292	Up	[63]
	circFndc3b	Down	[64]
	circRNA Ttc3	Down	[65]
	circHipk3	Down	[66, 67]
	ciRS-7	Up	[68]
	circNfix	Up	[69]
	circPostn	Up	[70]
	circPAN3	Up	[71]
	circ_0060745	Up	[72, 73]
	circFASTKD1	Up	[74]
Myocardial ischemia/reperfusion injury	circNCX1	Up	[75]
	circTLK1	Up	[76]
	ACR	Down	[77]
	circ-0068566	Down	[78]
	circ-ZNF609	Up	[79]
	circFoxo3	Up	[80]
Dilated cardiomyopathy	circCHD7	Up	[81]
	circATXN10	Up	[81]
	circDNA6JC	Down	[81]
	circNPPA	Up	[81]
	circMYH6	Up	[81]
	circMYH7	Up	[81]
	circFOXO3	Up	[49]
	circ-ZNF609	Up	[82]
	circAMOTL1	Down	[83]
	circFhod3	Up	[84]
	circPAN3	Up	[85]
Cardiac fibrosis	circNLGN	Up	[86, 87]
	circNCX1	Up	[75]
	circSlc8a1	Up	[88]
	circRNA_000203	Up	[89]
	circRNA_010567	Up	[90]
	circRNA_010567	Up	[90]
	circUBE3A	Up	[47]

cated in coronary heart disease and cardiomyocyte hypertrophy [11, 12]. Collectively, these findings underscore the close relationship between alterations in circRNA composition and abundance and various cardiovascular pathologies. In the following section, we will briefly review specific examples of circRNAs reported to be involved in the regulation of cardiovascular diseases (Table 17.1).

4.1 Pathological Cardiac Hypertrophy

Pathological cardiac hypertrophy is characterized by the abnormal enlargement of myocardial cells, typically stemming from prolonged stress or pathological conditions, which ultimately leads to compromised cardiac function. Within the field of cardiac hypertrophy research, circRNAs have received comparatively less attention than miRNAs and long noncoding RNAs (lncRNAs). Initial profiling of circRNAs in cardiac tissues unveiled 575 distinct circRNA species, many of which originated from genomic loci associated with cardiovascular diseases and exhibited significant biological correlations with cardiac hypertrophy [56]. The limited number of differentially expressed circRNAs between human decompensated and healthy cardiac tissues, as well as between sham-operated and transverse aortic constriction (TAC)-induced murine cardiac tissues, may be attributed to limitations in sequencing depth or the prolonged half-life and inherent stability of circRNAs.

The heart-related circular RNA (HRCR), identified as the first circRNA with a functional role in cardiac hypertrophy, operates as an miRNA sponge to inhibit miRNA activity in murine hypertrophy models induced by isoproterenol (ISO) or TAC [58]. HRCR exerts its anti-hypertrophic effects by upregulating activity-regulated cytoskeleton-associated protein (ARC) expression in mice through competitive binding with miR-233 [58]. Furthermore, the upregulation of circSLC8A1-1 has been implicated in hypertrophy progression via the sequestration of miR-133, a key regulator of myocardial hypertrophy [59]. Suppression of circSLC8A1-1 expression using adeno-associated virus type 9 (AAV9) has been shown to ameliorate hypertrophy in stress overload models, while elevated circSLC8A1-1 levels have been linked to HF development [59]. Moreover, researchers have explored the role of circRNA_000203 in angiotensin II (AngII)-induced hypertrophy. Notably, circRNA_000203 expression was found to be elevated in both AngII-injected murine myocardium and AngII-treated neonatal mouse ventricular cardiomyocytes (NMVCs) [60]. Using in vivo transgenic mouse models (tg-circRNA_000203), it was determined that circRNA_000203 overexpression exacerbated AngII-induced hypertrophy. This exacerbation is mediated by circRNA_000203 acting as a sponge for miR-26b-5p and miR-140-3p, thereby reducing their inhibition of the hypertrophic reactive transcription factor GATA-binding protein 4 (Gata4). Furthermore, circRNA_000203 expression was found to contribute to the exacerbation of AngII-induced hypertrophy through activation of the nuclear factor kappa B (NF-κB) signaling pathway.

4.2 Myocardial Infarction

Myocardial infarction (MI), commonly known as a heart attack, occurs when there is a sudden blockage of blood flow to a portion of the heart muscle, leading to tissue damage or cell death due to insufficient oxygen supply. Emerging evidence indicates that circRNAs play a role in the pathogenesis of acute myocardial infarction (AMI), with certain circRNAs exhibiting cardioprotective effects in MI models. Among these, myocardial infarction-related circular RNA (MICRA) has attracted attention due to its association with HF in MI patients [61]. Vausort et al. reported that MICRA levels are reduced in MI patients, with lower MICRA expression associated with an increased risk of left ventricular dysfunction [62]. Zhou Q et al. reported that cZNF292, when combined with clinical information (CM) such as age, gender, body mass index, heart rate, and diastolic blood pressure, can effectively diagnosis AMI [63]. Additionally, circRNA derived from the Fndc3b gene showed reduced expression following MI in murine models [64]. Notably, the presence of circFndc3b in cardiac endothelial cells enhanced endothelial function and conferred protection against cardiomyocyte apoptosis [64]. Furthermore, Cai et al. demonstrated the cardioprotective effects of circRNA Ttc3 through miR-15b sequestration in a rodent MI model [65]. CircHipk3, known for its involvement in various physiological and pathological functions, was found to have a significant impact on cardiac regeneration following AMI [66]. Elevated circHipk3 expression in neonatal mouse hearts suppressed cardiomyocyte proliferation when inhibited, attributed to enhanced acetylation and stability of the Notch1 intracellular domain (N1ICD) [67]. Furthermore, circHipk3 functions as a miR-133a sponge, activating connective tissue growth factor (CTGF) expression and promoting endothelial cell activation [66].

Conversely, some circRNAs have been found to exacerbate MI. Among them, Cdr1as (or ciRS-7) has received considerable attention for its capacity to worsen MI via the miR-7a/specificity protein 1 (SP1)/poly(ADP-ribose) polymerase (PARP) axis [68]. Another notable circRNA, circNfix, is classified as a superenhancer-associated circular RNA (SE-circRNA) [69]. Superenhancers consist of clusters of active enhancers that interact with key transcription factors [91]. Researchers observed increased circNfix expression in adult rodent cardiac tissues. CircNfix facilitates the interaction between Y-box binding protein 1 (Ybx1) and neural precursor cell expressed developmentally downregulated gene 4-like (Nedd4l), thereby promoting the ubiquitination-mediated degradation of Ybx1, which suppresses cardiomyocyte proliferation, enlarges the infarct size, and impairs cardiac regeneration [69]. CircPostn has been demonstrated to be upregulated in both plasma samples as well as in mouse models and patients suffering from MI [70]. Additionally, circPostn exacerbates myocardial injury and remodeling by modulating the miR-96-5p/BCL2 interacting protein 3 (BNIP3) axis [70]. CircPAN3, derived from PAN3 genes, is known to contribute to stem cell renewal and drug resistance [92, 93]. Li et al. discovered that circPAN3 is involved in MI-induced myocardial fibrosis via the miR-221/forkhead box O3 (FoxO3)/autophagy-related protein 7 (ATG7) axis [71]. Additionally, expression of circ_0060745 was found to be elevated in the infarcted myocardium of mice with AMI, particularly in cardiac fibroblasts (CF) [72]. Increased expression of circ_0060745 contributes to AMI development by modulating NF-κB activation and expanding MI size [73]. Gao et al. investigated the role of circFASTKD1 in angiogenesis following MI [74]. Their findings suggest that downregulation of circFASTKD1 enhances the viability of human cardiac microvascular endothelial cells (HCMECs), while its overexpression in human umbilical vein endothelial cells (HUVECs) reduces cell viability and impairs angiogenesis [74]. Elevated circFASTKD1 expression inhibits the migratory capacity and angiogenic tube formation of vascular endothelial cells, indicating a potential inhibitory effect on post-MI angiogenesis [74].

4.3 Myocardial Ischemia-Reperfusion Injury

Myocardial ischemia-reperfusion injury (MIRI) is a prevalent cardiovascular disorder characterized by a cascade of inflammatory responses, oxidative stress, and cellular apoptosis [94]. CircRNAs play a pivotal role in myocardial ischemia-reperfusion (I/R) injury by modulating pathophysiological processes through mechanisms such as regulating autophagy and influencing cell apoptosis [95]. CircNCX1, derived from the sodium/calcium exchanger 1 (NCX1) gene transcript, is widely expressed in cardiac tissue. Li et al. demonstrated that circNCX1 regulates the activity of the proapoptotic gene cell death-inducing protein 1 (CDIP1) by inhibiting miR-133a-3p, thereby promoting cardiomyocyte apoptosis induced by I/R injury [75]. Additionally, circTLK1, an upregulated circRNA identified in an MIRI mouse model, exacerbates myocardial injury and cardiomyocyte apoptosis during MIRI by targeting the TNF signaling pathway mediated by miR-214/ripk1 axis [76]. Furthermore, Zhou et al. identified an autophagy-related circular RNA (ACR) that attenuates cardiomyocyte autophagy and protects the heart from I/R injury by targeting PTEN-induced kinase 1 (PINK1) [77]. Another study by Hou et al. found that circ-0068566 is significantly downregulated in cardiomyocytes after myocardial I/R and mitigates cardiomyocyte apoptosis by targeting the hsa-miR-6322/Poly (ADP-ribose) polymerase-2 (PARP2) pathway [78].

Additionally, circular RNA MICRA, also known as circ-ZNF609, has been reported by Wang et al. to be upregulated during myocardial I/R remodeling. Knockdown of circ-ZNF609 protects against acute I/R injury and alleviates left ventricular dysfunction after I/R remodeling by modulating RNA N6-methyladenosine-modified YAP expression [79]. Moreover, mitochondria serve as the CENTRAL regulators of apoptosis, with the apoptotic mitochondrial pathway involving increased mitochondrial outer membrane permeability in response to various stimuli, leading to the release of cytochrome C into the cytoplasm. This release triggers the formation of the apoptosome by binding with apoptotic protease activating factor-1, which subsequently activates caspases and initiates a cascade reaction culminating in cell apoptosis [96]. Notably, studies have shown that circFoxo3 is significantly upregulated during myocardial I/R injury. Its overexpression can lead to mitochondrial dysfunction and substantial upregulation of apoptosis-related gene expression. Conversely, downregulation of circFoxo3 has been found to inhibit cardiomyocyte apoptosis and improve heart function by interacting with Foxo3, thereby reducing I/R injury [80].

4.4 Dilated Cardiomyopathy

Dilated cardiomyopathy (DCM) is characterized by the enlargement of the left or both ventricles of the heart, along with a reduction in contractile function. Its etiology involves a combination of genetic and nongenetic factors, with numerous genes and alleles contributing to its development [97, 98]. Extensive studies on dysregulated circRNAs in DCM have revealed a range of these molecules primarily originating from genes associated with DCM mutations. Comparative analyses with control groups have shown significant dysregulation of circRNAs derived from chromodomain helicase DNA-binding protein 7 (CHD7), ataxin 10 (ATXN10), and DnaJ heat shock protein family (Hsp40) member C6 (DNAJ6C) in DCM patients. Specifically, circ-CHD7 and circ-ATXN10 are upregulated, whereas circ-DNA6JC is downregulated. These findings introduce new therapeutic avenues, as circRNAs potentially relevant to the disease exhibit distinct characteristics [81]. Based on research findings, an RNA sequencing (RNA-seq) analysis of left ventricular tissues from individuals with DCM and healthy controls revealed a substantial presence of circRNAs in both normal and DCM-afflicted hearts. The study observed a notable increase in natriuretic peptide A (NPPA) expression levels in DCM samples along with a high abundance of circRNAs derived from the myosin heavy chain 6 (MYH6) and myosin heavy chain 7 (MYH7) genes. These circRNAs are widely conserved and prevalent in the cardiac tissues across various species. Given the pivotal roles of these genes in maintaining cardiovascular health and their involvement in cardiovascular disorders, the significance of their circRNAs is emphasized due to their crucial roles in the pathophysiology of cardiac conditions, including DCM [99].

Doxorubicin (DOX), a chemotherapy medication, induces chronic cardiac toxicity resembling the characteristics of DCM [100]. In a cohort study, differential expression of circRNAs was noted in young and aged hearts, with circFOXO3 being highly expressed in aged hearts [49]. CircFOXO3 exacerbates DOX-induced cardiomyopathy by intensifying cellular senescence through binding to antiaging proteins [49]. Circ-ZNF609 is significantly upregulated in adult cardiomyocytes of DOX-treated mice, and its suppression alleviates DOX-induced cardiotoxicity by upregulating FTO [82]. Conversely, circAMOTL1, which is highly expressed in neonatal hearts, protects against DOX-induced cardiomyopathy, activating AKT for cardiac repair [83]. Research on the RNA-binding protein QKI provides additional evidence for the involvement of circRNAs in DCM [84]. Elevated levels of Qki5 prevent cardiomyocyte apoptosis and heart enlargement in DOX-induced cardiomyopathy by regulating circRNAs, including those originating from TTN, formin homology 2 domain containing 3 (Fhod3), and striatin (Strn3) [84]. Notably, the suppression of TTN-derived circRNAs renders cardiac cells more susceptible to DOX-induced toxicity [84]. Additionally, circPAN3, which is negatively regulated by miR-31-5p and targets QKI, contributes to DOX-induced cardiac toxicity [85].

4.5 Cardiac Fibrosis

Cardiac fibrosis refers to the excessive accumulation of fibrous tissue within the heart muscle, which can result from various factors, including chronic inflammation, hypertension, and heart disease. Over time, this buildup of fibrous tissue can lead to stiffening of the heart muscle, impairing its ability to pump blood efficiently and contributing to heart failure and other cardiovascular complications [101]. Accumulating evidence suggests the involvement of circRNAs in cardiac fibrosis [102]. CircNLGN is highly expressed in cardiac tissues, where it activates and upregulates inflammatory cytokines through interactions with Histone H2A family member X (H2AX), thereby exacerbating the adverse effects of doxorubicin-induced cardiac fibrosis [86]. Additionally, circNLGN interacts with Lamin B1 (LMNB1) through its unique 9-amino acid motif, enhancing nuclear targeting, which leads to abnormal collagen deposition, increased proliferation of CFs, and decreased viability of cardiomyocytes [87]. CircNCX1, a circular transcript derived from the ncx1 gene, has been implicated in the regulation of myocardial fibrosis by acting as a sponge for miR133a-3p. In murine models, inhibition of circNCX1 using short hairpin RNA (shRNA) resulted in a significant reduction in collagen deposition and improved cardiac function following myocardial I/R injury [75].

Another study utilizing luciferase assays and reciprocal pull-down experiments confirmed the functional interaction between circSlc8a1 and miR-133. Previous research has shown that miR-133a can prevent excessive collagen expression induced by diabetes-induced myocardial fibrosis by targeting COL4A1, ERK1/2, and SMAD-2 phosphorylation [88]. Consequently, circNCX1 and circSlc8a1 may serve as novel therapeutic targets for the prevention and treatment of myocardial fibrosis by acting as miR-133 sponges. Additionally, circRNA microarray analysis conducted on myocardial tissue from diabetic db/db mice revealed significant upregulation of circRNA_000203, circRNA_010567, and circRNA_010567 in both diabetic myocardium and CFs treated with Angiotensin II. Overexpression of circRNA_000203 inhibits the antifibrotic effects of miR-26b-5p by interfering with its interaction with the 3′ UTR of connective tissue growth factor (CTGF) and Col1a2. This interference leads to increased expression of Col1a2, Col3a1, and α-SMA in CFs [89]. circRNA_010567, on the other hand, promotes the expression of fibrosis-associated proteins in CFs by targeting miR-141 [90]. Specifically, circRNA_010567 enhances the expression of fibrosis-associated proteins through the miR-141/TGFβ1 axis [90].

Moreover, extracellular circRNAs act as messengers facilitating interactions among various cell types during myocardial remodeling. For instance, small extracellular vesicle-derived circUBE3A from M2 macrophages enhances CF activity, promoting their migration and inducing phenotypic changes mediated by the miR-138-5p/RhoC signaling pathway, thereby exacerbating cardiac fibrosis following AMI [47]. These findings underscore the crucial role of circRNAs in modulating cardiac fibrosis processes.

5 Clinical Prospects

In the modern era, an increasing array of risk factors contributes to the onset and progression of CVDs. The pivotal role of early diagnosis and accurate prognosis forecasting cannot be overstated in the context of effective treatment and mortality reduction. Currently, a multitude of diagnostic modalities are employed for cardiovascular conditions, including blood pressure monitoring, dynamic electrocardiography, cardiovascular ultrasonography, magnetic resonance imaging, and computed tomography scanning. However, the implementation of these diagnostic methods necessitates substantial resources and sophisticated equipment. Furthermore, despite their widespread use, achieving precise detection of CVD onset and progression remains a challenge. As a result, certain blood biomarkers, notably atrial natriuretic peptide (ANP), creatine kinase (CK), creatine kinase-MB (CK-MB), and cardiac troponin (cTn), are widely recognized as sensitive indicators of CVDs [103]. In comparison, circRNAs, owing to their abundance, evolutionary conservation, and stable circular configuration, are undergoing extensive investigation as potential diagnostic and prognostic biomarkers in the field of CVDs.

5.1 Myocardial Infarction and Heart Failure

CircRNAs have shown promise as biomarkers for MI and HF, with 29 upregulated and 34 downregulated circRNAs identified as potential indicators through transcriptome sequencing and quantitative polymerase chain reaction (qPCR) analysis [104]. A study found that circRNA MICRA, which originates from exon 1 of the ZNF609 gene, is absent in peripheral blood cells and serum samples [105, 106]. MICRA, consisting of 874 nucleotides, is associated with HF progression following MI and has predictive value for left ventricular remodeling after AMI [61]. Reduced MICRA levels categorize patients into groups with decreased ejection fraction, whereas elevated levels forecast preserved ejection fraction after 4 months, indicating its potential as a novel biomarker for AMI prognosis [61, 62]. Additionally, cZNF292 combined with clinical information (CM), including age, gender, body mass index, heart rate, and diastolic blood pressure, can effectively diagnosis acute myocardial infarction (AMI). In the validation cohort, CM + cZNF292 was able to distinguish between AMI and non-AMI patients, unstable angina and AMI patients, as well as acute coronary syndrome (ACS) and non-ACS patients. Therefore, circulating cZNF292 may serve as a potential biomarker for AMI [63].

5.2 Coronary Artery Disease

Coronary artery disease (CAD) poses a significant risk to cardiovascular health, being linked to elevated morbidity and mortality rates. Improved diagnostic and prognostic approaches are essential to minimizing its impact [107]. Early detection of CAD is critical for enhancing patient outcomes. CircRNA-284, which interacts with miR-221 and miR-222, has shown potential as a diagnostic biomarker due to its elevated levels in the serum of individuals with carotid atherosclerotic plaques and its association with plaque rupture and stroke risk. Additionally, differential expression of circRNAs in blood, such as hsa_circRNA11783-2 and hsa_circ_0124644, suggests their potential utility as diagnostic biomarkers for CAD.

5.3 Hypertension

Hypertension, characterized by elevated systolic (≥140 mmHg) and/or diastolic (≥90 mmHg) blood pressure, is a prevalent chronic condition and a major contributor to CVD, often impairing the function of vital organs [108–110]. Profiling and bioinformatic analyses have revealed dysregulation of circRNAs in hypertension patients, with hsa-circ-0005870 notably downregulated and associated with the cellular stress response, suggesting its potential as a novel indicator for hypertension.

5.4 Chronic Thromboembolic Pulmonary Hypertension

Chronic thromboembolic pulmonary hypertension (CTEPH) poses substantial health risks due to pulmonary artery obstruction and subsequent vascular remodeling [111–113]. Early identification of pulmonary hypertension is crucial for improving cardiac function and patient prognosis [114, 115]. Analysis of blood samples from CTEPH patients revealed significant alterations in 351 circRNAs, with 122 upregulated and 229 downregulated [116]. Notably, circRNA hsa_circ_0002062 was found to regulate 761 miRNAs, while hsa_circ_0022342 was associated with the regulation of 453 miRNAs. This indicates their potential involvement in pathways related to CTEPH, such as the erb-b2 receptor tyrosine kinase (ErbB) signaling pathway. Therefore, these dysregulated circRNAs may serve as promising biomarkers for diagnosing CTEPH [116–118].

6 Conclusion

The emerging field of circRNAs in heart diseases has demonstrated their significant potential as diagnostic biomarkers, therapeutic targets, and regulators of disease pathogenesis. Extensive research has implicated circRNAs in a range of cardiovascular disorders, including pathological cardiac hypertrophy, myocardial infarction, MIRI, DCM, and cardiac fibrosis. Their diverse functions, such as acting as miRNA sponges, modulating gene expression, interacting with RBPs, and even undergoing translation into peptides, emphasize their complex regulatory roles in cardiovascular biology. Furthermore, the identification of specific circRNAs associated with various disease states presents exciting opportunities for the development of novel therapeutic interventions and precision medicine approaches. Moving forward, continued exploration of the mechanistic underpinnings of circRNA-mediated regulation in heart diseases will be essential to fully elucidate their potential and translate these findings into clinical applications.

Acknowledgments This work was supported by the grants from National Natural Science Foundation of China (82020108002 and 82225005 to J.X., 82401851 to T.Y.), the grant from Science and Technology Commission of Shanghai Municipality (23410750100 to J.X.), the China Postdoctoral Science Foundation (2023 M742205 to T.Y.), and Postdoctoral Fellowship Program of CPSF (GZB20230398 to T.Y.).

Disclosures The authors declare no competing interests.

References

1. Matsuzawa Y, Funahashi T, Nakamura T (2011) The concept of metabolic syndrome: contribution of visceral fat accumulation and its molecular mechanism. J Atheroscler Thromb 18(8):629–639. https://doi.org/10.5551/jat.7922
2. Gong X, Wu G, Zeng C (2019) Role of circular RNAs in cardiovascular diseases. Exp Biol Med (Maywood) 244(2):73–82. https://doi.org/10.1177/1535370218822988
3. Sanger HL, Klotz G, Riesner D, Gross HJ, Kleinschmidt AK (1976) Viroids are single-stranded covalently closed circular RNA molecules existing as highly base-paired rod-like structures. Proc Natl Acad Sci USA 73(11):3852–3856. https://doi.org/10.1073/pnas.73.11.3852
4. Kos A, Dijkema R, Arnberg AC, van der Meide PH, Schellekens H (1986) The hepatitis delta (delta) virus possesses a circular RNA. Nature 323(6088):558–560. https://doi.org/10.1038/323558a0
5. Capel B, Swain A, Nicolis S, Hacker A, Walter M, Koopman P, Goodfellow P, Lovell-Badge R (1993) Circular transcripts of the testis-determining gene Sry in adult mouse testis. Cell 73(5):1019–1030. https://doi.org/10.1016/0092-8674(93)90279-y
6. Danan M, Schwartz S, Edelheit S, Sorek R (2012) Transcriptome-wide discovery of circular RNAs in Archaea. Nucleic Acids Res 40(7):3131–3142. https://doi.org/10.1093/nar/gkr1009
7. Wang PL, Bao Y, Yee MC, Barrett SP, Hogan GJ, Olsen MN, Dinneny JR, Brown PO, Salzman J (2014) Circular RNA is expressed across the eukaryotic tree of life. PLoS One 9(6):e90859. https://doi.org/10.1371/journal.pone.0090859
8. Memczak S, Jens M, Elefsinioti A, Torti F, Krueger J, Rybak A, Maier L, Mackowiak SD, Gregersen LH, Munschauer M, Loewer A, Ziebold U, Landthaler M, Kocks C, le Noble F, Rajewsky N (2013) Circular RNAs are a large class of animal RNAs with regulatory potency. Nature 495(7441):333–338. https://doi.org/10.1038/nature11928
9. Maass PG, Glazar P, Memczak S, Dittmar G, Hollfinger I, Schreyer L, Sauer AV, Toka O, Aiuti A, Luft FC, Rajewsky N (2017) A map of human circular RNAs in clinically relevant tissues. J Mol Med (Berl) 95(11):1179–1189. https://doi.org/10.1007/s00109-017-1582-9
10. Lei W, Feng T, Fang X, Yu Y, Yang J, Zhao ZA, Liu J, Shen Z, Deng W, Hu S (2018) Signature of circular RNAs in human induced pluripotent stem cells and derived cardiomyocytes. Stem Cell Res Ther 9(1):56. https://doi.org/10.1186/s13287-018-0793-5
11. Lin F, Zhao G, Chen Z, Wang X, Lv F, Zhang Y, Yang X, Liang W, Cai R, Li J, Li M, Zhang G (2019) circRNA-miRNA association for coronary heart disease. Mol Med Rep 19(4):2527–2536. https://doi.org/10.3892/mmr.2019.9905
12. Meng Z, Chen C, Cao H, Wang J, Shen E (2019) Whole transcriptome sequencing reveals biologically significant RNA markers and related regulating biological pathways in cardiomyocyte hypertrophy induced by high glucose. J Cell Biochem 120(1):1018–1027. https://doi.org/10.1002/jcb.27546
13. Cocquerelle C, Daubersies P, Majerus MA, Kerckaert JP, Bailleul B (1992) Splicing with inverted order of exons occurs proximal to large introns. EMBO J 11(3):1095–1098. https://doi.org/10.1002/j.1460-2075.1992.tb05148.x
14. Lu C, Shi X, Wang AY, Tao Y, Wang Z, Huang C, Qiao Y, Hu H, Liu L (2018) RNA-Seq profiling of circular RNAs in human laryngeal squamous cell carcinomas. Mol Cancer 17(1):86. https://doi.org/10.1186/s12943-018-0833-x
15. AbouHaidar MG, Venkataraman S, Golshani A, Liu B, Ahmad T (2014) Novel coding, translation, and gene expression of a replicating covalently closed circular RNA of 220 nt. Proc Natl Acad Sci USA 111(40):14542–14547. https://doi.org/10.1073/pnas.1402814111

16. Barrett SP, Wang PL, Salzman J (2015) Circular RNA biogenesis can proceed through an exon-containing lariat precursor. elife 4:e07540. https://doi.org/10.7554/eLife.07540
17. Starke S, Jost I, Rossbach O, Schneider T, Schreiner S, Hung LH, Bindereif A (2015) Exon circularization requires canonical splice signals. Cell Rep 10(1):103–111. https://doi.org/10.1016/j.celrep.2014.12.002
18. Wilusz JE (2015) Repetitive elements regulate circular RNA biogenesis. Mob Genet Elements 5(3):1–7. https://doi.org/10.1080/2159256X.2015.1045682
19. Liang D, Tatomer DC, Luo Z, Wu H, Yang L, Chen LL, Cherry S, Wilusz JE (2017) The output of protein-coding genes shifts to circular RNAs when the pre-mRNA processing machinery is limiting. Mol Cell 68(5):940–954.e943. https://doi.org/10.1016/j.molcel.2017.10.034
20. Yang L, Wilusz JE, Chen LL (2022) Biogenesis and regulatory roles of circular RNAs. Annu Rev Cell Dev Biol 38:263–289. https://doi.org/10.1146/annurev-cellbio-120420-125117
21. Ashwal-Fluss R, Meyer M, Pamudurti NR, Ivanov A, Bartok O, Hanan M, Evantal N, Memczak S, Rajewsky N, Kadener S (2014) circRNA biogenesis competes with pre-mRNA splicing. Mol Cell 56(1):55–66. https://doi.org/10.1016/j.molcel.2014.08.019
22. Zhang Y, Zhang XO, Chen T, Xiang JF, Yin QF, Xing YH, Zhu S, Yang L, Chen LL (2013) Circular intronic long noncoding RNAs. Mol Cell 51(6):792–806. https://doi.org/10.1016/j.molcel.2013.08.017
23. Malviya A, Bhuyan R (2023) The recent advancements in circRNA research: from biogenesis to therapeutic interventions. Pathol Res Pract 248:154697. https://doi.org/10.1016/j.prp.2023.154697
24. Daniel C, Behm M, Ohman M (2015) The role of Alu elements in the cis-regulation of RNA processing. Cell Mol Life Sci 72(21):4063–4076. https://doi.org/10.1007/s00018-015-1990-3
25. Szabo L, Morey R, Palpant NJ, Wang PL, Afari N, Jiang C, Parast MM, Murry CE, Laurent LC, Salzman J (2015) Statistically based splicing detection reveals neural enrichment and tissue-specific induction of circular RNA during human fetal development. Genome Biol 16(1):126. https://doi.org/10.1186/s13059-015-0690-5
26. Ebbesen KK, Hansen TB, Kjems J (2017) Insights into circular RNA biology. RNA Biol 14(8):1035–1045. https://doi.org/10.1080/15476286.2016.1271524
27. Das A, Sinha T, Shyamal S, Panda AC (2021) Emerging role of circular RNA-protein interactions. Noncoding RNA 7(3):48. https://doi.org/10.3390/ncrna7030048
28. Liang Y, Liu N, Yang L, Tang J, Wang Y, Mei M (2021) A brief review of circRNA biogenesis, detection, and function. Curr Genomics 22(7):485–495. https://doi.org/10.2174/1389202922666210331130722
29. Petkovic S, Muller S (2015) RNA circularization strategies in vivo and in vitro. Nucleic Acids Res 43(4):2454–2465. https://doi.org/10.1093/nar/gkv045
30. Yang L, Han B, Zhang Y, Bai Y, Chao J, Hu G, Yao H (2018) Engagement of circular RNA HECW2 in the nonautophagic role of ATG5 implicated in the endothelial-mesenchymal transition. Autophagy 14(3):404–418. https://doi.org/10.1080/15548627.2017.1414755
31. Zheng Q, Bao C, Guo W, Li S, Chen J, Chen B, Luo Y, Lyu D, Li Y, Shi G, Liang L, Gu J, He X, Huang S (2016) Circular RNA profiling reveals an abundant circHIPK3 that regulates cell growth by sponging multiple miRNAs. Nat Commun 7:11215. https://doi.org/10.1038/ncomms11215
32. Lin SP, Ye S, Long Y, Fan Y, Mao HF, Chen MT, Ma QJ (2016) Circular RNA expression alterations are involved in OGD/R-induced neuron injury. Biochem Biophys Res Commun 471(1):52–56. https://doi.org/10.1016/j.bbrc.2016.01.183
33. Veno MT, Hansen TB, Veno ST, Clausen BH, Grebing M, Finsen B, Holm IE, Kjems J (2015) Spatio-temporal regulation of circular RNA expression during porcine embryonic brain development. Genome Biol 16:245. https://doi.org/10.1186/s13059-015-0801-3
34. Lukiw WJ (2013) Circular RNA (circRNA) in Alzheimer's disease (AD). Front Genet 4:307. https://doi.org/10.3389/fgene.2013.00307
35. You X, Vlatkovic I, Babic A, Will T, Epstein I, Tushev G, Akbalik G, Wang M, Glock C, Quedenau C, Wang X, Hou J, Liu H, Sun W, Sambandan S, Chen T, Schuman EM, Chen W (2015) Neural circular RNAs are derived from synaptic genes and regulated by development and plasticity. Nat Neurosci 18(4):603–610. https://doi.org/10.1038/nn.3975
36. Misir S, Wu N, Yang BB (2022) Specific expression and functions of circular RNAs. Cell Death Differ 29(3):481–491. https://doi.org/10.1038/s41418-022-00948-7
37. Gan H, Lei Y, Yuan N, Tang K, Hao W, Ma Q, Wu M, Zhou X, Li X, Huang J, Deng L, Chen J (2021) Circular RNAs in depression: biogenesis, function, expression, and therapeutic potential. Biomed Pharmacother 137:111244. https://doi.org/10.1016/j.biopha.2021.111244
38. Saikishore R, Velmurugan P, Ranjithkumar D, Latha R, Sathiamoorthi T, Arun A, Ravi AV, Sivakumar S (2020) The circular RNA-miRNA axis: a special RNA signature regulatory transcriptome as a potential biomarker for OSCC. Mol Ther Nucleic Acids 22:352–361. https://doi.org/10.1016/j.omtn.2020.09.001
39. Zang J, Lu D, Xu A (2020) The interaction of circRNAs and RNA binding proteins: an important part of circRNA maintenance and function. J Neurosci Res 98(1):87–97. https://doi.org/10.1002/jnr.24356
40. Wang H, Gao X, Yu S, Wang W, Liu G, Jiang X, Sun D (2022) Circular RNAs regulate parental gene

expression: a new direction for molecular oncology research. Front Oncol 12:947775. https://doi.org/10.3389/fonc.2022.947775
41. Wu C, Wang S, Cao T, Huang T, Xu L, Wang J, Li Q, Wang Y, Qian L, Xu L, Xia Y, Huang X (2023) Newly discovered mechanisms that mediate tumorigenesis and tumour progression: circRNA-encoded proteins. J Cell Mol Med 27(12):1609–1620. https://doi.org/10.1111/jcmm.17751
42. Liu B, Li J, Cairns MJ (2014) Identifying miRNAs, targets and functions. Brief Bioinform 15(1):1–19. https://doi.org/10.1093/bib/bbs075
43. Krek A, Grun D, Poy MN, Wolf R, Rosenberg L, Epstein EJ, MacMenamin P, da Piedade I, Gunsalus KC, Stoffel M, Rajewsky N (2005) Combinatorial microRNA target predictions. Nat Genet 37(5):495–500. https://doi.org/10.1038/ng1536
44. Lewis BP, Burge CB, Bartel DP (2005) Conserved seed pairing, often flanked by adenosines, indicates that thousands of human genes are microRNA targets. Cell 120(1):15–20. https://doi.org/10.1016/j.cell.2004.12.035
45. Kristensen LS, Andersen MS, Stagsted LVW, Ebbesen KK, Hansen TB, Kjems J (2019) The biogenesis, biology and characterization of circular RNAs. Nat Rev Genet 20(11):675–691. https://doi.org/10.1038/s41576-019-0158-7
46. Yang ZG, Awan FM, Du WW, Zeng Y, Lyu J, Wu D, Gupta S, Yang W, Yang BB (2017) The circular RNA interacts with STAT3, increasing its nuclear translocation and wound repair by modulating Dnmt3a and miR-17 function. Mol Ther 25(9):2062–2074. https://doi.org/10.1016/j.ymthe.2017.05.022
47. Zhang L, Wang Y, Zhang Y, Zhao Y, Li P (2021) Pathogenic mechanisms and the potential clinical value of circFoxo3 in cancers. Mol Ther Nucleic Acids 23:908–917. https://doi.org/10.1016/j.omtn.2021.01.010
48. Du WW, Yang W, Liu E, Yang Z, Dhaliwal P, Yang BB (2016) Foxo3 circular RNA retards cell cycle progression via forming ternary complexes with p21 and CDK2. Nucleic Acids Res 44(6):2846–2858. https://doi.org/10.1093/nar/gkw027
49. Du WW, Yang W, Chen Y, Wu ZK, Foster FS, Yang Z, Li X, Yang BB (2017) Foxo3 circular RNA promotes cardiac senescence by modulating multiple factors associated with stress and senescence responses. Eur Heart J 38(18):1402–1412. https://doi.org/10.1093/eurheartj/ehw001
50. Sun Y, Qiu L, Chen J, Wang Y, Qian J, Huang L, Ma H (2020) Construction of circRNA-associated ceRNA network reveals novel biomarkers for esophageal cancer. Comput Math Methods Med 2020:7958362. https://doi.org/10.1155/2020/7958362
51. Li X, Zhang JL, Lei YN, Liu XQ, Xue W, Zhang Y, Nan F, Gao X, Zhang J, Wei J, Yang L, Chen LL (2021) Linking circular intronic RNA degradation and function in transcription by RNase H1. Sci China Life Sci 64(11):1795–1809. https://doi.org/10.1007/s11427-021-1993-6
52. Fontemaggi G, Turco C, Esposito G, Di Agostino S (2021) New molecular mechanisms and clinical impact of circRNAs in human cancer. Cancers (Basel) 13(13):3154. https://doi.org/10.3390/cancers13133154
53. Wang Y, Wang Z (2015) Efficient backsplicing produces translatable circular mRNAs. RNA 21(2):172–179. https://doi.org/10.1261/rna.048272.114
54. Legnini I, Di Timoteo G, Rossi F, Morlando M, Briganti F, Sthandier O, Fatica A, Santini T, Andronache A, Wade M, Laneve P, Rajewsky N, Bozzoni I (2017) Circ-ZNF609 is a circular RNA that can be translated and functions in myogenesis. Mol Cell 66(1):22–37.e29. https://doi.org/10.1016/j.molcel.2017.02.017
55. Tan WL, Lim BT, Anene-Nzelu CG, Ackers-Johnson M, Dashi A, See K, Tiang Z, Lee DP, Chua WW, Luu TD, Li PY, Richards AM, Foo RS (2017) A landscape of circular RNA expression in the human heart. Cardiovasc Res 113(3):298–309. https://doi.org/10.1093/cvr/cvw250
56. Jakobi T, Czaja-Hasse LF, Reinhardt R, Dieterich C (2016) Profiling and validation of the circular RNA repertoire in adult murine hearts. Genomics Proteomics Bioinformatics 14(4):216–223. https://doi.org/10.1016/j.gpb.2016.02.003
57. Liu H, Hu Y, Zhuang B, Yin J, Chen XH, Wang J, Li MM, Xu J, Wang XY, Yu ZB, Han SP (2018) Differential expression of CircRNAs in embryonic heart tissue associated with ventricular septal defect. Int J Med Sci 15(7):703–712. https://doi.org/10.7150/ijms.21660
58. Li Y, Liang Y, Zhu Y, Zhang Y, Bei Y (2018) Noncoding RNAs in cardiac hypertrophy. J Cardiovasc Transl Res 11(6):439–449. https://doi.org/10.1007/s12265-018-9797-x
59. Lim TB, Aliwarga E, Luu TDA, Li YP, Ng SL, Annadoray L, Sian S, Ackers-Johnson MA, Foo RS (2019) Targeting the highly abundant circular RNA circSlc8a1 in cardiomyocytes attenuates pressure overload induced hypertrophy. Cardiovasc Res 115(14):1998–2007. https://doi.org/10.1093/cvr/cvz130
60. Li H, Xu JD, Fang XH, Zhu JN, Yang J, Pan R, Yuan SJ, Zeng N, Yang ZZ, Yang H, Wang XP, Duan JZ, Wang S, Luo JF, Wu SL, Shan ZX (2020) Circular RNA circRNA_000203 aggravates cardiac hypertrophy via suppressing miR-26b-5p and miR-140-3p binding to Gata4. Cardiovasc Res 116(7):1323–1334. https://doi.org/10.1093/cvr/cvz215
61. Salgado-Somoza A, Zhang L, Vausort M, Devaux Y (2017) The circular RNA MICRA for risk stratification after myocardial infarction. Int J Cardiol Heart Vasc 17:33–36. https://doi.org/10.1016/j.ijcha.2017.11.001
62. Vausort M, Salgado-Somoza A, Zhang L, Leszek P, Scholz M, Teren A, Burkhardt R, Thiery J, Wagner DR, Devaux Y (2016) Myocardial infarction-associated circular RNA predicting left ventricular

dysfunction. J Am Coll Cardiol 68(11):1247–1248. https://doi.org/10.1016/j.jacc.2016.06.040
63. Zhou Q, Boeckel JN, Yao J, Zhao J, Bai Y, Lv Y, Hu M, Meng D, Xie Y, Yu P, Xi P, Xu J, Zhang Y, Dimmeler S, Xiao J (2023) Diagnosis of acute myocardial infarction using a combination of circulating circular RNA cZNF292 and clinical information based on machine learning. MedComm (2020) 4(3):e299. https://doi.org/10.1002/mco2.299
64. Garikipati VNS, Verma SK, Cheng Z, Liang D, Truongcao MM, Cimini M, Yue Y, Huang G, Wang C, Benedict C, Tang Y, Mallaredy V, Ibetti J, Grisanti L, Schumacher SM, Gao E, Rajan S, Wilusz JE, Goukassian D, Houser SR, Koch WJ, Kishore R (2019) Circular RNA CircFndc3b modulates cardiac repair after myocardial infarction via FUS/VEGF-A axis. Nat Commun 10(1):4317. https://doi.org/10.1038/s41467-019-11777-7
65. Cai L, Qi B, Wu X, Peng S, Zhou G, Wei Y, Xu J, Chen S, Liu S (2019) Circular RNA Ttc3 regulates cardiac function after myocardial infarction by sponging miR-15b. J Mol Cell Cardiol 130:10–22. https://doi.org/10.1016/j.yjmcc.2019.03.007
66. Si X, Zheng H, Wei G, Li M, Li W, Wang H, Guo H, Sun J, Li C, Zhong S, Liao W, Liao Y, Huang S, Bin J (2020) circRNA Hipk3 induces cardiac regeneration after myocardial infarction in mice by binding to Notch1 and miR-133a. Mol Ther Nucleic Acids 21:636–655. https://doi.org/10.1016/j.omtn.2020.06.024
67. Rai AK, Lee B, Hebbard C, Uchida S, Garikipati VNS (2021) Decoding the complexity of circular RNAs in cardiovascular disease. Pharmacol Res 171:105766. https://doi.org/10.1016/j.phrs.2021.105766
68. Geng HH, Li R, Su YM, Xiao J, Pan M, Cai XX, Ji XP (2016) The circular RNA Cdr1as promotes myocardial infarction by mediating the regulation of miR-7a on its target genes expression. PLoS One 11(3):e0151753. https://doi.org/10.1371/journal.pone.0151753
69. Huang S, Li X, Zheng H, Si X, Li B, Wei G, Li C, Chen Y, Chen Y, Liao W, Liao Y, Bin J (2019) Loss of super-enhancer-regulated circRNA Nfix induces cardiac regeneration after myocardial infarction in adult mice. Circulation 139(25):2857–2876. https://doi.org/10.1161/CIRCULATIONAHA.118.038361
70. Cheng N, Wang MY, Wu YB, Cui HM, Wei SX, Liu B, Wang R (2020) Circular RNA POSTN promotes myocardial infarction-induced myocardial injury and cardiac remodeling by regulating miR-96-5p/BNIP3 axis. Front Cell Dev Biol 8:618574. https://doi.org/10.3389/fcell.2020.618574
71. Li F, Long TY, Bi SS, Sheikh SA, Zhang CL (2020) circPAN3 exerts a profibrotic role via sponging miR-221 through FoxO3/ATG7-activated autophagy in a rat model of myocardial infarction. Life Sci 257:118015. https://doi.org/10.1016/j.lfs.2020.118015
72. Zhao Q, Li W, Pan W, Wang Z (2021) CircRNA 010567 plays a significant role in myocardial infarction via the regulation of the miRNA-141/DAPK1 axis. J Thorac Dis 13(4):2447–2459. https://doi.org/10.21037/jtd-21-212
73. Zhai C, Qian G, Wu H, Pan H, Xie S, Sun Z, Shao P, Tang G, Hu H, Zhang S (2020) Knockdown of circ_0060745 alleviates acute myocardial infarction by suppressing NF-kappaB activation. J Cell Mol Med 24(21):12401–12410. https://doi.org/10.1111/jcmm.15748
74. Gao WQ, Hu XM, Zhang Q, Yang L, Lv XZ, Chen S, Wu P, Duan DW, Lang YH, Ning M, Lai KG, Zhang ZY, Liang B, Bao JY, Wu HD, Li T (2020) Downregulation of circFASTKD1 ameliorates myocardial infarction by promoting angiogenesis. Aging (Albany NY) 13(3):3588–3604. https://doi.org/10.18632/aging.202305
75. Li M, Ding W, Tariq MA, Chang W, Zhang X, Xu W, Hou L, Wang Y, Wang J (2018) A circular transcript of ncx1 gene mediates ischemic myocardial injury by targeting miR-133a-3p. Theranostics 8(21):5855–5869. https://doi.org/10.7150/thno.27285
76. Song YF, Zhao L, Wang BC, Sun JJ, Hu JL, Zhu XL, Zhao J, Zheng DK, Ge ZW (2020) The circular RNA TLK1 exacerbates myocardial ischemia/reperfusion injury via targeting miR-214/RIPK1 through TNF signaling pathway. Free Radic Biol Med 155:69–80. https://doi.org/10.1016/j.freeradbiomed.2020.05.013
77. Zhou LY, Zhai M, Huang Y, Xu S, An T, Wang YH, Zhang RC, Liu CY, Dong YH, Wang M, Qian LL, Ponnusamy M, Zhang YH, Zhang J, Wang K (2019) The circular RNA ACR attenuates myocardial ischemia/reperfusion injury by suppressing autophagy via modulation of the Pink1/FAM65B pathway. Cell Death Differ 26(7):1299–1315. https://doi.org/10.1038/s41418-018-0206-4
78. Zhou HF, Xu LL, Xie B, Ding HG, Fang F, Fang Q (2020) Hsa-circ-0068566 inhibited the development of myocardial ischemia reperfusion injury by regulating hsa-miR-6322/PARP2 signal pathway. Eur Rev Med Pharmacol Sci 24(12):6980–6993. https://doi.org/10.26355/eurrev_202006_21690
79. Wang L, Yu P, Wang J, Xu G, Wang T, Feng J, Bei Y, Xu J, Wang H, Das S, Xiao J (2022) Downregulation of circ-ZNF609 promotes heart repair by modulating RNA N(6)-methyladenosine-modified yap expression. Research (Wash, DC) 2022:9825916. https://doi.org/10.34133/2022/9825916
80. Su Y, Zhu C, Wang B, Zheng H, McAlister V, Lacefield JC, Quan D, Mele T, Greasley A, Liu K, Zheng X (2021) Circular RNA Foxo3 in cardiac ischemia-reperfusion injury in heart transplantation: a new regulator and target. Am J Transplant 21(9):2992–3004. https://doi.org/10.1111/ajt.16475
81. Siede D, Rapti K, Gorska AA, Katus HA, Altmuller J, Boeckel JN, Meder B, Maack C, Volkers M, Muller OJ, Backs J, Dieterich C (2017) Identification of cir-

cular RNAs with host gene-independent expression in human model systems for cardiac differentiation and disease. J Mol Cell Cardiol 109:48–56. https://doi.org/10.1016/j.yjmcc.2017.06.015
82. Yu P, Wang J, Xu GE, Zhao X, Cui X, Feng J, Sun J, Wang T, Spanos M, Lehmann HI, Li G, Xu J, Wang L, Xiao J (2023) RNA m(6)A-regulated circ-ZNF609 suppression ameliorates doxorubicin-induced cardiotoxicity by upregulating FTO. JACC Basic Transl Sci 8(6):677–698. https://doi.org/10.1016/j.jacbts.2022.12.005
83. Zeng Y, Du WW, Wu Y, Yang Z, Awan FM, Li X, Yang W, Zhang C, Yang Q, Yee A, Chen Y, Yang F, Sun H, Huang R, Yee AJ, Li RK, Wu Z, Backx PH, Yang BB (2017) A circular RNA binds to and activates AKT phosphorylation and nuclear localization reducing apoptosis and enhancing cardiac repair. Theranostics 7(16):3842–3855. https://doi.org/10.7150/thno.19764
84. Gupta SK, Garg A, Bar C, Chatterjee S, Foinquinos A, Milting H, Streckfuss-Bomeke K, Fiedler J, Thum T (2018) Quaking inhibits doxorubicin-mediated cardiotoxicity through regulation of cardiac circular RNA expression. Circ Res 122(2):246–254. https://doi.org/10.1161/CIRCRESAHA.117.311335
85. Ji X, Ding W, Xu T, Zheng X, Zhang J, Liu M, Liu G, Wang J (2020) MicroRNA-31-5p attenuates doxorubicin-induced cardiotoxicity via quaking and circular RNA Pan3. J Mol Cell Cardiol 140:56–67. https://doi.org/10.1016/j.yjmcc.2020.02.009
86. Xu J, Du WW, Wu N, Li F, Li X, Xie Y, Wang S, Yang BB (2022) The circular RNA circNlgnmediates doxorubicin-inducedcardiac remodeling and fibrosis. Mol Ther Nucleic Acids 28:175–189. https://doi.org/10.1016/j.omtn.2022.03.007
87. Du WW, Xu J, Yang W, Wu N, Li F, Zhou L, Wang S, Li X, He AT, Du KY, Zeng K, Ma J, Lyu J, Zhang C, Zhou C, Maksimovic K, Yang BB (2021) A neuroligin isoform translated by circNlgn contributes to cardiac remodeling. Circ Res 129(5):568–582. https://doi.org/10.1161/CIRCRESAHA.120.318364
88. Chen S, Puthanveetil P, Feng B, Matkovich SJ, Dorn GW 2nd, Chakrabarti S (2014) Cardiac miR-133a overexpression prevents early cardiac fibrosis in diabetes. J Cell Mol Med 18(3):415–421. https://doi.org/10.1111/jcmm.12218
89. Tang CM, Zhang M, Huang L, Hu ZQ, Zhu JN, Xiao Z, Zhang Z, Lin QX, Zheng XL, Yang M, Wu SL, Cheng JD, Shan ZX (2017) CircRNA_000203 enhances the expression of fibrosis-associated genes by derepressing targets of miR-26b-5p, Col1a2 and CTGF, in cardiac fibroblasts. Sci Rep 7:40342. https://doi.org/10.1038/srep40342
90. Zhou B, Yu JW (2017) A novel identified circular RNA, circRNA_010567, promotes myocardial fibrosis via suppressing miR-141 by targeting TGF-beta1. Biochem Biophys Res Commun 487(4):769–775. https://doi.org/10.1016/j.bbrc.2017.04.044
91. Whyte WA, Orlando DA, Hnisz D, Abraham BJ, Lin CY, Kagey MH, Rahl PB, Lee TI, Young RA (2013) Master transcription factors and mediator establish super-enhancers at key cell identity genes. Cell 153(2):307–319. https://doi.org/10.1016/j.cell.2013.03.035
92. Zhu P, Zhu X, Wu J, He L, Lu T, Wang Y, Liu B, Ye B, Sun L, Fan D, Wang J, Yang L, Qin X, Du Y, Li C, He L, Ren W, Wu X, Tian Y, Fan Z (2019) IL-13 secreted by ILC2s promotes the self-renewal of intestinal stem cells through circular RNA circPan3. Nat Immunol 20(2):183–194. https://doi.org/10.1038/s41590-018-0297-6
93. Shang J, Chen WM, Liu S, Wang ZH, Wei TN, Chen ZZ, Wu WB (2019) CircPAN3 contributes to drug resistance in acute myeloid leukemia through regulation of autophagy. Leuk Res 85:106198. https://doi.org/10.1016/j.leukres.2019.106198
94. Algoet M, Janssens S, Himmelreich U, Gsell W, Pusovnik M, Van den Eynde J, Oosterlinck W (2023) Myocardial ischemia-reperfusion injury and the influence of inflammation. Trends Cardiovasc Med 33(6):357–366. https://doi.org/10.1016/j.tcm.2022.02.005
95. Chen KY, Liu Z, Lu JH, Yang SY, Hu XY, Liang GY (2024) The function of circular RNAs in myocardial ischemia-reperfusion injury: underlying mechanisms and therapeutic advancement. Cardiovasc Drugs Ther. https://doi.org/10.1007/s10557-024-07557-1
96. Du J, Song D, Li J, Li Y, Li B, Li L (2022) Paeonol triggers apoptosis in HeLa cervical cancer cells: the role of mitochondria-related caspase pathway. Psychopharmacology 239(7):2083–2092. https://doi.org/10.1007/s00213-021-05811-0
97. McNally EM, Mestroni L (2017) Dilated cardiomyopathy: genetic determinants and mechanisms. Circ Res 121(7):731–748. https://doi.org/10.1161/CIRCRESAHA.116.309396
98. Hanselmann A, Veltmann C, Bauersachs J, Berliner D (2020) Dilated cardiomyopathies and noncompaction cardiomyopathy. Herz 45(3):212–220. https://doi.org/10.1007/s00059-020-04903-5
99. Dong K, He X, Su H, Fulton DJR, Zhou J (2020) Genomic analysis of circular RNAs in heart. BMC Med Genet 13(1):167. https://doi.org/10.1186/s12920-020-00817-7
100. Chatterjee K, Zhang J, Honbo N, Karliner JS (2010) Doxorubicin cardiomyopathy. Cardiology 115(2):155–162. https://doi.org/10.1159/000265166
101. Maruyama K, Imanaka-Yoshida K (2022) The pathogenesis of cardiac fibrosis: a review of recent progress. Int J Mol Sci 23(5). https://doi.org/10.3390/ijms23052617
102. Yousefi F, Soltani BM (2021) Circular RNAs as potential theranostics in the cardiac fibrosis. Heart Fail Rev 26(1):195–203. https://doi.org/10.1007/s10741-019-09908-9
103. Heil B, Tang WH (2015) Biomarkers: their potential in the diagnosis and treatment of heart failure. Cleve Clin J Med 82(12 Suppl 2):S28–S35. https://doi.org/10.3949/ccjm.82.s2.05

104. Wu HJ, Zhang CY, Zhang S, Chang M, Wang HY (2016) Microarray expression profile of circular RNAs in heart tissue of mice with myocardial infarction-induced heart failure. Cell Physiol Biochem 39(1):205–216. https://doi.org/10.1159/000445617
105. Yang Y, Yu T, Jiang S, Zhang Y, Li M, Tang N, Ponnusamy M, Wang JX, Li PF (2017) miRNAs as potential therapeutic targets and diagnostic biomarkers for cardiovascular disease with a particular focus on WO2010091204. Expert Opin Ther Pat 27(9):1021–1029. https://doi.org/10.1080/13543776.2017.1344217
106. Zhu H, Fan GC (2013) Whether circulating miRNAs or miRNA-carriers serve as biomarkers for acute myocardial infarction. J Biomark Drug Dev 1(1):1000e103. https://doi.org/10.4172/jbdd.1000e103
107. Burd CE, Jeck WR, Liu Y, Sanoff HK, Wang Z, Sharpless NE (2010) Expression of linear and novel circular forms of an INK4/ARF-associated non-coding RNA correlates with atherosclerosis risk. PLoS Genet 6(12):e1001233. https://doi.org/10.1371/journal.pgen.1001233
108. Hamamura K, Yanagida M, Ishikawa H, Banzai M, Yoshitake H, Nonaka D, Tanaka K, Sakuraba M, Miyakuni Y, Takamori K, Nojima M, Yoshida K, Fujiwara H, Takeda S, Araki Y (2018) Quantitative measurement of a candidate serum biomarker peptide derived from alpha2-HS-glycoprotein, and a preliminary trial of multidimensional peptide analysis in females with pregnancy-induced hypertension. Ann Clin Biochem 55(2):287–295. https://doi.org/10.1177/0004563217717748
109. Richter MJ, Schermuly R, Seeger W, Rao Y, Ghofrani HA, Gall H (2016) Relevance of angiopoietin-2 and soluble P-selectin levels in patients with pulmonary arterial hypertension receiving combination therapy with oral treprostinil: a FREEDOM-C2 biomarker substudy. Pulm Circ 6(4):516–523. https://doi.org/10.1086/688671
110. Smukowska-Gorynia A, Tomaszewska I, Malaczynska-Rajpold K, Marcinkowska J, Komosa A, Janus M, Olasinska-Wisniewska A, Slawek S, Araszkiewicz A, Jankiewicz S, Mularek-Kubzdela T (2018) Red blood cells distribution width as a potential prognostic biomarker in patients with pulmonary arterial hypertension and chronic thromboembolic pulmonary hypertension. Heart Lung Circ 27(7):842–848. https://doi.org/10.1016/j.hlc.2017.08.007
111. Ghofrani HA, Simonneau G, D'Armini AM, Fedullo P, Howard LS, Jais X, Jenkins DP, Jing ZC, Madani MM, Martin N, Mayer E, Papadakis K, Richard D, Kim NH, Investigators MS (2017) Macitentan for the treatment of inoperable chronic thromboembolic pulmonary hypertension (MERIT-1): results from the multicentre, phase 2, randomised, double-blind, placebo-controlled study. Lancet Respir Med 5(10):785–794. https://doi.org/10.1016/S2213-2600(17)30305-3
112. Teerapuncharoen K, Bag R (2022) Chronic thromboembolic pulmonary hypertension. Lung 200(3):283–299. https://doi.org/10.1007/s00408-022-00539-w
113. Miwa H, Tanabe N, Jujo T, Kato F, Anazawa R, Yamamoto K, Naito A, Kasai H, Nishimura R, Suda R, Sugiura T, Sakao S, Ishida K, Masuda M, Tatsumi K (2018) Long-term outcome of chronic thromboembolic pulmonary hypertension at a single Japanese pulmonary endarterectomy center. Circ J 82(5):1428–1436. https://doi.org/10.1253/circj.CJ-17-1242
114. Ghofrani HA, D'Armini AM, Grimminger F, Hoeper MM, Jansa P, Kim NH, Mayer E, Simonneau G, Wilkins MR, Fritsch A, Neuser D, Weimann G, Wang C, Group C-S (2013) Riociguat for the treatment of chronic thromboembolic pulmonary hypertension. N Engl J Med 369(4):319–329. https://doi.org/10.1056/NEJMoa1209657
115. Sulica R, Fenton R, Cefali F (2015) Early observations on the use of riociguat in a large, metropolitan pulmonary arterial hypertension/chronic thromboembolic pulmonary hypertension treatment center. Cardiol Ther 4(2):209–218. https://doi.org/10.1007/s40119-015-0046-y
116. Miao R, Wang Y, Wan J, Leng D, Gong J, Li J, Liang Y, Zhai Z, Yang Y (2017) Microarray expression profile of circular RNAs in chronic thromboembolic pulmonary hypertension. Medicine (Baltimore) 96(27):e7354. https://doi.org/10.1097/MD.0000000000007354
117. Kamath-Rayne BD, Saal H, Lang S, Habli M (2013) Recurrent severe oligohydramnios and fetal pulmonary hypoplasia associated with ErbB4 mutation. Obstet Gynecol 121(2 Pt 2 Suppl 1):499–501. https://doi.org/10.1097/aog.0b013e31827c60e1
118. Chair SY, Chan JYW, Waye MMY, Liu T, Law BMH, Chien WT (2021) Exploration of potential genetic biomarkers for heart failure: a systematic review. Int J Environ Res Public Health 18(11):5904. https://doi.org/10.3390/ijerph18115904

Circular RNAs and Cancers

18

Mohammad Mahdi Heidari,
Mohammad Saeid Moeinfar, Mahdi Rafiyan,
Parsa Sepehri, Ameneh Jafari, Arash Raisi,
Fatemeh Aslanbeigi, Ashkan Bahrami,
Maryam Nasr Azadani, Neda Rahimian,
Maryam Mahjoubin-Tehran, and Hamed Mirzaei

Abstract

Circular RNAs (circRNAs) are a recently discovered type of endogenous RNA molecules that are formed through the back-splicing process of exons in precursor messenger RNAs (pre-mRNAs), resulting in a unique covalently closed circular structure. CircRNAs are just beginning to be acknowledged for their roles as microRNA sponges and protein scaffolds, their ability to regulate transcription and splicing, and their occasional use as templates for polypeptide synthesis. However, the exploration of their functions is still in its early stages. circRNAs have essential functions in a wide range of different cellular functions. The abnormal presence of circRNAs in tumors contributes to continuous cell proliferation and growth, enhances invasion capabilities, and evades the process of cell aging and death. This indicates the possibility of using them as markers and targets for treatment in a clinical setting. This article provides a thorough examination of current advancements in the study of circular RNAs (circRNAs) in cancer, focusing particularly on their molecular mechanisms and potential impact in a clinical setting.

M. M. Heidari
Department of Pediatric, School of Medicine, Iran University of Medical Sciences, Tehran, Iran

Student Research Committee, Kashan University of Medical Sciences, Kashan, Iran

School of Medicine, Kashan University of Medical Sciences, Kashan, Iran
e-mail: Heidari.mo@iums.ac.ir

M. S. Moeinfar · M. Rafiyan · P. Sepehri · A. Raisi
F. Aslanbeigi · A. Bahrami · M. Nasr Azadani
Student Research Committee, Kashan University of Medical Sciences, Kashan, Iran

School of Medicine, Kashan University of Medical Sciences, Kashan, Iran

A. Jafari
Chronic Respiratory Diseases Research Center, NRITLD, Shahid Beheshti University of Medical Sciences, Tehran, Iran

N. Rahimian (✉)
Endocrine Research Center, Institute of Endocrinology and Metabolism, Iran University of Medical Sciences (IUMS), Tehran, Iran

Department of Internal Medicine, School of Medicine, Firoozgar Hospital, Iran University of Medical Sciences, Tehran, Iran
e-mail: Rahimian.n@iums.ac.ir

M. Mahjoubin-Tehran (✉)
School of Pharmacy, Mashhad University of Medical Sciences, Mashhad, Iran

H. Mirzaei (✉)
Research Center for Biochemistry and Nutrition in Metabolic Diseases, Institute for Basic Sciences, Kashan University of Medical Sciences, Kashan, Iran

J. Xiao (ed.), *Advances in Circular RNAs*, Advances in Experimental Medicine and Biology 1485,
https://doi.org/10.1007/978-981-96-9428-0_18

Keywords

Cancer · Noncoding RNAs · Circular RNAs

1 Introduction

Circular RNAs, also known as circRNAs, are a type of RNA molecule that is circular in structure and does not have distinct start and end points or poly (A) tails. The discovery of circular RNAs (circRNAs) in plant-based viruses dates back to 1976, marking a span of 40 years since their initial detection [1]. In 1979, it was discovered that eukaryotes produce an RNA splicing product known as circRNAs, and in 1986, they were also identified in humans after being triggered by hepatitis delta virus [2, 3]. Initially, circRNAs were thought to be useless compounds resulting from errors in RNA splicing, and therefore, they have not received enough recognition in the scientific community. During the year 2012, Salzman and colleagues conducted a study that identified the prevalence of circRNA in mammalian cells, concluding that over 10% of active genes have the capacity to generate circRNAs [4]. Hansen et al. and Memczak et al. published findings in 2013 revealing that CDR1as (also referred to as ciRS-7), a type of circular transcript found in connection with cerebellar degeneration-related protein 1 antisense RNA, can act as a miRNA sponge for miR-7 [5, 6]. These pieces of writing have revolutionized circRNAs into a central subject of study and emerging leaders in the noncoding RNA realm.

CircRNAs have become increasingly prevalent in eukaryotic cells and have been observed to have dynamic expression patterns in different developmental stages and under various physiological circumstances, thanks to the advancements in high-throughput deep RNA sequencing and bioinformatics technology [7, 8]. Numerous studies have conclusively proven that circRNAs play a significant role in the development of numerous human illnesses, such as neurodegenerative disorders, heart conditions, Alzheimer's disease, joint degeneration, diabetes, silicosis, and cancer [9–15].

Indeed, circRNAs have been found to have integral functions in promoting the progression, spread, maintenance of stemness, and resistance to treatment in cancer [16, 17]. CircRNA, an integral part of the RNA interaction system, has been established as highly prevalent, enduring, varied, and lasting [7].

New research indicates that circRNAs have diverse roles, including acting as natural RNAs that absorb miRNAs, controlling the activity of parent genes, influencing the process of alternative splicing, managing interactions between RNA and proteins, and functioning as supportive structures in the formation of protein complexes [18, 19].

2 Circular RNAs and GI Cancers

Hepatocellular carcinoma (HCC) is now a major cancer worldwide, causing a significant number of deaths due to its high mortality rate. The rate of occurrence of liver cancer is more substantial among males compared to females, and it is linked to certain risk factors such as hepatitis B and C, smoking, diabetes, and infections from Aspergillus [20]. Research has shown that numerous circRNAs have the potential to serve as biomarkers for detecting HCC and can also act as targets for hindering tumor progression. One instance is the detection of hsa-circ-0004018, which is particularly responsive to alpha-fetoprotein, while circ-0016788 has the potential to be utilized as a tailored therapy for individuals with HCC [21]. In the same way, circZKSCAN1 and circPCNX have the potential to serve as indications of liver cancer, whereas hsa-circRNA-100084 behaves as a regulator for hsa-miR-23A-5P, enhancing the expression of IGF2 [22, 23]. The transfer of circ-0051443 from healthy cells to cancer cells through exosomes leads to the exertion of tumor-suppressing effects, effectively prohibiting the growth of HCC [24]. The gene circ-0091579 plays a crucial role in preventing the growth and spread of HCC by specifically targeting miRNA-490-3p. Conversely, circABCB10 functions as a "sponge" for miR-670-3p, resulting in an increase in HMG20A levels and ultimately facilitating HCC progression. Nonetheless, the suppression of circABCB10 results in the hindering of tumor cell migration and spread. Further exploration of circRNAs' functions in liver cancer is essential in order to gain a deeper understanding of their impact and potential for treatment. Many patients diagnosed with liver cancer frequently encounter difficulties with drug resistance. Various studies have revealed that circRNAs play a significant role in the development of drug resistance in these individuals. One instance is the connection between circRNA-101237 and the sensitivity of liver cancer patients to cisplatin treatment. Elevated levels of circRNA101237 have been observed in the blood of hepatocellular carcinoma (HCC) patients who do not respond to cisplatin treatment and in Huh7 cells. Those with a high amount of this circRNA often have an unfavorable outlook and a low chance of survival. The ability of circRNA_101237 to act as a stand-alone indicator for forecasting the outcome of individuals with liver cancer is notable [25]. Hsa-circ-u0006294, hsa-circ-u0035944, and hsa-circ-u0084663 are capable of activating the Wnt/β-catenin pathway, which leads to the inhibition of HCC growth. Additionally, in cases of HCC, hsa-circ-u0006294 and hsa-circ-u0035944 have been identified as downregulated and have been shown to play a role in sorafenib resistance [26]. Similarly, in hepatocellular carcinoma (HCC), the expression of circ-0003418, which hinders the Wnt/β-catenin signaling pathway, is reduced; however, this decrease enhances the susceptibility of HCC to cisplatin. Conversely, the suppression of circ-0003418 leads to the proliferation and advancement of HCC [27]. The level of circFBXO11 is increased in HCC, leading to the enhancement of both disease advancement and resistance to oxaliplatin by activating the miR-605/FOXO3/ABCB1 pathway [28].

CircRNAs have been the subject of significant research interest due to their involvement in tumorigenesis and various other diseases. There is a growing body of evidence suggesting a correlation between irregularities in circRNAs and the development of colorectal cancer (refer to Table 18.1 and Fig. 18.1). For example, it was found that circHERC4 levels were significantly increased in colorectal cancer (CRC) tissues and were also closely linked to both the spread of cancer cells to nearby lymph nodes and the progression of the tumor [29]. On the contrary, circ-PLCE1 expression was decreased in colorectal cancer (CRC) tissue samples and was found to be correlated with lower survival rates and more advanced clinical stages [30]. It is intriguing to note that the levels of circRNAs exhibit variation among various forms of cancer, suggesting unique biological functions. For instance, circH-IPK3 showed a notable increase in expression levels in tissues and cell lines of gastric cancer, HCC, breast cancer, CRC, and lung cancer, while displaying a decrease in expression in bladder cancer [31–34]. Many research studies have documented the irregular expression of circRNAs in different types of cancers; however, the underlying reasons for their dysregulation are mostly unidentified and require additional exploration [35].

It is possible that the disordered expression of circular RNA in colorectal cancer is a result of irregular expression of host genes, such as chromosomal amplification, depletion, or translocation. CircPRKCI is a circular RNA that is found in the 3q26.2 region, which has shown to have proto-oncogenic properties in various types of cancers such as lung cancer, glioma, and esophageal cancer [36]. As a result, the increased expression of circPRKCI in cancer cells could potentially be attributed to the amplification of PRKCI. Fusion-circular RNAs, also known as F-circular RNAs, are produced as a result of chromosomal translocations that are associated with cancer in the genes of the host [37]. For instance, in non-small cell lung cancer, cell migration was enhanced by two new circular RNAs (F-circSR1 and F-circSR2) produced by the oncogenic SLC34A2-ROS1 fusion gene [38]. Moreover, F-circEA derived from the fusion of oncogenic EML4-ALK promoted the movement and invasion of cells in lung cancer [39]. The combination of genes creates operative proteins that play a role in the development of cancer.

Numerous experiments have revealed that RNase L can eliminate circRNAs, particularly

Table 18.1 Various circular RNAs in GI cancers

Type of cancer	Circular RNAs	Expression	Target	References
Cholangiocarcinoma	CircNFIB	Down	MEK1/ERK signaling	[80]
	circACTN4	Up	YBX1	[81]
	cGGNBP2	Up	cGGNBP2-184aa	[82]
	CircRNA MBOAT2	Up	PTBP1 FASN	[83]
	CircZNF215	Up	PTEN/AKT pathway	[84]
	circ-CCAC1	Up	Yin Yang 1 (YY1)	[85]
	Circular RNA Circ0021205	Up	MiR-204-5p/RAB22A Axis	[86]
	hsa_circ_0001649	Down		[87]
	Circ_0005230	Up	miR-1238 and miR-1299	[88]
	Circ-SMARCA5	Down		[89]
	circEIF3C	Up	miR-34a-5p/B7–H4 axis	[90]
	circ-ZNF609	Up	miR-432-5p LRRC1 YY1 and eIF4A3	[91]
	circLTBP2	Up	miR-338-3p TGFβ	[92]
	circDNM3OS	Up	miR-145-5p MORC2	[93]
	circSLCO1B3	Up	HOXC8 and PD-L1	[94]
	CircRNA SMARCA5	Down	miRNA-95-3p and TNF receptor-associated factor 3 gene (TRAF3)	[95]
	Circ-LAMP1	Up	miR-556-5p- and miR-567-mediated YY1 activation	[96]
	CircPCNXL2	Up	STRAP	[97]
	CircSETD3 (hsa_circ_0000567)	Down	microRNA-421	[98]
	Circ-RAPGEF5	Up	miR-3185 SAE1	[99]
	Circ_0007534	Up	DDX3X	[100]
	CDR1as	Up	MicroRNA 641	[101]
	cPKM	Up	STMN1/TGFB1 axis	[102]
	Circ_0000284	Up	miR-152-3p-mediated PDK1 expression	[103]
	circHMGCS1-016	Up	CD73 and GAL-8	[104]
	circ_0059961	Down	miR-629-5p/SFRP2 axis	[105]
	Hsa_circ_0019054	Up	miR-340-5p	[106]
	circ_0000591	Up	miR-326/TLR4/MyD88/IL6 axis	[107]
	circRTN4IP1	Up	miR-541-5p HIF1A	[108]
Esophageal cancer	CircBCAR3	Up	miR-27a-3p	[109]
	circCYP24A1	Up	PKM2	[110]
	Circular RNA LPAR3	Up	microRNA-198	[111]
	circular RNA AKT3	Down	miR-17-5p	[112]
	circPRKCI	Up	miR-186-5p/PARP9 axis	[113]
	hsa_circ_0000554	Up	miR-485-5p/fermitin family members 1 axis	[114]
	circ_0014715	Up		[115]
	circ_0086414	Down	miR-1290	[116]
	circRAD23B	Up	miR-5095 PARP2 and AKT2	[117]
	circ_0006168	Up	miR-384/RBBP7 axis S6K/S6 pathway	[118]

(continued)

Table 18.1 (continued)

Type of cancer	Circular RNAs	Expression	Target	References
	hsa_circ_0000277	Up	miR-873-5p/SOX4 axis	[119]
	hsa_circ_0000277	Up	miR-4766-5p/LAMA1 axis	[120]
	Circ_0001093	Up	miR-579-3p/glutaminase axis	[121]
	circ-TNRC6B	Down	miR-452-5p/DAG1 axis	[122]
	hsa_circ_0000700	Up	miR-1229	[123]
	circFNDC3B	Up	?	[124]
	circ-LRP6	Up	miR-182 Myc	[125]
	circ_0023984	Up	miR-1294/PI3K/Akt/c-Myc pathway	[126]
	circLARP4	Down	PTEN/PI3K/AKT pathway	[127]
	circPOLR1C	Up	miR-361-3p apoptosis- and EMT-related genes	[128]
	circ-Foxo3	Down	miR-23a/PTEN axis	[55]
	circ_0001273	Up	miR-622/SLC1A5 signaling axis	[129]
	circ-AGFG	Up	microRNA-497-5p/solute carrier family 1 member 5 axis	[130]
	CircPVT1	Up	MiR-30a-5p/FZD3 axis	[131]
	CircRNA-0008717	Up	miR-203/Slug	[132]
	hsa_circ_0003823	Up	miR-607/CRISP3 axis	[116]
	circTRPS1-2	Down	Reducing the production of ribosomes	[133]
	CircRNA VIM	Down	miR-124/PD-L1 axis	[134]
	Circular FAT1	Down		[135]
	hsa_circ_ RNA0023397	Down	miR-106b	[136]
	CircGOT1	Up	GOT1	[137]
	circ_0003340	Down	miR-564/TPX2	[138]
	has_circ_0067934	Up		[139]
	circRUNX1	Up	miR-449b-5p/FOXP3 axis	[140]
	Hsa_circ_0006948	Up	miR-490-3p/HMGA2 axis	[52]
	circ_0001944	Up	miR-338-5p PDK1	[141]
	circ_0072088	Up	miR-377	[142]
	Circ-POSTN	Up	miR-876-5p/FYN axis	[143]
	circ-FIRRE	Up	GLI2 mRNA HNRNPC	[144]
	circ_0006948	Up	microRNA-3612/LASP1 Axis	[145]
	Circ_0058063	Up	miR-377-3p/HOXA1 axis	[146]
	circ-231	Up		[147]
	Circ-PRMT5	Up	miR-203	[148]
	Circ_0024108	Up	miR-488-3p and USP14	[149]
	Hsa_circ_0004370	Up	miR-1294/LASP1 pathway	[150]
	circMMP1	Up	MiR-671-5p	[151]
	hsa_circ_0000977	Up	miR-874-3p	[152]
	circ_0001946	Down	miR-1290/SOX6	[153]
	Circular RNA ciRS-7	Up	miR-7/KLF4 and NF-κB signals	[58]
	circFIG 4	Up	miR-493-5p/E2F3 axis	[154]
	CircATIC	Down	miR-10-3p RHCG	[155]
	Circ_0003855	Up	miR-622/FLOT1	[156]
	circHIPK3	Up	microRNA-124 Serine/threonine kinase 3	[157]

(continued)

Table 18.1 (continued)

Type of cancer	Circular RNAs	Expression	Target	References
	Hsa_circ_0001615	Up	miR-142-5p/β-catenin	[158]
	Circ-SFMBT2	Up	miR-107-dependent regulation of SLC1A5	[159]
	CircLPAR3	Up	miR-873-5p/LDHA axis	[160]
	hsa_circ_0000654	Up	miR-149-5p/IL-6/STAT3 pathway	[53]
	Hsa_circ_0046534	Up	miR-339-5p MMP2	[161]
	Circ_0005231	Up	miR-383-5p KIAA0101	[162]
	Circ-TTC17	Up		[163]
	circSSPO	Up	micrRNA-6820-5p-mediated KLK8 and PKD1	[164]
	circNTRK2	Up	miR-140-3p NRIP1	[165]
	Circ_0023984	Up	miR-433-3p/REV3L Axis	[166]
	Circ_0003340	Up	miR-940/PRKAA1 axis	[167]
	hsa_circ_0004771	Up	miR-339-5p/CDC25A axis	[51]
	circUBE4B	Up	MAPK/ERK signaling	[168]
	Circ_0003340	Up	miR-874-3p ENAH	[169]
Colorectal cancer	Circ3823	Up	TCF7	[170]
	hsa_circ_0008285	Down	PTEN	[171]
	circRNA 100,146	Up	HMGA2	[172]
	circ_0000467	Up	EN2	[173]
	circDENND4C	Up	GLUT1	[174]
	circMETTL15	Up	ESCO2	[175]
	circCTNNA1	Up	FOXM1	[176]
	circSPARC	Up	STAT	[177]
	Hsa_circ_0001666	Down	PCDH10	[178]
	circ_0007142	Up	SGK1	[179]
	circ_0065378	Down	TUSC1	[180]
	Hsa_circ_0006732	Up	RAB3D	[181]
	circ-FARSA	Up	LASP1	[182]
	circN4BP2L2	Up	CXCR4	[183]
	circ-ERBIN	Up	4EBP-1	[184]
	circNOX4	Up	CKS1B	[185]
Gastric cancer	circDLG1	Up	CXCL12	[186]
	circEIF4G3	Down	δ-catenin ubiquitin SIK1	[187]
	circNRIP1	Up	AKT1/mTOR pathway	[188]
	hsa_circ_0004872	Down	Smad4/ADAR1	[189]
	circCUL2	Up	ROCK2	[190]
	circ_0000039	Up	DEK	[191]
	circAFF2	Up	ANTXR 1	[192]
	CircCACTIN	Up	TGFBR1	[193]
	hsa_circ_0009172	Down	NTRK3	[194]
	circ-ITCH	Down	Klotho	[195]
	hsa-circ-0000670	Up	SIX4	[196]
	Circ_0081146	Up	HMGB1	[197]
	circATXN7	Up	ENTPD4	[198]
	circFGD4	Down	APC/β-catenin signaling pathway	[199]
	circNHSL1	Up	YWHAZ	[200]
	circ-MAT2B	Up	HIF-1α	[201]

(continued)

Table 18.1 (continued)

Type of cancer	Circular RNAs	Expression	Target	References
Pancreatic cancer	circ_0066147	Up	E2F2	[202]
	hsa_circ_0006117	Up	KRAS/MAPK	[203]
	ciRS-7	Up	EGFR/STAT3	[75]
	circ-LDLRAD3	Up	PTN	[72]
	CDR1as	Up	E2F3	[204]
	circSFMBT1	Up	PAK1	[205]
	Circ_03955	Up	HIF-1α	[206]
	Circ-STIL	Up	AQP3	[207]
	hsa_circ_0050102	Up	NPSR1	[208]
	circBFAR	Up	MET/Akt	[68]
	CircSEC24A	Up	TGFBR2	[209]
	Circ-MBOAT2	Up	GOT1	[210]
	circNFIB1	Down	PIK3R1/VEGF-C	[211]
	hsa_circ_0071036	Up	miR-489	[212]
	circ_0001666	Up	SOX4	[213]
	hsa_ circRNA_001587	Up	SLC4A4	[214]
	hsa_ circRNA_104348	Up	RTKN2 Wnt/β-catenin pathway	[215]
	circ_UBXN7	Up		[216]
	CircGPR137B	Down	FTO	[217]
	CircRPN2	Down	ENO1	[218]
	circ_0091579	Up	CASC3	[219]
	circ_0001459	Up	IGF1R	[220]
	circ_0008934	Up	TMTC3	[221]
	has-circ-0001141	Down	miR-184	[222]
	circFBXO11	Up	FOXO3/ABCB1	[28]
	circZNF566	Up	TDO2	[223]
	Circ_0008043	Up	RAB21	[224]
	hsa_circ_0005397	Up	PDK2	[225]
	hsa_circ_0091579	Up	H3F3B	[226]
	circZFR	Up	HMGA2	[227]
	Circ-ZEB1	Up	PIK3CA	[228]

after the discharge of extracellular vesicles, viral invasion, or poly I: C stimulation. The stimulation of RNase L can result in the breakdown of circRNAs, leading to a rise in the phosphorylation of protein kinase R (PKR). CircRNAs may also be expelled from cells through exocytosis, in addition to undergoing degradation. In addition, the presence of circRNAs within exosomes may potentially play a role in the process of cellular communication, highlighting the necessity for further research into the processes of circRNA degradation and transportation outside of cells. Moreover, circular RNAs (circRNAs) are plentiful within the cytoplasm and have the ability to be carried from the cytoplasm to exosomes.

New research has revealed that alterations in RNA at the m6A location have been connected to the breakdown, interpretation, and manifestation of circular RNA in cancerous cells [40]. One of the most common changes observed in eukaryotic noncoding RNAs is the addition of m6A, which impacts various functions of RNA such as its stability, splicing, export, translation, and degradation. These alterations have a significant impact on the biological processes involved. The MALAT1 gene undergoes significant methylation at the m6A site, which results in the inhibition of local RNA structure and promotes the interaction between heterogeneous nuclear ribonucleoprotein C (HNRNPC) and two specific

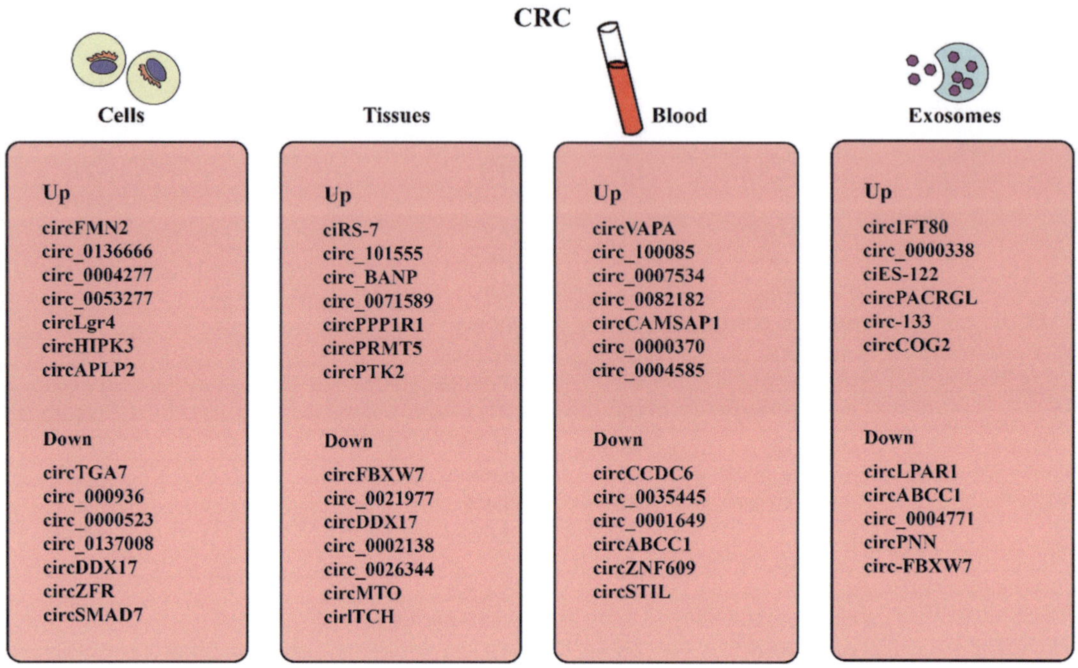

Fig. 18.1 Circular RNAs (circRNAs) display abnormal levels in various sources including blood/serum, cells, colorectal cancer (CRC) tissues, and exosomes derived from CRC patients. The illustration emphasizes the dysregulations of circRNAs, with "Up" indicating an increase and "Down" indicating a decrease in expression

residues of the gene through an "m6A switch" process [40]. The modification of m6A led to a quicker movement of circNSUN2 to the cytoplasm, ultimately enhancing the durability of HMGA2 mRNA. This, in turn, triggered the spread of CRC by producing a complex of circNSUN2, IGF2BP2, and HMGA2 in the cytoplasm [41]. Lately, circRNAs that have undergone m6A modification have been detected through the application of specific methylation patterns based on cell types. The process of m6A modification attracts the proteins YTHDF3 and eIF4G2 in order to control the production of proteins from circular RNAs. ALKBH5 and METTL3 have the ability to impact the production of circRNAs by modifying the levels of m6A. As an illustration, the levels of m6A in circRNA-SORE were increased in cells that were resistant to sorafenib. However, they were reduced when the process of m6A modification was inhibited [42]. The most recent discoveries revealed that METTL3-mediated m6A modification played a key role in stabilizing the expression of circCUX1 [43]. Therefore, when m6A sites are lost or m6A is removed from circRNAs, the methylation levels of these circRNAs may decrease, leading to the disruption of their proper functioning. In addition, the relaxation of restrictions on pivotal elements during the breakdown of circRNAs could potentially lead to atypical levels of circRNA functioning.

Gastric cancer (GC) is the predominant form of tumor within the digestive system, with a staggering 1 million new cases reported in 2018 and an estimated 783,000 fatalities attributed to it [44]. These numbers show that GC is the fifth most commonly diagnosed form of cancer and ranks as the third main factor of death caused by cancer [45]. Because of the significant number of deaths attributed to GC, it is imperative to urgently discover viable molecular indicators for detecting it in its initial stages, as well as potential targets for its treatment.

The function of circHIPK3 is to act as a sponge for miR-124 and miR-29, leading to a decrease in the expression of these microRNAs.

This, in turn, results in an increase in the proliferation of gastric cancer cells. In this procedure, the circHIPK3-miRNA-124/miRNA-29b axes cause the mRNA transcription levels of three specific genes (COL1A1, COL4A1, and CDK6) to increase in GC cells. The potential outcome of GC's survival can be foreseen by identifying the levels of expression for both COL1A1 and COL4A1. These results demonstrate that not just circHIPK3, but also COL1A1 and COL4A1, can act as predictive indicators for the lifespan of individuals with GC. This enhances our comprehension of the genetic processes involved in GC [46].

According to previous studies, miR-206 has been found to act as an inhibitory microRNA in gastric cancer (GC). When it is overexpressed, it can prevent the expression of CXCR4. Circ_0056618, a gene overexpressed in gastric cancer, acts as a competitive inhibitor of miR-206 in gastric cancer. This inhibits the growth and spread of cancer cells by increasing the levels of CXCR4. As a result, the combination of circ_0056618, miR-206, and CXCR4 effectively increases the likelihood of survival for individuals diagnosed with GC, as they work together to restrain the growth and spread of GC cells [47].

Circular RNAs, known as circRNAs, possess the ability to act as tumor-suppressing genes. In particular, circLARP4, which is originated from the ninth and tenth exons of the LARP4 gene, exhibits high levels of expression in gastric cancer. By serving as a sponge for miR-424, this process can suppress DNA synthesis, as well as the growth and invasion of cells within the body. This is achieved through the activation of the LATS1 and YAP genes within GC. Therefore, it can be concluded that circLARP4 plays a crucial role in inhibiting tumor growth in GC through its control over the miR-424/LATS1/YAP signaling pathway. The presence of circLARP4 can function as a standalone indicator for predicting the 5-year overall survival rate in individuals with GC, as well as those receiving chemotherapy treatment [48].

Esophageal cancer (EC) is a highly destructive form of cancer, boasting a survival rate of only 15–20% within a span of 5 years. The eighth most commonly occurring tumor globally, EC ranks as the sixth top cause of deaths attributed to cancer. Because of its unfavorable outlook and limited chances of survival, it is especially crucial to pinpoint dependable biomarkers in order to detect it at an early stage [49].

According to a research investigation, there are a total of 1045 circRNAs that have shown an increase and 1032 that have shown a decrease in expression in EC. Numerous studies support the notion that circRNAs play a crucial role in controlling gene transcription in EC [50]. An illustration of this can be seen in the case of hsa-cric-0004771, which plays a role in the promotion of esophageal squamous cell carcinoma by functioning through the miR-339-5P/CDC25A pathway. However, by suppressing the expression of hsa-cric-0004771, the growth and movement of ESCC is hindered. Nevertheless, the detailed explanations for how it works are yet to be fully understood [51]. The expression of circUBAP2 is increased in esophageal squamous cell carcinoma (ESCC) and functions as a competitive inhibitor of miR-422a. Reducing the presence of circUBAP2 effectively hinders the advancement of ESCC and hinders the dissemination of cancerous cells (54). On a similar note, the increased expression of hsa-circ-0006948 in ESCC facilitates the proliferation and maturation of cancer cells by means of the miR-490-3p/HMGA2 pathway [52]. Likewise, the expression of circ-0000654 is increased in esophageal squamous cell carcinoma (ESCC), leading to enhanced cell growth and movement through the regulatory pathway of miR-149-5p/IL-6/STAT3. In contrast, inhibiting circ-0000654 has the ability to hinder tumor advancement [53]. One of the circRNAs that is increased in ESCC is circGSK3. Its role in promoting the proliferation of tumor cells is achieved through the activation of β-catenin signaling. This is strongly associated with a negative prognosis [54]. The expression of irc-Foxo3 is notably reduced in ESCC and this actively influences the miR-23a/PTEN pathway to prevent the growth and progression of ESCC [55]. A study revealed a noteworthy increase in circPVT1 expression in EC, proposing its involvement in the growth of TE-10 tumor cells through the regulation of miR-4663 [56].

The overexpression of ciRS-7 in ESCC is significantly higher than normal and contributes to the enhancement of tumor cell proliferation. This gene has the potential to function as an oncogene, accelerating the growth of tumor cells by acting as a sponge for miR-876-5p. Further research by Meng et al. has revealed that ciRS-7 plays a role in hindering autophagy in ESCC triggered by a lack of nutrients or the use of rapamycin as an immunosuppressant [57]. Huang and colleagues firmly established that ciRS-7 plays a significant role in enhancing the ability of ESCC cells to migrate and invade through the facilitation of miR-7/KLF4 and NF-κB signaling pathways [58]. Suppression of KLF-4 in ESCC leads to a decrease in the invasive capacity of ciRS-7 [57, 58]. The development of drug resistance in EC is influenced by a variety of factors including cell growth, spread to other parts of the body, and levels of glycolytic enzymes. Additionally, circRNAs, lncRNAs, miRNAs, and the enzyme ACSS2 may play a role in this process. Out of the previously mentioned elements, lncRNAs have been extensively examined. One instance is the impact of ACSS2 on the intake of cisplatin in ESCC, which shields cancer cells from the effects of cisplatin [59]. On the other hand, there is currently a lack of studies exploring the connection between circRNAs and drug resistance in EC. However, previous research has shown that circRNA-001275 contributes to the activation of Wnt7a, a member of the Wnt family, leading to the development of resistance against cisplatin. This effect is achieved through the inhibition of miR-370-3p [60].

Gallbladder cancer is the prevailing cancerous tumor of the biliary system, and it has a 5-year survival rate of only 5–10% [61, 62]. Gallbladder tumors are identified either unexpectedly or through symptoms displayed by the patient [63]. Typically, individuals suffering from gallbladder cancer and displaying symptoms are found at a progressed stage of the disease. As a result, the vast majority (90%) of gallbladder cancers are identified at regional or metastatic levels during diagnosis. Surgical removal is the widely accepted and only effective way to cure gallbladder cancer. Alternative treatments, including chemoradiotherapy, have shown little success in most instances when surgical resection is not performed [64]. Extensive investigation is needed to uncover possible biomarkers that can aid in the early identification and prediction of outcomes for gallbladder cancer, as it has a significant rate of recurrence, morbidity, and mortality.

New research has shown definitively that circHIPK3 has a crucial impact in both colon cancer and gallbladder cancer, emphasizing its significance. The expression of this gene is increased in the tissues of individuals with gallbladder cancer and can control the reduction of miR-124 and the increase of ROCK1-CDK6 by acting as a miRNA sponge, ultimately supporting the growth of gallbladder cancer cells [65]. circFOXP1 (hsa_circ_0008234), which originates from the section of the FOXP1 gene that contains exons, is increased in both gallbladder cancer tissues and cells. By acting as a sponge for miR-370, it can facilitate the expression of PKLR in GBC cells, ultimately leading to the advancement of the Warburg effect in the progression of GBC. The impact entails stimulating the growth, movement, and invasion of cells, as well as preventing cell death in GBC [66].

Out of all types of cancers, pancreatic cancer (PC) has the seventh highest rate of deaths. The prevalence of PC is more common among males compared to females, and it is particularly widespread in North America, Australia, and Europe [67]. Many circRNAs have been identified as being associated with prostate cancer (PC) and could potentially serve as targets for therapy. One instance is the case of circBFAR (hsa-circ-0009065), which increases in expression in pancreatic ductal adenocarcinoma (PDAC). It functions as a sponge for miR-34b-5p and influences the MET/PI3K/Akt signaling pathway [68]. The gene circFOXK2 is also increased in levels in pancreatic ductal adenocarcinoma (PDAC) and facilitates the production of ANK1, GDNF, and PAX6 by functioning as a reservoir for miR-942, ultimately stimulating the growth of cancerous cells. In accordance with this, the reduction of circFOXK2 resulted in hindrance of tumor advancement [69]. The circ-ADAM9 gene exhibits significant expression levels in prostate cancer (PC) and plays a crucial role in activating the ERK/VEGF signaling pathway through the regulation of miR-217 [70]. In the context of PDAC, irc-ASH2L is markedly overexpressed and functions as a reservoir for miR-34a, facilitating the growth of cancerous cells and advancing the development of tumors through its control of Notch1 [71]. The circ-LDLRAD3 gene exhibits its increased expression in PC and plays a role in controlling cell growth through the miR-137-3p/PTN pathway. The prognosis of PC is impacted and the invasion of tumor cells is prevented by its suppression [72]. The expression of circ-0000977 is elevated in pancreatic cancer, and it functions as a competitive inhibitor for miR-153. The attenuation of circ-0000977 demonstrates the diminishing impact of HIF1A and ADAM10 on pancreatic cancer cells in a low oxygen environment [73]. The expression of hsa-circRNA-0007334 is increased in pancreatic ductal adenocarcinoma (PDAC) and it acts as a competitor for binding with hsa-miR-144-3p [74]. In the same way as in EC, the significant increase in ciRS-7 levels contributes to the progression of PDAC by functioning as a sponge for miR-7 and controlling the activity of the EGFR/STAT3 signaling pathway. Downregulation of circular RNA sponge for miR-7 suppresses the growth and infiltration of cancerous cells [75]. The gene circ-0030235 shows increased levels in PDAC and functions as a miRNA sponge for both miR-1253 and miR-1294. Higher expression of this gene is associated with a lower prognosis in PDAC [76]. The expression of circZMYM2 is significantly increased in pancreatic cancer. It functions as a sponge for miR-335-5p and controls the activity of the downstream oncogene JMJD2C. Downregulation of circZMYM2 promotes the suppression of tumor development, while the increased expression of circ-0007534 in PDAC functions as a decoy for miR-625 and miR-892b, stimulating the proliferation and infiltration of cancerous cells [77]. The expression of circ-IARS is enhanced in pancreatic ductal adenocarcinoma (PDAC), where it functions as a sponge for miR-122, leading to increased RhoA activity and F-actin expression. This results in a decrease in ZO-1 expression and ultimately promotes the formation and spread of tumors by heightening the permeability of endothelial cell monolayers, as well as facilitating invasion and metastasis [78]. Additionally, it can be noted that circ-PDE8A is increased in pancreatic ductal adenocarcinoma (PDAC) and serves as a buffer for miR-338, controlling the levels of MACC1 and promoting the proliferation of malignant cells through activation of the MACC/MET/ERK or AKT signaling pathways [79]. Table 18.1 lists various circular RNAs in GI cancers.

3 Circular RNAs and Gynecological Cancers

Cervical cancer (CC) is ranked as the fourth most prevalent type of cancer among women across the globe, causing over 300,000 fatalities annually [229]. It is widely known that high-risk types, specifically 16 and 18, of the human papillomavirus (HPV) are the primary cause of CC [229]. Extensive research has confirmed that a significant amount of circRNAs exhibit varying levels of expression in CC tissues when compared to healthy cervical tissues, functioning as either oncogenes or tumor suppressors in the development and advancement of CC [230–243]. A comprehensive list of all the circular RNAs (circRNAs) that exhibit abnormal expression in CC is presented in Table 18.2. Additionally, their roles and functions in CC are visually depicted in Fig. 18.2. Numerous investigations have exhibited that circRNAs act as miRNA reservoirs, thereby exerting control over various facets of cervical cancer advancement by modulating a multitude of signaling pathways (**see** Fig. 18.2). The research findings clearly indicate that the networks formed by circRNA, miRNA, and mRNA as well as the presence of a circRNA that encodes the HPV E7 oncoprotein have significant impacts on the regulation of cervical cancer development and evolution.

A variety of circular RNAs, specifically circ0007534, circEIF4G2, circ0075341, circ0005576, circ0000515, circSLC26A4, and circAMOTL1, have been previously identified as contributors to the development of CC by elevat-

Table 18.2 Various circular RNAs in gynecological cancers

Type of cancer	Circular RNAs	Expression	Target	References
Cervical cancer	circRNA_0000285	Up	FUS	[251]
	circ0001955	Up	NCAPG2	[290]
	hsa_circ_0043280	Down	PAQR3	[291]
	circCDK6	Down	miR-449a	[292]
	circLMO1	Down	ACSL4	[293]
	circ_0000119	Up	PAK2	[294]
	circEPSTI1	Up	SLC7A11	[295]
	circ_0001823	Up	RAB8A	[296]
	Circ_0003221	Up	CPEB4	[297]
	circRNA_101996	Up	TRIM37	[298]
	circ_0011385	Up	SOX4	[299]
	circSLC26A4	Up	HOXA7	[244]
	hsa_circ_0011385	Up	PRDX6	[300]
	circ_0000326	Down	CDK4	[301]
	hsa_circ_0003204	Up	MAPK pathway	[302]
	circUBAP2	Up	SOX4	[303]
Endometrial cancer	hsa_circ_0011324	Up	Rapamycin kinase (mTOR)	[304]
	Circ-IFT80	Up	FAM98A	[305]
	hsa_circ_0023404	Up	MAPK1	[306]
	CircESRP1	Up	CPEB4	[307]
	CircSMAD2	Up	MFGE8	[308]
	circ_0007534	Up	ZEB2	[309]
	hsa_circ_0005797	Up	Catenin delta 1 signaling	[310]
	hsa_circRNA_0001776	Down	LRIG2	[311]
	CircZNF124	Up	SLC7A5	[312]
	circ_0000043	Up	CTNND1	[313]
	Circ_PUM1	Up	NOTCH3	[288]
	hsa_circ_0002577	Up	IGF1R/PI3K/Akt pathway	[314]
	circSLC6A6	Up	PI4KB/Hedgehog axis	[315]
	CircWDR26	Up	MSH2	[316]
	Hsa_circ_0000437	Down	CDKN1B	[317]
	Circ_0002577	Up	MACC1	[318]
Ovarian cancer	hsa_circ_0007444	Down	DICER1	[319]
	Circular PVT1	Up	FOXM1	[320]
	Circular RNA S-7	Up	ZEB1 and MDM2	[321]
	circMTO1	Down	KLF15	[322]
	hsa_circ_0000144	Up	ELK3	[323]
	hsa_circ_0008732	Down	FBXW7	[324]
	circEXOC6B	Down	RUS1	[325]
	CircEXOC6B	Down	FOXO3	[326]
	circCELSR1	Up	BRD4	[327]
	hsa_circ_0078607	Down	Fas signaling pathway	[328]
	Circular RNA-9119	Down	microRNA-21-5p-PTEN-Akt pathway	[329]
	Circ_0026123	Up	RAB1A	[330]
	CircATRNL1	Down	Smad4	[331]
	CircCRIM1	Up	ZEB2	[332]
	Hsa_circ_0001756	Up	EGFR/MAPK signaling pathway	[333]
	Circ_0013958	Up	PLXNB2	[334]
	circRNA-UBAP2	Up	PRPF8	[335]

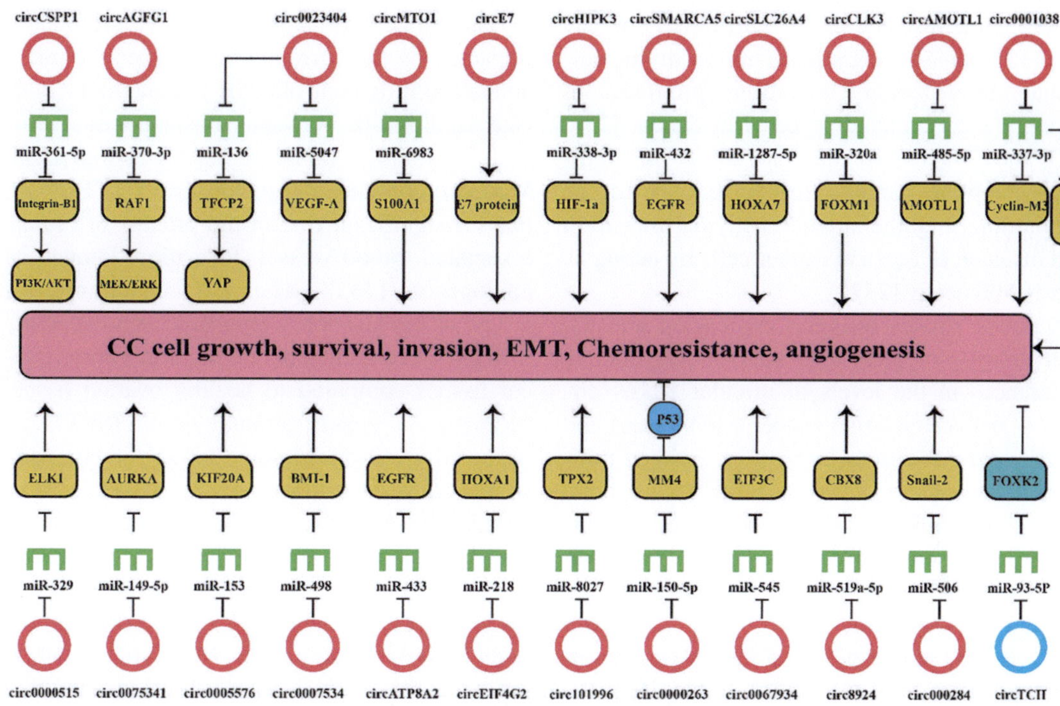

Fig. 18.2 The intricate networks involving circRNA, miRNA, and mRNA, as well as a specific circRNA that encodes the HPV E7 oncoprotein, play significant roles in regulating the progression, formation of new blood vessels, and resistance to chemotherapy in CC. Oncogenic circRNAs are represented by red circles, while tumor-suppressive ones are represented by green circles (This figure is adapted from [289])

ing the expression of targeted downstream mRNAs, including HOXA1, AURKA, ELK1, BMI-1, KIF20A, HOXA7, and AMOTL1, through their role as miRNA sponges [244–250]. Moreover, it was observed that CC tissues displayed elevated levels of circRNAs, including circ0000285, circ0018289, and circMYBL2. This increase in circRNA expression was positively correlated with enhanced invasion and metastasis of CC cells [236, 251, 252].

The PI3K/AKT signaling pathway plays a crucial role in regulating numerous cellular functions, such as proliferation, growth, invasion, and chemoresistance. It is a frequently altered pathway in various types of human cancers, including CC, contributing to the development and progression of the disease [253, 254]. Integrin β1 has the ability to trigger the onset and advancement of cancer by activating various downstream pathways, such as the PI3K/AKT signaling pathway [255]. CircCSPP1 promotes cervical cancer cell proliferation and migration by facilitating the activation of the PI3K/AKT pathway through the upregulation of integrin β1 expression, achieved by inhibiting miR-361-5p via a sponge effect [256]. The cascade of RAF/MEK/ERK is a crucial factor in the advancement and growth of cancer [257]. In a recent research report, it was proposed that there was an increase in the expression of circAGFG1 in cell lines of cervical cancer when compared to healthy cervical cells. The study revealed that this increase led to the promotion of cell growth and movement in cervical cancer cells by boosting the expression of RAF1 and activating the MEK/ERK pathway, accomplished by acting as a sponge for miR-370-3p, which suppresses RAF1 [258]. Epithelial–mesenchymal transition (EMT) is a natural process that involves the transformation of epithelial cells into mesenchymal cells, resulting in cancer cells that possess aggressive traits such as invasiveness, characteristics of cancer stem cells, and increased ability to withstand anticancer therapies [259]. Through various mechanisms, recent experimental research has demonstrated that circular RNAs have a significant impact on the epithelial–mesenchymal transition program in human cancers, including cervical cancer [237, 240, 260–264]. As an example, circ000284 indirectly amplifies Snail-2 levels, leading to the facilitation of epithelial–mesenchymal transition and invasion of cervical cancer cells by acting as a miR-506 decoy [237].

Various studies on gene expression and the analysis of biological data have shown notable differences in the levels of circular RNA (circRNA) between ovarian cancer (OC) tissues and healthy ovarian tissues, as well as between miliary and nonmiliary OC types. Furthermore, there were significant contrasts in the levels of circRNA in the serum samples of OC patients compared to healthy individuals, as well as within primary and metastatic sites of OC and between OC cells and noncancerous human ovarian surface epithelial cells [265–270]. Table 18.2 displays the various circRNAs and their corresponding mechanisms observed in OC, while Fig. 18.3 illustrates the biological functions they serve. Certain circular RNAs have the ability to activate their inhibitory effects on tumor growth in ovarian cancer cells as shown in Table 18.2 and Fig. 18.3 [271–279]. Based on these findings, it appears that the most frequently used approach for controlling the cancerous characteristics of ovarian cancer cells by circRNAs is through the use of the "miRNA" sponge mechanism It is noteworthy to mention that a study suggested an alternative mechanism in which circITCH could potentially impede the growth of ovarian cancer cells by reducing the expression of lncRNA HULC. However, the specific relationship between circITCH and HULC is yet to be established [273].

The abnormal activation of the Wnt/β-catenin pathway is a critical element in fueling the capacity of cancer stem cells to regenerate, metastasize, and withstand chemotherapy in ovarian cancer [280]. By absorbing miR-9, circPLEKHM3 plays a vital role in inhibiting cancer growth by enhancing the production of BRCA1, DNAJB6, and KLF4. This causes the PI3K/AKT and Wnt/β-catenin signaling pathways to become inactive in cells of ovarian cancer [278].

The process of angiogenesis is profoundly impacted by the surrounding microenvironment and is carefully controlled by maintaining a delicate equilibrium between substances that promote and inhibit blood vessel formation [281]. Vascular endothelial growth factor (VEGF-A) plays a significant role in the process of tumor-associated blood vessel formation known as angiogenesis [282]. The expression of circRhoC, a type of circular RNA obtained from the RhoC mRNA, was significantly elevated in ovarian cancer tissues compared to healthy ovarian tissue. Moreover, a greater amount of circRhoC was observed in advanced stage tumors in comparison to early stage tumors [283]. Knockdown of circular RNA RhoC using short hairpin RNA greatly reduced the ability of cells to migrate and invade in laboratory settings, and also impeded the spread of ovarian cancer cells throughout the body in live animal models [283]. After conducting a detailed examination of the mechanisms involved, the researchers concluded that circRhoC acts as a reservoir for miR-302e, leading to an increase in the expression of VEGF-A. This could potentially facilitate the development of new blood vessels and spread of ovarian cancer [283].

There is a scarce amount of research encompassing circRNAs in the context of EC that has been published [284–288]. The composition pattern of circular RNAs (circRNAs) was observed to be changed in the tissues of EC as compared to the surrounding healthy tissues [284]. In a similar manner, a separate research project discovered 62,167 circRNAs that exhibited varying levels of expression in grade 3 EC tissues in comparison to the neighboring healthy tissues [285]. A recent investigation found a strong connection between RNA-binding protein QKI and circRNAs, as well as a negative correlation between QKI and the effectiveness of certain miRNAs. This suggests the possibility of a regulatory feedback loop within EC involving QKI, miRNAs and circRNAs [286]. In a separate research, a discovery was made of a select group of circRNAs that exhibited varying levels of expression. Specifically, 209 circRNAs showed an increase in expression, while 66 circRNAs exhibited a decrease in expression, all of which were found

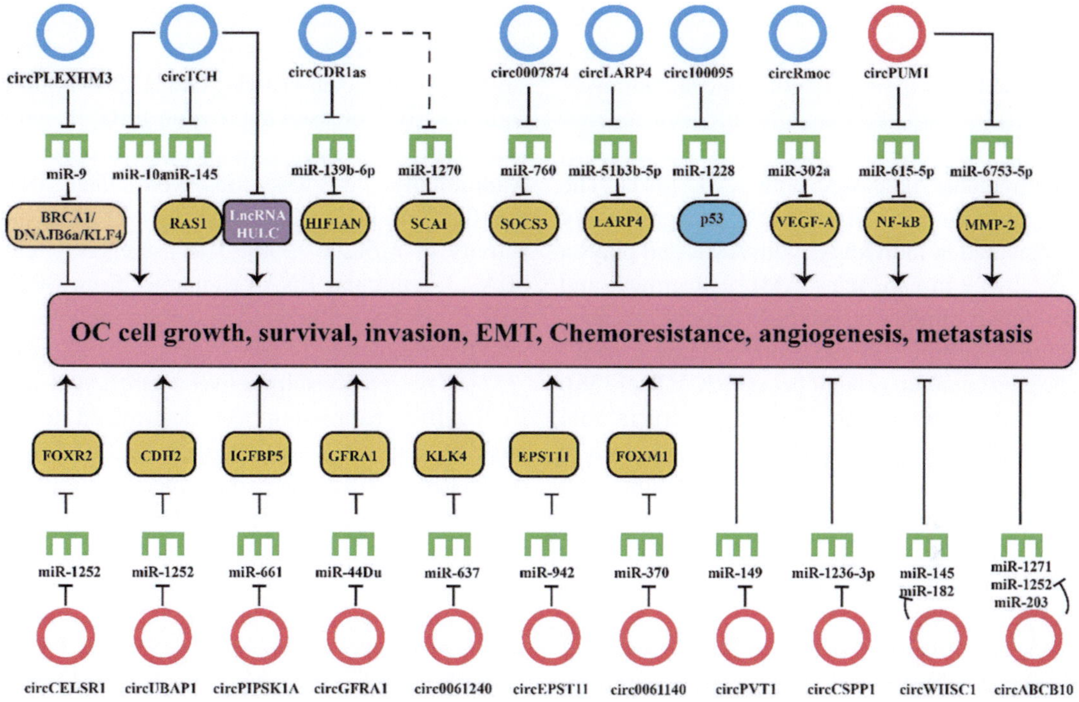

Fig. 18.3 The networks formed by circRNAs, miRNAs, and mRNAs are crucial in facilitating the progression of ovarian cancer, as well as regulating angiogenesis and resistance to chemotherapy. The red circles represent oncogenic circRNAs, while the green circles represent circRNAs that suppress tumor growth (This figure is adapted from [289])

in extracellular vesicles extracted from the serum of individuals with EC, as opposed to a group of healthy individuals used as a control [287]. CircPUM1 exhibited a pronounced increase in expression levels in both EC cell lines and EC tissues. Its role in facilitating the progression of EC was attributed to its regulation of the miR-136/NOTCH3 axis [288]. Table 18.2 lists various circular RNAs in gynecological cancers.

4 Circular RNAs and Hematological Cancers

Acute myeloid leukemia (AML) is a medical condition that results from the uncontrolled growth and spread of abnormal myeloid cells in the bone marrow [336]. The most common form of short-term leukemia in individuals of adult age is known as acute myelogenous leukemia (AML) [337]. Despite the considerable advancements in the diagnosis and treatment approaches for AML, the outlook for this condition remains unfavorable [338]. With the advancement of analytical techniques, many circRNAs have been identified to be abnormally expressed in AML. A research investigation examining 273 circular RNA molecules that showed elevated levels in acute myeloid leukemia (AML) determined that circ-00004136 was notably increased in pediatric cases of AML, contributing to the growth of leukemic cells by acting as a decoy for miR-142 [339]. Circ-RNF13 showed a considerable increase in expression levels among AML patients of adult age and was found to suppress AML cell proliferation, movement, and invasion by modulating the expression of miRNA-1224-5p [340]. The overexpression of miRNA-1224-5p induces premature cell death and halts the progression of AML cells in the G1 stage of the cell cycle [340]. The levels of circ-vimentin have been observed to greatly increase and show a strong connection with levels of leukocytes and French-American-British (FAB) classifications. In addition, an increase in circ-vimentin levels is seen as a separate and unfavorable factor for substantially reduced overall survival (OS) and leukemia-free survival (LFS) in individuals diagnosed with AML [341]. The study revealed that circ-RAD18 is significantly upregulated in individuals with AML and plays a crucial role in enhancing AML cell growth and facilitating tumor advancement through its interaction with miR-206 and protein kinase CAMP-activated catalytic subunit beta (PRKACB) [338]. The protein kinase PRKACB plays a crucial role in a multitude of physiological processes, such as cellular proliferation, genetic regulation, and metabolic activity [338, 342]. The direct impact of circRAD18 is to suppress the activity of miR-206, whereas miR-206 has the effect of inhibiting PRKACB [338].

Chronic myeloid leukemia (CML) is a type of blood cancer that occurs when there is a reciprocal exchange of genetic material between chromosomes 9 and 22 [343, 344]. As a result of this translocation, the breakpoint cluster region (BCR) and Abelson (ABL) gene are both formed [343, 345]. Imatinib is the primary medication used as the initial treatment for CML [346]. The development of drug resistance poses a significant challenge to effectively treating CML [347]. Elevated levels of circ-100053 and circ-0080145 have been found to be linked to the development of imatinib resistance in CML [348, 349].

Individuals who have elevated levels of circ-100053 are at a greater risk of developing a resistance to imatinib compared to those with lower levels, ultimately resulting in a unfavorable outlook for their condition [348]. In addition, circ-100053 is linked to the presence of the BCR/ABL mutation [348]. The levels of circ-0080145 expression were notably elevated in individuals who demonstrated resistance to imatinib, as compared to those who were responsive to the medication. Knocking down circ-0080145 significantly enhanced the sensitivity of individuals with chronic myeloid leukemia (CML) to imatinib, suppressed resistance to imatinib and cell growth, and triggered programmed cell death in imatinib-resistant CML cells by targeting miR-326 (specifically the miR-326 binding site in circ-0080145) and the protein alpha 1 (a gene targeted by miR-326) [349]. The results of the study revealed that hsa-circ-0080145 exhibited significantly increased expression in CML cells that had a strong ability to proliferate. Additionally, it was observed that hsa-circ-0080145 had a direct inhibitory effect on the activity of miR296 [350]. The presence of circBA9.3, a circular RNA originating from BCR-ABL1, has been identified in certain individuals who have shown resistance to imatinib treatment [351]. The relationship between circBA9.3 and BCR-ABL1 expression was determined to be positive, and the detection of BCR-ABL1 was associated with a negative outcome. Additionally, studies have revealed that circBA9.3 can impede the apoptosis of CML cells and facilitate the production of BCR-ABL1 protein [351].

Chronic lymphocytic leukemia (CLL) is characterized by the uncontrolled growth of CD5+ monoclonal B cells, leading to a cancerous state [352, 353]. The expression of circ-CBFB was notably elevated in CLL, leading to an acceleration of CLL cell proliferation and a suppression of CLL cell apoptosis [354]. Circular RNA CBFB is responsible for the enhancement of FZD3 gene expression by hindering the action of miR-607 through the stimulation of the Wnt/β-catenin pathway, ultimately expediting the progression of chronic lymphocytic leukemia [354]. High levels of mc-COX2, a circular RNA derived from the mitochondrial genome, have been shown to have a negative correlation with prognosis. Conversely, reduced mc-COX2 levels have been linked to impaired mitochondrial function, decreased cell proliferation, and the initiation of apoptosis [355]. The plasma form of circ-RPL15 was discovered to actively facilitate the advancement of CLL [356].

The occurrence of chromosomal translocations can lead to the formation of fusion genes that have oncogenic properties, playing a significant role in the onset of various tumors and being a major contributor to leukemia [37]. Guarnerio and colleagues provided evidence that f-circRNAs, which arise from chromosomal rearrangements in leukemia, have a crucial role in driving cell growth and disease advancement. These f-circRNAs sustain the survival of leukemic cells and contribute to their resistance to chemotherapy, ultimately leading to decreased effectiveness of treatment [37]. Moreover, a deeper understanding of the source of fusion gene chaperone circular RNAs (FP-circular RNAs) has shown that these molecules may act as cancer-causing genes in leukemias caused by chromosomal translocations [357]. An illustration of this can be seen in how CircAF4, which is discovered within MLL leukemia, has its origins in the fusion of the MLL gene with AF4. This leads to the development of leukemia and simultaneously suppresses the activity of the MLL-AF4 gene by acting as a sponge for the miR-128-3p molecule [358]. In conclusion, both FP-CircRNAs and f-circRNAs have potential as a focus for diagnosing and treating chromosomal translocation leukemia. Fig. 18.4 showcases the mechanisms and roles of circRNAs in this type of cancer.

There are different categories of lymphomas, which can lead to different ways of presenting themselves clinically, treatment choices, and expected outcomes depending on the classification [359]. T cell lymphoblastic lymphoma (T-LBL) is a cancerous growth originating from the excessive multiplication of malignant thymocytes, resulting in a significant accumulation of undeveloped T cells in the mediastinum and lymphoid tissues [359]. A recent discovery showed that in T-LBL tissues, there is an increased expression of circ-LAMP1. Additionally, circ-LAMP1 has been observed to regulate the expression of discoidin domain receptor tyrosine kinase 2 (DDR2) by suppressing miR-615-5p, and it also has the ability to promote DDR2 expression and hinder T-LBL cell apoptosis through the use of the miR-615-5p/DDR2 pathway [360].

Diffuse large B cell lymphoma (DLBCL), a prevalent type of lymphoma in adults, remains incurable in about 30% of cases [361–363]. This ailment displays extreme aggressiveness, arising either from within or outside of the lymph nodes, and is also triggered by the conversion of other cancers like CLL [362, 364, 365]. Circ-APC is a circular RNA that arises from the reverse splicing of APC exon 7 to exon 14 and functions to sequester and reduce the levels of miR-888 or physically interact with the APC promoter in order to enhance the expression of APC [366]. Circ-APC has the potential to serve as a biomarker for both the diagnosis and prognosis of DLBCL [366]. Blocking APC activity leads to a reduction in the accumulation of β-catenin in the cell's nucleus, effectively repressing the Wnt/β--catenin pathway and impeding the growth of

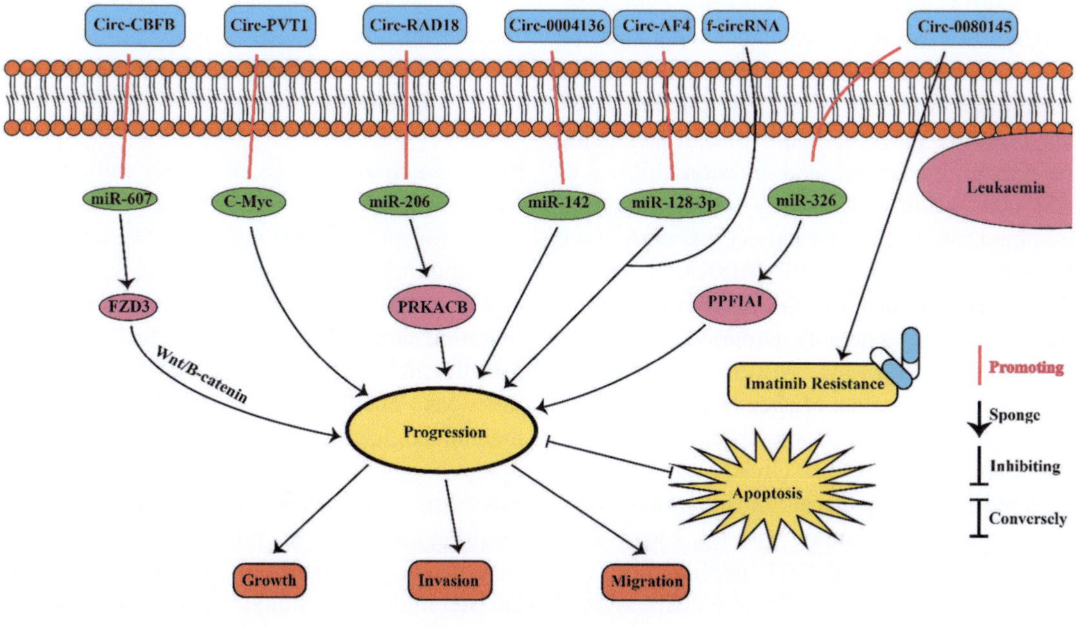

Fig. 18.4 The regulatory processes and roles of circular RNA in leukemia

DLBCL [367]. The movement and growth of DLBCL cells is enhanced by circCFL1 [368]. By utilizing the dual-luciferase reporter gene technique, it was validated that circCFL1 acts as a competing endogenous RNA for miR-107, thereby decreasing the levels of miR-107. This decrease in miR-107 results in the downregulation of its target gene HMGB1, as miR-107 normally binds to and inhibits HMGB1 expression [368]. HMGB1 intensifies the process of tumor invasion and spread and expedites the growth of tumors [369]. circCFL1 plays a critical role in controlling the concentration of HMGB1 through the mediation of mir-107, thereby facilitating the process of cell migration and proliferation in DLBCL [368].

Constituting a mere portion of 1–2% of all tumors globally, it also leads to approximately 2% of deaths caused by cancer [370]. Circular RNAs (circRNAs) have a significant role in regulating activities in MM. Some specific circRNAs exhibit increased levels in MM. Recent studies have demonstrated that circ-CDYL is significantly upregulated in both MM cells and plasma [371]. The absorption of miR-1180 and subsequent increase in YAP expression by circ-CDYL plays a critical role in promoting the proliferation of MM cells [371]. The study revealed that there was an excessive expression of circ_0000142 in individuals suffering from MM, and this was strongly associated with elevated levels of International Staging System (ISS) and Durie-Salmon stages [372]. The circular RNA circ_0000142 played a major role in enhancing the growth, movement, and invasion abilities of MM cells while concurrently suppressing programmed cell death. This activity was carried out through the control of miR-610/AKT3 expression [372]. The level of hsa-circRNA-101237 in the bone marrow is notably increased in individuals diagnosed with MM, as well as in MM cell lines and BTZ-resistant MM cells, particularly in those with relapsed or treatment-resistant disease [373]. The presence of hsa-circRNA-101237 was observed to notably elevate in individuals diagnosed with MM presenting with t, t, 13q14 deletion, P53 deletion, and 1q21 amplification, and its levels correlated with the prognosis, overall survival (OS), and progression-free survival (PFS) in MM patients [373].

The overexpression of circ_0007841 was observed in both MM cell lines and patients with MM. This resulted in the stimulation of the PI3K/AKT signaling cascade through the miR-338-3p/bromodomain containing 4 (BRD4) axis [374]. The miRNA, circ_0007841, stimulated the progression of cell division, proliferation, and spread while hindering the programmed cell death, apoptosis, in MM cells through its interaction with the miR-338-3p/BRD4 signaling pathway [374]. The destructive actions of myeloma cells are intensified by the release of exosomes from mesenchymal stromal cells (MSCs) through the process of circ_0007841 [374]. The circular non-coding RNA hsa-circ-0007841 was discovered to have significantly higher expression levels in multiple myeloma cells that were resistant to drugs. This was linked to abnormalities in the chromosomes, specifically in the 1q21 region, translocation of chromosomes 4 and 14, as well as mutations in the ATR and IRF4 genes [375, 376]. The circular RNA hsa-circ-0007841 is responsible for developing a resistance to doxorubicin in multiple myeloma cells by stimulating the ATP-binding cassette transporter G2 (ABCG2) [376]. Table 18.3 lists various circular RNAs in hematological cancers.

5 Circular RNAs and Other Cancers

The main characteristics of cancer cells involve their ability to continuously and persistently divide for an extended duration. The controlled regulation of healthy tissues plays a vital role in initiating and advancing the process of cell proliferation and differentiation by facilitating the growth and release of signals that promote these functions. The balance of cellular numbers is maintained, leading to the appropriate shape and functioning of robust tissues. Nevertheless, in cancerous cells, these signals are disturbed, resulting in unrestricted proliferation and replication [391]. Furthermore, cancer cells have devised multiple techniques that either hinder or

Table 18.3 Various circular RNAs in hematological cancers

Type of cancer	Circular RNAs	Expression	Target	References
Leukemia	circ_0040823	Down	PTEN	[377]
	circ_0005774	Down	circ_0005774/miR-192-5p/ ULK1 ceRNA pathway	[378]
	circPVT1	Up	CXCR4	[379]
	circCRIM1	Down	F-box	[380]
	Hsa-circ_0003420	Up	IGF2BP1	[381]
Lymphoma	ZDHHC11	Up	miR-150	[382]
	PVT1	Down	–	[383]
	Circ_0008012	Up	IKKα:IKKβ:IKKγ	[384]
	circKIF4A	Up	miR-1231	[385]
	circ_0000877	Up	miR-370-3p/MAP4K4/ Hippo	[386]
	circ_0094100	Down	miR-217/ATP1B1	[387]
	circCTNNA1	Up	miR-34a	[388]
	circZNF609	Up	miR-153	[389]
	circ_0003645	Up	miR-335-5p/NFIB	[390]

facilitate their ability to escape programmed cell death. In specific unusual circumstances, especially in instances of cancer, cells lose the ability to go through planned cell death via apoptosis, leading to uncontrolled proliferation. Cells have numerous ways to prevent programmed cell death, and one of these strategies involves creating anti-apoptotic molecules [392].

Recent studies have revealed that circRNAs have an impact on the proliferation, cellular division, and apoptosis of prostate cancer cells. In their study, Shan and colleagues discovered that by suppressing circFMN2, there was a significant decrease in the advancement of prostate cancer (PCa) in living organisms. They also found that this led to a slowdown in the cell cycle and induced programmed cell death in PCa cells, all through the regulation of the miR-1238/LHX2 pathway. Mao et al.'s research indicates a notable connection between circPDHX and Gleason score, pathological T stage, and overall survival among individuals diagnosed with prostate cancer (PCa). Moreover, research has demonstrated that circPDHX prompts cell reproduction in controlled laboratory studies and aids in the development of tumors in animal models [393]. Liu and team reported that there was an upregulation of circHIPK3 in prostate cancer, which promoted G2/M transition by functioning as a miR-338-3p decoy [394]. In their study, Deng et al. stated that when circ_0088233 was suppressed, it led to a reduction in cell proliferation and caused a pause in the G1 stage, along with promoting apoptosis. This was accomplished through a targeted interaction with hsa-miR-185-3p [395]. In the analysis by Zhang et al., it was found that a reduction in circ_0057553 ultimately led to a decline in the overall survival rate of cells and a rise in the occurrence of programmed cell death [396]. According to the study conducted by Huang et al., by inhibiting circABCC4, there was a decrease in tumor growth in living subjects and disruption of cell cycle progression in laboratory experiments through the regulation of the miR-1182-FOXP4 pathway [397].

A multitude of circular RNAs played a role in promoting melanoma by causing cells to progress and tumors to grow larger as a result of their heightened levels. In the case of circular RNA circ_0083444, this molecule is formed from the circularization of specific exons of the MTUS1 gene. In the context of conjunctival melanoma, it functions as a sponge for two microRNAs, miR-622 and miR-1208, as indicated by a bioinformatics tool. Some possible routes that may assist the advancement of melanoma include the ErbB, MAPK, and Wnt signaling pathways [398].

In a separate study, researchers noted a rise in the amount of circ_0084043 being produced in melanoma cells [399]. Through the analysis of bioinformatics data and luciferase reporter assays, circ_0084043 was determined to function as an enhancer of Snail expression by interacting with miR-153-3p. As a result, there was a rise in the spread, invasion, and movement of melanoma cells. Cotransfection of the miR-153-3p inhibitor effectively reversed the impact of depleting circ_0084043 in melanoma cells. Blocking the function of circ_0084043 led to a notable decrease in the growth of tumors in the melanoma model, specifically during the time period of day 12–21, compared to the group that did not receive this treatment. In the year 2020, Chen Z. and other researchers presented further proof to validate the idea that circ_0084043 has a vital impact on the progression of melanoma. Their findings showed that circ_0084043 acts as a literal absorber for miR-429, which results in an upsurge in the levels of tribbles homolog 2 (TRIB2) and facilitates its role in promoting cancerous effects [400]. Moreover, the inhibition of circ_0084043 impaired the triggering of the Wnt/β-catenin pathway by controlling the miR-429/TRIB2 axis. Jin and colleagues discovered that an increase in circ_0085533 (also known as circ-MYC) significantly enhanced the rate of growth, glycolysis, production of lactate, and activity of lactate dehydrogenase A (LDHA) in MEL-CV cells by attaching to miR-1236 [401]. The use of Western blotting and luciferase activity assay demonstrated that miR-1236 has the ability to bind to the 3′ UTR of the LDHA gene and decrease the expression of LDHA. Additionally, it has been observed that the circ-MYC molecule is stringently regulated by the transcriptional factor c-Myc and the splicing factor SRSF1, indicating the crucial influence of the c-Myc-SRSF1 combination on promoting melanoma growth. Circular RNAs play a crucial role in the development and advancement of kidney cancer. By conducting RNA sequencing experiments, researchers have identified approximately 2000 circular RNAs (circRNAs) that exhibit differential expression in renal cell carcinoma (RCC) [402]. These circular RNAs play a critical role in various cellular processes within renal cells, including but not limited to cell growth, spread, transformation into different cell types, programmed cell death, and metabolic activities (as shown in Fig. 18.5). Studies indicate that disrupted circRNA levels may be linked to excessive growth of kidney cancer cells and disruption of proper regulation of the cell cycle. For instance, when the level of circ000926 expression rises in cancerous cells in the kidneys, it has been observed to coincide with a negative prognosis for the patient [403]. Inhibition of Circ 000926 blocks the development of xenografts and decreases the proliferation of renal cancer cells. By directly obstructing the action of miR-411, it acts as a sponge for the miRNA and boosts the levels of CDH2. The growth of ACHN and 769-P kidney cancer cells is lowered when the activity of CircEGLN3, which is highly abundant in renal cancer, is inhibited [404]. CircEGNL3 acts as a miRNA decoy for miR-1224-3p, thereby leading to higher levels of HMGXB3 expression. This, in turn, counters the inhibitory impact of miR-1224-3p on renal cancer, and facilitates enhanced cell cycle progression and metastatic potential. In distinct investigations, it was observed that the levels of circSCARB1 were elevated in different cell lines and tissue of RCC [405]. CircSCARB1's binding to miR-510-5p and SDC3, which is a direct target of miR-510-5p, promotes cell growth suppression observed when circSCARB1 is suppressed. When SDC3 is simultaneously transfected, it only has a partial effect on controlling cell growth, while transfection with a miR-510-5p mimic results in reduced expression of SDC3 and decreased cellular expansion. Additionally, an excessive amount of circMTO1 slows down the proliferation of 786-O and A498 renal cancer cells, while the inhibition of circMTO1 promotes the growth of OS-RC-2 and SN12C renal cancer cells [406]. CircMTO1 effectively absorbs both miR-223 and miR-9, leading to a decrease in their levels. Suppressing circMTO1 causes an increase in RCC cell proliferation as it reduces the amount of LMX1A, a direct target of miR-9. Through transfection with a miR-9 mimic, previous studies have revealed that miR-9 plays a crucial role in regulating LMX1A expression in RCC. Overexpression of a miR-9 inhibitor and

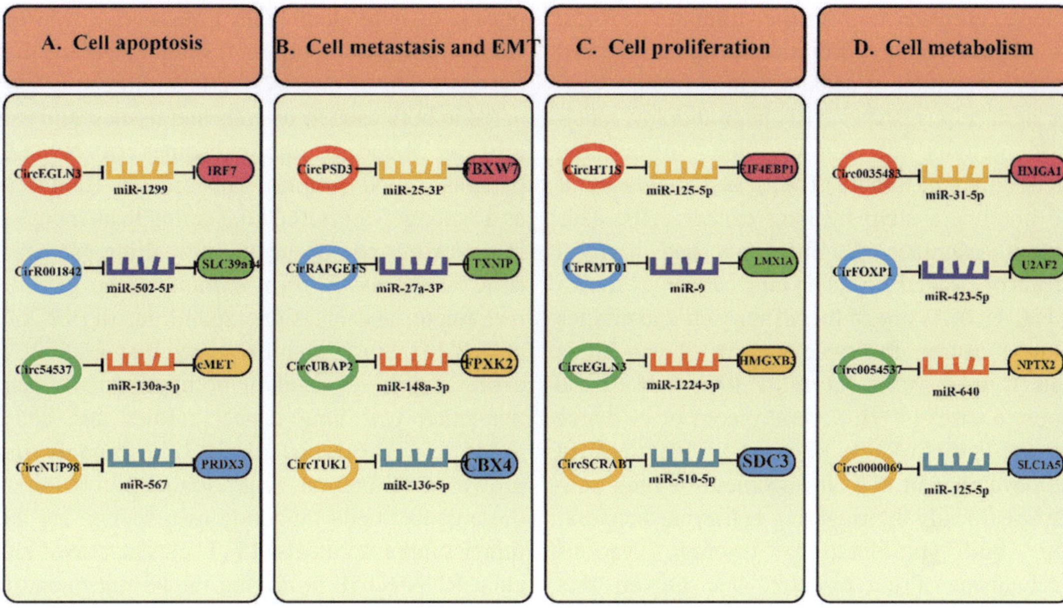

Fig. 18.5 CircRNAs have been found to play a crucial role in RCC by inhibiting miRNAs and their related proteins. (**a**) The circular RNAs Circ0054537, CircEGLN3, Circ001842, and CircNUP98 all impact the process of apoptosis in renal cancer cells. (**b**) Circular RNAs (CircRNAs) are involved in the process of epithelial–mesenchymal transition (EMT) and the spread of cancer to other parts of the body (metastasis) in renal cell carcinoma (RCC). This includes the newly discovered CircPSD3, CircUBAP2, CircRAPGEF5, and CircTLK1. (**c**) Circular RNAs are involved in regulating the cell cycle and growth of renal carcinoma cells. Specifically, CircSCARB1, CircMTO1, CircEGLN3, CircCHT15, and CircMTO1 have all been shown to play a role in influencing the proliferation of renal cancer cells. (**d**) Circular RNAs (CircRNAs) have a significant impact on the activity of renal cancer cells. Specifically, Circ0035483, Circ0000069, CircFOXP1, and Circ0054537 play a crucial role in altering the metabolic processes of renal cancer cells (This figure is adapted from [435])

LMX1A counteracts the tumor-promoting effect of circMTO1 silencing. In the same way, circTLK1 was identified as a potential circular RNA derived from the TLK1 gene through a large-scale analysis of circRNAs in kidney cancer cell lines using high-throughput sequencing. The cells in RCC possess a significant amount of CircTLK1, and when it is suppressed, the growth of the cells is also hindered. According to data from GSE100186 and GSE137836 sequencing, individuals with advanced TNM stage and elevated levels of circSDHC in their tissues have an unfavorable result in their prognosis for RCC [407].

Many circular RNAs have been identified to have abnormal expression levels in lung cancer cells, with either promoting or inhibiting effects on tumor growth. Numerous studies have provided overwhelming proof that there exist a large number of circRNAs with varied *expression* levels in both tissue and plasma samples of individuals with lung cancer. In their study, Chen and colleagues utilized high-throughput sequencing of plasmatic exosomes to examine the expression of circular RNAs in individuals with initial-stage lung adenocarcinoma. They identified 182 exosomal circRNAs that were differentially expressed, with 105 showing increased expression and 78 showing decreased expression compared to the control group [408]. Through the utilization of advanced sequencing technology, Zhang and colleagues discovered a total of 35 different circular RNAs that displayed abnormal expression in tissues affected by small-cell lung cancer. Out of all of these, there were 5 circular RNAs that showed a notable increase, while 30 circular RNAs showed a significant decrease [39]. The functions of imbalanced circular RNAs in lung cancer encompass the stimulation of cell growth, movement, infiltration, programmed cell death, replication of cells, potential of lung cancer stem cells, resistance to chemotherapy, metabolic activity of tumors, the surrounding environment of tumors (TME), and avoidance of the immune system by lung cancer cells. This section examines the functions and harmful effects of circular RNAs on lung cancer, as shown in Fig. 18.6. As one of the primary characteristics of lung cancer, the rapid growth of cancerous cells is responsible for more than half of the reported cases [409]. Several pieces of evidence suggest that circRNAs play a significant role in the development and advancement of lung cancer, specifically by triggering cell cycle acceleration and proliferation through various mechanisms. Prior research has proven that F-circEA that originates from the EML4-ALK fusion gene leads to increased levels of F-circEA and enhances the migratory and invasive capacity of lung cells. F-circEA has the potential to serve as an innovative liquid biopsy indicator for identifying lung cancer as it can be found in the plasma of individuals with EML4-ALK translocation [410]. Furthermore, it should be noted that the detection of circSATB2 is significantly elevated in both cancerous cells and tissues, and can also be found in the exosomes of NSCLC patients' blood samples. This circular RNA acts as a "sponge" for miR-326, leading to an increase in the expression of the actin-bundling protein 1 (FSCN1). This, in turn, stimulates the growth, movement, and invasion capabilities of NSCLC cells [411]. Studies have revealed that circBIRC6 expression is elevated in both primary human non-small cell lung cancer tissues and cells. When circBIRC6 is silenced, it hinders the growth, proliferation, migration, and invasion of these cancer cells by acting as a sponge for the tumor suppressor miR-145 [412]. The role of circular RNA NT5E in driving the advancement of human non-small cell lung cancer (NSCLC) cells is achieved through its activity as a sponge for miR-134. Similarly, circular RNA ARHGAP10 promotes the growth of human NSCLC cells by activating the miR-150-5p/GLUT1 pathway [413, 414].

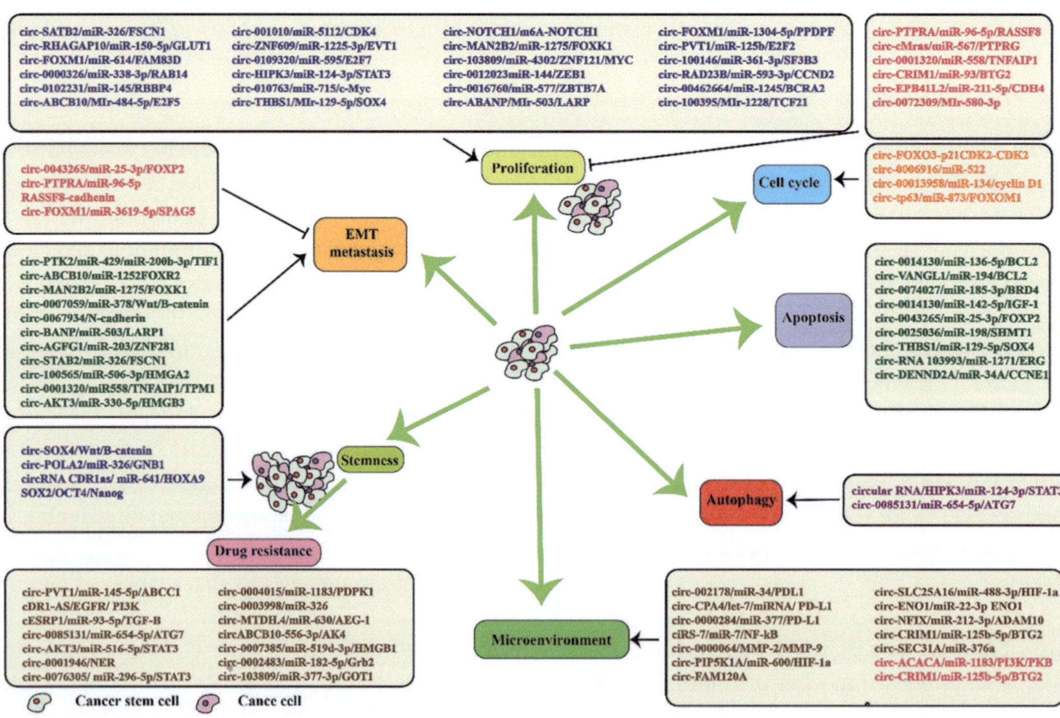

Fig. 18.6 The role of circRNAs in the core features of lung cancer (This figure is adapted from [436])

Circular RNAs (circRNAs) have the ability to either enhance or hinder the advancement of breast cancer by acting as sponge molecules for miRNAs, producing different proteins, and governing the behavior of their parent genes. These alterations are apparent in the control of tumor development, spread, and other clinicopathological characteristics, as seen through changes in proliferation, migration, and invasion both on and off the laboratory setting. Furthermore, an illustration of the various instances, classifications of BC, and associated clinical and pathological characteristics are also shown. In a research project that involved 120 individuals who had been diagnosed with breast cancer, it was found that there was a direct connection between the expression of hsa_circ_0005273 and factors such as lymph node metastasis, tumor size, TNM stage, and distant metastasis. However, there was no significant relationship between hsa_circ_0005273 expression and age in this cohort [415]. A study involving 60 samples of triple-negative breast cancer (TNBC) and their corresponding noncancerous tissues revealed that circSEPT9 was significantly associated with advanced T, N, and TNM stages [416]. A group of 240 BC tissues indicated that elevated levels of circACTN4 were associated with increasing levels of T, N, and TNM stages [417]. According to research, the level of expression of circAGFG1 is linked to both a negative prognosis and the tumor (T) and lymph node (N) stages in the TNM staging system. Similarly, circNFIC has been found to be reduced in breast cancer (BC) and has a negative correlation with the spread of cancer to lymph nodes ($P = 0.021$) and a poor prognosis ($P = 0.002$) [418].

According to a research led by Yang and colleagues, they found that circUVRAG was greatly increased in both BC tissues and cell lines. When this particular molecule was suppressed, there was a significant decrease in cell growth, which was facilitated by the activation of miR-223. This subsequently led to the suppression of FGFR2 [419] (Fig. 18.7). CircRGNEF is a circular RNA that contains two exons derived from the RGNEF gene. This molecule has been found to influence the development of breast cancer cells by acting as a sponge for miR-548, which leads to an increase in the levels of KIF2C [420]. In addition, malfunction of the circRNA-based Tgf-β2/smad3 signaling pathway was also involved in the advancement of BC. To illustrate, it was discovered by Su and colleagues that circRIP2 played a vital role in the advancement of breast cancer by utilizing the Tgf-β2/smad3 signaling mechanism through absorbing miR-1305 [421]. In the same manner, Mao and colleagues proved that an RNA circular structure called hsa_circ_0068871 was increased in breast cancer and facilitated the growth and death of cancer cells by stimulating the STAT3 pathway, with the help of FGFR3, through its ability to act as a sponge for miR-181a-5p [422]. By absorbing miR-145-5p, it was shown that the increased levels of circCEP128 promoted breast cancer cell growth and hindered programmed cell death by affecting the expression of SOX11 [423]. It was reported that the expression of CircDOCK1 was notably higher in BC tissue, and reducing its levels significantly hindered the advancement of EJ-m3 and 5673 BC cell lines by upregulating miR-132-3p expression [424]. The overexpression of has_circ_0068307 significantly increased and facilitated the advancement of breast cancer cells by activating the miR-147/c-Myc pathway [425]. Moreover, circ_0008532 was found to enhance the proliferation of BC cells by acting as a sponge for miR-155-5p and miR-330-5p, resulting in elevated levels of MTGR1 [426].

Studies conducted recently have revealed that certain circular RNAs (circRNAs) contribute to promoting the advancement of glioma. Cancer cells, in contrast to regular cells, exhibit uncontrolled and speedy growth. Numerous studies have demonstrated that circRNAs play a crucial role in sustaining the proliferative stimulus of glioma cells. A variant of oncogenic E-cadherin, encoded by circular RNA, is crucial in preserving the oncogenic properties of glioma stem cells. This is achieved by activating EGFR through binding to the EGFR CR2 domain, regardless of the presence of EGF [427]. Moreover, glioma exhibited elevated levels of circCPA4, which acted as a sponge for miR-760 and indirectly

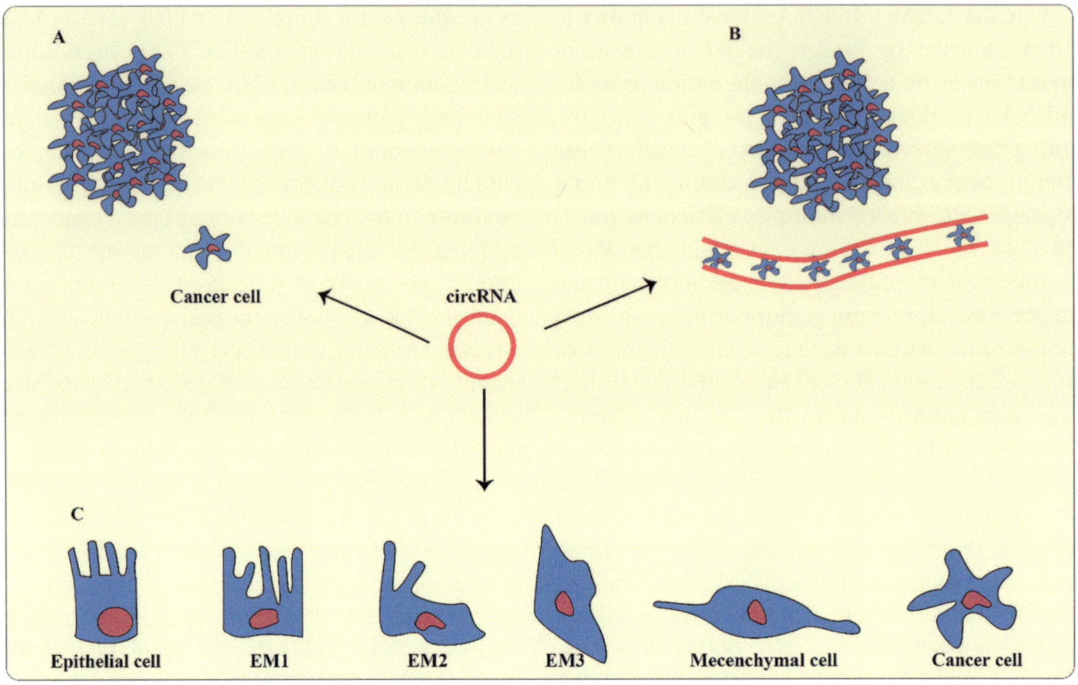

Fig. 18.7 Circular RNAs control the processes of cellular growth, movement, and penetration while also influencing the transition of cells between epithelial and mesenchymal states known as epithelial–mesenchymal transition (EMT). (**a**) The functions of circular RNAs (circRNAs) in the process of cell proliferation. (**b**) The involvement of circular RNAs in the process of cell migration and invasion. (**c**) The functions of circular RNAs in controlling epithelial-to-mesenchymal transition

controlled the expression of MEF2D. The inhibition of circCPA4 resulted in the suppression of tumor cell growth, indicating that circCPA4 plays a beneficial role in the development of malignant cell characteristics [428]. CircRNAs have a noteworthy role in suppressing or impeding the apoptosis of glioma cells. Specifically, circRNA MAPK4 acts as a sponge for miR-125a-3p and utilizes the p38/MAPK axis to hinder cell apoptosis in glioma [429]. According to a different study, the level of circRNA PIP5K1A was significantly elevated in samples of glioma compared to normal samples. This led to the activation of TCF12 and PI3K/AKT pathways by acting as a sponge for miR-515-5p by circRNA PIP5K1AAs a result of circRNA PIP5K1A being regulated, the detrimental actions of glioma were intensified and apoptosis was triggered [430]. The amount of research focused on circRNAs functioning as tumor suppressors in glioma is significantly lower than the number of studies on circRNAs functioning as oncogenes. Additionally, the expression of CDR1as was found to be substantially elevated in the brain compared to other normal organ tissues. The levels of CDR1as were reduced in glioma cases when compared to normal brain tissue. Significantly, it was discovered that CDR1as exerts a suppressive role in the development of tumors by disrupting the p53/MDM2 complex. This results in reduced MDM2-mediated ubiquitination, leading to the stabilization of p53 through CDR1as [431]. In a separate study, it was discovered that circ_SFMBT2 had a detrimental effect on glioma cells and was able to counteract the tumor-promoting effects induced by mir-182-5p in glioma by acting as a mir-182-5p sponge [432]. Furthermore, the presence of circBRAF was reduced in cases of glioma, and it was found to inhibit harmful actions of glioma by controlling the miR-1290/FBXW7 pathway [433]. The amount of hsa_circ_0001017 in glioma tissue was significantly decreased, and overexpression of hsa_circ_0001017 inhibited tumor growth by controlling the let-7g-3p/NDST3 signaling pathway [434]. Table 18.4 lists various circular RNAs in other cancers.

6 CircRNAs as Biomarkers in Cancer

The characteristics of circRNAs discussed in earlier sections (including durability, adherence to tradition, widespread occurrence, and precision) suggest that circRNAs hold promise as useful markers for cancer prognosis and diagnosis. In recent years, there have been multiple research investigations showing that circRNAs can be consistently detected and found in significant amounts in different bodily fluids, including saliva, plasma, serum, and exosomes. This characteristic makes circRNAs highly suitable as potential noninvasive biomarkers for cancer through liquid biopsy methods [537]. Studies have revealed that there is a higher level of expression of circ-ZEB1.33 in human liver cancer tissues compared to normal tissues, as well as in serum samples from patients with liver cancer in comparison to healthy individuals. The levels of this circRNA in both tissue and serum have been found to be associated with various stages of TNM and the overall survival of liver cancer patients, indicating that circ-ZEB1.33 could potentially be a useful biomarker in predicting the prognosis of HCC [538]. The expression of Hsa_circ_0000190 was reduced in both gastric cancer tissues and plasma samples [539]. The degree to which it was expressed was greatly linked to the size of the tumor, the spread of the cancer to distant areas, the spread of the cancer to lymph nodes, the TNM stage and levels of CA19-9. The AUC values for hsa_circ_0000190 in tissues and plasma were 0.75 and 0.60, respectively. However, when the two were combined, the AUC increased to 0.775. Additionally, the sensitivity and specificity of this combination were 0.712 and 0.750, respectively. In a different research, it was found that hsa_circ_0000745 had low expression levels in both gastric cancer tissues and plasma samples [540]. The presence of hsa_circ_0000745 in GC tissues was found to be linked to the level of tumor differentiation, while its level in plasma was associated with the tumor-node metastasis stage. The AUC for hsa_circ_0000745 in plasma was 0.683; however, when paired with carcinoembryogenic antigen (CEA) levels, the AUC rose to 0.775. This indicates that hsa_circ_0000745 in plasma, along with CEA levels, has strong potential as a diagnostic marker for gastric cancer. In their study, Zhao and team utilized microarray analysis to examine the presence of circRNA in the saliva of individuals with oral squamous cell carcinoma, in comparison to those without the disease. They identified a total of 20 circRNAs that were expressed at lower levels and 12 circRNAs that were expressed at higher levels in the saliva of patients with oral squamous cell carcinoma compared to healthy controls [541]. Two circRNAs, hsa_circ_0001874 and hsa_circ_0001971, were found to be upregulated and had an AUC of 0.863 and 0.845, respectively. The combination of these two circular RNAs, hsa_circ_0001874 and hsa_circ_0001971, was found to have an AUC of 0.922, indicating strong potential as a diagnostic biomarker for oral squamous cell carcinoma (OSCC). A risk score based on these two circRNAs was also able to effectively differentiate between patients with OSCC and those with oral leukoplakia, with an AUC of 0.863. This suggests that hsa_circ_0001874 and hsa_circ_0001971 found in saliva may have promising potential for diagnosing OSCC. Furthermore, recent research has indicated that circular RNAs (circRNAs) are highly concentrated and resilient in exosomes, tiny membrane-bound structures released by cancerous cells into the surrounding bodily fluids. Chen and team exposed that in comparison to healthy individuals, the circPRMT5 was notably concentrated in both exosomes found in the serum and urine of those affected with urothelial carcinoma [542]. High levels of circPRMT5 detected in both serum and urinary exosomes were found to have a strong correlation with lymph node metastasis and aggressive tumor progression, indicating that circPRMT5 could potentially serve as a valuable prognostic indicator for urothelial carcinoma. Furthermore, it was discov-

Table 18.4 Various circular RNAs in other cancers

Type of cancer	Circular RNAs	Expression	Target	References
Prostate cancer	CircSMARCC1	Up	miR-1322/CCL20/CCR6 signaling	[437]
	RNA EPHA3	Down	BMP2	[438]
	CircDHRS3	Down	MEIS2	[439]
	circPDHX	Up	MiR-378a-3p IGF1R	[393]
	circROBO1	Up	PGK1	[440]
	circMID1	Up	IGF1R and YTHDC2	[441]
	Circular RNA_LARP4	Down	FOXO3A	[442]
	circSMARCA5	Down	TIMP3	[443]
	circARHGEF28	Down	miR-671-5p/LGALS3BP/ NF-κB axis	[444]
	circHIPK3	Up	MTDH	[445]
	Circ_0076305	Up	PGK1	[446]
	circ_0057558	Down	miR-206/USP33/c-Myc Axis	[447]
	circFOXO3	Up	SLC25A15	[448]
	circMBOAT2	Up	mTOR	[449]
	irc_0085494	Up	NRBP1	[450]
Melanoma	circ0001591	Up	ROCK1/PI3K/AKT	[451]
	Hsa_circ_0103232	Up	RAB3D	[452]
	circITCH	Down	TFCP2	[453]
	circ_0020710	Up	CXCL12	[454]
	hsa_circ_0062270	Up	CDC45	[455]
	Circ_0016418	Up	GLS	[456]
	circ_0079593	Up	ABHD2	[457]
	Circ-0079593	Up	EGFR	[458]
	CircFAT1	Up	SLC7A11	[459]
	Circ_0081054	Up	RAB9A	[460]
	hsa_circ_0025039	Up	CDK4	[461]
	Circ-GLI1	Up	?	[462]
	Circ-FOXM1	Up	FLOT2	[463]
	Circ_0119872	Up	G3BP1	[464]
	Circ_0084043	Up	KLF3	[465]
Kidney (renal cell) cancer	circTXNDC11	Up	MAPK/ERK pathway	[466]
	CircPPAP2B	Up	CYP1B1	[467]
	circVAMP3	Up		[468]
	CircNTNG1	Down	HOXA5	[469]
	circ_0001368	Down	LATS2	[470]
	Hsa_circ_0065217	Up	ALPK2	[471]
	circ-EGLN3	Up	IRF7	[472]
	circSDHC	Up	CDKN3/E2F1 axis	[407]
	circMYLK	Up	VEGFC	[473]
	circ_0054537	Up	NPTX2	[474]
	circ-ABCB10	Up		[475]
	circKL	Down	FBXW7	[476]
	hsa_circ_0015004	Down	RAC1	[477]
	circDVL1	Down	PCDH7	[478]
	Cir-CCDC66	Up	HGF/c-Met	[479]

(continued)

Table 18.4 (continued)

Type of cancer	Circular RNAs	Expression	Target	References
Lung cancer	circTAB2	Down	GLIS2	[480]
	circNDUFB2	Down	IGF2BPs	[481]
	CircVAPA	Up	miR-377-3p and miR-494-3p/IGF1R/AKT axis	[482]
	CircHERC1	Up	HMGB1	[483]
	circ_103820	Down	LATS2 and SOCS6	[484]
	hsa_circ_0012673	Up	miR-320a/LIMK18521 axis	[485]
	circMAN2B2	Up	FOXK1	[486]
	circPIP5K1A	Up	HIF-1α	[487]
	hsa_circ_0030998	Down	miR-558	[488]
	Circ_0010235	Up	PDL1	[489]
	Circ_0001421	Up	CDCA3	[490]
	Circ_0120376	Up	CEP55	[491]
	Circ_0001777	Down	3'UTR	[492]
	hsa_circ_0050386	Down	SRSF3/FN1 axis	[493]
	hsa_circ_0077837	Up	Phosphatase and tensin homolog	[494]
Breast cancer	circRNF20	Up	miR-487a/HIF-1α/HK2	[495]
	hsa_circ_001783	Up	miR-200c-3p	[496]
	circBCBM1	Up	BRD4	[497]
	circSEPT9	Up	LIF/Stat3 signaling	[416]
	hsa_circ_0025202	Down	FOXO3a	[498]
	circACTN4	Up		[417]
	circRNA-0001283	Down	HIPK3	[499]
	circIKBKB	Up	NF-κB	[500]
	hsa_circ_0067842	Up	HuR/CMTM6/PD-L1 axis	[501]
	Circular RNA_0006014	Up	NTRK2 and PIK3/AKT pathway	[502]
	CircNR3C2	Down	HRD1	[503]
	circZFAND6	Up	FASN	[504]
	circHIPK3	Up	MiR-326	[505]
	circTP63	Up	FOXM1	[506]
Bladder cancer	CircPTK2	Up	PABPC1/SETDB1	[507]
	CircXRN2	Down	Histone lactylation Hippo pathway	[508]
	circ-FOXO3	Down	TGFBR2	[509]
	CircLONP2	Up	YAP1	[510]
	circSEMA5A	Up	ENO1	[511]
	circTAF4B	Up	TGFA	[512]
	RBPMS	Down	RAI2	[513]
	ACVR2A	Down	EYA4	[514]
	circGLIS3	Up	SKP1/Cyclin D1	[515]
	circ_0000629	Down	CDC73	[516]
	circMTO1	Down		[517]
	circKDM4C	Up	ZEB1	[518]
	circ_0000326	Up	microRNA-338-3p/ETS Proto-Oncogene 1/phosphoinositide-3 kinase/Akt pathway	[519]
	circRGNEF	Up	KIF2C	[420]
	CircZNF609	Up	CDC25B	[520]
	circRNA_103809	Up	FBXL18	[521]

(continued)

Table 18.4 (continued)

Type of cancer	Circular RNAs	Expression	Target	References
Glioblastoma	circASAP1	Up	NRAS/MEK1/ERK1-2 signaling	[522]
	CircLRFN5	Down	PRRX2/GCH1	[523]
	circVPS18	Up	BCAT1	[524]
	hsa_circ_0008344	Up		[525]
	circ-0010117	Up	SPEN	[526]
	Hsa_circ_0072309	Down		[527]
	circLGMN	Up	LGMN	[528]
	hsa_circ_0075323	Up		[529]
	circSERPINE2	Up	BCL2	[530]
	hsa_circ_0019421	Up	ROS1	[531]
	circMMD	Up	FUBP1/FIR/DVL1 and miR-15b-5p/FZD6 signaling	[532]
	circPOSTN	Up	STC1	[533]
	Circ_0060055	Up	API5	[534]
	circ_0074027	Up	IL17RD	[535]
	circ_0000741	Up	TRIM14	[536]

ered that several other circular RNAs, including hsa_circ_0006633, hsa_circ_0000520, hsa_circ_0000673, hsa_circ_0001017, hsa_circ_0061276, circ-TTC17, circ-LDLRAD3, hsa_circ_0001785, hsa_circ_0001445, hsa_circ_0000181, hsa_circ_0013958, and hsa_circ_0000285, were also present in plasma, serum or exosomes and had the ability to differentiate cancer patients from healthy individuals, making them potentially valuable biomarkers for cancer [163, 543–551].

7 Conclusion

The scientific community's perception of circRNAs has shifted significantly as our comprehension of them has grown. They are now seen as more than just random splicing errors and instead are recognized as a distinct group of RNA molecules that are physically stable and play a role in various biological processes. Circular RNAs (circRNAs), which are found in various biological systems, are typically formed through the reverse splicing of pre-mRNA. CircRNAs and cancer have recently become a subject of focus in research. An extensive amount of circRNAs are not functioning properly and have an impact on controlling the progression of cancer. Numerous instances of studies investigating the functions of circRNAs in relation to cancer have been presented previously. To sum up, circRNAs have the ability to engage in different signaling pathways and impact a range of cellular functions such as cell growth, movement, invasiveness, programmed cell death, cellular recycling, and resistance to medications, among others. The potential of this ability has motivated scientists to contemplate the potential therapeutic applications of directing attention toward circRNAs and the pathways they are linked to. Moreover, scientists have also attempted to artificially create circRNAs for the purpose of treating diseases. One instance involved utilizing a synthesized circRNA containing eight sites that can bind to miRNA-122 in order to compete with it. This miRNA is crucial for the hepatitis C virus (HCV) to replicate, so by doing this, the spread of HCV is hindered. This approach was as effective as the anti-miR medication miravirsen. In this research, the concept of using engineered circRNAs for treating specific diseases was suggested, similar to how targeted drugs are used. However, this notion still has a long way to go before it can be applied in clinical settings. CircRNAs possess a stable architecture, conservative sequence, and distinct expression profile, rendering them valuable biomarkers for cancer. These circular RNAs hold immense promise in various clinical arenas, including disease detection, prognostication, and therapeutic monitoring. However, a majority of previous research has consisted of individual institution-based and retrospective projects. Our understanding of circRNAs is in its early stages and the process by which they are formed is not yet fully comprehended. Although certain circRNAs have been actively studied and described, our comprehension of circRNAs remains inadequate. Aside from their roles as miRNA sponges, protein scaffolds, translation templates, and transcriptional regulators, there are likely more functions of miRNAs that have yet to be uncovered. Currently, it is impossible to determine the extent to which these mechanisms work together and the relative significance of each one. Furthermore, the transportation and breakdown processes of circRNAs remain largely unknown. As a result, there is still a significant amount of work that needs to be done. The identification of circRNAs has undeniably added depth to our understanding of RNA regulatory networks and has provided innovative pathways for the creation of practical diagnostic and prognostic tools as well as potential treatment targets for cancer.

References

1. Sanger HL, Klotz G, Riesner D, Gross HJ, Kleinschmidt AK (1976) Viroids are single-stranded covalently closed circular RNA molecules existing as highly base-paired rod-like structures. Proc Natl Acad Sci USA 73(11):3852–3856. https://doi.org/10.1073/pnas.73.11.3852
2. Hsu MT, Coca-Prados M (1979) Electron microscopic evidence for the circular form of RNA in the cytoplasm of eukaryotic cells. Nature 280(5720):339–340. https://doi.org/10.1038/280339a0
3. Kos A, Dijkema R, Arnberg AC, van der Meide PH, Schellekens H (1986) The hepatitis delta (delta) virus possesses a circular RNA. Nature 323(6088):558–560. https://doi.org/10.1038/323558a0
4. Salzman J, Gawad C, Wang PL, Lacayo N, Brown PO (2012) Circular RNAs are the predominant transcript isoform from hundreds of human genes in diverse cell types. PLoS One 7(2):e30733. https://doi.org/10.1371/journal.pone.0030733
5. Hansen TB, Jensen TI, Clausen BH, Bramsen JB, Finsen B, Damgaard CK, Kjems J (2013) Natural RNA circles function as efficient microRNA sponges. Nature 495(7441):384–388. https://doi.org/10.1038/nature11993
6. Memczak S, Jens M, Elefsinioti A, Torti F, Krueger J, Rybak A, Maier L, Mackowiak SD, Gregersen LH, Munschauer M, Loewer A, Ziebold U, Landthaler M, Kocks C, le Noble F, Rajewsky N (2013) Circular RNAs are a large class of animal RNAs with regulatory potency. Nature 495(7441):333–338. https://doi.org/10.1038/nature11928
7. Jeck WR, Sorrentino JA, Wang K, Slevin MK, Burd CE, Liu J, Marzluff WF, Sharpless NE (2013) Circular RNAs are abundant, conserved, and associated with ALU repeats. RNA 19(2):141–157. https://doi.org/10.1261/rna.035667.112
8. Salzman J, Chen RE, Olsen MN, Wang PL, Brown PO (2013) Cell-type specific features of circular RNA expression. PLoS Genet 9(9):e1003777. https://doi.org/10.1371/journal.pgen.1003777
9. Bai Y, Zhang Y, Han B, Yang L, Chen X, Huang R, Wu F, Chao J, Liu P, Hu G, Zhang JH, Yao H (2018) Circular RNA DLGAP4 ameliorates ischemic stroke outcomes by targeting miR-143 to regulate endothelial-mesenchymal transition associated with blood-brain barrier integrity. J Neurosci 38(1):32–50. https://doi.org/10.1523/jneurosci.1348-17.2017
10. Holdt LM, Stahringer A, Sass K, Pichler G, Kulak NA, Wilfert W, Kohlmaier A, Herbst A, Northoff BH, Nicolaou A, Gäbel G, Beutner F, Scholz M, Thiery J, Musunuru K, Krohn K, Mann M, Teupser D (2016) Circular non-coding RNA ANRIL modulates ribosomal RNA maturation and atherosclerosis in humans. Nat Commun 7:12429. https://doi.org/10.1038/ncomms12429
11. Zhao Y, Alexandrov PN, Jaber V, Lukiw WJ (2016) Deficiency in the ubiquitin conjugating enzyme UBE2A in Alzheimer's disease (AD) is linked to deficits in a natural circular miRNA-7 sponge (circRNA; ciRS-7). Genes (Basel) 7(12):116. https://doi.org/10.3390/genes7120116
12. Liu Q, Zhang X, Hu X, Dai L, Fu X, Zhang J, Ao Y (2016) Circular RNA related to the chondrocyte ECM regulates MMP13 expression by functioning as a MiR-136 'sponge' in human cartilage degradation. Sci Rep 6:22572. https://doi.org/10.1038/srep22572
13. Zhao Z, Li X, Jian D, Hao P, Rao L, Li M (2017) Hsa_circ_0054633 in peripheral blood can be used as a diagnostic biomarker of pre-diabetes and type 2 diabetes mellitus. Acta Diabetol 54(3):237–245. https://doi.org/10.1007/s00592-016-0943-0
14. Zhou Z, Jiang R, Yang X, Guo H, Fang S, Zhang Y, Cheng Y, Wang J, Yao H, Chao J (2018) circRNA mediates silica-induced macrophage activation via HECTD1/ZC3H12A-dependent ubiquitination.

Theranostics 8(2):575–592. https://doi.org/10.7150/thno.21648
15. Geng Y, Jiang J, Wu C (2018) Function and clinical significance of circRNAs in solid tumors. J Hematol Oncol 11(1):98. https://doi.org/10.1186/s13045-018-0643-z
16. Guarnerio J, Bezzi M, Jeong JC, Paffenholz SV, Berry K, Naldini MM, Lo-Coco F, Tay Y, Beck AH, Pandolfi PP (2016) Oncogenic role of fusion-circRNAs derived from cancer-associated chromosomal translocations. Cell 166(4):1055–1056. https://doi.org/10.1016/j.cell.2016.07.035
17. Liang HF, Zhang XZ, Liu BG, Jia GT, Li WL (2017) Circular RNA circ-ABCB10 promotes breast cancer proliferation and progression through sponging miR-1271. Am J Cancer Res 7(7):1566–1576
18. Zhang Y, Liang W, Zhang P, Chen J, Qian H, Zhang X, Xu W (2017) Circular RNAs: emerging cancer biomarkers and targets. J Exp Clin Cancer Res 36:1–13
19. Zhang M, Xin Y (2018) Circular RNAs: a new frontier for cancer diagnosis and therapy. J Hematol Oncol 11:1–9
20. Hu J, Li P, Song Y, Ge YX, Meng XM, Huang C, Li J, Xu T (2018) Progress and prospects of circular RNAs in Hepatocellular carcinoma: novel insights into their function. J Cell Physiol 233(6):4408–4422. https://doi.org/10.1002/jcp.26154
21. Cheng F, Wang L, Zhang J (2020) Circular RNA 0016788 displays as a biomarker for tumor progression and poor prognosis in surgical hepatocellular carcinoma patients. J Clin Lab Anal 34(7):e23300. https://doi.org/10.1002/jcla.23300
22. Yao Z, Luo J, Hu K, Lin J, Huang H, Wang Q, Zhang P, Xiong Z, He C, Huang Z, Liu B, Yang Y (2017) ZKSCAN1 gene and its related circular RNA (circZKSCAN1) both inhibit hepatocellular carcinoma cell growth, migration, and invasion but through different signaling pathways. Mol Oncol 11(4):422–437. https://doi.org/10.1002/1878-0261.12045
23. Yang J, Li Y, Yu Z, Zhou Y, Tu J, Lou J, Wang Y (2020) Circular RNA Circ100084 functions as sponge of miR-23a-5p to regulate IGF2 expression in hepatocellular carcinoma. Mol Med Rep 21(6):2395–2404. https://doi.org/10.3892/mmr.2020.11069
24. Chen W, Quan Y, Fan S, Wang H, Liang J, Huang L, Chen L, Liu Q, He P, Ye Y (2020) Exosome-transmitted circular RNA hsa_circ_0051443 suppresses hepatocellular carcinoma progression. Cancer Lett 475:119–128. https://doi.org/10.1016/j.canlet.2020.01.022
25. Zhou S, Wei J, Wang Y, Liu X (2020) Cisplatin resistance-associated circRNA_101237 serves as a prognostic biomarker in hepatocellular carcinoma. Exp Ther Med 19(4):2733–2740. https://doi.org/10.3892/etm.2020.8526
26. Wu MY, Tang YP, Liu JJ, Liang R, Luo XL (2020) Global transcriptomic study of circRNAs expression profile in sorafenib resistant hepatocellular carcinoma cells. J Cancer 11(10):2993–3001. https://doi.org/10.7150/jca.39854
27. Chen H, Liu S, Li M, Huang P, Li X (2019) circ_0003418 inhibits tumorigenesis and cisplatin chemoresistance through Wnt/β-catenin pathway in hepatocellular carcinoma. Onco Targets Ther 12:9539–9549. https://doi.org/10.2147/ott.S229507
28. Li J, Qin X, Wu R, Wan L, Zhang L, Liu R (2020) Circular RNA circFBXO11 modulates hepatocellular carcinoma progress and oxaliplatin resistance through miR-605/FOXO3/ABCB1 axis. J Cell Mol Med 24(9):5152–5161. https://doi.org/10.1111/jcmm.15162
29. He J, Chu Z, Lai W, Lan Q, Zeng Y, Lu D, Jin S, Xu H, Su P, Yin D, Chu Z, Liu L (2021) Circular RNA circHERC4 as a novel oncogenic driver to promote tumor metastasis via the miR-556-5p/CTBP2/E-cadherin axis in colorectal cancer. J Hematol Oncol 14(1):194. https://doi.org/10.1186/s13045-021-01210-2
30. Liang ZX, Liu HS, Xiong L, Yang X, Wang FW, Zeng ZW, He XW, Wu XR, Lan P (2021) A novel NF-κB regulator encoded by circPLCE1 inhibits colorectal carcinoma progression by promoting RPS3 ubiquitin-dependent degradation. Mol Cancer 20(1):103. https://doi.org/10.1186/s12943-021-01404-9
31. Yang D, Hu Z, Zhang Y, Zhang X, Xu J, Fu H, Zhu Z, Feng D, Cai Q (2021) CircHIPK3 promotes the tumorigenesis and development of gastric cancer through miR-637/AKT1 pathway. Front Oncol 11:637761. https://doi.org/10.3389/fonc.2021.637761
32. Chen ZG, Zhao HJ, Lin L, Liu JB, Bai JZ, Wang GS (2020) Circular RNA CirCHIPK3 promotes cell proliferation and invasion of breast cancer by sponging miR-193a/HMGB1/PI3K/AKT axis. Thorac Cancer 11(9):2660–2671. https://doi.org/10.1111/1759-7714.13603
33. Chen G, Shi Y, Liu M, Sun J (2018) circHIPK3 regulates cell proliferation and migration by sponging miR-124 and regulating AQP3 expression in hepatocellular carcinoma. Cell Death Dis 9(2):175. https://doi.org/10.1038/s41419-017-0204-3
34. Li Y, Zheng F, Xiao X, Xie F, Tao D, Huang C, Liu D, Wang M, Wang L, Zeng F, Jiang G (2017) CircHIPK3 sponges miR-558 to suppress heparanase expression in bladder cancer cells. EMBO Rep 18(9):1646–1659. https://doi.org/10.15252/embr.201643581
35. Kristensen LS, Jakobsen T, Hager H, Kjems J (2022) The emerging roles of circRNAs in cancer and oncology. Nat Rev Clin Oncol 19(3):188–206. https://doi.org/10.1038/s41571-021-00585-y
36. Qiu M, Xia W, Chen R, Wang S, Xu Y, Ma Z, Xu W, Zhang E, Wang J, Fang T, Hu J, Dong G, Yin R, Wang J, Xu L (2018) The circular RNA circPRKCI promotes tumor growth in lung adenocarcinoma. Cancer Res 78(11):2839–2851. https://doi.org/10.1158/0008-5472.Can-17-2808

37. Guarnerio J, Bezzi M, Jeong JC, Paffenholz SV, Berry K, Naldini MM, Lo-Coco F, Tay Y, Beck AH, Pandolfi PP (2016) Oncogenic role of fusion-circRNAs derived from cancer-associated chromosomal translocations. Cell 165(2):289–302. https://doi.org/10.1016/j.cell.2016.03.020
38. Wu K, Liao X, Gong Y, He J, Zhou JK, Tan S, Pu W, Huang C, Wei YQ, Peng Y (2019) Circular RNA F-circSR derived from SLC34A2-ROS1 fusion gene promotes cell migration in non-small cell lung cancer. Mol Cancer 18(1):98. https://doi.org/10.1186/s12943-019-1028-9
39. Tan S, Gou Q, Pu W, Guo C, Yang Y, Wu K, Liu Y, Liu L, Wei YQ, Peng Y (2018) Circular RNA F-circEA produced from EML4-ALK fusion gene as a novel liquid biopsy biomarker for non-small cell lung cancer. Cell Res 28(6):693–695. https://doi.org/10.1038/s41422-018-0033-7
40. Zhang L, Hou C, Chen C, Guo Y, Yuan W, Yin D, Liu J, Sun Z (2020) The role of N(6)-methyladenosine (m(6)A) modification in the regulation of circRNAs. Mol Cancer 19(1):105. https://doi.org/10.1186/s12943-020-01224-3
41. Chen RX, Chen X, Xia LP, Zhang JX, Pan ZZ, Ma XD, Han K, Chen JW, Judde JG, Deas O, Wang F, Ma NF, Guan X, Yun JP, Wang FW, Xu RH, Dan X (2019) N(6)-methyladenosine modification of circNSUN2 facilitates cytoplasmic export and stabilizes HMGA2 to promote colorectal liver metastasis. Nat Commun 10(1):4695. https://doi.org/10.1038/s41467-019-12651-2
42. Xu J, Wan Z, Tang M, Lin Z, Jiang S, Ji L, Gorshkov K, Mao Q, Xia S, Cen D, Zheng J, Liang X, Cai X (2020) N(6)-methyladenosine-modified CircRNA-SORE sustains sorafenib resistance in hepatocellular carcinoma by regulating β-catenin signaling. Mol Cancer 19(1):163. https://doi.org/10.1186/s12943-020-01281-8
43. Wu P, Fang X, Liu Y, Tang Y, Wang W, Li X, Fan Y (2021) N6-methyladenosine modification of circCUX1 confers radioresistance of hypopharyngeal squamous cell carcinoma through caspase1 pathway. Cell Death Dis 12(4):298. https://doi.org/10.1038/s41419-021-03558-2
44. Sun HD, Xu ZP, Sun ZQ, Zhu B, Wang Q, Zhou J, Jin H, Zhao A, Tang WW, Cao XF (2018) Down-regulation of circPVRL3 promotes the proliferation and migration of gastric cancer cells. Sci Rep 8(1):10111. https://doi.org/10.1038/s41598-018-27837-9
45. Bray F, Ferlay J, Soerjomataram I, Siegel RL, Torre LA, Jemal A (2018) Global cancer statistics 2018: GLOBOCAN estimates of incidence and mortality worldwide for 36 cancers in 185 countries. CA Cancer J Clin 68(6):394–424. https://doi.org/10.3322/caac.21492
46. Cheng J, Zhuo H, Xu M, Wang L, Xu H, Peng J, Hou J, Lin L, Cai J (2018) Regulatory network of circRNA-miRNA-mRNA contributes to the histological classification and disease progression in gastric cancer. J Transl Med 16(1):216. https://doi.org/10.1186/s12967-018-1582-8
47. Li H, Yao G, Feng B, Lu X, Fan Y (2018) Circ_0056618 and CXCR4 act as competing endogenous in gastric cancer by regulating miR-206. J Cell Biochem 119(11):9543–9551. https://doi.org/10.1002/jcb.27271
48. Zhang J, Liu H, Hou L, Wang G, Zhang R, Huang Y, Chen X, Zhu J (2017) Circular RNA_LARP4 inhibits cell proliferation and invasion of gastric cancer by sponging miR-424-5p and regulating LATS1 expression. Mol Cancer 16(1):151. https://doi.org/10.1186/s12943-017-0719-3
49. Naeli P, Pourhanifeh MH, Karimzadeh MR, Shabaninejad Z, Movahedpour A, Tarrahimofrad H, Mirzaei HR, Bafrani HH, Savardashtaki A, Mirzaei H, Hamblin MR (2020) Circular RNAs and gastrointestinal cancers: Epigenetic regulators with a prognostic and therapeutic role. Crit Rev Oncol Hematol 145:102854. https://doi.org/10.1016/j.critrevonc.2019.102854
50. Fan L, Cao Q, Liu J, Zhang J, Li B (2019) Circular RNA profiling and its potential for esophageal squamous cell cancer diagnosis and prognosis. Mol Cancer 18(1):16. https://doi.org/10.1186/s12943-018-0936-4
51. Huang E, Fu J, Yu Q, Xie P, Yang Z, Ji H, Wang L, Luo G, Zhang Y, Li K (2020) CircRNA hsa_circ_0004771 promotes esophageal squamous cell cancer progression via miR-339-5p/CDC25A axis. Epigenomics 12(7):587–603. https://doi.org/10.2217/epi-2019-0404
52. Pan Z, Lin J, Wu D, He X, Wang W, Hu X, Zhang L, Wang M (2019) Hsa_circ_0006948 enhances cancer progression and epithelial-mesenchymal transition through the miR-490-3p/HMGA2 axis in esophageal squamous cell carcinoma. Aging 11(24):11937–11954. https://doi.org/10.18632/aging.102519
53. Xu Z, Tie X, Li N, Yi Z, Shen F, Zhang Y (2020) Circular RNA hsa_circ_0000654 promotes esophageal squamous cell carcinoma progression by regulating the miR-149-5p/IL-6/STAT3 pathway. IUBMB Life 72(3):426–439. https://doi.org/10.1002/iub.2202
54. Hu X, Wu D, He X, Zhao H, He Z, Lin J, Wang K, Wang W, Pan Z, Lin H, Wang M (2019) circGSK3β promotes metastasis in esophageal squamous cell carcinoma by augmenting β-catenin signaling. Mol Cancer 18(1):160. https://doi.org/10.1186/s12943-019-1095-y
55. Xing Y, Zha WJ, Li XM, Li H, Gao F, Ye T, Du WQ, Liu YC (2020) Circular RNA circ-Foxo3 inhibits esophageal squamous cell cancer progression via the miR-23a/PTEN axis. J Cell Biochem 121(3):2595–2605. https://doi.org/10.1002/jcb.29481
56. Zhong R, Chen Z, Mo T, Li Z, Zhang P (2019) Potential role of circPVT1 as a proliferative factor and treatment target in esophageal carcinoma.

Cancer Cell Int 19:267. https://doi.org/10.1186/s12935-019-0985-9
57. Meng L, Liu S, Ding P, Chang S, Sang M (2020) Circular RNA ciRS-7 inhibits autophagy of ESCC cells by functioning as miR-1299 sponge to target EGFR signaling. J Cell Biochem 121(2):1039–1049. https://doi.org/10.1002/jcb.29339
58. Huang H, Wei L, Qin T, Yang N, Li Z, Xu Z (2019) Circular RNA ciRS-7 triggers the migration and invasion of esophageal squamous cell carcinoma via miR-7/KLF4 and NF-κB signals. Cancer Biol Ther 20(1):73–80. https://doi.org/10.1080/15384047.2018.1507254
59. Mi L, Zhou Y, Wu D, Tao Q, Wang X, Zhu H, Gao X, Wang J, Ling R, Deng J, Mao C, Chen D (2019) ACSS2/AMPK/PCNA pathway-driven proliferation and chemoresistance of esophageal squamous carcinoma cells under nutrient stress. Mol Med Rep 20(6):5286–5296. https://doi.org/10.3892/mmr.2019.10735
60. Zou FW, Yang SZ, Li WY, Liu CY, Liu XH, Hu CH, Liu ZH, Xu S (2020) circRNA_001275 upregulates Wnt7a expression by competitively sponging miR-370-3p to promote cisplatin resistance in esophageal cancer. Int J Oncol 57(1):151–160. https://doi.org/10.3892/ijo.2020.5050
61. Kim Y, Bang SS, Jee S, Park S, Shin SJ, Jang K (2020) Prevalence and clinicopathological significance of MET overexpression and gene amplification in patients with gallbladder carcinoma. Cancer Res Treat 52(2):481–491. https://doi.org/10.4143/crt.2019.370
62. McNamara MG, Metran-Nascente C, Knox JJ (2013) State-of-the-art in the management of locally advanced and metastatic gallbladder cancer. Curr Opin Oncol 25(4):425–431. https://doi.org/10.1097/CCO.0b013e3283620fd8
63. Sung YN, Song M, Lee JH, Song KB, Hwang DW, Ahn CS, Hwang S, Hong SM (2020) Validation of the 8th edition of the American Joint Committee on cancer staging system for gallbladder cancer and implications for the follow-up of patients without node dissection. Cancer Res Treat 52(2):455–468. https://doi.org/10.4143/crt.2019.271
64. Kakaei F, Beheshtirouy S, Nejatollahi SM, Zarrintan S, Mafi MR (2015) Surgical treatment of gallbladder carcinoma: a critical review. Updat Surg 67(4):339–351. https://doi.org/10.1007/s13304-015-0328-x
65. Kai D, Yannian L, Yitian C, Dinghao G, Xin Z, Wu J (2018) Circular RNA HIPK3 promotes gallbladder cancer cell growth by sponging microRNA-124. Biochem Biophys Res Commun 503(2):863–869. https://doi.org/10.1016/j.bbrc.2018.06.088
66. Wang S, Zhang Y, Cai Q, Ma M, Jin LY, Weng M, Zhou D, Tang Z, Wang JD, Quan Z (2019) Circular RNA FOXP1 promotes tumor progression and Warburg effect in gallbladder cancer by regulating PKLR expression. Mol Cancer 18(1):145. https://doi.org/10.1186/s12943-019-1078-z
67. Arnold M, Abnet CC, Neale RE, Vignat J, Giovannucci EL, McGlynn KA, Bray F (2020) Global burden of 5 major types of gastrointestinal cancer. Gastroenterology 159(1):335–349.e315. https://doi.org/10.1053/j.gastro.2020.02.068
68. Guo X, Zhou Q, Su D, Luo Y, Fu Z, Huang L, Li Z, Jiang D, Kong Y, Li Z, Chen R, Chen C (2020) Circular RNA circBFAR promotes the progression of pancreatic ductal adenocarcinoma via the miR-34b-5p/MET/Akt axis. Mol Cancer 19(1):83. https://doi.org/10.1186/s12943-020-01196-4
69. Wong CH, Lou UK, Li Y, Chan SL, Tong JH, To KF, Chen Y (2020) CircFOXK2 promotes growth and metastasis of pancreatic ductal adenocarcinoma by complexing with RNA-binding proteins and sponging MiR-942. Cancer Res 80(11):2138–2149. https://doi.org/10.1158/0008-5472.Can-19-3268
70. Xing C, Ye H, Wang W, Sun M, Zhang J, Zhao Z, Jiang G (2019) Circular RNA ADAM9 facilitates the malignant behaviours of pancreatic cancer by sponging miR-217 and upregulating PRSS3 expression. Artif Cells Nanomed Biotechnol 47(1):3920–3928. https://doi.org/10.1080/21691401.2019.1671856
71. Chen Y, Li Z, Zhang M, Wang B, Ye J, Zhang Y, Tang D, Ma D, Jin W, Li X, Wang S (2019) Circ-ASH2L promotes tumor progression by sponging miR-34a to regulate Notch1 in pancreatic ductal adenocarcinoma. J Exp Clin Cancer Res 38(1):466. https://doi.org/10.1186/s13046-019-1436-0
72. Yao J, Zhang C, Chen Y, Gao S (2019) Downregulation of circular RNA circ-LDLRAD3 suppresses pancreatic cancer progression through miR-137-3p/PTN axis. Life Sci 239:116871. https://doi.org/10.1016/j.lfs.2019.116871
73. Ou ZL, Luo Z, Wei W, Liang S, Gao TL, Lu YB (2019) Hypoxia-induced shedding of MICA and HIF1A-mediated immune escape of pancreatic cancer cells from NK cells: role of circ_0000977/miR-153 axis. RNA Biol 16(11):1592–1603. https://doi.org/10.1080/15476286.2019.1649585
74. Yang J, Cong X, Ren M, Sun H, Liu T, Chen G, Wang Q, Li Z, Yu S, Yang Q (2019) Circular RNA hsa_circRNA_0007334 is predicted to promote MMP7 and COL1A1 expression by functioning as a miRNA sponge in pancreatic ductal adenocarcinoma. J Oncol 2019:7630894. https://doi.org/10.1155/2019/7630894
75. Liu L, Liu FB, Huang M, Xie K, Xie QS, Liu CH, Shen MJ, Huang Q (2019) Circular RNA ciRS-7 promotes the proliferation and metastasis of pancreatic cancer by regulating miR-7-mediated EGFR/STAT3 signaling pathway. Hepatobiliary Pancreat Dis Int 18(6):580–586. https://doi.org/10.1016/j.hbpd.2019.03.003
76. Xu Y, Yao Y, Gao P, Cui Y (2019) Upregulated circular RNA circ_0030235 predicts unfavorable prognosis in pancreatic ductal adenocarcinoma and facilitates cell progression by sponging miR-1253 and miR-1294. Biochem Biophys Res Commun 509(1):138–142. https://doi.org/10.1016/j.bbrc.2018.12.088

77. Hao L, Rong W, Bai L, Cui H, Zhang S, Li Y, Chen D, Meng X (2019) Upregulated circular RNA circ_0007534 indicates an unfavorable prognosis in pancreatic ductal adenocarcinoma and regulates cell proliferation, apoptosis, and invasion by sponging miR-625 and miR-892b. J Cell Biochem 120(3):3780–3789. https://doi.org/10.1002/jcb.27658
78. Li J, Li Z, Jiang P, Peng M, Zhang X, Chen K, Liu H, Bi H, Liu X, Li X (2018) Circular RNA IARS (circ-IARS) secreted by pancreatic cancer cells and located within exosomes regulates endothelial monolayer permeability to promote tumor metastasis. J Exp Clin Cancer Res 37(1):177. https://doi.org/10.1186/s13046-018-0822-3
79. Li Z, Yanfang W, Li J, Jiang P, Peng T, Chen K, Zhao X, Zhang Y, Zhen P, Zhu J, Li X (2018) Tumor-released exosomal circular RNA PDE8A promotes invasive growth via the miR-338/MACC1/MET pathway in pancreatic cancer. Cancer Lett 432:237–250. https://doi.org/10.1016/j.canlet.2018.04.035
80. Du J, Lan T, Liao H, Feng X, Chen X, Liao W, Hou G, Xu L, Feng Q, Xie K, Liao M, Chen X, Huang J, Yuan K, Zeng Y (2022) CircNFIB inhibits tumor growth and metastasis through suppressing MEK1/ERK signaling in intrahepatic cholangiocarcinoma. Mol Cancer 21(1):18. https://doi.org/10.1186/s12943-021-01482-9
81. Chen Q, Wang H, Li Z, Li F, Liang L, Zou Y, Shen H, Li J, Xia Y, Cheng Z, Yang T, Wang K, Shen F (2022) Circular RNA ACTN4 promotes intrahepatic cholangiocarcinoma progression by recruiting YBX1 to initiate FZD7 transcription. J Hepatol 76(1):135–147. https://doi.org/10.1016/j.jhep.2021.08.027
82. Li H, Lan T, Liu H, Liu C, Dai J, Xu L, Cai Y, Hou G, Xie K, Liao M, Li J, Huang J, Yuan K, Wang G, Zeng Y, Wu H (2022) IL-6-induced cGGNBP2 encodes a protein to promote cell growth and metastasis in intrahepatic cholangiocarcinoma. Hepatology (Baltimore, Md) 75(6):1402–1419. https://doi.org/10.1002/hep.32232
83. Yu X, Tong H, Chen J, Tang C, Wang S, Si Y, Wang S, Tang Z (2023) CircRNA MBOAT2 promotes intrahepatic cholangiocarcinoma progression and lipid metabolism reprogramming by stabilizing PTBP1 to facilitate FASN mRNA cytoplasmic export. Cell Death Dis 14(1):20. https://doi.org/10.1038/s41419-022-05540-y
84. Liao W, Du J, Li L, Wu X, Chen X, Feng Q, Xu L, Chen X, Liao M, Huang J, Yuan K, Zeng Y (2023) CircZNF215 promotes tumor growth and metastasis through inactivation of the PTEN/AKT pathway in intrahepatic cholangiocarcinoma. J Exp Clin Cancer Res 42(1):125. https://doi.org/10.1186/s13046-023-02699-w
85. Xu Y, Leng K, Yao Y, Kang P, Liao G, Han Y, Shi G, Ji D, Huang P, Zheng W, Li Z, Li J, Huang L, Yu L, Zhou Y, Jiang X, Wang H, Li C, Su Z, Tai S, Zhong X, Wang Z, Cui Y (2021) A circular RNA, cholangiocarcinoma-associated circular RNA 1, contributes to cholangiocarcinoma progression, induces angiogenesis, and disrupts vascular endothelial barriers. Hepatology (Baltimore, Md) 73(4):1419–1435. https://doi.org/10.1002/hep.31493
86. Tu J, Chen W, Zheng L, Fang S, Zhang D, Kong C, Yang Y, Qiu R, Zhao Z, Lu C, Lu X, Ji J (2021) Circular RNA Circ0021205 promotes cholangiocarcinoma progression through MiR-204-5p/RAB22A axis. Front Cell Dev Biol 9:653207. https://doi.org/10.3389/fcell.2021.653207
87. Xu Y, Yao Y, Zhong X, Leng K, Qin W, Qu L, Cui Y, Jiang X (2018) Downregulated circular RNA hsa_circ_0001649 regulates proliferation, migration and invasion in cholangiocarcinoma cells. Biochem Biophys Res Commun 496(2):455–461. https://doi.org/10.1016/j.bbrc.2018.01.077
88. Xu Y, Yao Y, Liu Y, Wang Z, Hu Z, Su Z, Li C, Wang H, Jiang X, Kang P, Sun D, Zhong X, Cui Y (2019) Elevation of circular RNA circ_0005230 facilitates cell growth and metastasis via sponging miR-1238 and miR-1299 in cholangiocarcinoma. Aging 11(7):1907–1917. https://doi.org/10.18632/aging.101872
89. Lu Q, Fang T (2020) Circular RNA SMARCA5 correlates with favorable clinical tumor features and prognosis, and increases chemotherapy sensitivity in intrahepatic cholangiocarcinoma. J Clin Lab Anal 34(4):e23138. https://doi.org/10.1002/jcla.23138
90. Zhong X, Ji C, Ren D, Ke A, Yang Z (2023) Circular RNA circEIF3C promotes intrahepatic cholangiocarcinoma progression and immune evasion via the miR-34a-5p/B7-H4 axis. Genes Dis 10(2):370–372. https://doi.org/10.1016/j.gendis.2022.05.005
91. Guan C, Liu L, Zhao Y, Zhang X, Liu G, Wang H, Gao X, Zhong X, Jiang X (2021) YY1 and eIF4A3 are mediators of the cell proliferation, migration and invasion in cholangiocarcinoma promoted by circ-ZNF609 by targeting miR-432-5p to regulate LRRC1. Aging 13(23):25195–25212. https://doi.org/10.18632/aging.203735
92. Louis C, Ferlier T, Leroux R, Pineau R, Desoteux M, Papoutsoglou P, Leclerc D, Angenard G, Vaquero J, Macias RIR, Edeline J, Coulouarn C (2023) TGFβ-induced circLTBP2 predicts a poor prognosis in intrahepatic cholangiocarcinoma and mediates gemcitabine resistance by sponging miR-338-3p. JHEP Rep 5(12):100900. https://doi.org/10.1016/j.jhepr.2023.100900
93. Su Y, Yu T, Wang Y, Huang X, Wei X (2021) Circular RNA circDNM3OS functions as a miR-145-5p sponge to accelerate cholangiocarcinoma growth and glutamine metabolism by upregulating MORC2. Onco Targets Ther 14:1117–1129. https://doi.org/10.2147/ott.s289241
94. Li J, Xu X, Xu K, Zhou X, Wu K, Yao Y, Liu Z, Chen C, Wang L, Sun Z, Jiao D, Han X (2024) N6-methyladenosine-modified circSLCO1B3 promotes intrahepatic cholangiocarcinoma progression via regulating HOXC8 and PD-L1. J Exp Clin Cancer Res 43(1):119. https://doi.org/10.1186/s13046-024-03006-x

95. Wang G, Gao X, Sun Z, He T, Huang C, Li S, Long H (2023) Circular RNA SMARCA5 inhibits cholangiocarcinoma via microRNA-95-3p/tumor necrosis factor receptor associated factor 3 axis. Anti-Cancer Drugs 34(9):1002–1009. https://doi.org/10.1097/cad.0000000000001487
96. Xu Y, Gao P, Wang Z, Su Z, Liao G, Han Y, Cui Y, Yao Y, Zhong X (2021) Circ-LAMP1 contributes to the growth and metastasis of cholangiocarcinoma via miR-556-5p and miR-567 mediated YY1 activation. J Cell Mol Med 25(7):3226–3238. https://doi.org/10.1111/jcmm.16392
97. Liu S, Wang Y, Wang T, Shi K, Fan S, Li C, Chen R, Wang J, Jiang W, Zhang Y, Chen Y, Xu X, Yu Y, Li C, Li X (2024) CircPCNXL2 promotes tumor growth and metastasis by interacting with STRAP to regulate ERK signaling in intrahepatic cholangiocarcinoma. Mol Cancer 23(1):35. https://doi.org/10.1186/s12943-024-01950-y
98. Xiong W, Zhang A, Xiao X, Liu W (2022) CircSETD3 (hsa_circ_0000567) inhibits proliferation and induces apoptosis in cholangiocarcinoma cells via downregulation of microRNA-421 expression. Bioengineered 13(4):10191–10201. https://doi.org/10.1080/21655979.2022.2061283
99. Zheng J, Wang Y, Tao L, Cai J, Shen Z, Liu Y, Pan H, Li S, Ruan Y, Chen T, Ye Z, Lin K, Sun Y, Xu J, Liang X (2023) Circ-RAPGEF5 promotes intrahepatic cholangiocarcinoma progression by stabilizing SAE1 to facilitate SUMOylation. J Exp Clin Cancer Res 42(1):239. https://doi.org/10.1186/s13046-023-02813-y
100. Liu S, Guan C, Sha X, Gao X, Zou X, Yang C, Zhang X, Zhong X, Jiang X (2024) Circ_0007534 promotes cholangiocarcinoma stemness and resistance to anoikis through DDX3X-mediated positive feedback regulation of parental gene DDX42. Cell Signal 118:111141. https://doi.org/10.1016/j.cellsig.2024.111141
101. Li D, Tang Z, Gao Z, Shen P, Liu Z, Dang X (2020) Circular RNA CDR1as exerts oncogenic properties partially through regulating MicroRNA 641 in cholangiocarcinoma. Mol Cel Biol 40(15):e00042. https://doi.org/10.1128/mcb.00042-20
102. Chen ZW, Kang FP, Xie CK, Liao CY, Li G, Wu YD, Lin HY, Zhu SC, Hu JF, Lin CF, Huang Y, Tian YF, Huang L, Wang ZW, Chen S (2023) A novel trojan horse nanotherapy strategy targeting the cPKM-STMN1/TGFB1 axis for effective treatment of intrahepatic cholangiocarcinoma. Adv Sci (Weinheim, Baden-Wurttemberg, Germany) 10(32):e2303814. https://doi.org/10.1002/advs.202303814
103. Sun J, Feng M, Zou H, Mao Y, Yu W (2023) Circ_0000284 facilitates the growth, metastasis and glycolysis of intrahepatic cholangiocarcinoma through miR-152-3p-mediated PDK1 expression. Histol Histopathol 38(10):1129–1143. https://doi.org/10.14670/hh-18-544
104. Xu YP, Dong ZN, Wang SW, Zheng YM, Zhang C, Zhou YQ, Zhao YJ, Zhao Y, Wang F, Peng R, Tang MC, Bai DS, Huang XY, Guo CY (2021) circHMGCS1-016 reshapes immune environment by sponging miR-1236-3p to regulate CD73 and GAL-8 expression in intrahepatic cholangiocarcinoma. J Exp Clin Cancer Res 40(1):290. https://doi.org/10.1186/s13046-021-02095-2
105. Zhang X, Zhao Y, Wang W, Yu S, Liu L, Sun D, Li W, Jiang X (2022) Upregulation of circ_0059961 suppresses cholangiocarcinoma development by modulating miR-629-5p/SFRP2 axis. Pathol Res Pract 234:153901. https://doi.org/10.1016/j.prp.2022.153901
106. Tang J, Tang R, Gu P, Han J, Huang W, Xue F (2022) Hsa_circ_0019054 up-regulates HIF1A through sequestering miR-340-5p to promote the tumorigenesis of intrahepatic cholangiocarcinoma. Hum Exp Toxicol 41:9603271221126494. https://doi.org/10.1177/09603271221126494
107. Xiao F, Xu F, Zhang H, Shuai X (2022) Circ_0000591 served as endogenous RNA for miR-326 to promote progression of cholangiocarcinoma via the TLR4/MyD88/IL6 axis. Biochem Biophys Res Commun 600:101–108. https://doi.org/10.1016/j.bbrc.2022.02.073
108. Tang J, Wang R, Tang R, Gu P, Han J, Huang W (2022) CircRTN4IP1 regulates the malignant progression of intrahepatic cholangiocarcinoma by sponging miR-541-5p to induce HIF1A production. Pathol Res Pract 230:153732. https://doi.org/10.1016/j.prp.2021.153732
109. Xi Y, Shen Y, Wu D, Zhang J, Lin C, Wang L, Yu C, Yu B, Shen W (2022) CircBCAR3 accelerates esophageal cancer tumorigenesis and metastasis via sponging miR-27a-3p. Mol Cancer 21(1):145. https://doi.org/10.1186/s12943-022-01615-8
110. Gu L, Sang Y, Nan X, Zheng Y, Liu F, Meng L, Sang M, Shan B (2022) circCYP24A1 facilitates esophageal squamous cell carcinoma progression through binding PKM2 to regulate NF-κB-induced CCL5 secretion. Mol Cancer 21(1):217. https://doi.org/10.1186/s12943-022-01686-7
111. Shi Y, Fang N, Li Y, Guo Z, Jiang W, He Y, Ma Z, Chen Y (2020) Circular RNA LPAR3 sponges microRNA-198 to facilitate esophageal cancer migration, invasion, and metastasis. Cancer Sci 111(8):2824–2836. https://doi.org/10.1111/cas.14511
112. Zang HL, Ji FJ, Ju HY, Tian XF (2021) Circular RNA AKT3 governs malignant behaviors of esophageal cancer cells by sponging miR-17-5p. World J Gastroenterol 27(3):240–254. https://doi.org/10.3748/wjg.v27.i3.240
113. Ma Y, Zhang D, Wu H, Li P, Zhao W, Yang X, Xing X, Li S, Li J (2020) Circular RNA PRKCI silencing represses esophageal cancer progression and elevates cell radiosensitivity through regulating

the miR-186-5p/PARP9 axis. Life Sci 259:118168. https://doi.org/10.1016/j.lfs.2020.118168
114. Wang J, Chen Y, Wu R, Lin Y (2021) Circular RNA hsa_circ_0000554 promotes progression and elevates radioresistance through the miR-485-5p/fermitin family members 1 axis in esophageal cancer. Anti-Cancer Drugs 32(4):405–416. https://doi.org/10.1097/cad.0000000000001007
115. Zhang H, Ju L, Hu P, Ye J, Yang C, Huang J (2021) Circular RNA 0014715 facilitates cell proliferation and inhibits apoptosis in esophageal squamous cell carcinoma. Cancer Manag Res 13:4735–4749. https://doi.org/10.2147/cmar.s314882
116. Jiang Q, Wang H, Yuan D, Qian X, Ma X, Yan M, Xing W (2022) Circular_0086414 induces SPARC like 1 (SPARCL1) production to inhibit esophageal cancer cell proliferation, invasion and glycolysis and induce cell apoptosis by sponging miR-1290. Bioengineered 13(5):12099–12114. https://doi.org/10.1080/21655979.2022.2073114
117. Lan X, Liu X, Sun J, Yuan Q, Li J (2019) CircRAD23B facilitates proliferation and invasion of esophageal cancer cells by sponging miR-5095. Biochem Biophys Res Commun 516(2):357–364. https://doi.org/10.1016/j.bbrc.2019.06.044
118. Xie ZF, Li HT, Xie SH, Ma M (2020) Circular RNA hsa_circ_0006168 contributes to cell proliferation, migration and invasion in esophageal cancer by regulating miR-384/RBBP7 axis via activation of S6K/S6 pathway. Eur Rev Med Pharmacol Sci 24(1):151–163. https://doi.org/10.26355/eurrev_202001_19906
119. Cheng J, Zhang R, Yan M, Li Y (2022) Circular RNA hsa_circ_0000277 promotes tumor progression and DDP resistance in esophageal squamous cell carcinoma. BMC Cancer 22(1):238. https://doi.org/10.1186/s12885-022-09241-9
120. Zhou PL, Wu Z, Zhang W, Xu M, Ren J, Zhang Q, Sun Z, Han X (2021) Circular RNA hsa_circ_0000277 sequesters miR-4766-5p to upregulate LAMA1 and promote esophageal carcinoma progression. Cell Death Dis 12(7):676. https://doi.org/10.1038/s41419-021-03911-5
121. Qian CJ, Tong YY, Wang YC, Teng XS, Yao J (2022) Circ_0001093 promotes glutamine metabolism and cancer progression of esophageal squamous cell carcinoma by targeting miR-579-3p/glutaminase axis. J Bioenerg Biomembr 54(2):119–134. https://doi.org/10.1007/s10863-022-09935-6
122. Xu R, Ding P, Zhao X, Li Z, Liu F, Gu L, Zheng Y, Sang M, Meng L (2023) Circular RNA circ-TNRC6B inhibits the proliferation and invasion of esophageal squamous cell carcinoma cells by regulating the miR-452-5p/DAG1 axis. Mol Oncol 17(7):1437–1452. https://doi.org/10.1002/1878-0261.13432
123. Fang J, Ji WH, Wang FZ, Xie TM, Wang L, Fu ZF, Wang Z, Yan FQ, Shen QL, Ye ZM (2021) Circular RNA hsa_circ_0000700 promotes cell proliferation and migration in esophageal squamous cell carcinoma by sponging miR-1229. J Cancer 12(9):2610–2623. https://doi.org/10.7150/jca.47112
124. Luo G, Li R, Li Z (2018) CircRNA circFNDC3B promotes esophageal cancer progression via cell proliferation, apoptosis, and migration regulation. Int J Clin Exp Pathol 11(8):4188–4196
125. Wang J, Zhu W, Tao G, Wang W (2020) Circular RNA circ-LRP6 facilitates Myc-driven tumorigenesis in esophageal squamous cell cancer. Bioengineered 11(1):932–938. https://doi.org/10.1080/21655979.2020.1809922
126. Liang W, Wang C, Wang J, Zhang M (2022) Hsa_circ_0023984 regulates cell proliferation, migration, and invasion in esophageal squamous cancer via regulating miR-1294/PI3K/Akt/c-Myc pathway. Appl Biochem Biotechnol 194(9):1–16. https://doi.org/10.1007/s12010-022-03935-3
127. Chen Z, Yao N, Gu H, Song Y, Ye Z, Li L, Lu P, Shao Q (2020) Circular RNA_LARP4 sponges miR-1323 and hampers progression of esophageal squamous cell carcinoma through modulating PTEN/PI3K/AKT pathway. Dig Dis Sci 65(8):2272–2283. https://doi.org/10.1007/s10620-019-05973-0
128. Fang Y, Yin J, Shen Y, Wang H, Tang H, Chen X (2022) circPOLR1C promotes the development of esophageal cancer by adsorbing miR-361-3p and regulating cancer cell apoptosis and metastasis. J Oncol 2022:9124142. https://doi.org/10.1155/2022/9124142
129. Wu B, Chen Y, Chen Y, Xie X, Liang H, Peng F, Che W (2022) Circ_0001273 downregulation inhibits the growth, migration and glutamine metabolism of esophageal cancer cells via targeting the miR-622/SLC1A5 signaling axis. Thorac Cancer 13(12):1795–1805. https://doi.org/10.1111/1759-7714.14458
130. Chen S, Liu Z, Feng Q, Zhou J, Huang J, Yu J (2023) Circular RNA circ-AGFG1 contributes to esophageal squamous cell carcinoma progression and glutamine catabolism by targeting microRNA-497-5p/solute carrier family 1 member 5 axis. Anti-Cancer Drugs 34(2):195–206. https://doi.org/10.1097/cad.0000000000001400
131. Yao W, Wang J, Meng F, Zhu Z, Jia X, Xu L, Zhang Q, Wei L (2021) Circular RNA CircPVT1 inhibits 5-fluorouracil chemosensitivity by regulating ferroptosis through MiR-30a-5p/FZD3 axis in esophageal cancer cells. Front Oncol 11:780938. https://doi.org/10.3389/fonc.2021.780938
132. Wang T, Wang J, Ren W, Chen S, Cheng YF, Zhang XM (2020) CircRNA-0008717 promotes cell proliferation, migration, and invasion by regulating miR-203/Slug in esophageal cancer cells. Ann Transl Med 8(16):999. https://doi.org/10.21037/atm-20-5205
133. Zhao R, Chen P, Qu C, Liang J, Cheng Y, Sun Z, Tian H (2023) Circular RNA circTRPS1-2 inhibits the proliferation and migration of esophageal squamous cell carcinoma by reducing the production of ribosomes. Cell Death Dis 9(1):5. https://doi.org/10.1038/s41420-023-01300-9

134. Gao C, Xu YJ, Qi L, Bao YF, Zhang L, Zheng L (2022) CircRNA VIM silence synergizes with sevoflurane to inhibit immune escape and multiple oncogenic activities of esophageal cancer by simultaneously regulating miR-124/PD-L1 axis. Cell Biol Toxicol 38(5):825–845. https://doi.org/10.1007/s10565-021-09613-0
135. Takaki W, Konishi H, Shoda K, Arita T, Kataoka S, Shibamoto J, Furuke H, Takabatake K, Shimizu H, Komatsu S, Shiozaki A, Fujiwara H, Masuda K, Otsuji E (2021) Significance of circular FAT1 as a prognostic factor and tumor suppressor for esophageal squamous cell carcinoma. Ann Surg Oncol 28(13):8508–8518. https://doi.org/10.1245/s10434-021-10089-9
136. Bagheri R, Rashidlamir A, Ashtary-Larky D, Wong A, Alipour M, Motevalli MS, Chebbi A, Laher I, Zouhal H (2020) Does green tea extract enhance the anti-inflammatory effects of exercise on fat loss? Br J Clin Pharmacol 86(4):753–762. https://doi.org/10.1111/bcp.14176
137. Zhou S, Guo Z, Lv X, Zhang X (2022) CircGOT1 promotes cell proliferation, mobility, and glycolysis-mediated cisplatin resistance via inhibiting its host gene GOT1 in esophageal squamous cell cancer. Cell Cycle (Georgetown, Tex) 21(3):247–260. https://doi.org/10.1080/15384101.2021.2015671
138. Hou Y, Liu H, Pan W (2020) Knockdown of circ_0003340 induces cell apoptosis, inhibits invasion and proliferation through miR-564/TPX2 in esophageal cancer cells. Exp Cell Res 394(2):112142. https://doi.org/10.1016/j.yexcr.2020.112142
139. Xia W, Qiu M, Chen R, Wang S, Leng X, Wang J, Xu Y, Hu J, Dong G, Xu PL, Yin R (2016) Circular RNA has_circ_0067934 is upregulated in esophageal squamous cell carcinoma and promoted proliferation. Sci Rep 6:35576. https://doi.org/10.1038/srep35576
140. Wang C, Zhou M, Zhu P, Ju C, Sheng J, Du D, Wan J, Yin H, Xing Y, Li H, He J, He F (2022) IGF2BP2-induced circRUNX1 facilitates the growth and metastasis of esophageal squamous cell carcinoma through miR-449b-5p/FOXP3 axis. J Exp Clin Cancer Res 41(1):347. https://doi.org/10.1186/s13046-022-02550-8
141. Wang J, Yao W, Li J, Zhang Q, Wei L (2024) Circ_0001944 depletion inhibits glycolysis and esophageal cancer progression by binding to miR-338-5p to reduce PDK1 expression. J Bioenerg Biomembr 56(1):73–85. https://doi.org/10.1007/s10863-023-09988-1
142. Fang N, Shi Y, Fan Y, Long T, Shu Y, Zhou J (2020) Circ_0072088 promotes proliferation, migration, and invasion of esophageal squamous cell cancer by absorbing miR-377. J Oncol 2020:8967126. https://doi.org/10.1155/2020/8967126
143. Chu A, Sun C, Liu Z, Liu S, Li M, Song R, Gan L, Wang Y, Fan R (2024) Circ-POSTN promotes the progression and reduces radiosensitivity in esophageal cancer by regulating the miR-876-5p/FYN axis. Thorac Cancer 15(13):1082–1094. https://doi.org/10.1111/1759-7714.15273
144. Zhou Y, Xue X, Luo J, Li P, Xiao Z, Zhang W, Zhou J, Li P, Zhao J, Ge H, Tian Z, Zhao X (2023) Circular RNA circ-FIRRE interacts with HNRNPC to promote esophageal squamous cell carcinoma progression by stabilizing GLI2 mRNA. Cancer Sci 114(9):3608–3622. https://doi.org/10.1111/cas.15899
145. Tang R, Zhou Q, Xu Q, Lu L, Zhou Y (2022) Circular RNA circ_0006948 promotes esophageal squamous cell carcinoma progression by regulating microRNA-3612/LASP1 axis. Dig Dis Sci 67(6):2158–2172. https://doi.org/10.1007/s10620-021-07057-4
146. Chen L, Luo C, Xu Y, Hu J, Chen H (2023) Circ_0058063 regulates the development of esophageal cancer through miR-377-3p/HOXA1 axis. Anti-Cancer Drugs 34(4):495–506. https://doi.org/10.1097/cad.0000000000001454
147. Huang GW, Yang TQ, Chen QQ, Liu XM, Xie LH, Huang W, Chen XL, Geng YQ, Gu J (2024) circular RNA circ-231 promotes protein biogenesis of TPI1 and PRDX6 through mediating the interaction of eIF4A3 with STAU1 to facilitate unwinding of secondary structure in 5' UTR, enhancing progression of human esophageal squamous cell carcinoma (ESCC). J Cancer 15(9):2518–2537. https://doi.org/10.7150/jca.92578
148. Zhang LW, Wang B, Yang JX, Yang H (2020) Circ-PRMT5 stimulates migration in esophageal cancer by binding miR-203. Eur Rev Med Pharmacol Sci 24(19):9965–9972. https://doi.org/10.26355/eurrev_202010_23209
149. Hu T, Peng H, Yang F, Zhang F, He J (2023) Circ_0024108 promotes the progression of esophageal cancer cells. Gen Thorac Cardiovasc Surg 71(7):418–431. https://doi.org/10.1007/s11748-023-01909-8
150. Zhang Z, Lin W, Gao L, Chen K, Yang C, Zhuang L, Peng S, Kang M, Lin J (2019) Hsa_circ_0004370 promotes esophageal cancer progression through miR-1294/LASP1 pathway. Biosci Rep 39(5):BSR20182377. https://doi.org/10.1042/bsr20182377
151. Li R, Chai L, Lei L, Guo R, Wen X (2023) MiR-671-5p sponging activity of circMMP1 promotes esophageal squamous cancer progression. Thorac Cancer 14(29):2924–2933. https://doi.org/10.1111/1759-7714.15078
152. Li N, Wu J, Hu B, Lu H, Gao J, Zhu L, Zheng D (2022) Upregulation of hsa_circ_0000977 participates in esophageal squamous cancer progression by sponging miR-874-3p. J Clin Lab Anal 36(6):e24458. https://doi.org/10.1002/jcla.24458
153. Wang J, Yao W, Li J, Zhang Q, Wei L (2022) Identification of a novel circ_0001946/miR-1290/SOX6 ceRNA network in esophageal squamous cell

cancer. Thorac Cancer 13(9):1299–1310. https://doi.org/10.1111/1759-7714.14381
154. Huang Z, Wang C, Zhao X (2022) circFIG 4 drives the carcinogenesis and metastasis of esophagus cancer via the miR-493-5p/E2F3 axis. Thorac Cancer 13(6):783–794. https://doi.org/10.1111/1759-7714.14321
155. Zhang K, Fan R, Zhao D, Liu P, Yang Z, Liu J, Zhang S, Rao S, Wang Y, Wan L (2022) CircATIC inhibits esophageal carcinoma progression and promotes radiosensitivity by elevating RHCG through sponging miR-10-3p. Thorac Cancer 13(7):934–946. https://doi.org/10.1111/1759-7714.14326
156. Tian J, Hu X, Zhang X (2024) Circ_0003855 involvement of esophageal cancer progression through miR-622/FLOT1. Oncol Res 32(5):925–931. https://doi.org/10.32604/or.2023.042447
157. Yao D, Lin S, Chen S, Wang Z (2022) circHIPK3 regulates cell proliferation and migration by sponging microRNA-124 and regulating serine/threonine kinase 3 expression in esophageal squamous cell carcinoma. Bioengineered 13(4):9767–9780. https://doi.org/10.1080/21655979.2022.2060776
158. Dai Y, Xu Q, Xia M, Chen C, Xiong X, Yang X, Wang W (2024) Hsa_circ_0001615 downregulation inhibits esophageal cancer development through miR-142-5p/β-catenin. PeerJ 12:e17089. https://doi.org/10.7717/peerj.17089
159. Chang Z, Fu Y, Jia Y, Gao M, Song L, Zhang W, Zhao R, Qin Y (2021) Circ-SFMBT2 drives the malignant phenotypes of esophageal cancer by the miR-107-dependent regulation of SLC1A5. Cancer Cell Int 21(1):495. https://doi.org/10.1186/s12935-021-02156-8
160. Cheng Y, Ma Z, Liu S, Yang X, Li S (2022) CircLPAR3 knockdown suppresses esophageal squamous cell carcinoma cell oncogenic phenotypes and Warburg effect through miR-873-5p/LDHA axis. Hum Exp Toxicol 41:9603271221143695. https://doi.org/10.1177/09603271221143695
161. Xu T, Hu Y, Zhao Y, Qi Y, Zhang S, Li P (2023) Hsa_circ_0046534 accelerates esophageal squamous cell carcinoma proliferation and metastasis via regulating MMP2 expression by sponging miR-339-5p. Cell Signal 112:110906. https://doi.org/10.1016/j.cellsig.2023.110906
162. Sun Z, Zhang S, Zhang N, Wang J, Wang J, Liu J (2022) Circ_0005231 promotes the progression of esophageal squamous cell carcinoma via sponging miR-383-5p and regulating KIAA0101. Thorac Cancer 13(12):1751–1762. https://doi.org/10.1111/1759-7714.14450
163. Wang Q, Zhang Q, Sun H, Tang W, Yang L, Xu Z, Liu Z, Jin H, Cao X (2019) Circ-TTC17 promotes proliferation and migration of esophageal squamous cell carcinoma. Dig Dis Sci 64(3):751–758. https://doi.org/10.1007/s10620-018-5382-z
164. Luo Q, Li J, Miao H, Su S, Chen Y, Xu C, Zhao C, Huang J, Ling K, Lin C, Yan H, Zhang S (2023) circSSPO boosts growth of esophageal squamous cell carcinoma through upregulation of micrRNA-6820-5p-mediated KLK8 and PKD1 expression. Cell Biol Toxicol 39(6):3219–3234. https://doi.org/10.1007/s10565-023-09828-3
165. Chen X, Jiang J, Zhao Y, Wang X, Zhang C, Zhuan L, Zhang D, Zheng Y (2020) Circular RNA circNTRK2 facilitates the progression of esophageal squamous cell carcinoma through up-regulating NRIP1 expression via miR-140-3p. J Exp Clin Cancer Res 39(1):133. https://doi.org/10.1186/s13046-020-01640-9
166. Li T, Li S (2022) Circ_0023984 facilitates esophageal squamous cell carcinoma progression by regulating miR-433-3p/REV3L axis. Dig Dis Sci 67(3):892–903. https://doi.org/10.1007/s10620-021-06916-4
167. Guan X, Guan X, Wang Y, Lan T, Cheng T, Cui Y, Xu H (2022) Circ_0003340 downregulation mitigates esophageal squamous cell carcinoma progression by targeting miR-940/PRKAA1 axis. Thorac Cancer 13(8):1164–1175. https://doi.org/10.1111/1759-7714.14377
168. Lyu Y, Tan B, Li L, Liang R, Lei K, Wang K, Wu D, Lin H, Wang M (2023) A novel protein encoded by circUBE4B promotes progression of esophageal squamous cell carcinoma by augmenting MAPK/ERK signaling. Cell Death Dis 14(6):346. https://doi.org/10.1038/s41419-023-05865-2
169. Wang J, Zhao N, Peng S, Zhang T (2023) Circ_0003340 regulates the expression of ENAH to affect the development of esophageal cancer through miR-874-3p. Thorac Cancer 14(9):815–826. https://doi.org/10.1111/1759-7714.14812
170. Guo Y, Guo Y, Chen C, Fan D, Wu X, Zhao L, Shao B, Sun Z, Ji Z (2021) Circ3823 contributes to growth, metastasis and angiogenesis of colorectal cancer: involvement of miR-30c-5p/TCF7 axis. Mol Cancer 20(1):93. https://doi.org/10.1186/s12943-021-01372-0
171. Wang J, Luo J, Liu G, Li X (2020) Circular RNA hsa_circ_0008285 inhibits colorectal cancer cell proliferation and migration via the miR-382-5p/PTEN axis. Biochem Biophys Res Commun 527(2):503–510. https://doi.org/10.1016/j.bbrc.2020.03.165
172. Liu K, Mou Y, Shi X, Liu T, Chen Z, Zuo X (2021) Circular RNA 100146 promotes colorectal cancer progression by the MicroRNA 149/HMGA2 axis. Mol Cell Biol 41(2):e00445-20. https://doi.org/10.1128/mcb.00445-20
173. Xie L, Pan Z (2021) Circular RNA circ_0000467 regulates colorectal cancer development via miR-382-5p/EN2 axis. Bioengineered 12(1):886–897. https://doi.org/10.1080/21655979.2021.1889130
174. Zhang ZJ, Zhang YH, Qin XJ, Wang YX, Fu J (2020) Circular RNA circDENND4C facilitates proliferation, migration and glycolysis of colorectal cancer cells through miR-760/GLUT1 axis. Eur Rev Med Pharmacol Sci 24(5):2387–2400. https://doi.org/10.26355/eurrev_202003_20506
175. Guo F, Luo Y, Ye G, Tang W (2023) Circular RNA METTL15/miR-374a-5p/ESCO2 axis induces colorectal cancer development. Acta Biochim

Pol 70(3):541–549. https://doi.org/10.18388/abp.2020_6470
176. Chen P, Yao Y, Yang N, Gong L, Kong Y, Wu A (2020) Circular RNA circCTNNA1 promotes colorectal cancer progression by sponging miR-149-5p and regulating FOXM1 expression. Cell Death Dis 11(7):557. https://doi.org/10.1038/s41419-020-02757-7
177. Wang J, Zhang Y, Song H, Yin H, Jiang T, Xu Y, Liu L, Wang H, Gao H, Wang R, Song J (2021) The circular RNA circSPARC enhances the migration and proliferation of colorectal cancer by regulating the JAK/STAT pathway. Mol Cancer 20(1):81. https://doi.org/10.1186/s12943-021-01375-x
178. Zhou J, Wang L, Sun Q, Chen R, Zhang C, Yang P, Tan Y, Peng C, Wang T, Jin C, Ji J, Jin K, Sun Y (2021) Hsa_circ_0001666 suppresses the progression of colorectal cancer through the miR-576-5p/PCDH10 axis. Clin Transl Med 11(11):e565. https://doi.org/10.1002/ctm2.565
179. Wen T, Wu H, Zhang L, Li K, Xiao X, Zhang L, Zhang Y (2021) Circular RNA circ_0007142 regulates cell proliferation, apoptosis, migration and invasion via miR-455-5p/SGK1 axis in colorectal cancer. Anti-Cancer Drugs 32(1):22–33. https://doi.org/10.1097/cad.0000000000000992
180. Yan D, Liu W, Liu Y, Zhu X (2022) Circular RNA circ_0065378 upregulates tumor suppressor candidate 1 by competitively binding with miR-4701-5p to alleviate colorectal cancer progression. J Gastroenterol Hepatol 37(6):1107–1118. https://doi.org/10.1111/jgh.15862
181. Yang T, Sun J, Wang W, Li D, Yang X, Jia A, Ma Y, Fan Z (2022) Hsa_circ_0006732 regulates colorectal cancer cell proliferation, invasion and EMT by miR-127-5p/RAB3D axis. Mol Cell Biochem 477(12):2751–2760. https://doi.org/10.1007/s11010-022-04458-5
182. Lu C, Fu L, Qian X, Dou L, Cang S (2020) Knockdown of circular RNA circ-FARSA restricts colorectal cancer cell growth through regulation of miR-330-5p/LASP1 axis. Arch Biochem Biophys 689:108434. https://doi.org/10.1016/j.abb.2020.108434
183. Yang KD, Wang Y, Zhang F, Luo BH, Feng DY, Zeng ZJ (2022) CircN4BP2L2 promotes colorectal cancer growth and metastasis through regulation of the miR-340-5p/CXCR4 axis. Lab Investig 102(1):38–47. https://doi.org/10.1038/s41374-021-00632-3
184. Chen LY, Wang L, Ren YX, Pang Z, Liu Y, Sun XD, Tu J, Zhi Z, Qin Y, Sun LN, Li JM (2020) The circular RNA circ-ERBIN promotes growth and metastasis of colorectal cancer by miR-125a-5p and miR-138-5p/4EBP-1 mediated cap-independent HIF-1α translation. Mol Cancer 19(1):164. https://doi.org/10.1186/s12943-020-01272-9
185. Wang X, Tao G, Huang D, Liang S, Zheng D (2020) Circular RNA NOX4 promotes the development of colorectal cancer via the microRNA-485-5p/CKS1B axis. Oncol Rep 44(5):2009–2020. https://doi.org/10.3892/or.2020.7758
186. Chen DL, Sheng H, Zhang DS, Jin Y, Zhao BT, Chen N, Song K, Xu RH (2021) The circular RNA circDLG1 promotes gastric cancer progression and anti-PD-1 resistance through the regulation of CXCL12 by sponging miR-141-3p. Mol Cancer 20(1):166. https://doi.org/10.1186/s12943-021-01475-8
187. Zang X, Jiang J, Gu J, Chen Y, Wang M, Zhang Y, Fu M, Shi H, Cai H, Qian H, Xu W, Zhang X (2022) Circular RNA EIF4G3 suppresses gastric cancer progression through inhibition of β-catenin by promoting δ-catenin ubiquitin degradation and upregulating SIK1. Mol Cancer 21(1):141. https://doi.org/10.1186/s12943-022-01606-9
188. Zhang X, Wang S, Wang H, Cao J, Huang X, Chen Z, Xu P, Sun G, Xu J, Lv J, Xu Z (2019) Circular RNA circNRIP1 acts as a microRNA-149-5p sponge to promote gastric cancer progression via the AKT1/mTOR pathway. Mol Cancer 18(1):20. https://doi.org/10.1186/s12943-018-0935-5
189. Ma C, Wang X, Yang F, Zang Y, Liu J, Wang X, Xu X, Li W, Jia J, Liu Z (2020) Circular RNA hsa_circ_0004872 inhibits gastric cancer progression via the miR-224/Smad4/ADAR1 successive regulatory circuit. Mol Cancer 19(1):157. https://doi.org/10.1186/s12943-020-01268-5
190. Peng L, Sang H, Wei S, Li Y, Jin D, Zhu X, Li X, Dang Y, Zhang G (2020) circCUL2 regulates gastric cancer malignant transformation and cisplatin resistance by modulating autophagy activation via miR-142-3p/ROCK2. Mol Cancer 19(1):156. https://doi.org/10.1186/s12943-020-01270-x
191. Fan D, Wang C, Wang D, Zhang N, Yi T (2021) Circular RNA circ_0000039 enhances gastric cancer progression through miR-1292-5p/DEK axis. Cancer Biomarkers 30(2):167–177. https://doi.org/10.3233/cbm-201754
192. Bu X, Chen Z, Zhang A, Zhou X, Zhang X, Yuan H, Zhang Y, Yin C, Yan Y (2021) Circular RNA circAFF2 accelerates gastric cancer development by activating miR-6894-5p and regulating ANTXR 1 expression. Clin Res Hepatol Gastroenterol 45(3):101671. https://doi.org/10.1016/j.clinre.2021.101671
193. Zhang L, Song X, Chen X, Wang Q, Zheng X, Wu C, Jiang J (2019) Circular RNA CircCACTIN promotes gastric cancer progression by sponging MiR-331-3p and regulating TGFBR1 expression. Int J Biol Sci 15(5):1091–1103. https://doi.org/10.7150/ijbs.31533
194. Wang H, Wang N, Zheng X, Wu L, Fan C, Li X, Ye K, Han S (2021) Circular RNA hsa_circ_0009172 suppresses gastric cancer by regulation of microRNA-485-3p-mediated NTRK3. Cancer Gene Ther 28(12):1312–1324. https://doi.org/10.1038/s41417-020-00280-7
195. Wang Y, Wang H, Zheng R, Wu P, Sun Z, Chen J, Zhang L, Zhang C, Qian H, Jiang J, Xu W (2021) Circular RNA ITCH suppresses metastasis of gastric cancer via regulating miR-199a-5p/Klotho axis. Cell

Cycle (Georgetown, Tex) 20(5–6):522–536. https://doi.org/10.1080/15384101.2021.1878327
196. Liu P, Cai S, Li N (2020) Circular RNA-hsa-circ-0000670 promotes gastric cancer progression through the microRNA-384/SIX4 axis. Exp Cell Res 394(2):112141. https://doi.org/10.1016/j.yexcr.2020.112141
197. Xu Q, Liao B, Hu S, Zhou Y, Xia W (2021) Circular RNA 0081146 facilitates the progression of gastric cancer by sponging miR-144 and up-regulating HMGB1. Biotechnol Lett 43(4):767–779. https://doi.org/10.1007/s10529-020-03058-x
198. Zhang Z, Wu H, Chen Z, Li G, Liu B (2020) Circular RNA ATXN7 promotes the development of gastric cancer through sponging miR-4319 and regulating ENTPD4. Cancer Cell Int 20:25. https://doi.org/10.1186/s12935-020-1106-5
199. Dai X, Liu J, Guo X, Cheng A, Deng X, Guo L, Wang Z (2020) Circular RNA circFGD4 suppresses gastric cancer progression via modulating miR-532-3p/APC/β-catenin signalling pathway. Clin Sci (London, England: 1979) 134(13):1821–1839. https://doi.org/10.1042/cs20191043
200. Hui C, Tian L, He X (2020) Circular RNA circNHSL1 contributes to gastric cancer progression through the miR-149-5p/YWHAZ axis. Cancer Manag Res 12:7117–7130. https://doi.org/10.2147/cmar.s253152
201. Liu J, Liu H, Zeng Q, Xu P, Liu M, Yang N (2020) Circular RNA circ-MAT2B facilitates glycolysis and growth of gastric cancer through regulating the miR-515-5p/HIF-1α axis. Cancer Cell Int 20:171. https://doi.org/10.1186/s12935-020-01256-1
202. Zhang J, Zhang Z (2021) Mechanisms of circular RNA circ_0066147 on pancreatic cancer progression. Open Life Sci 16(1):495–510. https://doi.org/10.1515/biol-2021-0047
203. Liu T, Zhou L, He Z, Chen Y, Jiang X, Xu J, Jiang J (2021) Circular RNA hsa_circ_0006117 facilitates pancreatic cancer progression by regulating the miR-96-5p/KRAS/MAPK signaling pathway. J Oncol 2021:9213205. https://doi.org/10.1155/2021/9213205
204. Xiong X, Feng J, Yang X, Li H, Shi Q, Tao J, Chang J (2021) Circular RNA CDR1as promotes tumor progression by regulating miR-432-5p/E2F3 axis in pancreatic cancer. Cancer Cell Int 21(1):112. https://doi.org/10.1186/s12935-021-01812-3
205. Xu S, Lei SL, Liu KJ, Yi SG, Yang ZL, Yao HL (2021) circSFMBT1 promotes pancreatic cancer growth and metastasis via targeting miR-330-5p/PAK1 axis. Cancer Gene Ther 28(3–4):234–249. https://doi.org/10.1038/s41417-020-00215-2
206. Liu A, Xu J (2021) Circ_03955 promotes pancreatic cancer tumorigenesis and Warburg effect by targeting the miR-3662/HIF-1α axis. Clin Transl Oncol 23(9):1905–1914. https://doi.org/10.1007/s12094-021-02599-5
207. Liu J, He X, Zou Y, Wang K (2022) Circular RNA Circ-STIL contributes to cell growth and metastasis in hepatocellular carcinoma via regulating miR-345-5p/AQP3 axis. Dig Dis Sci 67(6):2269–2282. https://doi.org/10.1007/s10620-021-07054-7
208. Hua S, Gao J, Li T, Wang M, You L, Chen G, Han X, Liao Q (2021) The promoting effects of hsa_circ_0050102 in pancreatic cancer and the molecular mechanism by targeting miR-1182/NPSR1. Carcinogenesis 42(3):471–480. https://doi.org/10.1093/carcin/bgaa130
209. Chen Y, Xu S, Liu X, Jiang X, Jiang J (2021) CircSEC24A upregulates TGFBR2 expression to accelerate pancreatic cancer proliferation and migration via sponging to miR-606. Cancer Cell Int 21(1):671. https://doi.org/10.1186/s12935-021-02392-y
210. Zhou X, Liu K, Cui J, Xiong J, Wu H, Peng T, Guo Y (2021) Circ-MBOAT2 knockdown represses tumor progression and glutamine catabolism by miR-433-3p/GOT1 axis in pancreatic cancer. J Exp Clin Cancer Res 40(1):124. https://doi.org/10.1186/s13046-021-01894-x
211. Kong Y, Li Y, Luo Y, Zhu J, Zheng H, Gao B, Guo X, Li Z, Chen R, Chen C (2020) circNFIB1 inhibits lymphangiogenesis and lymphatic metastasis via the miR-486-5p/PIK3R1/VEGF-C axis in pancreatic cancer. Mol Cancer 19(1):82. https://doi.org/10.1186/s12943-020-01205-6
212. Han X, Fang Y, Chen P, Xu Y, Zhou W, Rong Y, Li JA, Chen W, Lou W (2021) Upregulated circRNA hsa_circ_0071036 promotes tumourigenesis of pancreatic cancer by sponging miR-489 and predicts unfavorable characteristics and prognosis. Cell Cycle (Georgetown, Tex) 20(4):369–382. https://doi.org/10.1080/15384101.2021.1874684
213. Zhang R, Zhu W, Ma C, Ai K (2021) Silencing of circRNA circ_0001666 represses EMT in pancreatic cancer through upregulating miR-1251 and downregulating SOX4. Front Mol Biosci 8:684866. https://doi.org/10.3389/fmolb.2021.684866
214. Zhang X, Tan P, Zhuang Y, Du L (2020) hsa_circRNA_001587 upregulates SLC4A4 expression to inhibit migration, invasion, and angiogenesis of pancreatic cancer cells via binding to microRNA-223. Am J Physiol Gastrointest Liver Physiol 319(6):G703–g717. https://doi.org/10.1152/ajpgi.00118.2020
215. Huang G, Liang M, Liu H, Huang J, Li P, Wang C, Zhang Y, Lin Y, Jiang X (2020) CircRNA hsa_circRNA_104348 promotes hepatocellular carcinoma progression through modulating miR-187-3p/RTKN2 axis and activating Wnt/β-catenin pathway. Cell Death Dis 11(12):1065. https://doi.org/10.1038/s41419-020-03276-1
216. Yao JN, Zhang XX, Zhang YZ, Li YL, Wang CF, Zhang LF (2020) Circular RNA-UBXN7 promotes proliferation, migration and suppresses apoptosis in hepatocellular cancer. Zhonghua gan zang bing za zhi/Zhonghua ganzangbing zazhi/Chin J Hepatol 28(5):421–427. https://doi.org/10.3760/cma.j.cn501113-20200305-00086

217. Liu L, Gu M, Ma J, Wang Y, Li M, Wang H, Yin X, Li X (2022) CircGPR137B/miR-4739/FTO feedback loop suppresses tumorigenesis and metastasis of hepatocellular carcinoma. Mol Cancer 21(1):149. https://doi.org/10.1186/s12943-022-01619-4
218. Li J, Hu ZQ, Yu SY, Mao L, Zhou ZJ, Wang PC, Gong Y, Su S, Zhou J, Fan J, Zhou SL, Huang XW (2022) CircRPN2 inhibits aerobic glycolysis and metastasis in hepatocellular carcinoma. Cancer Res 82(6):1055–1069. https://doi.org/10.1158/0008-5472.can-21-1259
219. Liu W, Yin C, Liu Y (2021) Circular RNA circ_0091579 promotes hepatocellular carcinoma proliferation, migration, invasion, and glycolysis through miR-490-5p/CASC3 axis. Cancer Biother Radiopharm 36(10):863–878. https://doi.org/10.1089/cbr.2019.3472
220. Shen D, Zhao H, Zeng P, Ge M, Shrestha S, Zhao W (2022) Circular RNA circ_0001459 accelerates hepatocellular carcinoma progression via the miR-6165/IGF1R axis. Ann N Y Acad Sci 1512(1):46–60. https://doi.org/10.1111/nyas.14753
221. Li JX, Wang JJ, Deng ZF, Zheng H, Yang CM, Yuan Y, Yang C, Gu FF, Wu WQ, Qiao GL, Ma LJ (2022) Circular RNA circ_0008934 promotes hepatocellular carcinoma growth and metastasis through modulating miR-1305/TMTC3 axis. Hum Cell 35(2):498–510. https://doi.org/10.1007/s13577-021-00657-2
222. Guo X, Wang Z, Deng X, Lu Y, Huang X, Lin J, Lan X, Su Q, Wang C (2022) Circular RNA CircITCH (has-circ-0001141) suppresses hepatocellular carcinoma (HCC) progression by sponging miR-184. Cell Cycle (Georgetown, Tex) 21(15):1557–1577. https://doi.org/10.1080/15384101.2022.2057633
223. Li S, Weng J, Song F, Li L, Xiao C, Yang W, Xu J (2020) Circular RNA circZNF566 promotes hepatocellular carcinoma progression by sponging miR-4738-3p and regulating TDO2 expression. Cell Death Dis 11(6):452. https://doi.org/10.1038/s41419-020-2616-8
224. Zhang K, Fang T, Zhao D, Cen F, Yan X, Jin X (2022) Circular RNA Circ_0008043 promotes the proliferation and metastasis of hepatocellular carcinoma cells by regulating the microRNA (miR)-326/RAB21 axis. Bioengineered 13(3):6600–6614. https://doi.org/10.1080/21655979.2022.2044260
225. Gong J, Du C, Sun N, Xiao X, Wu H (2021) Circular RNA hsa_circ_0005397 promotes hepatocellular carcinoma progression by regulating the miR-326/PDK2 axis. J Gene Med 23(6):e3332. https://doi.org/10.1002/jgm.3332
226. Chen Y, Song S, Zhang L, Zhang Y (2021) Circular RNA hsa_circ_0091579 facilitates the Warburg effect and malignancy of hepatocellular carcinoma cells via the miR-624/H3F3B axis. Clin Transl Oncol 23(11):2280–2292. https://doi.org/10.1007/s12094-021-02627-4
227. Xu R, Yin S, Zheng M, Pei X, Ji X (2021) Circular RNA circZFR promotes hepatocellular carcinoma progression by regulating miR-375/HMGA2 axis. Dig Dis Sci 66(12):4361–4373. https://doi.org/10.1007/s10620-020-06805-2
228. Liu W, Zheng L, Zhang R, Hou P, Wang J, Wu L, Li J (2022) Circ-ZEB1 promotes PIK3CA expression by silencing miR-199a-3p and affects the proliferation and apoptosis of hepatocellular carcinoma. Mol Cancer 21(1):72. https://doi.org/10.1186/s12943-022-01529-5
229. Cohen PA, Jhingran A, Oaknin A, Denny L (2019) Cervical cancer. Lancet 393(10167):169–182. https://doi.org/10.1016/s0140-6736(18)32470-x
230. Li S, Teng S, Xu J, Su G, Zhang Y, Zhao J, Zhang S, Wang H, Qin W, Lu ZJ, Guo Y, Zhu Q, Wang D (2019) Microarray is an efficient tool for circRNA profiling. Brief Bioinform 20(4):1420–1433. https://doi.org/10.1093/bib/bby006
231. Zheng SR, Zhang HR, Zhang ZF, Lai SY, Huang LJ, Liu J, Bai X, Ding K, Zhou JY (2018) Human papillomavirus 16 E7 oncoprotein alters the expression profiles of circular RNAs in Caski cells. J Cancer 9(20):3755–3764. https://doi.org/10.7150/jca.24253
232. Yu D, Li Y, Ming Z, Wang H, Dong Z, Qiu L, Wang T (2018) Comprehensive circular RNA expression profile in radiation-treated HeLa cells and analysis of radioresistance-related circRNAs. PeerJ 6:e5011. https://doi.org/10.7717/peerj.5011
233. Wang H, Zhao Y, Chen M, Cui J (2017) Identification of novel long non-coding and circular RNAs in human papillomavirus-mediated cervical cancer. Front Microbiol 8:1720. https://doi.org/10.3389/fmicb.2017.01720
234. Yi Y, Liu Y, Wu W, Wu K, Zhang W (2019) Reconstruction and analysis of circRNA-miRNA-mRNA network in the pathology of cervical cancer. Oncol Rep 41(4):2209–2225. https://doi.org/10.3892/or.2019.7028
235. Di Liddo A, de Oliveira Freitas Machado C, Fischer S, Ebersberger S, Heumüller AW, Weigand JE, Müller-McNicoll M, Zarnack K (2019) A combined computational pipeline to detect circular RNAs in human cancer cells under hypoxic stress. J Mol Cell Biol 11(10):829–844. https://doi.org/10.1093/jmcb/mjz094
236. Gao YL, Zhang MY, Xu B, Han LJ, Lan SF, Chen J, Dong YJ, Cao LL (2017) Circular RNA expression profiles reveal that hsa_circ_0018289 is up-regulated in cervical cancer and promotes the tumorigenesis. Oncotarget 8(49):86625–86633. https://doi.org/10.18632/oncotarget.21257
237. Ma HB, Yao YN, Yu JJ, Chen XX, Li HF (2018) Extensive profiling of circular RNAs and the potential regulatory role of circRNA-000284 in cell proliferation and invasion of cervical cancer via sponging miR-506. Am J Transl Res 10(2):592–604
238. Zhang J, Zhao X, Zhang J, Zheng X, Li F (2018) Circular RNA hsa_circ_0023404 exerts an onco-

genic role in cervical cancer through regulating miR-136/TFCP2/YAP pathway. Biochem Biophys Res Commun 501(2):428–433. https://doi.org/10.1016/j.bbrc.2018.05.006

239. Liu J, Wang D, Long Z, Liu J, Li W (2018) CircRNA8924 promotes cervical cancer cell proliferation, migration and invasion by competitively binding to MiR-518d-5p /519-5p family and modulating the expression of CBX8. Cell Physiol Biochem 48(1):173–184. https://doi.org/10.1159/000491716
240. Hu C, Wang Y, Li A, Zhang J, Xue F, Zhu L (2019) Overexpressed circ_0067934 acts as an oncogene to facilitate cervical cancer progression via the miR-545/EIF3C axis. J Cell Physiol 234(6):9225–9232. https://doi.org/10.1002/jcp.27601
241. Cai H, Zhang P, Xu M, Yan L, Liu N, Wu X (2019) Circular RNA hsa_circ_0000263 participates in cervical cancer development by regulating target gene of miR-150-5p. J Cell Physiol 234(7):11391–11400. https://doi.org/10.1002/jcp.27796
242. Tian JDC, Liang L (2018) Involvement of circular RNA SMARCA5/microRNA-620 axis in the regulation of cervical cancer cell proliferation, invasion and migration. Eur Rev Med Pharmacol Sci 22(24):8589–8598. https://doi.org/10.26355/eurrev_201812_16622
243. Song T, Xu A, Zhang Z, Gao F, Zhao L, Chen X, Gao J, Kong X (2019) CircRNA hsa_circRNA_101996 increases cervical cancer proliferation and invasion through activating TPX2 expression by restraining miR-8075. J Cell Physiol 234(8):14296–14305. https://doi.org/10.1002/jcp.28128
244. Ji F, Du R, Chen T, Zhang M, Zhu Y, Luo X, Ding Y (2020) Circular RNA circSLC26A4 accelerates cervical cancer progression via miR-1287-5p/HOXA7 axis. Mol Ther Nucleic Acids 19:413–420. https://doi.org/10.1016/j.omtn.2019.11.032
245. Ou R, Lv J, Zhang Q, Lin F, Zhu L, Huang F, Li X, Li T, Zhao L, Ren Y, Xu Y (2020) circAMOTL1 motivates AMOTL1 expression to facilitate cervical cancer growth. Mol Ther Nucleic Acids 19:50–60. https://doi.org/10.1016/j.omtn.2019.09.022
246. Tang Q, Chen Z, Zhao L, Xu H (2019) Circular RNA hsa_circ_0000515 acts as a miR-326 sponge to promote cervical cancer progression through up-regulation of ELK1. Aging (Albany NY) 11(22):9982–9999. https://doi.org/10.18632/aging.102356
247. Shao S, Wang C, Wang S, Zhang H, Zhang Y (2020) Hsa_circ_0075341 is up-regulated and exerts oncogenic properties by sponging miR-149-5p in cervical cancer. Biomed Pharmacother 121:109582. https://doi.org/10.1016/j.biopha.2019.109582
248. Ma H, Tian T, Liu X, Xia M, Chen C, Mai L, Xie S, Yu L (2019) Upregulated circ_0005576 facilitates cervical cancer progression via the miR-153/KIF20A axis. Biomed Pharmacother 118:109311. https://doi.org/10.1016/j.biopha.2019.109311
249. Rong X, Gao W, Yang X, Guo J (2019) Downregulation of hsa_circ_0007534 restricts the proliferation and invasion of cervical cancer through regulating miR-498/BMI-1 signaling. Life Sci 235:116785. https://doi.org/10.1016/j.lfs.2019.116785
250. Mao Y, Zhang L, Li Y (2019) circEIF4G2 modulates the malignant features of cervical cancer via the miR-218/HOXA1 pathway. Mol Med Rep 19(5):3714–3722. https://doi.org/10.3892/mmr.2019.10032
251. Chen RX, Liu HL, Yang LL, Kang FH, Xin LP, Huang LR, Guo QF, Wang YL (2019) Circular RNA circRNA_0000285 promotes cervical cancer development by regulating FUS. Eur Rev Med Pharmacol Sci 23(20):8771–8778. https://doi.org/10.26355/eurrev_201910_19271
252. Wang J, Li H, Liang Z (2019) circ-MYBL2 serves as a sponge for miR-361-3p promoting cervical cancer cells proliferation and invasion. Onco Targets Ther 12:9957–9964. https://doi.org/10.2147/ott.S218976
253. Shi X, Wang J, Lei Y, Cong C, Tan D, Zhou X (2019) Research progress on the PI3K/AKT signaling pathway in gynecological cancer (Review). Mol Med Rep 19(6):4529–4535. https://doi.org/10.3892/mmr.2019.10121
254. Bahrami A, Hasanzadeh M, Hassanian SM, ShahidSales S, Ghayour-Mobarhan M, Ferns GA, Avan A (2017) The potential value of the PI3K/Akt/mTOR signaling pathway for assessing prognosis in cervical cancer and as a target for therapy. J Cell Biochem 118(12):4163–4169. https://doi.org/10.1002/jcb.26118
255. Cruz da Silva E, Dontenwill M, Choulier L, Lehmann M (2019) Role of integrins in resistance to therapies targeting growth factor receptors in cancer. Cancers (Basel) 11(5):692. https://doi.org/10.3390/cancers11050692
256. Yang W, Xie T (2020) Hsa_circ_CSPP1/MiR-361-5p/ITGB1 regulates proliferation and migration of cervical cancer (CC) by modulating the PI3K-Akt signaling pathway. Reprod Sci 27(1):132–144. https://doi.org/10.1007/s43032-019-00008-5
257. Dhillon AS, Hagan S, Rath O, Kolch W (2007) MAP kinase signalling pathways in cancer. Oncogene 26(22):3279–3290. https://doi.org/10.1038/sj.onc.1210421
258. Wu F, Zhou J (2019) CircAGFG1 promotes cervical cancer progression via miR-370-3p/RAF1 signaling. BMC Cancer 19(1):1067. https://doi.org/10.1186/s12885-019-6269-x
259. Lu W, Kang Y (2019) Epithelial-mesenchymal plasticity in cancer progression and metastasis. Dev Cell 49(3):361–374. https://doi.org/10.1016/j.devcel.2019.04.010
260. Shang BQ, Li ML, Quan HY, Hou PF, Li ZW, Chu SF, Zheng JN, Bai J (2019) Functional roles of circular RNAs during epithelial-to-mesenchymal transition. Mol Cancer 18(1):138. https://doi.org/10.1186/s12943-019-1071-6
261. Qian W, Huang T, Feng W (2020) Circular RNA HIPK3 promotes EMT of cervical cancer through

sponging miR-338-3p to up-regulate HIF-1α. Cancer Manag Res 12:177–187. https://doi.org/10.2147/cmar.S232235
262. Zhou Y, Shen L, Wang YZ, Zhou CC (2020) The potential of ciRS-7 for predicting onset and prognosis of cervical cancer. Neoplasma 67(2):312–322. https://doi.org/10.4149/neo_2019_190415N334
263. Hong H, Zhu H, Zhao S, Wang K, Zhang N, Tian Y, Li Y, Wang Y, Lv X, Wei T, Liu Y, Fan S, Liu Y, Li Y, Cai A, Jin S, Qin Q, Li H (2019) The novel circCLK3/miR-320a/FoxM1 axis promotes cervical cancer progression. Cell Death Dis 10(12):950. https://doi.org/10.1038/s41419-019-2183-z
264. Jiao J, Zhang T, Jiao X, Huang T, Zhao L, Ma D, Cui B (2020) hsa_circ_0000745 promotes cervical cancer by increasing cell proliferation, migration, and invasion. J Cell Physiol 235(2):1287–1295. https://doi.org/10.1002/jcp.29045
265. Teng F, Xu J, Zhang M, Liu S, Gu Y, Zhang M, Wang X, Ni J, Qian B, Shen R, Jia X (2019) Comprehensive circular RNA expression profiles and the tumor-suppressive function of circHIPK3 in ovarian cancer. Int J Biochem Cell Biol 112:8–17. https://doi.org/10.1016/j.biocel.2019.04.011
266. Gao Y, Zhang C, Liu Y, Wang M (2019) Circular RNA profiling reveals circRNA1656 as a novel biomarker in high grade serous ovarian cancer. Biosci Trends 13(2):204–211. https://doi.org/10.5582/bst.2019.01021
267. Bachmayr-Heyda A, Auer K, Sukhbaatar N, Aust S, Deycmar S, Reiner AT, Polterauer S, Dekan S, Pils D (2016) Small RNAs and the competing endogenous RNA network in high grade serous ovarian cancer tumor spread. Oncotarget 7(26):39640–39653. https://doi.org/10.18632/oncotarget.9243
268. Wang J, Wu A, Yang B, Zhu X, Teng Y, Ai Z (2020) Profiling and bioinformatics analyses reveal differential circular RNA expression in ovarian cancer. Gene 724:144150. https://doi.org/10.1016/j.gene.2019.144150
269. Ahmed I, Karedath T, Andrews SS, Al-Azwani IK, Mohamoud YA, Querleu D, Rafii A, Malek JA (2016) Altered expression pattern of circular RNAs in primary and metastatic sites of epithelial ovarian carcinoma. Oncotarget 7(24):36366–36381. https://doi.org/10.18632/oncotarget.8917
270. Bachmayr-Heyda A, Reiner AT, Auer K, Sukhbaatar N, Aust S, Bachleitner-Hofmann T, Mesteri I, Grunt TW, Zeillinger R, Pils D (2015) Correlation of circular RNA abundance with proliferation—exemplified with colorectal and ovarian cancer, idiopathic lung fibrosis, and normal human tissues. Sci Rep 5:8057. https://doi.org/10.1038/srep08057
271. Hu J, Wang L, Chen J, Gao H, Zhao W, Huang Y, Jiang T, Zhou J, Chen Y (2018) The circular RNA circ-ITCH suppresses ovarian carcinoma progression through targeting miR-145/RASA1 signaling. Biochem Biophys Res Commun 505(1):222–228. https://doi.org/10.1016/j.bbrc.2018.09.060
272. Luo L, Gao YQ, Sun XF (2018) Circular RNA ITCH suppresses proliferation and promotes apoptosis in human epithelial ovarian cancer cells by sponging miR-10a-α. Eur Rev Med Pharmacol Sci 22(23):8119–8126. https://doi.org/10.26355/eurrev_201812_16503
273. Yan H, Xiang H, Sun B, Feng F, Chen P (2020) Circular RNA-ITCH inhibits the proliferation of ovarian carcinoma by downregulating lncRNA HULC. Reprod Sci 27(1):375–379. https://doi.org/10.1007/s43032-019-00049-w
274. Luo L, Gao Y, Sun X (2018) Circ-ITCH correlates with small tumor size, decreased FIGO stage and prolonged overall survival, and it inhibits cells proliferation while promotes cells apoptosis in epithelial ovarian cancer. Cancer Biomark 23(4):505–513. https://doi.org/10.3233/cbm-181609
275. Chen H, Mao M, Jiang J, Zhu D, Li P (2019) Circular RNA CDR1as acts as a sponge of miR-135b-5p to suppress ovarian cancer progression. Onco Targets Ther 12:3869–3879. https://doi.org/10.2147/ott.S207938
276. Lin W, Ye H, You K, Chen L (2020) Up-regulation of circ_LARP4 suppresses cell proliferation and migration in ovarian cancer by regulating miR-513b-5p/LARP4 axis. Cancer Cell Int 20:5. https://doi.org/10.1186/s12935-019-1071-z
277. Li L, Yu P, Zhang P, Wu H, Chen Q, Li S, Wang Y (2020) Upregulation of hsa_circ_0007874 suppresses the progression of ovarian cancer by regulating the miR-760/SOCS3 pathway. Cancer Med 9(7):2491–2499. https://doi.org/10.1002/cam4.2866
278. Zhang L, Zhou Q, Qiu Q, Hou L, Wu M, Li J, Li X, Lu B, Cheng X, Liu P, Lu W, Lu Y (2019) CircPLEKHM3 acts as a tumor suppressor through regulation of the miR-9/BRCA1/DNAJB6/KLF4/AKT1 axis in ovarian cancer. Mol Cancer 18(1):144. https://doi.org/10.1186/s12943-019-1080-5
279. Li X, Lin S, Mo Z, Jiang J, Tang H, Wu C, Song J (2020) CircRNA_100395 inhibits cell proliferation and metastasis in ovarian cancer via regulating miR-1228/p53/epithelial-mesenchymal transition (EMT) axis. J Cancer 11(3):599–609. https://doi.org/10.7150/jca.35041
280. Nguyen VHL, Hough R, Bernaudo S, Peng C (2019) Wnt/β-catenin signalling in ovarian cancer: insights into its hyperactivation and function in tumorigenesis. J Ovarian Res 12(1):122. https://doi.org/10.1186/s13048-019-0596-z
281. Krill LS, Tewari KS (2015) Exploring the therapeutic rationale for angiogenesis blockade in cervical cancer. Clin Ther 37(1):9–19. https://doi.org/10.1016/j.clinthera.2014.11.012
282. Chung AS, Lee J, Ferrara N (2010) Targeting the tumour vasculature: insights from physiological angiogenesis. Nat Rev Cancer 10(7):505–514. https://doi.org/10.1038/nrc2868
283. Wang LL, Zong ZH, Liu Y, Guan X, Chen S, Zhao Y (2019) CircRhoC promotes tumorigenicity and

progression in ovarian cancer by functioning as a miR-302e sponge to positively regulate VEGFA. J Cell Mol Med 23(12):8472–8481. https://doi.org/10.1111/jcmm.14736

284. Chen BJ, Byrne FL, Takenaka K, Modesitt SC, Olzomer EM, Mills JD, Farrell R, Hoehn KL, Janitz M (2018) Analysis of the circular RNA transcriptome in endometrial cancer. Oncotarget 9(5):5786–5796. https://doi.org/10.18632/oncotarget.23534
285. Ye F, Tang QL, Ma F, Cai L, Chen M, Ran XX, Wang XY, Jiang XF (2019) Analysis of the circular RNA transcriptome in the grade 3 endometrial cancer. Cancer Manag Res 11:6215–6227
286. Dou Y, Kawaler EA, Cui Zhou D, Gritsenko MA, Huang C, Blumenberg L, Karpova A, Petyuk VA, Savage SR, Satpathy S, Liu W, Wu Y, Tsai CF, Wen B, Li Z, Cao S, Moon J, Shi Z, Cornwell M, Wyczalkowski MA, Chu RK, Vasaikar S, Zhou H, Gao Q, Moore RJ, Li K, Sethuraman S, Monroe ME, Zhao R, Heiman D, Krug K, Clauser K, Kothadia R, Maruvka Y, Pico AR, Oliphant AE, Hoskins EL, Pugh SL, Beecroft SJI, Adams DW, Jarman JC, Kong A, Chang HY, Reva B, Liao Y, Rykunov D, Colaprico A, Chen XS, Czekański A, Jędryka M, Matkowski R, Wiznerowicz M, Hiltke T, Boja E, Kinsinger CR, Mesri M, Robles AI, Rodriguez H, Mutch D, Fuh K, Ellis MJ, DeLair D, Thiagarajan M, Mani DR, Getz G, Noble M, Nesvizhskii AI, Wang P, Anderson ML, Levine DA, Smith RD, Payne SH, Ruggles KV, Rodland KD, Ding L, Zhang B, Liu T, Fenyö D (2020) Proteogenomic characterization of endometrial carcinoma. Cell 180(4):729–748.e726. https://doi.org/10.1016/j.cell.2020.01.026
287. Xu H, Gong Z, Shen Y, Fang Y, Zhong S (2018) Circular RNA expression in extracellular vesicles isolated from serum of patients with endometrial cancer. Epigenomics 10(2):187–197. https://doi.org/10.2217/epi-2017-0109
288. Zong ZH, Liu Y, Chen S, Zhao Y (2020) Circ_PUM1 promotes the development of endometrial cancer by targeting the miR-136/NOTCH3 pathway. J Cell Mol Med 24(7):4127–4135. https://doi.org/10.1111/jcmm.15069
289. Dong P, Xu D, Xiong Y, Yue J, Ihira K, Konno Y, Watari H (2020) The expression, functions and mechanisms of circular RNAs in gynecological cancers. Cancer 12(6):1472. https://doi.org/10.3390/cancers12061472
290. Wang W, Luo H, Chang J, Yang X, Zhang X, Zhang Q, Li Y, Zhao Y, Liu J, Zou B, Hao M (2023) Circular RNA circ0001955 promotes cervical cancer tumorigenesis and metastasis via the miR-188-3p/NCAPG2 axis. J Transl Med 21(1):356. https://doi.org/10.1186/s12967-023-04194-4
291. Zhang C, Liu P, Huang J, Liao Y, Pan C, Liu J, Du Q, Liu T, Shang C, Ooi S, Chen R, Xia M, Jiang H, Xu M, Zou Q, Zhou Y, Huang H, Pan Y, Yuan L, Wang W, Yao S (2021) Circular RNA hsa_circ_0043280 inhibits cervical cancer tumor growth and metastasis via miR-203a-3p/PAQR3 axis. Cell Death Dis 12(10):888. https://doi.org/10.1038/s41419-021-04193-7
292. Zhong P, Guo A, Wang L, Lin X, Feng M (2022) Circular RNA CDK6 suppresses cervical cancer proliferation and metastasis by sponging miR-449a. Bioengineered 13(3):4885–4897. https://doi.org/10.1080/21655979.2022.2036898
293. Ou R, Lu S, Wang L, Wang Y, Lv M, Li T, Xu Y, Lu J, Ge RS (2022) Circular RNA circLMO1 suppresses cervical cancer growth and metastasis by triggering miR-4291/ACSL4-mediated ferroptosis. Front Oncol 12:858598. https://doi.org/10.3389/fonc.2022.858598
294. Zhang J, Chu G, Zheng L, Liu J, He J (2023) Circular RNA circ_0000119 promotes cervical cancer cell growth and migration via miR-433-3p/PAK2 axis. J Appl Genet 64(3):531–543. https://doi.org/10.1007/s13353-023-00772-w
295. Wu P, Li C, Ye DM, Yu K, Li Y, Tang H, Xu G, Yi S, Zhang Z (2021) Circular RNA circEPSTI1 accelerates cervical cancer progression via miR-375/409-3P/515-5p-SLC7A11 axis. Aging 13(3):4663–4673. https://doi.org/10.18632/aging.202518
296. Ji H, Hu N (2022) Circular RNA 0001823 aggravates the growth and metastasis of the cervical cancer cells through modulating the microRNA-613/RAB8A axis. Bioengineered 13(4):10335–10349. https://doi.org/10.1080/21655979.2022.2063665
297. Xie H, Wang J, Wang B (2021) Circular RNA Circ_0003221 promotes cervical cancer progression by regulating miR-758-3p/CPEB4 axis. Cancer Manag Res 13:5337–5350. https://doi.org/10.2147/cmar.s311242
298. Song TF, Xu AL, Chen XH, Gao JY, Gao F, Kong XC (2021) Circular RNA circRNA_101996 promoted cervical cancer development by regulating miR-1236-3p/TRIM37 axis. Kaohsiung J Med Sci 37(7):547–561. https://doi.org/10.1002/kjm2.12378
299. Xu AL, Wang WS, Zhao MY, Sun JN, Chen XR, Hou JQ (2021) Circular RNA circ_0011385 promotes cervical cancer progression through competitively binding to miR-149-5p and up-regulating SOX4 expression. Kaohsiung J Med Sci 37(12):1058–1068. https://doi.org/10.1002/kjm2.12432
300. Hu L, Zhou M, Xue L, Zhang J (2022) Circular RNA hsa_circ_0011385 contributes to cervical cancer progression through sequestering miR-149-5p and increasing PRDX6 expression. Reprod Biol 22(2):100619. https://doi.org/10.1016/j.repbio.2022.100619
301. Wang Z, Ren C, Yang L, Zhang X, Liu J, Zhu Y, Jiang D (2021) Silencing of circular RNA_0000326 inhibits cervical cancer cell proliferation, migration and invasion by boosting microRNA-338-3p-dependent down-regulation of CDK4. Aging 13(6):9119–9134. https://doi.org/10.18632/aging.103711
302. Huang XB, Song KJ, Chen GB, Liu R, Jiang ZF, He YL (2020) Circular RNA hsa_circ_0003204 promotes cervical cancer cell proliferation, migration, and invasion by regulating MAPK pathway. Cancer

Biol Ther 21(10):972–982. https://doi.org/10.1080/15384047.2020.1824513
303. Meng L, Jia X, Yu W, Wang C, Chen J, Liu F (2020) Circular RNA UBAP2 contributes to tumor growth and metastasis of cervical cancer via modulating miR-361-3p/SOX4 axis. Cancer Cell Int 20:357. https://doi.org/10.1186/s12935-020-01436-z
304. Liu D, Bi X, Yang Y (2022) Circular RNA hsa_circ_0011324 is involved in endometrial cancer progression and the evolution of its mechanism. Bioengineered 13(3):7485–7499. https://doi.org/10.1080/21655979.2022.2049026
305. Wang N, Guo Y, Song L, Tong T, Fan X (2022) Circular RNA intraflagellar transport 80 facilitates endometrial cancer progression through modulating miR-545-3p/FAM98A signaling. J Gynecol Oncol 33(1):e2. https://doi.org/10.3802/jgo.2022.33.e2
306. Chen Z, Huang M, You J, Lin Y, Huang Q, He C (2022) Circular RNA hsa_circ_0023404 promotes the proliferation, migration and invasion in endometrial cancer cells through regulating miR-217/MAPK1 axis. Eur J Med Res 27(1):242. https://doi.org/10.1186/s40001-022-00866-x
307. Shi R, Zhang W, Zhang J, Yu Z, An L, Zhao R, Zhou X, Wang Z, Wei S, Wang H (2022) CircESRP1 enhances metastasis and epithelial-mesenchymal transition in endometrial cancer via the miR-874-3p/CPEB4 axis. J Transl Med 20(1):139. https://doi.org/10.1186/s12967-022-03334-6
308. Wu Y, Wang F, Shi J, Guo X, Li F (2023) CircSMAD2 accelerates endometrial cancer cell proliferation and metastasis by regulating the miR-1277-5p/MFGE8 axis. J Gynecol Oncol 34(2):e19. https://doi.org/10.3802/jgo.2023.34.e19
309. Yi H, Han Y, Li S (2022) Oncogenic circular RNA circ_0007534 contributes to paclitaxel resistance in endometrial cancer by sponging miR-625 and promoting ZEB2 expression. Front Oncol 12:985470. https://doi.org/10.3389/fonc.2022.985470
310. Liu Y, Yuan H, He T (2022) Downregulated circular RNA hsa_circ_0005797 inhibits endometrial cancer by modulating microRNA-298/Catenin delta 1 signaling. Bioengineered 13(3):4634–4645. https://doi.org/10.1080/21655979.2021.2013113
311. Jia Y, Liu M, Wang S (2020) CircRNA hsa_circRNA_0001776 inhibits proliferation and promotes apoptosis in endometrial cancer via downregulating LRIG2 by sponging miR-182. Cancer Cell Int 20:412. https://doi.org/10.1186/s12935-020-01437-y
312. Shu L, Peng Y, Zhong L, Feng X, Qiao L, Yi Y (2021) CircZNF124 regulates cell proliferation, leucine uptake, migration and invasion by miR-199b-5p/SLC7A5 pathway in endometrial cancer. Immun Inflamm Dis 9(4):1291–1305. https://doi.org/10.1002/iid3.477
313. Wei D, Tian M, Fan W, Zhong X, Wang S, Chen Y, Zhang S (2021) Circular RNA circ_0000043 promotes endometrial carcinoma progression by regulating miR-1271-5p/CTNND1 axis. Arch Gynecol Obstet 303(4):1075–1087. https://doi.org/10.1007/s00404-020-05849-z
314. Wang Y, Yin L, Sun X (2020) CircRNA hsa_circ_0002577 accelerates endometrial cancer progression through activating IGF1R/PI3K/Akt pathway. J Exp Clin Cancer Res 39(1):169. https://doi.org/10.1186/s13046-020-01679-8
315. Hu J, Peng X, Du W, Huang Y, Zhang C, Zhang X (2021) circSLC6A6 sponges miR-497-5p to promote endometrial cancer progression via the PI4KB/Hedgehog axis. J Immunol Res 2021:5512391. https://doi.org/10.1155/2021/5512391
316. Lei TX, He DJ, Cao J, Lv WG (2022) CircWDR26 regulates endometrial carcinoma progression via miR-212-3p-mediated typing genes MSH2. Eur J Med Res 27(1):135. https://doi.org/10.1186/s40001-022-00755-3
317. Li X, Liu Y (2022) Hsa_circ_0000437 inhibits the development of endometrial carcinoma through miR-626/CDKN1B axis. Protein Pept Lett 29(7):611–620. https://doi.org/10.2174/0929866529666220622125016
318. Wang W, Jin W, Liu X, Zheng L (2022) Circ_0002577/miR-126-5p/MACC1 axis promotes endometrial carcinoma progression by regulation of proliferation, migration, invasion, and apoptosis of endometrial carcinoma cells. Arch Gynecol Obstet 306(2):481–491. https://doi.org/10.1007/s00404-022-06412-8
319. Zhang M, Sun Y, Xu H, Shi Y, Shen R, Teng F, Xu J, Jia X (2023) Circular RNA hsa_circ_0007444 inhibits ovarian cancer progression through miR-23a-3p/DICER1 axis. Acta Biochim Biophys Sin 55(4):574–586. https://doi.org/10.3724/abbs.2023052
320. Li M, Chi C, Zhou L, Chen Y, Tang X (2021) Circular PVT1 regulates cell proliferation and invasion via miR-149-5p/FOXM1 axis in ovarian cancer. J Cancer 12(2):611–621. https://doi.org/10.7150/jca.52234
321. Zhang F, Xu Y, Ye W, Jiang J, Wu C (2020) Circular RNA S-7 promotes ovarian cancer EMT via sponging miR-641 to up-regulate ZEB1 and MDM2. Biosci Rep 40(7):BSR20200825. https://doi.org/10.1042/bsr20200825
322. Wang N, Cao QX, Tian J, Ren L, Cheng HL, Yang SQ (2020) Circular RNA MTO1 inhibits the proliferation and invasion of ovarian cancer cells through the miR-182-5p/KLF15 axis. Cell Transplant 29:963689720943613. https://doi.org/10.1177/0963689720943613
323. Wu D, Liu J, Yu L, Wu S, Qiu X (2022) Circular RNA hsa_circ_0000144 aggravates ovarian cancer progression by regulating ELK3 via sponging miR-610. J Ovarian Res 15(1):113. https://doi.org/10.1186/s13048-022-01048-3
324. Liu T, Yuan L, Zou X (2022) Circular RNA circBNC2 (hsa_circ_0008732) inhibits the progression of ovarian cancer through microRNA-223-3p/

FBXW7 axis. J Ovarian Res 15(1):95. https://doi.org/10.1186/s13048-022-01025-w
325. Chen M, Hou P, Zhou M, Ren Q, Wang X, Huang L, Hui S, Yi L, Mi M (2020) Resveratrol attenuates high-fat diet-induced non-alcoholic steatohepatitis by maintaining gut barrier integrity and inhibiting gut inflammation through regulation of the endocannabinoid system. Clin Nutr (Edinburgh, Scotland) 39(4):1264–1275. https://doi.org/10.1016/j.clnu.2019.05.020
326. Zheng Y, Li Z, Yang S, Wang Y, Luan Z (2022) CircEXOC6B suppresses the proliferation and motility and sensitizes ovarian cancer cells to paclitaxel through miR-376c-3p/FOXO3 axis. Cancer Biother Radiopharm 37(9):802–814. https://doi.org/10.1089/cbr.2020.3739
327. Zeng XY, Yuan J, Wang C, Zeng D, Yong JH, Jiang XY, Lan H, Xiao SS (2020) circCELSR1 facilitates ovarian cancer proliferation and metastasis by sponging miR-598 to activate BRD4 signals. Mol Med (Cambridge, Mass) 26(1):70. https://doi.org/10.1186/s10020-020-00194-y
328. Zhang N, Jin Y, Hu Q, Cheng S, Wang C, Yang Z, Wang Y (2020) Circular RNA hsa_circ_0078607 suppresses ovarian cancer progression by regulating miR-518a-5p/Fas signaling pathway. J Ovarian Res 13(1):64. https://doi.org/10.1186/s13048-020-00664-1
329. Gong J, Xu X, Zhang X, Zhou Y (2020) Circular RNA-9119 suppresses in ovarian cancer cell viability via targeting the microRNA-21-5p-PTEN-Akt pathway. Aging 12(14):14314–14328. https://doi.org/10.18632/aging.103470
330. Wei L, He W, Zhao H, Zhao P (2022) Circ_0026123 promotes cisplatin resistance and progression of ovarian cancer by upregulating RAB1A through sequestering miR-543. Anti-Cancer Drugs 33(10):1069–1080. https://doi.org/10.1097/cad.0000000000001373
331. Wang J, Li Y, Zhou JH, Shen FR, Shi X, Chen YG (2021) CircATRNL1 activates Smad4 signaling to inhibit angiogenesis and ovarian cancer metastasis via miR-378. Mol Oncol 15(4):1217–1233. https://doi.org/10.1002/1878-0261.12893
332. Du Y, Liu X, Zhang S, Chen S, Guan X, Li Q, Chen X, Zhao Y (2021) CircCRIM1 promotes ovarian cancer progression by working as ceRNAs of CRIM1 and targeting miR-383-5p/ZEB2 axis. Reprod Biol Endocrinol 19(1):176. https://doi.org/10.1186/s12958-021-00857-3
333. Ji J, Li C, Wang J, Wang L, Huang H, Li Y, Fang J (2022) Hsa_circ_0001756 promotes ovarian cancer progression through regulating IGF2BP2-mediated RAB5A expression and the EGFR/MAPK signaling pathway. Cell Cycle (Georgetown, Tex) 21(7):685–696. https://doi.org/10.1080/15384101.2021.2010166
334. Liang Y, Meng K, Qiu R (2021) Circular RNA Circ_0013958 functions as a tumor promoter in ovarian cancer by regulating miR-637/PLXNB2 axis. Front Genet 12:644451. https://doi.org/10.3389/fgene.2021.644451
335. Xu Q, Deng B, Li M, Chen Y, Zhuan L (2020) circRNA-UBAP2 promotes the proliferation and inhibits apoptosis of ovarian cancer though miR-382-5p/PRPF8 axis. J Ovarian Res 13(1):81. https://doi.org/10.1186/s13048-020-00685-w
336. Yi L, Zhou L, Luo J, Yang Q (2021) Circ-PTK2 promotes the proliferation and suppressed the apoptosis of acute myeloid leukemia cells through targeting miR-330-5p/FOXM1 axis. Blood Cells Mol Dis 86:102506. https://doi.org/10.1016/j.bcmd.2020.102506
337. Prada-Arismendy J, Arroyave JC, Röthlisberger S (2017) Molecular biomarkers in acute myeloid leukemia. Blood Rev 31(1):63–76. https://doi.org/10.1016/j.blre.2016.08.005
338. Wang Y, Guo T, Liu Q, Xie X (2020) CircRAD18 accelerates the progression of acute myeloid leukemia by modulation of miR-206/PRKACB axis. Cancer Manag Res 12:10887–10896. https://doi.org/10.2147/cmar.S277432
339. Yuan DM, Ma J, Fang WB (2019) Identification of non-coding RNA regulatory networks in pediatric acute myeloid leukemia reveals circ-0004136 could promote cell proliferation by sponging miR-142. Eur Rev Med Pharmacol Sci 23(21):9251–9258. https://doi.org/10.26355/eurrev_201911_19417
340. Zhang R, Li Y, Wang H, Zhu K, Zhang G (2020) The regulation of circRNA RNF13/miRNA-1224-5p axis promotes the malignant evolution in acute myeloid leukemia. Biomed Res Int 2020:5654380. https://doi.org/10.1155/2020/5654380
341. Yi YY, Yi J, Zhu X, Zhang J, Zhou J, Tang X, Lin J, Wang P, Deng ZQ (2019) Circular RNA of vimentin expression as a valuable predictor for acute myeloid leukemia development and prognosis. J Cell Physiol 234(4):3711–3719. https://doi.org/10.1002/jcp.27145
342. Wu DM, Wen X, Han XR, Wang S, Wang YJ, Shen M, Fan SH, Zhang ZF, Shan Q, Li MQ, Hu B, Chen GQ, Lu J, Zheng YL (2018) Role of circular RNA DLEU2 in human acute myeloid leukemia. Mol Cell Biol 38(20):e00259-18. https://doi.org/10.1128/mcb.00259-18
343. Baccarani M, Castagnetti F, Gugliotta G, Rosti G, Soverini S, Albeer A, Pfirrmann M (2019) The proportion of different BCR-ABL1 transcript types in chronic myeloid leukemia. An international overview. Leukemia 33(5):1173–1183. https://doi.org/10.1038/s41375-018-0341-4
344. Agarwal P, Isringhausen S, Li H, Paterson AJ, He J, Gomariz Á, Nagasawa T, Nombela-Arrieta C, Bhatia R (2019) Mesenchymal niche-specific expression of Cxcl12 controls quiescence of treatment-resistant leukemia stem cells. Cell Stem Cell 24(5):769–784.e766. https://doi.org/10.1016/j.stem.2019.02.018
345. Asnafi AA, Deris Zayeri Z, Shahrabi S, Zibara K, Vosughi T (2019) Chronic myeloid leukemia with complex karyotypes: prognosis and therapeutic

approaches. J Cell Physiol 234(5):5798–5806. https://doi.org/10.1002/jcp.27505
346. Vener C, Banzi R, Ambrogi F, Ferrero A, Saglio G, Pravettoni G, Sant M (2020) First-line imatinib vs second- and third-generation TKIs for chronic-phase CML: a systematic review and meta-analysis. Blood Adv 4(12):2723–2735. https://doi.org/10.1182/bloodadvances.2019001329
347. Lu Z, Xu N, He B, Pan C, Lan Y, Zhou H, Liu X (2017) Inhibition of autophagy enhances the selective anti-cancer activity of tigecycline to overcome drug resistance in the treatment of chronic myeloid leukemia. J Exp Clin Cancer Res 36(1):43. https://doi.org/10.1186/s13046-017-0512-6
348. Ping L, Jian-Jun C, Chu-Shu L, Guang-Hua L, Ming Z (2019) High circ_100053 predicts a poor outcome for chronic myeloid leukemia and is involved in imatinib resistance. Oncol Res. https://doi.org/10.3727/096504018x15412701483326
349. Che H, Ding H, Jia X (2020) circ_0080145 enhances imatinib resistance of chronic myeloid leukemia by regulating miR-326/PPFIA1 axis. Cancer Biother Radiopharm. https://doi.org/10.1089/cbr.2020.3600
350. Liu J, Kong F, Lou S, Yang D, Gu L (2018) Global identification of circular RNAs in chronic myeloid leukemia reveals hsa_circ_0080145 regulates cell proliferation by sponging miR-29b. Biochem Biophys Res Commun 504(4):660–665. https://doi.org/10.1016/j.bbrc.2018.08.154
351. Pan Y, Lou J, Wang H, An N, Chen H, Zhang Q, Du X (2018) CircBA9.3 supports the survival of leukaemic cells by up-regulating c-ABL1 or BCR-ABL1 protein levels. Blood Cells Mol Dis 73:38–44. https://doi.org/10.1016/j.bcmd.2018.09.002
352. Hallek M (2017) Chronic lymphocytic leukemia: 2017 update on diagnosis, risk stratification, and treatment. Am J Hematol 92(9):946–965. https://doi.org/10.1002/ajh.24826
353. Hallek M, Shanafelt TD, Eichhorst B (2018) Chronic lymphocytic leukaemia. Lancet 391(10129):1524–1537. https://doi.org/10.1016/s0140-6736(18)30422-7
354. Xia L, Wu L, Bao J, Li Q, Chen X, Xia H, Xia R (2018) Circular RNA circ-CBFB promotes proliferation and inhibits apoptosis in chronic lymphocytic leukemia through regulating miR-607/FZD3/Wnt/β-catenin pathway. Biochem Biophys Res Commun 503(1):385–390. https://doi.org/10.1016/j.bbrc.2018.06.045
355. Wu Z, Sun H, Wang C, Liu W, Liu M, Zhu Y, Xu W, Jin H, Li J (2020) Mitochondrial genome-derived circRNA mc-COX2 functions as an oncogene in chronic lymphocytic leukemia. Mol Ther Nucleic Acids 20:801–811. https://doi.org/10.1016/j.omtn.2020.04.017
356. Wu Z, Sun H, Liu W, Zhu H, Fu J, Yang C, Fan L, Wang L, Liu Y, Xu W, Li J, Jin H (2020) Circ-RPL15: a plasma circular RNA as novel oncogenic driver to promote progression of chronic lymphocytic leukemia. Leukemia 34(3):919–923. https://doi.org/10.1038/s41375-019-0594-6
357. Lin Z, Long F, Zhao M, Zhang X, Yang M (2020) The role of circular RNAs in hematological malignancies. Genomics 112(6):4000–4008
358. Huang W, Fang K, Chen TQ, Zeng ZC, Sun YM, Han C, Sun LY, Chen ZH, Yang QQ, Pan Q, Luo XQ, Wang WT, Chen YQ (2019) circRNA circAF4 functions as an oncogene to regulate MLL-AF4 fusion protein expression and inhibit MLL leukemia progression. J Hematol Oncol 12(1):103. https://doi.org/10.1186/s13045-019-0800-z
359. Kroeze E, Loeffen JLC, Poort VM, Meijerink JPP (2020) T-cell lymphoblastic lymphoma and leukemia: different diseases from a common premalignant progenitor? Blood Adv 4(14):3466–3473. https://doi.org/10.1182/bloodadvances.2020001822
360. Deng L, Liu G, Zheng C, Zhang L, Kang Y, Yang F (2019) Circ-LAMP1 promotes T-cell lymphoblastic lymphoma progression via acting as a ceRNA for miR-615-5p to regulate DDR2 expression. Gene 701:146–151. https://doi.org/10.1016/j.gene.2019.03.052
361. Wang L, Li LR, Young KH (2020) New agents and regimens for diffuse large B cell lymphoma. J Hematol Oncol 13(1):175. https://doi.org/10.1186/s13045-020-01011-z
362. Pasqualucci L, Dalla-Favera R (2018) Genetics of diffuse large B-cell lymphoma. Blood 131(21):2307–2319. https://doi.org/10.1182/blood-2017-11-764332
363. Li S, Young KH, Medeiros LJ (2018) Diffuse large B-cell lymphoma. Pathology 50(1):74–87. https://doi.org/10.1016/j.pathol.2017.09.006
364. He MY, Kridel R (2021) Treatment resistance in diffuse large B-cell lymphoma. Leukemia 35(8):2151–2165. https://doi.org/10.1038/s41375-021-01285-3
365. Ennishi D, Hsi ED, Steidl C, Scott DW (2020) Toward a new molecular taxonomy of diffuse large B-cell lymphoma. Cancer Discov 10(9):1267–1281. https://doi.org/10.1158/2159-8290.Cd-20-0174
366. Hu Y, Zhao Y, Shi C, Ren P, Wei B, Guo Y, Ma J (2019) A circular RNA from APC inhibits the proliferation of diffuse large B-cell lymphoma by inactivating Wnt/β-catenin signaling via interacting with TET1 and miR-888. Aging (Albany NY) 11(19):8068–8084. https://doi.org/10.18632/aging.102122
367. Fodde R, Brabletz T (2007) Wnt/beta-catenin signaling in cancer stemness and malignant behavior. Curr Opin Cell Biol 19(2):150–158. https://doi.org/10.1016/j.ceb.2007.02.007
368. Chen X, Xie X, Zhou W (2020) CircCFL1/MiR-107 axis targeting HMGB1 promotes the malignant progression of diffuse large B-cell lymphoma tumors. Cancer Manag Res 12:9351–9362. https://doi.org/10.2147/cmar.S263222
369. Sims GP, Rowe DC, Rietdijk ST, Herbst R, Coyle AJ (2010) HMGB1 and RAGE in inflammation and

cancer. Annu Rev Immunol 28:367–388. https://doi.org/10.1146/annurev.immunol.021908.132603
370. Poh C, Keegan T, Rosenberg AS (2021) Second primary malignancies in multiple myeloma: a review. Blood Rev 46:100757. https://doi.org/10.1016/j.blre.2020.100757
371. Chen F, Wang X, Fu S, Wang S, Fu Y, Zhang J, Liu Z (2020) Circular RNA circ-CDYL sponges miR-1180 to elevate yes-associated protein in multiple myeloma. Exp Biol Med (Maywood) 245(11):925–932. https://doi.org/10.1177/1535370220918191
372. Liu F, Wang YL, Wei JM, Huang ZD (2021) Upregulation of circ_0000142 promotes multiple myeloma progression by adsorbing miR-610 and upregulating AKT3 expression. J Biochem 169(3):327–336. https://doi.org/10.1093/jb/mvaa106
373. Liu X, Tang H, Liu J, Wang X (2020) hsa_circRNA_101237: a novel diagnostic and prognostic biomarker and potential therapeutic target for multiple myeloma. Cancer Manag Res 12:2109–2118. https://doi.org/10.2147/cmar.S241089
374. Wang Y, Lin Q, Song C, Ma R, Li X (2020) Circ_0007841 promotes the progression of multiple myeloma through targeting miR-338-3p/BRD4 signaling cascade. Cancer Cell Int 20:383. https://doi.org/10.1186/s12935-020-01475-6
375. Gao M, Li C, Xiao H, Dong H, Jiang S, Fu Y, Gong L (2019) hsa_circ_0007841: a novel potential biomarker and drug resistance for multiple myeloma. Front Oncol 9:1261. https://doi.org/10.3389/fonc.2019.01261
376. Song Y, Hu N, Song X, Yang J (2020) Hsa_Circ_0007841 enhances multiple myeloma chemotherapy resistance through upregulating ABCG2. Technol Cancer Res Treat 19:1533033820928371. https://doi.org/10.1177/1533033820928371
377. Wang N, Yang B, Jin J, He Y, Wu X, Yang Y, Zhou W, He Z (2022) Circular RNA circ_0040823 inhibits the proliferation of acute myeloid leukemia cells and induces apoptosis by regulating miR-516b/PTEN. J Gene Med 24(3):e3404. https://doi.org/10.1002/jgm.3404
378. Li Q, Luan Q, Zhu H, Zhao Y, Ji J, Wu F, Yan J (2021) Circular RNA circ_0005774 contributes to proliferation and suppresses apoptosis of acute myeloid leukemia cells via circ_0005774/miR-192-5p/ULK1 ceRNA pathway. Biochem Biophys Res Commun 551:78–85. https://doi.org/10.1016/j.bbrc.2021.02.058
379. Sheng XF, Hong LL, Fan L, Zhang Y, Chen KL, Mu J, Shen SY, Zhuang HF (2023) Circular RNA PVT1 regulates cell proliferation, migration, and apoptosis by stabilizing c-Myc and downstream target CXCR4 expression in acute myeloid leukemia. Turk J Hematol 40(2):82–91. https://doi.org/10.4274/tjh.galenos.2023.2022.0435
380. Wang L, Liang Y, Mao Q, Xia W, Chen B, Shen H, Xu L, Jiang F, Dong G (2019) Circular RNA circ-CRIM1 inhibits invasion and metastasis in lung adenocarcinoma through the microRNA (miR)-182/miR-93-leukemia inhibitory factor receptor pathway. Cancer Sci 110(9):2960–2972. https://doi.org/10.1111/cas.14131
381. Lin G, Fei Y, Zhang Y (2021) Hsa-circ_0003420 induces apoptosis in acute myeloid leukemia stem cells and impairs stem cell properties. Immunopharmacol Immunotoxicol 43(5):622–631. https://doi.org/10.1080/08923973.2021.1963272
382. Liu Y, Zhao X, Seitz A, Hooijsma AA, Ravanbakhsh R, Sheveleva S, de Jong D, Koerts J, Dzikiewicz-Krawczyk A, van den Berg A, Ziel-Swier L, Kluiver J (2024) Circular ZDHHC11 supports Burkitt lymphoma growth independent of its miR-150 binding capacity. Sci Rep 14(1):8730. https://doi.org/10.1038/s41598-024-59443-3
383. Zhao X, van den Berg A, Winkle M, Koerts J, Seitz A, de Jong D, Rutgers B, van der Sluis T, Bakker E, Kluiver J (2023) Proliferation-promoting roles of linear and circular PVT1 are independent of their ability to bind miRNAs in B-cell lymphoma. Int J Biol Macromol 253(Pt 2):126744. https://doi.org/10.1016/j.ijbiomac.2023.126744
384. Tang YL, Su JY, Luo JS, Zhang LD, Zheng LM, Liang C, Wang LN, Li Y, Fan Z, Huang DP, Sun P, Luo Z, Qi NH, Lan JJ, Zhang XL, Huang LB, Luo XQ (2023) Gene expression network and Circ_0008012 promote progression in MLL/AF4 positive acute lymphoblastic leukemia. Recent Pat Anticancer Drug Discov 18(4):538–548. https://doi.org/10.2174/1574892818666221207115016
385. He R, Wen W, Fu B, Zhu R, Chen G, Bai S, Cao X, Wang H (2022) CircKIF4A is a prognostic factor and modulator of natural killer/T-cell lymphoma progression. Cancer 14(19):4950. https://doi.org/10.3390/cancers14194950
386. Zhan C, Zhou H, Zhang W, Si C (2022) Hsa_circ_0000877 facilitates the progression of diffuse large B-cell lymphoma by miR-370-3p/mitogen-activated protein kinase kinase kinase kinase 4/Hippo pathway. Anti-Cancer Drugs 33(10):1091–1102. https://doi.org/10.1097/cad.0000000000001366
387. Cao J, Huang S, Li X (2022) Rapamycin inhibits the progression of human acute myeloid leukemia by regulating the circ_0094100/miR-217/ATP1B1 axis. Exp Hematol 112–113:60–69.e62. https://doi.org/10.1016/j.exphem.2022.07.298
388. Lu C, Song C, Han Q, Jing X (2022) CircCTNNA1 is upregulated in mantle cell lymphoma and predicts poor survival by sponging miR-34a to increase cell proliferation. Mediterr J Hematol Infect Dis 14(1):e2022047. https://doi.org/10.4084/mjhid.2022.047
389. Yang CS, Lou Y, Ke QP, Xu XJ, Zhang Y (2022) Mechanism of circZNF609 targeting miR-153 to regulate the proliferation and apoptosis of diffuse large B-cell lymphoma. Zhonghua zhong liu za zhi [Chin J Oncol] 44(3):238–245. https://doi.org/10.3760/cma.j.cn112152-20200723-00677
390. Lin D, Wang Y, Lei L, Lin C (2022) Circ_0003645 serves as miR-335-5p sponge to promote the bio-

logical process of diffuse large B-cell lymphoma by upregulating NFIB. Autoimmunity 55(2):127–135. https://doi.org/10.1080/08916934.2021.2023863
391. Hanahan D, Weinberg RA (2011) Hallmarks of cancer: the next generation. Cell 144(5):646–674. https://doi.org/10.1016/j.cell.2011.02.013
392. Mohammad RM, Muqbil I, Lowe L, Yedjou C, Hsu HY, Lin LT, Siegelin MD, Fimognari C, Kumar NB, Dou QP, Yang H, Samadi AK, Russo GL, Spagnuolo C, Ray SK, Chakrabarti M, Morre JD, Coley HM, Honoki K, Fujii H, Georgakilas AG, Amedei A, Niccolai E, Amin A, Ashraf SS, Helferich WG, Yang X, Boosani CS, Guha G, Bhakta D, Ciriolo MR, Aquilano K, Chen S, Mohammed SI, Keith WN, Bilsland A, Halicka D, Nowsheen S, Azmi AS (2015) Broad targeting of resistance to apoptosis in cancer. Semin Cancer Biol 35(Suppl):S78–S103. https://doi.org/10.1016/j.semcancer.2015.03.001
393. Mao Y, Li W, Hua B, Gu X, Pan W, Chen Q, Xu B, Lu C, Wang Z (2020) Circular RNA_PDHX promotes the proliferation and invasion of prostate cancer by sponging MiR-378a-3p. Front Cell Dev Biol 8:602707. https://doi.org/10.3389/fcell.2020.602707
394. Liu F, Fan Y, Ou L, Li T, Fan J, Duan L, Yang J, Luo C, Wu X (2020) CircHIPK3 facilitates the G2/M transition in prostate cancer cells by sponging miR-338-3p. Onco Targets Ther 13:4545–4558. https://doi.org/10.2147/ott.S242482
395. Deng ZH, Yu GS, Deng KL, Feng ZH, Huang Q, Pan B, Deng JZ (2020) Hsa_circ_0088233 alleviates proliferation, migration, and invasion of prostate cancer by targeting hsa-miR-185-3p. Front Cell Dev Biol 8:528155. https://doi.org/10.3389/fcell.2020.528155
396. Zhang PF, Gao C, Huang XY, Lu JC, Guo XJ, Shi GM, Cai JB, Ke AW (2020) Cancer cell-derived exosomal circUHRF1 induces natural killer cell exhaustion and may cause resistance to anti-PD1 therapy in hepatocellular carcinoma. Mol Cancer 19(1):110. https://doi.org/10.1186/s12943-020-01222-5
397. Huang E, Chen X, Yuan Y (2019) Downregulated circular RNA itchy E3 ubiquitin protein ligase correlates with advanced pathologic T stage, high lymph node metastasis risk and poor survivals in prostate cancer patients. Cancer Biomark 26(1):41–50. https://doi.org/10.3233/cbm-182111
398. Shang Q, Li Y, Wang H, Ge S, Jia R (2019) Altered expression profile of circular RNAs in conjunctival melanoma. Epigenomics 11(7):787–804. https://doi.org/10.2217/epi-2019-0029
399. Luan W, Shi Y, Zhou Z, Xia Y, Wang J (2018) circRNA_0084043 promote malignant melanoma progression via miR-153-3p/Snail axis. Biochem Biophys Res Commun 502(1):22–29. https://doi.org/10.1016/j.bbrc.2018.05.114
400. Chen Z, Chen J, Wa Q, He M, Wang X, Zhou J, Cen Y (2020) Knockdown of circ_0084043 suppresses the development of human melanoma cells through miR-429/tribbles homolog 2 axis and Wnt/β-catenin pathway. Life Sci 243:117323. https://doi.org/10.1016/j.lfs.2020.117323
401. Jin C, Dong D, Yang Z, Xia R, Tao S, Piao M (2020) CircMYC regulates glycolysis and cell proliferation in melanoma. Cell Biochem Biophys 78(1):77–88. https://doi.org/10.1007/s12013-019-00895-0
402. Huang Z, Ding Y, Zhang L, He S, Jia Z, Gu C, Wang T, Li H, Li X, Jin Z, Ding Y, Yang J (2020) Upregulated circPDK1 promotes RCC cell migration and invasion by regulating the miR-377-3P-NOTCH1 axis in renal cell carcinoma. Onco Targets Ther 13:11237–11252. https://doi.org/10.2147/ott.S280434
403. Zhang D, Yang XJ, Luo QD, Fu DL, Li ZL, Zhang P, Chong T (2019) Down-regulation of circular RNA_000926 attenuates renal cell carcinoma progression through miRNA-411-dependent CDH2 inhibition. Am J Pathol 189(12):2469–2486. https://doi.org/10.1016/j.ajpath.2019.06.016
404. Zhang G, Wang J, Tan W, Han X, Han B, Wang H, Xia Y, Sun Y, Li H (2021) Circular RNA EGLN3 silencing represses renal cell carcinoma progression through the miR-1224-3p/HMGXB3 axis. Acta Histochem 123(6):151752. https://doi.org/10.1016/j.acthis.2021.151752
405. Sun J, Pan S, Cui H, Li H (2020) CircRNA SCARB1 promotes renal cell carcinoma progression via Mir-510-5p/SDC3 axis. Curr Cancer Drug Targets 20(6):461–470. https://doi.org/10.2174/1568009620666200409130032
406. Li K, Wan CL, Guo Y (2020) Circular RNA circMTO1 suppresses RCC cancer cell progression via miR9/LMX1A axis. Technol Cancer Res Treat 19:1533033820914286. https://doi.org/10.1177/1533033820914286
407. Cen J, Liang Y, Huang Y, Pan Y, Shu G, Zheng Z, Liao X, Zhou M, Chen D, Fang Y, Chen W, Luo J, Zhang J (2021) Circular RNA circSDHC serves as a sponge for miR-127-3p to promote the proliferation and metastasis of renal cell carcinoma via the CDKN3/E2F1 axis. Mol Cancer 20(1):19. https://doi.org/10.1186/s12943-021-01314-w
408. Hirsch FR, Scagliotti GV, Mulshine JL, Kwon R, Curran WJ Jr, Wu YL, Paz-Ares L (2017) Lung cancer: current therapies and new targeted treatments. Lancet 389(10066):299–311. https://doi.org/10.1016/s0140-6736(16)30958-8
409. Thompson JC, Hwang WT, Davis C, Deshpande C, Jeffries S, Rajpurohit Y, Krishna V, Smirnov D, Verona R, Lorenzi MV, Langer CJ, Albelda SM (2020) Gene signatures of tumor inflammation and epithelial-to-mesenchymal transition (EMT) predict responses to immune checkpoint blockade in lung cancer with high accuracy. Lung Cancer 139:1–8. https://doi.org/10.1016/j.lungcan.2019.10.012
410. Yuan Z, Yuan Z, Hasnat M, Zhang H, Liang P, Sun L, Jiang Z, Zhang L (2020) A new perspective of triptolide-associated hepatotoxicity: the relevance of NF-κB and NF-κB-mediated cellular FLICE-

inhibitory protein. Acta Pharm Sin B 10(5):861–877. https://doi.org/10.1016/j.apsb.2020.02.009
411. Zhang N, Nan A, Chen L, Li X, Jia Y, Qiu M, Dai X, Zhou H, Zhu J, Zhang H, Jiang Y (2020) Circular RNA circSATB2 promotes progression of non-small cell lung cancer cells. Mol Cancer 19(1):101. https://doi.org/10.1186/s12943-020-01221-6
412. Jin M, Shi C, Yang C, Liu J, Huang G (2019) Upregulated circRNA ARHGAP10 predicts an unfavorable prognosis in NSCLC through regulation of the miR-150-5p/GLUT-1 axis. Mol Ther Nucleic Acids 18:219–231. https://doi.org/10.1016/j.omtn.2019.08.016
413. Yu C, Cheng Z, Cui S, Mao X, Li B, Fu Y, Wang H, Jin H, Ye Q, Zhao X, Jiang L, Qin W (2020) circFOXM1 promotes proliferation of non-small cell lung carcinoma cells by acting as a ceRNA to upregulate FAM83D. J Exp Clin Cancer Res 39(1):55. https://doi.org/10.1186/s13046-020-01555-5
414. Xiong Q, Liu B, Ding M, Zhou J, Yang C, Chen Y (2020) Hypoxia and cancer related pathology. Cancer Lett 486:1–7. https://doi.org/10.1016/j.canlet.2020.05.002
415. Wang X, Ji C, Hu J, Deng X, Zheng W, Yu Y, Hua K, Zhou X, Fang L (2021) Hsa_circ_0005273 facilitates breast cancer tumorigenesis by regulating YAP1-hippo signaling pathway. J Exp Clin Cancer Res 40(1):29. https://doi.org/10.1186/s13046-021-01830-z
416. Zheng X, Huang M, Xing L, Yang R, Wang X, Jiang R, Zhang L, Chen J (2020) The circRNA circSEPT9 mediated by E2F1 and EIF4A3 facilitates the carcinogenesis and development of triple-negative breast cancer. Mol Cancer 19(1):73. https://doi.org/10.1186/s12943-020-01183-9
417. Wang X, Xing L, Yang R, Chen H, Wang M, Jiang R, Zhang L, Chen J (2021) The circACTN4 interacts with FUBP1 to promote tumorigenesis and progression of breast cancer by regulating the expression of proto-oncogene MYC. Mol Cancer 20(1):91. https://doi.org/10.1186/s12943-021-01383-x
418. Xu G, Ye D, Zhao Q, He R, Ma W, Li Y, Tang S, Zhou Z, Li X, Zhang Z (2020) circNFIC suppresses breast cancer progression by sponging miR-658. J Cancer 11(14):4222–4229. https://doi.org/10.7150/jca.38830
419. Liu M, Wang Q, Shen J, Yang BB, Ding X (2019) Circbank: a comprehensive database for circRNA with standard nomenclature. RNA Biol 16(7):899–905. https://doi.org/10.1080/15476286.2019.1600395
420. Yang C, Li Q, Chen X, Zhang Z, Mou Z, Ye F, Jin S, Jun X, Tang F, Jiang H (2020) Circular RNA circRGNEF promotes bladder cancer progression via miR-548/KIF2C axis regulation. Aging (Albany NY) 12(8):6865–6879. https://doi.org/10.18632/aging.103047
421. Su Y, Feng W, Shi J, Chen L, Huang J, Lin T (2020) circRIP2 accelerates bladder cancer progression via miR-1305/Tgf-β2/smad3 pathway. Mol Cancer 19(1):23. https://doi.org/10.1186/s12943-019-1129-5
422. Mao W, Huang X, Wang L, Zhang Z, Liu M, Li Y, Luo M, Yao X, Fan J, Geng J (2019) Circular RNA hsa_circ_0068871 regulates FGFR3 expression and activates STAT3 by targeting miR-181a-5p to promote bladder cancer progression. J Exp Clin Cancer Res 38(1):169. https://doi.org/10.1186/s13046-019-1136-9
423. Wu Z, Huang W, Wang X, Wang T, Chen Y, Chen B, Liu R, Bai P, Xing J (2018) Circular RNA CEP128 acts as a sponge of miR-145-5p in promoting the bladder cancer progression via regulating SOX11. Mol Med 24(1):40. https://doi.org/10.1186/s10020-018-0039-0
424. Liu P, Li X, Guo X, Chen J, Li C, Chen M, Liu L, Zhang X, Zu X (2019) Circular RNA DOCK1 promotes bladder carcinoma progression via modulating circDOCK1/hsa-miR-132-3p/Sox5 signalling pathway. Cell Prolif 52(4):e12614. https://doi.org/10.1111/cpr.12614
425. Chen Q, Yin Q, Mao Y, Zhang Z, Wu S, Cheng Z, Chen X, Xu H, Jin S, Jiang H, Yang C (2020) Hsa_circ_0068307 mediates bladder cancer stem cell-like properties via miR-147/c-Myc axis regulation. Cancer Cell Int 20:151. https://doi.org/10.1186/s12935-020-01235-6
426. Chen L, Yang X, Zhao J, Xiong M, Almaraihah R, Chen Z, Hou T (2020) Circ_0008532 promotes bladder cancer progression by regulation of the miR-155-5p/miR-330-5p/MTGR1 axis. J Exp Clin Cancer Res 39(1):94. https://doi.org/10.1186/s13046-020-01592-0
427. Gao X, Xia X, Li F, Zhang M, Zhou H, Wu X, Zhong J, Zhao Z, Zhao K, Liu D, Xiao F, Xu Q, Jiang T, Li B, Cheng SY, Zhang N (2021) Circular RNA-encoded oncogenic E-cadherin variant promotes glioblastoma tumorigenicity through activation of EGFR-STAT3 signalling. Nat Cell Biol 23(3):278–291. https://doi.org/10.1038/s41556-021-00639-4
428. Zhang Y, Cai Z, Liang J, Chai E, Lu A, Shang Y (2020) CircCPA4 promotes the malignant phenotypes in glioma via miR-760/MEF2D axis. Neurochem Res 45(12):2903–2913. https://doi.org/10.1007/s11064-020-03139-3
429. He J, Huang Z, He M, Liao J, Zhang Q, Wang S, Xie L, Ouyang L, Koeffler HP, Yin D, Liu A (2020) Circular RNA MAPK4 (circ-MAPK4) inhibits cell apoptosis via MAPK signaling pathway by sponging miR-125a-3p in gliomas. Mol Cancer 19(1):17. https://doi.org/10.1186/s12943-019-1120-1
430. Zheng K, Xie H, Wu W, Wen X, Zeng Z, Shi Y (2021) CircRNA PIP5K1A promotes the progression of glioma through upregulation of the TCF12/PI3K/AKT pathway by sponging miR-515-5p. Cancer Cell Int 21(1):27. https://doi.org/10.1186/s12935-020-01699-6
431. Lou J, Hao Y, Lin K, Lyu Y, Chen M, Wang H, Zou D, Jiang X, Wang R, Jin D, Lam EW, Shao S, Liu Q, Yan J, Wang X, Chen P, Zhang B, Jin B (2020) Circular

RNA CDR1as disrupts the p53/MDM2 complex to inhibit Gliomagenesis. Mol Cancer 19(1):138. https://doi.org/10.1186/s12943-020-01253-y

432. Zhang S, Qin W, Yang S, Guan N, Sui X, Guo W (2020) Circular RNA SFMBT2 inhibits the proliferation and metastasis of glioma cells through Mir-182-5p/Mtss1 pathway. Technol Cancer Res Treat 19:1533033820945799. https://doi.org/10.1177/1533033820945799
433. Zhang J, Chen Z, Liu X, Yang C, Xie D (2021) Gain of circBRAF represses glioma progression by regulating miR-1290/FBXW7 axis. Neurochem Res 46(5):1203–1213. https://doi.org/10.1007/s11064-021-03259-4
434. Li T, Li Y, Zhang W, Zhang J (2021) Hsa_circ_0001017 inhibits proliferation and metastasis via regulating the let-7g-3p/NDST3 axis in glioma. Folia Neuropathol 59(2):174–188. https://doi.org/10.5114/fn.2021.107115
435. Huang C, Esfani Sarafraz P, Enayati P, Mortazavi Mamaghani E, Babakhanzadeh E, Nazari M (2023) Circular RNAs in renal cell carcinoma: from mechanistic to clinical perspective. Cancer Cell Int 23(1):288. https://doi.org/10.1186/s12935-023-03128-w
436. Ren W, Yuan Y, Peng J, Mutti L, Jiang X (2022) The function and clinical implication of circular RNAs in lung cancer. Front Oncol 12:862602. https://doi.org/10.3389/fonc.2022.862602
437. Xie T, Fu DJ, Li ZM, Lv DJ, Song XL, Yu YZ, Wang C, Li KJ, Zhai B, Wu J, Feng NH, Zhao SC (2022) CircSMARCC1 facilitates tumor progression by disrupting the crosstalk between prostate cancer cells and tumor-associated macrophages via miR-1322/CCL20/CCR6 signaling. Mol Cancer 21(1):173. https://doi.org/10.1186/s12943-022-01630-9
438. Feng H, Deng Z, Peng W, Wei X, Liu J, Wang T (2023) Circular RNA EPHA3 suppresses progression and metastasis in prostate cancer through the miR-513a-3p/BMP2 axis. J Transl Med 21(1):288. https://doi.org/10.1186/s12967-023-04132-4
439. Dai X, Chen X, Chen W, Ou Y, Chen Y, Wu S, Zhou Q, Yang C, Zhang L, Jiang H (2023) CircDHRS3 inhibits prostate cancer cell proliferation and metastasis through the circDHRS3/miR-421/MEIS2 axis. Epigenetics 18(1):2178802. https://doi.org/10.1080/15592294.2023.2178802
440. Zhou Z, Qin J, Song C, Wu T, Quan Q, Zhang Y, Zou Y, Liu L, Tang H, Zhao J (2023) circROBO1 promotes prostate cancer growth and enzalutamide resistance via accelerating glycolysis. J Cancer 14(13):2574–2584. https://doi.org/10.7150/jca.86940
441. Ding Y, Wang M, Yang J (2022) Circular RNA midline-1 (circMID1) promotes proliferation, migration, invasion and glycolysis in prostate cancer. Bioengineered 13(3):6293–6308. https://doi.org/10.1080/21655979.2022.2037367
442. Weng XD, Yan T, Liu CL (2020) Circular RNA_LARP4 inhibits cell migration and invasion of prostate cancer by targeting FOXO3A. Eur Rev Med Pharmacol Sci 24(10):5303–5309. https://doi.org/10.26355/eurrev_202005_21312
443. Xie X, Sun FK, Huang X, Wang CH, Dai J, Zhao JP, Fang C, He W (2021) A circular RNA, circSMARCA5, inhibits prostate cancer proliferative, migrative, and invasive capabilities via the miR-181b-5p/miR-17-3p-TIMP3 axis. Aging 13(15):19908–19919. https://doi.org/10.18632/aging.203408
444. Guo K, Shi J, Tang Z, Lai C, Liu C, Li K, Li Z, Xu K (2023) Circular RNA circARHGEF28 inhibited the progression of prostate cancer via the miR-671-5p/LGALS3BP/NF-κB axis. Cancer Sci 114(7):2907–2919. https://doi.org/10.1111/cas.15820
445. Liu DC, Song LL, Li XZ, Liang Q, Zhang ZG, Han CH (2021) Circular RNA circHIPK3 modulates prostate cancer progression via targeting miR-448/MTDH signaling. Clin Transl Oncol 23(12):2497–2506. https://doi.org/10.1007/s12094-021-02650-5
446. Ding X, Sun J, Zhang X (2022) Circ_0076305 facilitates prostate cancer development via sponging miR-411-5p and regulating PGK1. Andrologia 54(6):e14406. https://doi.org/10.1111/and.14406
447. Ding T, Zhu Y, Jin H, Zhang P, Guo J, Zheng J (2021) Circular RNA circ_0057558 controls prostate cancer cell proliferation through regulating miR-206/USP33/c-Myc axis. Front Cell Dev Biol 9:644397. https://doi.org/10.3389/fcell.2021.644397
448. Kong Z, Wan X, Lu Y, Zhang Y, Huang Y, Xu Y, Liu Y, Zhao P, Xiang X, Li L, Li Y (2020) Circular RNA circFOXO3 promotes prostate cancer progression through sponging miR-29a-3p. J Cell Mol Med 24(1):799–813. https://doi.org/10.1111/jcmm.14791
449. Shi J, Liu C, Chen C, Guo K, Tang Z, Luo Y, Chen L, Su Y, Xu K (2020) Circular RNA circMBOAT2 promotes prostate cancer progression via a miR-1271-5p/mTOR axis. Aging 12(13):13255–13280. https://doi.org/10.18632/aging.103432
450. Cao C, Sun G, Le K, Xu Q, Liu C (2023) The circular RNA Circ_0085494 regulates prostate cancer progression through NRBP1/miR-497-5p axis. Biochem Genet 61(5):1775–1790. https://doi.org/10.1007/s10528-023-10341-9
451. Yin D, Wei G, Yang F, Sun X (2021) Circular RNA has circ 0001591 promoted cell proliferation and metastasis of human melanoma via ROCK1/PI3K/AKT by targeting miR-431-5p. Hum Exp Toxicol 40(2):310–324. https://doi.org/10.1177/0960327120950014
452. Lin X, Zhong L, Wang N, Chu X, Liu B (2022) Hsa_circ_0103232 promotes melanoma cells proliferation and invasion via targeting miR-661/RAB3D. Cell Cycle (Georgetown, Tex) 21(17):1811–1826. https://doi.org/10.1080/15384101.2022.2072636

453. Zhang J, Cai Y, Sheng S, Zhao C, Jiang B (2022) circITCH suppresses cell proliferation and metastasis through miR-660/TFCP2 pathway in melanoma. Cancer Med 11(12):2405–2413. https://doi.org/10.1002/cam4.4627
454. Wei CY, Zhu MX, Lu NH, Liu JQ, Yang YW, Zhang Y, Shi YD, Feng ZH, Li JX, Qi FZ, Gu JY (2020) Circular RNA circ_0020710 drives tumor progression and immune evasion by regulating the miR-370-3p/CXCL12 axis in melanoma. Mol Cancer 19(1):84. https://doi.org/10.1186/s12943-020-01191-9
455. Hao T, Yang Y, He J, Bai J, Zheng Y, Luo Z (2022) Knockdown of circular RNA hsa_circ_0062270 suppresses the progression of melanoma via downregulation of CDC45. Histol Histopathol 37(4):373–383. https://doi.org/10.14670/hh-18-412
456. Lu R, Zhang X, Li X, Wan X (2020) Circ_0016418 promotes melanoma development and glutamine catabolism by regulating the miR-605-5p/GLS axis. Int J Clin Exp Pathol 13(7):1791–1801
457. Zhao F, Jia Z, Feng Y, Li Z, Feng J (2021) Circular RNA circ_0079593 enhances malignant melanoma progression by the regulation of the miR-573/ABHD2 axis. J Dermatol Sci 102(1):7–15. https://doi.org/10.1016/j.jdermsci.2021.01.008
458. Hao YL, Wang XQ (2021) Circ-0079593 promotes proliferation and migration of melanoma cells by sponging microRNA-433 and elevating EGFR expression. Eur Rev Med Pharmacol Sci 25(2):779–786. https://doi.org/10.26355/eurrev_202101_24639
459. Lu T, Yang D, Li X (2023) CircFAT1 promotes the proliferation and invasion of malignant melanoma through miR375-SLC7A11 signal axis. Anti Cancer Agents Med Chem 23(20):2200–2208. https://doi.org/10.2174/1871520623666230609163916
460. Li X, Kong Y, Li H, Xu M, Jiang M, Sun W, Xu S (2023) Circ_0081054 facilitates melanoma development via sponging miR-637 and regulating RAB9A. Skin Res Technol 29(5):e13313. https://doi.org/10.1111/srt.13313
461. Bian D, Wu Y, Song G (2018) Novel circular RNA, hsa_circ_0025039 promotes cell growth, invasion and glucose metabolism in malignant melanoma via the miR-198/CDK4 axis. Biomed Pharmacother 108:165–176. https://doi.org/10.1016/j.biopha.2018.08.152
462. Chen J, Zhou X, Yang J, Sun Q, Liu Y, Li N, Zhang Z, Xu H (2020) Circ-GLI1 promotes metastasis in melanoma through interacting with p70S6K2 to activate Hedgehog/GLI1 and Wnt/β-catenin pathways and upregulate Cyr61. Cell Death Dis 11(7):596. https://doi.org/10.1038/s41419-020-02799-x
463. Tian S, Han G, Lu L, Meng X (2020) Circ-FOXM1 contributes to cell proliferation, invasion, and glycolysis and represses apoptosis in melanoma by regulating miR-143-3p/FLOT2 axis. World J Surg Oncol 18(1):56. https://doi.org/10.1186/s12957-020-01832-9
464. Liu S, Chen L, Chen H, Xu K, Peng X, Zhang M (2021) Circ_0119872 promotes uveal melanoma development by regulating the miR-622/G3BP1 axis and downstream signalling pathways. J Exp Clin Cancer Res 40(1):66. https://doi.org/10.1186/s13046-021-01833-w
465. Wu S, Tang Y, Liu W (2020) Circ_0084043 promotes cell proliferation and glycolysis but blocks cell apoptosis in melanoma via circ_0084043-miR-31-KLF3 axis. Open Life Sci 15(1):774–786. https://doi.org/10.1515/biol-2020-0071
466. Yang CY, Wang J, Zhang JQ, Cai HM (2021) Human circular RNA hsa_circRNA_101705 (circTXNDC11) regulates renal cancer progression by regulating MAPK/ERK pathway. Bioengineered 12(1):4432–4441. https://doi.org/10.1080/21655979.2021.1955579
467. Zheng Z, Zeng X, Zhu Y, Leng M, Zhang Z, Wang Q, Liu X, Zeng S, Xiao Y, Hu C, Pang S, Wang T, Xu B, Peng P, Li F, Tan W (2024) CircPPAP2B controls metastasis of clear cell renal cell carcinoma via HNRNPC-dependent alternative splicing and targeting the miR-182-5p/CYP1B1 axis. Mol Cancer 23(1):4. https://doi.org/10.1186/s12943-023-01912-w
468. Li J, Zhang Q, Guan Y, Liao D, Jiang D, Xiong H, Zhan H, Pang J (2022) Circular RNA circVAMP3 promotes aerobic glycolysis and proliferation by regulating LDHA in renal cell carcinoma. Cell Death Dis 13(5):443. https://doi.org/10.1038/s41419-022-04863-0
469. Liang Y, Cen J, Huang Y, Fang Y, Wang Y, Shu G, Pan Y, Huang K, Dong J, Zhou M, Xu Y, Luo J, Liu M, Zhang J (2022) CircNTNG1 inhibits renal cell carcinoma progression via HOXA5-mediated epigenetic silencing of Slug. Mol Cancer 21(1):224. https://doi.org/10.1186/s12943-022-01694-7
470. Chen L, Wu D, Ding T (2020) Circular RNA circ_0001368 inhibited growth and invasion in renal cell carcinoma by sponging miR-492 and targeting LATS2. Gene 753:144781. https://doi.org/10.1016/j.gene.2020.144781
471. Yan JS, Chen Q, Li YL, Gao YQ (2021) Hsa_circ_0065217 promotes growth and metastasis of renal cancer through regulating the miR-214-3p-ALPK2 axis. Cell Cycle (Georgetown, Tex) 20(23):2519–2530. https://doi.org/10.1080/15384101.2021.1991123
472. Lin L, Cai J (2020) Circular RNA circ-EGLN3 promotes renal cell carcinoma proliferation and aggressiveness via miR-1299-mediated IRF7 activation. J Cell Biochem 121(11):4377–4385. https://doi.org/10.1002/jcb.29620
473. Li J, Huang C, Zou Y, Yu J, Gui Y (2020) Circular RNA MYLK promotes tumour growth and metastasis via modulating miR-513a-5p/VEGFC signalling in renal cell carcinoma. J Cell Mol Med 24(12):6609–6621. https://doi.org/10.1111/jcmm.15308

474. Pei L, Lv X, Jia G, Tan X, Li M, Zhang A (2021) Silencing circular RNA circ_0054537 and upregulating microRNA-640 suppress malignant progression of renal cell carcinoma via regulating neuronal pentraxin-2 (NPTX2). Bioengineered 12(1):8279–8295. https://doi.org/10.1080/21655979.2021.1984002
475. Huang Y, Zhang Y, Jia L, Liu C, Xu F (2019) Circular RNA ABCB10 promotes tumor progression and correlates with pejorative prognosis in clear cell renal cell carcinoma. Int J Biol Markers 34(2):176–183. https://doi.org/10.1177/1724600819842279
476. Cao J, Yu U, Li L, Yuan X, Chen S, Xu H, Yi M, Liu S (2022) circKL inhibits the growth and metastasis of kidney cancer by sponging miR-182-5p and upregulating FBXW7. Oncol Rep 47(4):75. https://doi.org/10.3892/or.2022.8286
477. Zhang D, Yang X, Luo Q, Fu D, Li H, Zhang P, Tie C (2023) Circular RNA CSPP1 motivates renal cell carcinoma carcinogenesis and the Warburg effect by targeting RAC1 through microRNA-493-5p. Acta Biochim Pol 70(3):693–701. https://doi.org/10.18388/abp.2020_6299
478. Wang Y, Zhang Y, Su X, Qiu Q, Yuan Y, Weng C, Zou S, Tian Y, Han W, Liu P, Guo X, Mao J, Fu X, Wang P, Lin W (2022) Circular RNA circDVL1 inhibits clear cell renal cell carcinoma progression through the miR-412-3p/PCDH7 axis. Int J Biol Sci 18(4):1491–1507. https://doi.org/10.7150/ijbs.69351
479. Yang J, Yang L, Li S, Hu N (2020) HGF/c-Met promote renal carcinoma cancer stem cells enrichment through upregulation of Cir-CCDC66. Technol Cancer Res Treat 19:1533033819901114. https://doi.org/10.1177/1533033819901114
480. Man W, Cui Y, Li J, Li Y, Jin J, Jin Y, Wu X, Zhong R, Li X, Yao H, Lin Y, Jiang L, Wang Y (2023) circTAB2 inhibits lung cancer proliferation, migration and invasion by sponging miR-3142 to upregulate GLIS2. Apoptosis 28(3–4):471–484. https://doi.org/10.1007/s10495-022-01805-1
481. Li B, Zhu L, Lu C, Wang C, Wang H, Jin H, Ma X, Cheng Z, Yu C, Wang S, Zuo Q, Zhou Y, Wang J, Yang C, Lv Y, Jiang L, Qin W (2021) circNDUFB2 inhibits non-small cell lung cancer progression via destabilizing IGF2BPs and activating anti-tumor immunity. Nat Commun 12(1):295. https://doi.org/10.1038/s41467-020-20527-z
482. Hua J, Wang X, Ma L, Li J, Cao G, Zhang S, Lin W (2022) CircVAPA promotes small cell lung cancer progression by modulating the miR-377-3p and miR-494-3p/IGF1R/AKT axis. Mol Cancer 21(1):123. https://doi.org/10.1186/s12943-022-01595-9
483. Cui Y, Wu X, Jin J, Man W, Li J, Li X, Li Y, Yao H, Zhong R, Chen S, Wu J, Zhu T, Lin Y, Xu J, Wang Y (2023) CircHERC1 promotes non-small cell lung cancer cell progression by sequestering FOXO1 in the cytoplasm and regulating the miR-142-3p-HMGB1 axis. Mol Cancer 22(1):179. https://doi.org/10.1186/s12943-023-01888-7
484. Chi Y, Zheng W, Bao G, Wu L, He X, Gan R, Shen Y, Yin X, Jin M (2021) Circular RNA circ_103820 suppresses lung cancer tumorigenesis by sponging miR-200b-3p to release LATS2 and SOCS6. Cell Death Dis 12(2):185. https://doi.org/10.1038/s41419-021-03472-7
485. Qin H, Liu J, Du ZH, Hu R, Yu YK, Wang QA (2020) Circular RNA hsa_circ_0012673 facilitates lung cancer cell proliferation and invasion via miR-320a/LIMK18521 axis. Eur Rev Med Pharmacol Sci 24(4):1841–1852. https://doi.org/10.26355/eurrev_202002_20362
486. Ma X, Yang X, Bao W, Li S, Liang S, Sun Y, Zhao Y, Wang J, Zhao C (2018) Circular RNA circMAN2B2 facilitates lung cancer cell proliferation and invasion via miR-1275/FOXK1 axis. Biochem Biophys Res Commun 498(4):1009–1015. https://doi.org/10.1016/j.bbrc.2018.03.105
487. Chi Y, Luo Q, Song Y, Yang F, Wang Y, Jin M, Zhang D (2019) Circular RNA circPIP5K1A promotes non-small cell lung cancer proliferation and metastasis through miR-600/HIF-1α regulation. J Cell Biochem 120(11):19019–19030. https://doi.org/10.1002/jcb.29225
488. Li X, Feng Y, Yang B, Xiao T, Ren H, Yu X, Li L, Li M, Zhang W (2021) A novel circular RNA, hsa_circ_0030998 suppresses lung cancer tumorigenesis and Taxol resistance by sponging miR-558. Mol Oncol 15(8):2235–2248. https://doi.org/10.1002/1878-0261.12852
489. Zhao J, Yan W, Huang W, Li Y (2022) Circ_0010235 facilitates lung cancer development and immune escape by regulating miR-636/PDL1 axis. Thorac Cancer 13(7):965–976. https://doi.org/10.1111/1759-7714.14338
490. Zhang K, Hu H, Xu J, Qiu L, Chen H, Jiang X, Jiang Y (2020) Circ_0001421 facilitates glycolysis and lung cancer development by regulating miR-4677-3p/CDCA3. Diagn Pathol 15(1):133. https://doi.org/10.1186/s13000-020-01048-1
491. Du T, Yi S, Wang Y, Zhao Q, Ma P, Jiang W (2022) Circular RNA_0120376 regulates microRNA-148b-3 and centrosomal protein 55 to promote non-small cell lung cancer development. Bioengineered 13(5):11844–11855. https://doi.org/10.1080/21655979.2022.2052647
492. Zhu L, Liu Y, Tang H, Wang P (2023) Circular RNA Circ_0001777 suppresses lung adenocarcinoma progression in vitro and in vivo. Biochem Genet 61(2):704–724. https://doi.org/10.1007/s10528-022-10284-7
493. Chen J, Rao B, Huang Z, Xie C, Yu Y, Yang B, Wu D, Wang D, Qiu F, Zhou Y, Deng Y, Lu J (2024) Circular RNA hsa_circ_0050386 suppresses non-small cell lung cancer progression via regulating the SRSF3/FN1 axis. J Transl Med 22(1):47. https://doi.org/10.1186/s12967-023-04812-1
494. Li D, Lv H, Gao H, Wang Z, Wang D, Tian K, Li L (2022) Circular RNA hsa_circ_0077837 is upregulated in non-small cell lung cancer to downregulate

phosphatase and tensin homolog through methylation. Bioengineered 13(3):6711–6718. https://doi.org/10.1080/21655979.2022.2025707

495. Cao L, Wang M, Dong Y, Xu B, Chen J, Ding Y, Qiu S, Li L, Karamfilova Zaharieva E, Zhou X, Xu Y (2020) Circular RNA circRNF20 promotes breast cancer tumorigenesis and Warburg effect through miR-487a/HIF-1α/HK2. Cell Death Dis 11(2):145. https://doi.org/10.1038/s41419-020-2336-0
496. Liu Z, Zhou Y, Liang G, Ling Y, Tan W, Tan L, Andrews R, Zhong W, Zhang X, Song E, Gong C (2019) Circular RNA hsa_circ_001783 regulates breast cancer progression via sponging miR-200c-3p. Cell Death Dis 10(2):55. https://doi.org/10.1038/s41419-018-1287-1
497. Fu B, Liu W, Zhu C, Li P, Wang L, Pan L, Li K, Cai P, Meng M, Wang Y, Zhang A, Tang W, An M (2021) Circular RNA circBCBM1 promotes breast cancer brain metastasis by modulating miR-125a/BRD4 axis. Int J Biol Sci 17(12):3104–3117. https://doi.org/10.7150/ijbs.58916
498. Sang Y, Chen B, Song X, Li Y, Liang Y, Han D, Zhang N, Zhang H, Liu Y, Chen T, Li C, Wang L, Zhao W, Yang Q (2019) circRNA_0025202 regulates tamoxifen sensitivity and tumor progression via regulating the miR-182-5p/FOXO3a axis in breast cancer. Mol Ther 27(9):1638–1652. https://doi.org/10.1016/j.ymthe.2019.05.011
499. Hu Y, Guo F, Zhu H, Tan X, Zhu X, Liu X, Zhang W, Yang Q, Jiang Y (2020) Circular RNA-0001283 suppresses breast cancer proliferation and invasion via MiR-187/HIPK3 axis. Med Sci Monit 26:e921502. https://doi.org/10.12659/msm.921502
500. Xu Y, Zhang S, Liao X, Li M, Chen S, Li X, Wu X, Yang M, Tang M, Hu Y, Li Z, Yu R, Huang M, Song L, Li J (2021) Circular RNA circIKBKB promotes breast cancer bone metastasis through sustaining NF-κB/bone remodeling factors signaling. Mol Cancer 20(1):98. https://doi.org/10.1186/s12943-021-01394-8
501. Li J, Dong X, Kong X, Wang Y, Li Y, Tong Y, Zhao W, Duan W, Li P, Wang Y, Wang C (2023) Circular RNA hsa_circ_0067842 facilitates tumor metastasis and immune escape in breast cancer through HuR/CMTM6/PD-L1 axis. Biol Direct 18(1):48. https://doi.org/10.1186/s13062-023-00397-3
502. Zhou X, Jian W, Luo Q, Zheng W, Deng X, Wang X, Borkhuu O, Ji C, Li D, Fang L (2022) Circular RNA_0006014 promotes breast cancer progression through sponging miR-885-3p to regulate NTRK2 and PIK3/AKT pathway. Aging 14(7):3105–3128. https://doi.org/10.18632/aging.203996
503. Fan Y, Wang J, Jin W, Sun Y, Xu Y, Wang Y, Liang X, Su D (2021) CircNR3C2 promotes HRD1-mediated tumor-suppressive effect via sponging miR-513a-3p in triple-negative breast cancer. Mol Cancer 20(1):25. https://doi.org/10.1186/s12943-021-01321-x
504. Huang X, Tan W, Liu Z, Fu X, Li Z, Lai S, Li Q, Zhong X, Qu F, Zhang H, Li H (2023) EIF4A3-induced circZFAND6 promotes breast cancer proliferation and metastasis through the miR-647/FASN axis. Life Sci 324:121745. https://doi.org/10.1016/j.lfs.2023.121745
505. Luo N, Liu S, Li X, Hu Y, Zhang K (2021) Circular RNA circHIPK3 promotes breast cancer progression via sponging MiR-326. Cell Cycle (Georgetown, Tex) 20(13):1320–1333. https://doi.org/10.1080/15384101.2021.1939476
506. Deng Y, Xia J, Xu YE (2021) Circular RNA circTP63 enhances estrogen receptor-positive breast cancer progression and malignant behaviors through the miR-873-3p/FOXM1 axis. Anti-Cancer Drugs 32(1):44–52. https://doi.org/10.1097/cad.0000000000001010
507. Meng X, Xiao W, Sun J, Li W, Yuan H, Yu T, Zhang X, Dong W (2023) CircPTK2/PABPC1/SETDB1 axis promotes EMT-mediated tumor metastasis and gemcitabine resistance in bladder cancer. Cancer Lett 554:216023. https://doi.org/10.1016/j.canlet.2022.216023
508. Xie B, Lin J, Chen X, Zhou X, Zhang Y, Fan M, Xiang J, He N, Hu Z, Wang F (2023) CircXRN2 suppresses tumor progression driven by histone lactylation through activating the Hippo pathway in human bladder cancer. Mol Cancer 22(1):151. https://doi.org/10.1186/s12943-023-01856-1
509. Li Y, Qiao L, Zang Y, Ni W, Xu Z (2020) Circular RNA FOXO3 suppresses bladder cancer progression and metastasis by regulating MiR-9-5p/TGFBR2. Cancer Manag Res 12:5049–5056. https://doi.org/10.2147/cmar.S253412
510. Zhang X, Xiong H, Zhao Y, Lin S, Huang X, Lin C, Mao S, Chen D (2022) Circular RNA LONP2 regulates proliferation, invasion, and apoptosis of bladder cancer cells by sponging microRNA-584-5p. Bioengineered 13(4):8823–8835. https://doi.org/10.1080/21655979.2022.2054753
511. Wang L, Li H, Qiao Q, Ge Y, Ma L, Wang Q (2020) Circular RNA circSEMA5A promotes bladder cancer progression by upregulating ENO1 and SEMA5A expression. Aging 12(21):21674–21686. https://doi.org/10.18632/aging.103971
512. Zhang X, Li X, Fu X, Yu M, Qin G, Chen G, Huang C (2021) Circular RNA TAF4B promotes bladder cancer progression by sponging miR-1298-5p and regulating TGFA expression. Front Oncol 11:643362. https://doi.org/10.3389/fonc.2021.643362
513. Yang C, Mou Z, Zhang Z, Wu S, Zhou Q, Chen Y, Gong J, Xu C, Ou Y, Chen X, Dai X, Jiang H (2021) Circular RNA RBPMS inhibits bladder cancer progression via miR-330-3p/RAI2 regulation. Mol Ther Nucleic Acids 23:872–886. https://doi.org/10.1016/j.omtn.2021.01.009
514. Dong W, Bi J, Liu H, Yan D, He Q, Zhou Q, Wang Q, Xie R, Su Y, Yang M, Lin T, Huang J (2019) Circular RNA ACVR2A suppresses bladder cancer cells proliferation and metastasis through miR-626/EYA4 axis. Mol Cancer 18(1):95. https://doi.org/10.1186/s12943-019-1025-z
515. Wu S, Yang J, Xu H, Wang X, Zhang R, Lu W, Yang J, Li X, Chen S, Zou Y, Nan A (2022) Circular RNA circGLIS3 promotes bladder cancer proliferation

via the miR-1273f/SKP1/Cyclin D1 axis. Cell Biol Toxicol 38(1):129–146. https://doi.org/10.1007/s10565-021-09591-3

516. Wang J, Luo J, Wu X, Gao Z (2021) Circular RNA_0000629 suppresses bladder cancer progression mediating MicroRNA-1290/CDC73. Cancer Manag Res 13:2701–2715. https://doi.org/10.2147/cmar.s292863
517. Li Y, Wan B, Liu L, Zhou L, Zeng Q (2019) Circular RNA circMTO1 suppresses bladder cancer metastasis by sponging miR-221 and inhibiting epithelial-to-mesenchymal transition. Biochem Biophys Res Commun 508(4):991–996. https://doi.org/10.1016/j.bbrc.2018.12.046
518. Ma X, Ying Y, Sun J, Xie H, Li J, He L, Wang W, Chen S, Shen H, Yi J, Luo J, Wang X, Zheng X, Liu B, Xie L (2021) circKDM4C enhances bladder cancer invasion and metastasis through miR-200bc-3p/ZEB1 axis. Cell Death Dis 7(1):365. https://doi.org/10.1038/s41420-021-00712-9
519. Chen Y, Wang D, Shu T, Sun K, Zhao J, Wang M, Huang Y, Wang P, Zheng H, Cai Z, Yang Z (2021) Circular RNA_0000326 promotes bladder cancer progression via microRNA-338-3p/ETS Proto-Oncogene 1/phosphoinositide-3 kinase/Akt pathway. Bioengineered 12(2):11410–11422. https://doi.org/10.1080/21655979.2021.2008738
520. Feng D, Lv J, Li K, Cao Q, Han J, Yu H, Cheng Y, Zhuang J, Cai L, Yang H, Yang X, Lu Q (2023) CircZNF609 promotes bladder cancer progression and inhibits cisplatin sensitivity via miR-1200/CDC25B pathway. Cell Biol Toxicol 39(5):1–18. https://doi.org/10.1007/s10565-022-09715-3
521. Huang W, Lu Y, Wang F, Huang X, Yu Z (2020) Circular RNA circRNA_103809 accelerates bladder cancer progression and enhances chemo-resistance by activation of miR-516a-5p/FBXL18 axis. Cancer Manag Res 12:7561–7568. https://doi.org/10.2147/cmar.s263083
522. Wei Y, Lu C, Zhou P, Zhao L, Lyu X, Yin J, Shi Z, You Y (2021) EIF4A3-induced circular RNA ASAP1 promotes tumorigenesis and temozolomide resistance of glioblastoma via NRAS/MEK1/ERK1-2 signaling. Neuro-Oncology 23(4):611–624. https://doi.org/10.1093/neuonc/noaa214
523. Jiang Y, Zhao J, Li R, Liu Y, Zhou L, Wang C, Lv C, Gao L, Cui D (2022) CircLRFN5 inhibits the progression of glioblastoma via PRRX2/GCH1 mediated ferroptosis. J Exp Clin Cancer Res 41(1):307. https://doi.org/10.1186/s13046-022-02518-8
524. Huang Q, Li W, Huang Y, Chen Q, Wei W (2022) Circular RNA VPS18 promotes glioblastoma progression by regulating miR-1229-3p/BCAT1 axis. Neurotox Res 40(5):1138–1151. https://doi.org/10.1007/s12640-022-00530-6
525. Zhou J, Wang H, Chu J, Huang Q, Li G, Yan Y, Xu T, Chen J, Wang Y (2018) Circular RNA hsa_circ_0008344 regulates glioblastoma cell proliferation, migration, invasion, and apoptosis. J Clin Lab Anal 32(7):e22454. https://doi.org/10.1002/jcla.22454
526. Yang X, Liu Y, Zhou X, Chen K, Xu J, Xu S (2022) Circular RNA 0010117 promotes aggressive glioblastoma behavior by regulating the miRNA-6779-5p/SPEN axis. Transl Oncol 25:101515. https://doi.org/10.1016/j.tranon.2022.101515
527. Yuan F, Sun Q, Xu Y, Zhang H, Deng G, Yang J, Zhang S, Liu J, Liu B, Chen Q (2021) Hsa_circ_0072309 inhibits proliferation and invasion of glioblastoma. Pathol Res Pract 222:153433. https://doi.org/10.1016/j.prp.2021.153433
528. Chen B, Wang M, Huang R, Liao K, Wang T, Yang R, Zhang W, Shi Z, Ren L, Lv Q, Ma C, Lin Y, Qiu Y (2021) Circular RNA circLGMN facilitates glioblastoma progression by targeting miR-127-3p/LGMN axis. Cancer Lett 522:225–237. https://doi.org/10.1016/j.canlet.2021.09.030
529. Zhang W, Shi Z, Chen S, Shen S, Tu S, Yang J, Qiu Y, Lin Y, Dai X (2023) Circular RNA hsa_circ_0075323 promotes glioblastoma cells proliferation and invasion via regulation of autophagy. Cell Div 18(1):1. https://doi.org/10.1186/s13008-023-00084-9
530. Li D, Li L, Chen X, Yang W, Cao Y (2021) Circular RNA SERPINE2 promotes development of glioblastoma by regulating the miR-361-3p/miR-324-5p/BCL2 signaling pathway. Mol Ther Oncolytics 22:483–494. https://doi.org/10.1016/j.omto.2021.07.010
531. Zhu F, Cheng C, Qin H, Wang H, Yu H (2020) A novel circular RNA circENTPD7 contributes to glioblastoma progression by targeting ROS1. Cancer Cell Int 20:118. https://doi.org/10.1186/s12935-020-01208-9
532. Xu S, Luo C, Chen D, Tang L, Cheng Q, Chen L, Liu Z (2023) circMMD reduction following tumor treating fields inhibits glioblastoma progression through FUBP1/FIR/DVL1 and miR-15b-5p/FZD6 signaling. J Exp Clin Cancer Res 42(1):64. https://doi.org/10.1186/s13046-023-02642-z
533. Long N, Xu X, Lin H, Lv Y, Zou S, Cao H, Chen X, Zhao Y, Qi X, Yang H, Liu J, Chu L (2022) Circular RNA circPOSTN promotes neovascularization by regulating miR-219a-2-3p/STC1 axis and stimulating the secretion of VEGFA in glioblastoma. Cell Death Dis 8(1):349. https://doi.org/10.1038/s41420-022-01136-9
534. Yuan J, Liu Z, Liu J, Fan R (2022) Circ_0060055 promotes the growth, invasion, and radioresistance of glioblastoma by targeting MiR-197-3p/API5 axis. Neurotox Res 40(5):1292–1303. https://doi.org/10.1007/s12640-022-00548-w
535. Qian L, Guan J, Wu Y, Wang Q (2019) Upregulated circular RNA circ_0074027 promotes glioblastoma cell growth and invasion by regulating miR-518a-5p/IL17RD signaling pathway. Biochem Biophys Res Commun 510(4):515–519. https://doi.org/10.1016/j.bbrc.2019.01.140
536. Meng L, Wang Y, Tu Q, Zhu Y, Dai X, Yang J (2023) Circular RNA circ_0000741/miR-379-5p/TRIM14 signaling axis promotes HDAC inhibitor (SAHA) tolerance in glioblastoma. Metab Brain

Dis 38(4):1351–1364. https://doi.org/10.1007/s11011-023-01184-9
537. Cui X, Wang J, Guo Z, Li M, Li M, Liu S, Liu H, Li W, Yin X, Tao J, Xu W (2018) Emerging function and potential diagnostic value of circular RNAs in cancer. Mol Cancer 17(1):123. https://doi.org/10.1186/s12943-018-0877-y
538. Gong Y, Mao J, Wu D, Wang X, Li L, Zhu L, Song R (2018) Circ-ZEB1.33 promotes the proliferation of human HCC by sponging miR-200a-3p and upregulating CDK6. Cancer Cell Int 18:116. https://doi.org/10.1186/s12935-018-0602-3
539. Chen S, Li T, Zhao Q, Xiao B, Guo J (2017) Using circular RNA hsa_circ_0000190 as a new biomarker in the diagnosis of gastric cancer. Clin Chim Acta 466:167–171. https://doi.org/10.1016/j.cca.2017.01.025
540. Huang M, He YR, Liang LC, Huang Q, Zhu ZQ (2017) Circular RNA hsa_circ_0000745 may serve as a diagnostic marker for gastric cancer. World J Gastroenterol 23(34):6330–6338. https://doi.org/10.3748/wjg.v23.i34.6330
541. Zhao SY, Wang J, Ouyang SB, Huang ZK, Liao L (2018) Salivary circular RNAs Hsa_Circ_0001874 and Hsa_Circ_0001971 as novel biomarkers for the diagnosis of oral squamous cell carcinoma. Cell Physiol Biochem 47(6):2511–2521. https://doi.org/10.1159/000491624
542. Chen X, Chen RX, Wei WS, Li YH, Feng ZH, Tan L, Chen JW, Yuan GJ, Chen SL, Guo SJ, Xiao KH, Liu ZW, Luo JH, Zhou FJ, Xie D (2018) PRMT5 circular RNA promotes metastasis of urothelial carcinoma of the bladder through sponging miR-30c to induce epithelial-mesenchymal transition. Clin Cancer Res 24(24):6319–6330. https://doi.org/10.1158/1078-0432.Ccr-18-1270
543. Lu R, Shao Y, Ye G, Xiao B, Guo J (2017) Low expression of hsa_circ_0006633 in human gastric cancer and its clinical significances. Tumour Biol 39(6):1010428317704175. https://doi.org/10.1177/1010428317704175
544. Sun H, Tang W, Rong D, Jin H, Fu K, Zhang W, Liu Z, Cao H, Cao X (2018) Hsa_circ_0000520, a potential new circular RNA biomarker, is involved in gastric carcinoma. Cancer Biomark 21(2):299–306. https://doi.org/10.3233/cbm-170379
545. Li T, Shao Y, Fu L, Xie Y, Zhu L, Sun W, Yu R, Xiao B, Guo J (2018) Plasma circular RNA profiling of patients with gastric cancer and their droplet digital RT-PCR detection. J Mol Med (Berl) 96(1):85–96. https://doi.org/10.1007/s00109-017-1600-y
546. Yang F, Liu DY, Guo JT, Ge N, Zhu P, Liu X, Wang S, Wang GX, Sun SY (2017) Circular RNA circ-LDLRAD3 as a biomarker in diagnosis of pancreatic cancer. World J Gastroenterol 23(47):8345–8354. https://doi.org/10.3748/wjg.v23.i47.8345
547. Yin WB, Yan MG, Fang X, Guo JJ, Xiong W, Zhang RP (2018) Circulating circular RNA hsa_circ_0001785 acts as a diagnostic biomarker for breast cancer detection. Clin Chim Acta 487:363–368. https://doi.org/10.1016/j.cca.2017.10.011
548. Zhang X, Zhou H, Jing W, Luo P, Qiu S, Liu X, Zhu M, Liang C, Yu M, Tu J (2018) The circular RNA hsa_circ_0001445 regulates the proliferation and migration of hepatocellular carcinoma and may serve as a diagnostic biomarker. Dis Markers 2018:3073467. https://doi.org/10.1155/2018/3073467
549. Zhao Q, Chen S, Li T, Xiao B, Zhang X (2018) Clinical values of circular RNA 0000181 in the screening of gastric cancer. J Clin Lab Anal 32(4):e22333. https://doi.org/10.1002/jcla.22333
550. Chi BJ, Zhao DM, Liu L, Yin XZ, Wang FF, Bi S, Gui SL, Zhou SB, Qin WB, Wu DM, Wang SQ (2019) Downregulation of hsa_circ_0000285 serves as a prognostic biomarker for bladder cancer and is involved in cisplatin resistance. Neoplasma 66(2):197–202. https://doi.org/10.4149/neo_2018_180318N185
551. Shuai M, Hong J, Huang D, Zhang X, Tian Y (2018) Upregulation of circRNA_0000285 serves as a prognostic biomarker for nasopharyngeal carcinoma and is involved in radiosensitivity. Oncol Lett 16(5):6495–6501. https://doi.org/10.3892/ol.2018.9471

Circular RNAs in Organ Fibrosis 19

Prabhat Ranjan, Roshan Kumar Dutta,
and Suresh Kumar Verma

Abstract

Fibrosis is an irreversible process that occurs in all major organs in the body; however, excessive fibrosis results in organ dysfunction and possible death. This process involves robust proliferation of fibroblasts and, in turn, excessive accumulation of extracellular matrix (ECM) within and around damaged tissues, leading to abnormal architectural remodeling (Cannito S, Novo E, Parola M, Adv Drug Deliv Rev 121:57–84, 2017; Kendall RT, Feghali-Bostwick CA, Front Pharmacol 5:123, 2014). The fibrotic process can be triggered and initiated by various fibrotic factors such as transforming growth factor-β (TGF-β) (Travis MA, Sheppard D, Annu Rev Immunol 32:51–82, 2014), members of the connective tissue growth factor (CTGF) family as well as the platelet-derived growth factor (PDGF) family (Wynn TA, J Pathol 214(2):199–210, 2008). Despite advanced modern medicine, there has been no satisfactory progress in inhibiting or reversing the progression of fibrosis. Therefore, research continues to uncover advanced antifibrotic therapeutics. Among the many known factors involved in the development and progression of organ fibrosis, circular RNAs (circRNAs) have recently drawn attention as having an essential role in organ fibrosis. Their critical involvement in fibrosis opens a new avenue of promising drug therapy. This chapter summarizes the current research progress on the biological function and underlying molecular mechanism(s) of circRNAs in regulating organ fibrosis, including heart, lung, and kidney. Furthermore, we discuss the contribution of circRNAs to antifibrotic therapy and what its future holds.

Keywords

CircRNA · Organ fibrosis · miRNA sponge · Small extracellular vesicles · m6A RNA methylation

1 Role of Circular RNAs in Organ Fibrosis

CircRNAs are expressed in a wide range of living species, from flies to humans, and play a vital role in the pathogenesis of human diseases [5]. Primarily, they are present in the cytoplasm, while a few circRNAs reside in the nucleus [6]. These are circular structures formed during pre-mRNA splicing by ligating exons or introns. They emerged as a unique member of a gene

P. Ranjan · R. K. Dutta · S. K. Verma (✉)
Division of Cardiovascular Disease, Department of Medicine, The University of Alabama at Birmingham, Birmingham, AL, USA
e-mail: sverma@uabmc.edu

J. Xiao (ed.), *Advances in Circular RNAs*, Advances in Experimental Medicine and Biology 1485,
https://doi.org/10.1007/978-981-96-9428-0_19

regulatory network that either sequesters other genes or promotes their splicing and expression. Unlike other linear RNAs, they lack 5′-caps and 3′-poly(A) tails, therefore relatively stable, being resistant to exonuclease degradation [7]. These regulatory RNAs usually enhance interactions among RNA, DNA, and proteins and establish important biological networks [6, 8, 9]. CircRNAs adopt different regulatory mechanisms, such as binding with specific miRNAs, and inhibit their functions by sequestering their sequence or binding complementary sequences through their response element [10, 11]. So, it prevents miRNAs from binding to their target genes and stops their inhibitory processes, resulting in the target genes being upregulated [5]. Since circRNAs sponge various miRNAs and proteins, their roles are highly diverse [12]. Due to its presence in the cytoplasm and open reading frame, several circRNAs, such as circ-ZNF609, circ-Foxo3, etc., translate functional proteins [13, 14]. Given its extensive coverage and numerous unknowns, there remains significant potential for discovery in the role of circularRNA in biological systems.

2 Biogenesis, Transportation, and Degradation

2.1 RNAs and Its Types

RNA, the product of transcription originating from DNA, is crucial in further downstream processes, including translation into physiologically active proteins. RNA is classified into two main categories: coding RNAs and noncoding RNAs. Coding RNAs, primarily represented by messenger RNA (mRNA), are translated to yield functional proteins. Other coding RNAs, such as transfer RNA (tRNA) and ribosomal RNA (rRNA), facilitate translation.

Noncoding RNAs (ncRNAs) encompass various types, including small nuclear RNA (snRNA), small nucleolar RNA (snoRNA), Piwi-interacting RNA (piRNA), microRNA (miRNA), long noncoding RNA (lncRNA), and circular RNA (CircRNA). Depending on their localization and interactions with DNA, RNA, and proteins, these ncRNAs can modulate RNA stability and translation, interfere with signaling pathways, and ultimately affect gene expression in diverse biological and pathophysiological contexts [15, 16]. This chapter focused on circular RNA (circRNA) and its functions in organ fibrosis.

2.2 Historical Background of CircRNA Discovery

The initial discovery of circRNA occurred during the study of potato spindle tuber disease (PSTD), a virus-led condition characterized by a single-stranded, closed-ring RNA genome [17]. The structural confirmation of circRNA was achieved using electron microscopy in the cytoplasm of eukaryotic cells, specifically in mouse testis, which revealed the expression of Sry circRNA [18–20]. The advent of next-generation sequencing post-2012 solidified the existence of circRNA. Using innovative techniques, Salzman et al. identified circular transcripts formed by exon rearrangements in human blood cells, leukemic cells, and HeLa cell lines [21].

2.3 Circular RNA: Structure and Stability

Circular RNA (circRNA) is a noncoding RNA molecule prevalent in eukaryotes. Unlike most RNA, which has a linear structure with 5′ caps and 3′ poly(A) tails, circRNA forms a covalently closed loop without these features, making it exceptionally stable and resistant to degradation by RNA exonucleases [22]. CircRNAs have diverse origins and tissue-specific expressions, playing significant roles in organismal growth and development. CircRNAs act as miRNA sponges due to their abundance of miRNA-binding sites, thereby impeding the inhibitory effects of miRNAs and influencing the expression of their target genes [23].

2.4 CircRNA in Human Physiology and Disease

Research has established circRNA as an active participant in various human diseases, including cardiovascular diseases, diabetes mellitus, and cancer. CircRNAs also affect plant diseases, biological tissue development, and cell aging. Their specific expression patterns in different biological samples make circRNAs a promising biomarker for disease diagnosis and tissue development assessment [21, 24, 25].

Understanding the mechanisms by which circRNAs contribute to disease pathology could pave the way for novel therapeutic interventions. As research progresses, the significance of circRNAs in human physiology and their potential in medical applications will become increasingly apparent.

2.5 Biogenesis

2.5.1 Introduction to CircRNA Diversity

CircRNAs exhibit diverse origins, characterized by their stability, evolutionary conservation, tissue specificity, subcellular localization, and developmental stage-specific expression. These RNA molecules are primarily generated through a process known as back-splicing, where the 3′ end of an RNA sequence is joined to the 5′ end during the splicing of precursor mRNA, forming a circular structure. Back-splicing, a type of alternative splicing, differs mechanistically from linear alternative splicing [26].

2.5.2 Regulation of CircRNA Synthesis

The circRNA synthesis is regulated by both cis-acting and trans-acting elements, although the exact mechanisms remain not fully understood. Research indicates that inhibiting splicing and spliceosome components alters circRNA biogenesis [27, 28]. This suggests that alterations in pre-mRNA processing may divert mRNAs to alternative pathways, resulting in back-splicing and the production of circRNAs [29].

Exons and introns are initially part of the mRNA during the transcription process. Introns are subsequently removed to create mature mRNA for protein translation. Splicing is a fundamental cellular process, and back-splicing involves looping intron sequences flanking downstream splice-donor sites and upstream splice-acceptor sites. This looping, mediated by base pairing between inverted repeat elements such as Alu elements [30–32] or RNA-binding protein (RBP) dimerization [23, 33, 34], brings these sites closer together.

2.5.3 Origins of CircRNA

CircRNAs originate from five primary sources based on gene structure annotation:

Exon-Only Circular RNA (Exon CircRNA): Formed solely from exons [21]

Intron-Only Circular RNA (Circular Intronic RNA): Derived exclusively from introns [35]

Exon–Intron Circular RNA (EIciRNA): Resulting from back-splicing of upstream exons with retained introns [36]

Circular RNA from Fusion Genes (f-CircRNA): Formed through gene fusions [37, 38]

Read-Through Circular RNA (rt-CircRNA): Generated by polymerase II read-through [39]

2.5.4 Mechanisms of CircRNA Synthesis

Three primary mechanisms drive circRNA synthesis through back-splicing:

1. *Flanking Intron Reverse Complementary Pairing:* Introns flanking exon pairs complement each other, enabling the 5′ splice donor site of the precursor mRNA to connect directly with the 3′ splice acceptor site, resulting in circRNA synthesis. This direct back-splicing model is facilitated by intron pairs and is aided by reverse complementary matches (RCMs), intronic complementary sequences (ICSs), and Alu elements, which are abundant in flanking introns and facilitate exon circularization [40–42].
2. *RNA-Binding Protein-Driven Circularization:* RNA-binding proteins (RBPs) facilitate

tissue-specific circRNA generation by interacting with specific motifs within flanking intron sequences. For example, in *Drosophila melanogaster* and humans, circular RNA from the second exon of the MBL/MBNL1 gene relies on specific binding motifs in adjacent introns, with heterogeneous nuclear ribonucleoprotein L (HNRNPL) promoting CircRNA formation by binding to these sequences [27, 43–45].

3. *Lasso-Driven Cyclization:* During GU/AG splicing of the mRNA precursor, exon skipping can lead to the formation of a lasso intermediate containing both intronic and exonic sequences. This intermediate subsequently undergoes back-splicing to generate circRNA. Recent research in plants such as *Arabidopsis thaliana*, *Solanum lycopersicum*, and *Zea mays* indicates that circRNAs formed through the lasso mechanism are prevalent [40, 46, 47]. Despite significant advancements, the precise mechanisms underlying CircRNA biogenesis remain largely unclear, warranting further investigation [48].

2.6 Localization and Transportation of CircRNA

2.6.1 Introduction to CircRNA Localization

CircRNAs, products of back-splicing from pre-mRNA, are primarily synthesized in the nucleus but exhibit notable abundance in the cytoplasm of eukaryotic cells [24, 49, 50]. While their role as miRNA sponges in disease contexts is well documented, recent interest has focused on direct interactions between circRNAs and mRNA or proteins, influencing their localization and transportation.

2.6.2 Cytoplasmic Localization and Functional Implications

In the cytoplasm, circRNAs engage in diverse interactions that regulate protein localization and function. For instance, circFoxo3, highly expressed in aged heart tissues, binds antisenescence and antistress proteins such as ID1, E2F1, FAK, and HIF1α. By retaining these proteins in the cytoplasm, circFoxo3 promotes cellular senescence and impacts cardiac function [51, 52]. Similarly, circAmotl1 facilitates the nuclear translocation of Stat3, enhancing wound repair, and interacts with c-myc to promote tumorigenesis by preventing its cytoplasmic translocation [53]. Conversely, circDnmt1 inhibits the translocation of p53 and AUF1, while circAGO2 facilitates the cytoplasmic translocation of HuR, processes implicated in cancer progression [52].

2.6.3 Mechanisms of CircRNA Transport Across Cellular Compartments

The nuclear export of circular RNA (circRNA) selectively requires Ran-GTP, exportin-2, and IGF2BP1. It has been seen that enhancing the nuclear Ran-GTP gradient through depletion or chemical inhibition of the major protein exporter CRM1 selectively increases the nuclear export of circRNAs. In contrast, reducing nuclear Ran-GTP gradient blocks circRNA export. Unlike Ran-GTP, exportin-2 depletion specifically inhibited the nuclear export of circRNA. Further analysis of nuclear circRNA-binding proteins indicates that the interaction between IGF2BP1 and circRNA is enhanced by Ran-GTP. The formation of circRNA export complex in the nucleus is primarily promoted by Ran-GTP through its interactions with exportin-2, circRNA, and IGF2BP1. These findings demonstrate that adaptors such as IGF2BP1, which bind directly to circular RNAs, recruit Ran-GTP and exportin-2 to facilitate circRNA export via a mechanism analogous to protein export rather than mRNA export [54].

2.6.4 CircRNAs in Small Extracellular Vesicles (sEVs)

In addition to their cellular roles, circRNAs are enriched and stable in extracellular vesicles (sEVs), which is also known as exosomes [55]. The circRNAs containing sEVs play significant roles in intercellular communication and disease pathology. The primary mechanisms regulating circRNA sorting into sEVs include:

1. *Competitive Regulation by lncRNAs:*
   Long noncoding RNAs (lncRNAs) play diverse regulatory roles in gene expression and cellular processes. One such role involves the regulation of circular RNA (circRNA) sorting. For example, expression analysis of 17 lncRNAs and 31 circRNAs in biopsy and serum exosome samples from patients with colorectal cancer demonstrated the ability to protect tumor cells from TUG1 tumor suppressor activity by secreting lncRNAs via exosomes [56].
2. *miRNA Sponges:*
   Circular RNAs can act as microRNA (miRNA) sponges, meaning they can bind to and sequester miRNAs and prevent their binding with respective mRNAs. Interestingly, exosomal circCDR1as has been identified in liver cancer cells to sponge miR-7. When miR-7 is overexpressed in these cells, the level of exosomal circCDR1as decreases, while its intracellular levels increase. Thus, circRNAs can modulate miRNA activity both inside the cell and in the extracellular environment [55].
3. *RNA-Binding Protein Recognition:*
   RNA-binding proteins (RBPs) are crucial for various aspects of RNA metabolism, including the sorting of circRNAs. In certain cells, such as DKs-8 cells, specific RBPs bind to circFAT1 and regulate its inclusion in exosomes. This selective binding by RBPs underscores the role of protein–RNA interactions in the sorting mechanism of circRNAs [57].
4. *Size Preference:*
   The length of circRNAs can influence their preferential export via exosomes. Studies have found that exosomes tend to release relatively shorter circRNAs. Specifically, the average length of circRNAs retained within cells is approximately 459 nucleotides, whereas those found in exosomes average around 435 nucleotides. This size preference suggests that shorter circRNAs are more likely to be exported through exosomes [58]. lncRNA, miRNA sponging, RBP recognition, and size preference are key factors influencing the sorting and release of circRNAs via exosomes. These mechanisms contribute to the intricate regulation of circRNA dynamics within and outside the cell, highlighting the sophisticated nature of RNA-based regulation in cellular processes. Despite these insights, the precise mechanisms governing circRNA sorting into extracellular vesicles, particularly exosomes, remain an area of active research. Understanding these mechanisms could illuminate new avenues for therapeutic interventions and biomarker development in various diseases.

2.7 Degradation of CircRNA

2.7.1 Resilience to Degradation and Mechanisms of Breakdown

Circular RNAs (circRNAs), defined by their unique closed-loop structure, demonstrate increased stability compared to linear RNAs, primarily due to their resistance to exonuclease degradation. However, despite this inherent stability, circRNAs are susceptible to degradation processes involving endonucleases and a subsequent cascade of exonucleases or endonucleases [59]. The precise mechanisms governing these degradation pathways remain partially understood, warranting further investigation.

Endonuclease Recognition and m6A Modification:

Recent studies have shed light on the role of N6-methylation of adenosine (m6A) in circRNA degradation [59]. HRSP12, an m6A reader protein, can recognize this modification. HRSP12 binds to the methylated circRNA and recruits the RNase P/MRP endonuclease complex. This complex facilitates the initial cleavage of the circRNA, thereby initiating its degradation [59]. This mechanism underscores the significance of m6A modifications in regulating circRNA stability and degradation.

RNA Interference and Exogenous Factors:

Synthetic RNA interference methods, such as short hairpin RNAs (shRNAs) or small interfering RNAs (siRNAs), are commonly used to study circRNA degradation. These methods mimic natural RNA interference pathways to target and degrade specific circRNAs. For instance, miR-671 targets binding sites on circRNA CDR1as, leading to its AGO2-mediated degradation [60, 61]. Additionally, exogenous factors such as exposure to poly(I), a synthetic analog of double-stranded RNA (dsRNA), or infection with certain viruses can activate RNase L. RNase L is an endoribonuclease that degrades viral and cellular RNAs, including circRNAs, thereby contributing to the breakdown of these molecules [61].

RNase L Activation:

The activation of RNase L is particularly relevant in pathological conditions such as systemic lupus erythematosus (SLE). In SLE, the immune system is hyperactivated, leading to the spontaneous activation of RNase L in peripheral blood mononuclear cells (PBMCs). This activation is mediated by protein kinase R (PKR), which is itself activated by the formation of intradouble-stranded RNA (dsRNA) duplexes, a structure that many circRNAs can form. Once activated, RNase L degrades circRNAs globally, contributing to the overall RNA degradation observed in these cells [61, 62].

Complex Secondary Structures and Helicase Activity:

CircRNAs often possess intricate secondary structures that require specialized mechanisms for degradation. The ATP-dependent RNA helicase UPF1 and the endonuclease G3BP1 can recognize and bind to these highly structured regions. UPF1 unwinds these regions, allowing G3BP1 to cleave the circRNA at specific sites, thereby facilitating its degradation [52].

Exosomal CircRNAs:

Beyond intracellular degradation, circRNAs can be released from cells via exocytosis or taken up by cells through endocytosis, processes often mediated by exosomes. As described in the preceding section, exosomes are small extracellular vesicles that facilitate intercellular communication by transporting various biomolecules, including circRNAs. Notably, circRNAs are found abundantly in exosomes; in some cases, their levels in exosomes exceed those within cells [58, 62]. Exosomal circRNAs can bind to proteins and play significant roles in intercellular communication. Degradation of exosomal circRNAs is thought to involve renal and hepatic metabolism, although the precise mechanisms are still under investigation. Despite these advances, much remains to be elucidated regarding the degradation mechanisms of circRNAs and their functional roles in cellular signaling. Further research is needed to fully understand these processes and their implications for cell biology and disease.

3 Biological Function and Regulatory Roles of CircRNAs

CircRNAs are involved in regulating various cellular pathways and affect physiologies. Figure 19.1 depicts circRNA biogenesis and overall biological functions, especially in organ fibrosis.

3.1 Biological Functions of CircRNAs

1. *Conservation and Noncoding Functions:* CircRNAs exhibit enhanced nucleotide conservation at the third position of codons across the closely related species, indicating significant noncoding functions and enhanced stability [25]. Many circRNAs act as miRNA sponges, regulating gene expression by sequestering miRNAs and influencing their availability to target mRNAs [25, 49, 63].
2. *Interaction with RNA-Binding Proteins (RBPs):* CircRNAs interact with various RBPs, serving as protein sponges, enhancing protein functions, acting as scaffolds for enzyme–substrate complexes, and recruiting proteins to specific cellular locations. These interactions contribute to diverse regulatory

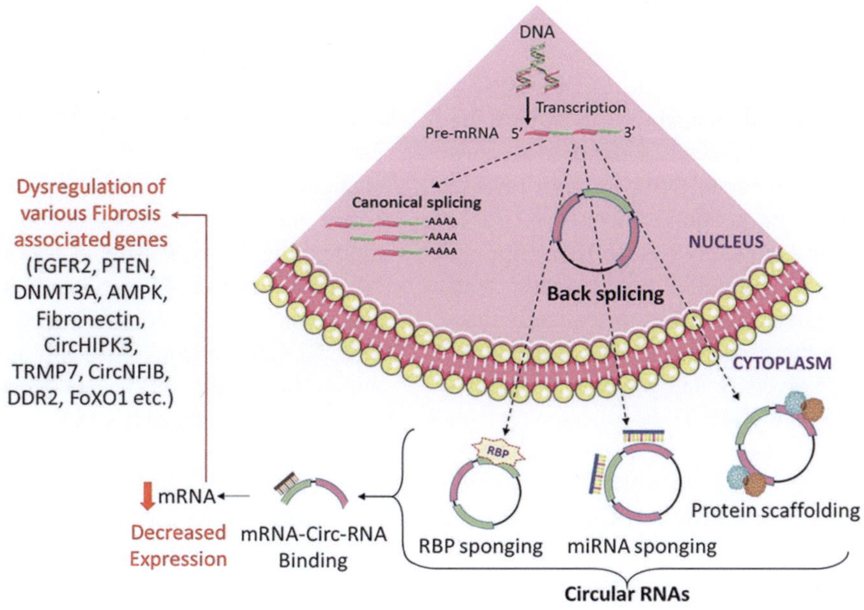

Fig. 19.1 Pre-mRNA after transcription undergoes splicing and gives rise to either linear or circular RNAs. CircRNAs interact with RNA-binding proteins, protein scaffolds, and various miRNAs that usually regulate several gene pathways. This figure shows that circRNAs dysregulate multiple genes involved in fibrosis pathways

roles in cellular processes [14, 27, 35, 36, 64–68].

3. *Cap-Independent Translation:* While predominantly noncoding, specific circRNAs undergo cap-independent translation under certain conditions, facilitated by motifs like m6A RNA methylation and interactions with translation-related proteins. This capability underscores their potential functional diversity beyond canonical noncoding roles [69–74].
4. *Regulation of Gene Expression:* CircRNAs regulate host gene expression through various mechanisms. For instance, they can bind to DNA loci of host genes, forming RNA heteroduplex R-loop structures that modulate transcription and alternative splicing, influencing phenotypic outcomes such as flowering in plants [75, 76]. Additionally, circRNAs interact with U1 snRNA to form complexes that enhance RNA Pol II transcription, illustrating their role in transcriptional regulation [36, 77].
5. *m6A Motifs and Translational Activation:* In human cells, circRNAs contain numerous m6A motifs that can activate their translational function. These motifs are recognized by specific proteins, facilitating the initiation of circRNA translation, thereby expanding their functional repertoire beyond traditional non-coding roles [71].
6. *Ribosome Association and Enhanced Translation:* Certain circRNAs in *Drosophila melanogaster* utilize the promoter regions of host genes to associate with ribosomes. This association enhances translation efficiency, suggesting a direct role of circRNAs in the regulation of protein synthesis [70].
7. *In Vitro Cyclized RNA and Protein Expression:* Recent studies have demonstrated that high concentrations of *in vitro* cyclized RNA can induce protein expression in mice. This highlights the potential of circRNAs as effective tools for protein expression, offering novel

applications in biotechnology and therapeutic research [78].

8. *Induction of Pseudogenes by CircRNAs:* CircRNAs have the capability to induce the formation of pseudogenes. Like pseudogenes derived from linear mRNAs through reverse transcription and genomic integration, stable circRNAs can undergo reverse transcription. This process generates pseudogenes with reversed sequences that are subsequently integrated into the genome, thereby contributing to genomic diversity and potentially impacting gene regulation [79].

3.2 Regulatory Roles of CircRNAs

CircRNAs, beyond their linear counterparts, have emerged as crucial players in gene regulation through their miRNA sponge functions. Here's an overview of their roles and implications:

1. *Multiple Binding Sites for miRNAs:* CircRNAs possess multiple binding sites, enabling miRNAs to sequester and inhibit their activity. This function has been observed across various plant species, including rice, wheat, tomato, tangerine, and Arabidopsis [80, 81].
2. *Competing Endogenous RNAs (ceRNAs):* CircRNAs containing multiple miRNAs binding sites can act as ceRNAs. It has the potential to competitively bind miRNAs and regulate the expression of their target mRNAs/genes. This regulatory mechanism is influenced by the stoichiometric relationship between circRNAs and miRNAs, impacting gene expression networks [82, 83]. CiRS-7 as a prototypical example, ciRS-7 (circular RNA sponge for miR-7), is extensively studied and highly expressed in various tissues, especially in the brain. It contains over 70 conserved binding sites for miR-7, suggesting a potent role in regulating miR-7 activity across different cellular contexts [49]. Studies have shown that deletion of ciRS-7 in mice leads to downregulation of miR-7 levels, impacting the expression of miR-7 target genes and influencing behaviors associated with neuropsychiatric disorders [84, 85].
3. *Role in Development and Disease:* Various circRNAs, including circHIPK3, circBIRC6, circCCDC66, and circPVT1, act as miRNA sponges in cancers. They can function as either tumor suppressors or oncogenes by modulating the expression of oncogenic miRNAs that target vital genes involved in cancer progression [63, 86, 87]. For instance, circCCDC66 sponges miR-33b and miR-93, which target the MYC oncogene, thereby influencing tumorigenesis [87].

3.3 Individual CircRNAs Function Through Interactions with Proteins

Circular RNAs (circRNAs) are integral players in cellular regulation, surpassing their established role as microRNA sponges to serve diverse functions such as protein sponges, scaffolds for protein complexes, and facilitators of protein localization within cells. A prominent example is circMbl, found in both *Drosophila melanogaster* and humans. circMbl, derived from the muscleblind (mbl) gene in flies and its homolog MBNL1 in humans, acts as a protein sponge by binding mbl and MBNL1 proteins, respectively. This interaction regulates circMbl biogenesis by looping nascent RNA, forming an autoregulatory loop where excess mbl/MBNL1 proteins reduce linear mRNA production, influencing splicing dynamics and gene expression modulation [27]. In human cervical carcinoma cells, circPABPN1 illustrates another facet of circRNA function as a protein sponge. By binding to Hu-antigen R (HUR), circPABPN1 represses the translation of nuclear poly(A) binding protein 1 (PABPN1) mRNA, impacting cellular processes influenced by PABPN1 and implicating circRNAs in cancer-related mechanisms [65]. Additionally, circANRIL functions as a protein sponge and regulator of nucleolar stress in human vascular smooth muscle cells and macrophages. Its interaction with pescadillo homolog 1 (PES1) disrupts pre-rRNA processing and ribosome biogenesis, leading to p53 activation and potentially contributing to atheroprotection through modulation of cell proliferation and apoptosis in atherosclerotic plaques [88].

Beyond their role as protein sponges, circRNAs like circ-Amotl1 and circ-Foxo3 function as crucial protein scaffolds. circ-Amotl1 facilitates interaction between 3-phosphoinositide-dependent protein kinase 1 (PDK1) and AKT1, promoting AKT1 phosphorylation and nuclear translocation associated with cardioprotective effects in doxorubicin-induced cardiomyopathy model [66]. In contrast, circ-Foxo3 regulates mutant p53 stability by enhancing its MDM2-mediated ubiquitylation and degradation, influencing cellular responses to stress and DNA damage [14, 69]. Furthermore, circRNAs such as FECR1 highlight their role in protein recruitment. FECR1 recruits ten-eleven translocation methylcytosine dioxygenase 1 (TET1) to the promoter region of its host gene FLI1, facilitating CpG demethylation and active transcription of FLI1. This regulatory mechanism impacts hematopoietic and endothelial cell biology, underscoring the intricate roles circRNAs play in gene expression regulation [68].

3.4 Cooperative Actions of CircRNAs Interacting with Proteins

The binding of specific proteins to numerous circRNAs within the cytoplasm forms a molecular reservoir that facilitates rapid responses to extracellular stimuli, particularly during viral infections [61, 89]. Notably, antiviral proteins NF90/NF110 prefer binding circRNAs over their linear counterparts due to the high stability of circRNAs, thereby establishing them as reservoirs for NF90/NF110 [89].

Furthermore, many endogenous circRNAs form intramolecular RNA duplexes capable of binding to the nucleic acid receptor PKR, which possesses antiviral properties. Upon viral infection, these circRNAs are cleaved by RNase L, an endoribonuclease activated by 2′,5′-linked oligonucleotides generated by oligoadenylate synthetase upon binding to viral double-stranded RNA [90]. This cleavage releases and activates PKR molecules, contributing to the antiviral response [61].

Viral infection triggers a global regulation of circRNAs, leading to increased nuclear export of NF90/NF110 and activation of RNase L, collectively resulting in a marked reduction in circRNA abundance [89]. These findings underscore the multifaceted roles of circRNAs beyond their traditional function as miRNA sponges. In addition to their involvement in antiviral responses, circRNAs interact with RNA-binding proteins (RBPs) to modulate protein functions, participate in cap-independent translation, and regulate gene expression through diverse mechanisms. A comprehensive understanding of these roles is pivotal for unraveling the contributions of circRNAs to biological processes and exploring their potential as therapeutic targets in various diseases. Ongoing research promises to unveil further complexities and expand our understanding of circRNA functions in cellular and molecular biology.

4 Mechanism of Organ Fibrosis

Fibrosis is a pathological deviation from the normal physiological wound healing process, leading to tissue scarring. It can affect various organs, such as the heart, kidneys, liver, lungs, and skin. Common factors triggering fibrotic signaling across organs include TGF-β, PDGF, and danger-associated molecular patterns (DAMPs). This section of this chapter aims to provide insights into the role of circRNAs in the progression of fibrosis across different organs.

4.1 Role of CircRNA in Cardiac Fibrosis

Several pathological conditions such as high blood pressure, diabetes, inflammation, and myocardial infarction lead to cardiac fibrosis, a process of robust fibroblast proliferation and secretion of extracellular matrix (ECM) [91].

Cardiac fibrosis is primarily caused by irregular extracellular matrix production and degradation that induces structural and functional abnormalities in cardiac cells and impairs heart muscle function. This leads to cardiac stiffness in myocardial infarction and, thus, ventricular arrhythmias. The causes of fibrosis in various pathological conditions have not been fully revealed, and many new mechanisms are still being regularly elucidated. Depending on the pathological manifestations, there are two types of fibrosis, i.e., reactive and replacement. Accumulating evidence suggests the role of circRNA in regulating cardiac fibrosis via several mechanisms. Zhu et al. (2019) found that myocardial infarction significantly reduces circNFIB expression levels and promotes cardiac fibrosis [92]. At molecular level, they found that miR-433 that was associated with fibroblast proliferation via antizyme inhibitor 1 (AZIN1) and c-Jun N-terminal kinase 1 (JNK1) pathways and overexpressing circNFIB sequestered miR-433 and thereby attenuated cardiac fibrosis. Another study demonstrates that circHIPK3 reduces the expression of fibrosis-associated genes such as α-SMA and COL1A1 and reduces cardiac fibrosis by sponging miR-29b-3p [93]. In contrast, the same fibrotic genes are upregulated by myo9A gene transcribed circRNA_000203, which does so by sequestering miR-26b-5p and disrupting its binding with the 3′ UTRs of Col1a2 and CTGF [94]. Another study showed that circRNA_010567 binds with miR-141 and enhances cardiac fibrosis by modulating TGF-β1 expression in diabetic mouse heart and cardiac fibroblasts, and inhibition of circ_010567 inhibited the expression of Col1, Col3, and α-SMA [95]. Heart-related circRNA (HRCR) mm9-circ-012559 has been found to target miR-223 and prevent isoproterenol-induced cardiac hypertrophy, interstitial fibrosis, and heart failure in mice, as proven by overexpression of HRCR [96]. Recently, Guo J et al. (2022) found that the protein encoded from circ_0036176 of myosin IXA inhibits cardiac proliferation by suppressing retinoblastoma (RB) tumor suppressor gene pathway. In contrast, the binding of miR-218-5p to circ_0036176 reverses this process [97]. Various chemotherapy agents are also known to play an important role in inducing cardiac fibrosis. For example, doxorubicin causes cardiac hypertrophy, fibrosis, myopathy as well as cardiotoxicity. Xu J et al. demonstrated that circRNA (circN1gn-N1gn173) is abundantly found in the heart and causes doxorubicin-induced cardiac inflammation and fibrosis through H2AX [98]. Further, lamininB1 binds to circNIgn and increases its localization in the nucleus, ultimately leading to increased collagen deposition and myocyte death [99]. In addition, some studies also suggest that many circRNA also exhibit a protective effect against doxorubicin-induced cardiotoxicity. Circ-INSR is such a highly conserved noncoding RNA that is derived from the gene encoding insulin receptor and interacts with mitochondrial single-stranded DNA-binding protein 1 (SSBP1) to stabilize mtDNA and thereby improves mitochondrial function and protects against doxorubicin-mediated cardiotoxicity [100]. Beyond all these, many circRNAs are packaged into small extracellular vesicles (sEVs) and undergo m6A modification that diversely regulates their turnover and may influence cardiac fibrogenesis [12]. One such example is circUbe3a, which is derived from M2 macrophage-secreted sEVs after acute myocardial infarction and exacerbates cardiac fibrosis via miR-138-5p/RhoC signaling pathway [101]. CircRNA has been known to affect the Hippo signaling pathway. CircYap binds to tropomyosin-4 (TPM4) and subsequently decreases actin polymerization and cardiac fibrosis [102]. circZNF609 also shows regulatory effects on the YAP pathway by interfering with YTHDF recognition, m6A reader protein, and downregulation of circZNF609, which exhibits cardioprotection [103]. Recently, it has been shown that circslc8a1, abundantly present in mice hearts, significantly decreased after 8–12 weeks of transverse aortic constriction surgery to develop therapy. They found that silencing cardiac-specific circslc8a1 induces cardiac hypertrophy and fibrosis, and its overexpression abolished that effect [104]. In addition, Ni et al. also identified significantly upregulated circHIPK3 expression in cardiac fibroblasts in mice treated with angiotensin-II. Further, this group showed that circHIPK3/miR-29b-3p axis positively regulates fibrosis-associated genes (Col1a1, Col3a1, and α-SMA) expression and ECM contents [93]. These and many other studies have highlighted the diverse roles of circRNAs in cardiac fibrosis. Thus, regulation of circRNAs could be a potential therapeutic strategy for treating cardiac fibrosis.

4.2 Contribution of CircRNA in Pulmonary Fibrosis

CircRNA has played an important role in several diseases. In this section, we are discussing how it causes pulmonary fibrosis. Like cardiac fibrosis, the lung also exhibits fibrotic remodeling, known as pulmonary fibrosis. Pulmonary fibrosis is a fatal progressive disease, which is usually reflected in an age-dependent manner [105]. There are several forms of pulmonary fibrosis, such as cystic fibrosis, silicosis, idiopathic pulmonary fibrosis (IPF), and chronic obstructive pulmonary disease (COPD). Multiple previous studies have shown the role of circRNA in regulating pulmonary fibrosis. For example, circHIPK3 promotes fibrogenesis in IPF patients via promoting fibroblast proliferation and enhancing the transition of fibroblast-to-myofibroblast in the lung [93]. CircHipk3 is upregulated and promotes forkhead box K2 expression by targeting miR-30a-3p in TGF-β1-activated fibroblasts; however, its silencing reduces fibroblast activation, proliferation, and glycolysis [106]. Furthermore, another investigation showed that circHipk3 sponge miR-338-3p promotes fibroblast-to-myofibroblast transition by increasing the expression of SRY-related HMGB4 and collagen 1a1 [107]. Moreover, other studies revealed that some other circRNAs, such as circANKRD42, regulate the biochemical signals and further mechanical stiffness in lung fibrosis by translocating YAP1 into the nucleus through the miR-324-5p/AJUBA and miR-136-5p axes [108]. In contrast, circRNATADA2A is known to delay pulmonary fibrosis. CircRNATADA2A binds to miR-526b and miR-203 and then increases the expression of caveolin (Cav)-1 and Cav-2, inhibiting the proliferation and activation of fibroblasts [109]. As we discussed, the role of m6A modification in regulating cardiac fibrosis is also involved in pulmonary fibrosis. It has been reported that methyltransferase-like 3 (METTL3) targets highly m6A-modified m6AcircRNAs (hsa_circ_0000672 and hsa_circ_0005654) and confirms their fibrogenic role in silicosis mouse models [110]. Chu et al. found that the expression of circHECTD1 is inversely associated with HECTD1 gene expression, which mediates autophagy in fibroblasts exposed to silica, thereby controlling fibrogenesis [111]. Gain- and loss-of-function experiments demonstrated that miR-29b-2-5p expression inhibits the proliferation and migration of lung fibroblasts through either circ949 or circ057 [112].

Recently, the role of circ_PWWP2A in TGF-β-induced lung fibroblast proliferation through miR-27b-3p/GATA3 axis and BLM-induced pulmonary fibrosis in mice was established [113]. Li Xuan et al. (2024) also confirmed that TGFβ treatment activates human fetal lung fibroblast 1 (HFL1 cells) and fibrosis by upregulating circ_0035796 and L1CAM expression, while decreased miR-150-5p expression was observed [114]. Further, Wu Tao et al. observed that (circ) RNA_0001861 upregulation rescues fibrosis by modulating the miR-296-5p/Bcl-2 binding component 3 axis [115]. Furthermore, Shen M. et al. (2023) demonstrated that dysregulated circ_0007535 promotes TGF-β1-induced pulmonary fibrosis by increasing the absorption of miR-18a-5p to upregulate TGF-β1-receptor 1 [116]. Gan W. et al. (2024) highlighted the role of exosome-derived circRNAs. They examined 11 dysregulated circRNAs from IPF patient-derived blood exosomes, of which hsa_circ_0044226, hsa_circ_0004099, and hsa_circ_0008898 were significantly increased compared to volunteer controls [117].

4.3 Role of CircRNA in Renal Fibrosis

Chronic kidney disease (CKD) impacts over 10% of the global population, with renal fibrosis representing a common final stage in the progression of numerous kidney disorders. Renal fibrosis is characterized by its complexity, irreversibility, and detrimental impact on the kidney and various organs, such as the cardiovascular, skeletal, and digestive systems. This pathological process begins at specific focal sites following injurious stimuli, creating a specialized microenvironment that activates fibroblasts and promotes extracellular matrix (ECM) deposition [118].

ECM–cell interactions play a crucial role in integrating various fibrogenic signals and trigger intracellular pathways that facilitate ECM component production and assembly, perpetuating a detrimental cycle. Understanding the intricate cellular and molecular events underlying this process is vital for enhancing antifibrotic therapies for renal fibrosis. Emerging research highlights the significant role of circular RNAs (circRNAs) in the pathogenesis of renal fibrosis [119, 120]. For instance, circ_0000064 is notably upregulated in mesangial cells under high glucose-induced conditions and in kidneys of diabetic nephropathy patients and promotes proliferation, inflammation, and ECM accumulation by targeting miR-424-5p and enhancing WNT2B expression [121].

Renal fibrosis, a common consequence of chronic kidney disease (CKD), is exacerbated by various circular RNAs (circRNAs), which offer potential as biomarkers and therapeutic targets. For example, circ_0000064 is upregulated in the serum of diabetic nephropathy patients and high glucose (HG)-induced HK-2 cells. Silencing circ_0000064 mitigates HG-induced HK-2 cell injury by targeting the miR-532-3p/ROCK1 axis [122]. Conversely, the overexpression of circHIPK3 worsens renal tubule-interstitial fibrosis via miR-30a sponging and TGF-β1 signaling activation in HK-2 cells [123]. Furthermore, circHIPK3 displays diverse expression and mechanisms across kidney diseases; it is downregulated in HG-induced HK-2 cells, where it alleviates HG toxicity by targeting miR-326 and miR-487a-3p to increase SIRT1 expression [124]. In diabetic nephropathy, circHIPK3 is significantly upregulated and promotes cell proliferation and profibrotic markers (TGF-β1, Collagen 1A, fibronectin) by sponging miR-185 [125]. G2/M cell cycle arrest is a key factor in renal fibrosis progression. Interestingly, CircBNC2, induced by DEAH-box helicase 9, promotes CDK1/cyclin B1 complex formation, combating G2/M arrest and reducing fibrosis in ischemic reperfusion injury [126]. CircPTPN14 enhances MYC transcription and fibrotic gene expression through its scaffold role, exacerbating renal fibrosis [126].

CircRNAs are also present in urinary exosomes, making them promising biomarkers and therapeutic targets. Cao et al. identified exosomal circRNA profiles in urine samples from CKD patients, highlighting hsa_circ_0036649, which is significantly decreased in patients with renal fibrosis and correlates with kidney function and fibrosis degree [127, 128]. In addition, exosomal circRNA, hsa_circ_0008925, is elevated in renal fibrosis patients and associated with serum creatinine, blood urea nitrogen, eGFR, cystatin C, and pathological scores for tubulointerstitial fibrosis and glomerular sclerosis [128]. Circ_DLGAP4, found in HG-treated mesangial cells and diabetic kidney disease patients, promotes mesangial cell proliferation and fibrosis by sponging miR-143 and activating the Erb-b2 receptor tyrosine kinase 3/NF-κB/MMP-2 axis [129].

Multiple studies highlight the critical role of miR-29b-3p in inhibiting fibrosis by targeting genes such as COL1A1 and COL3A1 (PMID: 30967276). CircRNA-HIPK3 targets miR-29b-3p, thereby modulating cardiac fibroblast activity and liver fibrosis [130]. Additionally, WISP1, implicated in fibrosis across various organs, including the kidneys [131], promotes renal fibrosis by upregulating collagen I and α-SMA [132]. In an elegant study, authors demonstrated that knockdown of circRNA_33702 attenuates TGF-β1-induced renal fibrosis by downregulating WISP1. This regulatory effect is reversed by a miR-29b-3p inhibitor, implicating circRNA_33702 in fibrosis modulation through the miR-29b-3p/WISP1 axis. These findings were validated in unilateral ureteral obstruction (UUO) mouse models, where CircRNA_33702 inhibition ameliorated renal fibrosis [133]. These discoveries underscore the pivotal role of circRNAs in renal fibrosis and suggest their potential utility as biomarkers in clinical biospecimens such as blood and urine from patients with kidney diseases.

5 Therapeutic Potential and Clinical Challenges

The clinical manifestations of progressive fibrotic disease are influenced by multiple factors, including aging, individual genetics, external exposures such as smoking, silica, or radiation, and autoimmune comorbidities [134]. Fibrosis imposes a significant public health burden, with substantial socioeconomic costs and detrimental effects on quality of life. Treating progressive fibrosis remains challenging due to limited treatment options. Currently, anti-inflammatory drugs and immunosuppressants are primarily used, but there is a pressing need for more specific therapies. In earlier sections, this chapter discussed the potential of circular RNAs (circRNAs) as game changers in regulating cell proliferation, cell apoptosis, and inflammatory responses during major organ injuries. However, most studies on circRNAs have been conducted *in vitro* settings, not in patient samples. Only a few studies have explored the role of circRNAs in human patients. Additionally, current research has primarily focused on individual circRNAs. More dynamic studies are needed to investigate interactions between multiple circRNA species, which may resemble specific disease pathologies more closely. To advance the use of circRNAs in clinical settings, robust, standardized protocols will be required to handle large data sets in the context of specific diseases. This will enable the use of more precise circRNAs or their antisense counterparts in clinical applications.

Acknowledgments The authors acknowledge the Department of Medicine, University of Alabama at Birmingham, for providing necessary resources.

Author Contributions PR and SKV designed the layout. All authors reviewed the literature and wrote the chapter. SKV has edited the manuscript in its final form. All authors read and approved the final manuscript.

Conflicts of Interest No competing interests are declared.

Funding This study was supported by the National Institutes of Health (HL135060), American Heart Association transformational project award (958292), UAB CCVC/UMC pilot grant to SKV, and American Heart Association Career Development Grant (23CDA1047979) to PR, and American Heart Association Post-doctoral grant to RKD (24POST1194627).

References

1. Cannito S, Novo E, Parola M (2017) Therapeutic pro-fibrogenic signaling pathways in fibroblasts. Adv Drug Deliv Rev 121:57–84. https://doi.org/10.1016/j.addr.2017.05.017
2. Kendall RT, Feghali-Bostwick CA (2014) Fibroblasts in fibrosis: novel roles and mediators. Front Pharmacol 5:123. https://doi.org/10.3389/fphar.2014.00123
3. Travis MA, Sheppard D (2014) TGF-beta activation and function in immunity. Annu Rev Immunol 32:51–82. https://doi.org/10.1146/annurev-immunol-032713-120257
4. Wynn TA (2008) Cellular and molecular mechanisms of fibrosis. J Pathol 214(2):199–210. https://doi.org/10.1002/path.2277
5. Altesha MA, Ni T, Khan A, Liu K, Zheng X (2019) Circular RNA in cardiovascular disease. J Cell Physiol 234(5):5588–5600. https://doi.org/10.1002/jcp.27384
6. Zhang Y, Liang W, Zhang P, Chen J, Qian H, Zhang X, Xu W (2017) Circular RNAs: emerging cancer biomarkers and targets. J Exp Clin Cancer Res 36(1):152. https://doi.org/10.1186/s13046-017-0624-z
7. Kristensen LS, Jakobsen T, Hager H, Kjems J (2022) The emerging roles of circRNAs in cancer and oncology. Nat Rev Clin Oncol 19(3):188–206. https://doi.org/10.1038/s41571-021-00585-y
8. Wang Y, Wang Z (2015) Efficient backsplicing produces translatable circular mRNAs. RNA 21(2):172–179. https://doi.org/10.1261/rna.048272.114
9. Liu J, Liu T, Wang X, He A (2017) Circles reshaping the RNA world: from waste to treasure.

Mol Cancer 16(1):58. https://doi.org/10.1186/s12943-017-0630-y
10. Tay Y, Rinn J, Pandolfi PP (2014) The multilayered complexity of ceRNA crosstalk and competition. Nature 505(7483):344–352. https://doi.org/10.1038/nature12986
11. van Rossum D, Verheijen BM, Pasterkamp RJ (2016) Circular RNAs: novel regulators of neuronal development. Front Mol Neurosci 9:74. https://doi.org/10.3389/fnmol.2016.00074
12. Wei L, Liu L, Bai M, Ning X, Sun S (2023) CircRNAs: versatile players and new targets in organ fibrosis. Cell Commun Signal 21(1):90. https://doi.org/10.1186/s12964-023-01051-1
13. Granados-Riveron JT, Aquino-Jarquin G (2016) The complexity of the translation ability of circRNAs. Biochim Biophys Acta 1859(10):1245–1251. https://doi.org/10.1016/j.bbagrm.2016.07.009
14. Du WW, Yang W, Liu E, Yang Z, Dhaliwal P, Yang BB (2016) Foxo3 circular RNA retards cell cycle progression via forming ternary complexes with p21 and CDK2. Nucleic Acids Res 44(6):2846–2858. https://doi.org/10.1093/nar/gkw027
15. Statello L, Guo CJ, Chen LL, Huarte M (2021) Gene regulation by long non-coding RNAs and its biological functions. Nat Rev Mol Cell Biol 22(2):96–118. https://doi.org/10.1038/s41580-020-00315-9
16. Nemeth K, Bayraktar R, Ferracin M, Calin GA (2024) Non-coding RNAs in disease: from mechanisms to therapeutics. Nat Rev Genet 25(3):211–232. https://doi.org/10.1038/s41576-023-00662-1
17. Diener TO (1971) Potato spindle tuber "virus". IV. A replicating, low molecular weight RNA. Virology 45(2):411–428. https://doi.org/10.1016/0042-6822(71)90342-4
18. Sanger HL, Klotz G, Riesner D, Gross HJ, Kleinschmidt AK (1976) Viroids are single-stranded covalently closed circular RNA molecules existing as highly base-paired rod-like structures. Proc Natl Acad Sci USA 73(11):3852–3856. https://doi.org/10.1073/pnas.73.11.3852
19. Hsu MT, Coca-Prados M (1979) Electron microscopic evidence for the circular form of RNA in the cytoplasm of eukaryotic cells. Nature 280(5720):339–340. https://doi.org/10.1038/280339a0
20. Cao QP, Gaudette MF, Robinson DH, Crain WR (1995) Expression of the mouse testis-determining gene Sry in male preimplantation embryos. Mol Reprod Dev 40(2):196–204. https://doi.org/10.1002/mrd.1080400208
21. Salzman J, Gawad C, Wang PL, Lacayo N, Brown PO (2012) Circular RNAs are the predominant transcript isoform from hundreds of human genes in diverse cell types. PLoS One 7(2):e30733. https://doi.org/10.1371/journal.pone.0030733
22. Nigro JM, Cho KR, Fearon ER, Kern SE, Ruppert JM, Oliner JD, Kinzler KW, Vogelstein B (1991) Scrambled exons. Cell 64(3):607–613. https://doi.org/10.1016/0092-8674(91)90244-s
23. Kristensen LS, Andersen MS, Stagsted LVW, Ebbesen KK, Hansen TB, Kjems J (2019) The biogenesis, biology and characterization of circular RNAs. Nat Rev Genet 20(11):675–691. https://doi.org/10.1038/s41576-019-0158-7
24. Jeck WR, Sorrentino JA, Wang K, Slevin MK, Burd CE, Liu J, Marzluff WF, Sharpless NE (2013) Circular RNAs are abundant, conserved, and associated with ALU repeats. RNA 19(2):141–157. https://doi.org/10.1261/rna.035667.112
25. Memczak S, Jens M, Elefsinioti A, Torti F, Krueger J, Rybak A, Maier L, Mackowiak SD, Gregersen LH, Munschauer M, Loewer A, Ziebold U, Landthaler M, Kocks C, le Noble F, Rajewsky N (2013) Circular RNAs are a large class of animal RNAs with regulatory potency. Nature 495(7441):333–338. https://doi.org/10.1038/nature11928
26. Fu XD, Ares M Jr (2014) Context-dependent control of alternative splicing by RNA-binding proteins. Nat Rev Genet 15(10):689–701. https://doi.org/10.1038/nrg3778
27. Ashwal-Fluss R, Meyer M, Pamudurti NR, Ivanov A, Bartok O, Hanan M, Evantal N, Memczak S, Rajewsky N, Kadener S (2014) circRNA biogenesis competes with pre-mRNA splicing. Mol Cell 56(1):55–66. https://doi.org/10.1016/j.molcel.2014.08.019
28. Starke S, Jost I, Rossbach O, Schneider T, Schreiner S, Hung LH, Bindereif A (2015) Exon circularization requires canonical splice signals. Cell Rep 10(1):103–111. https://doi.org/10.1016/j.celrep.2014.12.002
29. Liang D, Tatomer DC, Luo Z, Wu H, Yang L, Chen LL, Cherry S, Wilusz JE (2017) The output of protein-coding genes shifts to circular RNAs when the pre-mRNA processing machinery is limiting. Mol Cell 68(5):940–954.e943. https://doi.org/10.1016/j.molcel.2017.10.034
30. Ivanov A, Memczak S, Wyler E, Torti F, Porath HT, Orejuela MR, Piechotta M, Levanon EY, Landthaler M, Dieterich C, Rajewsky N (2015) Analysis of intron sequences reveals hallmarks of circular RNA biogenesis in animals. Cell Rep 10(2):170–177. https://doi.org/10.1016/j.celrep.2014.12.019
31. Zhang XO, Wang HB, Zhang Y, Lu X, Chen LL, Yang L (2014) Complementary sequence-mediated exon circularization. Cell 159(1):134–147. https://doi.org/10.1016/j.cell.2014.09.001
32. Kelly S, Greenman C, Cook PR, Papantonis A (2015) Exon skipping is correlated with exon circularization. J Mol Biol 427(15):2414–2417. https://doi.org/10.1016/j.jmb.2015.02.018
33. Conn SJ, Pillman KA, Toubia J, Conn VM, Salmanidis M, Phillips CA, Roslan S, Schreiber AW, Gregory PA, Goodall GJ (2015) The RNA binding protein quaking regulates formation of circRNAs. Cell 160(6):1125–1134. https://doi.org/10.1016/j.cell.2015.02.014
34. Kramer MC, Liang D, Tatomer DC, Gold B, March ZM, Cherry S, Wilusz JE (2015) Combinatorial

control of Drosophila circular RNA expression by intronic repeats, hnRNPs, and SR proteins. Genes Dev 29(20):2168–2182. https://doi.org/10.1101/gad.270421.115
35. Zhang Y, Zhang XO, Chen T, Xiang JF, Yin QF, Xing YH, Zhu S, Yang L, Chen LL (2013) Circular intronic long noncoding RNAs. Mol Cell 51(6):792–806. https://doi.org/10.1016/j.molcel.2013.08.017
36. Li Z, Huang C, Bao C, Chen L, Lin M, Wang X, Zhong G, Yu B, Hu W, Dai L, Zhu P, Chang Z, Wu Q, Zhao Y, Jia Y, Xu P, Liu H, Shan G (2015) Exon-intron circular RNAs regulate transcription in the nucleus. Nat Struct Mol Biol 22(3):256–264. https://doi.org/10.1038/nsmb.2959
37. Guarnerio J, Bezzi M, Jeong JC, Paffenholz SV, Berry K, Naldini MM, Lo-Coco F, Tay Y, Beck AH, Pandolfi PP (2016) Oncogenic role of fusion-circRNAs derived from cancer-associated chromosomal translocations. Cell 165(2):289–302. https://doi.org/10.1016/j.cell.2016.03.020
38. Tan S, Sun D, Pu W, Gou Q, Guo C, Gong Y, Li J, Wei YQ, Liu L, Zhao Y, Peng Y (2018) Circular RNA F-circEA-2a derived from EML4-ALK fusion gene promotes cell migration and invasion in non-small cell lung cancer. Mol Cancer 17(1):138. https://doi.org/10.1186/s12943-018-0887-9
39. Peng L, Yuan XQ, Li GC (2015) The emerging landscape of circular RNA ciRS-7 in cancer (Review). Oncol Rep 33(6):2669–2674. https://doi.org/10.3892/or.2015.3904
40. Chen LL, Yang L (2015) Regulation of circRNA biogenesis. RNA Biol 12(4):381–388. https://doi.org/10.1080/15476286.2015.1020271
41. Ebbesen KK, Kjems J, Hansen TB (2016) Circular RNAs: identification, biogenesis and function. Biochim Biophys Acta 1859(1):163–168. https://doi.org/10.1016/j.bbagrm.2015.07.007
42. Liang D, Wilusz JE (2014) Short intronic repeat sequences facilitate circular RNA production. Genes Dev 28(20):2233–2247. https://doi.org/10.1101/gad.251926.114
43. Hafner M, Landthaler M, Burger L, Khorshid M, Hausser J, Berninger P, Rothballer A, Ascano M Jr, Jungkamp AC, Munschauer M, Ulrich A, Wardle GS, Dewell S, Zavolan M, Tuschl T (2010) Transcriptome-wide identification of RNA-binding protein and microRNA target sites by PAR-CLIP. Cell 141(1):129–141. https://doi.org/10.1016/j.cell.2010.03.009
44. Ji P, Wu W, Chen S, Zheng Y, Zhou L, Zhang J, Cheng H, Yan J, Zhang S, Yang P, Zhao F (2019) Expanded expression landscape and prioritization of circular RNAs in mammals. Cell Rep 26(12):3444–3460.e3445. https://doi.org/10.1016/j.celrep.2019.02.078
45. Fei T, Chen Y, Xiao T, Li W, Cato L, Zhang P, Cotter MB, Bowden M, Lis RT, Zhao SG, Wu Q, Feng FY, Loda M, He HH, Liu XS, Brown M (2017) Genome-wide CRISPR screen identifies HNRNPL as a prostate cancer dependency regulating RNA splicing. Proc Natl Acad Sci USA 114(26):E5207–E5215. https://doi.org/10.1073/pnas.1617467114
46. Zhang X, Zhang Y, Wang T, Li Z, Cheng J, Ge H, Tang Q, Chen K, Liu L, Lu C, Guo J, Zheng B, Zheng Y (2019) A comprehensive map of intron branchpoints and lariat RNAs in plants. Plant Cell 31(5):956–973. https://doi.org/10.1105/tpc.18.00711
47. Ye CY, Zhang X, Chu Q, Liu C, Yu Y, Jiang W, Zhu QH, Fan L, Guo L (2017) Full-length sequence assembly reveals circular RNAs with diverse non-GT/AG splicing signals in rice. RNA Biol 14(8):1055–1063. https://doi.org/10.1080/15476286.2016.1245268
48. Liang Y, Liu N, Yang L, Tang J, Wang Y, Mei M (2021) A brief review of circRNA biogenesis, detection, and function. Curr Genomics 22(7):485–495. https://doi.org/10.2174/1389202922666210331130722
49. Hansen TB, Jensen TI, Clausen BH, Bramsen JB, Finsen B, Damgaard CK, Kjems J (2013) Natural RNA circles function as efficient microRNA sponges. Nature 495(7441):384–388. https://doi.org/10.1038/nature11993
50. Errichelli L, Dini Modigliani S, Laneve P, Colantoni A, Legnini I, Capauto D, Rosa A, De Santis R, Scarfo R, Peruzzi G, Lu L, Caffarelli E, Shneider NA, Morlando M, Bozzoni I (2017) FUS affects circular RNA expression in murine embryonic stem cell-derived motor neurons. Nat Commun 8:14741. https://doi.org/10.1038/ncomms14741
51. Du WW, Yang W, Chen Y, Wu ZK, Foster FS, Yang Z, Li X, Yang BB (2017) Foxo3 circular RNA promotes cardiac senescence by modulating multiple factors associated with stress and senescence responses. Eur Heart J 38(18):1402–1412. https://doi.org/10.1093/eurheartj/ehw001
52. Yang Q, Li F, He AT, Yang BB (2021) Circular RNAs: expression, localization, and therapeutic potentials. Mol Ther 29(5):1683–1702. https://doi.org/10.1016/j.ymthe.2021.01.018
53. Wu D, Chen S, Huang D, Huang Z, Zhen N, Zhou Z, Chen J (2024) circ-Amotl1 in extracellular vesicles derived from ADSCs improves wound healing by upregulating SPARC translation. Regen Ther 25:290–301. https://doi.org/10.1016/j.reth.2024.01.005
54. Ngo LH, Bert AG, Dredge BK, Williams T, Murphy V, Li W, Hamilton WB, Carey KT, Toubia J, Pillman KA, Liu D, Desogus J, Chao JA, Deans AJ, Goodall GJ, Wickramasinghe VO (2024) Nuclear export of circular RNA. Nature 627(8002):212–220. https://doi.org/10.1038/s41586-024-07060-5
55. Li Y, Zheng Q, Bao C, Li S, Guo W, Zhao J, Chen D, Gu J, He X, Huang S (2015) Circular RNA is enriched and stable in exosomes: a promising biomarker for cancer diagnosis. Cell Res 25(8):981–984. https://doi.org/10.1038/cr.2015.82
56. Barbagallo C, Brex D, Caponnetto A, Cirnigliaro M, Scalia M, Magnano A, Caltabiano R, Barbagallo D, Biondi A, Cappellani A, Basile F, Di Pietro C,

Purrello M, Ragusa M (2018) LncRNA UCA1, upregulated in CRC biopsies and downregulated in serum exosomes, controls mRNA expression by RNA-RNA interactions. Mol Ther Nucleic Acids 12:229–241. https://doi.org/10.1016/j.omtn.2018.05.009
57. Dou Y, Cha DJ, Franklin JL, Higginbotham JN, Jeppesen DK, Weaver AM, Prasad N, Levy S, Coffey RJ, Patton JG, Zhang B (2016) Circular RNAs are down-regulated in KRAS mutant colon cancer cells and can be transferred to exosomes. Sci Rep 6:37982. https://doi.org/10.1038/srep37982
58. Preusser C, Hung LH, Schneider T, Schreiner S, Hardt M, Moebus A, Santoso S, Bindereif A (2018) Selective release of circRNAs in platelet-derived extracellular vesicles. J Extracell Vesicles 7(1):1424473. https://doi.org/10.1080/20013078.2018.1424473
59. Park OH, Ha H, Lee Y, Boo SH, Kwon DH, Song HK, Kim YK (2019) Endoribonucleolytic cleavage of m(6)A-containing RNAs by RNase P/MRP complex. Mol Cell 74(3):494–507.e498. https://doi.org/10.1016/j.molcel.2019.02.034
60. Hansen TB, Wiklund ED, Bramsen JB, Villadsen SB, Statham AL, Clark SJ, Kjems J (2011) miRNA-dependent gene silencing involving Ago2-mediated cleavage of a circular antisense RNA. EMBO J 30(21):4414–4422. https://doi.org/10.1038/emboj.2011.359
61. Liu CX, Li X, Nan F, Jiang S, Gao X, Guo SK, Xue W, Cui Y, Dong K, Ding H, Qu B, Zhou Z, Shen N, Yang L, Chen LL (2019) Structure and degradation of circular RNAs regulate PKR activation in innate immunity. Cell 177(4):865–880.e821. https://doi.org/10.1016/j.cell.2019.03.046
62. Lin H, Yu J, Gu X, Ge S, Fan X (2021) Novel insights into exosomal circular RNAs: redefining intercellular communication in cancer biology. Clin Transl Med 11(12):e636. https://doi.org/10.1002/ctm2.636
63. Zheng Q, Bao C, Guo W, Li S, Chen J, Chen B, Luo Y, Lyu D, Li Y, Shi G, Liang L, Gu J, He X, Huang S (2016) Circular RNA profiling reveals an abundant circHIPK3 that regulates cell growth by sponging multiple miRNAs. Nat Commun 7:11215. https://doi.org/10.1038/ncomms11215
64. Dudekula DB, Panda AC, Grammatikakis I, De S, Abdelmohsen K, Gorospe M (2016) CircInteractome: a web tool for exploring circular RNAs and their interacting proteins and microRNAs. RNA Biol 13(1):34–42. https://doi.org/10.1080/15476286.2015.1128065
65. Abdelmohsen K, Panda AC, Munk R, Grammatikakis I, Dudekula DB, De S, Kim J, Noh JH, Kim KM, Martindale JL, Gorospe M (2017) Identification of HuR target circular RNAs uncovers suppression of PABPN1 translation by CircPABPN1. RNA Biol 14(3):361–369. https://doi.org/10.1080/15476286.2017.1279788
66. Zeng Y, Du WW, Wu Y, Yang Z, Awan FM, Li X, Yang W, Zhang C, Yang Q, Yee A, Chen Y, Yang F, Sun H, Huang R, Yee AJ, Li RK, Wu Z, Backx PH, Yang BB (2017) A circular RNA binds to and activates AKT phosphorylation and nuclear localization reducing apoptosis and enhancing cardiac repair. Theranostics 7(16):3842–3855. https://doi.org/10.7150/thno.19764
67. Du WW, Fang L, Yang W, Wu N, Awan FM, Yang Z, Yang BB (2017) Induction of tumor apoptosis through a circular RNA enhancing Foxo3 activity. Cell Death Differ 24(2):357–370. https://doi.org/10.1038/cdd.2016.133
68. Chen N, Zhao G, Yan X, Lv Z, Yin H, Zhang S, Song W, Li X, Li L, Du Z, Jia L, Zhou L, Li W, Hoffman AR, Hu JF, Cui J (2018) A novel FLI1 exonic circular RNA promotes metastasis in breast cancer by coordinately regulating TET1 and DNMT1. Genome Biol 19(1):218. https://doi.org/10.1186/s13059-018-1594-y
69. Legnini I, Di Timoteo G, Rossi F, Morlando M, Briganti F, Sthandier O, Fatica A, Santini T, Andronache A, Wade M, Laneve P, Rajewsky N, Bozzoni I (2017) Circ-ZNF609 is a circular rna that can be translated and functions in myogenesis. Mol Cell 66(1):22–37.e29. https://doi.org/10.1016/j.molcel.2017.02.017
70. Pamudurti NR, Bartok O, Jens M, Ashwal-Fluss R, Stottmeister C, Ruhe L, Hanan M, Wyler E, Perez-Hernandez D, Ramberger E, Shenzis S, Samson M, Dittmar G, Landthaler M, Chekulaeva M, Rajewsky N, Kadener S (2017) Translation of CircRNAs. Mol Cell 66(1):9–21.e27. https://doi.org/10.1016/j.molcel.2017.02.021
71. Yang Y, Fan X, Mao M, Song X, Wu P, Zhang Y, Jin Y, Yang Y, Chen LL, Wang Y, Wong CC, Xiao X, Wang Z (2017) Extensive translation of circular RNAs driven by N(6)-methyladenosine. Cell Res 27(5):626–641. https://doi.org/10.1038/cr.2017.31
72. Yang Y, Gao X, Zhang M, Yan S, Sun C, Xiao F, Huang N, Yang X, Zhao K, Zhou H, Huang S, Xie B, Zhang N (2018) Novel role of FBXW7 circular RNA in repressing glioma tumorigenesis. J Natl Cancer Inst 110(3):304–315. https://doi.org/10.1093/jnci/djx166
73. Zhang M, Huang N, Yang X, Luo J, Yan S, Xiao F, Chen W, Gao X, Zhao K, Zhou H, Li Z, Ming L, Xie B, Zhang N (2018) A novel protein encoded by the circular form of the SHPRH gene suppresses glioma tumorigenesis. Oncogene 37(13):1805–1814. https://doi.org/10.1038/s41388-017-0019-9
74. Zhang M, Zhao K, Xu X, Yang Y, Yan S, Wei P, Liu H, Xu J, Xiao F, Zhou H, Yang X, Huang N, Liu J, He K, Xie K, Zhang G, Huang S, Zhang N (2018) A peptide encoded by circular form of LINC-PINT suppresses oncogenic transcriptional elongation in glioblastoma. Nat Commun 9(1):4475. https://doi.org/10.1038/s41467-018-06862-2
75. Conn VM, Hugouvieux V, Nayak A, Conos SA, Capovilla G, Cildir G, Jourdain A, Tergaonkar V, Schmid M, Zubieta C, Conn SJ (2017) A circRNA from SEPALLATA3 regulates splicing of its cog-

nate mRNA through R-loop formation. Nat Plants 3:17053. https://doi.org/10.1038/nplants.2017.53
76. Wang Y, Gao Y, Zhang H, Wang H, Liu X, Xu X, Zhang Z, Kohnen MV, Hu K, Wang H, Xi F, Zhao L, Lin C, Gu L (2019) Genome-wide profiling of circular RNAs in the rapidly growing shoots of moso bamboo (*Phyllostachys edulis*). Plant Cell Physiol 60(6):1354–1373. https://doi.org/10.1093/pcp/pcz043
77. Lu T, Cui L, Zhou Y, Zhu C, Fan D, Gong H, Zhao Q, Zhou C, Zhao Y, Lu D, Luo J, Wang Y, Tian Q, Feng Q, Huang T, Han B (2015) Transcriptome-wide investigation of circular RNAs in rice. RNA 21(12):2076–2087. https://doi.org/10.1261/rna.052282.115
78. Wesselhoeft RA, Kowalski PS, Parker-Hale FC, Huang Y, Bisaria N, Anderson DG (2019) RNA circularization diminishes immunogenicity and can extend translation duration in vivo. Mol Cell 74(3):508–520.e504. https://doi.org/10.1016/j.molcel.2019.02.015
79. Dong R, Zhang XO, Zhang Y, Ma XK, Chen LL, Yang L (2016) CircRNA-derived pseudogenes. Cell Res 26(6):747–750. https://doi.org/10.1038/cr.2016.42
80. Mathews DH, Sabina J, Zuker M, Turner DH (1999) Expanded sequence dependence of thermodynamic parameters improves prediction of RNA secondary structure. J Mol Biol 288(5):911–940. https://doi.org/10.1006/jmbi.1999.2700
81. Su H, Tao T, Yang Z, Kang X, Zhang X, Kang D, Wu S, Li C (2019) Circular RNA cTFRC acts as the sponge of MicroRNA-107 to promote bladder carcinoma progression. Mol Cancer 18(1):27. https://doi.org/10.1186/s12943-019-0951-0
82. Denzler R, Agarwal V, Stefano J, Bartel DP, Stoffel M (2014) Assessing the ceRNA hypothesis with quantitative measurements of miRNA and target abundance. Mol Cell 54(5):766–776. https://doi.org/10.1016/j.molcel.2014.03.045
83. Thomson DW, Dinger ME (2016) Endogenous microRNA sponges: evidence and controversy. Nat Rev Genet 17(5):272–283. https://doi.org/10.1038/nrg.2016.20
84. Weng W, Wei Q, Toden S, Yoshida K, Nagasaka T, Fujiwara T, Cai S, Qin H, Ma Y, Goel A (2017) Circular RNA ciRS-7-A promising prognostic biomarker and a potential therapeutic target in colorectal cancer. Clin Cancer Res 23(14):3918–3928. https://doi.org/10.1158/1078-0432.CCR-16-2541
85. Kleaveland B, Shi CY, Stefano J, Bartel DP (2018) A network of noncoding regulatory RNAs acts in the mammalian brain. Cell 174(2):350–362.e317. https://doi.org/10.1016/j.cell.2018.05.022
86. Kristensen LS, Hansen TB, Veno MT, Kjems J (2018) Circular RNAs in cancer: opportunities and challenges in the field. Oncogene 37(5):555–565. https://doi.org/10.1038/onc.2017.361
87. Hsiao KY, Lin YC, Gupta SK, Chang N, Yen L, Sun HS, Tsai SJ (2017) Noncoding effects of circular RNA CCDC66 promote colon cancer growth and metastasis. Cancer Res 77(9):2339–2350. https://doi.org/10.1158/0008-5472.CAN-16-1883
88. Holdt LM, Stahringer A, Sass K, Pichler G, Kulak NA, Wilfert W, Kohlmaier A, Herbst A, Northoff BH, Nicolaou A, Gabel G, Beutner F, Scholz M, Thiery J, Musunuru K, Krohn K, Mann M, Teupser D (2016) Circular non-coding RNA ANRIL modulates ribosomal RNA maturation and atherosclerosis in humans. Nat Commun 7:12429. https://doi.org/10.1038/ncomms12429
89. Li X, Liu CX, Xue W, Zhang Y, Jiang S, Yin QF, Wei J, Yao RW, Yang L, Chen LL (2017) Coordinated circRNA biogenesis and function with NF90/NF110 in viral infection. Mol Cell 67(2):214–227.e217. https://doi.org/10.1016/j.molcel.2017.05.023
90. Schlee M, Hartmann G (2016) Discriminating self from non-self in nucleic acid sensing. Nat Rev Immunol 16(9):566–580. https://doi.org/10.1038/nri.2016.78
91. Ma ZG, Yuan YP, Wu HM, Zhang X, Tang QZ (2018) Cardiac fibrosis: new insights into the pathogenesis. Int J Biol Sci 14(12):1645–1657. https://doi.org/10.7150/ijbs.28103
92. Zhu Y, Pan W, Yang T, Meng X, Jiang Z, Tao L, Wang L (2019) Upregulation of circular RNA CircNFIB attenuates cardiac fibrosis by sponging miR-433. Front Genet 10:564. https://doi.org/10.3389/fgene.2019.00564
93. Ni H, Li W, Zhuge Y, Xu S, Wang Y, Chen Y, Shen G, Wang F (2019) Inhibition of circHIPK3 prevents angiotensin II-induced cardiac fibrosis by sponging miR-29b-3p. Int J Cardiol 292:188–196. https://doi.org/10.1016/j.ijcard.2019.04.006
94. Tang CM, Zhang M, Huang L, Hu ZQ, Zhu JN, Xiao Z, Zhang Z, Lin QX, Zheng XL, Yang M, Wu SL, Cheng JD, Shan ZX (2017) CircRNA_000203 enhances the expression of fibrosis-associated genes by derepressing targets of miR-26b-5p, Col1a2 and CTGF, in cardiac fibroblasts. Sci Rep 7:40342. https://doi.org/10.1038/srep40342
95. Zhou B, Yu JW (2017) A novel identified circular RNA, circRNA_010567, promotes myocardial fibrosis via suppressing miR-141 by targeting TGF-beta1. Biochem Biophys Res Commun 487(4):769–775. https://doi.org/10.1016/j.bbrc.2017.04.044
96. Wang K, Long B, Liu F, Wang JX, Liu CY, Zhao B, Zhou LY, Sun T, Wang M, Yu T, Gong Y, Liu J, Dong YH, Li N, Li PF (2016) A circular RNA protects the heart from pathological hypertrophy and heart failure by targeting miR-223. Eur Heart J 37(33):2602–2611. https://doi.org/10.1093/eurheartj/ehv713
97. Guo J, Chen LW, Huang ZQ, Guo JS, Li H, Shan Y, Chen ZR, Yan YM, Zhu JN, Guo HM, Fang XH, Shan ZX (2022) Suppression of the inhibitory effect of circ_0036176-translated Myo9a-208 on cardiac fibroblast proliferation by miR-218-5p. J Cardiovasc Transl Res 15(3):548–559. https://doi.org/10.1007/s12265-022-10228-x

98. Xu J, Du WW, Wu N, Li F, Li X, Xie Y, Wang S, Yang BB (2022) The circular RNA circNlgnmediates doxorubicin-inducedcardiac remodeling and fibrosis. Mol Ther Nucleic Acids 28:175–189. https://doi.org/10.1016/j.omtn.2022.03.007
99. Du WW, Xu J, Yang W, Wu N, Li F, Zhou L, Wang S, Li X, He AT, Du KY, Zeng K, Ma J, Lyu J, Zhang C, Zhou C, Maksimovic K, Yang BB (2021) A neuroligin isoform translated by circNlgn contributes to cardiac remodeling. Circ Res 129(5):568–582. https://doi.org/10.1161/CIRCRESAHA.120.318364
100. Lu D, Chatterjee S, Xiao K, Riedel I, Huang CK, Costa A, Cushman S, Neufeldt D, Rode L, Schmidt A, Juchem M, Leonardy J, Buchler G, Blume J, Gern OL, Kalinke U, Wen Tan WL, Foo R, Vink A, van Laake LW, van der Meer P, Bar C, Thum T (2022) A circular RNA derived from the insulin receptor locus protects against doxorubicin-induced cardiotoxicity. Eur Heart J 43(42):4496–4511. https://doi.org/10.1093/eurheartj/ehac337
101. Wang Y, Li C, Zhao R, Qiu Z, Shen C, Wang Z, Liu W, Zhang W, Ge J, Shi B (2021) CircUbe3a from M2 macrophage-derived small extracellular vesicles mediates myocardial fibrosis after acute myocardial infarction. Theranostics 11(13):6315–6333. https://doi.org/10.7150/thno.52843
102. Wu N, Xu J, Du WW, Li X, Awan FM, Li F, Misir S, Eshaghi E, Lyu J, Zhou L, Zeng K, Adil A, Wang S, Yang BB (2021) YAP circular RNA, circYap, attenuates cardiac fibrosis via binding with tropomyosin-4 and gamma-actin decreasing actin polymerization. Mol Ther 29(3):1138–1150. https://doi.org/10.1016/j.ymthe.2020.12.004
103. Wang L, Yu P, Wang J, Xu G, Wang T, Feng J, Bei Y, Xu J, Wang H, Das S, Xiao J (2022) Downregulation of circ-ZNF609 promotes heart repair by modulating RNA N(6)-methyladenosine-modified yap expression. Research (Wash D C) 2022:9825916. https://doi.org/10.34133/2022/9825916
104. Wu N, Li F, Yang W, Du WW, Awan FM, Zhang C, Lyu J, Misir S, Zeng K, Eshaghi E, Yang BB (2023) Silencing mouse circular RNA circSlc8a1 by circular antisense cA-circSlc8a1 induces cardiac hepatopathy. Mol Ther 31(6):1688–1704. https://doi.org/10.1016/j.ymthe.2022.10.005
105. Marcon A, Schievano E, Fedeli U (2021) Mortality associated with idiopathic pulmonary fibrosis in Northeastern Italy, 2008-2020: a multiple cause of death analysis. Int J Environ Res Public Health 18(14):7249. https://doi.org/10.3390/ijerph18147249
106. Xu Q, Cheng D, Li G, Liu Y, Li P, Sun W, Ma D, Ni C (2021) CircHIPK3 regulates pulmonary fibrosis by facilitating glycolysis in miR-30a-3p/FOXK2-dependent manner. Int J Biol Sci 17(9):2294–2307. https://doi.org/10.7150/ijbs.57915
107. Zhang JX, Lu J, Xie H, Wang DP, Ni HE, Zhu Y, Ren LH, Meng XX, Wang RL (2019) circHIPK3 regulates lung fibroblast-to-myofibroblast transition by functioning as a competing endogenous RNA. Cell Death Dis 10(3):182. https://doi.org/10.1038/s41419-019-1430-7
108. Xu P, Zhang J, Wang M, Liu B, Li R, Li H, Zhai N, Liu W, Lv C, Song X (2022) hnRNPL-activated circANKRD42 back-splicing and circANKRD42-mediated crosstalk of mechanical stiffness and biochemical signal in lung fibrosis. Mol Ther 30(6):2370–2387. https://doi.org/10.1016/j.ymthe.2022.01.045
109. Li J, Li P, Zhang G, Qin P, Zhang D, Zhao W (2020) CircRNA TADA2A relieves idiopathic pulmonary fibrosis by inhibiting proliferation and activation of fibroblasts. Cell Death Dis 11(7):553. https://doi.org/10.1038/s41419-020-02747-9
110. Wang S, Luo W, Huang J, Chen M, Ding J, Cheng Y, Zhang W, Fang S, Wang J, Chao J (2022) The combined effects of circular RNA methylation promote pulmonary fibrosis. Am J Respir Cell Mol Biol 66(5):510–523. https://doi.org/10.1165/rcmb.2021-0379OC
111. Chu H, Wang W, Luo W, Zhang W, Cheng Y, Huang J, Wang J, Dai X, Fang S, Chao J (2019) CircHECTD1 mediates pulmonary fibroblast activation via HECTD1. Ther Adv Chronic Dis 10:2040622319891558. https://doi.org/10.1177/2040622319891558
112. Li C, Wang Z, Zhang J, Zhao X, Xu P, Liu X, Li M, Lv C, Song X (2019) Crosstalk of mRNA, miRNA, lncRNA, and circRNA and their regulatory pattern in pulmonary fibrosis. Mol Ther Nucleic Acids 18:204–218. https://doi.org/10.1016/j.omtn.2019.08.018
113. Su L, Nian Y, Zhu T (2023) Circ_PWWP2A promotes lung fibroblast proliferation and fibrosis via the miR-27b-3p/GATA3 axis, thereby aggravating idiopathic pulmonary fibrosis. Acta Biochim Pol 70(3):525–532. https://doi.org/10.18388/abp.2020_6459
114. Li J, Chen X, Zhang B, Wang C (2023) Circ_0035796 depletion inhibits transforming growth factor-beta1-induced pulmonary fibrosis in a miR-150-5p/L1CAM-dependent manner. Autoimmunity 56(1):2250099. https://doi.org/10.1080/08916934.2023.2250099
115. Wu T, Wu S, Jiao H, Feng J, Zeng X (2023) Overexpression of hsa_circ_0001861 inhibits pulmonary fibrosis through targeting miR-296-5p/BCL-2 binding component 3 axis. Eur J Histochem 67(4):3839. https://doi.org/10.4081/ejh.2023.3839
116. Shen M, Wang X, Chang X, Li Z, Jiang N, Han Z, Liu X (2023) Circ_0007535 upregulates TGFBR1 to promote pulmonary fibrosis in TGF-beta1-treated lung fibroblasts via sequestering miR-18a-5p. Autoimmunity 56(1):2259128. https://doi.org/10.1080/08916934.2023.2259128
117. Gan W, Song W, Gao Y, Zheng X, Wang F, Zhang Z, Zen K, Liang H, Yan X (2024) Exosomal circRNAs in the plasma serve as novel biomarkers for IPF diagnosis and progression prediction. J Transl Med 22(1):264. https://doi.org/10.1186/s12967-024-05034-9

118. Li L, Fu H, Liu Y (2022) The fibrogenic niche in kidney fibrosis: components and mechanisms. Nat Rev Nephrol 18(9):545–557. https://doi.org/10.1038/s41581-022-00590-z
119. Tu C, Wang L, Wei L, Jiang Z (2022) The role of circular RNA in diabetic nephropathy. Int J Med Sci 19(5):916–923. https://doi.org/10.7150/ijms.71648
120. van Zonneveld AJ, Kolling M, Bijkerk R, Lorenzen JM (2021) Circular RNAs in kidney disease and cancer. Nat Rev Nephrol 17(12):814–826. https://doi.org/10.1038/s41581-021-00465-9
121. Li J, Min Y, Zhao Q (2022) Circ_0000064 knockdown attenuates high glucose-induced proliferation, inflammation and extracellular matrix deposition of mesangial cells through miR-424-5p-mediated WNT2B inhibition in cell models of diabetic nephropathy. Clin Exp Nephrol 26(10):943–954. https://doi.org/10.1007/s10157-022-02241-w
122. Wang H, Huang S, Hu T, Fei S, Zhang H (2022) Circ_0000064 promotes high glucose-induced renal tubular epithelial cells injury to facilitate diabetic nephropathy progression through miR-532-3p/ROCK1 axis. BMC Endocr Disord 22(1):67. https://doi.org/10.1186/s12902-022-00968-x
123. Wu Y, Luan J, Jiao C, Zhang S, Ma C, Zhang Y, Fu J, Lai EY, Kopp JB, Pi J, Zhou H (2021) circHIPK3 exacerbates folic acid-induced renal tubulointerstitial fibrosis by sponging miR-30a. Front Physiol 12:715567. https://doi.org/10.3389/fphys.2021.715567
124. Zhuang L, Wang Z, Hu X, Yang Q, Pei X, Jin G (2021) CircHIPK3 alleviates high glucose toxicity to human renal tubular epithelial HK-2 cells through regulation of miR-326/miR-487a-3p/SIRT1. Diabetes Metab Syndr Obes 14:729–740. https://doi.org/10.2147/DMSO.S289624
125. Liu R, Zhang M, Ge Y (2021) Circular RNA HIPK3 exacerbates diabetic nephropathy and promotes proliferation by sponging miR-185. Gene 765:145065. https://doi.org/10.1016/j.gene.2020.145065
126. Wang P, Huang Z, Peng Y, Li H, Lin T, Zhao Y, Hu Z, Zhou Z, Zhou W, Liu Y, Hou FF (2022) Circular RNA circBNC2 inhibits epithelial cell G2-M arrest to prevent fibrotic maladaptive repair. Nat Commun 13(1):6502. https://doi.org/10.1038/s41467-022-34287-5
127. Cao Y, Shi Y, Yang Y, Wu Z, Peng N, Xiao J, Dou F, Xu J, Pei W, Fu C, Chen P, Wang Y (2022) Urinary exosomes derived circRNAs as biomarkers for chronic renal fibrosis. Ann Med 54(1):1966–1976. https://doi.org/10.1080/07853890.2022.2098374
128. Cao Y, Shi Y, Wang Y, Yang Y, Guo W, Zhang C, Pei W, Fu C (2022) Exosomal hsa_circ_0008925 from urine is related to chronic renal fibrosis. Dis Markers 2022:1899282. https://doi.org/10.1155/2022/1899282
129. Bai S, Xiong X, Tang B, Ji T, Li X, Qu X, Li W (2020) Exosomal circ_DLGAP4 promotes diabetic kidney disease progression by sponging miR-143 and targeting ERBB3/NF-kappaB/MMP-2 axis. Cell Death Dis 11(11):1008. https://doi.org/10.1038/s41419-020-03169-3
130. Yang R, Xu X, Li H, Chen J, Xiang X, Dong Z, Zhang D (2017) p53 induces miR199a-3p to suppress SOCS7 for STAT3 activation and renal fibrosis in UUO. Sci Rep 7:43409. https://doi.org/10.1038/srep43409
131. CORRIGENDUM (2021) WNT1-inducible signaling protein-1 mediates TGF-beta1-induced renal fibrosis in tubular epithelial cells and unilateral ureteral obstruction mouse models via autophagy. J Cell Physiol 236(12):8253. https://doi.org/10.1002/jcp.30469
132. Chen YZ, Sun DQ, Zheng Y, Zheng GK, Chen RQ, Lin M, Huang LF, Huang C, Song D, Wu BQ (2019) WISP1 silencing confers protection against epithelial-mesenchymal transition of renal tubular epithelial cells in rats via inactivation of the wnt/beta-catenin signaling pathway in uremia. J Cell Physiol 234(6):9673–9686. https://doi.org/10.1002/jcp.27654
133. Ai K, Yi L, Wang Y, Li Y (2023) CircRNA_33702 promotes renal fibrosis by targeting the miR-29b-3p/WNT1-inducible signaling pathway protein 1 pathway. J Pharmacol Exp Ther 384(1):61–71. https://doi.org/10.1124/jpet.122.001280
134. Moss BJ, Ryter SW, Rosas IO (2022) Pathogenic mechanisms underlying idiopathic pulmonary fibrosis. Annu Rev Pathol 17:515–546. https://doi.org/10.1146/annurev-pathol-042320-030240

Circular RNAs in Metabolic Diseases

20

Luisa Sophie Rajcsanyi, Triinu Peters, and Anke Hinney

Abstract

Prevalence rates of metabolic diseases, such as obesity, the metabolic syndrome (MetS), non-alcoholic fatty liver disease (NAFLD) and diabetes mellitus (DM), have been rising continuously. So far, the underlying disease mechanisms are still not fully understood. However, over the last years, evidence emerged that circRNAs may be regulators of metabolic diseases and related mechanisms. Several studies have been conducted analysing the impact of circRNAs in obesity, NAFLD and DM, while solely few investigated circRNAs pertaining to the MetS. Here, we highlight circRNAs described to be relevant for the respective disease pathogenesis.

Keywords

CircRNAs · Obesity · NAFLD · MetS · Diabetes mellitus

L. S. Rajcsanyi (✉) · T. Peters · A. Hinney
Section of Molecular Genetics in Mental Disorders, LVR University Hospital, University Hospital Essen, Essen, Germany

Center for Translational Neuro- and Behavioural Sciences, University Hospital Essen, Essen, Germany

Institute of Sex- and Gender-Sensitive Medicine, University Hospital Essen, Essen, Germany
e-mail: luisa.rajcsanyi@uni-due.de

1 Introduction

A dysregulated metabolism can cause complex and highly heterogeneous metabolic diseases [1], like obesity, metabolic syndrome (MetS), non-alcoholic fatty liver disease (NAFLD) and diabetes mellitus (DM). All these conditions constitute a substantial health risk, particularly given their rising prevalence rates [2]. Hence, understanding the pathophysiology of these diseases as well as links between them is crucial. In recent years, evidence emerged that circular RNAs (circRNAs) may be major regulators of metabolic diseases and related mechanisms [3, 4].

First reported in 1976 [5] and initially thought to be a splicing by-product [6, 7], circRNAs are covalently closed RNAs exhibiting a differential expression between tissues, cell types and developmental stages [7–9]. So far, excessive research highlighted circRNAs' functionalities in indirectly regulating gene expression by sponging microRNAs (miRNAs) [8, 10–12] and interacting with proteins like the RNA polymerase II [13]. Despite their lack of a 5'-cap structure [7, 14], several circRNAs have been reported to be translated based on the presence of internal ribosome entry sites or N6-methyladenosine (m6A) modifications [15–17].

In this chapter, we will present circRNAs reported to be relevant for the pathogenesis of obesity, MetS, NAFLD and DM in human tissues, or rodent models.

J. Xiao (ed.), *Advances in Circular RNAs*, Advances in Experimental Medicine and Biology 1485,
https://doi.org/10.1007/978-981-96-9428-0_20

2 Obesity

Obesity is a chronic disease defined as an elevated accumulation of adipose tissue resulting in a BMI > 30 kg/m^2 [18]. Its prevalence has risen since 1975 [19]. According to the World Health Organization (WHO), more than 650 million individuals over the age of 18 years had obesity in 2016, representing about 13% of the global adult population [20]. Projections suggest that by 2035, nearly 2 billion people, including both adults and children, will have obesity [21]. Individuals with obesity frequently experience a variety of comorbidities like type 2 diabetes mellitus (T2DM) [22], cardiovascular diseases [23, 24] and certain types of cancer [25–27], frequently leading to a premature death [28–32]. The number of life years lost due to obesity and its respective comorbidities can be as high as 13 years for men aged 20–30 years having a BMI greater than 45 kg/m^2 [31].

The complex aetiology of obesity involves both environmental and genetic factors, contributing to a prolonged energy imbalance leading to an accumulation of adipose tissue [19, 33, 34]. Genetic factors account for approximately 50% of the body weight variance [33, 35]. Given this underlying genetic basis of obesity, it is typically classified into monogenic and polygenic forms. Monogenic obesity is caused by mutations in single genes, particularly involved in the leptin–melanocortin pathway. Conversely, polygenic obesity is triggered by small cumulative effects of multiple genetic variants in several genes and is the predominant form of obesity detected in the general population [33, 34].

2.1 CircRNAs in Obesity

Since the discovery of circRNAs as functional modalities [7], their roles in obesity and its related mechanisms, like adipogenesis, have been extensively analysed (see Table 20.1). Several studies performed expression profiling in human subcutaneous (SAT) [36, 37] or visceral adipose tissue (VAT) [37–40] as well as murine cells and tissues [11, 37, 41–45]. The investigation of circRNA profiles in human tissue (excluding human cell lines) has predominantly been performed through RNA-sequencing approaches [37, 38, 40] and circRNA microarrays [39, 46].

One microarray analysis revealed 244 differentially expressed circRNAs in VAT of patients with obesity ($n = 3$; BMI > 28 kg/m^2) compared to lean individuals ($n = 3$) [39]. The expression of the most highly expressed circRNA, circSAMD4A, correlated with the BMI ($r = 0.212$, $p < 0.001$) and a poor prognosis for a non-remission one year after bariatric surgery (odds ratio = 13.795, 95% CI = [2.25, 84.62], $p = 0.005$) in 60 patients with obesity [39]. In vitro analyses showed that a knockdown of circSAMD4A inhibited the differentiation of VAT-isolated preadipocytes, while in an in vivo model analysing high-fat diet (HFD) fed mice, a knockdown led to a decreased food intake, lowered body fat and elevated sensitivity to insulin as well as improved glucose tolerance. Mechanistically, circSAMD4A binds to miR-138-5p and ultimately regulates the expression of *Ezh2* [39].

By profiling circRNAs' expression via whole transcriptome RNA-sequencing in eight VAT samples of patients with obesity (BMI ≥ 28 kg/m^2) or lean control individuals (BMI 18.5 –23.9 kg/m^2), one circRNA impacting FTO was detected [40]. Notably, *FTO* is a risk locus relevant for body weight regulation already detected in the first respective genome-wide association study (GWAS) and had been confirmed in every subsequent GWAS to date [47–50]. The respective *FTO*-impacting circRNA, circMAPK9, was higher expressed in VAT of individuals with obesity ($n = 70$) than in VAT of lean individuals ($n = 20$; $p < 0.05$) [40]. Silencing circMAPK9 reduced triglyceride (TG) levels, formation of lipid droplets and the expression of *PPARγ*, *CEBPα* and *FABP4*. MiR-1322, which was targeted by circMAPK9, is involved in the process of adipocyte maturation by affecting FTO. The expression levels of circMAPK9 and *FTO* were positively correlated in VAT of 70 patients with obesity ($r = 0.36$, $p = 0.002$) [40].

Another study performing whole transcriptome RNA-sequencing uncovered expression profiles in VAT of six patients with obesity

Table 20.1 Circular RNAs implicated in obesity

CircRNAs	Reported functionalities	References
circArhgap5-2 and human homologue circARHGAP5-1	Crucial regulators of adipogenesis	[37]
circTshz2-1	Crucial regulator of adipogenesis	[37]
circNrxn2	Facilitates browning of WAT in HFD mice via miR-103/*Fgf10* axis	[43]
circARF3	Mitigates inflammation in adipose tissue by promoting mitophagy	[44]
circSAMD4A	Regulates preadipocyte differentiation via miR-138-5p/*Ezh2* axis	[39]
hsa_circ_0017650	Upregulated in adipocytes compared to HPA-v cells	[46]
hsa_circRNA9227-1	Upregulated in adipocytes compared to HPA-v cells	[46]
circTshz2-2	Affects spatial memory in obese mice by regulating *Bdnf* transcription	[60]
circOgdh	Mediates brown adipocyte lipolysis via miR-34a-5p/*Atgl* axis	[11]
circORC5	Acts via miR-197-5p/*TNFRSF10D*	[38]
circNTRK2	Acts via miR-760/*LAT*	[38]
circMAPK9	Promotes adipogenesis via miR-1322/*FTO* axis	[40]
circRNA1709	Might regulate expression of *PTEN*	[42]
mmu_circ_0001874	Affects adipogenesis and lipid accumulation via miR-24-3p/*Igf2* and binds to *Igf2bp2/Ucp1* regulating thermogenesis	[45]
hsa_circ_0046367	Associated with paediatric obesity	[51]
hsa_circ_0000284		

circRNA circular RNA, *HFD* high-fat diet, *HPA-v* human preadipocytes–visceral

(BMI $\geq$ 32.5 kg/m^2) and four lean controls (BMI $\leq$ 20 kg/m^2). In total, 25 circRNAs were upregulated, while 27 exhibited a downregulation. Three circRNAs that were most likely associated with inflammation were analysed in a subsequent quantitative PCR (qPCR). Of these, only two (circORC5 and circNTRK2) were found to be downregulated in VAT of patients with obesity ($n = 30$) compared to VAT of lean individuals ($n = 30$; $p < 0.001$) [38].

While studies pertaining circRNAs in obesity predominantly analysed adult individuals, merely one study investigated circRNAs putatively implicated in paediatric obesity [51]. Here, candidate circRNAs were selected based on a literature search identifying circRNAs related to metabolic diseases. Four circRNAs were subsequently screened in peripheral blood leukocytes of children with obesity ($n = 30$) and children with normal weight ($n = 30$). As the expression of two circRNAs (hsa_circ_0000284 and hsa_circ_0046367) differed between the analysed groups, those were investigated in a larger cohort of children with obesity ($n = 75$) and controls ($n = 75$). Hereby, hsa_circ_0000284 was upregulated ($p = 0.036$), while hsa_circ_0046367 was downregulated ($p = 0.004$) in children with obesity [51].

Impacts of single-nucleotide polymorphisms (SNPs) on circRNAs have been implied for various phenotypes [52, 53]. For BMI, one study analysed the positional overlap of circRNAs derived from various publicly available databases and SNPs identified in a GWAS for BMI [54]. It was determined that BMI-associated SNPs (genome-wide threshold $p < 5 \times 10^{-8}$) are more frequently located on circRNAs than non-significant variants ($p \geq 5 \times 10^{-8}$). This enrichment was further evident for a GWAS for anorexia nervosa (AN) [54], a psychiatric disorder characterised, among other symptoms, by a markedly reduced body weight [55, 56]. Additionally, an in vitro analysis of a BMI-associated SNP (rs4752856) located on hsa_circ_0022025 was performed in individuals heterozygous for this SNP, revealing increased levels of hsa_circ_0022025 derived from the BMI-increasing allele of rs4752856 [54].

Several studies analysed circRNAs related to white adipose tissue (WAT) browning, a reversible process in which beige adipocytes are formed upon a prolonged cold exposure [57, 58]. For instance, circNrxn2 was reported to facilitate this process and further act as a miR-103 sponge eventually targeting *Fgf10* in murine adipose tissue. In HFD-fed mice housed at 4 °C for 12 h, this cold stimulation led to a reduced size of adipocytes and an upregulation of circNrxn2 and *Fgf10*, while miR-103 levels were diminished. Besides, circNrxn2 facilitated mRNA and protein levels of UCP1, hinting at a circNrxn2-induced WAT browning. The identified axis of circNrxn2-miR-103-*Fgf10* also promoted M2 macrophage polarisation. This as well as browning of WAT were facilitated by circNrxn2's activation of the PPARγ signalling pathway [43]. Similarly, several circRNAs detected to be differentially expressed in adipose tissue of human and mice revealed a more than two-fold upregulation during adipogenesis in murine WAT under HFD conditions ($n = 10$; $p < 0.05$) [37]. Two murine circRNAs, circTshz2-1 and circArhgap5-2, and the human homologue of the latter, circARHGAP5-1, were found to be crucial regulators of adipocyte differentiation [37].

CircOgdh, which is highly expressed in murine brown adipose tissue (BAT), exhibited a distinct expression during adipocyte differentiation [11]. Here, an upregulation at early stages and a subsequent decrease on day 6 of differentiation were reported. Overexpression of circOgdh facilitated adipocyte proliferation, while reducing lipid deposition in brown adipocytes. Functionally, circOgdh binds miR-34a-5p regulating the expression of a key lipolysis gene, *Atgl*, enhancing brown adipocyte lipolysis and inhibiting lipid droplet accumulation [11].

A recent analysis initially focused on the role of miR-24-3p in obesity [45]. Hereby, an overexpression of this miRNA prevented weight gain in an HFD-induced obesity mouse model by reducing lipid accumulation. In a subsequent investigation of circRNAs, mmu_circ_0001874 was detected to sponge miR-24-3p targeting *Igf2*. A knockdown of this respective circRNA resulted in a diminished TG accumulation in differentiated adipocytes as well as hepatocytes and led to anti-inflammatory effects both on fat and liver tissue in vivo. Mmu_circ_0001874 further binds to *Igf2bp2* and thus interacts indirectly with *Ucp1*, a crucial thermogenic regulator [45].

Further, one circRNA (circARF3) implicated in obesity-related inflammation has been investigated through analyses of miR-103 which was upregulated in adipose tissue of diet-induced obesity (DIO) mice [44]. MiR-103 is a positive regulator of inflammation in adipose tissue by regulating TRAF3. An overexpression of this gene attenuated the miR-103-induced inflammation by facilitating mitophagy. CircARF3 was found to block the effects of miR-103 on TRAF3 and thus inhibits adipose tissue inflammation [44].

Obesity as well as HFD in mice can impair cognitive functions [59]. One study detected no histological changes in the hippocampus [41], a relevant brain structure for memory [59], but observed 56 dysregulated circRNAs [41]. One specific circRNA, circTshz2-2, was neuron-specific and regulated the neuronal cell cycle and spatial memory in the brain. Binding to YY1 transcriptional complex suppressed the transcription of *Bdnf*, inducing proteins related to cell cycle and shortened neurite length [60].

3 Metabolic Syndrome

Originally known as syndrome X [61], the MetS represents a cluster of risk factors including abdominal obesity, hypertension and dyslipidaemia [62, 63]. To date, multiple definitions are in use, and no universally accepted criteria are available [62]. Depending on the applied definition, prevalence rates can vary substantially. Accordingly, between 12.5% and 31.4% of the global population are affected [64]. Since the 1980s up to 2012, the prevalence of MetS has increased more than 35% in adults from the United States [63]. Even 3% and 5% of children and adolescents have MetS, respectively [65].

Multiple comorbidities have been linked to MetS. For instance, patients with MetS have an elevated risk of cardiovascular diseases as well as a generally higher risk for a premature death [66].

The pathophysiology of MetS is still not fully elucidated. Yet, besides genetic and epigenetic factors, insulin resistance (IR), chronic inflammatory states and neurohormonal activation are known to be major driver for MetS development [63].

3.1 CircRNAs in MetS

Unlike for other metabolic diseases, less is known about the relevance of circRNAs in MetS (see Table 20.2). So far, only two circRNAs (circRNF111 and hsa_circH19) have been reported to be relevant. For instance, circRNF111 was downregulated in serum and urine of patients with MetS ($n = 40$) compared to healthy individuals ($n = 40$) [67]. These levels of circRNF111 were further associated with IR and TG content in urine and serum (urine—IR: $r = -0.23$, $p = 0.039$; urine—TG: $r = -0.29$, $p = 0.01$; serum—IR: $r = -0.28$, $p = 0.012$; serum—TG: $r = -0.29$, $p = 0.01$). Studying human adipose-derived stem cells (hADSCs) demonstrated that a knockdown of circRNF111 inhibited glucose uptake as well as phosphorylation of Akt, while concomitantly increasing lipid droplet deposition. Administration of sh-circRNF111 in HFD-fed mice suppressed circRNF111's expression and resulted in a decreased glucose tolerance. Mechanistically, circRNF111 sponges miR-143-3p, which targets *IGF2R,* and thereby alleviating sensitivity to insulin and lipid deposition. These findings hint at a protective role of circRNF111 in the progression of MetS [67].

The other circRNA implicated in MetS, hsa_circH19, was upregulated in serum samples of patients with MetS ($n = 50$) compared to non-MetS samples ($n = 50$). Binary regression analysis indicated hsa_circH19's potential as a risk factor for MetS independent of sex, age or BMI. Knockdown of hsa_circH19 led to the activation of lipogenesis as the expression of genes like *PPARγ* and *SREBP1c* was upregulated upon circular knockdown. Bioinformatic analyses revealed multiple bindings sites of hsa_circH19 with PTBP1, which facilitates cleavage and nuclear translocation of SREBP1 [68].

Table 20.2 CircRNAs implicated in MetS

CircRNA	Reported functionalities	References
circRNF111	Affects insulin resistance and lipid deposition via miR-143-3p/*IGF2R* axis	[67]
hsa_circH19	Involved in differentiation of hADSCs by interacting with PTBP1	[68]

hADSCs human adipose-derived stem cells

4 Non-Alcoholic Fatty Liver Disease

NAFLD is the most common cause of chronic liver diseases and is characterised by hepatic fat accumulation, which is not due to factors like excessive alcohol consumption [69, 70]. Globally, prevalence rates of NAFLD have risen substantially in the last decades. Approximately 1 billion individuals are affected, representing about 25% of the global population [71]. Diagnosis is typically based on histological examination and the presence of steatosis in more than 5% of hepatocytes [70]. Yet, the term NAFLD covers a wide spectrum of conditions ranging from the rather benign non-alcoholic fatty liver (NAFL) to the severe form of non-alcoholic steatohepatitis (NASH). The latter can eventually progress into advanced liver diseases such as cirrhosis or hepatocellular carcinoma [69, 70].

The pathogenesis of NAFLD is based on two hypotheses. The traditional 'two-hit' hypothesis has been replaced by the 'multi-hit' hypothesis [72, 73]. Initially, the hepatic fat accumulation was attributed to a primary hit caused by factors like a sedentary lifestyle, HFD, obesity or IR, rendering the liver more susceptible to additional impairments. Secondary hits might, for instance, be oxidative stress or mitochondrial dysfunction [72, 73]. Conversely, the 'multi-hit' hypothesis considers the collective impact of various factors and processes encompassing genetic predisposition, various stressors, epigenetic effects and dietary factors [72, 73].

4.1 CircRNAs in Non-Alcoholic Fatty Liver Disease

In recent years, numerous studies have explored circRNAs in the context of NAFLD (see Table 20.3). Similar to the other metabolic diseases, expression profiles of circRNAs in various cell lines and murine tissues [74–85] as well as samples of patients with NAFLD or even NASH [77, 80, 86–88] have been identified.

One of the initial studies, conducted in 2017, examined a hepatic steatosis cell model using the human hepatocarcinoma cell line HepG2 revealing 357 dysregulated circRNAs [74]. Hereby, the circRNA_021412/miR-1972/*LPIN1* cascade was detected. LPIN1, a coactivator of transcription factors like PPARα, highlighted the potential relevance of this cascade to hepatic steatosis [74]. Another circRNA, namely circRNA_0046367, was also investigated in a similar hepatic steatosis cell model of HepG2 cells. Hereby, circRNA_0046367, which was acting as a miR-34a sponge, exhibited a loss of expression ($p < 0.01$), while concomitantly being a pivotal regulator of *PPARα*, ultimately ameliorating hepatic steatosis [86]. In samples of patients with NAFLD ($n = 5$), a hepatic downregulation of circRNA_0046367 and *PPARα* as well as increased level of miR-34a ($p < 0.01$) in comparison to non-steatosis controls ($n = 3$) were reported. MiR-34a was further targeted by an additional NAFLD-linked circRNA (circRNA_0046366) which regulated *PPARα* expression and restored lipid metabolism [75]. Several other circRNAs, including circScd1 [78], circLDLR [82], circPI4KB [83], circPTK2 [76], circ_0048179 [81] and circRNA_0001805 [77], were also reported to be relevant for lipid droplet formation and accumulation, indicating a potential function in modulating NAFLD development. For instance, overexpression of circLDLR reduced TG and cholesterol levels as well as mitigating accumulation of lipid droplets in hepatocytes treated with oleic acid/palmitic acid to establish a NAFLD cell model [82]. Binding to miR-667-5p regulated the expression of *SIRT1*. In turn, a knockdown of *SIRT1* induced the development of NAFLD [82]. CircPI4KB, a circRNA downregulated in liver tissues, and upregulated in serum of mice fed a western diet, was found to function as a miR-122 sponge. Overexpression of circPI4KB reduced intra-hepatocyte miR-122 level suppressing its lipid deposition protective effect [83].

By studying the roles of circ_0057558 and miR-206 in NAFLD using an in vitro model with long-chain free fatty acid (FFA) treated liver cells and an HFD in vivo model, it was determined that circ_0057558 binds miR-206 activating ROCK1/AMPK signalling which in turn contributes to the suppression of lipogenesis and TG secretion. Injections of miR-206 mimics ameliorated NAFLD progression in vivo [84].

Yet, only few studies have investigated the role of circRNAs in the more severe form of NAFLD, NASH. For instance, the expression profile of circRNAs in primary fibroblasts either isolated from patients with NASH cirrhosis ($n = 4$) or individuals without NAFLD ($n = 4$) revealed 15 up- and 11 downregulated circRNAs in NASH fibroblasts, respectively [80]. Downregulated circRNAs, like the steatohepatitis-associated circRNA ATP5B Regulator (SCAR), generally derived from the mitochondrial genome. SCAR acts via PGC-1α, binds to ATP5B and suppresses mitochondrial ROS output and activation of fibroblasts. When comparing clinical samples of 31 patients with simple steatosis, 28 patients with NASH but with no or just mild fibrosis, 13 patients with NASH cirrhosis and 34 individuals without NAFLD, an SCAR expression decline from simple steatosis to NASH cirrhosis was detected [80].

Another circRNA, circRNA_002581, was investigated in an FFA-induced NASH-like cell model as well as methionine and choline deficiency diet-induced NASH mouse model [79]. Overexpression of circRNA_002581 revealed an alleviation of the effects of miR-122 on its target, *CPEB1*, which is part of the CPEB1-PREN-AMPK-mTOR pathway associated with autophagy. In contrast, a knockdown of circRNA_002581 diminished lipid droplet accumulation and reduced levels of pro-inflammatory cytokines [79].

Given that the characteristics of metabolic diseases are predominantly overlapping, one study

Table 20.3 Circular RNAs involved in NAFLD

CircRNAs	Reported functionalities	References
circRNA_021412	Involved in hepatic steatosis via miR-1972/*LPIN1* axis	[74]
circRNA_0046367	Involved in hepatic steatosis via miR-34a/*PPARα*	[86]
circRNA_0046366	Restores hepatic steatosis via miR-34a/*PPARα*	[75]
circScd1	Affects degree of steatosis in NAFLD and facilitates fatty liver disease via JAK2/STAT5 signalling	[78]
hsa_circ_0048179	Suppresses lipid deposition via elevating *GPX4* activation through miR-188-3p sponging	[81]
SCAR	Attenuates mitochondrial ROS levels and fibroblast activation	[80]
circRNA_002581	Involved in NASH pathogenesis by regulating autophagy	[79]
circ_0057558	Promotes lipogenesis and TG secretion via miR-206/ROCK1/AMPK axis	[84]
circRNA_0001805	Mitigates NAFLD progression by targeting miR-106a-5p/miR-320a and *ABCA1*/*CPT1*	[77]
circPTK2	Involved in NAFLD development via miR-200c/SIK2/PI3K/Akt axis	[76]
circRNA608	Mediates mitophagy of hepatic stellate cells via miR-222/PINK1 axis	[85]
circLDLR	Regulates autophagy via miR-667-5p/*SIRT1* axis	[82]
circPI4KB	Binds miR-122 and diminishes its protective effects on lipid deposition	[83]
hsa_circ_0004535	Improves lipid droplet fusion and cell morphology via miR-1827/CASP8 axis	[87]

NAFLD non-alcoholic fatty liver disease, *NASH* non-alcoholic steatohepatitis, *ROS* reactive oxygen species, *TG* triglycerides analysed circRNAs in patients with T2DM and NAFLD [87]. RNA-sequencing of three controls, three patients with T2DM and three patients with T2DM and NAFLD, revealed 62 and 75 differentially expressed circRNAs in patients with T2DM and patients with T2DM and NAFLD, respectively. Ten circRNAs were dysregulated in both groups of patients with T2DM (with and without NAFLD). Via network construction and analyses, CASP8 and hsa_circ_0004535 were identified as candidates for further investigations. In comparison to controls, both were downregulated in blood of patients with T2DM regardless of NAFLD diagnosis. Hsa_circ_0004535 binds to CASP8 and miR-1827. Overexpression of hsa_circ_0004535 and CASP8 improved cell structure and lipid droplet fusion [87].

5 Diabetes Mellitus

Hyperglycaemia caused by alterations in insulin secretion and action characterises diabetes mellitus, which may ultimately have adverse effects on heart, vasculature, eyes, kidneys and nerves [89]. Two types of DM are distinguished: type 1 diabetes mellitus (T1DM) and T2DM. While T1DM is casued by β-cell destruction, typically with a predominant onset in childhood and adolescence, T2DM is triggered by a defect in insulin secretion alongside IR [90, 91]. In 2021, 529 million individuals worldwide had diabetes, with T2DM, accounting for about 96% of cases [90, 92]. Of note, the prevalence of T2DM is slightly higher in males than in females, with men developing T2DM at a younger age and lower BMI [92, 93]. Projections hint that by 2050, approximately 1.31 billion people will be living with either type of diabetes [92].

Hypertension, hyperlipidaemia, obesity as well as cardiovascular diseases are the predominant reported diabetic comorbidities [94, 95]. About 30% of patients with T2DM present with three or more comorbidities at the time of diagnosis. Ten years later, up to 60% will have at least three comorbidities [94]. A substantial proportion, up to 65%, of deaths in individuals with diabetes are linked to cardiovascular diseases [91]. More than half of the reported disability-adjusted life years (DALY) were attributed to an elevated BMI [92]. Overall, diabetes contributed to 1.5 million deaths in 2019, with 48% of these occurring before the age of 70 [96].

Pathogenesis of DM is based on a dysregulation of the interplay between insulin action and secretion resulting in elevated blood glucose levels. In early stages of diabetes, decreased insulin sensitivity causes the β-cells to hyperfunction to prevent hyperglycaemia. Over time, however, this increased insulin secretion is unable to counteract the reduced sensitivity. Further, β-cell function is reduced, causing a deficiency in insulin and eventually leading to hyperglycaemia [97].

5.1 CircRNAs in Diabetes Mellitus

The pathogenesis of DM is predominantly characterised by a loss or dysfunction of β-cells causing insufficient insulin secretion [89–91]. CircRNAs have been previously reported to regulate β-cell function and proliferation as well as insulin secretion (see Tables 20.4 and 20.5). One of the first respective studies identified numerous abundant circRNAs in human islet cells, with many of those being conserved in murine islet cells [98]. In this cell type of diabetic mice deficient for the leptin receptor (*db/db*), expression levels of circHIPK3 and ciRS-7/CDR1as were found to be reduced. A mimicking of this decreased expression in islets cells of wild-type animals led to impaired insulin secretion and reduced β-cell proliferation [98]. CiRS-7/CDR1as is known to be a potent miR-7 sponge [10]. In contrast, circHIPK3 binds to miR-124-3p and miR-338-3p in MIN6B1 cells, thereby regulating the expression of genes pivotal for β-cells like *Slc2a2, Akt1* and *Mtpn* [98]. Both circRNAs can serve as effective biomarkers, with circHIPK3 distinguishing between patients with T2DM and controls (AUC = 0.665, 95% CI = [0.589, 0.752], p = 0.001) and between patients with T2DM and pre-diabetic patients with a sensitivity of 50% and specificity of 75% (AUC = 0.613, 95% CI = [0.524, 0.697],

Table 20.4 CircRNAs involved in type 2 diabetes mellitus

circRNA	Reported functionalities	References
ciRS-7/CDR1as	Affects insulin secretion and content in islet cells	[123]
	Affects insulin secretion and β-cell proliferation	[98]
	Increased expression in pre-diabetic patients compared to controls	[99]
circHIPK3	Involved in insulin secretion, β-cell proliferation and apoptosis	[98]
	Increased expression in patients with T2DM compared to controls and pre-diabetic patients	[99]
	Regulates insulin resistance and hyperglycaemia via miR-192-5p/*FOXO1* axis	[100]
circANKRD36	Expression positively correlated with glucose and glycosylated haemoglobin	[101]
	Involved in insulin resistance and inflammation via miR-145/XBP1 axis	[120]
hsa_ circ_0068087	Involved in inflammation via miR-192/TLR4/NF-κB/NLRP3 axis	[111]
circ_0009024	Negatively correlated with glucose, glycosylated haemoglobin, IL-6, and TNF-α	[104]
hsa_circ_ CCNB1	Positively correlated with glucose, glycosylated haemoglobin, IL-6, and TNF-α	[104
circ-Tulp4	Affects β-cell proliferation via miR-7222-3p/*Soat1/Cyclin D1* axis	[115]]
circCIRBP	Reduced expression in islet cells of individuals with T2DM compared to controls	[103]
circCAMSAP1	Higher expression in islet cells of individuals with T2DM and lower expression in blood of patients with overt diabetes compared to controls	[103]
circRPH3AL	Higher expression in islet cells of individuals with T2DM compared to controls	[103]
circZKSCAN1	Higher expression in islet cells of individuals with T2DM compared to controls	[103]
ci-Ins2	Regulates transcription of insulin secretion-related genes by interacting with RBP TDP-43	[116]
circBPTF	Knockdown protects HUVECs against high glucose induced inflammatory injuries and oxidative stress via miR-384/LIN28B axis	[112]
hsa_ circ_0054633	Decreased in patients with T2DM receiving insulin treatment, negatively correlated with C peptide and positively correlated with IL-17 and TNF-α	[109]
	Increased in patients with T2DM and expression was correlated with fasting blood glucose and haemoglobin A1c	[108]
	Highly expressed in serum of patients with DM and further affects cell proliferation, apoptosis and insulin secretion in HG-treated NES2Y cells	[124]
	Expression increased in patients with T2DM compared to pre-diabetic individuals and controls	[102]
circLRP6	Involved in HG-induced proliferation, oxidative stress and inflammation in mesangial cells via miR-205/HMGB1 axis and activation of TLR4/ NF-κB pathway	[117]
	Involved in cell viability, apoptosis, and insulin secretion via miR-9-5p/PRMT1 axis	[118]
circ_0000712	High expression in *db/db* diabetic nephropathy mice and involved in apoptosis, inflammation, oxidative stress and fibrosis	[122]
Hsa_ circ_0071336	Downregulated in patients with T2DM compared to controls and acts as a miR-93-5p sponge regulating *GLUT4*	[110]
Hsa_ circ_0115355	Reduced expression in patients with T2DM compared to controls and acts via miR-145/*SIRT1* axis	[105]
circPIP5K1A	Upregulated in patients with T2DM compared to controls and regulates inflammation, oxidative damage and apoptosis in INS-1 β-cells	[106]
	Upregulated in T2DM rats, involved in IR and inhibits inflammation via miR-552-3p/ENO1 axis	[107]
circ_0000491	Involved in apoptosis, inflammation, oxidative stress and fibrosis in an in vitro model via miR-455-3p/Hmgb1 axis	[121]
circGlis3	Increased expression in islet cells of obese male mice, involved in insulin secretion and obesity-induced apoptosis via miR-124-3p/*NeuroD1/Creb1* axis	[113]
	Involved in β-cell dysfunction, acts via hnRNPF/*Sirt1* axis and encodes protein regulating GLIS3's transcriptional activity	[114]

HG high glucose, *HUVEC* human umbilical vein endothelial cells, *RBP* RNA-binding protein, *TDP-43* TAR DNA-binding protein 43 kDa

Table 20.5 CircRNAs involved in type 1 diabetes mellitus

circRNA	Reported functionalities	References
circPPM1F	Upregulated in children with T1DM and regulates M1 macrophage activation	[125]
hsa_circ_0060450	Upregulated in patients with T1DM and involved in inflammation via miR-199-5p/SHP2 axis	[126]

p = 0.023) [99]. Further, ciRS-7/CDR1as was expressed in higher levels in pre-diabetic individuals (n = 60) compared to controls (n = 69, p = 0.004) and was positively correlated with the BMI (r = 0.216, p = 0.002). Positive correlations of circHIPK3 and ciRS-7/CDR1as with fasting blood glucose (r = 0.22, p = 0.002; r = 0.18, p = 0.011) in patients with T2DM (n = 70) have been reported as well [99]. Additionally, circHIPK3 may act via the miR-192-5p/*FOXO1* axis contributing to IR and hyperglycaemia in HepG2 cells [100].

Over the last years, several studies have investigated expression profiles of circRNAs in patients with T2DM yielding in candidate circular transcripts that might be relevant for the disease [101–103]. For instance, circular expression profiles of five patients with T2DM revealed 220 differentially expressed circular transcripts. Of these, circANKRD36 was the most upregulated circRNA. In patients with T2DM (n = 38), its expression level was higher than in 40 healthy controls (p = 0.02). Further, circANKRD36' expression levels correlated positively with glucose (r = 0.325, p = 0.0047), glycosylated haemoglobin (r = 0.3171, p = 0.0056) and IL-6 (r = 0.393, p = 0.031) [101]. Another study analysed the expression profiles of circRNAs in pooled samples of five human islets [103]. In total, 2619 circRNAs were found to be expressed with the highest expressions found for circCAMSAP1, circCIRBP, circRPH3AL, circRHOBTB3 and circZKSCAN1. Analysing a confirmation study group of 50 controls and 20 additional patients with T2DM, three (circCAMSAP1, circCIRBP and circRPH3AL) of these five circRNAs were found to be upregulated following correction for multiple testing (p < 0.001). A nominal association of circZKSCAN1 with T2DM status was detected (p = 0.03). In a human ß-cell line (EndoC-ßH1), the expression of circCIRBP and circRPH3AL changed following a palmitate treatment. CircCIRBP's expression increased (p = 0.021), whereas the expression of circRPHAL3 decreased after treatment (p = 0.022). Analysing the five most highly expressed circRNAs in peripheral blood of 133 non-diabetic controls, 106 samples of patients with overt diabetes and 46 patients with impaired glucose tolerance revealed a nominal association of circCAMSAP1 (p < 0.05) with diabetes status [103].

Additional analyses identified circRNAs that may correlate with diabetes status like circ_CCNB1 and hsa_circ_0009024 [104], hsa_circ_0115355 [105], circPIP5K1A [106, 107], hsa_circ_0054633 [102, 108, 109] as well as hsa_circ_0071336 [110]. For instance, in blood of patients with T2DM (n=30), hsa_circ_0071336 was downregulated when compared to healthy controls (n = 30) [110]. This circRNA allowed the discrimination between patients with T2DM (n = 112) and those with an impaired fasting glucose (n = 98) from controls (n = 103), as a low expression was an indicator for T2DM, and IR. Hsa_circ_0071336 functions through miR-93-5p/*GLUT4* axis [110]. Further, in serum of patients with T2DM (n = 58), circPIP5K1A showed increased expression levels, while miR-552-3p was downregulated. Suppression of circPIP5K1A caused decreased inflammatory rate, oxidative damage, and apoptosis of glycolipid toxicity-stimulated INS-1E cells [106]. Similarly, circPIP5K1A downregulation led to reduced glucose and lipid levels, and suppressed inflammation in T2DM rats [107].

Additionally, hsa_circ_0068087 was reported to be upregulated in patients with DM [102] as well as in human umbilical vein endothelial cells (HUVEC) following a glucose increase [111]. A knockdown of hsa_circ_0068087 caused a suppression of inflammation and HUVEC dysfunction due to high glucose (HG) levels by inhibiting the TLR4/NF-κB/NLRP3 inflammasome signalling pathway [111]. Similarly, circBPTF was upregulated in HG-induced HUVECs. A knockdown of this circRNA resulted in an elevated cell viability and suppressed cell apoptosis, as well as the release of proinflammatory cytokines and oxidative stress. Mechanistically, circBPTF targets miR-384, which in turn affects LIN28B [112].

By profiling circRNA expressions in mouse models, circGlis3 was the most abundant circRNAs in islet cells of HFD-fed and leptin-deficient $LEP^{ob/ob}$ mice [113]. It was upregulated in the pancreas of HFD-fed mice in comparison to mice fed a normal chow diet ($p < 0.0001$), while the respective mRNA ($p = 0.0239$) and protein ($p < 0.0001$) levels were downregulated. In serum of patients with impaired glucose tolerance ($n = 21$), dynamic expression levels were determined. Interestingly, circGlis3's expression was elevated in serum of patients with obesity and moderate diabetes having compensated ß-cell function. In the decompensation stage, circGlis3 levels were reduced. Additionally, the expression of circGlis3 was positively correlated with the BMI (<30 kg/m^2). Mechanistically, circGlis3 upregulated the expression of *NeuroD1* and *Creb1*, in turn promoting insulin secretion by sponging miR-124-3p. It further decreases apoptosis by interacting with SCOTIN, a pro-apoptotic factor [113]. Another study showed that circGlis3's biogenesis was regulated by CELF1 and revealed circGlis3's dual mechanism to impair β-cell function. Firstly, binding of circGlis3 to hnRNPF leads to the suppression of hnRNPF's nuclear translocation, which in turn inhibits *Sirt1* transcription. Secondly, circGlis3 encodes a 348 amino acid protein interacting with GLIS3 itself suppressing its transcriptional activity [114].

One highly expressed circRNA in murine pancreatic islet cells is circ-Tulp4 [115]. It is regulated by non-esterified fatty acid (NEFA) in Min6 cells indicating a potential role of circ-Tul4 in β-cell lipotoxicity and development of T2DM. Circ-Tulp4 is crucial for the maintenance of β-cell function and mediated β-cell proliferation by targeting miR-7222-3p, which in turn affected *Soat1/Cyclin D1* [115].

The intronic circRNA ci-Ins2 showed a diminished expression in both, the diabetic and severely obese *db/db* mice as well as the diabetic but non-obese Goto-Kakizaki rats, whereas no expression change was observed in the leptin deficient $LEP^{ob/ob}$ mice [116]. By interacting with the RNA-binding protein TAR DNA-binding protein 43 kDa (TDP-43), ci-Ins2 regulated the transcription of key genes for insulin secretion [116]. Analyses of in vitro models revealed that silencing circLRP6, which was previously associated with cell injury induced by high glucose levels and reported to sponge miR-205 [117], caused an increased cell viability and insulin secretion, while decreasing apoptosis via circLRP6/miR-9-5p/PRMT1 axis [118].

Besides their role in DM pathogenesis, circRNAs have been implicated in IR [119]. For instance, in rats with T2DM, high levels of circANKRD36 were detected [120]. Following a depletion of circANKRD36, blood glucose and insulin levels decreased, while adiponectin levels increased improving IR. Further, silencing of circANKRD36 led to diminished inflammation and cell apoptosis in pancreatic tissue of T2DM rats. Mechanistically, circANKRD36 binds miR-145, eventually regulating *XBP1* expression [120].

In vitro and in vivo analyses revealed that circ_0000491 binds miR-455-3p, thereby being an indirect regulator of*Hmgb1*. A circRNA knockdown led to the inhibition of miR-445-3p's protective role against cell injury and caused HG-induced apoptosis, inflammation, oxidative stress and fibrosis in vitro [121]. Further, circ_0000712 knockdown inhibited cell injury by targeting miR-879-5p which subsequently mediates expression of SOX6 [122].

Studies analysing circRNAs in regard to T1DM are scarce (see Table 20.5). Yet, a circRNA (circPPM1F), identified to be differentially expressed in a microarray analysis ($n = 4$ patients with T1DM and $n = 4$ controls), was confirmed to be upregulated in children with T1DM ($n = 43$; $p < 0.001$) [125]. CircPPM1F was positively correlated with IL-6 ($r = 0.643$, $p < 0.001$), IL-1ß ($r = 0.543$, $p < 0.001$) and TNF-α ($r = 0.4721$, $p < 0.01$) and found to be predominantly expressed in monocytes, but not T and B cells. As M1 macrophage genes, like *IL-1ß* and *TNF-α,* were downregulated and phosphorylation of p65 was reduced ($p < 0.05$) upon circPPM1F knockdown in THP1-derived macrophages, this circRNA was implied to enhance M1 macrophage activation via elevating NF-κB signalling. CircPPM1F's ability to facilitate M1 macrophage activation was based on its suppression of PPM1F translation. Its macrophage activation was able to induce pancreatic ß-cell apoptosis via the MAPK pathway in vitro. Similarly, circPPM1F increased pancreatic injury through the activation of MAPK and STAT3 pathways in mice. The expression of circPPM1F was found to be regulated by the RNA-binding proteins EIF4A3 and FUS [125]. Additionally, hsa_circ_0060450, deriving from the *MYBL2* gene, was found to be differentially expressed in patients with T1DM ($n = 3$) by analysing expression profiles using a microarray. It could sponge miR-199a-5p, enhancing its target gene *SHP2*'s expression and consequently inhibits JAK-STAT pathway leading to the inhibition of macrophage-mediated inflammation [126].

6 Limitations of Studies Analysing CircRNAs in Metabolic Diseases

In recent years, numerous studies have focussed on circRNAs in the context of metabolic diseases. However, caution is warranted when interpreting these findings. The small sample sizes, lack of replication, the neglect of *p*-value correction for multiple testing as well as the analysis of homogenous populations raise questions regarding the robustness of the reported results. As these limitations are inherent in many studies [51, 75, 99, 109], it is important to critically review published studies and the reported findings.

A common limitation across these studies is the insufficient demographic data provided for human participants, particularly in terms of sex. Information about the number of female and male participants is sporadically given and sometimes moved to the supplementary materials. Moreover, studies examining expression profiles generally offer phenotypic or demographic data only for the overall study group. Solely, Fang et al. [101] included phenotypic data, including sex, specifically for the samples analysed within RNA-sequencing. Further, despite certain studies providing information on the inclusion of both sexes, sex-specific analyses are rarely performed [e.g. 40, 51, 67, 86, 103]. Only one study explored the expression levels of a circRNA (hsa_circ_0054633) in a sex-specific manner. Even though no sex-specific expression was determined [102], this analysis offers novel insights into the analysed circRNA and highlights the significance of considering these factors.

The reported findings primarily pertain to populations recruited in China, raising uncertainties regarding the replicability of the results in other populations or potential variations in expression profiles. Further, studies on circRNAs in obesity, often use varying BMI threshold to define obesity. For instance, one study defined obesity if the BMI exceeded 32.5 kg/m^2 [38], while others used a BMI cut-off of 28 kg/m^2 [39, 40]. Methodological flaws extend to animal and cell culture studies, with frequent omission of the number of animals included, a predominant focus on male mice and a lack of information regarding the sex of the analysed cell lines [e.g. 75, 100, 112].

The widespread neglect in correcting for multiple testing amplifies the risk of false-positive results. Only few studies, such as one by Haque and colleagues [103], distinguished between nominal and corrected *p*-values. Exploration of correlations between circRNA expression and various disease-related parameters predominantly appears to be weak to moderate, emphasising the need for larger study groups to validate and strengthen the detected associations.

For the identification of circRNAs, several studies apply diverse algorithms, with only few studies using multiple algorithms to minimise the risk of false positives. Adopting multiple algorithms aligns with recent best practices in circRNA research [127].

Addressing these widespread issues is crucial for advancing our understanding of circRNAs in metabolic diseases. Existing publications outlining best practices in circRNA research [127] should be employed when planning and conducting studies. Future research must strive for comprehensive demographic reporting, the analyses of diverse populations, sex-specific evaluation of findings, and a rigorous statistical analysis, including corrections for multiple testing. Overcoming these challenges will enhance reliability, validity, and clinical relevance of circRNA research, paving the way for its potential diagnostic and therapeutic applications in metabolic diseases.

7 Conclusion and Outlook

Since the recognition of circRNAs as functionally relevant RNA species, several studies have been conducted on their roles in metabolic diseases. Certain metabolic diseases, such as diabetes mellitus, have already been well studied, while the metabolic syndrome, for instance, still requires further intensive research in this regard. In general, several circRNAs have been reported to be relevant for metabolic diseases and might even act as biomarkers. Yet, since many studies refer to homogenous populations without comprehensive phenotypic data, often do not correct for the number of tests performed and a large proportion of these transcripts have only been reported by one study, caution is advised when interpreting individual data and studies. Thus, replication studies including larger and more diverse cohorts are urgently needed. Further, due to their specific expression in various tissues and cell types8, 9, it is important to use and analyse easily accessible biomaterials (e.g., blood) for biomarker identification, in addition to the already extensively studied disease-specific tissues, such as adipose tissue and β-cells. Particularly, as circRNAs are repeatedly associated with therapeutic applications [128, 129], ongoing investigation of circRNAs in metabolic diseases is needed.

References

1. Clemente-Suarez VJ, Martin-Rodriguez A, Redondo-Florez L, Lopez-Mora C, Yanez-Sepulveda R, Tornero-Aguilera JF (2023) New insights and potential therapeutic interventions in metabolic diseases. Int J Mol Sci 24(13):10672. https://doi.org/10.3390/ijms241310672
2. Chew NWS, Ng CH, Tan DJH, Kong G, Lin C, Chin YH, Lim WH, Huang DQ, Quek J, Fu CE, Xiao J, Syn N, Foo R, Khoo CM, Wang JW, Dimitriadis GK, Young DY, Siddiqui MS, Lam CSP, Wang Y, Figtree GA, Chan MY, Cummings DE, Noureddin M, Wong VW, Ma RCW, Mantzoros CS, Sanyal A, Muthiah MD (2023) The global burden of metabolic disease: data from 2000 to 2019. Cell Metab 35(3):414–428.e413. https://doi.org/10.1016/j.cmet.2023.02.003
3. Wu YL, Li HF, Chen HH, Lin H (2022) Emergent roles of circular RNAs in metabolism and metabolic disorders. Int J Mol Sci 23(3):1032. https://doi.org/10.3390/ijms23031032
4. Dandare A, Khan MJ, Naeem A, Liaquat A (2023) Clinical relevance of circulating non-coding RNAs in metabolic diseases: emphasis on obesity, diabetes, cardiovascular diseases and metabolic syndrome. Genes Dis 10(6):2393–2413. https://doi.org/10.1016/j.gendis.2022.05.022
5. Sanger HL, Klotz G, Riesner D, Gross HJ, Kleinschmidt AK (1976) Viroids are single-stranded covalently closed circular RNA molecules existing as highly base-paired rod-like structures. Proc Natl Acad Sci USA 73(11):3852–3856. https://doi.org/10.1073/pnas.73.11.3852
6. Barrett SP, Salzman J (2016) Circular RNAs: analysis, expression and potential functions. Development 143(11):1838–1847. https://doi.org/10.1242/dev.128074
7. Salzman J, Gawad C, Wang PL, Lacayo N, Brown PO (2012) Circular RNAs are the predominant transcript isoform from hundreds of human genes in diverse cell types. PLoS One 7(2):e30733. https://doi.org/10.1371/journal.pone.0030733
8. Memczak S, Jens M, Elefsinioti A, Torti F, Krueger J, Rybak A, Maier L, Mackowiak SD, Gregersen LH, Munschauer M, Loewer A, Ziebold U, Landthaler M, Kocks C, le Noble F, Rajewsky N (2013) Circular RNAs are a large class of animal RNAs with regulatory potency. Nature 495(7441):333–338. https://doi.org/10.1038/nature11928
9. Salzman J, Chen RE, Olsen MN, Wang PL, Brown PO (2013) Cell-type specific features of circular RNA expression. PLoS Genet 9(9):e1003777. https://doi.org/10.1371/journal.pgen.1003777
10. Hansen TB, Jensen TI, Clausen BH, Bramsen JB, Finsen B, Damgaard CK, Kjems J (2013) Natural RNA circles function as efficient microRNA sponges. Nature 495(7441):384–388. https://doi.org/10.1038/nature11993

11. Liu K, Liu X, Deng Y, Li Z, Tang A (2022) CircRNA-mediated regulation of brown adipose tissue adipogenesis. Front Nutr 9:926024. https://doi.org/10.3389/fnut.2022.926024
12. Piwecka M, Glazar P, Hernandez-Miranda LR, Memczak S, Wolf SA, Rybak-Wolf A, Filipchyk A, Klironomos F, Cerda Jara CA, Fenske P, Trimbuch T, Zywitza V, Plass M, Schreyer L, Ayoub S, Kocks C, Kuhn R, Rosenmund C, Birchmeier C, Rajewsky N (2017) Loss of a mammalian circular RNA locus causes miRNA deregulation and affects brain function. Science 357(6357):eaam8526. https://doi.org/10.1126/science.aam8526
13. Li Z, Huang C, Bao C, Chen L, Lin M, Wang X, Zhong G, Yu B, Hu W, Dai L, Zhu P, Chang Z, Wu Q, Zhao Y, Jia Y, Xu P, Liu H, Shan G (2015) Exon-intron circular RNAs regulate transcription in the nucleus. Nat Struct Mol Biol 22(3):256–264. https://doi.org/10.1038/nsmb.2959
14. Jeck WR, Sorrentino JA, Wang K, Slevin MK, Burd CE, Liu J, Marzluff WF, Sharpless NE (2013) Circular RNAs are abundant, conserved, and associated with ALU repeats. RNA 19(2):141–157. https://doi.org/10.1261/rna.035667.112
15. Legnini I, Di Timoteo G, Rossi F, Morlando M, Briganti F, Sthandier O, Fatica A, Santini T, Andronache A, Wade M, Laneve P, Rajewsky N, Bozzoni I (2017) Circ-ZNF609 is a circular RNA that can be translated and functions in myogenesis. Mol Cell 66(1):22–37.e29. https://doi.org/10.1016/j.molcel.2017.02.017
16. Pamudurti NR, Bartok O, Jens M, Ashwal-Fluss R, Stottmeister C, Ruhe L, Hanan M, Wyler E, Perez-Hernandez D, Ramberger E, Shenzis S, Samson M, Dittmar G, Landthaler M, Chekulaeva M, Rajewsky N, Kadener S (2017) Translation of CircRNAs. Mol Cell 66(1):9–21.e27. https://doi.org/10.1016/j.molcel.2017.02.021
17. Yang Y, Fan X, Mao M, Song X, Wu P, Zhang Y, Jin Y, Yang Y, Chen LL, Wang Y, Wong CC, Xiao X, Wang Z (2017) Extensive translation of circular RNAs driven by N(6)-methyladenosine. Cell Res 27(5):626–641. https://doi.org/10.1038/cr.2017.31
18. World Health Organization W (2000) Obesity: preventing and managing the global epidemic: report of a WHO consultation, World Health Organization technical report series 894. World Health Organisation
19. Blüher M (2019) Obesity: global epidemiology and pathogenesis. Nat Rev Endocrinol 15(5):288–298. https://doi.org/10.1038/s41574-019-0176-8
20. World Health Organization W (n.d.) Obesity and overweight. https://www.who.int/news-room/fact-sheets/detail/obesity-and-overweight. Accessed 25 May 2023
21. World Obesity Federation (2023) World Obesity Atlas 2023. World Obesity Federation
22. Schnurr TM, Jakupovic H, Carrasquilla GD, Angquist L, Grarup N, Sorensen TIA, Tjonneland A, Overvad K, Pedersen O, Hansen T, Kilpelainen TO (2020) Obesity, unfavourable lifestyle and genetic risk of type 2 diabetes: a case-cohort study. Diabetologia 63(7):1324–1332. https://doi.org/10.1007/s00125-020-05140-5
23. Cercato C, Fonseca FA (2019) Cardiovascular risk and obesity. Diabetol Metab Syndr 11:74. https://doi.org/10.1186/s13098-019-0468-0
24. Powell-Wiley TM, Poirier P, Burke LE, Despres JP, Gordon-Larsen P, Lavie CJ, Lear SA, Ndumele CE, Neeland IJ, Sanders P, St-Onge MP, American Heart Association Council on L, Cardiometabolic H, Council on C, Stroke N, Council on Clinical C, Council on E, Prevention, Stroke C (2021) Obesity and cardiovascular disease: a scientific statement from the American Heart Association. Circulation 143(21):e984–e1010. https://doi.org/10.1161/CIR.0000000000000973
25. Avgerinos KI, Spyrou N, Mantzoros CS, Dalamaga M (2019) Obesity and cancer risk: emerging biological mechanisms and perspectives. Metabolism 92:121–135. https://doi.org/10.1016/j.metabol.2018.11.001
26. Iyengar NM, Gucalp A, Dannenberg AJ, Hudis CA (2016) Obesity and cancer mechanisms: tumor microenvironment and inflammation. J Clin Oncol 34(35):4270–4276. https://doi.org/10.1200/JCO.2016.67.4283
27. Petrelli F, Cortellini A, Indini A, Tomasello G, Ghidini M, Nigro O, Salati M, Dottorini L, Iaculli A, Varricchio A, Rampulla V, Barni S, Cabiddu M, Bossi A, Ghidini A, Zaniboni A (2021) Association of obesity with survival outcomes in patients with cancer: a systematic review and meta-analysis. JAMA Netw Open 4(3):e213520. https://doi.org/10.1001/jamanetworkopen.2021.3520
28. Berrington de Gonzalez A, Hartge P, Cerhan JR, Flint AJ, Hannan L, MacInnis RJ, Moore SC, Tobias GS, Anton-Culver H, Freeman LB, Beeson WL, Clipp SL, English DR, Folsom AR, Freedman DM, Giles G, Hakansson N, Henderson KD, Hoffman-Bolton J, Hoppin JA, Koenig KL, Lee IM, Linet MS, Park Y, Pocobelli G, Schatzkin A, Sesso HD, Weiderpass E, Willcox BJ, Wolk A, Zeleniuch-Jacquotte A, Willett WC, Thun MJ (2010) Body-mass index and mortality among 1.46 million white adults. N Engl J Med 363(23):2211–2219. https://doi.org/10.1056/NEJMoa1000367
29. Bhaskaran K, Dos-Santos-Silva I, Leon DA, Douglas IJ, Smeeth L (2018) Association of BMI with overall and cause-specific mortality: a population-based cohort study of 3.6 million adults in the UK. Lancet Diabetes Endocrinol 6(12):944–953. https://doi.org/10.1016/S2213-8587(18)30288-2
30. Chen C, Ye Y, Zhang Y, Pan XF, Pan A (2019) Weight change across adulthood in relation to all cause and cause specific mortality: prospective cohort study. BMJ 367:l5584. https://doi.org/10.1136/bmj.l5584
31. Fontaine KR, Redden DT, Wang C, Westfall AO, Allison DB (2003) Years of life lost due to obesity. JAMA 289(2):187–193. https://doi.org/10.1001/jama.289.2.187

32. Visaria A, Setoguchi S (2023) Body mass index and all-cause mortality in a 21st century U.S. population: a National Health Interview Survey analysis. PLoS One 18(7):e0287218. https://doi.org/10.1371/journal.pone.0287218
33. Hinney A, Korner A, Fischer-Posovszky P (2022) The promise of new anti-obesity therapies arising from knowledge of genetic obesity traits. Nat Rev Endocrinol 18(10):623–637. https://doi.org/10.1038/s41574-022-00716-0
34. Loos RJF, Yeo GSH (2021) The genetics of obesity: from discovery to biology. Nat Rev Genet 23:120–133. https://doi.org/10.1038/s41576-021-00414-z
35. Bouchard C (2021) Genetics of obesity: what we have learned over decades of research. Obesity (Silver Spring) 29(5):802–820. https://doi.org/10.1002/oby.23116
36. Yi X, Wan X, Khan MA, Sun X, Wang Z, Chen K, Peng L (2024) Expression analysis of circRNAs in human adipogenesis. Diabetes Metab Syndr Obes 17:45–54. https://doi.org/10.2147/DMSO.S381603
37. Arcinas C, Tan W, Fang W, Desai TP, Teh DCS, Degirmenci U, Xu D, Foo R, Sun L (2019) Adipose circular RNAs exhibit dynamic regulation in obesity and functional role in adipogenesis. Nat Metab 1(7):688–703. https://doi.org/10.1038/s42255-019-0078-z
38. Li G, Zhang H, Cao K, Zhai Z, Wang Z, Yang L, Han J (2022) Transcriptome of visceral adipose tissue identifies an inflammation-related ceRNA network that regulates obesity. Mol Cell Biochem 477(4):1095–1106. https://doi.org/10.1007/s11010-022-04362-y
39. Liu Y, Liu H, Li Y, Mao R, Yang H, Zhang Y, Zhang Y, Guo P, Zhan D, Zhang T (2020) Circular RNA SAMD4A controls adipogenesis in obesity through the miR-138-5p/EZH2 axis. Theranostics 10(10):4705–4719. https://doi.org/10.7150/thno.42417
40. Chen S, Song P, Wang Y, Wang Z, Xue J, Jiang Y, Zhou Y, Zhao J, Tang L (2023) CircMAPK9 promotes adipogenesis through modulating hsa-miR-1322/FTO axis in obesity. iScience 26(10):107756. https://doi.org/10.1016/j.isci.2023.107756
41. Niu Y, Chang P, Liu T, Shen X, Zhao H, Zhang M, Lei S, Chen B, Yu J (2022) Obese mice induced by high-fat diet have differential expression of circular RNAs involved in endoplasmic reticulum stress and neuronal synaptic plasticity of hippocampus leading to obesity-associated cognitive impairment. Front Mol Neurosci 15:1000482. https://doi.org/10.3389/fnmol.2022.1000482
42. Zhang X, Gu S, Shen S, Luo T, Zhao H, Liu S, Feng J, Yang M, Yi L, Fan Z, Liu Y, Han R (2023) Identification of circular RNA profiles in the liver of diet-induced obese mice and construction of the ceRNA network. Genes (Basel) 14(3). https://doi.org/10.3390/genes14030688
43. Zhang T, Zhang Z, Xia T, Liu C, Sun C (2019) circNrxn2 promoted WAT browning via sponging miR-103 to relieve its inhibition of FGF10 in HFD mice. Mol Ther Nucleic Acids 17:551–562. https://doi.org/10.1016/j.omtn.2019.06.019
44. Zhang Z, Zhang T, Feng R, Huang H, Xia T, Sun C (2019) circARF3 alleviates mitophagy-mediated inflammation by targeting miR-103/TRAF3 in mouse adipose tissue. Mol Ther Nucleic Acids 14:192–203. https://doi.org/10.1016/j.omtn.2018.11.014
45. Shao J, Wang M, Zhang A, Liu Z, Jiang G, Tang T, Wang J, Jia X, Lai S (2023) Interference of a mammalian circRNA regulates lipid metabolism reprogramming by targeting miR-24-3p/Igf2/PI3K-AKT-mTOR and Igf2bp2/Ucp1 axis. Cell Mol Life Sci 80(9):252. https://doi.org/10.1007/s00018-023-04899-1
46. Sun W, Sun X, Chu W, Yu S, Dong F, Xu G (2020) CircRNA expression profiles in human visceral preadipocytes and adipocytes. Mol Med Rep 21(2):815–821. https://doi.org/10.3892/mmr.2019.10886
47. Hinney A, Nguyen TT, Scherag A, Friedel S, Bronner G, Muller TD, Grallert H, Illig T, Wichmann HE, Rief W, Schafer H, Hebebrand J (2007) Genome wide association (GWA) study for early onset extreme obesity supports the role of fat mass and obesity associated gene (FTO) variants. PLoS One 2(12):e1361. https://doi.org/10.1371/journal.pone.0001361
48. Locke AE, Kahali B, Berndt SI, Justice AE, Pers TH, Day FR, Powell C, Vedantam S, Buchkovich ML, Yang J, Croteau-Chonka DC, Esko T, Fall T, Ferreira T, Gustafsson S, Kutalik Z, Luan J, Magi R, Randall JC, Winkler TW, Wood AR, Workalemahu T, Faul JD, Smith JA, Zhao JH, Zhao W, Chen J, Fehrmann R, Hedman AK, Karjalainen J, Schmidt EM, Absher D, Amin N, Anderson D, Beekman M, Bolton JL, Bragg-Gresham JL, Buyske S, Demirkan A, Deng G, Ehret GB, Feenstra B, Feitosa MF, Fischer K, Goel A, Gong J, Jackson AU, Kanoni S, Kleber ME, Kristiansson K, Lim U, Lotay V, Mangino M, Leach IM, Medina-Gomez C, Medland SE, Nalls MA, Palmer CD, Pasko D, Pechlivanis S, Peters MJ, Prokopenko I, Shungin D, Stancakova A, Strawbridge RJ, Sung YJ, Tanaka T, Teumer A, Trompet S, van der Laan SW, van Setten J, Van Vliet-Ostaptchouk JV, Wang Z, Yengo L, Zhang W, Isaacs A, Albrecht E, Arnlov J, Arscott GM, Attwood AP, Bandinelli S, Barrett A, Bas IN, Bellis C, Bennett AJ, Berne C, Blagieva R, Bluher M, Bohringer S, Bonnycastle LL, Bottcher Y, Boyd HA, Bruinenberg M, Caspersen IH, Chen YI, Clarke R, Daw EW, de Craen AJM, Delgado G, Dimitriou M, Doney ASF, Eklund N, Estrada K, Eury E, Folkersen L, Fraser RM, Garcia ME, Geller F, Giedraitis V, Gigante B, Go AS, Golay A, Goodall AH, Gordon SD, Gorski M, Grabe HJ, Grallert H, Grammer TB, Grassler J, Gronberg H, Groves CJ, Gusto G, Haessler J, Hall P, Haller T, Hallmans G, Hartman CA, Hassinen

M, Hayward C, Heard-Costa NL, Helmer Q, Hengstenberg C, Holmen O, Hottenga JJ, James AL, Jeff JM, Johansson A, Jolley J, Juliusdottir T, Kinnunen L, Koenig W, Koskenvuo M, Kratzer W, Laitinen J, Lamina C, Leander K, Lee NR, Lichtner P, Lind L, Lindstrom J, Lo KS, Lobbens S, Lorbeer R, Lu Y, Mach F, Magnusson PKE, Mahajan A, McArdle WL, McLachlan S, Menni C, Merger S, Mihailov E, Milani L, Moayyeri A, Monda KL, Morken MA, Mulas A, Muller G, Muller-Nurasyid M, Musk AW, Nagaraja R, Nothen MM, Nolte IM, Pilz S, Rayner NW, Renstrom F, Rettig R, Ried JS, Ripke S, Robertson NR, Rose LM, Sanna S, Scharnagl H, Scholtens S, Schumacher FR, Scott WR, Seufferlein T, Shi J, Smith AV, Smolonska J, Stanton AV, Steinthorsdottir V, Stirrups K, Stringham HM, Sundstrom J, Swertz MA, Swift AJ, Syvanen AC, Tan ST, Tayo BO, Thorand B, Thorleifsson G, Tyrer JP, Uh HW, Vandenput L, Verhulst FC, Vermeulen SH, Verweij N, Vonk JM, Waite LL, Warren HR, Waterworth D, Weedon MN, Wilkens LR, Willenborg C, Wilsgaard T, Wojczynski MK, Wong A, Wright AF, Zhang Q, LifeLines Cohort S, Brennan EP, Choi M, Dastani Z, Drong AW, Eriksson P, Franco-Cereceda A, Gadin JR, Gharavi AG, Goddard ME, Handsaker RE, Huang J, Karpe F, Kathiresan S, Keildson S, Kiryluk K, Kubo M, Lee JY, Liang L, Lifton RP, Ma B, McCarroll SA, McKnight AJ, Min JL, Moffatt MF, Montgomery GW, Murabito JM, Nicholson G, Nyholt DR, Okada Y, Perry JRB, Dorajoo R, Reinmaa E, Salem RM, Sandholm N, Scott RA, Stolk L, Takahashi A, Tanaka T, van 't Hooft FM, Vinkhuyzen AAE, Westra HJ, Zheng W, Zondervan KT, Consortium AD, Group A-BW, Consortium CAD, Consortium CK, Glgc, Icbp, Investigators M, Mu TC, Consortium MI, Consortium P, ReproGen C, Consortium G, International Endogene C, Heath AC, Arveiler D, Bakker SJL, Beilby J, Bergman RN, Blangero J, Bovet P, Campbell H, Caulfield MJ, Cesana G, Chakravarti A, Chasman DI, Chines PS, Collins FS, Crawford DC, Cupples LA, Cusi D, Danesh J, de Faire U, den Ruijter HM, Dominiczak AF, Erbel R, Erdmann J, Eriksson JG, Farrall M, Felix SB, Ferrannini E, Ferrieres J, Ford I, Forouhi NG, Forrester T, Franco OH, Gansevoort RT, Gejman PV, Gieger C, Gottesman O, Gudnason V, Gyllensten U, Hall AS, Harris TB, Hattersley AT, Hicks AA, Hindorff LA, Hingorani AD, Hofman A, Homuth G, Hovingh GK, Humphries SE, Hunt SC, Hypponen E, Illig T, Jacobs KB, Jarvelin MR, Jockel KH, Johansen B, Jousilahti P, Jukema JW, Jula AM, Kaprio J, Kastelein JJP, Keinanen-Kiukaanniemi SM, Kiemeney LA, Knekt P, Kooner JS, Kooperberg C, Kovacs P, Kraja AT, Kumari M, Kuusisto J, Lakka TA, Langenberg C, Marchand LL, Lehtimaki T, Lyssenko V, Mannisto S, Marette A, Matise TC, McKenzie CA, McKnight B, Moll FL, Morris AD, Morris AP, Murray JC, Nelis M, Ohlsson C, Oldehinkel AJ, Ong KK, Madden PAF, Pasterkamp G, Peden JF, Peters A, Postma DS, Pramstaller PP, Price JF, Qi L, Raitakari OT, Rankinen T, Rao DC, Rice TK, Ridker PM, Rioux JD, Ritchie MD, Rudan I, Salomaa V, Samani NJ, Saramies J, Sarzynski MA, Schunkert H, Schwarz PEH, Sever P, Shuldiner AR, Sinisalo J, Stolk RP, Strauch K, Tonjes A, Tregouet DA, Tremblay A, Tremoli E, Virtamo J, Vohl MC, Volker U, Waeber G, Willemsen G, Witteman JC, Zillikens MC, Adair LS, Amouyel P, Asselbergs FW, Assimes TL, Bochud M, Boehm BO, Boerwinkle E, Bornstein SR, Bottinger EP, Bouchard C, Cauchi S, Chambers JC, Chanock SJ, Cooper RS, de Bakker PIW, Dedoussis G, Ferrucci L, Franks PW, Froguel P, Groop LC, Haiman CA, Hamsten A, Hui J, Hunter DJ, Hveem K, Kaplan RC, Kivimaki M, Kuh D, Laakso M, Liu Y, Martin NG, Marz W, Melbye M, Metspalu A, Moebus S, Munroe PB, Njolstad I, Oostra BA, Palmer CNA, Pedersen NL, Perola M, Perusse L, Peters U, Power C, Quertermous T, Rauramaa R, Rivadeneira F, Saaristo TE, Saleheen D, Sattar N, Schadt EE, Schlessinger D, Slagboom PE, Snieder H, Spector TD, Thorsteinsdottir U, Stumvoll M, Tuomilehto J, Uitterlinden AG, Uusitupa M, van der Harst P, Walker M, Wallaschofski H, Wareham NJ, Watkins H, Weir DR, Wichmann HE, Wilson JF, Zanen P, Borecki IB, Deloukas P, Fox CS, Heid IM, O'Connell JR, Strachan DP, Stefansson K, van Duijn CM, Abecasis GR, Franke L, Frayling TM, McCarthy MI, Visscher PM, Scherag A, Willer CJ, Boehnke M, Mohlke KL, Lindgren CM, Beckmann JS, Barroso I, North KE, Ingelsson E, Hirschhorn JN, Loos RJF, Speliotes EK (2015) Genetic studies of body mass index yield new insights for obesity biology. Nature 518(7538):197–206. https://doi.org/10.1038/nature14177

49. Yengo L, Sidorenko J, Kemper KE, Zheng Z, Wood AR, Weedon MN, Frayling TM, Hirschhorn J, Yang J, Visscher PM, Consortium G (2018) Meta-analysis of genome-wide association studies for height and body mass index in approximately 700000 individuals of European ancestry. Hum Mol Genet 27(20):3641–3649. https://doi.org/10.1093/hmg/ddy271
50. Frayling TM, Timpson NJ, Weedon MN, Zeggini E, Freathy RM, Lindgren CM, Perry JR, Elliott KS, Lango H, Rayner NW, Shields B, Harries LW, Barrett JC, Ellard S, Groves CJ, Knight B, Patch AM, Ness AR, Ebrahim S, Lawlor DA, Ring SM, Ben-Shlomo Y, Jarvelin MR, Sovio U, Bennett AJ, Melzer D, Ferrucci L, Loos RJ, Barroso I, Wareham NJ, Karpe F, Owen KR, Cardon LR, Walker M, Hitman GA, Palmer CN, Doney AS, Morris AD, Smith GD, Hattersley AT, McCarthy MI (2007) A common variant in the FTO gene is associated with body mass index and predisposes to childhood and adult obesity. Science 316(5826):889–894. https://doi.org/10.1126/science.1141634
51. Li G, Xu X, Yang L, Cai Y, Sun Y, Guo J, Lin Y, Hu Y, Chen M, Li H, Wu S (2023) Exploring the association between circRNA expression and pediatric

obesity based on a case-control study and related bioinformatics analysis. BMC Pediatr 23(1):561. https://doi.org/10.1186/s12887-023-04261-1
52. Paraboschi EM, Cardamone G, Solda G, Duga S, Asselta R (2018) Interpreting non-coding genetic variation in multiple sclerosis genome-wide associated regions. Front Genet 9:647. https://doi.org/10.3389/fgene.2018.00647
53. Zhou YL, Wu WP, Cheng J, Liang LL, Cen JM, Chen C, Liu X, Xiong XD (2020) CircFOXO3 rs12196996, a polymorphism at the gene flanking intron, is associated with circFOXO3 levels and the risk of coronary artery disease. Aging (Albany NY) 12(13):13076–13089. https://doi.org/10.18632/aging.103398
54. Rajcsanyi LS, Diebels I, Pastoors L, Kanber D, Peters T, Volckmar AL, Zheng Y, Grosse M, Dieterich C, Hebebrand J, Kaiser FJ, Horsthemke B, Hinney A (2022) Evidence for correlations between BMI-associated SNPs and circRNAs. Sci Rep 12(1):12643. https://doi.org/10.1038/s41598-022-16495-7
55. Hebebrand J, Hinney A, Antel J (2023) Could leptin substitution therapy potentially terminate entrapment in anorexia nervosa? Nat Rev Endocrinol 19(8):435–436. https://doi.org/10.1038/s41574-023-00863-y
56. Bulik CM, Carroll IM, Mehler P (2021) Reframing anorexia nervosa as a metabo-psychiatric disorder. Trends Endocrinol Metab 32(10):752–761. https://doi.org/10.1016/j.tem.2021.07.010
57. Sakers A, De Siqueira MK, Seale P, Villanueva CJ (2022) Adipose-tissue plasticity in health and disease. Cell 185(3):419–446. https://doi.org/10.1016/j.cell.2021.12.016
58. Tews D, Wabitsch M (2022) Brown adipose tissue in children and its metabolic function. Horm Res Paediatr 95(2):104–111. https://doi.org/10.1159/000518353
59. Dye L, Boyle NB, Champ C, Lawton C (2017) The relationship between obesity and cognitive health and decline. Proc Nutr Soc 76(4):443–454. https://doi.org/10.1017/S0029665117002014
60. Yoon G, Lim YH, Jo D, Ryu J, Song J, Kim YK (2021) Obesity-linked circular RNA circTshz2-2 regulates the neuronal cell cycle and spatial memory in the brain. Mol Psychiatry 26(11):6350–6364. https://doi.org/10.1038/s41380-021-01303-x
61. Reaven GM (1988) Banting lecture 1988. Role of insulin resistance in human disease. Diabetes 37(12):1595–1607. https://doi.org/10.2337/diab.37.12.1595
62. McCracken E, Monaghan M, Sreenivasan S (2018) Pathophysiology of the metabolic syndrome. Clin Dermatol 36(1):14–20. https://doi.org/10.1016/j.clindermatol.2017.09.004
63. Fahed G, Aoun L, Bou Zerdan M, Allam S, Bou Zerdan M, Bouferraa Y, Assi HI (2022) Metabolic syndrome: updates on pathophysiology and management in 2021. Int J Mol Sci 23(2). https://doi.org/10.3390/ijms23020786
64. Noubiap JJ, Nansseu JR, Lontchi-Yimagou E, Nkeck JR, Nyaga UF, Ngouo AT, Tounouga DN, Tianyi FL, Foka AJ, Ndoadoumgue AL, Bigna JJ (2022) Geographic distribution of metabolic syndrome and its components in the general adult population: a meta-analysis of global data from 28 million individuals. Diabetes Res Clin Pract 188:109924. https://doi.org/10.1016/j.diabres.2022.109924
65. Noubiap JJ, Nansseu JR, Lontchi-Yimagou E, Nkeck JR, Nyaga UF, Ngouo AT, Tounouga DN, Tianyi FL, Foka AJ, Ndoadoumgue AL, Bigna JJ (2022) Global, regional, and country estimates of metabolic syndrome burden in children and adolescents in 2020: a systematic review and modelling analysis. Lancet Child Adolesc Health 6(3):158–170. https://doi.org/10.1016/S2352-4642(21)00374-6
66. Guembe MJ, Fernandez-Lazaro CI, Sayon-Orea C, Toledo E, Moreno-Iribas C, Investigators RS (2020) Risk for cardiovascular disease associated with metabolic syndrome and its components: a 13-year prospective study in the RIVANA cohort. Cardiovasc Diabetol 19(1):195. https://doi.org/10.1186/s12933-020-01166-6
67. Lin X, Du Y, Lu W, Gui W, Sun S, Zhu Y, Wang G, Eserberg DT, Zheng F, Zhou J, Wu F, Li H (2021) CircRNF111 protects against insulin resistance and lipid deposition via regulating miR-143-3p/IGF2R axis in metabolic syndrome. Front Cell Dev Biol 9:663148. https://doi.org/10.3389/fcell.2021.663148
68. Zhu Y, Gui W, Lin X, Li H (2020) Knock-down of circular RNA H19 induces human adipose-derived stem cells adipogenic differentiation via a mechanism involving the polypyrimidine tract-binding protein 1. Exp Cell Res 387(2):111753. https://doi.org/10.1016/j.yexcr.2019.111753
69. Pouwels S, Sakran N, Graham Y, Leal A, Pintar T, Yang W, Kassir R, Singhal R, Mahawar K, Ramnarain D (2022) Non-alcoholic fatty liver disease (NAFLD): a review of pathophysiology, clinical management and effects of weight loss. BMC Endocr Disord 22(1):63. https://doi.org/10.1186/s12902-022-00980-1
70. Loomba R, Friedman SL, Shulman GI (2021) Mechanisms and disease consequences of nonalcoholic fatty liver disease. Cell 184(10):2537–2564. https://doi.org/10.1016/j.cell.2021.04.015
71. Younossi Z, Tacke F, Arrese M, Chander Sharma B, Mostafa I, Bugianesi E, Wai-Sun Wong V, Yilmaz Y, George J, Fan J, Vos MB (2019) Global perspectives on nonalcoholic fatty liver disease and nonalcoholic steatohepatitis. Hepatology 69(6):2672–2682. https://doi.org/10.1002/hep.30251
72. Buzzetti E, Pinzani M, Tsochatzis EA (2016) The multiple-hit pathogenesis of non-alcoholic fatty liver disease (NAFLD). Metabolism 65(8):1038–1048. https://doi.org/10.1016/j.metabol.2015.12.012
73. Hao L, Chai Z (2023) Nonalcoholic fatty liver disease: pathogenesis and intervention. Precision

Nutr 2(1):e00029. https://doi.org/10.1097/pn9.0000000000000029
74. Guo XY, He CX, Wang YQ, Sun C, Li GM, Su Q, Pan Q, Fan JG (2017) Circular RNA profiling and bioinformatic modeling identify its regulatory role in hepatic steatosis. Biomed Res Int 2017:5936171. https://doi.org/10.1155/2017/5936171
75. Guo XY, Sun F, Chen JN, Wang YQ, Pan Q, Fan JG (2018) circRNA_0046366 inhibits hepatocellular steatosis by normalization of PPAR signaling. World J Gastroenterol 24(3):323–337. https://doi.org/10.3748/wjg.v24.i3.323
76. Li Y, Cen CQ, Liu B, Zhou L, Huang XM, Liu GY (2022) Overexpression of circ PTK2 suppresses the progression of nonalcoholic fatty liver disease via the miR-200c/SIK2/PI3K/Akt axis. Kaohsiung J Med Sci 38(9):869–878. https://doi.org/10.1002/kjm2.12568
77. Li J, Qi J, Tang Y, Liu H, Zhou K, Dai Z, Yuan L, Sun C (2021) A nanodrug system overexpressed circRNA_0001805 alleviates nonalcoholic fatty liver disease via miR-106a-5p/miR-320a and ABCA1/CPT1 axis. J Nanobiotechnol 19(1):363. https://doi.org/10.1186/s12951-021-01108-8
78. Li P, Shan K, Liu Y, Zhang Y, Xu L, Xu L (2019) CircScd1 promotes fatty liver disease via the Janus kinase 2/signal transducer and activator of transcription 5 pathway. Dig Dis Sci 64(1):113–122. https://doi.org/10.1007/s10620-018-5290-2
79. Jin X, Gao J, Zheng R, Yu M, Ren Y, Yan T, Huang Y, Li Y (2020) Antagonizing circRNA_002581-miR-122-CPEB1 axis alleviates NASH through restoring PTEN-AMPK-mTOR pathway regulated autophagy. Cell Death Dis 11(2):123. https://doi.org/10.1038/s41419-020-2293-7
80. Zhao Q, Liu J, Deng H, Ma R, Liao JY, Liang H, Hu J, Li J, Guo Z, Cai J, Xu X, Gao Z, Su S (2020) Targeting mitochondria-located circRNA SCAR alleviates NASH via reducing mROS output. Cell 183(1):76–93.e22. https://doi.org/10.1016/j.cell.2020.08.009
81. Yang W, Zhao J, Zhao Y, Li W, Zhao L, Ren Y, Ou R, Xu Y (2020) Hsa_circ_0048179 attenuates free fatty acid-induced steatosis via hsa_circ_0048179/miR-188-3p/GPX4 signaling. Aging (Albany NY) 12(23):23996–24008. https://doi.org/10.18632/aging.104081
82. Yuan X, Li Y, Wen S, Xu C, Wang C, He Y, Zhou L (2022) CircLDLR acts as a sponge for miR-667-5p to regulate SIRT1 expression in non-alcoholic fatty liver disease. Lipids Health Dis 21(1):127. https://doi.org/10.1186/s12944-022-01740-9
83. Liu CH, Jiang W, Zeng Q, Wu D, Li H, Zhou L, Bai L, Tang H (2023) CircRNA-PI4KB induces hepatic lipid deposition in non-alcoholic fatty liver disease by transporting miRNA-122 to extra-hepatocytes. Int J Mol Sci 24(2):1297. https://doi.org/10.3390/ijms24021297
84. Chen X, Tan QQ, Tan XR, Li SJ, Zhang XX (2021) Circ_0057558 promotes nonalcoholic fatty liver disease by regulating ROCK1/AMPK signaling through targeting miR-206. Cell Death Dis 12(9):809. https://doi.org/10.1038/s41419-021-04090-z
85. Xu ZX, Li JZ, Li Q, Xu MY, Li HY (2022) CircRNA608-microRNA222-PINK1 axis regulates the mitophagy of hepatic stellate cells in NASH related fibrosis. Biochem Biophys Res Commun 610:35–42. https://doi.org/10.1016/j.bbrc.2022.04.008
86. Guo XY, Chen JN, Sun F, Wang YQ, Pan Q, Fan JG (2017) circRNA_0046367 prevents hepatoxicity of lipid peroxidation: an inhibitory role against hepatic steatosis. Oxidative Med Cell Longev 2017:3960197. https://doi.org/10.1155/2017/3960197
87. Li M, Zeng A, Tang X, Xu H, Xiong W, Guo Y (2023) Circ_0004535/miR-1827/CASP8 network involved in type 2 diabetes mellitus with nonalcoholic fatty liver disease. Sci Rep 13(1):19807. https://doi.org/10.1038/s41598-023-47189-3
88. Du C, Shen L, Ma Z, Du J, Jin S (2021) Bioinformatic analysis of crosstalk between circRNA, miRNA, and target gene network in NAFLD. Front Genet 12:671523. https://doi.org/10.3389/fgene.2021.671523
89. Galicia-Garcia U, Benito-Vicente A, Jebari S, Larrea-Sebal A, Siddiqi H, Uribe KB, Ostolaza H, Martin C (2020) Pathophysiology of type 2 diabetes mellitus. Int J Mol Sci 21(17):6275. https://doi.org/10.3390/ijms21176275
90. Leslie RD, Ma RCW, Franks PW, Nadeau KJ, Pearson ER, Redondo MJ (2023) Understanding diabetes heterogeneity: key steps towards precision medicine in diabetes. Lancet Diabetes Endocrinol 11(11):848–860. https://doi.org/10.1016/S2213-8587(23)00159-6
91. Sakran N, Graham Y, Pintar T, Yang W, Kassir R, Willigendael EM, Singhal R, Kooreman ZE, Ramnarain D, Mahawar K, Parmar C, Madhok B, Pouwels S (2022) The many faces of diabetes. Is there a need for re-classification? A narrative review. BMC Endocr Disord 22(1):9. https://doi.org/10.1186/s12902-021-00927-y
92. GBD Diabetes Collaborators (2023) Global, regional, and national burden of diabetes from 1990 to 2021, with projections of prevalence to 2050: a systematic analysis for the Global Burden of Disease Study 2021. Lancet 402(10397):203–234. https://doi.org/10.1016/S0140-6736(23)01301-6
93. Kautzky-Willer A, Leutner M, Harreiter J (2023) Sex differences in type 2 diabetes. Diabetologia 66(6):986–1002. https://doi.org/10.1007/s00125-023-05891-x
94. Pearson-Stuttard J, Holloway S, Polya R, Sloan R, Zhang L, Gregg EW, Harrison K, Elvidge J, Jonsson P, Porter T (2022) Variations in comorbidity burden in people with type 2 diabetes over disease duration: a population-based analysis of real world evidence. EClinicalMedicine 52:101584. https://doi.org/10.1016/j.eclinm.2022.101584

95. Iglay K, Hannachi H, Engel SS, Li X, O'Connell D, Moore LM, Rajpathak S (2021) Comorbidities in type 2 diabetes patients with and without atherosclerotic cardiovascular disease: a retrospective database analysis. Curr Med Res Opin 37(5):743–751. https://doi.org/10.1080/03007995.2021.1895736
96. Antar SA, Ashour NA, Sharaky M, Khattab M, Ashour NA, Zaid RT, Roh EJ, Elkamhawy A, Al-Karmalawy AA (2023) Diabetes mellitus: classification, mediators, and complications; A gate to identify potential targets for the development of new effective treatments. Biomed Pharmacother 168:115734. https://doi.org/10.1016/j.biopha.2023.115734
97. Banday MZ, Sameer AS, Nissar S (2020) Pathophysiology of diabetes: an overview. Avicenna J Med 10(4):174–188. https://doi.org/10.4103/ajm.ajm_53_20
98. Stoll L, Sobel J, Rodriguez-Trejo A, Guay C, Lee K, Veno MT, Kjems J, Laybutt DR, Regazzi R (2018) Circular RNAs as novel regulators of beta-cell functions in normal and disease conditions. Mol Metab 9:69–83. https://doi.org/10.1016/j.molmet.2018.01.010
99. Rezaeinejad F, Mirzaei A, Khalvati B, Sabz G, Alipoor B (2022) Circulating expression levels of CircHIPK3 and CDR1as circular-RNAs in type 2 diabetes patients. Mol Biol Rep 49(1):131–138. https://doi.org/10.1007/s11033-021-06850-8
100. Cai H, Jiang Z, Yang X, Lin J, Cai Q, Li X (2020) Circular RNA HIPK3 contributes to hyperglycemia and insulin homeostasis by sponging miR-192-5p and upregulating transcription factor forkhead box O1. Endocr J 67(4):397–408. https://doi.org/10.1507/endocrj.EJ19-0271
101. Fang Y, Wang X, Li W, Han J, Jin J, Su F, Zhang J, Huang W, Xiao F, Pan Q, Zou L (2018) Screening of circular RNAs and validation of circANKRD36 associated with inflammation in patients with type 2 diabetes mellitus. Int J Mol Med 42(4):1865–1874. https://doi.org/10.3892/ijmm.2018.3783
102. Zhao Z, Li X, Jian D, Hao P, Rao L, Li M (2017) Hsa_circ_0054633 in peripheral blood can be used as a diagnostic biomarker of pre-diabetes and type 2 diabetes mellitus. Acta Diabetol 54(3):237–245. https://doi.org/10.1007/s00592-016-0943-0
103. Haque S, Ames RM, Moore K, Lee BP, Jeffery N, Harries LW (2020) Islet-expressed circular RNAs are associated with type 2 diabetes status in human primary islets and in peripheral blood. BMC Med Genet 13(1):64. https://doi.org/10.1186/s12920-020-0713-2
104. Chen X, Yin J, Zhang F, Xiao T, Zhao M (2020) has_circ_CCNB1 and has_circ_0009024 function as potential biomarkers for the diagnosis of type 2 diabetes mellitus. J Clin Lab Anal 34(10):e23439. https://doi.org/10.1002/jcla.23439
105. Dai Y, Ma X, Zhang J, Yu S, Zhu Y, Wang J (2022) hsa_circ_0115355 promotes pancreatic beta-cell function in patients with type 2 diabetes through the miR-145/SIRT1 axis. J Clin Lab Anal 36(8):e24583. https://doi.org/10.1002/jcla.24583
106. Ren L (2022) Circular RNA PIP5K1A act as microRNA-552-3p sponge to regulates inflammation, oxidative damage in glucolipotoxicity-induced pancreatic INS-1 beta-cells via Janus kinase 1. Bioengineered 13(3):5724–5736. https://doi.org/10.1080/21655979.2021.2022076
107. Song G, Zhang Y, Jiang Y, Zhang H, Gu W, Xu X, Yao J, Chen Z (2023) Circular RNA PIP5K1A promotes glucose and lipid metabolism disorders and inflammation in type 2 diabetes mellitus. Mol Biotechnol 66(12):3549–3558. https://doi.org/10.1007/s12033-023-00954-1
108. Liang H, Hou L, Wang Q, Zhou X, Sha L, Xu L, Lu X (2021) Serum hsa_circ_0054633 is elevated and correlated with clinical features in type 2 diabetes mellitus. Ann Clin Lab Sci 51(1):90–96
109. Dai H, Hu F, Yang X, Hu P, Chu Y, Bu S (2021) Hsa_circ_0054633 association of C peptide is related to IL-17 and TNF-alpha in patients with diabetes mellitus receiving insulin treatment. J Clin Lab Anal 35(8):e23856. https://doi.org/10.1002/jcla.23856
110. Yan YX, Dong J, Li YL, Lu YK, Yang K, Wang T, Zhang X, Xiao HB (2022) CircRNA hsa_circ_0071336 is associated with type 2 diabetes through targeting the miR-93-5p/GLUT4 axis. FASEB J 36(5):e22324. https://doi.org/10.1096/fj.202200149RR
111. Cheng J, Liu Q, Hu N, Zheng F, Zhang X, Ni Y, Liu J (2019) Downregulation of hsa_circ_0068087 ameliorates TLR4/NF-kappaB/NLRP3 inflammasome-mediated inflammation and endothelial cell dysfunction in high glucose conditioned by sponging miR-197. Gene 709:1–7. https://doi.org/10.1016/j.gene.2019.05.012
112. Zhang W, Sui Y (2020) CircBPTF knockdown ameliorates high glucose-induced inflammatory injuries and oxidative stress by targeting the miR-384/LIN28B axis in human umbilical vein endothelial cells. Mol Cell Biochem 471(1–2):101–111. https://doi.org/10.1007/s11010-020-03770-2
113. Liu Y, Yang Y, Xu C, Liu J, Chen J, Li G, Huang B, Pan Y, Zhang Y, Wei Q, Pandol SJ, Zhang F, Li L, Jin L (2023) Circular RNA circGlis3 protects against islet beta-cell dysfunction and apoptosis in obesity. Nat Commun 14(1):351. https://doi.org/10.1038/s41467-023-35998-z
114. Xiong L, Gong Y, Liu H, Huang L, Zeng Z, Zheng X, Li W, Liang Z, Kang L (2024) circGlis3 promotes beta-cell dysfunction by binding to heterogeneous nuclear ribonucleoprotein F and encoding Glis3-348aa protein. iScience 27(1):108680. https://doi.org/10.1016/j.isci.2023.108680
115. Wu L, Xiong L, Li J, Peng Z, Zhang L, Shi P, Gong Y, Xiao H (2020) Circ-Tulp4 promotes beta-cell adaptation to lipotoxicity by regulating soat1 expression. J Mol Endocrinol 65(4):149–161. https://doi.org/10.1530/JME-20-0079

116. Stoll L, Rodriguez-Trejo A, Guay C, Brozzi F, Bayazit MB, Gattesco S, Menoud V, Sobel J, Marques AC, Veno MT, Esguerra JLS, Barghouth M, Suleiman M, Marselli L, Kjems J, Eliasson L, Renstrom E, Bouzakri K, Pinget M, Marchetti P, Regazzi R (2020) A circular RNA generated from an intron of the insulin gene controls insulin secretion. Nat Commun 11(1):5611. https://doi.org/10.1038/s41467-020-19381-w
117. Chen B, Li Y, Liu Y, Xu Z (2019) circLRP6 regulates high glucose-induced proliferation, oxidative stress, ECM accumulation, and inflammation in mesangial cells. J Cell Physiol 234(11):21249–21259. https://doi.org/10.1002/jcp.28730
118. Ma J, Wu Y, He Y (2021) Silencing circRNA LRP6 down-regulates PRMT1 to improve the streptozocin-induced pancreatic beta-cell injury and insulin secretion by sponging miR-9-5p. J Bioenerg Biomembr 53(3):333–342. https://doi.org/10.1007/s10863-021-09895-3
119. Zeng Y, Zheng Z, Liu F, Yi G (2021) Circular RNAs in metabolism and metabolic disorders. Obes Rev 22(7):e13220. https://doi.org/10.1111/obr.13220
120. Lu J, Pang L, Zhang B, Gong Z, Song C (2021) Silencing circANKRD36 inhibits streptozotocin-induced insulin resistance and inflammation in diabetic rats by targeting miR-145 via XBP1. Inflamm Res 70(6):695–704. https://doi.org/10.1007/s00011-021-01467-w
121. Wang J, Yang S, Li W, Zhao M, Li K (2022) Circ_0000491 promotes apoptosis, inflammation, oxidative stress, and fibrosis in high glucose-induced mesangial cells by regulating miR-455-3p/Hmgb1 axis. Nephron 146(1):72–83. https://doi.org/10.1159/000516870
122. Zhao L, Chen H, Zeng Y, Yang K, Zhang R, Li Z, Yang T, Ruan H (2021) Circular RNA circ_0000712 regulates high glucose-induced apoptosis, inflammation, oxidative stress, and fibrosis in (DN) by targeting the miR-879-5p/SOX6 axis. Endocr J 68(10):1155–1164. https://doi.org/10.1507/endocrj.EJ20-0739
123. Xu H, Guo S, Li W, Yu P (2015) The circular RNA Cdr1as, via miR-7 and its targets, regulates insulin transcription and secretion in islet cells. Sci Rep 5:12453. https://doi.org/10.1038/srep12453
124. Sun R, Xue W, Zhao J (2021) Hsa_circ_0054633 mediates apoptosis and insulin secretion in human pancreatic beta cells through miR-409-3p/caspase-8 axis. Diabetes Res Clin Pract 176:108837. https://doi.org/10.1016/j.diabres.2021.108837
125. Zhang C, Han X, Yang L, Fu J, Sun C, Huang S, Xiao W, Gao Y, Liang Q, Wang X, Luo F, Lu W, Zhou Y (2020) Circular RNA circPPM1F modulates M1 macrophage activation and pancreatic islet inflammation in type 1 diabetes mellitus. Theranostics 10(24):10908–10924. https://doi.org/10.7150/thno.48264
126. Yang L, Han X, Zhang C, Sun C, Huang S, Xiao W, Gao Y, Liang Q, Luo F, Lu W, Fu J, Zhou Y (2020) Hsa_circ_0060450 negatively regulates type I interferon-induced inflammation by serving as miR-199a-5p sponge in type 1 diabetes mellitus. Front Immunol 11:576903. https://doi.org/10.3389/fimmu.2020.576903
127. Nielsen AF, Bindereif A, Bozzoni I, Hanan M, Hansen TB, Irimia M, Kadener S, Kristensen LS, Legnini I, Morlando M, Jarlstad Olesen MT, Pasterkamp RJ, Preibisch S, Rajewsky N, Suenkel C, Kjems J (2022) Best practice standards for circular RNA research. Nat Methods 19(10):1208–1220. https://doi.org/10.1038/s41592-022-01487-2
128. Liu X, Zhang Y, Zhou S, Dain L, Mei L, Zhu G (2022) Circular RNA: an emerging frontier in RNA therapeutic targets, RNA therapeutics, and mRNA vaccines. J Control Release 348:84–94. https://doi.org/10.1016/j.jconrel.2022.05.043
129. Niu D, Wu Y, Lian J (2023) Circular RNA vaccine in disease prevention and treatment. Signal Transduct Target Ther 8(1):341. https://doi.org/10.1038/s41392-023-01561-x

Circular RNAs in Liver Diseases

21

Archittapon Nokkeaw and Chaiyaboot Ariyachet

Abstract

Circular RNAs (circRNAs), a class of recently discovered noncoding RNAs, exhibit a distinctive feature in comparison to other noncoding RNAs, such as microRNAs (miRNAs) and long noncoding RNAs (lncRNAs). They are characterized by the formation of a covalent bond between the 3′ and 5′ ends through a process known as back-splicing, rendering them remarkably resistant to degradation by exonucleases, particularly RNase R. This inherent stability contributes to their prolonged expression compared to their linear counterparts. Recent investigations have unveiled the pivotal role of circRNAs in the regulation of gene expression through various mechanisms, including the modulation of host gene splicing, acting as miRNA sponges, interacting with proteins, and even demonstrating translational capabilities to produce proteins. These diverse functions position circRNAs as key players in the complex landscape of the pathogenesis of various diseases, including liver disease. Unraveling the intricate involvement of circRNAs in these liver disorders provides valuable insights into their diagnostic and therapeutic potential. This chapter aims to comprehensively examine the current body of literature, consolidating our understanding of the complex regulatory networks orchestrated by circRNAs in liver pathophysiology. Additionally, it strives to identify potential avenues for therapeutic intervention and diagnostic innovation, thus contributing to the burgeoning field of precision medicine in liver disease management.

Keywords

Circular RNA · Liver diseases · Metabolic-associated fatty liver disease · Hepatocellular carcinoma · miRNA sponge

A. Nokkeaw
Department of Biochemistry, Faculty of Medicine, Chulalongkorn University, Bangkok, Thailand

Center of Excellence in Hepatitis and Liver Cancer, Faculty of Medicine, Chulalongkorn University, Bangkok, Thailand

Medical Biochemistry Program, Department of Biochemistry, Faculty of Medicine, Chulalongkorn University, Bangkok, Thailand

C. Ariyachet (✉)
Department of Biochemistry, Faculty of Medicine, Chulalongkorn University, Bangkok, Thailand

Center of Excellence in Hepatitis and Liver Cancer, Faculty of Medicine, Chulalongkorn University, Bangkok, Thailand
e-mail: Chaiyaboot.A@chula.ac.th

J. Xiao (ed.), *Advances in Circular RNAs*, Advances in Experimental Medicine and Biology 1485,
https://doi.org/10.1007/978-981-96-9428-0_21

1 Introduction

Liver diseases are considered the major global health burden, responsible for approximately two million fatalities globally each year [1]. Liver diseases mostly affect individuals with a history of alcohol use and obesity, who are particularly susceptible to liver-related complications. This fact further amplifies the gravity of this health issue due to the higher rate of alcohol consumption and obesity among individuals [2]. As we confront this formidable challenge, an imperative arises to delve deeper into the molecular intricacies that underpin liver pathophysiology.

In recent years, a distinctive class of noncoding RNAs (ncRNAs), termed circular RNAs (circRNAs), has emerged as a focal point of investigation about their association with the pathogenesis of liver diseases [3]. Characterized by their covalently closed loop structures, circRNAs exhibit distinct features that set them apart from their linear RNA counterparts [4]. The increasing amount of research in this field has brought to light the intricate involvement of circRNAs in various cellular processes, emphasizing their pivotal roles in the regulation of gene expression [5].

CircRNAs have been implicated in a multitude of functions, ranging from the modulation of gene splicing to serving as molecular sponges for microRNAs (miRNAs), interacting with proteins, and being translated into proteins [5]. These functions can affect biological processes. Moreover, the stability conferred by their closed-loop structure, rendering them resistant to exonucleases [6], further underscores their potential significance in the dynamic panorama of liver pathology.

This comprehensive literature review is intended to explore and consolidate the current state of knowledge regarding the roles of circRNAs in liver diseases. By synthesizing findings from diverse investigations, our aim is to unravel the mechanisms through which circRNAs contribute to the pathophysiology of liver diseases. Moreover, the chapter seeks to identify potential diagnostic and therapeutic implications of circRNAs, thereby advancing our understanding of novel medicine approaches in the management of liver disorders. In doing so, we aspire to contribute to the collective effort aimed at mitigating the global impact of liver diseases through novel interventions.

2 Liver Diseases: Global Health Burden

Liver diseases pose a significant global health burden, accounting for two million deaths annually and representing 4% of all deaths worldwide. This translates to one out of every 25 fatalities worldwide [7]. Chronic liver disease prevails globally, transcending considerations of age, gender, geographic location, or ethnicity [8].

Cirrhosis is a clinical symptom and complication of various liver disorders characterized by fibrosis and the architectural deformation of the liver with the creation of regenerative nodules [9]. The global age-standardized incidence rate for cirrhosis is 25.35 per 100,000 population. Consequently, it is estimated that approximately 2.05 million people worldwide are affected by cirrhosis [10]. The burden of advanced liver disease is expected to increase exponentially due to the dramatic rise in the incidence of metabolic risk factors worldwide, especially obesity in minors and juveniles, which pose a threat to global health [7].

In conclusion, liver diseases constitute a substantial global health burden. The escalating prevalence of these ailments emphasizes the critical need for a concerted global effort to formulate and execute effective strategies aimed at preventing, managing, and treating these conditions. Through a nuanced comprehension of the epidemiology and impact of liver diseases, tailored interventions and public health policies can be developed to alleviate their global burden. This collective endeavor will ultimately contribute to enhancing global health outcomes and alleviating the burden imposed by liver diseases.

3 Circular RNAs: A Molecular Perspective

CircRNAs, a subclass of ncRNAs characterized by the formation of covalently closed continuous loop structures devoid of the conventional terminal 5′ and 3′ ends [11], were once excluded from the scope of research because they were viewed as a low-abundant RNA and a result of splicing error without any biological function for a long time [12]. However, as high-throughput sequencing technology and bioinformatics have progressed, circRNAs have been discovered to be prevalent in eukaryotic cells, suggesting their significance in the regulation of cellular processes [13].

CircRNAs are characterized by their stability, high conservation, and tissue-specific expression [14]. They are involved in various physiological and pathological processes by regulating the expression level of genes and the activity of proteins [15]. CircRNAs have an essential function in tissue growth and development, as well as regulation of the cell cycle, apoptosis, stress, senescence, and inflammatory responses [16, 17].

CircRNAs can originate from various genetic structures, with a single gene location producing diverse types of circRNAs. These circular transcripts may encompass single exons or, more commonly, multiple exons derived from protein-coding sequences, constituting what is known as exonic circRNAs (ecircRNAs) [18]. Alternatively, certain circRNAs may exclusively consist of introns, labeled as circular intronic RNAs (ciRNAs) [19]. Moreover, circRNAs can contain both introns and exons, forming exon–intron circRNAs (EIciRNAs) [20]. Notably, circRNAs may also stem from the intron region of transfer RNA (tRNA) genes, designated as tRNA intronic circular RNAs (tricRNAs) [21].

The formation of circRNAs commonly transpires via the back-splicing of precursor mRNAs (pre-mRNAs), where a downstream 5′ splice-donor becomes covalently connected to an upstream 3′ splice-acceptor [22]. Additionally, splicing intermediates, including lariat precursors of pre-mRNAs resulting from exon-skipping or intronic lariat precursors, contribute to circRNA generation [23]. Despite the inherent unfavorable nature of back-splicing [24], cells employ *cis*- and *trans*-acting elements to facilitate this process [25]. Reverse complementary sequences, such as Alu elements, play pivotal roles in intron-pairing-driven circularization [26] (Fig. 21.1a). Simultaneously, RNA-binding proteins (RBPs) also act as trans-acting elements, fostering circularization through the proximity of splice sites or the stabilization of intronic RNA pairs [27] (Fig. 21.1a). Intron-pairing-driven and RBP-mediated circularization gives rise to EIciRNAs, which can undergo internal splicing to remove the intronic sequence, resulting in ecircRNAs [28]. In addition to intron-pairing-driven and RBP-mediated circularization, lariat-driven circularization plays a crucial role in the generation of EIciRNAs (Fig. 21.1b). Following this route, EIciRNAs undergo internal splicing processes to excise the intronic sequence, leading to the formation of ecircRNAs [20, 23]. Additionally, lariat-driven circularization supports the generation of ciRNAs [29]. Lastly, tricRNAs originate during pre-tRNA splicing through the recognition and processing of the bulge-helix-bulge (BHB) motif by the tRNA splicing endonuclease complex [30] (Fig. 21.1c).

Recent studies have shed light on the diverse functions of circRNAs, including (1) serving as miRNA sponges, (2) enhancing protein function, (3) inhibiting the splicing of linear RNA, (4) regulating gene expression by acting as protein sponges, and (5) functioning as protein scaffolds [31–33]. Furthermore, emerging research suggests that circRNAs may encode proteins, prompting inquiries about their classification as ncRNAs [34].

CircRNAs have emerged as significant contributors to the complex landscape of pathogenesis across a spectrum of disorders. Their involvement extends to various human diseases, encompassing conditions such as cancer [35], neurodegenerative disorders [36], and liver diseases [37]. In the realm of liver diseases, circRNAs exhibit distinct expression patterns in conditions like hepatocellular carcinoma (HCC), chronic hepatitis B (CHB), hepatic steatosis, and hepatoblastoma [37]. This observation under-

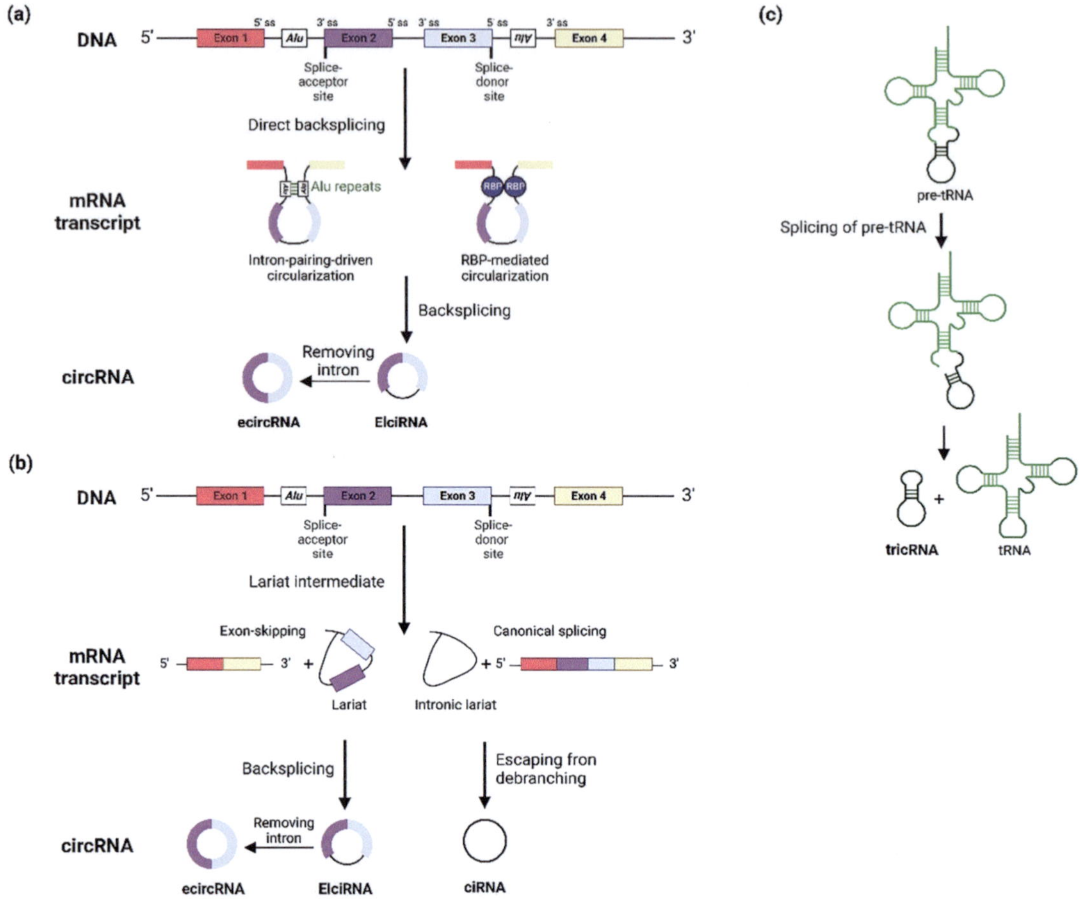

Fig. 21.1 Mechanisms of circRNA biogenesis. (**a**) Circular RNA (circRNA) biogenesis by direct backsplicing is driven by inverted repeat elements in intron-pairing-driven circularization and RNA-binding proteins (RBPs) in RBP-mediated circularization, resulting in exon–intron circRNAs (EIcircRNAs) or exonic circRNAs (ecircRNAs). (**b**) CircRNAs can also be formed through a process called "lariat-driven circularization" by forming lariat intermediates that are generated by exon-skipping or by escaping the debranching of intronic lariat. These processes generate exon-containing circRNAs (EIcircRNAs and ecircRNAs) or circular intronic RNA (ciRNA), respectively. (**c**) tRNA intronic circRNAs (tricRNAs) are produced during the splicing of pre-tRNA to form tricRNAs and mature tRNAs. (Created with BioRender.com)

scores the potential of circRNAs as promising biomarkers or therapeutic targets in the context of liver diseases.

4 Implications of Circular RNAs in Liver Diseases

The molecular intricacies of liver diseases are still not fully elucidated and require extensive research. Recently, within the realm of liver diseases, a multitude of circRNAs have come to light, participating in crucial processes related to liver development and diseases. In the ensuing section, we will provide a concise overview of specific instances of circRNAs implicated in the documented mechanisms of regulation within the liver system.

4.1 Alcohol-Associated Liver Disease

Alcoholic liver disease (ALD) emerges as a significant consequence of prolonged alcohol consumption, posing a substantial global health concern due to the accumulation of fats, inflammation, and scarring within the liver. The spectrum of ALD encompasses various liver disorders, including fatty liver, alcoholic hepatitis, cirrhosis, and hepatocellular carcinoma, with each stage carrying distinct clinical implications that underscore the complexity of its pathophysiology [38].

As a prominent contributor to liver-related morbidity and mortality worldwide, ALD necessitates a thorough investigation to elucidate its pathogenesis, progression, and potential therapeutic interventions [39]. In the United States, ALD accounts for 24% of all liver transplants, solidifying its position as a primary cause for such procedures [40]. This statistic underscores the critical need for a comprehensive understanding of ALD to inform clinical management strategies and develop effective therapeutic interventions.

Recent investigations have unveiled the participation of circRNAs in various facets of ALD. Specifically, circRNA_1639 exhibited an upregulation in Kupffer cells extracted from the livers of mice subjected to alcohol feeding. Circ_1639 has been implicated in the induction of ALD by activating the nuclear factor-kappa B (NF-κB) signaling pathway through the circRNA_1639/miR-122/TNFRSF13C axis. TNFRSF13C, known for its pivotal role in inflammatory responses, becomes activated through this axis, contributing to the pathogenesis of ALD [41].

Furthermore, both mm9_circ_018725 and mou_circ_1657 were identified to be upregulated in a mouse model of alcohol-induced liver disease [42, 43], indicating their potential involvement in advancing the disease progression. Mm9_circ_018725 is suggested to contribute to the promotion of proinflammatory factor secretion by macrophages, exacerbating the inflammatory milieu associated with ALD [42], possibly via miRNA sponging. The underlying mechanism of this circRNA needs further study. On the other hand, mou_circ_1657 appears to activate hepatic stellate cells (HSC), a known contributor to ALD, potentially through the sponge of miR-96-5p [43]. The findings collectively propose the significant roles of these circRNAs in the complex molecular landscape of ALD, shedding light on potential mechanisms for therapeutic exploration. The expression profile of circRNAs in patients diagnosed with ALD should be further investigated to confirm the role of circRNAs in ALD pathogenesis in humans.

4.2 Metabolic-Dysfunction-Associated Fatty Liver Disease (MAFLD)

Metabolic-dysfunction-associated fatty liver disease (MAFLD), or nonalcoholic fatty liver disease (NAFLD) in its previous terminology [44], is characterized by the excessive accumulation of fat in the liver, predominantly affecting individuals with minimal to no alcohol consumption [45]. MAFLD is particularly prevalent among those who are overweight or obese, with its global escalation attributed to unhealthy dietary patterns and sedentary lifestyles, especially in South America and Middle Eastern nations, and rapidly increasing in Asia. As the most prevalent form of chronic liver disease, MAFLD impacts approximately 25% of the global population [46].

Numerous cohort studies consistently associate MAFLD, especially in its advanced stages, with an elevated risk of all-cause mortality [47]. Notably, cardiovascular diseases (CVDs) emerge as the predominant causes of death in individuals with MAFLD, emphasizing a robust connection between MAFLD and an increased susceptibility to significant CVD events and additional cardiac complications [48]. In certain instances, MAFLD can progress to metabolic dysfunction-associated steatohepatitis (MASH), or, in an antiquated term, nonalcoholic steatohepatitis (NASH)—a severe manifestation characterized by liver inflammation and damage due to intracellular lipid deposition [44, 49]. MASH poses the risk of progressing to advanced liver fibrosis, known as cirrhosis, and potentially culminating in the development of liver cancer [49].

In the intricate landscape of MAFLD and MASH, miR-122, recognized as the most abundant liver-specific miRNA, plays a pivotal role in governing lipid metabolism and inflammatory processes. Dysregulation of miR-122 has been observed in individuals afflicted with MAFLD, establishing its critical involvement in the pathogenesis and progression of this liver disorder [50]. Target genes of miR-122 encompass a spectrum of processes integral to MAFLD, including hepatic lipid metabolism (e.g., Yin Yang 1) [51] and autophagy (e.g., CPEB1) [52].

In MASH, a decrease in hepatic miR-122 expression [53], accompanied by an increase in serum miR-122 levels [54], suggests a potential role for serum miR-122 as an extrahepatic marker for MASH [55]. Meanwhile, hepatic miR-122 might hold promise in therapeutic interventions for chronic liver diseases [56]. Consequently, circRNAs acting as sponges for hepatic miR-122 have been associated with the pathogenesis of MAFLD and MASH (Fig. 21.2). For example, circRNA_002581, identified as a miR-122 sponge, exhibits the potential to exacerbate MASH through the circRNA_002581-miR-122-CPEB1 regulatory axis. CPEB1 regulates autophagy through the PTEN–AMPK–mTOR pathway. This regulation occurs by decreasing PTEN levels and the phosphorylation of their target protein, AMPK, thereby resulting in the upregulation of phosphorylated mTOR and subsequently reducing autophagic function [52]. Autophagy, a crucial player in hepatic functions, exhibits impairment in the liver under conditions leading to the development of MASH, suggesting its potential as a novel therapeutic avenue for this disease [57]. Fascinatingly, circPI4KB has the capability to interact with miR-122, aiding in its extracellular transport and subsequently leading to reduced intracellular levels of miRNA-122. This reduction within hepatocytes contributes to the induction of lipid deposition [58]. These findings align with prior reports indicating a correlation between decreased miR-122 levels within the liver and an increase in circulating miR-122 levels [53, 54].

CircLDLR mitigates the progression of MAFLD by instigating autophagic flux via the miR-667-5p/SIRT1 axis. Sirtuin 1 (SIRT1), known for its role in autophagy regulation through NAD-dependent deacetylation of autophagy-related proteins, plays a pivotal role in metabolic regulation across various tissues. It is intricately involved in lipid metabolism, encompassing fatty acid synthesis, oxidation, and lipogenesis [59]. SIRT1 also promotes phosphorylation of AMPK through deacetylation of liver kinase B1 (LKB1), which in turn activates AMPK [60]. Moreover, SIRT1 can deacetylate and activate peroxisome proliferator-activated receptor gamma coactivator 1 alpha (PGC-1α), which further influenced the transcriptional activity of peroxisome proliferator-activated receptor α (PPARα), a key regulator of many lipid metabolisms involving genes [61]. Substantiated by accumulating evidence, SIRT1 is recognized for its functional impact on steering the development and advancement of MAFLD [62].

In both in vivo and in vitro models of MAFLD, circ_0057558 demonstrated an elevation in expression, whereas miR-206 exhibited a reduction [63]. MiR-206 was identified to directly bind to the 3′ untranslated region (3′-UTR) of ROCK1 [63], a hepatic inhibitor of AMPK activity upregulated in fatty liver disease [64]. This upregulation correlates with risk factors associated with MAFLD and insulin resistance, underscoring the pivotal role of the ROCK1/AMPK pathway in mediating cannabinoid-induced lipogenesis in the liver [64]. Circ_0057558 directly interacts with miR-206, alleviating the inhibition of ROCK1/AMPK signaling and promoting lipogenesis and triglyceride (TG) secretion, progressing MAFLD development [63].

CircACC1 intricately regulates lipid metabolism and energy homeostasis by promoting both fatty acid β-oxidation and glycolysis. Studies have demonstrated that circACC1 possesses the ability to bind to the regulatory β1 and γ1 subunits of AMPK, thereby facilitating the assembly, stabilization, and activation of the AMPK holoenzyme. Conversely, unassociated subunits undergo ubiquitination and subsequent proteasomal degradation. As a pivotal cellular energy status sensor, AMPK governs various metabolic processes, including fatty acid oxidation (FAO), glycolysis, and lipolysis [65]. AMPK can regulate β-oxidation by phosphorylating and inactivating

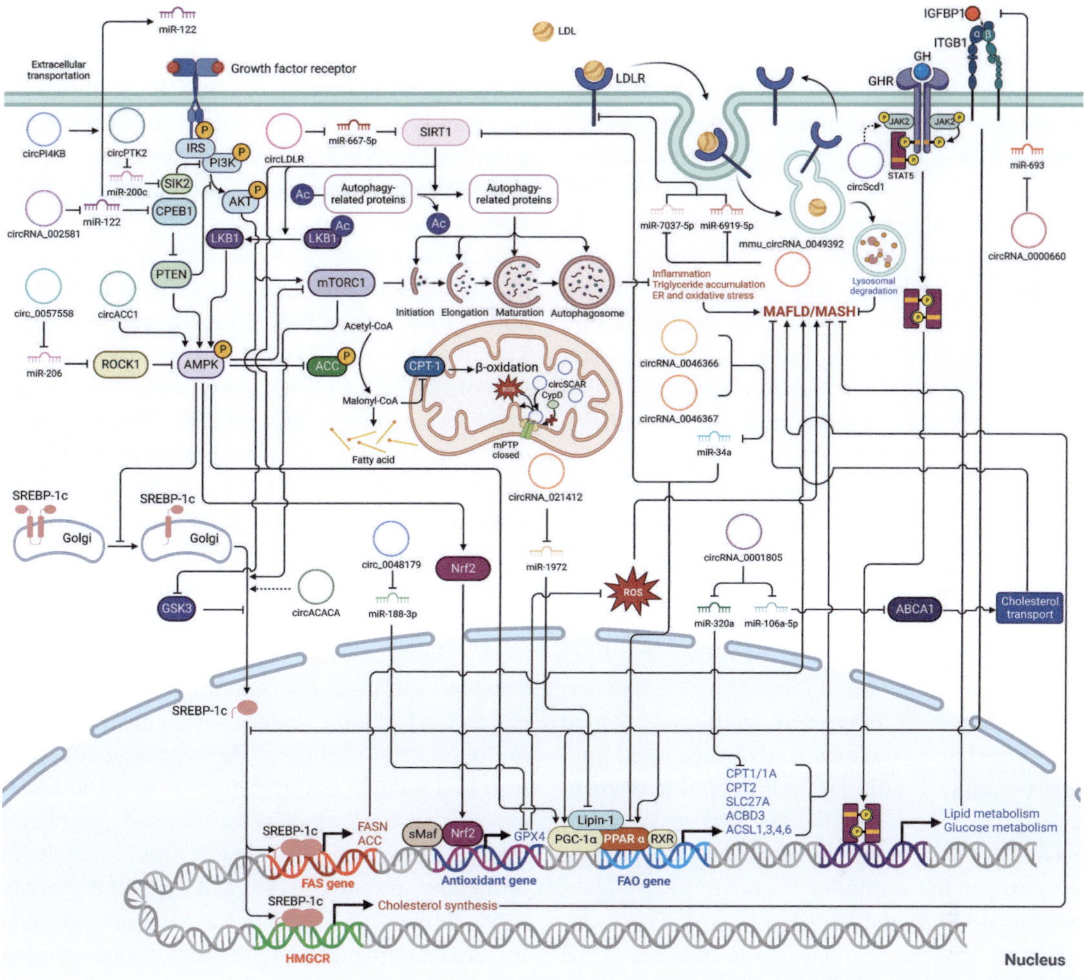

Fig. 21.2 Mechanisms of circRNAs in the development of MAFLD/MASH. Arrows represent activation, whereas bars symbolize inhibition (Created with BioRender.com)

acetyl-CoA carboxylase (ACC), which can carboxylate acetyl-CoA to generate malonyl-CoA. Malonyl-CoA further inhibits carnitine palmitoyltransferase I (CPT1), a rate-limiting enzyme in fatty acid oxidation. Thus, AMPK can suppress the fatty acid by inhibiting the production of malonyl-CoA, which can be used as a substrate to synthesize long-chain fatty acids, and increasing the lipid metabolism via CPT1 [66]. This complex interplay underscores the significance of circACC1 in orchestrating metabolic pathways through its regulatory interaction with AMPK subunits [65].

Notably, circACACA in chickens, originating from the ACC1 host gene similar to circACC1 but with different exons, also regulates lipid accumulation. In contrast to circACC1, circACACA, particularly in laying hens, exhibits a significant correlation with relative liver weight and triglyceride levels, contributing to increased lipid accumulation [65, 67]. The overexpression of circACACA further enhances the mRNA expression of sterol regulatory element-binding protein-1c (SREBP-1c) [67]. SREBP-1c, which is negatively regulated by AMPK, is capable of promoting the expression of fatty acid and cholesterol synthesis genes, including fatty acid synthase (FASN) [67, 68]; thus, AMPK could suppress the synthesis of fatty acid and cholesterol, which exacerbates the progression of

MAFLD [68]. SREBP-1c and FASN are found to be overexpressed in MALFD [69, 70]. Mulberry leaf extract has been observed to alleviate liver steatosis by downregulating the expression of circACACA [67]. This intriguing discovery prompts further investigation into the role of other circRNAs derived from ACC1, such as circACACA, in human lipid metabolism and explores the potential of mulberry leaf extract in reducing hepatic steatosis in humans. The mechanism by which circACACA regulates SREBP-1c and FASN is possibly via miRNA sponging. This hypothesis remains for future studies to confirm.

Both circRNA_0046366 and circRNA_0046367 can prevent hepatic steatosis by normalizing PPARα signaling via sponging miR-34a [71–73]. Mir-34a can also sponge SIRT1. MiR-34a silencing enhanced SIRT1 expression and reduced triglyceride accumulation [74]. PPARα plays a crucial role in promoting FAO while mitigating proinflammatory responses. Dysregulation in fatty acid synthesis coupled with a decline in FAO leads to uncontrolled lipid peroxidation, resulting in the generation of reactive oxygen species (ROS). This cascade of events contributes to inflammatory responses, thereby playing a significant role in the progression of MAFLD and MASH [75, 76]. The restoration of PPARα resulted in the transcriptional activation of genes linked to lipid metabolism, such as CPT1A, CPT2, solute-carrier family 27A (SLC27A), and acyl-CoA binding domain containing 3 (ACBD3), subsequently leading to the resolution of steatosis [71, 72]. Furthermore, CircRNA_021412 also plays a pivotal role in promoting fatty acid oxidation while concurrently inhibiting triglyceride synthesis. This regulatory function is achieved through the modulation of PPARα activation in hepatocytes, orchestrated by the miR-1972/lipin1 (LPIN1) pathway [77]. LPIN1 directly binds with PPARα and PGC-1α and induces the expression of PPARα, thus amplifying the PGC-1α/PPARα pathway [78].

The levels of hsa_circ_0048179 exhibit a significant reduction in the steatosis-induced HepG2 cell line. Elevating the expression of hsa_circ_0048179 proved effective in mitigating oleate/palmitate-induced steatosis through the hsa_circ_0048179/miR-188-3p/GPX4 signaling pathway [79]. GPX4, known for its role in safeguarding the liver from lipid peroxidation, played a pivotal role in this regulatory mechanism [80]. AMPK can phosphorylate nuclear factor E2-related factor 2 (Nrf2) and promote their nuclear translocation and activation, which affect downstream antioxidant GPX4 expression [81]. The inhibition of GPX4 intensified the progression of MASH, while activation of GPX4 ameliorated the severity of MASH [80]. This sequence of events underscores the intricate interplay of hsa_circ_0048179 and GPX4 in the modulation of steatosis and the potential therapeutic implications in the context of MASH.

Hepatic GH-JAK2-STAT5 signaling plays a central role in the progression of fatty liver disease. Disruption of this signaling pathway, achieved through the conditional deletion of either *Stat5* or *Jak2*, induces systemic alterations in lipid metabolism [82]. These genetic ablations lead to a notable accumulation of lipids in hepatocytes, accompanied by heightened peripheral lipolysis and increased hepatic lipid anabolism, ultimately promoting an MAFLD-like phenotype. Both STAT5 and JAK2 are indispensable for maintaining metabolic homeostasis in the liver, playing a critical role in averting liver steatosis through the regulation of lipogenic genes associated with fatty acid uptake and synthesis [82]. Hence, circScd1, with the capability to positively influence the expression levels of JAK2/STAT5, also impacts the extent of steatosis. The precise mechanism through which circScd1 modulates the expression levels of JAK2 and STAT5 remains undisclosed; nevertheless, it is conceivable that it acts as a sponge for miRNAs that exert negative regulation on the JAK2/STAT5 pathway. Alternatively, circScd1 might directly bind to proteins or undertake other functional roles [83].

Mmu_circRNA_0049392 is downregulated in an MAFLD mouse model, suggesting its potential role in the progression of the disease [84]. MiR-7037-5p and miR-6919-5p, which are known to regulate low-density lipoprotein receptor (LDLR), are predicted as mmu_circRNA_0049392 targets [84]. LDLR is accountable for the uptake of cholesterol-carrying lipoprotein particles into cells, subsequently undergoing degradation by lysosomes [85]. Consequently, the diminished expression of LDLR may render the liver susceptible to prolonged inflammation and damage when subjected to extended periods of high-fat, high-cholesterol (HFC) feeding [86]. Further studies are necessary to substantiate the role of mmu_circRNA_0049392, miR-7037-5p, miR-6919-5p, and the target genes of these miRNAs in the development of MAFLD.

Similar to mmu_circRNA_0049392, circRNA_0000660 plays a role in cellular lipid accumulation via the circRNA_0000660/miR-693/IGFBP1 regulatory axis [87]. Insulin-like growth factor binding protein 1 (IGFBP1) has the potential to mitigate lipid accumulation and inflammation in MAFLD by downregulating the expression of SREBP1 and their target genes involved in fatty acid synthesis processes. Additionally, IGFBP1 also promotes the expression of PPARα and PGC-1α and their target gene, CPT-1α, all of which are associated with the lipid oxidation pathway [88].

CircPTK2 displayed a notable downregulation in mice with MAFLD. The increased expression of circPTK2 exhibited a reversal of lipid accumulation in HepG2 cells treated with free fatty acids (FFA). By interacting with miR-200c, circPTK2 established a regulatory relationship, and SIK2 was identified as the direct target of miR-200c. Significantly, SIK2, serving as a negative regulator of the PI3K/Akt signaling pathway pivotal in the development of MAFLD [89]. The PI3K/Akt signaling pathway is involved in MAFLD by promoting mTOR activation and inactivating glycogen synthase kinase-3 (GSK3), a negative regulator of SREBP-1c, by promoting its proteasomal degradation. These events lead to the upregulation of lipogenic SREBP-1c [90–92].

Circ_0004535 actively participates in the pathological suppression of MAFLD in type 2 diabetes mellitus (T2DM) through its interaction with miR-1827. This interaction results in the upregulation of caspase 8 (CASP8) expression, consequently influencing the regulation of the Toll-like receptor (TLR) signaling pathway and apoptosis. CASP8, a pivotal factor, serves as a key suppressor of MAFLD with metabolic disorders and represents a therapeutic target for individuals with T2DM. Moreover, CASP8 is intricately linked to inflammatory pathways and the infiltration of immune cells in the context of T2DM. The multifaceted involvement of circ_0004535 in these processes underscores its significance in the complicated interplay of molecular mechanisms associated with MAFLD in T2DM [93].

Introducing a nanodrug system designed to selectively overexpress circRNA_0001805 in hepatocytes presents a promising avenue for mitigating MAFLD [94]. The expression of circRNA_0001805 was found to be notably diminished in primary hepatocytes challenged with FFA, in mice subjected to a high-fat diet (HFD), and in patients diagnosed with MAFLD. Its mechanism of action involved the targeting of miR-106a-5p and miR-320a, acting as upstream inhibitors of ATP-binding cassette subfamily A member 1 (ABCA1) and CPT1, respectively. Dysfunction in cholesterol outflow and fatty acid β-oxidation in the liver trigger hepatic lipid accumulation and a persistent inflammatory response, which aggravate MAFLD. This collaborative regulation of key factors contributed to the effective management of MAFLD progression [94].

While mammalian mitochondria possess compact genomes encoding a limited number of proteins, there exist a multitude of circular RNAs encoded by the mitochondrial genome, termed mecciRNAs [95]. These mecciRNAs have been identified and play crucial roles in various aspects of cellular physiology [96]. Among the myriad of mecciRNAs, several have been identified as being associated with MASH. Notably, hsa_circ_0089761 and hsa_circ_0089763 exhibit downregulation in MASH and are noteworthy for their significant miRNA-sponge capacity, a feature attributed to their extended sequence lengths.

Bioinformatic analysis revealed that hsa_circ_0089761 and hsa_circ_0089763 have the potential to target miR-642a-5p, miR-1248, miR-670-3p, and miR-1224-3p. This prediction gained further support as these miRNAs were found to be upregulated in an LPS-induced MASH model, confirming their association with the circRNAs mentioned [97]. Intriguingly, certain mecciRNAs function through mechanisms other than simply acting as miRNA sponges to regulate MASH. CircSCAR, derived from the mitochondrial DNA, exerts an inhibitory effect on mitochondrial reactive oxygen species (mROS) flux, thereby mitigating fibroblast activation in MASH. This is achieved through its direct interaction with ATP5B, a component of the ATP synthase within the mitochondrial permeability transition pore (mPTP) complex. CircSCAR effectively hinders the interaction between CypD and mPTP, consequently impeding the opening of mPTP and thereby reducing the output of mROS [98].

4.3 Viral Hepatitis

Viral hepatitis, characterized by liver inflammation induced by a virus, leads to liver damage. The primary culprits behind this condition are hepatitis B virus (HBV) and hepatitis C virus (HCV).

HBV, a viral infection with the potential for both acute and chronic manifestations, primarily targets the liver. According to the World Health Organization (WHO), approximately 296 million individuals were living with chronic hepatitis B infection in 2019, with an annual incidence of 1.5 million new infections. The year 2019 witnessed an estimated 820,000 deaths attributed to hepatitis B, primarily arising from complications like cirrhosis and hepatocellular carcinoma [99].

Circ_0000650 exhibits significant downregulation in CHB, suggesting its involvement in HBV-induced liver disease. The microRNA miR-6873-3p, known to target transforming growth factor-beta 2 (TGF-β2), is predicted to be sequestered by circ_0000650 [100]. Notably, TGF-β2, which undergoes substantial downregulation in CHB, exerts a suppressive effect on the expression of various hepatocyte-enriched transcription factors that can promote the transcription of the virus [101].

In contrast, levels of circ_0004812 exhibited upregulation in both CHB patients and HBV-infected hepatoma cells, identifying circ_0004812 as a potential target in CHB infection. Notably, knockdown of circ_0004812 leads to an increase in the expression of IFN-α and IFN-β, known inhibitors of HBV replication. Additionally, circ_0004812 was found to enhance the expression of follistatin-like protein 1 (FSTL1) by inhibiting miR-1287-5p. In circ_0004812 knockdown cells treated with an miR-1287-5p inhibitor, HBV DNA copies and the ratio of HBsAg to HBeAg demonstrated recovery. These findings collectively underscore the regulatory role of the circ_0004812/miR-1287-5p/FSTL1 axis in modulating HBV-induced immune suppression [102].

CircBACH1 plays a pivotal role in promoting HBV replication by acting as a sponge for miR-200a-3p, which, in turn, targets mitogen-activated protein kinase kinase kinase 2 (MAP3K2) [103]. As a member of the serine/threonine-protein kinase family, MAP3K2 actively participates in the intricate process of HBV replication [104]. The interaction between circBACH1, miR-200a-3p, and MAP3K2 underscores the multifaceted regulatory mechanisms at play during HBV replication.

While less prevalent than HBV infection, HCV infection remains a significant health concern. This virus can give rise to both acute and chronic hepatitis, manifesting across a spectrum of severity from a mild, transient illness to a severe, lifelong condition that may lead to liver cirrhosis and cancer. Globally, an estimated 58 million individuals live with chronic HCV infection, and approximately 1.5 million new infections emerge annually. Among these cases, there are around 3.2 million adolescents and children affected by chronic hepatitis C infection. The WHO reported an estimated 290,000 deaths attributed to hepatitis C in 2019, primarily resulting from cirrhosis and hepatocellular carcinoma [105]. Therefore, it becomes imperative to explore the involvement of circRNAs in the pathogenesis of HCV. In particular, cellular treatment with circPSD3 siRNAs appears to slightly augment the translation of viral RNA compared to cells subjected to control siRNAs. Although the precise mechanism through which circPSD3 operates in influencing HCV infection remains elusive, the findings substantiate the proposition that circPSD3 impacts HCV RNA abundance at a post-translational stage. Notably, circPSD3 has been established to facilitate HCV infection [106].

Interestingly, Coxsackie virus B3 (CVB3) has the potential to cause viral myocarditis, pericarditis, and other inflammation-related diseases, including hepatitis. In HeLa cells infected with CVB3, hsa_circ_0000367 or circSIAE exhibited a significant decrease. CircSIAE was found to inhibit the replication and proliferation of the CVB3 virus through the sponge-like adsorption of miR-331-3p, which targets thousand and one amino-acid kinase 2 (TAOK2) [107]. However, the role of circSIAE in hepatitis related to CVB3 infection requires further investigation, particularly using hepatocyte cells.

4.4 Autoimmune Liver Disease

Autoimmune diseases (ADs) constitute chronic pathologies arising from the breakdown of immunological tolerance toward self-antigens, leading to either systemic or organ-specific damage [108]. In instances where the immune system targets the liver, it gives rise to autoimmune liver disease. While the numerical impact of autoimmune liver diseases may not match the scale observed in liver conditions linked to alcohol abuse, hepatitis virus infections, or steatosis-related pathology, their health burden remains considerable. Prominent examples of autoimmune liver diseases include autoimmune hepatitis (AIH), primary biliary cirrhosis (PBC), and primary (autoimmune) sclerosing cholangitis (PSC) [109].

In the mouse model induced with AIH, differentially expressed circular RNAs (DECs) were identified through microarray analysis, with mmu_circ_0001520 found to be upregulated. This particular circRNA is suggested to function as a sponge for mmu-miR-193b-3p, which has the capability to target MAPK10, a signaling pathway involved in AIH [110]. Another study reveals that hsa_circ_402458, exhibiting upregulated expression in PBC, exerts a competitive binding effect on both miR-522-3p and miR-943. Earlier investigations have underscored the potential role of miR-522-3p in chronic inflammatory diseases, while miR-943 has been implicated in the onset and progression of PBC through its regulatory influence on the TGF-β signaling pathway, which was found to be associated with the development of PBC in a murine model [111].

4.5 Liver Fibrosis/Cirrhosis

Liver fibrosis, characterized by the pathological accumulation of extracellular matrix (ECM) proteins, notably collagen, is a common manifestation of various chronic liver diseases, encompassing those detailed in the preceding descriptions. The progression of liver fibrosis often culminates in cirrhosis, a severe stage associated with significant morbidity and mortality [112]. In 2019, the global tally of deaths attributed to cirrhosis reached 1.43 million, marking an 8.1% increase from the figures recorded in 2017 [113]. The distribution of cirrhosis-related deaths in 2017 revealed distinct etiological factors, with 31.5% among men attributed to HBV, 25.5% to HCV, 27.3% to ALD, 7.7% to MASH, and 8.0% to various other factors. Among women, the proportions of cirrhosis-related deaths were 24.0% for HBV, 26.7% for HCV, 20.6% for ALD, 11.3% for MASH, and 17.3% for other causes [114].

HSCs assume a central role in the initiation and advancement of liver fibrosis and cirrhosis. In instances of sustained liver damage, cells undergoing damage release proinflammatory or profibrotic factors, initiating specific signaling pathways that trigger the activation of HSCs. The activated HSCs, having undergone transdifferentiation, expedite the secretion and impede the degradation of ECM components. This culminates in the substantial production of ECM, exacerbating the progression of liver fibrosis [115, 116]. The continuous cycle of profibrotic signaling and ECM accumulation amplifies the severity of liver fibrosis, highlighting the critical role played by HSCs in this pathological cascade.

The primary fibrogenic pathway implicated in HSC activation and ECM synthesis is the TGF-β signaling pathway [117]. Numerous studies report circRNAs associated with HSC activation target miRNAs and proteins within the TGF-β pathway. One such circRNA is circPSD3 (mmu_circ_0001682). Acting as a miR-92b-3p sponge, circPSD3 promotes Smad7 expression [118], which, in turn, inhibits the TGF-β signaling pathway [119]. CircCREBBP functions as a miR-1291 sponge, enhancing LEFTY2 expression [120]. LEFTY2 inhibits the phosphorylation of Smad2/3, key molecules in the TGF-β pathway [121, 122]. Similarly, hsa_circ_0070963 (circSCLT1) acts as a miR-223-3p sponge, targeting the LEM domain containing 3 (LEMD3) [123], an inhibitor of Smad2/3 signaling in the TGF-β pathway [124]. Mmu_circ_34116 suppresses HSC activation by targeting the miR-22-3p/BMP7 axis [125], thereby inhibiting Smad3 activation and consequently antagonizing TGF-β signaling [126]. CircMTO1 interacts with miR-17-5p, upregulating Smad7 expression and negatively regulating the TGF-β signaling pathway [127]. Moreover, circMTO1 acts as a miR-181b-5p sponge, reducing HSC activation by enhancing PTEN activity, which is crucial as a negative regulator of the PI3K/Akt signaling pathway [128], another fibrogenic signaling pathway [129]. CircBNC2 plays an indirect antifibrotic role by safeguarding hepatocytes from injury-induced apoptosis. The circBNC2-translated protein (ctBNC2) facilitates CDK/cyclin complex translocation into the nucleus [130], preventing apoptosis and the release of TGF-β and damage-associated molecular patterns (DAMPs) from hepatocytes [131].

Certain circRNAs play an opposing role in modulating HSC activation via the TGF-β signaling pathway. Acting as profibrogenic circRNAs, circPWWP2A [132], circUBE2k [133], circTUBD1 [134, 135], and circCCNB1 [136] exert intricate regulatory effects. CircPWWP2A enhances TGF-β signaling by reducing miR-203 and miR-223 expression, leading to increased levels of FSTL1 and TLR4, respectively [132]. FSTL1 binds to TGF-β receptors [137], and TLR4 suppresses the expression of BAMBI, a negative regulator in the TGF-β pathway [138]. Overexpressed circUBE2K induces liver fibrosis by sponging miR-149-5p, leading to the upregulation of TGF-β2 expression [133]. CircTUBD1 interacts with miR-203a-3p and miR-146a-5p [134, 135], promoting TGF-β signaling by increasing Smad3 expression through miR-203a-3p sponging [134]. Additionally, circTUBD1 targets TLR4 through miR-146a-5p, enhancing TGF-β and NF-κB signaling pathways [135]. Lastly, circCCNB1 or hsa_circ_0072765 acts as a miR-197-3p sponge, leading to the negative regulation of miR-197-3p expression. The target gene of miR-197-3p is identified as TRPV3, a member of the Transient Receptor Potential (TRP) superfamily known to contribute to the progression of liver fibrosis [136]. Numerous circRNAs participate in fibrotic signaling pathways beyond TGF-β, also exhibiting roles as antifibrotic or profibrotic circRNAs by regulating HSC activation [139]. This highlights the significant involvement of circRNAs in the regulation of HSC activation and, consequently, in the development of liver fibrosis and cirrhosis.

4.6 Liver Cancer

Liver cancer stands as the fifth most prevalent cancer and the fourth leading cause of cancer-related mortality globally. The two primary types of primary liver cancers are HCC and intrahepatic cholangiocarcinoma (ICC). HCC constitutes over 80% of primary liver cancer cases worldwide [140]. Liver fibrosis emerges as a significant risk factor for HCC development, with up to 90% of cases arising in the context of a cir-

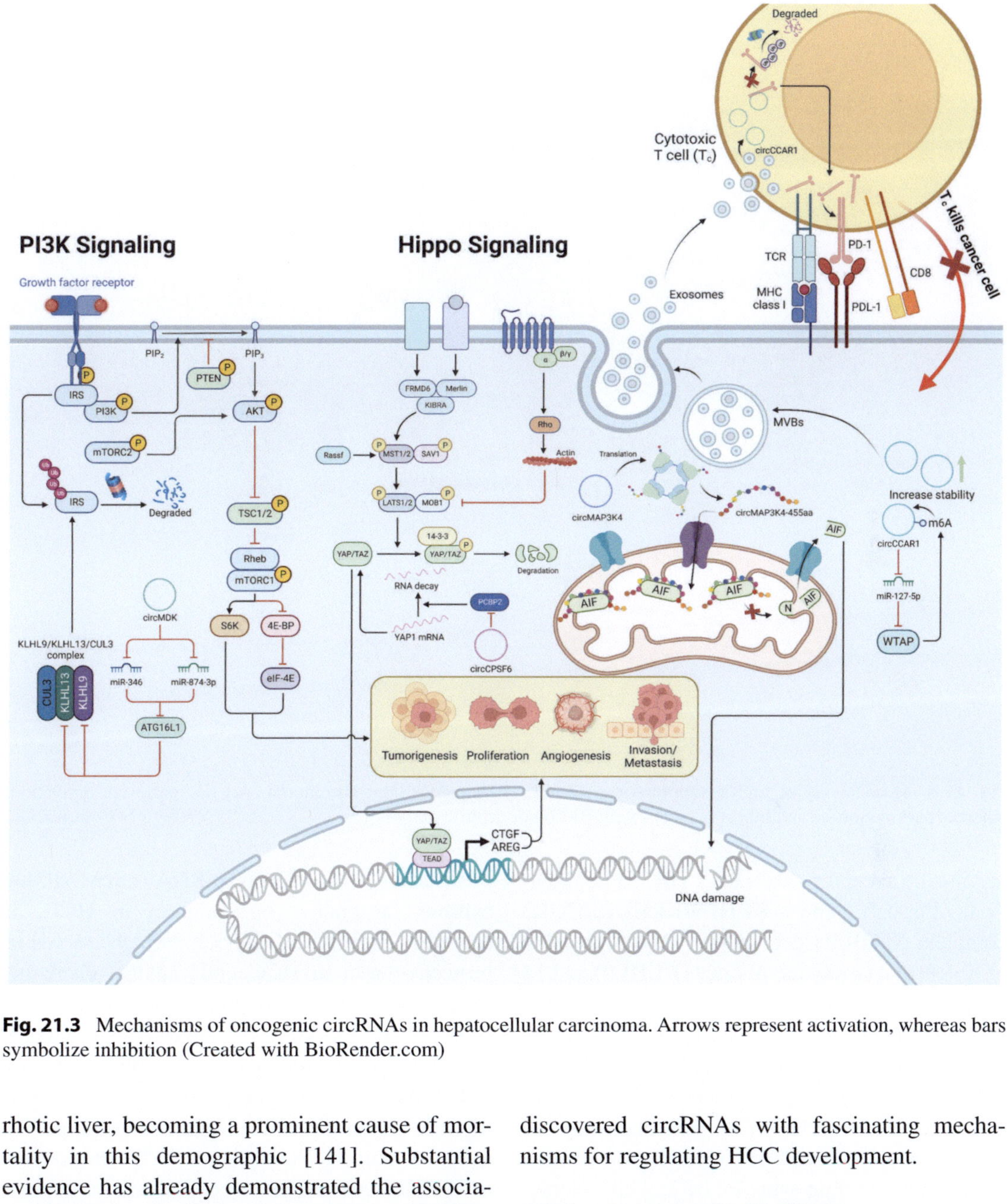

Fig. 21.3 Mechanisms of oncogenic circRNAs in hepatocellular carcinoma. Arrows represent activation, whereas bars symbolize inhibition (Created with BioRender.com)

rhotic liver, becoming a prominent cause of mortality in this demographic [141]. Substantial evidence has already demonstrated the association between circRNA dysregulation and HCC development. CircRNAs, in this context, can be categorized into oncogenic and tumor suppressors (Figs. 21.3 and 21.4). Currently, there is substantial research involving circRNAs that is considered to participate in cancer suppression or progression. Here we will list some of the recently discovered circRNAs with fascinating mechanisms for regulating HCC development.

4.6.1 Oncogenic CircRNAs

CircMDK facilitates the advancement of HCC through the miR-346/874-3p-ATG16L1 axis. Autophagy-related 16 Like 1 (ATG16L1) activates the PI3K/AKT/mTOR pathway, thereby enhancing cell growth, migration, invasion, and impeding apoptosis in HCC [142]. ATG16L1 regulates PI3K/Akt/mTOR signaling by inhibit-

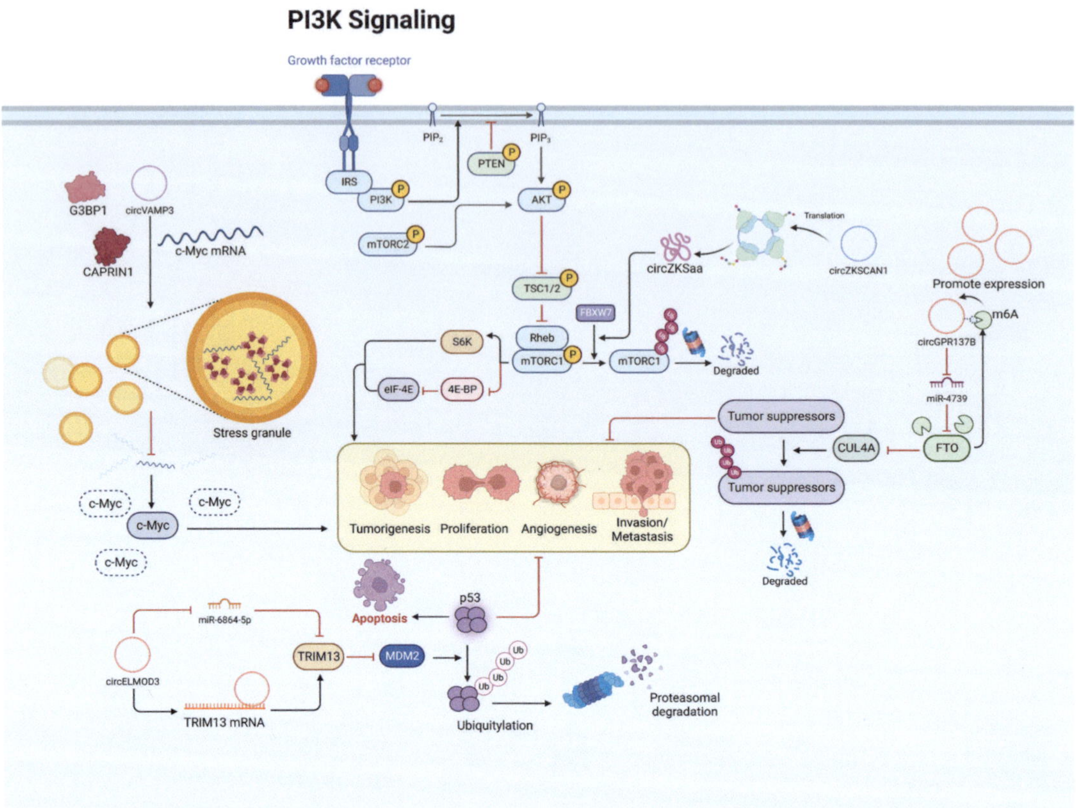

Fig. 21.4 Mechanisms of tumor suppressor circRNAs in hepatocellular carcinoma. Arrows represent activation, whereas bars symbolize inhibition (Created with BioRender.com)

ing insulin receptor substrate 1 (IRS1) degradation through the KLHL9/KLHL13/CUL3 complex. ATG16L1 can suppress the expression of Kelch-like family members 9 (KLHL9) and 13 (KLHL13). Both of which form an E3 ubiquitin ligase complex with cullin 3 (CUL3). These complexes promote the proteasomal degradation of IRS1. IRS1 can then promote the PI3K/Akt signaling pathway [143]. The tumorigenicity of HCC induced by circCPSF6 is facilitated through competitive binding with PCBP2. This interaction hinders the binding of PCBP2 to YAP1 mRNA, consequently maintaining the stability of YAP1 and activating the YAP signaling pathway. This activation, in turn, promotes the expression of YAP1 target genes, including connective tissue growth factor (CTGF) and amphiregulin (AREG), ultimately facilitating tumor proliferation and metastasis [144]. The circRNA circMAP3K4 exhibits significant upregulation in HCC. It encodes circMAP3K4-455aa, which serves a protective role in HCC cells against cisplatin-induced apoptosis. This protective mechanism is achieved by inhibiting the cleavage and nuclear translocation of the apoptosis-inducing factor (AIF). AIF, a mitochondrial protein, is released through N-terminal cleavage in response to apoptotic stimuli. Subsequently, AIF translocates to the nucleus, initiating cell death through a caspase-independent pathway. By impeding the cleavage of AIF, circMAP3K4-455aa prevents its translocation to the nucleus, consequently averting cell death in HCC cells experiencing stress. This inhibitory mechanism contributes to the evasion of apoptosis and promotes the progression of HCC [145].

4.6.2 Tumor Suppressor CircRNAs

CircGPR137B functions as a suppressor of HCC tumorigenesis and metastasis through the circGPR137B/miR-4739/FTO axis. Elevated miR-4739 expression correlates with advanced pathological stages and larger tumor sizes. CircGPR137B facilitates fat mass and obesity-associated protein (FTO) expression by acting as a sponge for miR-4739. The FTO protein translocates into the cellular nucleus, where it mediates m6A demethylation of circGPR137B, thereby enhancing its expression. This establishes a positive feedback loop that effectively suppresses tumor growth [146]. Many HCC patients usually exhibit higher levels of m6A mRNA modification. This further supports the antitumor role of FTO. CUL4A is among the genes that can be demethylated by FTO. The hypermethylation of CUL4A can increase their expression by enhancing mRNA translation. CUL4A E3 can form a ubiquitin ligase complex with other proteins [147]. These complexes can then mediate the ubiquitination and degradation of target proteins, many of which are tumor suppressor genes [148]. CircVAMP3 exerts a negative regulatory effect on the proliferation and metastasis of HCC. Through its interaction with CAPRIN1 and G3BP1, circVAMP3 induces the phase separation of CAPRIN1, facilitating stress granule formation. Acting as a molecular scaffolder, circVAMP3 promotes the sequestration of c-Myc mRNA within these stress granules, leading to the downregulation of the protein level of the Myc proto-oncogene protein [149]. CircZKSCAN1 encodes a secretory peptide known as circZKSaa. This peptide functions as a tumor suppressor within the PI3K/AKT/mTOR pathway by facilitating the interaction between FBXW7 and the mTOR protein. As a result, the mTOR protein undergoes ubiquitin-mediated degradation, influencing a pathway that is classically dysregulated and plays a crucial role in the pathogenesis of HCC [150]. CircELMOD3 suppressed the proliferation of HCC cells and induced apoptosis by functioning as a molecular sponge for miR-6864-5p. This action resulted in an upregulation of its downstream target gene, TRIM13. Moreover, circELMOD3 demonstrated an additional role in stabilizing the directly bound TRIM13 mRNA. Consequently, this stabilization led to the co-activation of the well-established tumor suppressor signaling pathway, p53 [151].

Interestingly, certain circRNAs exhibit functionality beyond their cells of origin, manifesting their effects in other cell types. Specifically, exosomal circRNAs assume a pivotal role, demonstrating diverse cellular functions, with some implicated in the development of various neoplasia [152]. Notably, the release of exosomal circCCAR1 by HCC cells contributes to immunosuppression by inducing $CD8^+$ T cell dysfunction through a circCCAR1/miR-127-5p/WTAP feedback loop, ultimately enhancing the growth and metastasis of HCC. Overexpression of WTAP elevates circCCAR1 levels via WTAP-mediated m6A modification, establishing a positive feedback loop that promotes the secretion of circCCAR1 from HCC cells. These released circCCAR1 exosomes are then internalized by $CD8^+$ T cells where, they prevent PD1 from ubiquitination-mediated degradation by directly binding to PD1. Consequently, PD1 overexpression in $CD8^+$ T cells diminishes their tumor-killing ability (Fig. 21.3) [153].

5 Conclusion and Future Perspectives

Numerous recent studies lend support to the hypothesis that circRNAs play integral roles in various cellular processes, associating them particularly with the pathogenesis of numerous diseases. Notably, liver-specific circRNAs have been implicated in the development of chronic liver diseases. The mechanisms through which circRNAs exert their effects include functioning as miRNA sponges, acting as protein sponges, enhancing protein function, and serving as protein scaffolds. Additionally, despite their classification as noncoding RNAs, some circRNAs exhibit functionality through the translation of peptides [32].

The implications of circRNAs in chronic liver diseases position them as potential biomarkers and therapeutic targets. As previously discussed, liver diseases are profoundly debilitating. Early diagnosis in patients can significantly mitigate mortality rates by providing timely warnings, enabling access to treatment before the disease advances into a more severe and challenging-to-treat stage. The conventional approach for diagnosing numerous liver diseases typically involves biopsy [154], a method fraught with various limitations such as invasiveness, high cost, poor acceptability, and the potential for sampling errors [155]. Because of these limitations, it is inappropriate for screening populations at risk [155]. Therefore, the utilization of circulating circRNAs associated with the pathogenesis of liver diseases emerges as a promising alternative for less invasive diagnostic tools. This innovative approach aims to enhance accessibility to early treatment options. CircRNAs exhibit abundant expression patterns specific to cell types and tissues, such as liver-specific circRNAs indicative of liver pathologic damage, which can be readily isolated from serum or plasma. A distinguishing feature lies in the resistance of circRNAs to degradation by RNase R exonuclease, rendering them more stable compared to their linear counterparts. This inherent stability positions circRNAs as promising serum markers for disease diagnosis and monitoring. Moreover, their predominant cytoplasmic localization facilitates easy detection in peripheral circulation through polymerase chain reactions, providing heightened sensitivity compared to protein biomarkers [156]. While the conceptual appeal of circRNAs as biomarkers for liver diseases is evident, additional investigations are imperative to delineate the specific circRNAs that can function as reliable biomarkers in diverse populations.

A noteworthy study has demonstrated the feasibility of utilizing circRNAs within nanoparticles as a drug delivery system for targeted delivery to the liver. Metal-organic frameworks (MOFs) emerge as promising nanocarriers due to their heightened specific surface area, porosity, modifiable pore dimensions, and facile chemical customization [157]. Glycyrrhizic acid (GA), known for its diverse range of biological advantages, including anti-inflammatory, antiviral, antitumor, antimicrobial, and hepatoprotective activities, is an ideal candidate for deployment as a nanocarrier [158]. However, owing to the poor solubility of MOFs constructed from GA and potential immune clearance affecting their bioavailability, encapsulation becomes imperative [94, 159]. To achieve effective liver-targeted delivery, red blood cell membranes modified with galactose were employed to selectively recognize asialoglycoprotein receptors (ASGPR), predominantly expressed on the surfaces of hepatocytes. This modification facilitates efficient liver-targeted delivery by providing camouflage against immune clearance and enhancing overall bioavailability [94, 160]. This study contributes to paving the way for future investigations into the therapeutic potential of circRNAs in humans. Interestingly, circRNAs could serve as therapeutic vaccines by utilizing those that encode cancer-specific cryptic antigenic peptides. This approach has the potential to induce antitumor immunity by stimulating adaptive immune responses, thereby suppressing tumor progression [161].

In summary, circRNAs have emerged as pivotal players in the epigenetic orchestration of the initiation and advancement of chronic liver diseases. The imperative for additional endeavors becomes evident, as there is an urgent requirement to advance circRNAs into novel diagnostic tools and therapeutic interventions for these conditions (Table 21.1).

Acknowledgments The authors were supported by Ratchadaphiseksomphot Matching Fund, Faculty of Medicine, Chulalongkorn University (RA-MF-31/65 and RA-MF-29/66), and Ratchadaphiseksomphot Fund, Graduate Affairs, Faculty of Medicine, Chulalongkorn University (GA66/077). A. Nokkaew was supported by the Second Century Fund (C2F), Chulalongkorn University, for the doctoral fellowship.

Competing Financial Interest The authors declare no competing financial interest.

Table 21.1 A list of circRNAs implicated in chronic liver disorders

CircRNAs	Diseases	Role	Actions	Mechanism	Refs
circRNA_1639	ALD	Promote	circRNA_1639/miR-122/TNFRSF13C	miRNA sponge	[41]
mm9_circ_018725		Promote	Mediates the apoptosis and inflammation	Possibly miRNA sponge	[42]
mou_circ_1657		Promote	mou_circ_1657/miR-96-5p	miRNA sponge	[43]
circRNA_002581	MAFLD/MASH	Promote	circRNA_002581/miR-122/CPEB1	miRNA sponge	[52]
circPI4KB		Promote	Promotes extracellular transport and subsequently leading to reduced intracellular levels of miRNA-122	miRNA sponge	[58]
circLDLR		Alleviate	circLDLR/miR-667-5p/SIRT1	miRNA sponge	[59]
circ_0057558		Promote	circ_0057558/miR-206/ROCK1/AMPK	miRNA sponge	[63]
circACC1		Alleviate	Binds to the regulatory β1 and γ1 subunits of AMPK, consequently promoting the assembly, stabilization, and activity of AMPK	Enhancer of protein function	[65]
circACACA		Promote	Promotes the expression of SREBP-1c and FASN	Possibly miRNA sponge	[67]
circRNA_0046366/7		Alleviate	circRNA_0046366-7/miR-34a/PPARα	miRNA sponge	[71, 72]
circRNA_021412		Alleviate	circRNA_021412/miR-1972/lipin1	miRNA sponge	[77]
circ_0048179		Alleviate	circ_0048179/miR-188-3p/GPX4	miRNA sponge	[79]
circScd1		Alleviate	Promotes JAK2/STAT5 pathway	Possibly miRNA sponge	[83]
mmu_circRNA_0049392		Alleviate	mmu_circRNA_0049392/miR-7037-5p, miR-6919-5p	miRNA sponge	[84]
circRNA_0000660		Alleviate	circRNA_0000660/miR_693/IGFBP1	miRNA sponge	[87]
circPTK2		Alleviate	circPTK2/miR-200c/SIK2/PI3K/Akt	miRNA sponge	[89]
circ_0004535		Alleviate	circ_0004535/miR-1827/CASP8	miRNA sponge	[93]
circRNA_0001805		Alleviate	circRNA_0001805/miR-106a-5p/ABCA1 circRNA_0001805/miR-320a/CPT1	miRNA sponge	[94]
circ_0089761/3		Alleviate	Target miR-642a-5p, miR-1248, miR-670-3p, and miR-1224-3p	miRNA sponge	[97]
circSCAR		Alleviate	Inhibits mROS flux by shuts down mPTP	Protein sponge	[98]
circ_0000650	Viral hepatitis	Alleviate	circ_0000650/miR-6873-3p/TGFβ2	miRNA sponge	[100]
circ_0004812		Promote	circ_0004812/miR-1287-5p/FSTL1	miRNA sponge	[102]
circBACH1		Promote	circBACH1/miR-200a-3p/MAP3K2	miRNA sponge	[103]

(continued)

Table 21.1 (continued)

CircRNAs	Diseases	Role	Actions	Mechanism	Refs
mmu_circ_0001520	Autoimmune	Promote	mmu_circ_0001520/miR-193b-3p/MAPK10	miRNA sponge	[110]
circ_402458		Promote	Targets miR-522-3p and miR-943	miRNA sponge	[111]
circPSD3	Fibrosis/ Cirrhosis	Alleviate	circPSD3/miR-92b-3p/Smad7	miRNA sponge	[118]
circCREBBP		Alleviate	circCREBBP/miR-1291/LEFTY2	miRNA sponge	[120]
circSCLT1		Alleviate	circSCLT1/miR-223-3p/LEMD3	miRNA sponge	[123]
mmu_circ_34116		Alleviate	mmu_circ_34116/miR-22-3p/BMP7/Smad3	miRNA sponge	[125]
circMTO1		Alleviate	circMTO1/miR-17-5p/Smad7 circMTO1/miR-181b-5p/PTEN	miRNA sponge	[127, 128]
circBNC2		Alleviate	Translates into ctBNC2 and prevent cell cycle arrest and the release of profibrotic factors	Protein template	[130]
circPWWP2A		Promote	circPWWP2A/miR-203/FSTL1 circPWWP2A/miR-223/TLR4	miRNA sponge	[132]
circUBE2k		Promote	circUBE2k/miR-149-5p/TGFβ2	miRNA sponge	[133]
circTUBD1		Promote	circTUBD1/miR-203a-3p/Smad3 circTUBD1/miR-146a-5p/TLR4	miRNA sponge	[134, 135]
circCCNB1		Promote	circCCNB1/miR-197-3p/TRPV3	miRNA sponge	[136]
circMDK	HCC	Promote	circMDK/miR-346, miR-874-3p/ATG16L1	miRNA sponge	[142]
circCPSF6		Promote	Activates YAP signaling pathway by hinders the binding of PCBP2 to YAP1 mRNA	Protein sponge	[144]
circMAP3K4		Promote	Encodes circMAP3K4-455aa which prevent cisplatin-induced apoptosis	Protein template	[145]
circGPR137B		Alleviate	circGPR137B/miR-4739/FTO	miRNA sponge	[146]
circVAMP3		Alleviate	Facilitates stress granule formation and downregulates the protein level of the Myc	Protein scaffold	[149]
circZKSCAN1		Alleviate	Encodes circZKSaa which facilitating the interaction between FBXW7 and the mTOR protein	Protein template	[150]
circELMOD3		Alleviate	circELMOD3/miR-6864-5p/TRIM13	miRNA sponge	[151]
circCCAR1		Promote	circCCAR1/miR-127-5p/WTAP	miRNA sponge	[153]

References

1. Asrani SK, Devarbhavi H, Eaton J, Kamath PS (2019) Burden of liver diseases in the world. J Hepatol 70(1):151–171. https://doi.org/10.1016/j.jhep.2018.09.014
2. Åberg F, Färkkilä M (2020) Drinking and obesity: alcoholic liver disease/nonalcoholic fatty liver disease interactions. Semin Liver Dis 40(2):154–162. https://doi.org/10.1055/s-0040-1701443
3. Song M, Xia L, Sun M, Yang C, Wang F (2018) Circular RNA in liver: health and diseases. Adv Exp Med Biol 1087:245–257. https://doi.org/10.1007/978-981-13-1426-1_20
4. Mumtaz PT, Taban Q, Dar MA, Mir S, Haq Z, Zargar SM, Shah RA, Ahmad SM (2020) Deep insights in circular RNAs: from biogenesis to therapeutics. Biol Proced Online 22(1):10. https://doi.org/10.1186/s12575-020-00122-8
5. Yu C-Y, Kuo H-C (2019) The emerging roles and functions of circular RNAs and their generation. J Biomed Sci 26(1):29. https://doi.org/10.1186/s12929-019-0523-z
6. Lasda E, Parker R (2014) Circular RNAs: diversity of form and function. RNA 20(12):1829–1842. https://doi.org/10.1261/rna.047126.114
7. Devarbhavi H, Asrani SK, Arab JP, Nartey YA, Pose E, Kamath PS (2023) Global burden of liver disease: 2023 update. J Hepatol 79(2):516–537. https://doi.org/10.1016/j.jhep.2023.03.017
8. Younossi ZM, Wong G, Anstee QM, Henry L (2023) The global burden of liver disease. Clin Gastroenterol Hepatol 21(8):1978–1991. https://doi.org/10.1016/j.cgh.2023.04.015
9. Nusrat S, Khan MS, Fazili J, Madhoun MF (2014) Cirrhosis and its complications: evidence based treatment. World J Gastroenterol 20(18):5442–5460. https://doi.org/10.3748/wjg.v20.i18.5442
10. Xiao S, Xie W, Zhang Y, Lei L, Pan Y (2023) Changing epidemiology of cirrhosis from 2010 to 2019: results from the Global Burden Disease study 2019. Ann Med 55(2):2252326. https://doi.org/10.1080/07853890.2023.2252326
11. Santer L, Bär C, Thum T (2019) Circular RNAs: a novel class of functional RNA molecules with a therapeutic perspective. Mol Ther 27(8):1350–1363. https://doi.org/10.1016/j.ymthe.2019.07.001
12. Geng Y, Jiang J, Wu C (2018) Function and clinical significance of circRNAs in solid tumors. J Hematol Oncol 11(1):98. https://doi.org/10.1186/s13045-018-0643-z
13. Chen L, Huang C, Wang X, Shan G (2015) Circular RNAs in eukaryotic cells. Curr Genomics 16(5):312–318. https://doi.org/10.2174/1389202916666150707161554
14. Fu L-Y, Wang S-W, Hu M-Y, Jiang Z-L, Shen L-L, Zhou Y-P, Guo J-M, Hu Y-R (2021) Circular RNAs in liver diseases: mechanisms and therapeutic targets. Life Sci 264:118707. https://doi.org/10.1016/j.lfs.2020.118707
15. Zhou M, Li S, Huang C (2024) Physiological and pathological functions of circular RNAs in the nervous system. Neural Regen Res 19(2):342–349. https://doi.org/10.4103/1673-5374.379017
16. Di Agostino S, Riccioli A, De Cesaris P, Fontemaggi G, Blandino G, Filippini A, Fazi F (2020) Circular RNAs in embryogenesis and cell differentiation with a focus on cancer development. Front Cell Dev Biol 8. https://doi.org/10.3389/fcell.2020.00389
17. Zhou W-Y, Cai Z-R, Liu J, Wang D-S, Ju H-Q, Xu R-H (2020) Circular RNA: metabolism, functions and interactions with proteins. Mol Cancer 19(1):172. https://doi.org/10.1186/s12943-020-01286-3
18. Guo JU, Agarwal V, Guo H, Bartel DP (2014) Expanded identification and characterization of mammalian circular RNAs. Genome Biol 15(7):409. https://doi.org/10.1186/s13059-014-0409-z
19. Zhang Y, Zhang XO, Chen T, Xiang JF, Yin QF, Xing YH, Zhu S, Yang L, Chen LL (2013) Circular intronic long noncoding RNAs. Mol Cell 51(6):792–806. https://doi.org/10.1016/j.molcel.2013.08.017
20. Li Z, Huang C, Bao C, Chen L, Lin M, Wang X, Zhong G, Yu B, Hu W, Dai L, Zhu P, Chang Z, Wu Q, Zhao Y, Jia Y, Xu P, Liu H, Shan G (2015) Exon-intron circular RNAs regulate transcription in the nucleus. Nat Struct Mol Biol 22(3):256–264. https://doi.org/10.1038/nsmb.2959
21. Noto JJ, Schmidt CA, Matera AG (2017) Engineering and expressing circular RNAs via tRNA splicing. RNA Biol 14(8):978–984. https://doi.org/10.1080/15476286.2017.1317911
22. Kelly S, Greenman C, Cook PR, Papantonis A (2015) Exon skipping is correlated with exon circularization. J Mol Biol 427(15):2414–2417. https://doi.org/10.1016/j.jmb.2015.02.018
23. Barrett SP, Wang PL, Salzman J (2015) Circular RNA biogenesis can proceed through an exon-containing lariat precursor. elife 4:e07540. https://doi.org/10.7554/eLife.07540
24. Enuka Y, Lauriola M, Feldman ME, Sas-Chen A, Ulitsky I, Yarden Y (2016) Circular RNAs are long-lived and display only minimal early alterations in response to a growth factor. Nucleic Acids Res 44(3):1370–1383. https://doi.org/10.1093/nar/gkv1367
25. Chen LL, Yang L (2015) Regulation of circRNA biogenesis. RNA Biol 12(4):381–388. https://doi.org/10.1080/15476286.2015.1020271
26. Jeck WR, Sorrentino JA, Wang K, Slevin MK, Burd CE, Liu J, Marzluff WF, Sharpless NE (2013) Circular RNAs are abundant, conserved, and associated with ALU repeats. RNA 19(2):141–157. https://doi.org/10.1261/rna.035667.112
27. Li X, Liu C-X, Xue W, Zhang Y, Jiang S, Yin Q-F, Wei J, Yao R-W, Yang L, Chen L-L (2017) Coordinated circRNA biogenesis and function with NF90/

NF110 in viral infection. Mol Cell 67(2):214–227. e217. https://doi.org/10.1016/j.molcel.2017.05.023
28. Wu J, Qi X, Liu L, Hu X, Liu J, Yang J, Yang J, Lu L, Zhang Z, Ma S, Li H, Yun X, Sun T, Wang Y, Wang Z, Liu Z, Zhao W (2019) Emerging epigenetic regulation of circular RNAs in human cancer. Mol Ther Nucleic Acids 16:589–596. https://doi.org/10.1016/j.omtn.2019.04.011
29. Qu S, Yang X, Li X, Wang J, Gao Y, Shang R, Sun W, Dou K, Li H (2015) Circular RNA: a new star of noncoding RNAs. Cancer Lett 365(2):141–148. https://doi.org/10.1016/j.canlet.2015.06.003
30. Schmidt CA, Giusto JD, Bao A, Hopper AK, Matera AG (2019) Molecular determinants of metazoan tricRNA biogenesis. Nucleic Acids Res 47(12):6452–6465. https://doi.org/10.1093/nar/gkz311
31. Verduci L, Tarcitano E, Strano S, Yarden Y, Blandino G (2021) CircRNAs: role in human diseases and potential use as biomarkers. Cell Death Dis 12(5):468. https://doi.org/10.1038/s41419-021-03743-3
32. Kristensen LS, Andersen MS, Stagsted LVW, Ebbesen KK, Hansen TB, Kjems J (2019) The biogenesis, biology and characterization of circular RNAs. Nat Rev Genet 20(11):675–691. https://doi.org/10.1038/s41576-019-0158-7
33. Conn VM, Hugouvieux V, Nayak A, Conos SA, Capovilla G, Cildir G, Jourdain A, Tergaonkar V, Schmid M, Zubieta C, Conn SJ (2017) A circRNA from SEPALLATA3 regulates splicing of its cognate mRNA through R-loop formation. Nat Plants 3:17053. https://doi.org/10.1038/nplants.2017.53
34. Schneider T, Bindereif A (2017) Circular RNAs: coding or noncoding? Cell Res 27(6):724–725. https://doi.org/10.1038/cr.2017.70
35. Su M, Xiao Y, Ma J, Tang Y, Tian B, Zhang Y, Li X, Wu Z, Yang D, Zhou Y, Wang H, Liao Q, Wang W (2019) Circular RNAs in cancer: emerging functions in hallmarks, stemness, resistance and roles as potential biomarkers. Mol Cancer 18(1):90. https://doi.org/10.1186/s12943-019-1002-6
36. Zhang Y, Zhao Y, Liu Y, Wang M, Yu W, Zhang L (2020) Exploring the regulatory roles of circular RNAs in Alzheimer's disease. Transl Neurodegener 9(1):35. https://doi.org/10.1186/s40035-020-00216-z
37. Wang P, Zhang Y, Deng L, Qu Z, Guo P, Liu L, Yu Z, Wang P, Liu N (2022) The function and regulation network mechanism of circRNA in liver diseases. Cancer Cell Int 22(1):141. https://doi.org/10.1186/s12935-022-02559-1
38. Osna NA, Donohue TM Jr, Kharbanda KK (2017) Alcoholic liver disease: pathogenesis and current management. Alcohol Res 38(2):147–161
39. Hernández-Évole H, Jiménez-Esquivel N, Pose E, Bataller R (2023) Alcohol-associated liver disease: epidemiology and management. Ann Hepatol. https://doi.org/10.1016/j.aohep.2023.101162
40. Martin EF (2018) Liver transplantation for alcoholic liver disease. Gastroenterol Hepatol (N Y) 14(9):532–535
41. Lu X, Liu Y, Xuan W, Ye J, Yao H, Huang C, Li J (2019) Circ_1639 induces cells inflammation responses by sponging miR-122 and regulating TNFRSF13C expression in alcoholic liver disease. Toxicol Lett 314:89–97. https://doi.org/10.1016/j.toxlet.2019.07.021
42. Meng H, Wang L, You H, Huang C, Li J (2019) Circular RNA expression profile of liver tissues in an EtOH-induced mouse model of alcoholic hepatitis. Eur J Pharmacol 862:172642. https://doi.org/10.1016/j.ejphar.2019.172642
43. Dou X, Feng L, Ying N, Ding Q, Song Q, Jiang F, Wang C, Li S (2020) RNA sequencing reveals a comprehensive circular RNA expression profile in a mouse model of alcoholic liver disease. Alcohol Clin Exp Res 44(2):415–422. https://doi.org/10.1111/acer.14265
44. Pipitone RM, Ciccioli C, Infantino G, La Mantia C, Parisi S, Tulone A, Pennisi G, Grimaudo S, Petta S (2023) MAFLD: a multisystem disease. Ther Adv Endocrinol Metab 14:20420188221145549. https://doi.org/10.1177/20420188221145549
45. Pouwels S, Sakran N, Graham Y, Leal A, Pintar T, Yang W, Kassir R, Singhal R, Mahawar K, Ramnarain D (2022) Non-alcoholic fatty liver disease (NAFLD): a review of pathophysiology, clinical management and effects of weight loss. BMC Endocr Disord 22(1):63. https://doi.org/10.1186/s12902-022-00980-1
46. Nakatsuka T, Tateishi R, Koike K (2022) Changing clinical management of NAFLD in Asia. Liver Int 42(9):1955–1968. https://doi.org/10.1111/liv.15046
47. Kim D, Ahmed A (2022) Nonalcoholic fatty liver disease in early life and all-cause and cause-specific mortality. Hepatobiliary Surg Nutr 11(2):317–319. https://doi.org/10.21037/hbsn-21-571
48. Misra VL, Khashab M, Chalasani N (2009) Nonalcoholic fatty liver disease and cardiovascular risk. Curr Gastroenterol Rep 11(1):50–55. https://doi.org/10.1007/s11894-009-0008-4
49. Pierantonelli I, Svegliati-Baroni G (2019) Nonalcoholic fatty liver disease: basic pathogenetic mechanisms in the progression from NAFLD to NASH. Transplantation 103(1):e1–e13. https://doi.org/10.1097/tp.0000000000002480
50. Bandiera S, Pfeffer S, Baumert TF, Zeisel MB (2015) miR-122—a key factor and therapeutic target in liver disease. J Hepatol 62(2):448–457. https://doi.org/10.1016/j.jhep.2014.10.004
51. Wu G-y, Rui C, Chen J-q, Sho E, Zhan S-s, Yuan X-w, Ding Y-t (2017) MicroRNA-122 inhibits lipid droplet formation and hepatic triglyceride accumulation via Yin Yang 1. Cell Physiol Biochem 44(4):1651–1664. https://doi.org/10.1159/000485765

52. Jin X, Gao J, Zheng R, Yu M, Ren Y, Yan T, Huang Y, Li Y (2020) Antagonizing circRNA_002581–miR-122–CPEB1 axis alleviates NASH through restoring PTEN–AMPK–mTOR pathway regulated autophagy. Cell Death Dis 11(2):123. https://doi.org/10.1038/s41419-020-2293-7
53. Cheung O, Puri P, Eicken C, Contos MJ, Mirshahi F, Maher JW, Kellum JM, Min H, Luketic VA, Sanyal AJ (2008) Nonalcoholic steatohepatitis is associated with altered hepatic MicroRNA expression. Hepatology 48(6):1810–1820. https://doi.org/10.1002/hep.22569
54. Yamada H, Suzuki K, Ichino N, Ando Y, Sawada A, Osakabe K, Sugimoto K, Ohashi K, Teradaira R, Inoue T, Hamajima N, Hashimoto S (2013) Associations between circulating microRNAs (miR-21, miR-34a, miR-122 and miR-451) and non-alcoholic fatty liver. Clin Chim Acta 424:99–103. https://doi.org/10.1016/j.cca.2013.05.021
55. Pirola CJ, Fernández Gianotti T, Castaño GO, Mallardi P, San Martino J, Mora Gonzalez Lopez Ledesma M, Flichman D, Mirshahi F, Sanyal AJ, Sookoian S (2015) Circulating microRNA signature in non-alcoholic fatty liver disease: from serum non-coding RNAs to liver histology and disease pathogenesis. Gut 64(5):800–812. https://doi.org/10.1136/gutjnl-2014-306996
56. Tsai WC, Hsu SD, Hsu CS, Lai TC, Chen SJ, Shen R, Huang Y, Chen HC, Lee CH, Tsai TF, Hsu MT, Wu JC, Huang HD, Shiao MS, Hsiao M, Tsou AP (2012) MicroRNA-122 plays a critical role in liver homeostasis and hepatocarcinogenesis. J Clin Invest 122(8):2884–2897. https://doi.org/10.1172/jci63455
57. Czaja MJ (2016) Function of autophagy in nonalcoholic fatty liver disease. Dig Dis Sci 61(5):1304–1313. https://doi.org/10.1007/s10620-015-4025-x
58. Liu CH, Jiang W, Zeng Q, Wu D, Li H, Zhou L, Bai L, Tang H (2023) CircRNA-PI4KB induces hepatic lipid deposition in non-alcoholic fatty liver disease by transporting miRNA-122 to extra-hepatocytes. Int J Mol Sci 24(2). https://doi.org/10.3390/ijms24021297
59. Yuan X, Li Y, Wen S, Xu C, Wang C, He Y, Zhou L (2022) CircLDLR acts as a sponge for miR-667-5p to regulate SIRT1 expression in non-alcoholic fatty liver disease. Lipids Health Dis 21(1):127. https://doi.org/10.1186/s12944-022-01740-9
60. Lan F, Cacicedo JM, Ruderman N, Ido Y (2008) SIRT1 modulation of the acetylation status, cytosolic localization, and activity of LKB1. Possible role in AMP-activated protein kinase activation. J Biol Chem 283(41):27628–27635. https://doi.org/10.1074/jbc.M805711200
61. Khan SA, Sathyanarayan A, Mashek MT, Ong KT, Wollaston-Hayden EE, Mashek DG (2015) ATGL-catalyzed lipolysis regulates SIRT1 to control PGC-1α/PPAR-α signaling. Diabetes 64(2):418–426. https://doi.org/10.2337/db14-0325
62. Ding RB, Bao J, Deng CX (2017) Emerging roles of SIRT1 in fatty liver diseases. Int J Biol Sci 13(7):852–867. https://doi.org/10.7150/ijbs.19370
63. Chen X, Tan Q-Q, Tan X-R, Li S-J, Zhang X-X (2021) Circ_0057558 promotes nonalcoholic fatty liver disease by regulating ROCK1/AMPK signaling through targeting miR-206. Cell Death Dis 12(9):809. https://doi.org/10.1038/s41419-021-04090-z
64. Huang H, Lee SH, Sousa-Lima I, Kim SS, Hwang WM, Dagon Y, Yang WM, Cho S, Kang MC, Seo JA, Shibata M, Cho H, Belew GD, Bhin J, Desai BN, Ryu MJ, Shong M, Li P, Meng H, Chung BH, Hwang D, Kim MS, Park KS, Macedo MP, White M, Jones J, Kim YB (2018) Rho-kinase/AMPK axis regulates hepatic lipogenesis during overnutrition. J Clin Invest 128(12):5335–5350. https://doi.org/10.1172/jci63562
65. Li Q, Wang Y, Wu S, Zhou Z, Ding X, Shi R, Thorne RF, Zhang XD, Hu W, Wu M (2019) CircACC1 regulates assembly and activation of AMPK complex under metabolic stress. Cell Metab 30(1):157–173.e157. https://doi.org/10.1016/j.cmet.2019.05.009
66. Hardie DG, Pan DA (2002) Regulation of fatty acid synthesis and oxidation by the AMP-activated protein kinase. Biochem Soc Trans 30(Pt 6):1064–1070. https://doi.org/10.1042/bst0301064
67. Chen F, Huang J, Zhao N, Jin F, Fan Q, Du E, Wei J (2023) Dietary Morus alba L. leaf supplementation improves hepatic lipid accumulation of laying hens via downregulating CircACACA. Poult Sci 102(11):103042. https://doi.org/10.1016/j.psj.2023.103042
68. Li Y, Xu S, Mihaylova MM, Zheng B, Hou X, Jiang B, Park O, Luo Z, Lefai E, Shyy JY, Gao B, Wierzbicki M, Verbeuren TJ, Shaw RJ, Cohen RA, Zang M (2011) AMPK phosphorylates and inhibits SREBP activity to attenuate hepatic steatosis and atherosclerosis in diet-induced insulin-resistant mice. Cell Metab 13(4):376–388. https://doi.org/10.1016/j.cmet.2011.03.009
69. Kohjima M, Higuchi N, Kato M, Kotoh K, Yoshimoto T, Fujino T, Yada M, Yada R, Harada N, Enjoji M, Takayanagi R, Nakamuta M (2008) SREBP-1c, regulated by the insulin and AMPK signaling pathways, plays a role in nonalcoholic fatty liver disease. Int J Mol Med 21(4):507–511. https://doi.org/10.3892/ijmm.21.4.507
70. Dorn C, Riener MO, Kirovski G, Saugspier M, Steib K, Weiss TS, Gäbele E, Kristiansen G, Hartmann A, Hellerbrand C (2010) Expression of fatty acid synthase in nonalcoholic fatty liver disease. Int J Clin Exp Pathol 3(5):505–514
71. Guo XY, Sun F, Chen JN, Wang YQ, Pan Q, Fan JG (2018) circRNA_0046366 inhibits hepatocellular steatosis by normalization of PPAR signaling. World J Gastroenterol 24(3):323–337. https://doi.org/10.3748/wjg.v24.i3.323
72. Guo XY, Chen JN, Sun F, Wang YQ, Pan Q, Fan JG (2017) circRNA_0046367 prevents hepatoxicity of lipid peroxidation: an inhibitory role against hepatic

steatosis. Oxidative Med Cell Longev 2017:3960197. https://doi.org/10.1155/2017/3960197

73. van Raalte DH, Li M, Pritchard PH, Wasan KM (2004) Peroxisome proliferator-activated receptor (PPAR)-alpha: a pharmacological target with a promising future. Pharm Res 21(9):1531–1538. https://doi.org/10.1023/b:pham.0000041444.06122.8d
74. Wang L, Sun M, Cao Y, Ma L, Shen Y, Velikanova AA, Li X, Sun C, Zhao Y (2020) miR-34a regulates lipid metabolism by targeting SIRT1 in non-alcoholic fatty liver disease with iron overload. Arch Biochem Biophys 695:108642. https://doi.org/10.1016/j.abb.2020.108642
75. Pawlak M, Lefebvre P, Staels B (2015) Molecular mechanism of PPARα action and its impact on lipid metabolism, inflammation and fibrosis in non-alcoholic fatty liver disease. J Hepatol 62(3):720–733. https://doi.org/10.1016/j.jhep.2014.10.039
76. Reddy JK, Rao MS (2006) Lipid metabolism and liver inflammation. II. Fatty liver disease and fatty acid oxidation. Am J Physiol Gastrointest Liver Physiol 290(5):G852–G858. https://doi.org/10.1152/ajpgi.00521.2005
77. Guo XY, He CX, Wang YQ, Sun C, Li GM, Su Q, Pan Q, Fan JG (2017) Circular RNA profiling and bioinformatic modeling identify its regulatory role in hepatic steatosis. Biomed Res Int 2017:5936171. https://doi.org/10.1155/2017/5936171
78. Finck BN, Gropler MC, Chen Z, Leone TC, Croce MA, Harris TE, Lawrence JC Jr, Kelly DP (2006) Lipin 1 is an inducible amplifier of the hepatic PGC-1alpha/PPARalpha regulatory pathway. Cell Metab 4(3):199–210. https://doi.org/10.1016/j.cmet.2006.08.005
79. Yang W, Zhao J, Zhao Y, Li W, Zhao L, Ren Y, Ou R, Xu Y (2020) Hsa_circ_0048179 attenuates free fatty acid-induced steatosis via hsa_circ_0048179/miR-188-3p/GPX4 signaling. Aging (Albany NY) 12(23):23996–24008. https://doi.org/10.18632/aging.104081
80. Zhang H, Zhang E, Hu H (2021) Role of ferroptosis in non-alcoholic fatty liver disease and its implications for therapeutic strategies. Biomedicines 9(11). https://doi.org/10.3390/biomedicines9111660
81. Zhang Y, Wu Q, Liu J, Zhang Z, Ma X, Zhang Y, Zhu J, Thring RW, Wu M, Gao Y, Tong H (2022) Sulforaphane alleviates high fat diet-induced insulin resistance via AMPK/Nrf2/GPx4 axis. Biomed Pharmacother 152:113273. https://doi.org/10.1016/j.biopha.2022.113273
82. Kaltenecker D, Themanns M, Mueller KM, Spirk K, Suske T, Merkel O, Kenner L, Luís A, Kozlov A, Haybaeck J, Müller M, Han X, Moriggl R (2019) Hepatic growth hormone – JAK2 – STAT5 signalling: metabolic function, non-alcoholic fatty liver disease and hepatocellular carcinoma progression. Cytokine 124:154569. https://doi.org/10.1016/j.cyto.2018.10.010
83. Li P, Shan K, Liu Y, Zhang Y, Xu L, Xu L (2019) CircScd1 promotes fatty liver disease via the Janus kinase 2/signal transducer and activator of transcription 5 pathway. Dig Dis Sci 64(1):113–122. https://doi.org/10.1007/s10620-018-5290-2
84. Yuan X, Diao J, Du A, Wen S, Zhou L, Pan Y (2020) Circular RNA expression profiles and features in NAFLD mice: a study using RNA-seq data. J Transl Med 18(1):476. https://doi.org/10.1186/s12967-020-02637-w
85. Jeon H, Blacklow SC (2005) Structure and physiologic function of the low-density lipoprotein receptor. Annu Rev Biochem 74(1):535–562. https://doi.org/10.1146/annurev.biochem.74.082803.133354
86. Bieghs V, Van Gorp PJ, Wouters K, Hendrikx T, Gijbels MJ, van Bilsen M, Bakker J, Binder CJ, Lütjohann D, Staels B, Hofker MH, Shiri-Sverdlov R (2012) LDL receptor knock-out mice are a physiological model particularly vulnerable to study the onset of inflammation in non-alcoholic fatty liver disease. PLoS One 7(1):e30668. https://doi.org/10.1371/journal.pone.0030668
87. Chen Q, Liu M, Luo Y, Yu H, Zhang J, Li D, He Q (2020) Maternal obesity alters circRNA expression and the potential role of mmu_circRNA_0000660 via sponging miR_693 in offspring liver at weaning age. Gene 731:144354. https://doi.org/10.1016/j.gene.2020.144354
88. Pan J, Cen L, Zhou T, Yu M, Chen X, Jiang W, Li Y, Yu C, Shen Z (2021) Insulin-like growth factor binding protein 1 ameliorates lipid accumulation and inflammation in nonalcoholic fatty liver disease. J Gastroenterol Hepatol 36(12):3438–3447. https://doi.org/10.1111/jgh.15627
89. Li Y, Cen C-Q, Liu B, Zhou L, Huang X-M, Liu G-Y (2022) Overexpression of circ PTK2 suppresses the progression of nonalcoholic fatty liver disease via the miR-200c/SIK2/PI3K/Akt axis. Kaohsiung J Med Sci 38(9):869–878. https://doi.org/10.1002/kjm2.12568
90. Matsuda S, Kobayashi M, Kitagishi Y (2013) Roles for PI3K/AKT/PTEN pathway in cell signaling of nonalcoholic fatty liver disease. ISRN Endocrinol 2013:472432. https://doi.org/10.1155/2013/472432
91. Feng J, Qiu S, Zhou S, Tan Y, Bai Y, Cao H, Guo J, Su Z (2022) mTOR: a potential new target in non-alcoholic fatty liver disease. Int J Mol Sci 23(16). https://doi.org/10.3390/ijms23169196
92. Dong Q, Giorgianni F, Beranova-Giorgianni S, Deng X, O'Meally RN, Bridges D, Park EA, Cole RN, Elam MB, Raghow R (2015) Glycogen synthase kinase-3-mediated phosphorylation of serine 73 targets sterol response element binding protein-1c (SREBP-1c) for proteasomal degradation. Biosci Rep 36(1):e00284. https://doi.org/10.1042/bsr20150234
93. Li M, Zeng A, Tang X, Xu H, Xiong W, Guo Y (2023) Circ_0004535/miR-1827/CASP8 network involved in type 2 diabetes mellitus with nonal-

coholic fatty liver disease. Sci Rep 13(1):19807. https://doi.org/10.1038/s41598-023-47189-3
94. Li J, Qi J, Tang Y, Liu H, Zhou K, Dai Z, Yuan L, Sun C (2021) A nanodrug system overexpressed circRNA_0001805 alleviates nonalcoholic fatty liver disease via miR-106a-5p/miR-320a and ABCA1/CPT1 axis. J Nanobiotechnol 19(1):363. https://doi.org/10.1186/s12951-021-01108-8
95. Liu X, Wang X, Li J, Hu S, Deng Y, Yin H, Bao X, Zhang QC, Wang G, Wang B, Shi Q, Shan G (2020) Identification of mecciRNAs and their roles in the mitochondrial entry of proteins. Sci China Life Sci 63(10):1429–1449. https://doi.org/10.1007/s11427-020-1631-9
96. Liu X, Shan G (2021) Mitochondria encoded noncoding RNAs in cell physiology. Front Cell Dev Biol 9. https://doi.org/10.3389/fcell.2021.713729
97. Liu B, Tian Y, He J, Gu Q, Jin B, Shen H, Li W, Shi L, Yu H, Shan G, Cai X (2022) The potential of mecciRNA in hepatic stellate cell to regulate progression of nonalcoholic hepatitis. J Transl Med 20(1):393. https://doi.org/10.1186/s12967-022-03595-1
98. Zhao Q, Liu J, Deng H, Ma R, Liao J-Y, Liang H, Hu J, Li J, Guo Z, Cai J, Xu X, Gao Z, Su S (2020) Targeting mitochondria-located circRNA SCAR alleviates NASH via reducing mROS output. Cell 183(1):76–93.e22. https://doi.org/10.1016/j.cell.2020.08.009
99. World Health Organization (2023) Hepatitis B. https://www.who.int/news-room/fact-sheets/detail/hepatitis-b
100. Zhou TC, Li X, Chen LJ, Fan JH, Lai X, Tang Y, Zhang L, Wei J (2018) Differential expression profile of hepatic circular RNAs in chronic hepatitis B. J Viral Hepat 25(11):1341–1351. https://doi.org/10.1111/jvh.12944
101. Xia Y, Cheng X, Li Y, Valdez K, Chen W, Liang TJ (2018) Hepatitis B virus deregulates the cell cycle to promote viral replication and a premalignant phenotype. J Virol 92(19). https://doi.org/10.1128/jvi.00722-18
102. Zhang L, Wang Z (2020) Circular RNA hsa_circ_0004812 impairs IFN-induced immune response by sponging miR-1287-5p to regulate FSTL1 in chronic hepatitis B. Virol J 17(1):40. https://doi.org/10.1186/s12985-020-01314-0
103. Du N, Li K, Wang Y, Song B, Zhou X, Duan S (2022) CircRNA circBACH1 facilitates hepatitis B virus replication and hepatoma development by regulating the miR-200a-3p/MAP3K2 axis. Histol Histopathol 37(9):863–877. https://doi.org/10.14670/hh-18-452
104. Chen S, Yang LEI, Pan A, Duan S, Li M, Li P, Huang J, Gao X, Huang X, Lin Y (2019) Inhibitory effect on the hepatitis B cells through the regulation of miR-122-MAP3K2 signal pathway. An Acad Bras Ciênc 91:e20180941
105. World Health Organization (2023) Hepatitis C. https://www.who.int/news-room/fact-sheets/detail/hepatitis-c
106. Chen TC, Tallo-Parra M, Cao QM, Kadener S, Böttcher R, Pérez-Vilaró G, Boonchuen P, Somboonwiwat K, Díez J, Sarnow P (2020) Host-derived circular RNAs display proviral activities in Hepatitis C virus-infected cells. PLoS Pathog 16(8):e1008346. https://doi.org/10.1371/journal.ppat.1008346
107. Yang Q, Li Y, Wang Y, Qiao X, Liu T, Wang H, Shen H (2021) The circRNA circSIAE inhibits replication of coxsackie virus B3 by targeting miR-331-3p and thousand and one amino-acid kinase 2. Front Cell Infect Microbiol 11:779919. https://doi.org/10.3389/fcimb.2021.779919
108. Smith DA, Germolec DR (1999) Introduction to immunology and autoimmunity. Environ Health Perspect 107 Suppl 5(Suppl 5):661–665. https://doi.org/10.1289/ehp.99107s5661
109. Invernizzi P, Mackay IR (2008) Autoimmune liver diseases. World J Gastroenterol 14(21):3290–3291. https://doi.org/10.3748/wjg.14.3290
110. Liu Y, Li Z, Hao J, Chen H, Hou T, Hao H (2020) Circular RNAs associated with a mouse model of concanavalin A-induced autoimmune hepatitis: preliminary screening and comprehensive functional analysis. FEBS Open Bio 10(11):2350–2362. https://doi.org/10.1002/2211-5463.12981
111. Zheng J, Li Z, Wang T, Zhao Y, Wang Y (2017) Microarray expression profile of circular RNAs in plasma from primary biliary cholangitis patients. Cell Physiol Biochem 44(4):1271–1281. https://doi.org/10.1159/000485487
112. Bataller R, Brenner DA (2005) Liver fibrosis. J Clin Invest 115(2):209–218. https://doi.org/10.1172/jci24282
113. GBD 2021 Diabetes Collaborators (2022) Global, regional, and national burden of hepatitis B, 1990-2019: a systematic analysis for the Global Burden of Disease Study 2019. Lancet Gastroenterol Hepatol 7(9):796–829. https://doi.org/10.1016/s2468-1253(22)00124-8
114. GBD 2017 Cirrhosis Collaborators (2020) The global, regional, and national burden of cirrhosis by cause in 195 countries and territories, 1990-2017: a systematic analysis for the Global Burden of Disease Study 2017. Lancet Gastroenterol Hepatol 5(3):245–266. https://doi.org/10.1016/s2468-1253(19)30349-8
115. Pei Q, Yi Q, Tang L (2023) Liver fibrosis resolution: from molecular mechanisms to therapeutic opportunities. Int J Mol Sci 24(11). https://doi.org/10.3390/ijms24119671
116. Chan Y-T, Wang N, Tan HY, Li S, Feng Y (2020) Targeting hepatic stellate cells for the treatment of liver fibrosis by natural products: is it the dawning of a new era? Front Pharmacol 11. https://doi.org/10.3389/fphar.2020.00548
117. Dewidar B, Meyer C, Dooley S, Meindl-Beinker AN (2019) TGF-β in hepatic stellate cell activation and liver fibrogenesis-updated 2019. Cells 8(11). https://doi.org/10.3390/cells8111419

118. Bu F-t, Zhu Y, Chen X, Wang A, Zhang Y-f, You H-m, Yang Y, Yang Y-r, Huang C, Li J (2021) Circular RNA circPSD3 alleviates hepatic fibrogenesis by regulating the miR-92b-3p/Smad7 axis. Mol Ther Nucleic Acids 23:847–862. https://doi.org/10.1016/j.omtn.2021.01.007
119. Yan X, Liu Z, Chen Y (2009) Regulation of TGF-beta signaling by Smad7. Acta Biochim Biophys Sin Shanghai 41(4):263–272. https://doi.org/10.1093/abbs/gmp018
120. Yang Y-R, Hu S, Bu F-T, Li H, Huang C, Meng X-M, Zhang L, Lv X-W, Li J (2021) Circular RNA CREBBP suppresses hepatic fibrosis via targeting the hsa-miR-1291/LEFTY2 axis. Front Pharmacol 12:741151
121. Yang Y-r, Bu F-t, Yang Y, Li H, Huang C, Meng X-m, Zhang L, Lv X-w, Li J (2020) LEFTY2 alleviates hepatic stellate cell activation and liver fibrosis by regulating the TGF-β1/Smad3 pathway. Mol Immunol 126:31–39. https://doi.org/10.1016/j.molimm.2020.07.012
122. Lin AH, Luo J, Mondshein LH, ten Dijke P, Vivien D, Contag CH, Wyss-Coray T (2005) Global analysis of Smad2/3-dependent TGF-β signaling in living mice reveals prominent tissue-specific responses to injury1. J Immunol 175(1):547–554. https://doi.org/10.4049/jimmunol.175.1.547
123. Ji D, Chen GF, Wang JC, Ji SH, Wu XW, Lu XJ, Chen JL, Li JT (2020) Hsa_circ_0070963 inhibits liver fibrosis via regulation of miR-223-3p and LEMD3. Aging (Albany NY) 12(2):1643–1655. https://doi.org/10.18632/aging.102705
124. Chambers DM, Moretti L, Zhang JJ, Cooper SW, Chambers DM, Santangelo PJ, Barker TH (2018) LEM domain–containing protein 3 antagonizes TGFβ–SMAD2/3 signaling in a stiffness-dependent manner in both the nucleus and cytosol. J Biol Chem 293(41):15867–15886. https://doi.org/10.1074/jbc.RA118.003658
125. Zhou Y, Lv X, Qu H, Zhao K, Fu L, Zhu L, Ye G, Guo J (2018) Preliminary screening and functional analysis of circular RNAs associated with hepatic stellate cell activation. Gene 677:317–323. https://doi.org/10.1016/j.gene.2018.08.052
126. Zou GL, Zuo S, Lu S, Hu RH, Lu YY, Yang J, Deng KS, Wu YT, Mu M, Zhu JJ, Zeng JZ, Zhang BF, Wu X, Zhao XK, Li HY (2019) Bone morphogenetic protein-7 represses hepatic stellate cell activation and liver fibrosis via regulation of TGF-β/Smad signaling pathway. World J Gastroenterol 25(30):4222–4234. https://doi.org/10.3748/wjg.v25.i30.4222
127. Wang W, Dong R, Guo Y, He J, Shao C, Yi P, Yu F, Gu D, Zheng J (2019) CircMTO1 inhibits liver fibrosis via regulation of miR-17-5p and Smad7. J Cell Mol Med 23(8):5486–5496. https://doi.org/10.1111/jcmm.14432
128. Jin H, Li C, Dong P, Huang J, Yu J, Zheng J (2020) Circular RNA cMTO1 promotes PTEN expression through sponging miR-181b-5p in liver fibrosis. Front Cell Dev Biol 8:714
129. Cai CX, Buddha H, Castelino-Prabhu S, Zhang Z, Britton RS, Bacon BR, Neuschwander-Tetri BA (2017) Activation of insulin-PI3K/Akt-p70S6K pathway in hepatic stellate cells contributes to fibrosis in nonalcoholic steatohepatitis. Dig Dis Sci 62(4):968–978. https://doi.org/10.1007/s10620-017-4470-9
130. Wang P, Huang Z, Peng Y, Li H, Lin T, Zhao Y, Hu Z, Zhou Z, Zhou W, Liu Y, Hou FF (2022) Circular RNA circBNC2 inhibits epithelial cell G2-M arrest to prevent fibrotic maladaptive repair. Nat Commun 13(1):6502. https://doi.org/10.1038/s41467-022-34287-5
131. Ignat S-R, Dinescu S, Hermenean A, Costache M (2020) Cellular interplay as a consequence of inflammatory signals leading to liver fibrosis development. Cells 9(2). https://doi.org/10.3390/cells9020461
132. Liu W, Feng R, Li X, Li D, Zhai W (2019) TGF-β- and lipopolysaccharide-induced upregulation of circular RNA PWWP2A promotes hepatic fibrosis via sponging miR-203 and miR-223. Aging (Albany NY) 11(21):9569–9580. https://doi.org/10.18632/aging.102405
133. Zhu S, Chen X, Wang J-N, Xu J-J, Wang A, Li J-J, Wu S, Wu Y-Y, Li X-F, Huang C, Li J (2021) Circular RNA circUbe2k promotes hepatic fibrosis via sponging miR-149-5p/TGF-β2 axis. FASEB J 35(6):e21622. https://doi.org/10.1096/fj.202002738R
134. Niu H, Zhang L, Wang B, Zhang GC, Liu J, Wu ZF, Du SS, Zeng ZC (2022) CircTUBD1 regulates radiation-induced liver fibrosis response via a circTUBD1/micro-203a-3p/Smad3 positive feedback loop. J Clin Transl Hepatol 10(4):680–691. https://doi.org/10.14218/jcth.2021.00511
135. Niu H, Zhang L, Chen YH, Yuan BY, Wu ZF, Cheng JC, Lin Q, Zeng ZC (2020) Circular RNA TUBD1 acts as the miR-146a-5p sponge to affect the viability and pro-inflammatory cytokine production of LX-2 cells through the TLR4 pathway. Radiat Res 193(4):383–393. https://doi.org/10.1667/rr15550.1
136. Jin Y, Guo X, Zhang R, Yan C (2023) Hsa_circ_0072765 knockdown inhibits proliferation, activation and migration in transforming growth factor-beta (TGF-β)-induced hepatic stellate cells (HSCs) by the miR-197-3p/TRPV3 axis. Histol Histopathol 38(11):1295–1306. https://doi.org/10.14670/hh-18-586
137. Shang H, Liu X, Guo H (2017) Knockdown of Fstl1 attenuates hepatic stellate cell activation through the TGF-β1/Smad3 signaling pathway. Mol Med Rep 16(5):7119–7123. https://doi.org/10.3892/mmr.2017.7445

138. Seki E, De Minicis S, Österreicher CH, Kluwe J, Osawa Y, Brenner DA, Schwabe RF (2007) TLR4 enhances TGF-β signaling and hepatic fibrosis. Nat Med 13(11):1324–1332. https://doi.org/10.1038/nm1663
139. Nokkeaw A, Thamjamrassri P, Tangkijvanich P, Ariyachet C (2023) Regulatory functions and mechanisms of circular RNAs in hepatic stellate cell activation and liver fibrosis. Cells 12(3). https://doi.org/10.3390/cells12030378
140. Chidambaranathan-Reghupaty S, Fisher PB, Sarkar D (2021) Hepatocellular carcinoma (HCC): epidemiology, etiology and molecular classification. Adv Cancer Res 149:1–61. https://doi.org/10.1016/bs.acr.2020.10.001
141. O'Rourke JM, Sagar VM, Shah T, Shetty S (2018) Carcinogenesis on the background of liver fibrosis: implications for the management of hepatocellular cancer. World J Gastroenterol 24(39):4436–4447. https://doi.org/10.3748/wjg.v24.i39.4436
142. Du A, Li S, Zhou Y, Disoma C, Liao Y, Zhang Y, Chen Z, Yang Q, Liu P, Liu S, Dong Z, Razzaq A, Tao S, Chen X, Liu Y, Xu L, Zhang Q, Li S, Peng J, Xia Z (2022) M6A-mediated upregulation of circMDK promotes tumorigenesis and acts as a nanotherapeutic target in hepatocellular carcinoma. Mol Cancer 21(1):109. https://doi.org/10.1186/s12943-022-01575-z
143. Frendo-Cumbo S, Jaldin-Fincati JR, Coyaud E, Laurent EMN, Townsend LK, Tan JMJ, Xavier RJ, Pillon NJ, Raught B, Wright DC, Brumell JH, Klip A (2019) Deficiency of the autophagy gene ATG16L1 induces insulin resistance through KLHL9/KLHL13/CUL3-mediated IRS1 degradation. J Biol Chem 294(44):16172–16185. https://doi.org/10.1074/jbc.RA119.009110
144. Chen Y, Ling Z, Cai X, Xu Y, Lv Z, Man D, Ge J, Yu C, Zhang D, Zhang Y, Xie H, Zhou L, Wu J, Zheng S (2021) Activation of YAP1 by N6-methyladenosine–modified circCPSF6 drives malignancy in hepatocellular carcinoma. Cancer Res 82(4):599–614. https://doi.org/10.1158/0008-5472.CAN-21-1628
145. Duan JL, Chen W, Xie JJ, Zhang ML, Nie RC, Liang H, Mei J, Han K, Xiang ZC, Wang FW, Teng K, Chen RX, Deng MH, Yin YX, Zhang N, Xie D, Cai MY (2022) A novel peptide encoded by N6-methyladenosine modified circMAP3K4 prevents apoptosis in hepatocellular carcinoma. Mol Cancer 21(1):93. https://doi.org/10.1186/s12943-022-01537-5
146. Liu L, Gu M, Ma J, Wang Y, Li M, Wang H, Yin X, Li X (2022) CircGPR137B/miR-4739/FTO feedback loop suppresses tumorigenesis and metastasis of hepatocellular carcinoma. Mol Cancer 21(1):149. https://doi.org/10.1186/s12943-022-01619-4
147. Mittenbühler MJ, Saedler K, Nolte H, Kern L, Zhou J, Qian SB, Meder L, Ullrich RT, Brüning JC, Wunderlich FT (2020) Hepatic FTO is dispensable for the regulation of metabolism but counteracts HCC development in vivo. Mol Metab 42:101085. https://doi.org/10.1016/j.molmet.2020.101085
148. Pan Y, Wang B, Yang X, Bai F, Xu Q, Li X, Gao L, Ma C, Liang X (2015) CUL4A facilitates hepatocarcinogenesis by promoting cell cycle progression and epithelial-mesenchymal transition. Sci Rep 5:17006. https://doi.org/10.1038/srep17006
149. Chen S, Cao X, Zhang J, Wu W, Zhang B, Zhao F (2022) circVAMP3 drives CAPRIN1 phase separation and inhibits hepatocellular carcinoma by suppressing c-Myc translation. Adv Sci (Weinh) 9(8):e2103817. https://doi.org/10.1002/advs.202103817
150. Song R, Ma S, Xu J, Ren X, Guo P, Liu H, Li P, Yin F, Liu M, Wang Q, Yu L, Liu J, Duan B, Rahman NA, Wołczyński S, Li G, Li X (2023) A novel polypeptide encoded by the circular RNA ZKSCAN1 suppresses HCC via degradation of mTOR. Mol Cancer 22(1):16. https://doi.org/10.1186/s12943-023-01719-9
151. Lai M, Liu M, Li D, Zhang R, Wang L, Li X, Chen S, Wu S, Xiao S, Wei L, Wang X, Nan A, Zeng X (2023) circELMOD3 increases and stabilizes TRIM13 by sponging miR-6864-5p and direct binding to inhibit HCC progression. iScience 26(10):107818. https://doi.org/10.1016/j.isci.2023.107818
152. Wang Y, Liu J, Ma J, Sun T, Zhou Q, Wang W, Wang G, Wu P, Wang H, Jiang L, Yuan W, Sun Z, Ming L (2019) Exosomal circRNAs: biogenesis, effect and application in human diseases. Mol Cancer 18(1):116. https://doi.org/10.1186/s12943-019-1041-z
153. Hu Z, Chen G, Zhao Y, Gao H, Li L, Yin Y, Jiang J, Wang L, Mang Y, Gao Y, Zhang S, Ran J, Li L (2023) Exosome-derived circCCAR1 promotes CD8 + T-cell dysfunction and anti-PD1 resistance in hepatocellular carcinoma. Mol Cancer 22(1):55. https://doi.org/10.1186/s12943-023-01759-1
154. Barrera-Saldaña HA, Fernández-Garza LE, Barrera-Barrera SA (2021) Liquid biopsy in chronic liver disease. Ann Hepatol 20. https://doi.org/10.1016/j.aohep.2020.03.008
155. Castera L, Friedrich-Rust M, Loomba R (2019) Noninvasive assessment of liver disease in patients with nonalcoholic fatty liver disease. Gastroenterology 156(5):1264–1281.e1264. https://doi.org/10.1053/j.gastro.2018.12.036
156. Zeng Q, Liu C-H, Ampuero J, Wu D, Jiang W, Zhou L, Li H, Bai L, Romero-Gómez M, Tang H (2024) Circular RNAs in non-alcoholic fatty liver disease: functions and clinical significance. RNA Biol 21(1):1–15. https://doi.org/10.1080/15476286.2023.2290769
157. Sun Y, Zheng L, Yang Y, Qian X, Fu T, Li X, Yang Z, Yan H, Cui C, Tan W (2020) Metal–organic framework nanocarriers for drug delivery in biomedical applications. Nanomicro Lett 12(1):103. https://doi.org/10.1007/s40820-020-00423-3
158. Nascimento MH, de Araújo DR (2022) Exploring the pharmacological potential of Glycyrrhizic acid: from therapeutic applications to trends in nanomedicine. Future Pharmacol 2(1):1–15. https://doi.org/10.3390/futurepharmacol2010001

159. Bernela M, Ahuja M, Thakur R (2016) Enhancement of anti-inflammatory activity of glycyrrhizic acid by encapsulation in chitosan-katira gum nanoparticles. Eur J Pharm Biopharm 105:141–147. https://doi.org/10.1016/j.ejpb.2016.06.003
160. Saraswat A, Vemana HP, Dukhande VV, Patel K (2022) Galactose-decorated liver tumor-specific nanoliposomes incorporating selective BRD4-targeted PROTAC for hepatocellular carcinoma therapy. Heliyon 8(1):e08702. https://doi.org/10.1016/j.heliyon.2021.e08702
161. Huang D, Zhu X, Ye S, Zhang J, Liao J, Zhang N, Zeng X, Wang J, Yang B, Zhang Y, Lao L, Chen J, Xin M, Nie Y, Saw PE, Su S, Song E (2023) Tumour circular RNAs elicit anti-tumour immunity by encoding cryptic peptides. Nature. https://doi.org/10.1038/s41586-023-06834-7

Circular RNAs in Respiratory Diseases

22

Sema Misir, Serap Ozer Yaman, Ceylan Hepokur, and Yuksel Aliyazicioglu

Abstract

Respiratory diseases are among the most severe diseases and are among the leading causes of death in the world. Early diagnosis and prognosis prediction, as well as effective treatment, are vital to reducing mortality rates. Recently, increasing reports have shown that circRNAs are associated with the occurrence and progression of cancerous and noncancerous respiratory diseases through different mechanisms. In this chapter, we summarize the critical roles of circRNAs in the occurrence and progression of respiratory diseases. We also discussed the opportunity to use these molecules as potential therapeutic targets for diagnosing, prognosing, and treating respiratory diseases.

Keywords

Biomarker · Circular RNA · Respiratory diseases · MicroRNA sponge

Abbreviations

circRNA	circular RNA
miRNA	microRNA
LC	Lung carcinoma
LLCC	Lung large cell carcinoma
NSCLC	Non-small cell lung cancer
LSCC	Laryngeal squamous cell carcinoma
NPC	Nasopharyngeal carcinoma
COPD	Chronic obstructive pulmonary disease
PH	Pulmonary hypertension
PAH	Pulmonary arterial hypertension
TB	Pulmonary tuberculosis
PBMC	Peripheral blood mononuclear cells

1 Introduction

Circular RNAs (circRNAs) are a class of endogenous, noncoding RNA molecules containing single-stranded transcripts that can covalently construct circular closed structures [1]. CircRNAs are extensively expressed in the human body in

S. Misir (✉) · C. Hepokur
Department of Biochemistry, Faculty of Pharmacy, Sivas Cumhuriyet University, Sivas, Turkey
e-mail: smisir@cumhuriyet.edu.tr; cozsoy@cumhuriyet.edu.tr

S. O. Yaman · Y. Aliyazicioglu
Department of Medical Biochemistry, Faculty of Medicine, Karadeniz Technical University, Trabzon, Turkey
e-mail: serapozer@ktu.edu.tr; yayazici@ktu.edu.tr

J. Xiao (ed.), *Advances in Circular RNAs*, Advances in Experimental Medicine and Biology 1485,
https://doi.org/10.1007/978-981-96-9428-0_22

tissue disease and developmental stage-specific manners and play a vital role in the pathogenesis and progression of human diseases [2]. During the development of various diseases such as cancers, cardiovascular diseases, and neurological diseases [3], circRNAs can act as microRNA (miRNA) sponges, interact with RNA-binding proteins (RBPs), regulate gene transcription, and even directly regulate gene expression through coding proteins [4]. Furthermore, circRNAs are considered attractive biomarker candidates for liquid biopsy of clinical samples, including whole blood, serum, saliva, urine, and cerebrospinal fluid, because of their exceptional stability and large abundance in bodily fluids and circulating extracellular vesicles [2]. Respiratory diseases are one of the most common medical conditions in the world and are a leading cause of death rates. Therefore, the importance of regulatory molecules such as circRNAs in diagnosing, progressing, and treating these diseases is increasing [5]. Recent studies have identified many circRNAs in respiratory diseases using circRNA microarrays or next-generation sequencing. Although the biological mechanisms of various circRNAs have been revealed in the pathological processes of some respiratory diseases, these molecular mechanisms remain largely unclear [6–8]. Therefore, more comprehensive studies are needed to explore the functions and mechanisms of these molecules [9].

This chapter discusses the critical roles of circRNAs in respiratory diseases, highlighting the underlying mechanisms of circRNAs. Additionally, focus is also placed on the current and future potential of circRNAs as novel biomarkers and therapeutic targets for diagnosis, prognosis, and treating respiratory diseases.

2 CircRNAs and Respiratory Diseases

According to many new studies conducted in recent years, circRNAs are closely related to the pathogenesis, development, and progression of various respiratory diseases, such as acute respiratory distress syndrome, pulmonary hypertension, pulmonary tuberculosis, silicosis, and respiratory cancers. Most studies have focused on the role of circRNAs in respiratory cancers, especially lung cancer [4, 9].

2.1 CircRNAs in Respiratory Cancers

As a regulatory molecule, circRNA has been stated in the formation and progression of common malignancies of the respiratory system, including lung carcinoma (LC), laryngeal squamous cell carcinoma (LSCC), and nasopharyngeal carcinoma (NPC). CircRNAs regulate tumor cells' proliferation, migration, invasion, and apoptosis mainly through the competitive endogenous RNA mechanism. Its unique structures can also play a vital role in diagnosing, prognosing, and treating respiratory system cancers [8]. Dysregulated circRNAs in common cancers of the respiratory system are shown in Fig. 22.1.

2.1.1 Lung Cancer

Lung cancer is one of the leading causes of cancer-related deaths and the second most common type of cancer worldwide [10]. It is essential to understand the mechanisms underlying the formation and development of lung cancer, which has the highest mortality rate due to malignancy, and to investigate new diagnostic and treatment targets [11]. Studies indicate that by controlling the circRNA–miRNA–mRNA pathways, circRNAs can either accelerate or slow the development and spread of lung cancer. They also play a role in several cellular functions, such as migration, differentiation, and proliferation [9]. The ability to escape from apoptosis and have superior proliferation ability are considered the main mechanisms of tumor formation [11]. It is stated that many circRNAs act as oncogenes and suppressors by acting on these mechanisms [10]. When circRNAs that have a suppressor effect on lung cancer are examined, Li et al. stated that upregulation of circ_0002483 inhibited NSCLC progression in vitro and in vivo and increased apoptosis by relieving FOXO3 suppression [12]. circITCH suppresses lung cancer progression by

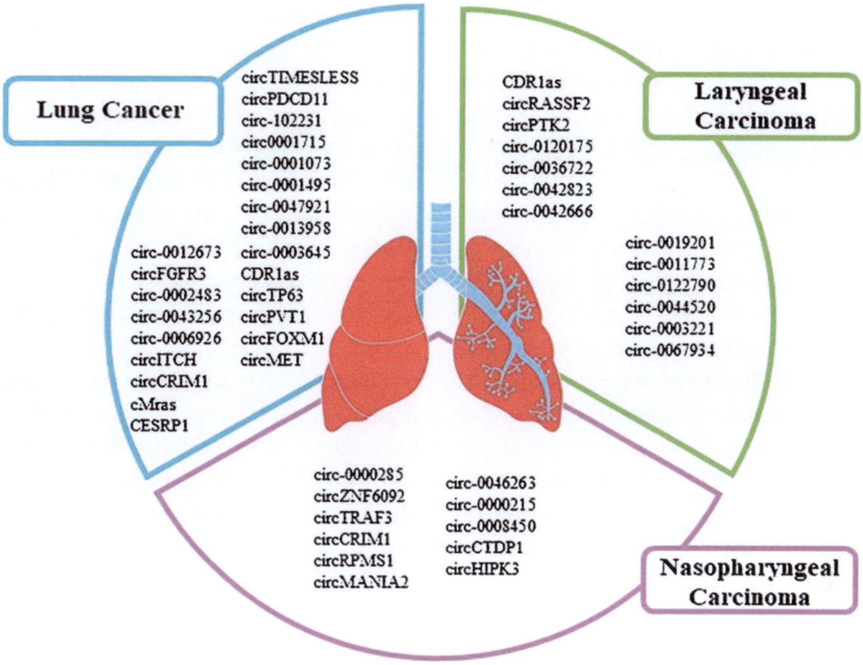

Fig. 22.1 Dysregulated circRNAs in common cancers of the respiratory system

targeting miR-7 and miR-214 and has been shown to play a protective role in lung cancer by inactivating the Wnt/β-catenin pathway [13]. Another circRNA, hsa_circ_0043256, indirectly affects ITCH expression by targeting miR-1252 and is a tumor suppressor in lung cancer [14]. Dai et al. reported that circ0006916 functioned as a tumor suppressor by blocking the cell cycle and cell proliferation and that its expression was markedly downregulated in A549 and 16HBE-T cells [6]. circCRIM1 and cMras have been stated to be expressed at low levels in lung adenocarcinoma tissues and cell lines [15, 16]. Overexpression of circCRIM1 has been shown to reduce the expression of the leukemia inhibitory factor receptor (LIFR) by targeting miR-93 and miR-182, thereby suppressing cell migration and invasion [15]. cMras has been reported to target miR-567 and downregulate G protein tyrosine phosphatase, receptor type G (PTPRG), to inhibit tumorigenesis [16]. Overexpression of cESRP1 has been shown to affect TGF-β signaling by regulating the miR-93-5p-Smad7/p21(CDKN1A) axis to arrest the cell cycle and induce cell apoptosis in large-cell lung cancer [17]. On the other hand, by encouraging the production of cell cycle and anti-apoptotic proteins, many oncogenic circRNAs maintain cell proliferation [11]. In a study to investigate the function of hsa_circ_0012673 in lung adenocarcinoma, it was stated that hsa_circ_0012673 targets miR-22, increases the expression of ErbB3, and increases the proliferation of cells [18]. It has been stated that circFGFR3 increases the expressions of Gal-1, p-AKT, and p-ERK1/2 by targeting miR-22-3p in NSCLC cells, thus supporting invasion and proliferation [19]. circ_0003645 has been reported to increase in both NSCLC cells and tissues and functions as an oncogene. circ_0003645 regulates TMEM14A expression by targeting miR-1179 and promotes tumor formation and progression by affecting cell growth, apoptosis, and metastatic properties [20]. In NSCLC, it has been shown that CDR1as, another circRNA with oncogenic properties, stimulates cell proliferation via miR-7/EGFR/CCNE1/PIK3CD signaling [7]. It has been stated that the expression of circP63 [21] and circ TIMELESS [22] is increased in lung squamous cell carcinoma (LUSC) and that these circRNAs support cell proliferation and migration. Another circRNA, circPVT1, has been shown to have increased expression in LUSC tissues and cell lines and to promote cell proliferation by attenuating the suppressive effects of miR-30d and miR-30e on their target cyclin F (CCNF) [23]. Early diagnosis significantly improves the prognosis of lung cancer and plays a crucial role in cancer treatment. It has been reported that circRNAs with aberrant expression in lung cancer may serve as promising biomarkers for diagnosis, prognosis, and therapy [10, 11]. For instance, circMET has been proposed as a novel, potential biomarker for prognosis and diagnosis in patients with non-small cell lung cancer [24]. Lui et al. reported that circ-FOXM1 was overexpressed in NSCLC tissues and may serve as a novel prognostic indicator and therapeutic target for NSCLC [25]. Another study revealed that circPDCD11 was overexpressed, especially in lung large cell carcinoma (LLCC), but not in other lung cancer subtypes, and increased plasma circPDCD11 levels could be used as a potential diagnostic and prognostic biomarker for LLCC [26]. Moreover, circ_102231 [27], circ_0001715 [28], hsa_circ_0001073 and hsa_circ_0001495 [29], circ_0047921 [30], and hsa_circ_0013958 [31] may serve as new diagnostic and prognostic biomarkers for lung adenocarcinoma. With all the information available, evaluating the functions and expression profiles of circRNAs in lung cancer may offer insights into the disease prognosis, early detection, and customized therapy.

2.1.2 Nasopharyngeal Carcinoma

Nasopharyngeal carcinoma is a common malignant tumor of the head and neck, and its incidence is characterized by different geographical distributions [32]. Radiation and radiotherapy are used to treat NPC, but the desired success is not at the desired level. Thus, many NPC patients face tumor recurrence or distant metastasis within a few years after treatment. The most effective way to reduce deaths from cancer is early diagnosis and early treatment. Therefore, it is essential to understand the molecular mechanisms and pathways underlying NPC and identify new biomarkers for diagnosis and prognosis [33]. Various studies conducted in recent years have revealed that circRNA is abnormally expressed in NPC and is associated with the proliferation, invasion, migration, and apoptosis of NPC [33, 34]. Ke et al. reported that circHIPK3 was highly expressed in NPC tissue samples and cell lines and that circHIPK3 promoted NPC cell proliferation, migration, and invasion. The circHIPK3/miR-4288/ELF3 axis plays a vital role in the development of NPC, providing a new perspective for the design of NPC intervention treatment targets [34]. Another study showed that circCTDP1 targets miR-320b and regulates NPC progression through HOXA10, and HOXA10 regulates TGFβ2 expression in NPC cells. It has been stated that circCTdP1, miR-320b, HOXA10, and TGFβ2 may become therapeutic targets for NPC [35]. It was revealed that circ-ZNF6092 expression increased in NPC and regulated the expression of oncogene Sp1 by targeting miR-150-5p, thus promoting the formation and development of NPC [36]. circTRAF3 plays an oncogene role in NPC. It has been stated that circTRAF3 promotes cell metastasis by targeting the miR-203a-3p/AKT3 axis, and the circTRAF3/miR-203a-3p/AKT3 pathway may be a potential therapeutic target for metastatic NPC [32]. Circ_0008450 silencing regulates the miR-577/CXCL9 axis and thus inhibits NPC cell proliferation and metastasis and facilitates apoptosis [37]. It has also been stated that in NPC, circCRIM1 supports metastasis [38], circ_0000215 supports cell proliferation and metastasis [39], and circ_0046263 exhibits an oncogenic character by regulating the miR-133a-5p/IGFBP3 axis [40]. Potential biomarkers for the diagnosis or prognosis of NPC have been demonstrated for circMAN1A2 [41], circ_0000285 [42], circRPMS1 [43], and circCRIM1 [38]. NPC-associated circRNAs have been shown to act through the ceRNA mechanism, thus playing a role in tumor inhibition or progression by regulating the expression of target genes. There are few studies on the impact of circRNAs on the formation and development of NPC through other molecular mechanisms, as well as on its diagnosis and treatment. Further studies in this field may enable a better understanding of NPC and the development of personalized treatment for patients with NPC.

2.1.3 Laryngeal Carcinoma

Laryngeal cancer is the second most common respiratory malignancy and accounts for 1% of cancer-related deaths worldwide [44, 45]. Laryngeal squamous cell carcinoma (LSCC) is the most common type of LC, with a high mortality rate due to its highly aggressive nature [46].

Although significant advances have been made in surgical techniques and adjuvant chemotherapy, laryngeal cancer is, unfortunately, one of the diseases with a decreasing survival rate. Therefore, identifying new biomarkers associated with laryngeal cancer tumorigenesis is critical for the early diagnosis and treatment of this disease [45]. Many studies have indicated that circRNAs have critical regulatory effects in the formation, development, and metastasis of LSCC [45, 47]. It has been implicated that CDR1as promotes LSCC growth and metastasis through regulation of the miR-7/CCNE1/PIK3CD signaling pathway, which functions as an oncogene [44]. In another study, circCORO1C was shown to have high expression in LSCC tissues and cells, leading to PBX3 accumulation by targeting let-7c-5p and increasing the proliferation, migration, and invasion of LSCC cells [47]. It was stated that circ_0044520 was highly expressed in LSCC tissues and cells. The circ_0044520/miR-338-3p/ROR2 regulatory network was established and revealed to be a promising candidate for developing new antitumor therapy [48]. Yang et al. indicated that circPTK2 expression was increased in LSCC cells and promoted malignant phenotypes in LSCC cells. It has been suggested that circPTK2 may provide new insight into therapeutic strategies for LSCC and that circPTK2 may be a potential diagnostic biomarker for LSCC [49]. LSCC cell proliferation, migration, and invasion were shown to be stimulated by circ_0067934 targeting miR-1324 [50], circ_0120175 targeting miR -330-3p/SLC7A11 axis [51], circCORO1C modulating the miR-654-3p/USP7 axis [52], and circ_0042823 modulating the miR-877-5p/FOXM1 axis [53]. Wei et al. reported that hsa_circ_0042666 expression was low in LSCC tumor tissue and would inhibit LSCC progression by regulating the miR-223/TGFBR3 axis in LSCC tumor tissue [46]. It was also suggested that circRASSF2 [54], circ_0019201, circ_0011773 and circ_0122790 [55], and hsa_circ_0036722 may be new clinical molecular markers for diagnosing LSCC patients [56]. Patients' quality of life and survival rate will increase with a better understanding of how LSCC develops and the development of effective molecular therapeutics, given the unfavorable prognosis and detrimental effects on LSCC patients' quality of life.

2.2 CircRNAs in Noncancerous Respiratory Diseases

Recently, dysregulated circRNA expression has been frequently identified in noncancerous respiratory diseases. Understanding these diseases' molecular causes and revealing relationships with circRNAs is essential. These molecules provide new information for new strategies for diagnosing, progressing, and treating noncancerous respiratory diseases [4, 9]. Dysregulated circRNAs in respiratory system diseases are shown in Fig. 22.2.

2.2.1 Chronic Obstructive Pulmonary Disease

Chronic obstructive pulmonary disease is a heterogeneous disease with high prevalence, morbidity, and mortality [57]. Studies have revealed that circRNAs contribute to the pathogenesis of COPD. Altered expression profiles in COPD patients are associated with inflammation, oxidative stress, and tissue restructuring. Understanding the mechanisms underlying COPD pathogenesis and revealing its relationship with circRNA may provide insight into new potential therapeutic targets [58]. It has been stated that circBBS9 expression increases significantly in COPD models and plays a role in the development of COPD by targeting miR-103a-3p, and circBBS9 may be considered a new target in treating COPD [59]. Miao et al. showed that circ_0040929 increased expression in serum exosomes from COPD patients. Additionally, in a cellular COPD model established with cigarette smoke extract (CSE) of human bronchial epithelial cells (16HBE), it was

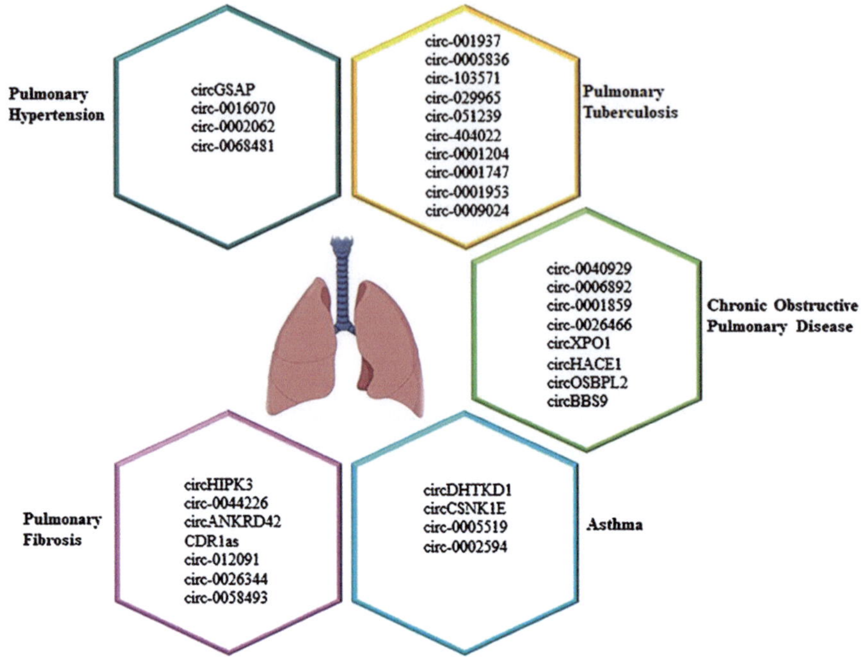

Fig. 22.2 Dysregulated circRNAs in respiratory system diseases

stated that circ_0040929 played a role in the pathogenesis of COPD by regulating the miR-515-5p/IGFBP3 axis [60]. In a cigarette smoke extract induced COPD model, circ_0006892 attenuated CSE-induced bronchial epithelial cell apoptosis and inflammatory response by regulating the miR-24/PHLPP2 axis. The results provide a new perspective on bronchial epithelial damage in COPD and suggest circ_0006892. It has been stated that it may be a promising target for COPD treatment [61]. Zhou et al. demonstrated that circHACE1 contributes to CSE-induced cell damage by regulating the miR-485-3p/TLR4 axis. It has been stated that silencing circHACE1 suppresses CSE-triggered apoptosis, inflammation, and oxidative stress in 16HBE cells [62]. In another study, circOSBPL2 contributed to the apoptosis, inflammation, and oxidative stress of human bronchial epithelial cells (HBEC) implicated in smoking-associated COPD via the miR-193a 5p/BRD4 axis. Therefore, circOSBPL2 may be a potential therapeutic strategy for future clinical intervention in COPD [63]. It has been stated that the expression of circXPO1 is increased in MLE12 cells treated with CSE and plays a role in the pathogenesis of COPD by regulating it through miR-23b-3p/TAB3 signaling [64]. Wang et al. stated that the expression of circ_0026466 increased in the blood of smokers with COPD and HBEC exposed to CSE. Dowregulation of circ_0026466 has alleviated CSE-induced survival and proliferation inhibition, induced apoptosis, oxidative stress, and inflammation in HBECs by regulating miR-153-3p/TRAF6/NF-κB axis [65]. The dysregulation of circRNAs in COPD patients' blood and lungs suggests that circRNAs may be considered potential diagnostic and therapeutic biomarkers for COPD disease. Many circRNAs have been identified as biomarkers of COPD, such as circ_0040929 [60], circRNA0001859 [66], and circHACE1 [62].

2.2.2 Pulmonary Hypertension

Pulmonary hypertension is a pathophysiological condition in which pulmonary artery pressure exceeds a certain threshold and is a significant cause of morbidity and mortality worldwide [9, 67]. Among the most common is pulmonary arterial hypertension (PAH), a specific subgroup of

PH in which the disease is characterized by primary pulmonary hypertension [67]. Recent studies have indicated that circRNAs play a role in initiating and developing PH. These studies revealed that the levels of circRNAs increased or decreased in different types of PH and revealed the functions of these molecules as miRNA sponges and biomarkers [68–70]. Sun et al. stated that the expression level of circGSAP was decreased in the lungs and plasma of COPD-PH patients. Ovarian expression of circGSAP has been shown to inhibit proliferation, apoptosis resistance, and G1/S transition in hypoxia-induced pulmonary microvascular endothelial cells (PMEC) [68]. Additionally, another study reported that circGSAP may be a biomarker for diagnosis and prognosis in patients with idiopathic pulmonary arterial hypertension (IPAH) [71]. It has been stated that hsa_circ_0016070 plays an essential role in the pathogenesis of pulmonary arterial hypertension (PAH). hsa_circ_0016070 has been demonstrated to be associated with vascular remodeling in PAH by promoting the proliferation of pulmonary artery smooth muscle cells (PASMCs) via miR-942/CCND1 [69], circ_0002062 is increased in pulmonary artery smooth muscle cells exposed to hypoxia, and silencing circ_0002062 inhibits cell proliferation and migration in these cells by regulating the miR-942 5p/CDK6 axis [72]. Additionally, another study showed that has_circ_0002062 may be a key circRNA for developing chronic thromboembolic pulmonary hypertension [73]. It has been stated that circRNA_0068481 may be a clinical biomarker for the diagnosis and prognosis of IPAH [70] and PAH [74]. Although there are limited circRNA studies on PH to date, these studies show that circRNAs have a critical impact on the pathogenesis of PH. Further function and mechanism studies are needed for PH.

2.2.3 Lung Tuberculosis

Pulmonary tuberculosis is caused by respiratory infections, mainly Mycobacterium tuberculosis, and is one of the most severe chronic diseases worldwide [75]. Current methods for the diagnosis of pulmonary tuberculosis have limited sensitivity and specificity [9]. The key to tuberculosis control is early, accurate diagnosis and fast treatment in order to limit the rising number of TB cases and offer successful therapy [76]. Increasingly, studies have shown that circRNAs play a regulatory role in the occurrence and development of various respiratory and infectious diseases through different mechanisms [75]. When compared to healthy controls, hsa_circRNA_001937, hsa_circRNA_009024, and hsa_circRNA_005086 were overexpressed in the study on PBMCs from TB patients, but hsa_circRNA_102101, hsa_circRNA_104964, and hsa_circRNA_104296 were dramatically reduced. According to ROC analysis, it is stated that hsa_circRNA_001937 (AUC = 0.873) can be used as a potential diagnostic biomarker of TB [77]. Zhuang et al. showed that hsa_circ_0005836 and has-circ-0009128 had significantly lower expression in peripheral blood mononuclear cells (PBMCs) of patients with active pulmonary TB compared with healthy controls. It also highlights that hsa_circ_0005836 may be a new potential biomarker for TB infection [76]. It was stated that the level of hsa_circRNA_103571 decreased significantly in TB plasma samples and could be used as a potential biomarker to diagnose active TB with an AUC value of 0.838 [78]. Another study identified three circRNAs (circRNA_029965, circRNA_051239, and circRNA_404022) that were significantly increased in the serum of active TB patients. With ROC analysis, the AUC value of these three circRNA combinations was found to be 0.992, and it was stated that these circRNAs could serve as ideal potential biomarkers for TB diagnosis [79]. Expression of hsa_circ_0001204 and hsa_circ_0001747 was significantly reduced in active TB plasma samples compared to healthy controls. ROC curve analysis showed that there was an AUC value of 0.928 to distinguish TB patients when hsa_circ_0001747 and hsa_circ_0001204 were used in combination, and this combination could be a new biomarker for the diagnosis of active TB [80]. Another study reported that hsa_circ_0001953 and hsa_circ_0009024 had high expression levels in TB patients, and this combination had an AUC value of 0.915 for the detection of TB [81]. The exact mechanisms of the effect of circRNAs on pulmonary TB are still largely unknown. The roles and mechanisms of circRNAs in other immune processes during Mtb infection need to be comprehensively investigated. In addition to being diagnostic biomarkers, circRNAs may be potential therapeutic targets or therapeutic strategies for tuberculosis.

2.2.4 Other Diseases

Pulmonary fibrosis is the general term used for various heterogeneous interstitial lung diseases. Many agents, such as microbial, therapeutic, and occupational/environmental toxins, can contribute to the initiation and progression of lung fibrosis [82]. Silicosis, the most common form of pulmonary fibrosis, is an occupational disease caused by inhalation of crystalline silica dust [83]. Additionally, idiopathic pulmonary fibrosis (IPF) is another frequently seen pulmonary fibrosis disease of unknown etiology [84]. Based on the complex pathology, understanding the mechanisms of pulmonary fibrosis is needed to develop new treatments for these diseases. Although increasing evidence suggests that circRNAs are involved as novel regulators in the pathological process of diseases, few studies have focused on the roles of circRNAs in pulmonary fibrosis. It has been shown that circHIPK3 can significantly increase FOXK2 expression by targeting miR-30a-3p, contributing to the progression of pulmonary fibrosis by regulating fibroblast glycolysis and activation. circHIPK3 may be a new and effective therapeutic in treating pulmonary fibrosis and other fibrotic diseases [82]. Yao et al. reported that CDR1as supports the EMT process of silica-induced pulmonary fibrosis by sponging miR-7 [85]. It has been shown that hsa_circ_0058493 may promote the development and progression of pulmonary fibrosis by regulating miR-3175/miR-2278 expression [84]. In addition, circRNA_0044226 [86], circRNA_012091 [87], circANKRD42 [88], and circRNA_0026344 [89] are involved in pulmonary fibrosis.

Asthma is a chronic lung disease caused by airway remodeling, inflammation, and genetic and environmental factors [90, 91]. Although incurable, regular check-ups and avoiding triggers are crucial to minimize its impact. CircRNAs have been implicated as potential regulators in asthma [58]. It was stated that hsa_circ_0005519 increased IL-13/IL-6 by regulating hsa-let-7a-5p in CD4+ T cells and could be used as a biomarker in classifying, diagnosing, and treating asthma [92]. Another study showed that the circCSNK1E/miR-34a-5p/VAMP2 axis regulates airway remodeling during asthma pathogenesis [93]. Quan et al. reported that circ_0002594 regulates airway remodeling by regulating the miR-139-5p/TRIM8 axis during the development of asthma [91]. In another study investigating circ_0002594, it was stated that the expression of circ_0002594 was increased in CD4+ T cells resulting from asthma and could be used in the diagnosis and treatment of Th2-mediated allergic asthma [94]. Additionally, the role of circDHTKD1 in asthma pathogenesis by regulating the miR-338-3p/ETS1 axis has been revealed. It has been shown that downregulation of circDHTKD1 reduces the expression levels of proinflammatory cytokines. Therefore, circDHTKD1 may play an inflammatory role in the progression of asthma [95].

3 Conclusion and Perspectives

circRNAs have an essential role in gene regulation and the capacity to modulate many molecular mechanisms. However, the molecular mechanisms underlying these molecules' physiological and pathological conditions remain largely unclear. In recent years, circRNAs have been shown to play a role in cancerous and noncancerous respiratory diseases. The stability and abundance of circRNAs in peripheral blood or other body fluids enable the use of these molecules as potential diagnostic and prognostic biomarkers in respiratory diseases. CircRNAs have shown high organ specificity in cancerous and noncancerous respiratory diseases. In diseases such as lung cancer, chronic obstructive pulmonary disease, pulmonary hypertension, and pulmonary tuberculosis, circRNAs have been identified as promising clinical biomarkers for early diagnosis and prognosis. However, their use in clinics requires more candidate circRNAs, sensitivity and reliability of circRNAs, circRNA diagnostic panels, and more clinical validation. One of the ultimate goals of medical research is to contribute to treatment. CircRNAs are being suggested as new therapeutic targets in many disorders, including respiratory ones. Controlling circRNA expression can provide therapeutic effects through modified chemical drugs and RNA interference constructs, which can increase the value of circRNA. In particular, one of the possible therapeutic effects is using RNAi approaches to destroy target protein-coding mRNAs, which are thought to be important in the pathogenesis of respiratory diseases, and creating a circRNA–miRNA–mRNA network for this use. Much advanced and in-depth research is needed in this field. Although there is no clinical study examining the therapeutic effect of circRNAs in respiratory diseases so far, their potential therapeutic value in different respiratory diseases, especially lung cancer, has been demonstrated based on cell and animal models. Considering all these results, improving our understanding of the role of circRNAs in the respiratory system will help provide new insights into disease mechanisms and reveal attractive therapeutic targets and diagnostic/prognostic biomarkers for respiratory disorders.

Conflicts of Interest The authors declare that there is no conflict of interest.

References

1. Wang Q, Sun Y, Zhao Q, Wu W, Wang L, Miao Y, Yuan P (2022) Circular RNAs in pulmonary hypertension: emerging biological concepts and potential mechanism. Anim Model Exp Med 5(1):38–47. https://doi.org/10.1002/ame2.12208
2. Han B, Chao J, Yao H (2018) Circular RNA and its mechanisms in disease: from the bench to the clinic. Pharmacol Ther 187:31–44. https://doi.org/10.1016/j.pharmthera.2018.01.010
3. Zhang Z, Yang T, Xiao J (2018) Circular RNAs: promising biomarkers for human diseases. EBioMedicine 34:267–274. https://doi.org/10.1016/j.ebiom.2018.07.036
4. Peng Z, Fang S, Jiang M, Zhao X, Zhou C, Gong Z (2020) Circular RNAs: regulatory functions in respiratory tract cancers. Clin Chim Acta 510:264–271. https://doi.org/10.1016/j.cca.2020.07.042
5. Hurgobin B, de Jong E, Bosco A (2018) Insights into respiratory disease through bioinformatics. Respirology 23(12):1117–1126. https://doi.org/10.1111/resp.13401
6. Dai X, Zhang N, Cheng Y, Yang T, Chen Y, Liu Z, Wang Z, Yang C, Jiang Y (2018) RNA-binding protein trinucleotide repeat-containing 6A regulates the formation of circular RNA circ0006916, with important functions in lung cancer cells. Carcinogenesis 9(8):981–992. https://doi.org/10.1093/carcin/bgy06
7. Zhang X, Yang D, Wei Y (2018) Overexpressed CDR1as functions as an oncogene to promote the tumor progression via miR-7 in non-small-cell lung cancer. Onco Targets Ther 11:3979–3987. https://doi.org/10.2147/OTT.S158316
8. Zhou Z, Jiang R, Yang X, Guo H, Fang S, Zhang Y, Cheng Y, Wang J, Yao H, Chao J (2018) circRNA mediates silica-induced macrophage activation via HECTD1/ZC3H12A-dependent ubiquitination. Theranostics 8(2):575–592. https://doi.org/10.7150/thno.21648
9. Wang J, Zhu M, Pan J, Chen C, Xia S, Song Y (2019) Circular RNAs: a rising star in respiratory diseases. Respir Res 20(1):1–10. https://doi.org/10.1186/s12931-018-0962-1
10. Dong H, Zhou J, Cheng Y, Wang M, Wang S, Xu H (2021) Biogenesis, functions, and role of circRNAs in lung cancer. Cancer Manag Res 13:6651–6671. https://doi.org/10.2147/CMAR.S324812
11. Li J, Zhang Q, Jiang D, Shao J, Li W, Wang C (2022) CircRNAs in lung cancer- role and clinical application. Cancer Lett 544:215810. https://doi.org/10.1016/j.canlet.2022.215810
12. Li X, Yang B, Ren H, Xiao T, Zhang L, Li L, Li M, Wang X, Zhou H, Zhang W (2019) Hsa_circ_0002483 inhibited the progression and enhanced the Taxol sensitivity of non-small cell lung cancer by targeting miR-182-5p. Cell Death Dis 10(12):953. https://doi.org/10.1038/s41419-019-2180-2
13. Wan L, Zhang L, Fan K, Sun QC, Wang JJ (2016) Circular RNA-ITCH suppresses lung cancer proliferation via inhibiting the Wnt/β-catenin pathway. Biomed Res Int 2016:1579490. https://doi.org/10.1155/2016/1579490
14. Tian F, Yu CT, Ye WD, Wang Q (2017) Cinnamaldehyde induces cell apoptosis mediated by a novel circular RNA hsa_circ_0043256 in non-small cell lung cancer. Biochem Biophys Res Commun 493(3):1260–1266. https://doi.org/10.1016/j.bbrc.2017.09.136
15. Wang L, Liang Y, Mao Q, Xia W, Chen B, Shen H, Xu L, Jiang F, Dong G (2019) Circular RNA circCRIM1 inhibits invasion and metastasis in lung adenocarcinoma through the microRNA (miR)-182/miR-93-leukemia inhibitory factor receptor pathway. Cancer Sci 110(9):2960–2972. https://doi.org/10.1111/cas.14131
16. Yu C, Tian F, Liu J, Su M, Wu M, Zhu X, Qian W (2019) Circular RNA cMras inhibits lung adenocar-

cinoma progression via modulating miR-567/PTPRG regulatory pathway. Cell Prolif 52(3):1–12. https://doi.org/10.1111/cpr.12610
17. Huang W, Yang Y, Wu J, Niu Y, Yao Y, Zhang J, Huang X, Liang S, Chen R, Chen S, Guo L (2020) Circular RNA cESRP1 sensitises small cell lung cancer cells to chemotherapy by sponging miR-93-5p to inhibit TGF-β signalling. Cell Death Differ 27(5):1709–1727. https://doi.org/10.1038/s41418-019-0455-x
18. Hu W, Bi ZY, Chen ZL, Xu S (2018) Emerging landscape of circular RNAs in lung cancer. Cancer Lett 427:18–27. https://doi.org/10.3892/ijo.2022.5469
19. Qiu BQ, Zhang PF, Xiong D, Xu JJ, Long X, Zhu SQ, Ye XD, Wu Y, Pei X, Zhang XM, Wu YB (2019) CircRNA fibroblast growth factor receptor 3 promotes tumor progression in non-small cell lung cancer by regulating galectin-1-AKT/ERK1/2 signaling. J Cell Physiol 234(7):11256–11264. https://doi.org/10.1002/jcp.27783
20. An J, Shi H, Zhang N, Song S (2019) Elevation of circular RNA circ_0003645 forecasts unfavorable prognosis and facilitates cell progression via miR-1179/TMEM14A pathway in non-small cell lung cancer. Biochem Biophys Res Commun 511(4):921–925. https://doi.org/10.1016/j.bbrc.2019.03.011
21. Cheng Z, Yu C, Cui S, Wang H, Jin H, Wang C, Li B, Qin M, Yang C, He J, Zuo Q, Wang S, Liu J, Ye W, Lv Y, Zhao F, Yao M, Jiang L, Qin W (2019) circTP63 functions as a ceRNA to promote lung squamous cell carcinoma progression by upregulating FOXM1. Nat Commun 10(1):3200. https://doi.org/10.1038/s41467-019-11162-4
22. Zhang W, Shi J, Cheng C, Wang H (2020) CircTIMELESS regulates the proliferation and invasion of lung squamous cell carcinoma cells via the miR-136-5p/ROCK1 axis. J Cell Physiol 235(9):5962–5971. https://doi.org/10.1002/jcp.29521
23. Shi J, Lv X, Zeng L, Zhong Y, Yuan J, Deng S, Liu B, Yuan B, Chen Y, Ming Z, Yang X, Fang P, Yang S, Chen G (2021) CircPVT1 promotes proliferation of lung squamous cell carcinoma by binding to miR-30d/e. J Exp Clin Cancer Res 40(1):1–15. https://doi.org/10.1186/s13046-021-01976-w
24. Pei X, Chen S, Long X, Zhu SQ, Qiu BQ, Lin K, Lu F, Xu J, Zhang PF, Wu YB (2020) circMET promotes NSCLC cell proliferation, metastasis, and immune evasion by regulating the miR-145-5p/CXCL3 axis. Aging (Albany NY) 12(13):13038–13058. https://doi.org/10.18632/aging.103392
25. Liu G, Shi H, Deng L, Zheng H, Kong W, Wen X, Bi H (2019) Circular RNA circ-FOXM1 facilitates cell progression as ceRNA to target PPDPF and MACC1 by sponging miR-1304-5p in non-small cell lung cancer. Biochem Biophys Res Commun 513(1):207–212. https://doi.org/10.1016/j.bbrc.2019.03.213
26. Ma T, Miao H, Xiong Y, Ma Y, Dong Z (2023) CircRNA circ-PDCD11 is highly expressed in lung large-cell carcinoma and predicts poor survival. Immunopharmacol Immunotoxicol 45(1):89–93. https://doi.org/10.1080/08923973.2022.2117628
27. Zong L, Sun Q, Zhang H, Chen Z, Deng Y, Li D, Zhang L (2018) Increased expression of circRNA_102231 in lung cancer and its clinical significance. Biomed Pharmacother 102:639–644. https://doi.org/10.1016/j.biopha.2018.03.084
28. Lu GJ, Cui J, Qian Q, Hou ZB, Xie HY, Hu W, Hao KK, Xia N, Zhang Y (2020) Overexpression of hsa_circ_0001715 is a potential diagnostic and prognostic biomarker in lung adenocarcinoma. Onco Targets Ther 13:10775–10783. https://doi.org/10.2147/OTT.S274932
29. Wang C, Tan S, Liu WR, Lei Q, Qiao W, Wu Y, Liu X, Cheng W, Wei YQ, Peng Y, Li W (2019) RNA-Seq profiling of circular RNA in human lung adenocarcinoma and squamous cell carcinoma. Mol Cancer 18(1):1–6. https://doi.org/10.1186/s12943-019-1061-8
30. Xian J, Su W, Liu L, Rao B, Lin M, Feng Y, Qiu F, Chen J, Zhou Q, Zhao Z, Lu J, Yang L (2020) Identification of three circular rna cargoes in serum exosomes as diagnostic biomarkers of non–small-cell lung cancer in the Chinese population. J Mol Diagn 22(8):1096–1108. https://doi.org/10.1016/j.jmoldx.2020.05.011
31. Zhu X, Wang X, Wei S, Chen Y, Chen Y, Fan X, Han S, Wu G (2017) Hsa_circ_ 0013958: a circular RNA and potential novel biomarker for lung adenocarcinoma. FEBSJ 284(14):2170–2182. https://doi.org/10.1111/febs.14132
32. Fang X, Huang W, Wu P, Zeng J, Li X (2021) CircRNA circTRAF3 promotes nasopharyngeal carcinoma metastasis through targeting miR-203a-3p/AKT3 axis. Pathol Res Pract 221(87):153438. https://doi.org/10.1016/j.prp.2021
33. Yang M, Huang W (2020) Circular RNAs in nasopharyngeal carcinoma. Clin Chim Acta 508(28):240–248. https://doi.org/10.1016/j.cca.2020.05.029
34. Ke Z, Xie F, Zheng C, Chen D (2019) CircHIPK3 promotes proliferation and invasion in nasopharyngeal carcinoma by abrogating miR-4288-induced ELF3 inhibition. J Cell Physiol 234(2):1699–1706. https://doi.org/10.1002/jcp.27041
35. Li H, You J, Xue H, Tan X, Chao C (2020) CircCTDP1 promotes nasopharyngeal carcinoma progression via a microRNA-320b/HOXA10/TGFβ2 pathway. Int J Mol Med 45(3):836–846. https://doi.org/10.3892/ijmm.2020.4467
36. Zhu L, Liu Y, Yang Y, Yin ZD (2019) CircRNA ZNF609 promotes growth and metastasis of nasopharyngeal carcinoma by competing with microRNA-150-5p. Eur Rev Med Pharmacol Sci 23(7):2817–2826. https://doi.org/10.26355/eurrev_201904_17558
37. Wei H, Liu D, Sun J, Mao Y, Zhao L, Zhu W, Xu G, Gao Z (2019) Circular RNA circ_0008450 upregulates CXCL9 expression by targeting miR-577 to regulate cell proliferation and invasion in nasopharyngeal carcinoma. Exp Mol Pathol 110:104288. https://doi.org/10.1016/j.yexmp.2019.104288

38. Hong X, Liu N, Liang Y, He Q, Yang X, Lei Y, Zhang P, Zhao Y, He S, Wang Y, Li J, Li Q, Ma J, Li Y (2020) Circular RNA CRIM1 functions as a ceRNA to promote nasopharyngeal carcinoma metastasis and docetaxel chemoresistance through upregulating FOXQ1. Mol Cancer 19(1):1–16. https://doi.org/10.1186/s12943-020-01149-x
39. Chen X, Xu W, Ma Z, Zhu J, Hu J, Li X, Zhu J, Hu J, Li X, Fu S (2021) Circ_0000215 exerts oncogenic function in nasopharyngeal carcinoma by targeting miR-512-5p. Front Cell Dev Biol 9:1–14. https://doi.org/10.3389/fcell.2021.688873
40. Yin L, Chen J, Ma C, Pei S, Du M, Zhang Y, Feng Y, Yin R, Bian X, He X, Feng J (2020) Hsa_circ_0046263 functions as a ceRNA to promote nasopharyngeal carcinoma progression by upregulating IGFBP3. Cell Death Dis 11(7):562. https://doi.org/10.1038/s41419-020-02785-3
41. Fan CM, Wang JP, Tang YY, Zhao J, He SY, Xiong F, Guo C, Xiang B, Zhou M, Li XL, Li Y, Li GY, Xiong W, Zeng ZY (2019) circMAN1A2 could serve as a novel serum biomarker for malignant tumors. Cancer Sci 110(7):2180–2188. https://doi.org/10.1111/cas.14034
42. Shuai M, Hong J, Huang D, Zhang X, Tian Y (2018) Upregulation of circRNA_0000285 serves as a prognostic biomarker for nasopharyngeal carcinoma and is involved in radiosensitivity. Oncol Lett 16(5):6495–6501. https://doi.org/10.3892/ol.2018.9471
43. Liu Q, Shuai M, Xia Y (2019) Knockdown of EBV-encoded circRNA circRPMS1 suppresses nasopharyngeal carcinoma cell proliferation and metastasis through sponging multiple miRNAs. Cancer Manag Res 11:8023–8031. https://doi.org/10.2147/CMAR.S218967
44. Zhang J, Hu H, Zhao Y, Zhao Y (2018) CDR1as is overexpressed in laryngeal squamous cell carcinoma to promote the tumour's progression via miR-7 signals. Cell Prolif 51(6):1–10. https://doi.org/10.1111/cpr.12521
45. Fan Y, Xia X, Zhu Y, Diao W, Zhu X, Gao Z, Chen X (2018) Circular RNA expression profile in laryngeal squamous cell carcinoma revealed by microarray. Cell Physiol Biochem 50(1):353–362. https://doi.org/10.1159/000494010
46. Wei Z, Chang K, Fan C (2019) Hsa_circ_0042666 inhibits proliferation and invasion via regulating miR-223/TGFBR3 axis in laryngeal squamous cell carcinoma. Biomed Pharmacother 119(288):109365. https://doi.org/10.1016/j.biopha.2019.109365
47. Wu Y, Zhang Y, Zheng X, Dai F, Lu Y, Dai L, Niu M, Guo H, Li W, Xue X, Bo Y, Guo Y, Qin J, Qin Y, Liu H, Zhang Y, Yang T, Li L, Zhang L, Hou R, Wen S, An C, Li H, Xu W, Gao W (2020) Circular RNA circCORO1C promotes laryngeal squamous cell carcinoma progression by modulating the let-7c-5p/PBX3 axis. Mol Cancer 19(1):1–18. https://doi.org/10.1186/s12943-023-01819-6
48. Yang H, Yu G, Wang Y, Guo X (2022) Circ_0044520 regulates the progression of laryngeal squamous cell carcinoma via the miR-338-3p/ROR2 axis. Histol Histopathol 37(6):513–526. https://doi.org/10.14670/HH-18-420
49. Yang Z, Jin J, Chang T (2021) CircPTK2 (hsa_circ_0003221) contributes to laryngeal squamous cell carcinoma by the miR-1278/YAP1 axis. J Oncol 2021. https://doi.org/10.1155/2021/2408384
50. Chu YL (2020) Circ-0067934 correlates with poor prognosis and promotes laryngeal squamous cell cancer progression by sponging miR-1324. Eur Rev Med Pharmacol Sci 24(8):4320–4327. https://doi.org/10.26355/eurrev_202004_21013
51. Fan D, Zhu Y (2022) Circ_0120175 promotes laryngeal squamous cell carcinoma development through up-regulating SLC7A11 by sponging miR-330-3p. J Mol Histol 53(2):159–171. https://doi.org/10.1007/s10735-022-10061-1
52. Zhang J, Li H, Li J, Ke S (2022) CircRNA CORO1C regulates miR-654-3p/USP7 axis to mediate laryngeal squamous cell carcinoma progression. Biochem Genet 60(5):1615–1629. https://doi.org/10.1007/s10528-021-10169-1
53. Wu T, Sun Y, Sun Z, Li S, Wang W, Yu B, Wang G (2021) Hsa_circ_0042823 accelerates cancer progression via miR-877-5p/FOXM1 axis in laryngeal squamous cell carcinoma. Ann Med 53(1):960–970. https://doi.org/10.1080/07853890.2021.1934725
54. Tian L, Cao J, Jiao H, Zhang J, Ren X, Liu X, Liu M, Sun Y (2019) CircRASSF2 promotes laryngeal squamous cell carcinoma progression by regulating the miR-302b-3p/IGF-1R axis. Clin Sci 133(9):1053–1066. https://doi.org/10.1042/CS20190110
55. Han J, Lin Q, Dong C (2021) Plasma cell-free circRNAs panel act as fingerprint predicts the occurrence of laryngeal squamous cell carcinoma. Aging (Albany NY) 13(13):17328–17336. https://doi.org/10.18632/aging.203215
56. Guo Y, Huang Q, Zheng J, Hsueh CY, Yuan X, Heng Y, Zhou L (2020) Diagnostic role of dysregulated circular RNA hsa_circ_0036722 in laryngeal squamous cell carcinoma. Onco Targets Ther 13:5709–5719. https://doi.org/10.2147/OTT.S231076
57. Zhong S, Chen C, Yang L, Jin M, Zeng Y, Zou GM, Zhang Q, Wang Y (2023) Identification of circRNA-associated ceRNA networks in peripheral blood mononuclear cells as potential biomarkers for chronic obstructive pulmonary disease. Biosci Rep 43(10):1–18. https://doi.org/10.1042/BSR20230005
58. Bhat AA, Gupta G, Goyal A, Thapa R, Hassan AW, Kazmi I, Alzarea SI, Kukreti N, Sekar M, Meenakshi DU, Singh SK, MacLoughlin R, Dua K (2023) Unwinding circular RNA's role in inflammatory pulmonary diseases. Naunyn Schmiedeberg's Arch Pharmacol 2. https://doi.org/10.1007/s00210-023-02809-7
59. Guo P, Lu J, Lei Y (2023) Significant role of circRNA BBS9 in chronic obstructive pulmonary

disease via miRNA-103a-3p/BCL2L13. BMC Pulm Med 23(1):1–9. https://doi.org/10.1186/s12890-023-02540-2

60. Miao Y, Wu J, Wu R, Wang E, Wang J (2022) Circ_0040929 serves as promising biomarker and potential target for chronic obstructive pulmonary disease. Int J COPD 17:2079–2092. https://doi.org/10.2147/COPD.S364553
61. Zhang C, Gu S, Kang X (2022) CircRNA circ_0006892 regulates miR-24/PHLPP2 axis to mitigate cigarette smoke extract-induced bronchial epithelial cell injury. Biotechnol Appl Biochem 69(2):735–748. https://doi.org/10.1002/bab.2148
62. Zhou F, Cao C, Chai H, Hong J, Zhu M (2021) Circ-HACE1 aggravates cigarette smoke extract-induced injury in human bronchial epithelial cells via regulating toll-like receptor 4 by sponging miR-485-3p. Int J COPD 16:1535–1547. https://doi.org/10.2147/COPD.S304859
63. Zheng C, Zhang Y, Zhao Y, Duan Y, Mu Q, Wang X (2021) Circ-OSBPL2 contributes to smoke-related chronic obstructive pulmonary disease by targeting mir-193a-5p/brd4 axis. Int J COPD 16:919–931. https://doi.org/10.2147/COPD.S298465
64. Du Y, Ding Y, Shi T, He W, Mei Z, Feng X, Zhang X, Jie Z (2022) Suppression of circXPO1 attenuates cigarette smoke-induced inflammation and cellular senescence of alveolar epithelial cells in chronic obstructive pulmonary disease. Int Immunopharmacol 111:109086. https://doi.org/10.1016/j.intimp.2022.109086
65. Wang C, Tang Y, Hou H, Su C, Gao Y, Yang X (2023) Circ_0026466 knockdown protects human bronchial epithelial cells from cigarette smoke extract–induced injury by promoting the miR-153-3p/Traf6/Nf-Kb pathway. Shock 60(1):121–129. https://doi.org/10.1097/SHK.0000000000002141
66. Chen S, Yao Y, Lu S, Chen J, Yang G, Tu L, Chen L (2020) CircRNA0001859, a new diagnostic and prognostic biomarkers for COPD and AECOPD. BMC Pulm Med 20(1):1–10. https://doi.org/10.1186/s12890-020-01333-1
67. Anderson JJ, Lau EM (2022) Pulmonary hypertension definition, classification, and epidemiology in Asia. JACC Asia 2(5):538–546. https://doi.org/10.1016/j.jacasi.2022.04.008
68. Sun Y, Wu W, Zhao Q, Jiang R, Li J, Wang L, Xia S, Liu M, Gong S, Liu J, Yuan P (2022) CircGSAP regulates the cell cycle of pulmonary microvascular endothelial cells via the miR-942-5p sponge in pulmonary hypertension. Front Cell Dev Biol 10:1–14. https://doi.org/10.3389/fcell.2022.967708
69. Zhou S, Jiang H, Li M, Wu P, Sun L, Liu Y, Zhu K, Zhang B, Sun G, Cao C, Wang R (2019) Circular RNA hsa_circ_0016070 is associated with pulmonary arterial hypertension by promoting PASMC proliferation. Mol Ther Nucleic Acids 18:275–284. https://doi.org/10.1016/j.omtn.2019.08.026
70. Zhang Y, Chen Y, Yao H, Lie Z, Chen G, Tan H, Zhou Y (2019) Elevated serum circ_0068481 levels as a potential diagnostic and prognostic indicator in idiopathic pulmonary arterial hypertension. Pulm Circ 9(4):1–9. https://doi.org/10.1177/2045894019888416
71. Yuan P, Wu WH, Gong SG, Jiang R, Zhao QH, Pudasaini B, Sun YY, Li JL, Liu JM, Wang L (2021) Impact of circGSAP in peripheral blood mononuclear cells on idiopathic pulmonary arterial hypertension. Am J Respir Crit Care Med 203(12):1579–1583. https://doi.org/10.1164/rccm.202005-2052LE
72. Wang Y, Tan X, Wu Y, Cao S, Lou Y, Zhang L, Hu F (2021) Hsa_circ_0002062 promotes the proliferation of pulmonary artery smooth muscle cells by regulating the hsa-miR-942-5p/CDK6 signaling pathway. Front Genet 2:1–13. https://doi.org/10.3389/fgene.2021.673229
73. Miao R, Wang Y, Wan J, Leng D, Gong J, Li J, Liang Y, Zhai Z, Yang Y (2017) Microarray expression profile of circular RNAs in chronic thromboembolic pulmonary hypertension. Med (United States) 96(27):1–8. https://doi.org/10.1097/MD.0000000000007354
74. Mei GH, Peng LZ (2021) Up-regulation of circRNA_0068481 promotes right ventricular hypertrophy in PAH patients via regulating miR-646/miR-570/miR-885. J Cell Mol Med 25(8):3735–3743. https://doi.org/10.1111/jcmm.16164
75. Yuan Q, Wen Z, Yang K, Zhang S, Zhang N, Song Y, Chen F (2022) Identification of key circRNAs related to pulmonary tuberculosis based on bioinformatics analysis. Biomed Res Int 2022:1717784. https://doi.org/10.1155/2022/1717784
76. Zhuang ZG, Zhang JA, Luo HL, Liu GB, Lu YB, Ge NH, Zheng BY, Li RX, Chen C, Wang X, Liu YQ, Liu FH, Zhou Y, XiZ C, Chen ZW, Xu JF (2017) The circular RNA of peripheral blood mononuclear cells: Hsa_circ_0005836 as a new diagnostic biomarker and therapeutic target of active pulmonary tuberculosis. Mol Immunol 90:264–272. https://doi.org/10.1016/j.molimm.2017.08.008
77. Huang ZK, Yao FY, Xu JQ, Deng Z, Su RG, Peng YP, Luo Q, Li JM (2018) Microarray expression profile of circular rnas in peripheral blood mononuclear cells from active tuberculosis patients. Cell Physiol Biochem 45(3):1230–1240. https://doi.org/10.1159/000487454
78. Yi Z, Gao K, Li R, Fu Y (2018) Dysregulated circRNAs in plasma from active tuberculosis patients. J Cell Mol Med 22(9):4076–4084. https://doi.org/10.1111/jcmm.13684
79. Liu H, Lu G, Wang W, Jiang X, Gu S, Wang J, Yan X, He F, Wang J (2020) A panel of circRNAs in the serum serves as biomarkers for mycobacterium tuberculosis infection. Front Microbiol 11:1–7. https://doi.org/10.3389/fmicb.2020.01215
80. Huang Z, Su R, Yao F, Peng Y, Luo Q, Li J (2018) Circulating circular RNAs hsa_circ_0001204 and hsa_circ_0001747 act as diagnostic biomarkers for active tuberculosis detection. Int J Clin Exp Pathol 11(2):586–594. eCollection 2018
81. Huang Z, Su R, Qing C, Peng Y, Luo Q, Li J (2018) Plasma circular RNAs hsa_circ_0001953 and hsa_

circ_0009024 as diagnostic biomarkers for active tuberculosis. Front Microbiol 9:586–594. https://doi.org/10.3389/fmicb.2018.02010
82. Xu Q, Cheng D, Li G, Liu Y, Li P, Sun W, Ma D, Ni C (2021) CircHIPK3 regulates pulmonary fibrosis by facilitating glycolysis in mir-30a-3p/foxk2-dependent manner. Int J Biol Sci 17(9):2294–2307. https://doi.org/10.7150/ijbs.57915
83. Cheng Z, Zhang Y, Zhao R, Zhou Y, Dong Y, Qiu A, Xu H, Liu Y, Zhang W, Chang Q, Chu M (2022) A novel circRNA-SNP may increase susceptibility to silicosis. Ecotoxicol Environ Saf 242:113855. https://doi.org/10.1016/j.ecoenv.2022.113855
84. Cheng Z, Zhang Y, Wu S, Zhao R, Yu Y, Zhou Y, Zhou Z, Dong Y, Qiu A, Xu H, Liu Y, Zhang W, Tian T, Wu Q, Gu H, Chu M (2022) Peripheral blood circular RNA hsa_circ_0058493 as a potential novel biomarker for silicosis and idiopathic pulmonary fibrosis. Ecotoxicol Environ Saf 236:11345. https://doi.org/10.1016/j.ecoenv.2022.113451
85. Yao W, Li Y, Han L, Ji X, Pan H, Liu Y, Yuan J, Yan W, Ni C (2018) The CDR1as/miR-7/TGFBR2 axis modulates EMT in silica-induced pulmonary fibrosis. Toxicol Sci 166(2):465–478. https://doi.org/10.1093/toxsci/kfy221
86. Qi F, Li Y, Yang X, Wu Y, Lin L, Liu X (2020) Hsa_circ_0044226 knockdown attenuates progression of pulmonary fibrosis by inhibiting CDC27. Aging (Albany NY) 12(14):14808–14818. https://doi.org/10.18632/aging.103543
87. Cheng Y, Luo W, Li Z, Cao M, Zhu Z, Han C, Dai X, Zhang W, Wang J, Yao H, Chao J (2019) CircRNA-012091/PPP1R13B-mediated lung fibrotic response in silicosis via endoplasmic reticulum stress and autophagy. Am J Respir Cell Mol Biol 61(3):380–391. https://doi.org/10.1165/rcmb.2019-0017OC
88. Xu P, Zhang J, Wang M, Liu B, Li R, Li H, Zhai N, Liu W, Lv C, Song X (2022) hnRNPL-activated circANKRD42 back-splicing and circANKRD42-mediated crosstalk of mechanical stiffness and biochemical signal in lung fibrosis. Mol Ther 30(6):2370–2387. https://doi.org/10.1016/j.ymthe.2022.01.045
89. Bai J, Deng J, Han Z, Cui Y, He R, Gu Y, Zhang Q (2021) CircRNA_0026344 via exosomal miR-21 regulation of Smad7 is involved in aberrant crosstalk of epithelium-fibroblasts during cigarette smoke-induced pulmonary fibrosis. Toxicol Lett 347:58–66. https://doi.org/10.1016/j.toxlet.2021.04.017
90. Sockrider M, Fussner L (2020) What is asthma? Am J Respir Crit Care Med 202(9):25–26. https://doi.org/10.1164/rccm.2029P25
91. Quan L, Ren G, Liu L, Huang W, Li M (2022) Circular RNA circ_0002594 regulates PDGF-BB-induced proliferation and migration of human airway smooth muscle cells via sponging miR-139-5p/TRIM8 in asthma. Autoimmunity 55(5):339–350. https://doi.org/10.1080/08916934.2022.2062596
92. Huang Z, Cao Y, Zhou M, Qi X, Fu B, Mou Y, Wu G, Xie J, Zhao J, Xiong W (2019) Hsa_circ_0005519 increases IL-13/IL-6 by regulating hsa-let-7a-5p in CD4+ T cells to affect asthma. Clin Exp Allergy 49(8):1116–1127. https://doi.org/10.1111/cea.13445
93. Ding L, Liu GL, Lu L, Ge L, Wang JY (2022) circ_CSNK1E modulates airway smooth muscle cells proliferation and migration via miR-34a-5p/VAMP2 axis in asthma. Cell Signal 95(321). https://doi.org/10.1016/j.cellsig.2022.110340
94. Huang Z, Fu B, Qi X, Xu Y, Mou Y, Zhou M, Cao Y, Wu G, Xie J, Zhao J, Wang Y, Xiong W (2021) Diagnostic and therapeutic value of Hsa_circ_0002594 for T helper 2-mediated allergic asthma. Int Arch Allergy Immunol 182(5):388–398. https://doi.org/10.1159/000511612
95. Qian F, He S, Yang X, Chen X, Zhao S, Wang J (2023) Circular RNA DHTKD1 targets miR-338-3p/ETS1 axis to regulate the inflammatory response in human bronchial epithelial cells. Exp Ther Med 26(1):1–13. https://doi.org/10.3892/etm.2023.12015

Circular RNAs in Brain Physiology and Neurologic Diseases

23

Mandana Amelimojarad
and Melika Amelimojarad

Abstract

In the past ten years, advancements in sequencing technology have allowed scientists to study the function and expression of noncoding RNAs (ncRNAs). One particular type of ncRNA, called circular RNAs (circRNAs), are closed loops that are found in large quantities and are evolutionarily conserved across different species. These circRNAs can be derived from both coding and noncoding transcripts and can regulate cellular proteins and sequester microRNAs. In the mammalian brain, circRNAs are more highly expressed compared to other tissues, suggesting their importance in brain function and pathology. However, there is still limited knowledge about circRNAs in brain cells. By gaining a better understanding of their role in neurological pathways, we can fill in the gaps and develop therapeutic approaches for future neurological diseases.

Keywords

CircRNAs · Noncoding RNA · Brain · Neurological disease

M. Amelimojarad (✉) · M. Amelimojarad
The First Affiliated Hospital, Dalian Medical University, Dalian, Liaoning, China

1 Introduction

Circular RNAs (circRNAs) have a unique circular structure that sets them apart from other noncoding RNAs, while circRNAs mainly consist of exonic sequences, and they can also contain noncoding sequences. circRNAs are formed through a process called back-splicing, involving the joining of exons, introns, or both exon–intron regions, and unlike linear RNAs, circRNAs lack a 5′ cap and a 3′ polyadenylated tail, making them more resistant to exonucleases [1]. Since RNA polymerase II transcription of protein-coding genes produces the majority of known circRNAs, their stability is expected to be 2.5–5 times greater than that of linear transcripts [2].

There is growing evidence that circRNAs, through binding to RBPs, sponging miRNAs, controlling mRNA expression, and protein translation, play significant roles in modifying biological processes [3]. CircRNAs are widely expressed in eukaryotic cells in a cell or tissue specificity plus developmental stage-specificity manner, and in some cases, they are generated more efficiently than their linear mRNAs [4]. Furthermore, it was observed that circRNAs are regulated independently from their linear counterparts. Previous studies have identified circRNAs in various bodily fluids such as blood, urine, cerebrospinal fluid, exosomes, and extracellular vesicles. Due to their presence in these fluids, circRNAs have been suggested as poten-

J. Xiao (ed.), *Advances in Circular RNAs*, Advances in Experimental Medicine and Biology 1485,
https://doi.org/10.1007/978-981-96-9428-0_23

tial biomarkers for cancer and a range of other diseases [5–7]. Several investigations conducted on mammalian cells have documented a negative correlation between the abundance of circRNAs and their corresponding linear mRNAs. This observation emphasizes that the majority of circRNAs are transcribed from the same strand as their linear counterparts, leading to competition during transcription. CircRNAs exhibit significant evolutionary conservation, with approximately 28% similarity between mouse and human, both in terms of their sequence and expression patterns. This conservation is particularly evident in the brain. The brain contains a higher concentration of circRNAs compared to other tissues, and the number of circRNAs steadily increases from embryonic to adult stages [8]. Over 100,000 circRNAs are expressed in the human brain, with an average of three circRNAs per gene. The majority of these circular RNAs are generated from distinct gene junctions [8, 9]. The presence of circRNAs in various biological processes and their involvement in human diseases, particularly neurological diseases, make them potential biomarkers and therapeutic targets [10].

2 CircRNAs in Brain and Brain Cancer

Advancements in bioinformatics algorithms have provided us with an increased probability of uncovering new circular RNAs (circRNAs) by rescuing numerous unmapped reads through chimeric mapping in a back-splicing order. These technological advancements have successfully identified robust associations between the misregulation of circRNAs and various neurodegenerative diseases [8, 11]. Research has demonstrated that circRNAs are present in large quantities and exhibit high expression levels in the central nervous system, particularly in nerve synapses. This is observed during the process of cellular differentiation as well as in response to bursts of electrical activity. Approximately 60% of circRNAs in the central nervous system are upregulated during development, with a significant emphasis on synaptogenesis. In contrast, only 2% of their linear isoforms display a similar trend, suggesting that circRNAs may play a role in regulating synaptic plasticity, neuronal development, aging, and human neuropsychiatric diseases [12, 13]. Brain cells have the remarkable ability to generate numerous circular RNAs (circRNAs), particularly in regions that linear mRNA expression lacks diversity. This phenomenon leads to the creation of a diverse transcriptome and a wide range of biological pathways, ultimately contributing to the performance of human brain cells. While circRNAs are predominantly found in the cytoplasm, a significant portion of them is localized to and enriched in synapses [14]. Furthermore, it has been observed that the levels of circRNAs are higher in the human fetal brain compared to other fetal tissue types or even in the adult human brain. This finding suggests the potential significance of circRNAs in brain development [15]. CircRNAs exhibit varying levels of expression across different types of brain cells. Notably, an increase in circRNA expression has been observed during the process of neuronal differentiation, particularly in forebrain neuron progenitor cells and the frontal cortex [16]. It is worth mentioning that the expression of circRNAs, when compared to their corresponding linear transcripts, tends to be lower and more cell-specific [17]. In addition to their presence in human brain tissue, circRNAs exhibit expression patterns that are specific to age and gender. Notably, circCD99, circPREX1, and circTSPAN15, as well as circEFCAB2, are predominantly expressed in males. On the other hand, circPCTP and circZNF484 show a higher expression in females [2]. Furthermore, circGRIA1, a conserved circRNA, demonstrates an age-related expression pattern in the prefrontal cortex and hippocampus of male rhesus macaques [18]. Recent research has revealed a fascinating connection between diet, vitamin D, and ncRNAs in the field of cancer therapy. Notably, studies have shed light on this important link [19, 20]. Furthermore, it has been observed that both diet and stress can have a profound impact on the expression profiles of circRNAs in the brain cortex. For instance, a high-fat diet–induced diabetes has been found to alter the levels of circRNAs in adult mice [21]. Additionally, significant alterations in the expression of circDYM have been observed in the blood of patients with chronic unpredictable stress and MDD [16]. These findings highlight the intricate relationship between diet, stress, and the expression of circRNAs, providing valuable insights for potential therapeutic interventions. Furthermore, an additional report presents the noteworthy alterations of hsa_circRNA_103636 in individuals with MDD following an 8-week course of antidepressant therapy [22]. The circular nature of circRNAs confers functional benefits such as enhanced axonal transport and an extended half-life ranging from 18.8 to 23.7 h, which is at least 2.5 times longer than the median half-life of their linear counterparts (4.0–7.4 h) [17].

A recent investigation on circRNA in the brain has revealed that numerous genes produce up to 36 neuronal circRNAs, with a significant portion originating from the synapse machinery and synaptic genes. This study is further supported by findings indicating that the level of circRNA is reduced in cultures comprising a substantial population of glial cells. This suggests that the observed increase in circRNA expression may be exclusive to neurons or potentially linked to interactions between neurons. These findings are documented in references [5].

The expression of circRNAs in various cellular compartments, such as synaptosomes, microdissected neuropil within hippocampal slices, or subcellular compartments like the cell body and dendrites, suggests their potential involvement in the translation of mRNA within neurons and their influence on mRNA levels [23]. This indicates that circRNAs play a crucial role in regulating the proper transport and translation of mRNAs in neurons, particularly through synaptic or dendritic content. Consequently, they have a significant impact on neuronal activity and are essential for adaptive responses to various stimuli [24]. It is intriguing to note that a majority of brain disorders are classified as synaptic circRNAs. These circRNAs play a role in altering mRNA translation and localization, thereby impacting synaptic dysfunction or synaptopathy. This early defect in numerous neurodegenerative diseases and brain tumors highlights the significance of circRNAs in these conditions [18, 25]. CircRNAs have been identified as crucial mediators in the progression of glioma, influencing various cellular processes such as proliferation, invasion, migration, and apoptosis. It has been discovered that a number of circRNAs, such as circSMARCA5, circMTO1, circHIPK3, circ007628, circTTBK2, and circMMP9, accelerate the development of gliomas by sequestering miRNAs and binding to their host genes. This information emphasizes the role that circRNAs play in the growth and progression of gliomas [26].

The circSMARCA5 molecule plays a role in promoting angiogenesis and the progression of glioma by regulating the mRNA of VEGF-A, which is known to be a powerful inducer of angiogenesis. It achieves this by interacting with the serine and arginine-rich splicing factor 1 (SRSF1) [27]. On the other hand, circZNF292 downregulates various components of the Wnt/β-catenin signaling pathway and induces hypoxia [28]. Another circular RNA, circ_002136 (also known as circCDK11A_001), appears to regulate the expression of SOX13, which in turn induces the expression of the spondin-2 gene. This gene is associated with tumor aggressiveness and contributes to the promotion of glioma angiogenesis [29]. Targeting HNF1β/Derlin-1 pathway prevented cell apoptosis and thereby promoted proliferation, migration, and invasion of glioma cells [30]. Other study indicates that the promotion of glioma invasion is facilitated by the sequestration of miR-143-3p, leading to an increase in integrin subunit beta 8 (ITGB8) expression, as observed in circ_0046701 [31]. We have summarized the list of circRNAs in brain pathology in Table 23.1.

3 CircRNA in Parkinson's Disease

Parkinson's disease (PD), which is the second most prevalent neurodegenerative disorder causing movement problems, affects around 1–2% of individuals who are 65 years old and older. Although the reasons behind the degeneration of dopaminergic neurons in PD brains are not well

Table 23.1 CircRNAs in brain pathobiology

Upregulated circRNA			
Gene	Function	Samples	References
circRNA_013636	Higher expression was related to PD	SAMP8 mice	[32]
circ-001372	Decreases neural apoptosis and inflammation through PIK3CA/Akt/NF-kappaB via miRNA-148b-3p	Sprague Dawley rats	[33]
circ_002136 (circCDK11A_001)	Sponges miR-138-5p and prevents the repression of SOX13	Glioma	[29]
circSMARCA5, circHIPK3, circMTO1, circTTBK2, circ007628, and circMMP9	Higher expression was related to angiogenesis and glioma progression	Glioma	[27]
circ-0000220	Higher expression was related to PD by promoting production of inflammatory cytokines	BV-2 microglial cells	[34]
circZNF292	Lower expression was related to glioma	Glioma cells	[28]
circ_002136	Regulate angiogenesis in glioma miR-1385p/SOX13	Glioma cells	[29]
circ-KIAA1586	Regulate pathological process of AD via miR-29b, miR-101, miR-15a	Glioma cells	[35]
circNF1–419	Regulate autophagy in AD via interaction with other proteins	AD mice	[36]
circRNA_0000950	Sponge miR-103 and regulate neuro-inflammation and apoptosis	PC12 cells	[37]
mmu_circ_017963	Sponge miR_7033-3p	SAMP8 AD mice	[32]
ciRS-7	Lower expression was related to AD with lower sponging of miR-7/UBE2A	AD human brain tissues	[38]
circRNA HDAC9	Lower expression was related to AD with lower sponging of miR-138/sirtuin1	AD patients and AD mice	[39]
mmu_circRNA_012180	Lower expression was related to AD	SAMP8 mice	[32]
circzip-2	Lower expression was related to PD	*Caenorhabditis elegans*	[40]
circDNAJC6	Higher expression was related to PD	Human brain	[5]

comprehended, it is believed to be influenced by various factors such as genetics, being male, exposure to environmental toxins, excessive oxidative stress, malfunctioning mitochondria, abnormal cell death, and imbalanced signaling pathways. Moreover, recent studies indicate that circular RNAs (circRNAs) may contribute to the progression of PD [41, 42]. In a recent study, researchers utilized the most up-to-date Parkinson's disease (PD) genome-wide association studies (GWAS) to examine 96 genes that exhibited the expression of one or more circRNAs (circular RNAs) after filtering. They also looked at 32 genes that expressed at least one validated circRNA out of a total of 109 genes. The findings from this study suggest that there is an enrichment of circRNA expression in loci associated with both PD and intellectual disability. One example of such a locus is the KANSL1 gene or DNAJC6, which has been identified as being linked to PD [43].

CircDNAJC6, which is generated from the specific location in the genome, exhibited high expression levels in dopamine neurons and moderate levels in pyramidal neurons, but no expression in nonneuronal cells. This enrichment of circRNA production in certain regions supports the involvement of cortical neurons and immune cells in the development of PD, extending beyond the typical degeneration of dopaminergic neurons [5, 42]. CircRNA has been found to play a role in PD by regulating the stability and transport of miRNAs, which in turn affects the expression of α-synuclein (α-Syn), oxidative stress, abnormal transcription, cell death, and autophagy [42]. For example, CDR1, a circRNA that is highly expressed in excitatory neurons, can regulate miR-7. It has been observed that PD patients have lower levels of miR-7 in the substantia nigra, which is associated with a higher accumulation of α-Syn in Lewy bodies, PD is a condition where the expression of α-Syn is affected, and one circRNA, known as circZip-2, has been identified as playing an indirect role in regulating α-Syn levels [44]. Interestingly, circZip-2 is derived from the Zip-2 gene, and its expression has been observed to decrease in individuals with PD [44]. Circum_0017204, Circum_0085869, Circum_0004381, and Circum_0090668 were identified as panels monitoring PD progression in the plasma samples obtained from PD patients with the capacity to distinguish between the late and early phases of PD [45]. Other represented panels include hsa_circRNA_101275, hsa_circRNA_038416, hsa_circRNA_103730, and hsa_circRNA_102850 in PD patients, as well as SLAIN1_circ_0000497, SLAIN2_circ_0126525, ANKRD12_circ_0000826, and PSEN1_circ_0003848 [46]. Both panels have proper ROC curves for PD diagnosis.

4 CircRNA in Alzheimer's Disease

The most common cause of dementia, affecting millions of individuals globally, is Alzheimer's disease (AD) [47]. The two histopathological hallmarks of AD are intracellular neurofibrillary tangles made of abnormally hyperphosphorylated Tau protein and extracellular senile plaques made of amyloid-β peptides [48]. Different studies reported the association of circRNAs with an increased risk of AD [49]. Moreover, circRNAs can regulate Aβ generation, microglia autophagy, and neuroinflammation in the brain. In this, while an RNA-seq data analysis measured cortical circRNA expression in people with and without AD, the findings demonstrated a strong correlation between circRNA expression and the degree of clinical dementia, and the severity of neuropathological impairment in AD [23]. Screening RNA from human brain regions, especially the MAPT locus which contributes to tau protein pathology, is related to producing many circRNAs via a back-splicing mechanism [35, 50]. Likewise, CDR1as, one of the first circRNAs to be identified and studied in brain pathobiology, has more than 70 traditional binding sites for miR-7 and is derived from the cerebellar degeneration-related protein 1 antisense transcript (CDR1AS) [51]. CDR1 downregulation results in upregulation of miR-7 and negative regulation of its targets, including UBE2A (ubiquitin protein ligase A). The importance of UBE2A is in eliminating amyloid peptides, which increase amyloid accumulation and senile plaque deposition. In addition, ciRS-7 can regulate β-secretase (BACE1) and APP protein levels [38]. Another circRNA related to Aβ generation is circRNA CWC27, which is significantly found to be lower in AD patients and mice models. CircCwc27 can attach to purine-rich element-binding protein A (Pur-α) and thereby decreases the expression of proteins such as phosphatase 1, regulatory inhibitor subunit1B (Ppp1r1b), amyloid precursor protein (APP), including dopamine receptor D1 (Drd1), and LIM homeobox 8 (Lhx8) and neurotrophic tyrosine kinase, receptor, type 1 (Ntrk1) and induces cognitive impairment [52] as shown in Fig. 23.1.

5 Conclusion

CircRNAs' stability, variety, and abundance in the brain make them excellent candidates for use as biomarkers. Their expression varies during development and under different medical circumstances. They have distinct expression patterns in diverse cells and organs. Furthermore, common laboratory procedures like as RT-PCR are reliable and noninvasive methods of detecting them from peripheral blood, plasma, and saliva. The main issue with circRNA research, meanwhile, is that most of it relies only on bioinformatics and lacks basic validation. Other issues include very low expression levels, an uneven stoichiometry between circRNAs and miRNAs, and a small binding area. Various neuronal cell types are a great option for researching and treating neurological diseases in the brain because of their pro-

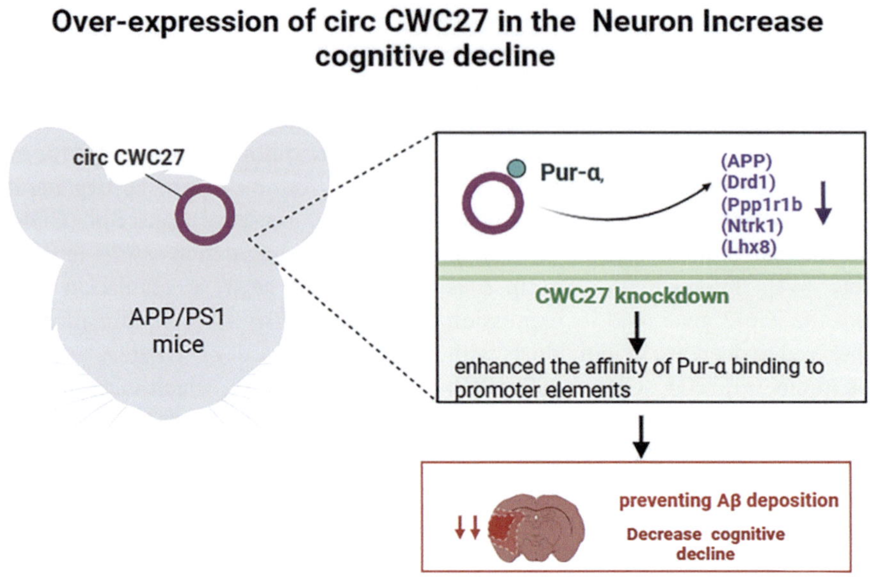

Fig. 23.1 Downregulation of circCwc27 inhibits Aβ plaque accumulation and cognitive decline by increasing our α-target protein transcription

pensity to release and transport circRNAs across the brain's network.

References

1. Zhang ZJ, Zhang YH, Qin XJ, Wang YX, Fu J (2020) Circular RNA circDENND4C facilitates proliferation, migration and glycolysis of colorectal cancer cells through miR-760/GLUT1 axis. Eur Rev Med Pharmacol Sci 24(5):2387–2400. https://doi.org/10.26355/eurrev_202003_20506
2. Sekar S, Liang WS (2019) Circular RNA expression and function in the brain. Noncoding RNA Res 4(1):23–29. https://doi.org/10.1016/J.NCRNA.2019.01.001
3. Mojarad MA, Mojarad MA, Nourbakhsh M (2022b) Circulating circular RNA ADAM9 a potential biomarker for human colorectal cancer. Gene Rep 26(1):101–516. https://doi.org/10.1016/J.GENREP.2022.101516
4. Gokool A, Anwar F, Voineagu I (2020) The landscape of circular RNA expression in the human brain. Biol Psychiatry 87(3):294–304. https://doi.org/10.1016/J.BIOPSYCH.2019.07.029
5. Dong X, Bai Y, Liao Z, Gritsch D, Liu X, Wang T, Borges-Monroy R, Ehrlich A, Serrano GE, Feany MB, Beach TG, Scherzer CR (2023) Circular RNAs in the human brain are tailored to neuron identity and neuropsychiatric disease. Nat Commun 14(1):1–12. https://doi.org/10.1038/s41467-023-40348-0
6. Tran AM, Chalbatani GM, Berland L, Cruz De los Santos M, Raj P, Jalali SA, Gharagouzloo E, Ivan C, Dragomir MP, Calin GA (2020) A new world of biomarkers and therapeutics for female reproductive system and breast cancers: circular RNAs. Front Cell Dev Biol 8(5):14–707. https://doi.org/10.3389/FCELL.2020.00050/BIBTEX
7. AmeliMojarad M, AmeliMojarad M, Pourmahdian A (2022) Circular RNA circ_0051620 sponges miR-338–3p and regulates ADAM17 to promote the gastric cancer progression. Pathol Res Pract 23(3):153–887. https://doi.org/10.1016/J.PRP.2022.153887
8. Hanan M, Soreq H, Kadener S (2017) CircRNAs in the brain. RNA Biol 14(8):1028. https://doi.org/10.1080/15476286.2016.1255398
9. Hanan M, Simchovitz A, Yayon N, Vaknine S, Cohen-Fultheim R, Karmon M, Madrer N, Rohrlich TM, Maman M, Bennett ER, Greenberg DS, Meshorer E, Levanon EY, Soreq H, Kadener S (2020) A Parkinson's disease CircRNAs resource reveals a link between circSLC8A1 and oxidative stress. EMBO Mol Med 12(9):e11942. https://doi.org/10.15252/EMMM.201911942
10. Mahmoudi E, Cairns MJ (2019) Circular RNAs are temporospatially regulated throughout development and ageing in the rat. Sci Rep 9(1):2564. https://doi.org/10.1038/S41598-019-38860-9
11. Mehta SL, Dempsey RJ, Vemuganti R (2020) Role of circular RNAs in brain development and CNS diseases. Prog Neurobiol 18(6):101–746. https://doi.org/10.1016/J.PNEUROBIO.2020.101746
12. Doxakis E (2022) Insights into the multifaceted role of circular RNAs: implications for Parkinson's disease pathogenesis and diagnosis. npj

Parkinson's Dis 8(1):1–9. https://doi.org/10.1038/s41531-021-00265-9
13. Xu K, Zhang Y, Li J (2021) Expression and function of circular RNAs in the mammalian brain. Cell Mol Life Sci 78(9):4189–4200. https://doi.org/10.1007/S00018-021-03780-3
14. Jiang MP, Xu WX, Hou JC, Xu Q, Wang DD, Tang JH (2021) The emerging role of the interactions between circular RNAs and RNA-binding proteins in common human cancers. J Cancer 12(17):5206–5219. https://doi.org/10.7150/JCA.58182
15. Xia S, Feng J, Lei L, Hu J, Xia L, Wang J, Xiang Y, Liu L, Zhong S, Han L, He C (2017) Comprehensive characterization of tissue-specific circular RNAs in the human and mouse genomes. Brief Bioinform 18(6):984–992. https://doi.org/10.1093/BIB/BBW081
16. Zhang Y, Xue W, Li X, Zhang J, Chen S, Zhang JL, Yang L, Chen LL (2016) The biogenesis of nascent circular RNAs. Cell Rep 15(3):611–624. https://doi.org/10.1016/j.celrep.2016.03.058
17. Enuka Y, Lauriola M, Feldman ME, Sas-Chen A, Ulitsky I, Yarden Y (2016) Circular RNAs are long-lived and display only minimal early alterations in response to a growth factor. Nucleic Acids Res 44(3):1370–1383. https://doi.org/10.1093/NAR/GKV1367
18. Xu K, Zhang Y, Xiong W, Zhang Z, Wang Z, Lv L, Liu C, Hu Z, Zheng YT, Lu L, Hu XT, Li J (2020) CircGRIA1 shows an age-related increase in male macaque brain and regulates synaptic plasticity and synaptogenesis. Nat Commun 11(1):1–15. https://doi.org/10.1038/s41467-020-17435-7
19. Bikle DD (2021) Vitamin D regulation of and by long non coding RNAs. Mol Cell Endocrinol 5(32):111–317. https://doi.org/10.1016/j.mce.2021.111317
20. Mojarad MA, Mojarad MA, Noourbakhsh M (2022a) Impact of vitamin D on expression of SIRT7 and CYP24A1 in human breast cancer cells. Res Vet Sci Med 2(1):31. https://doi.org/10.25259/RVSM_6_2021
21. Yoon G, Cho KA, Song J, Kim YK (2019) Transcriptomic analysis of high fat diet fed mouse brain cortex. Front Genet 10(1):83. https://doi.org/10.3389/FGENE.2019.00083
22. Cui X, Niu W, Kong L, He M, Jiang K, Chen S, Zhong A, Li W, Lu J, Zhang L (2016) hsa_circRNA_103636: potential novel diagnostic and therapeutic biomarker in major depressive disorder. Biomark Med 10(9):943–952. https://doi.org/10.2217/BMM-2016-0130
23. Westholm JO, Miura P, Olson S, Shenker S, Joseph B, Sanfilippo P, Celniker SE, Graveley BR, Lai EC (2014) Genome-wide analysis of drosophila circular RNAs reveals their structural and sequence properties and age-dependent neural accumulation. Cell Rep 9(5):1966–1980. https://doi.org/10.1016/J.CELREP.2014.10.062
24. Holt CE, Schuman EM (2013) The central dogma decentralized: new perspectives on RNA function and local translation in neurons. Neuron 80(3):648–657. https://doi.org/10.1016/J.NEURON.2013.10.036
25. Wu DP, Zhao YD, Yan QQ, Liu LL, Wei YS, Huang JL (2023) Circular RNAs: emerging players in brain aging and neurodegenerative diseases. J Pathol 259(1):1–9. https://doi.org/10.1002/PATH.6021
26. Han J, Zhao G, Ma X, Dong Q, Zhang H, Wang Y, Cui J (2018) CircRNA circ-BANP-mediated miR-503/LARP1 signaling contributes to lung cancer progression. Biochem Biophys Res Commun 503(4):2429–2435. https://doi.org/10.1016/j.bbrc.2018.06.172
27. Barbagallo D, Caponnetto A, Brex D, Mirabella F, Barbagallo C, Lauretta G, Morrone A, Certo F, Broggi G, Caltabiano R, Barbagallo GM, Spina-Purrello V, Ragusa M, Di Pietro C, Hansen TB, Purrello M (2019) CircSMARCA5 regulates VEGFA mRNA splicing and angiogenesis in glioblastoma multiforme through the binding of SRSF1. Cancers (Basel) 11(2):194. https://doi.org/10.3390/CANCERS11020194
28. Chen CH, Su YJ, Ding H, Duan J, Wang J (2020) Circular RNA ZNF292 affects proliferation and apoptosis of hepatocellular carcinoma cells by regulating Wnt/β-catenin pathway. Eur Rev Med Pharmacol Sci 24(23):12124–12130. https://doi.org/10.26355/EURREV_202012_24001
29. He Z, Ruan X, Liu X, Zheng J, Liu Y, Liu L, Ma J, Shao L, Wang D, Shen S, Yang C, Xue Y (2019) FUS/circ_002136/miR-138-5p/SOX13 feedback loop regulates angiogenesis in glioma. J Exp Clin Cancer Res 38(1):65. https://doi.org/10.1186/S13046-019-1065-7
30. Zheng J, Liu X, Xue Y, Gong W, Ma J, Xi Z, Que Z, Liu Y (2017) TTBK2 circular RNA promotes glioma malignancy by regulating miR-217/HNF1β/Derlin-1 pathway. J Hematol Oncol 10(1):52. https://doi.org/10.1186/S13045-017-0422-2
31. Li G, Yang H, Han K, Zhu D, Lun P, Zhao Y (2018) A novel circular RNA, hsa_circ_0046701, promotes carcinogenesis by increasing the expression of miR-142-3p target ITGB8 in glioma. Biochem Biophys Res Commun 498(1):254–261. https://doi.org/10.1016/J.BBRC.2018.01.076
32. Huang JL, Xu ZH, Yang SM, Yu C, Zhang F, Qin MC, Zhou Y, Zhong ZG, Wu DP (2018) Identification of differentially expressed profiles of Alzheimer's disease associated circular RNAs in a panax notoginseng saponins-treated Alzheimer's disease mouse model. Comput Struct Biotechnol J 1(6):523. https://doi.org/10.1016/J.CSBJ.2018.10.010
33. Wang M, Suo L, Yang S, Zhang W (2021) CircRNA 001372 reduces inflammation in propofol-induced neuroinflammation and neural apoptosis through PIK3CA/Akt/NF-κB by miRNA-148b-3p. J Investig Surg 34(11):1167–1177. https://doi.org/10.1080/08941939.2020.1771639
34. Li Y, Li Y, Li Y, Ashraf U, Ashraf U, Ashraf U, Chen Z, Chen Z, Chen Z, Zhou D, Zhou D, Zhou D, Imran M, Imran M, Imran M, Ye J, Ye J, Ye J, Chen H, Chen

H, Chen H, Cao S, Cao S, Cao S (2020) Genome-wide profiling of host-encoded circular RNAs highlights their potential role during the Japanese encephalitis virus-induced neuroinflammatory response. BMC Genomics 21(1):409. https://doi.org/10.1186/S12864-020-06822-5
35. Zhang Y, Yu F, Bao S, Sun J (2019) Systematic characterization of circular RNA-associated CeRNA network identified novel circRNA biomarkers in Alzheimer's disease. Front Bioeng Biotechnol 7(2):222. https://doi.org/10.3389/FBIOE.2019.00222/FULL
36. Diling C, Yinrui G, Longkai Q, Xiaocui T, Yadi L, Xin Y, Guoyan H, Ou S, Tianqiao Y, Dongdong W, Yizhen X, Yang BB, Qingping W (2019) Circular RNA NF1-419 enhances autophagy to ameliorate senile dementia by binding Dynamin-1 and Adaptor protein 2 B1 in AD-like mice. Aging (Albany NY) 11(24):12002. https://doi.org/10.18632/AGING.102529
37. Yang H, Wang H, Shang H, Chen X, Yang S, Qu Y, Ding J, Li X (2019) Circular RNA circ_0000950 promotes neuron apoptosis, suppresses neurite outgrowth and elevates inflammatory cytokines levels via directly sponging miR-103 in Alzheimer's disease. Cell Cycle 18(18):2197. https://doi.org/10.1080/15384101.2019.1629773
38. Zhao Y, Alexandrov PN, Jaber V, Lukiw WJ (2016) Deficiency in the ubiquitin conjugating enzyme UBE2A in Alzheimer's disease (AD) is linked to deficits in a natural circular miRNA-7 sponge (circRNA; ciRS-7). Genes (Basel) 7(12):116. https://doi.org/10.3390/GENES7120116
39. Lu Y, Tan L, Wang X (2019) Circular HDAC9/microRNA-138/Sirtuin-1 pathway mediates synaptic and amyloid precursor protein processing deficits in Alzheimer's disease. Neurosci Bull 35(5):877–888. https://doi.org/10.1007/S12264-019-00361-0
40. Kumar L, Shamsuzzama JP, Haque R, Shukla S, Nazir A (2018) Functional characterization of novel circular RNA molecule, circzip-2 and its synthesizing gene zip-2 in *C. elegans* model of Parkinson's disease. Mol Neurobiol 55(8):6914–6926. https://doi.org/10.1007/S12035-018-0903-5
41. Farrer MJ (2006) Genetics of Parkinson disease: paradigm shifts and future prospects. Nat Rev Genet 7(4):306–318. https://doi.org/10.1038/NRG1831
42. Liao J, Zhang Q, Huang J, He H, Lei J, Shen Y, Wang J, Xiao Y (2023) The emerging role of circular RNAs in Parkinson's disease. Front Neurosci 17(1):113–7363. https://doi.org/10.3389/FNINS.2023.1137363/BIBTEX
43. Roosen DA, Blauwendraat C, Cookson MR, Lewis PA (2019) DNAJC proteins and pathways to parkinsonism. FEBS J 286(16):3080–3094. https://doi.org/10.1111/FEBS.14936
44. Mehta SL, Chokkalla AK, Bathula S, Arruri V, Chelluboina B, Vemuganti R (2022) CDR1as regulates α-synuclein-mediated ischemic brain damage by controlling miR-7 availability. Mol Ther Nucleic Acids 3(1):57–67. https://doi.org/10.1016/J.OMTN.2022.11.022
45. Zhong L, Ju KJ, Chen A, Cao H (2021) Circulating CircRNAs panel acts as a biomarker for the early diagnosis and severity of Parkinson's disease. Front Aging Neurosci 1(3):684289. https://doi.org/10.3389/FNAGI.2021.684289
46. Xiao Y, Chen H, Liao J, Zhang Q, He H, Lei J, Huang J, Ouyang Q, Shen Y, Wang J (2022) The potential circular RNAs biomarker panel and regulatory networks of Parkinson's disease. Front Neurosci 1(16):893713. https://doi.org/10.3389/FNINS.2022.893713
47. Long JM, Holtzman DM (2019) Alzheimer disease: an update on pathobiology and treatment strategies. Cell 179(2):312–339. https://doi.org/10.1016/J.CELL.2019.09.001
48. Alhasan AA, Izuogu OG, Al-Balool HH, Steyn JS, Evans A, Colzani M, Ghevaert C, Mountford JC, Marenah L, Elliott DJ, Santibanez-Koref M, Jackson MS (2016) Circular RNA enrichment in platelets is a signature of transcriptome degradation. Blood 127(9):e1–e11. https://doi.org/10.1182/blood-2015-06-649434
49. Cochran KR, Veeraraghavan K, Kundu G, Mazan-Mamczarz K, Coletta C, Thambisetty M, Gorospe M, De S (2021) Systematic identification of circRNAs in Alzheimer's disease. Genes (Basel) 12(8):1258. https://doi.org/10.3390/GENES12081258
50. Welden JR, van Doorn J, Nelson PT, Stamm S (2018) The human MAPT locus generates circular RNAs. Biochim Biophys Acta (BBA) Mol Basis Dis 1864(9):2753–2760. https://doi.org/10.1016/J.BBADIS.2018.04.023
51. Song C, Zhang Y, Huang W, Shi J, Huang Q, Jiang M, Qiu Y, Wang T, Chen H, Wang H (2022) Circular RNA Cwc27 contributes to Alzheimer's disease pathogenesis by repressing Pur-α activity. Cell Death Differ 29(2):393–406. https://doi.org/10.1038/S41418-021-00865-1
52. Zou Y, Zheng S, Deng X, Yang A, Xie X, Tang H, Xie X (2019) The role of circular RNA CDR1as/ciRS-7 in regulating tumor microenvironment: a pan-cancer analysis. Biomol Ther 9(9):429. https://doi.org/10.3390/BIOM9090429

Circular RNAs in Immune-Related Diseases

24

Maria Laura Idda, Antonella Francesca Fara, and Matteo Floris

Abstract

CircRNAs are a novel class of noncoding RNAs characterized by a covalently closed loop structure lacking both a 3′ polyadenylation tail and a 5′ cap. Increasing evidence demonstrated that circRNAs can participate in several biological processes including immune system physiology and immune-related diseases. CircRNAs act on immune pathways through several molecular mechanisms including miRNAs and RBP sponges, protein scaffolds, and mRNA stability modulation via diverse signaling pathways including MAPK, JAK-STAT, and Wnt signaling pathway. Recently, circRNAs have also been studied as biomarkers for immune diseases as well as new potential therapeutic targets. Effects of dysregulated circRNA levels correlate, among others, with alterations in immune cell behavior, cytokines production, and immune response in general.

In this chapter, we describe new experimental progress in circRNAs that participates in the regulation of the immune system and in the pathogenesis of autoimmune diseases and infectious diseases.

Keywords

CircRNA · Immune response · Autoimmune · Infectious diseases

1 Introduction

The main role of the immune system is to monitor and defend host organisms against infecting pathogens, maintaining internal homeostasis through a balance between immune tolerance and appropriate immune response during immune surveillance. Indeed, immune surveillance represents a major component of the immune response; it is strongly related to specific features of the foreign antigen such as virus and bacterial [1, 2]. Thus, the immune system plays a dual role: (i) a functional immune system can quickly detect and respond to abnormal conditions in the body acting as a defense against microorganisms; (ii) a dysregulated response of the human immune system may lead to abnormal immune response to antigens aggravating the disease or leading to the onset of autoimmune diseases.

Physiologically, the immune response can be split into two branches: innate (general) and adaptive (specialized) immunity. *Innate immu-*

M. L. Idda (✉) · M. Floris
Department of Biomedical Science, University of Sassari, Sassari, Italy
e-mail: mlidda1@uniss.it; mfloris@uniss.it

A. F. Fara
Department of Medicine, Surgery and Pharmacy, University of Sassari, Sassari, Italy
e-mail: affara@uniss.it

J. Xiao (ed.), *Advances in Circular RNAs*, Advances in Experimental Medicine and Biology 1485,
https://doi.org/10.1007/978-981-96-9428-0_24

nity represents the first line of defense against pathogens, and it is characterized by a unique response to all foreign antigens; it gives rise to *adaptive immune response, which conducts* effector functions. The innate immune system is mediated by dendritic cells (DCs), macrophages, and natural killer cells (NKs). When the innate immune system is not able to destroy the germs, the adaptive immune system takes over the function of protecting the organism; the main players, T and B lymphocytes, target germs responsible for the infection, attacking them [2].

Recently, different lines of evidence have recognized circular RNAs (circRNAs) as key players of the immune responses with a primary role in the pathogenesis of autoimmune diseases, tumor immunity, and antimicrobial immunity. CircRNAs can participate in several biological processes related to physiological and pathological immune response by acting as miRNA sponges, protein interactors, and mRNA stability maintainers, thus participating in different steps of gene expression pathways [3–5].

The mechanisms involving circRNAs in immune-related pathways can be summarized in two major areas: (i) effects of circRNAs on immune-related signaling pathways, such as the MAPK signaling pathway, JAK-STAT signaling pathway, and Wnt signaling pathway, and (ii) the effects of circRNAs on immune cells and cytokine production. Furthermore, due to the circular structure, circRNAs are considered stable molecules protected from exonuclease degradation and can also be detected in several biological fluids such as serum, urine, and saliva; for these reasons, circRNAs possess the primary requirements to become potential biomarkers in the context of immune-related diseases [6, 7].

CircRNAs are single-stranded RNA molecules without 5′-terminal cap structure and 3′-terminal poly-A tail; however, endogenous circRNAs can form imperfect RNA duplexes (of around 16–26 bp) which act as inhibitors of the double-stranded RNA (dsRNA)-activated protein kinase (PKR); PKR is an enzyme with multiple functions, including regulation of innate immunity and antiviral defense [8]. Following viral infection or poly(I:C) stimulation (mimicking viral invasion), circRNAs are globally degraded by RNase L, a process required for PKR activation through phosphorylation. An increase in PKR phosphorylation and a circRNA reduction have been observed in peripheral blood mononuclear cells (PBMCs) derived from patients with systemic lupus erythematosus (SLE). Notably, overexpression of circRNA containing dsRNA structure in PBMCs or T cells purified from SLE patients can alleviate the effects of aberrant PKR activation, establishing a potential link between circRNAs and the pathogenesis of SLE [9].

Notably, several circRNA–miRNA–mRNA axes (ceRNA, competing endogenous RNAs) have been related to the onset, progression, and survival of patients affected by immune-related disease. For example, it has been observed that circRNA regulates immune-related mRNA through miRNA sponge during papillary thyroid cancer (PTC) and significantly influences immune regulation in patients affected by PTC. Among the most interesting results in the ceRNA subnetwork, circ_0016404-miR-1275-FGF1/CTSB/IL13RA1 and circ_0060055/circ_0038718-miR-150-3p CXCL14 axes were identified. Moreover, FGF1, PDGFC, CTSB, IL13RA1, TGFBR3, CXCL14, BMP7, LIFR, miR-125a-5p, miR-1275, miR-150-3p, miR-18b-5p, and miR-27b-3p were also found to have high diagnostic accuracy and may be potential novel diagnostic markers for PTC [10].

Epigenetic modifications also influence the heterogeneity of immune response at different levels. For example, N6-methyladenosine (m6A) RNA modification on human circRNAs, for example, affects innate immunity. The YT521-B homology domain family (YTHDF) protein YTHDF2 is a cytoplasmic m6A reader that recognizes and binds to m6A within the consensus RR(m6A)CH [11]. m6A reader YTHDF2 sequesters m6A-circRNA and is essential for suppression of innate immunity. Furthermore, only unmodified circRNA directly activates RNA pattern recognition receptor RIG-I. This activation occurs in the presence of lysine-63-linked polyubiquitin chains, leading to the filamentation of the adaptor protein mitochondrial antiviral-signaling protein (MAVS) and subsequent activation of the downstream transcription factor IRF3 (Interferon regulatory factor 3). These findings support a model in which circRNA modification translates nucleic acid chemical changes into organismal innate immunity [12]. In general, methylation of circRNA can be described as a mechanism affecting the biogenesis, transport, stability, translation, and degradation of circRNA.

To conclude, circRNAs were discovered as a new class of ncRNA that participate in the immune responses at different levels, affecting the processes of antimicrobial immunity, autoimmune diseases regulation, and immune response in presence of tumors. Here, we define the relationship between circRNAs and immune system, as well as the functions and mechanisms of their interactions during diseases' progression with focus on microorganisms' responses and autoimmunity; we will also highlight the clinical challenges in this field.

2 Regulation of Immune Cells

As described above, circRNAs achieve several functions in immune cell response; they have been detected in different types of immune cells, such as macrophages, B cells, and T cells. By modulating the activation, differentiation, or exhaustion of these populations, circRNAs may contribute to the physiological response of the immune system or in the development of various diseases. In the next section, we summarize recent insight into the role of circRNA in regulating immune cells (Table 24.1).

2.1 CircRNAs and Monocyte–Macrophage Activation

Growing evidence shows that circRNAs and macrophages have a key role in the progression and onset of several immune-mediated diseases, including the immune response associated with cancer.

Macrophages are present in different tissues and body cavities and are close to mucosal surfaces; they are a vital part of the innate immune system for host defense against many pathogens and cancer cells. Macrophage differentiation relies on various environmental conditions that determine their polarization. This polarization is tightly regulated, leading to either the classically triggered proinflammatory M1 response or the M2 immune response, which is mainly characterized by an anti-inflammatory pattern [36]. The levels of circRNAs can affect the activation of macrophages.

One of the first studies on this topic examined the function of a mouse circular RNA, *mcircRasGEF1B,* in mouse macrophages (for which an analog has been identified in humans, hcircRasGEF1B) and studied the effects of its silencing on transcriptome dynamics during lipopolysaccharide (LPS) response using an RNA sequencing approach (RNA-seq). Analysis of the RNA-seq data revealed that depletion of *mcircRasGEF1B* results in alteration of the expression of hundreds of genes. The most enriched functional categories were involved in cell cycle progression, mitotic division, metabolic activity, NF-κB and LPS signaling pathways, and macrophage activation, supporting the role of *mcircRasGEF1B* in the process of cellular response to LPS stimulation. Later, Lun Ng and colleagues observed that silencing of *mcircRasGEF1B* reduces LPS-induced Intercellular Adhesion Molecule 1 (ICAM-1) expression, and that *mcircRasGEF1B* regulates the stability of mature ICAM-1 mRNAs. These initial studies helped to functionally characterize circRNAs as a novel RNA species essential for properly calibrating immune responses [13, 14].

On the other hand, contrasting results are observed while analyzing different molecules. Zhang et al., for example, observed that circPPM1F participates in the activation of M1 macrophages differentiation in diabetic patients [15], while Whang and colleagues showed that circ_0110102 inhibits macrophage activation functioning through miR-580-5p sponge [16]. Thus, no general conclusions can be drawn, but each circRNA needs to be analyzed in a specific context.

Table 24.1 CircRNAs and their potential functions in immune cell regulation

CircRNA (hsa_)	miRNA target	References
Macrophage functions		
mcircRasGEF1B	ICAM-1	[13, 14]
circPPM1F	M1 macrophages differentiation	[15]
circ_0110102	miR-580-5p	[16]
circ_0005567	mir-492	[17]
circHIPK3	Macrophage inflammasome	[18–21]
circRNF19B	Macrophage inflammasome	
circ_0001490	Macrophage inflammasome	
circCdyl	Macrophage inflammasome	
circ_0000518	FUS/CaMKKβ/AMPK	[22]
circ_0043497/circ_0001204	Monocyte-derived macrophages modulation	[23]
circ_0050486	miR-1270	[24]
B cells		
circCFL1	miR-107	[25]
circ_OTUD7A	miR-431-5p	[26]
circ_0132266	miR-337-3p/PML axis	[27]
T cells		
circ_0045272	IL-2	[28]
circ_0001806	miR-126	[29]
circ_0012919	miR-125a-3p; CD4+ cells	[30]
circ_0005519	Let-7a-5p levels; CD4+ cells	[31]
circSPEF2	miR-16-5p/BACH2 axis; CD4+ cells	[32]
circIGF2BP3	PD-L1; CD8+ cells	[33]
circTRPS1	CD8+ T cell exhaustion	[34]
circRNA100783	CD28; CD8+ cells	[35]

CircRNAs can also play a role in directing the macrophage-mediated inflammatory response; for instance, circ_0005567 can lead to M2 macrophage polarization via the mir-492/ Suppressors of Cytokine Signaling 2 (SOCS2) axis, thus reducing chondrocyte apoptosis and possibly inhibiting osteoarthritis progression [17]. CircRNAs can also promote the M1 inflammatory response; in particular, circHIPK3, circRNF19B, circCdyl, and circ_0001490 were found to activate the macrophage inflammasome [18–21].

The M1/M2 polarization plays a key role in the context of several diseases, including multiple sclerosis (MS), an autoimmune disease (described in detail below), which mainly affects young women. Circ_0000518, which is upregulated in MS patients, was demonstrated to play an important role in the progression of MS with an unclear mechanism. To define the role of circ_0000518 in MS, human microglial clone 3 (HMC3) cells were treated with 100 ng/mL of LPS for 24 h followed by circ_0000518 silencing. LPS treatment promoted an apoptotic phenotype, increased the expression of M1 polarization markers, and inhibited M2 polarization marker expression; silencing of circ_0000518 reversed this effect. Furthermore, functional studies revealed that circ_0000518 promotes macrophage/microglial M1 polarization through the FUS/CaMKKβ/AMPK pathway, aggravating MS pathology [22].

Huang and colleagues characterize circRNA expression profiles in human monocyte-derived macrophages (MDMs) in response to *Mycobacterium tuberculosis* (TB) infection using microarray assay. Results revealed that many circRNAs were differentially expressed in human MDMs after TB infection; of these, 32 circRNAs were upregulated and 110 were downregulated.

Circ_0043497 and circ_0001204 were identified as effective diagnostic biomarkers for TB. This study demonstrated for the first time that circRNA alterations are involved in human MDMs response during TB infection, revealing new potential targets for TB diagnosis and treatment [23]. Furthermore, in the context of atherosclerosis (AS), a chronic inflammatory disease, Wang and colleagues used THP1 cells (a human monocytic cell line derived from an acute monocytic leukemia patient) to characterize the role of circ_0050486 in regulating monocyte functions. Circ_0050486 is significantly upregulated in THP1 stimulated with phorbol-12-myristate-13-acetate (PMA) and oxidized low-density lipoprotein (oxLDL) as compared to cells stimulated only with PMA. Moreover, the authors demonstrated that circ_0050486 silencing reduces IL-6 and TNF-α production and cell death rates in oxLDL-induced THP1 macrophages. They concluded that circ_0050486 knockdown inhibited inflammation and apoptosis by targeting miR-1270 in oxLDL-induced THP-1 macrophages [24].

3 B Cells

B lymphocytes (B cells) are a group of lymphocytes that express clonally diverse cell-surface immunoglobulin receptors; in humans, B lymphocytes are generated from pluripotent hematopoietic stem cells (HSCs) in the fetal liver, in the adult bone marrow (BM), and thymus. The B cell receptor (BCR), a membrane immunoglobulin, is essential for B cell development, differentiation, and survival. Depending on rearrangement of immunoglobulins (Ig) genes and expression of cell surface markers, the development of B cells is generally divided into four main stages: pro-B cells, pre-B cells, immature B cells, and mature B cells. The immature B cells move from the bone marrow to the spleen to differentiate into mature follicular (FO) B cells or marginal zone (MZ) B cells [37.

Several studies indicated that noncoding RNAs, including circRNAs, play multiple roles in the development of B lymphocytes. To deeply understand the role of circRNAs in B cell differentiation, Pan and colleagues performed an experimental analysis using B cell subsets purified from single-cell suspensions of mouse bone marrow. RNA-Seq was performed during B cell differentiation, and expression profiles of circRNAs, miRNAs, and mRNAs were analyzed at different stages. Interestingly, several circRNAs were specifically expressed in pro-B cells, pre-B cells, immature B cells, and mature B cells. The circRNA-associated ceRNA networks created in the sequential differentiation stages analyzed pointed out a potential role of circRNAs in these processes. This study represents the beginning of a new field exploring circRNA profiles, and circRNA–ceRNA networks in different B cell developmental stages strongly contribute to advancing research on B cell development and differentiation [38]. However, the relationship between circRNAs and human B-cell-mediated immunity remains to be explored and has been mostly assessed in the context of cancers, such as diffuse large B cell lymphoma (DLBCL). For example, upregulated circCFL1 acts as a sponge for miR-107, with subsequent upregulation of high mobility group box 1 (HMGB1) in DLBCL cells. Functional studies have then revealed that circCFL1 directly binds to miR-107, releasing HMGB1 from the inhibitory effects [25]. In another study, the silencing of the circ_OTUD7A has suppressed proliferation and metastasis generation in DLBCL, induced cell cycle arrest, and enhanced apoptosis. The mechanism works through the action of the circRNA as a miR-431-5p sponge, subsequently affecting the expression of FOXP1 (Forkhead box protein P1) protein [26]. The relevance of circRNAs on B cell biology has also been investigated in the context of leukemia; circ_0132266, for example, has been revealed to be downregulated in chronic lymphocytic leukemia, leading to enhancement of B cell viability influencing the miR-337-3p/PML axis activity [27]. B cell activity and correlation with circRNA were also analyzed in the context of autoimmune disease such as MS as deeply described in the MS section [39].

4 T Cells

T cells—key players in adaptive immunity—play a major role in the antigen response through several steps, including antigen recognition, activation, and infiltration. The major components of the T cell population are represented by the CD4+ and CD8+ T cell subpopulations.

Using high-throughput circRNA profiling in patients with SLE, 127 differentially expressed circRNAs in T cells were identified. Among them, circRNA circ_0045272 has been reported to be downregulated. Additionally, functional studies demonstrated that circ_0045272 silencing increases early apoptosis in Jurkat cells (an immortalized line of human T lymphocyte cells) and enhances IL-2 production [28]. In another high-throughput microarray study, downregulation of circ_0001806 was detected in patients affected by cryptococcal meningitis as compared to healthy controls. Using appropriate animal models, it was also demonstrated that circ_0001806 silencing increases the intensity of fungal infection and decreases survival of animals. Circ_0001806 has been shown to increase adrenomedullin level, decrease cell apoptosis, and reverse G1/S arrest in T cells through miR-126 sponge, thus demonstrating a key role in cell cycle regulation and apoptosis in T cells [29]. It is important to note that research focusing on the connection between circRNAs and CD4+ T cells mainly concentrates on SLE and asthma. For example, DNA methylation of CD11a and CD70 in CD4+ T cells from patients with SLE was associated with the downregulation of circ_0012919 [30]. Circ_0012919 has also been discovered to induce MDA5 expression in CD4+ T cells through miR-125a-3p downregulation [30]. Furthermore, circ_0002594 and circ_0005519 can affect asthma by regulating CD4+ T cells [31, 40]. Specifically, circ_0005519, another circRNA influencing T cell function is upregulated in CD4+ T cells of asthmatic patients and inversely correlated with let-7a-5p levels. Circ_0005519 levels, in CD4+ T cells, have also been correlated with nitric oxide and eosinophil ratio in the circulation of asthmatic patients participating in the study [31]. Changing context and exploring the potential mechanism of action of the circ_0128899 (circSPEF2) in DLBCL, Zhou and colleague found that circSPEF2 was a circRNA with abnormally downregulated expression in DLBCL. Upregulated expression of CircSPEF2 inhibits the proliferative activity and induces apoptosis of lymphoma cells in both in vitro and in vivo studies. It also increases CD4+ T cells and decreases the proportion of Treg cells in the tumor microenvironment. Mechanistically, circSPEF2 can influence DLBCL progression by modulating cellular proliferation and apoptosis, as well as the proportion of Treg and CD4 cells through the miR-16-5p/BACH2 axis [32].

m6A modification, one of the major posttranscriptional modifications of RNAs, participates in various aspects of RNA homeostasis [41]. m6A-modified circIGF2BP3 was found to inhibit CD8+ T cell responses and promote tumor immune evasion by promoting the deubiquitination of PD-L1 in lung cancer [33]. Moreover, studying bladder cancer microenvironment, exogenous circTRPS1 was proven to be related to CD8+ T cell exhaustion [34]. In addition, Chen et al. distinguished that the expression of circRNA100783 is influenced by time and CD28-related CD8(+) T-cell aging at the time of antigen exposure. Indeed, using circular RNA microarray data, they compared the circular RNA expression profiles between CD28(+) CD8(+) T cells and CD28(−)CD8(+) T cells isolated from healthy, elderly,or adult control subjects demonstrating that the circular RNA100783 may be involved in alternative splicing and phosphoprotein expression. Here, for the first time, circular RNA profiling was used to investigate CD8(+) T cell populations and the loss of CD28 expression during human ageing [35].

5 Autoimmune Diseases

With continuous research development, increasing evidence has emerged showing that circRNAs can interfere with biological processes of immune-related diseases. Here, we describe new experimental progress in circRNAs that partici-

Table 24.2 CircRNAs in immune-related diseases

circRNA (hsa_)	miRNA target	References
Multiple sclerosis		
circ_0005402, and circ_0035560	miR-155	[42, 43]
circ_0087862 and circ_0012077	Immunoglobulin G levels in CSF	[44]
circ_0043813	STAT3	[45, 46]
circRNA_101348 and circRNA_104361	AK2 and IKZP3—B cells	[39]
circINPP4B	miR-30a/Th17	[47]
Rheumatoid arthritis		
circRNA_09505	miR-6089	[48]
circ_0089172	IL-17A	[49]
circ0000175	miR-31-5p/GSDME	[50]
circ_0088036	miR-140-3p/SIRT 1 and miR-1263	[51, 52]
circ_0000479	miR-766	[53]
circ_0000396,	miR-574-5p/RSPO1	[54]
circ_0003353, circ_0005732, circ_0072428, circ_0091685	Immunity, inflammation, synovial invasion, and joint destruction	[55]
circ_0140271	Fatty acid metabolism pathways	[56]
Systemic lupus erythematosus		
circPOLR2A	PKR-IFN-β	[9]
circRTN4	Progression of lupus nephritis	[57]
circPTPN22	miR-4689—S1PR1	[58]
circLOC101928570	miR-150—c-myb/IL2RA pathway	[59]
circGARS	miR-19a	[60]
circ_0000479	Wnt signaling pathway	[61]
circ_0012919, circ_0006239 and circ_0002227	miR-125a-3p	[30]
Infectious diseases—SARS-CoV-2		
circ_0000479	miR-149-5p/RIG-I, IL-6 axis	[62]
Ppp1r10, C330019G07RiK	miR-124-3p	[63]
circ-3205	Spike (S) mRNA	[64]
Infectious diseases—other infection microorganism		
circ_001937	NF-kB pathway in TB	[23]
circ_0005836	Tuberculosis	[65]
circRasGEF1B	ICAM-1	[13]
circ_2858	miR-93-5p	[66]

pates in three major autoimmune diseases (multiple sclerosis, rheumatoid arthritis, and systemic lupus erythematosus), and in infectious diseases; a list of the circRNA described in this section is presented in Table 24.2. Tumor immunology is deeply analyzed elsewhere in this book.

6 Multiple Sclerosis

Multiple sclerosis (MS) can briefly be described as a chronic autoimmune and demyelinating disease of the central nervous system (CNS). Clinically, MS has been classified into four main categories based on the disease course: clinically isolated syndrome (CIS), relapsing-remitting MS (RRMS, involving 80% of patients), primary progressive MS (PPMS), and secondary progressive MS (SPMS) [67]. While the etiology of MS still needs to be fully understood, several lines of evidence strongly support a key role and interaction between genetic, environmental, and epigenetic factors [68]. Results from genome-wide association studies (GWAS) and genomic studies, in general, have identified a plethora of loci and genes potentially involved in predisposition to autoimmune diseases and in particular to MS, each of them generally characterized by a modest contribution to the disease risk [69–71]. Several studies have additionally revealed the implication of multiple genes influencing innate and adaptive immunity across the cellular components of the immune system, with enrichment for MS genes in both peripheral and brain-resident immune cells [69]. Recently, several epigenetic mechanisms that might be contributing to MS pathology have been identified. Among them, a key role for microRNA, long noncoding RNA and circular-RNA was documented by different studies [72, 73]. Noncoding RNA molecules are considered to be one of the most important regulators of gene expression, which might finally be implicated in controlling the production of key proteins by gene expression modulation at different level [74].

One of the first works on this topic was published some years ago by Iparraguirre and colleagues. Using RNA purified from blood samples of patients with RRMS, CIS, and healthy controls (HC), they identified a set of 406 circRNA differentially expressed in RRMS as compared to HC. Among them, circ_0005402 and circ_0035560 were validated and identified as circRNA downregulated in RRMS. Both circRNA are located on the chromosome 15 within the Annexin A2 (ANXA2) gene, recognized for playing a role in the blood–brain barrier (BBB) crossing and promotion of the miR-155 expression via STAT3 (Signal Transducer and Activator of Transcription 3). Interestingly, miR-155—a master regulator of inflammation—has been identified as a biomarker for MS [42]. Subsequently, the same authors analyzed the content of extracellular vesicles (EVs) purified from MS patients. Plasma-derived EVs were isolated from both MS patient and HC, and RNA-sequencing was used to identify differentially expressed linear and circRNAs. They found differences in the RNA type distribution, with circRNAs being enriched in EVs compared to leucocytes [43].

By analyzing the expression levels of circRNAs in the cerebrospinal fluid (CSF)-derived exosomes of patients with MS and other immune-mediated demyelinating diseases (e.g., Guillain–Barré syndrome, progressive multifocal leukoencephalitis and delayed hypoxic demyelinating encephalopathy), and comparing it to HC, He et al. identified more than 5000 differentially expressed circRNAs, of which 26 reached significant levels. Circ_0087862 and circ_0012077, significantly upregulated in MS as compared to HC, were further validated in an independent cohort. The study also discovered a potential connection between exosomal circ_0012077 expression levels and IgG levels in the CSF. Circ_0087862 or circ_0012077 were also suggested as a biomarker for immune-mediated demyelinating diseases with a diagnostic accuracy of 100% [44].

Searching for potential enrichments in circRNA elements at MS GWAS loci, with the aim to identify their possible role in disease susceptibility, Paraboschi et al. described a significant enrichment of circRNAs mapping the 73-linkage disequilibrium (LD) blocks containing MS-associated single nucleotide polymorphisms (SNPs) (a total of 482 circRNAs in MS-associated LD blocks as compared with the 194 ± 65 identified in the random sets of LD blocks).

The SNP rs2293152, for example, influences the STAT3 derived circ_0043813 expression levels [45]. STAT3 gene is a protein essential for the differentiation of the Th17 helper T cells, which have been implicated in a variety of autoimmune diseases [46]. In line with this study, Cardamone et al. proposed that MS-associated SNPs may influence the expression levels of circRNAs. Indeed, quantitative trait loci (QTL) analysis demonstrated a significant association between MS-associated loci and the expression of at least one circRNA, suggesting that MS-associated variants could impact MS pathogenesis through circRNA modulation. The identified circRNAs were associated with various immune functions, gene transcription, and chromatin regulation [75].

Moreover, Huang and colleagues discovered four circRNA/miRNA/mRNA networks impacting B cell proliferation and B cell receptor signaling pathways [76]. As already demonstrated by several studies [71]. B cell dysregulation seems to have a causative role in the pathogenesis of MS. To reach this result, the research group utilized bioinformatics methods examining the expression patterns of circRNAs, miRNAs, and mRNAs in MS patients from previously published data sets [72, 77]. Subsequently, Zurawska and colleagues identified 15 differentially expressed circRNAs, in RRMS when comparing relapse and remission patients; two of these circRNAs (circRNA_101348 and circRNA_104361) associated with inflammatory activity measured using magnetic resonance imaging (MRI) in RRMS patients. Additionally, the study found that the upregulation of these two circRNAs was linked to the upregulation of the adenylate kinase 2 (AK2) and the IKAROS family zinc finger 3 (IKZP3) mRNA expression, two genes involved in B cell functions [39]. By analyzing the role of circRNAs on T helper cell 17 (Th17) differentiation and MS, using transcriptome microarray analysis at different stages of EAE, Han and colleagues identified upregulation of circINPP4B during Th17 differentiation in vitro. Silencing of circINPP4B inhibited Th17 differentiation and alleviated EAE (animal model of MS), characterized by less demyelination and Th17 infiltration in the spinal cord. According to the authors, circINPP4B served as a sponge to miR-30a that has a key role in regulating Th17 differentiation. Additionally, circINPP4B levels were associated with the developing phases of clinical relapsing-remitting MS patients [47]. The findings listed in this section highlight the potential of circRNAs as a key regulator of immune-mediated demyelination diseases, supporting the relevance of continuing to study circRNA for early diagnosis and management of MS.

7 Rheumatoid Arthritis

Rheumatoid arthritis (RA) is a long-term autoimmune disease that primarily affects synovial joints and periarticular soft tissues. It is a systemic disease, meaning that over time it can affect other parts of the body, including the skin, lungs, heart, and eyes [78]. RA is estimated to affect more than 18 million people worldwide and is more common in women than men (with a female-to-male ratio of about 3:1) [79]. Disruptions in both the cellular and humoral immune systems trigger the distinctive production of autoantibodies, particularly rheumatoid factors (RFs) and other antibodies targeting post-translationally modified proteins (citrullination (ACPA), carbamylation (aCarP), and acetylation (AAPA)). These autoantibodies, alongside the migration of T and B lymphocytes into the synovial membrane, contribute to the pathogenesis of the disease [80]. The exact cause of RA is unknown, but it is thought to be a combination of genetic, epigenetic, and environmental factors. More than 150 loci associated with the risk of developing RA have been identified, with human leukocyte antigens (HLA) associations remaining the most robust [81]. Accumulating evidence has defined a critical role of noncoding RNAs in inflammation and autoimmune processes underlying RA pathogenesis, and during the past few years, the role of several circRNAs has drawn more and more attention in RA pathology [82, 83].

The expression level of several circRNA during RA has been assessed using different approaches. For example, the expression of circRNA_09505 was found to be upregulated in PBMCs isolated from patients with RA as compared with healthy controls using qPCR technology. Experimental validation revealed that circ_09505 can act as a miRNA sponge for miR-6089 in macrophages, and promote the production of TNF-α, IL-6, and IL-12 through ceRNA mechanism. Circ_09505 can also promote the expression of AKT1 (a direct target of miR-6089) by acting as a sponge, via IκBα/NF-κB signaling pathway in macrophages. Of note, silencing of circRNA_09505 significantly alleviates arthritis and inflammation in vivo in a collagen-induced arthritis (CIA) mouse model [48]. Circ_0089172 (an exonic circular RNA located on chromosome 9q34) is differentially expressed in RA patients and its downregulation decreased the proportion of Th17 cells and the transcript levels of IL-17A in vitro [49]. In another study, the effects of circular RNA circ_0000175 on pyroptosis—an inflammatory form of cell death—and inflammation were evaluated using serum and PBMCs purified from RA patients. Circ_0000175 can induce pyroptosis and trigger inflammatory response during RA through the miR-31-5p/GSDME axis [50]. Using a bioinformatics approach, the involvement of the circ_0088036/miR-140-3p/SIRT1 (silent information regulator 1) axis in rheumatoid arthritis (RA) was predicted. Indeed, circ_0088036 is significantly elevated in PBMCs and in fibroblast-like synoviocytes (FLSs) of RA patients as compared with FLSs of osteoarthritis patients, which has been used as reference for RA (OA). Functionally, upregulated circ_0088036 promoted the proliferation and migration of RA-FLSs promoting synovial pathogenesis [51].

Human fibroblast-like synoviocytes (HFLS) cells play a key role in RA by producing cytokines that perpetuate inflammation and produce proteases that contribute to cartilage destruction. Circ_0088036, for example, has been found to be significantly upregulated in sera and in HFLS cells of RA patients. Targeted silencing of circ_0088036 restrained proliferation, cell cycle progression, and inflammatory reaction promoting the apoptosis of HFLS-RA cells by inhibition of the NF-κB pathway and modulation of miR-1263 [52]. Circ_0000479 was also found to be upregulated in HFLS, and its knockdown suppressed pathogenic properties of HFLS during RA. Through bioinformatics analysis and screening, 18 miRNAs were identified that can bind to circ_0000479, and miR-766 was the most significantly upregulated after circ_0000479 knockdown. Circ_0000479 binds to miR-766 to promote RA progression via the FK506 binding protein 5 (FKBP5) [53]. In contrast, the expression level of circ_0000396 was downregulated in RA patients compared to healthy controls. Additionally, circ_0000396 overexpression was found to suppress the proliferation and inflammatory response triggering apoptosis of RA synovial fibroblasts (RASFs); functional analysis revealed that circ_0000396 restrained the proliferation and inflammation and induced the apoptosis of RASFs by mediating miR-574-5p/RSPO1 axis, providing a novel potential target for RA treatment [54].

Apart from the studies mentioned above—in which the function of a single circRNA in RA was specifically tested—only a few studies have extensively investigated all circRNAs for their potential role in RA. In a small study, using high-throughput sequencing of RNA extracted from PBMCs of three RA patients and three healthy controls, all circRNAs related to immunity and inflammation were investigated. Four differentially expressed circRNAs were verified by qPCR, circ_0003353, circ_0005732, circ_0072428, and circ_0091685. Further experimental validation confirmed a role for circ_0003353 in promoting immunity, inflammation, synovial invasion, and joint destruction. Consequently, circ_0003353 has been proposed as a biomarker of RA [55]. In another RNA sequencing study using PBMCs purified from 4 healthy controls (HC) and 4 RA patients, circ_0140271 was identified as a potential biomarker for RA. KEGG pathway enrichment analysis indicated circ_0140271 may act as a microRNA sponge and participate in fatty acid metabolism pathways [56].

8 Systemic Lupus Erythematosus

Systemic lupus erythematosus (SLE) is a chronic and systemic autoimmune disease characterized by a heterogeneous spectrum of clinical and serological signs generated mainly by the production of autoantibodies, activation of the complement cascade, and deposition of immune complexes. A significant aspect of SLE is the production of antinuclear antibodies (ANAs) targeting double-stranded DNA (dsDNA) as well as other nuclear autoantigens produced by uncleared apoptotic cells [84, 85]. SLE is characterized by high morbidity and mortality; thus, it is fundamental to properly identify the specific pathological mechanisms leading to disease onset and to identify accurate biomarkers for early diagnosis, prognosis, and effective therapy. The development of SLE affects the entire immune system [86], with complex interactions of genetics and environmental factors that jointly lead to immune dysregulation and diminished tolerance to self-antigens. NcRNAs, including circRNA, have also been identified as pathological factors for SLE representing key players in SLE pathogenesis [87].

As introduced above, an unexpected general role of circRNAs in autoimmune diseases was revealed by a study that demonstrated how imperfect endogenous circRNA duplexes act as inhibitors of the innate immunity-related PKR protein. Indeed, using SLE samples, it was shown how RNase L endonuclease, once activated, can globally degrade circRNAs. A consequence of the degradation during viral infection is the release of circRNA-associated PKR, which enables its activation. Furthermore, consistent with the role of circRNAs in suppressing PKR activation in normal cells in culture, SLE patients show reduced expression of circRNAs, associated with spontaneous RNase L activation and increased PKR activation. In addition, overexpression of circRNA containing dsRNA in PBMCs or SLE-derived T cells may reduce the aberrant PKR activation cascade, thus providing a link between circRNA and SLE: assays performed in PBMCs of three lupus patients showed that overexpression of circPOLR2A, for example, leads to reduced PKR activation and suppression of IFN-β and IFN-induced genes—a well-established signature of SLE [9].

Many other studies have investigated differential expression patterns of circular RNAs by comparing patients with SLE against healthy controls, using both RNA arrays and RNA seq. To explore whether deregulated circRNAs were involved in mesangial cell injury in lupus nephritis, human renal mesangial cells (HRMC), treated with lupus nephritis (LN) plasma or control plasma, were characterized by an Arraystar Human CircRNA Array; ten deregulated circRNAs were validated by qPCR. The circ_0054595 (named circRTN4) was notably upregulated in HRMCs exposed to LN plasma. Furthermore, circRTN4 was significantly increased in exosomes derived from monocytes of LN, and was also found upregulated in renal tissues, PBMCs and exosomes of LN patients. Overall, the finding suggested that circRTN4 may be systematically involved in the progression of LN [57]. RNA sequencing of PBMCs from patients with SLE and healthy individuals revealed 128 circRNAs that were significantly differentially expressed including 39 upregulated and 89 downregulated circRNAs in four new-onset SLE patients compared with three healthy controls [88]. Among the candidate circRNAs, particular interest has emerged in circPTPN22. The parental transcript of the circPTPN22, the protein tyrosine phosphatase nonreceptor 22 (PTPN22), has been involved in several autoimmune diseases including SLE and that is characterized by regulatory functions over T cells. Downregulation of circPTPN22 in patients with SLE was strongly and negatively correlated with SLE Disease Activity Index (SLE-DAI) scores. Follow-up studies also demonstrated that circPTPN22 acts as a miR-4689 sponge to regulate T cell activation by targeting S1PR1, another gene involved in autoimmune diseases that interacts with STAT3 [58].

Another study analyzed circRNA-seq data from SLE patients and healthy controls and identified the circLOC101928570 as significantly downregulated in SLE patients. CircLOC101928570 is derived from the LOC101928570 gene locus and is predicted to target miR-150, which plays an important role in the differentiation of T cells and B cells. The authors propose that circLOC101928570 acts as a miR-150 sponge to modulate the activation of immune inflammatory responses mediated by the c-myb/IL2RA pathway [59]. The IL2/IL2RA pathway which emerged from another study conducted by Li et al. and previously described in this chapter [28], represents a critical element for the maintenance of peripheral tolerance and has been early associated with autoimmunity [60].

In a subsequent analysis of SLE patient samples and healthy control samples, a significant positive correlation between the levels of circGARS and SLE-DAI (Systemic Lupus Erythematosus Disease Activity Index) was found. RNA pulldown assays suggested that circGARS acts as a miR-19a sponge and in combination with it regulates the expression of A20 (TNFAIP3), thus affecting the immune responses regulated by the NF-κB pathway. Likewise, circ-

GARS boosted the transcriptional levels in general by directly binding to m6A modification proteins [89].

Several other circRNAs have also been identified as potential biomarkers for SLE. Circ_0000479 (elsewhere cited for the association with RA and COVID-19) was found to be significantly upregulated in patients with SLE compared to healthy controls following next-generation sequencing (NGS) of PBMC purified from SLE patients. Circ_0000479 was predicted to regulate SLE progression by modulating metabolic pathways and the Wnt signaling pathway. These findings suggest that circ_0000479 may be a useful tool for diagnosing and for monitoring SLE progression [61]. Another study focused on the identification of biomarkers comparing the expression of circRNAs in CD4+ T cells purified from healthy controls and SLE patients. Circ_0012919, circ_0006239, and circ_0002227 were found to be upregulated and identified as candidate biomarkers for SLE. Circ_0012919 acts as a ceRNA for miR-125a-3p and regulates the expression of the DNA-methyltransferase 1 (DNMT1), CD70, and CD11a acting on DNA methylation mechanism [30]. Emerging indications recognize the involvement of circRNAs in SLE pathogenesis and may be used as innovative biomarkers in the future. Conversely, while existing data strongly support the potential use of circRNAs for SLE diagnosis and treatment, their application in clinical practice still needs to be properly evaluated.

9 Infectious Diseases and CircRNA

Infectious diseases are diseases driven by the action of protozoa, bacteria, viruses, etc. These pathogens can invade a host organism inducing alterations in the normal physiology of the host possibly leading to organism dysfunctions and death. CircRNAs can impact the regulatory network of infectious diseases and function as a biomarker; unfortunately, until now few studies have deeply analyzed this possibility and few circRNAs have been validated [90].

During viral infection, a release and activation of the circRNA-associated PKR protein is observed; it was also demonstrated that circRNAs transfection, independent of sequence, leads to activation of RIG-I (retinoic-acid-inducible gene-I). Due to the nature of circRNAs, recognition cannot be explained by the preference of RIG-I for 5′-triphosphate-containing dsRNA end. Indeed, it was found that, when circRNAs were synthesized in cells (following plasmids transfection), instead of being synthesized in vitro and transfected into cells, the immune stimulatory activity was dependent on the splicing mechanism. CircRNAs generated with endogenous introns and spliced by intracellular spliceosomes did not stimulate RIG-I. In opposition to it, circRNAs originated with self-splicing introns stimulated RIG-I. Additionally, mass spectrometry analysis showed that circRNAs generated by the host splicing (self) apparatus co-purify with multiple RNA-binding proteins (e.g., NF90) and splicing proteins, while self-spliced (nonself) circRNAs co-purify with few proteins. Using these data, they concluded that host spliceosomes mark endogenous circRNAs as "self" using a specific protein-binding partner that protects host circRNAs from RIG-I detection [91]. In this process, the RNA-binding proteins Nuclear Factor 90 and 110 (NF90/NF110) play a key role by stabilizing the secondary structure of intronic RNA, thereby promoting circRNA biogenesis. NF90/110, which have been involved in multiple immune pathways [92], can also act as global regulators of circRNA biogenesis by reducing their nuclear levels during viral infection [93].

The severe acute respiratory syndrome coronavirus 2 (SARS-CoV-2) strongly impacted life and health worldwide; thus, it is particularly important to identify new treatment technologies based on existing epidemic prevention measures such as drug development and vaccination [94]. CircRNAs represent a new option to better understand SARS-CoV-2 pathogenesis. Supporting a role of RIG-I pathway analyzed above, the expression levels of circ_0000479, RIG-I, and IL-6 were analyzed and found to be upregulated following COVID-19 infection which was correlated with decreased miR-149-5p expression.

The identified upregulation of circ_0000479, RIG-I, and IL-6, and downregulation of miR-149-5p, along with correlation studies, indicate that the circ_0000479/hsa-miR-149-5p/RIG-I, IL-6 axis could play a role in regulating the immune response against SARS-CoV-2 [62]. Furthermore, circRNA expression profile analysis was performed in human lung epithelial cells after COVID infection by several research groups. Arora et al. for example, identified a circRNA/lncRNA-miRNA–mRNA ceRNA network involving two circRNAs in SARS-CoV-2-infected cells; specifically, the predicted ceRNA network involved one miRNA (miR-124-3p), one mRNA (Ddx58), one lncRNA (Gm26917) and two circRNAs (Ppp1r10, C330019G07RiK) [63]. Using a genome-wide approach to study circRNAs in human lung epithelial cells infected with SARS-CoV-2 at different time points, Yang and colleagues identified a total of 6118 circRNAs, including 5641 known and 477 novel circRNAs; 42 circRNAs were significantly dysregulated, and of these, 17 were upregulated and 25 were downregulated following infection at multiple phases [95]. Finally, using nasopharyngeal swabs of SARS-CoV-2 infected or uninfected individuals the role of the circ-3205, previously predicted to be involved in SARS-CoV-2, was validated; indeed, the research group observed that circ-3205 was expressed only in infected samples and the expression level positively correlated with that of SARS-CoV-2 Spike (S) mRNA and viral load [64]. The integrative analysis performed in this section identified circRNAs and respective targets to enhance the understanding of molecular mechanisms associated with viral infection. Due to the relevance in our ages, here we focus on SARS-CoV-2 infection to underline the relevance of circRNA during viral infection; however, the literature provides multiple examples in which other viruses, such as CV-A16, Enterovirus A71 or Redondoviridae, have been correlated with circRNA levels [96–98].

Apart from viruses, circRNAs have been studied in many infectious diseases. For example, circ_001937 targets several miRNAs, thus influencing the functions of the NF-kB pathway and exerting a role in the immune responses during TB infection. CircRNA_001937 was correlated with TB severity and its expression significantly decreased after treatment [23]. Again, in the context of TB infection, the circ_0005836 has been identified as a potential novel biomarker for diagnosis and prognosis and a target for therapy of active pulmonary TB [65]. In another report, the LPS-inducible circRNA called circRasGEF1B was identified as a player in protecting cells against microbial infection by preserving the constancy of the mature mRNA of ICAM-1 in LPS-activated cells [13] and providing a new idea for antimicrobial infection therapies. Finally, Yang and colleagues identified a role for circ_2858 in regulating BBB (blood–brain barrier) permeability during the pathogenic process of *E. coli* meningitis and suggesting potential nucleic acid targets for future prevention and therapy of CNS infection induced by meningitis caused by *E. coli* [66]. In addition to bacteria and viruses, circRNAs can play key roles in many other infectious diseases, such as chlamydia infection [99] and Plasmodium falciparum malaria [100]; unfortunately, the clinical significance for all of them needs to be properly identified and clinical application of these results are still a long way from application.

10 Future Applications and Conclusions

CircRNAs are ncRNA characterized by unique features when compared to linear RNAs with regards to conformation, immunogenicity, and stability; importantly, circRNAs are more stable and less immunogenic than other classes of linear RNAs [101]. Their greater stability favors a possible role as biomarkers, while their reduced immunogenic action makes circRNAs superior to linear RNAs for practical applications. Thus, several efforts have been made to develop new circRNA-based methods able to interfere with cellular processes, modulation of immune responses, and translation of proteins; for example, noncoding aptamers, antisense RNAs, circular RNA templates able to sustain translation, modulators of innate immune responses, diagnostic and prognostic biomarkers for immune-related disease, and vaccine candidates are under development by different research and industrial groups [102]. As extensively described in this chapter, circRNAs are detectable components of exosomes and a key and stable component of all physiological body fluids (such as saliva, blood, and urine); these properties make circRNAs resistant to environmental fluctuations, rendering them potential biomarkers for autoimmune and infectious diseases. Unfortunately, until now, the clinical significance and possible clinical applications still need to be verified.

The chemically stable structure and reduced immunogenicity suggested that synthetic and viral-based translatable circular RNAs can be developed into circRNA-based therapies; indeed, the expression of relevant antigens using artificial circRNAs to trigger adaptive immune responses has already presented promising therapeutic effects in coronavirus disease 2019 (COVID-19) and hard-to-treat melanoma malignancies [103, 104]. Furthermore, unlike current mRNA vaccines, necessitating strict transportation and storage conditions, circRNA vaccines encapsulated with lipid nanoparticles can be safely kept for short periods at 4 °C and room temperature, thus facilitating vaccination campaigns. Furthermore, circRNAs can sponge miRNAs and other RNA molecules, reducing bioaccessibility and functions against their targets; this function has led to the introduction of in vitro *-generated* circular RNAs expression in specific sites, thus enabling a reduction in disease- or virus-related target activity. For example, SARS-CoV-2 viral replication was significantly suppressed in a cell culture model using a panel of antisense circRNAs designed to target the conserved 5′-UTR of the virus's genomic RNA and inhibiting virus replication by up to 90% as compared to the control [105].

Although progress has been made in understanding the involvement of circRNAs in immune response, the exact pathogenic mechanism leading to immune-related diseases still needs to be properly elucidated. CircRNAs could potentially serve as specific and stable biomarkers for immune diseases, therapeutic agents, and vaccine candidates, but more experimental studies are needed to properly define their protein-RNA target and mechanisms of regulation. However, even though circRNA studies are still at an initial stage in the prevention and treatment of immune-related diseases, we believe that with the development of new interdisciplinary studies and the emergence of advanced technology, there will be new and advanced opportunities for circRNAs to display their clinical talents in the future.

Competing Financial Interests The authors declare no competing financial interests.

References

1. Russell CD, Unger SA, Walton M, Schwarze J (2017) The human immune response to respiratory syncytial virus infection. Clin Microbiol Rev 30(2):481–502. https://doi.org/10.1128/CMR.00090-16
2. Chaplin DD (2010) Overview of the immune response. J Allergy Clin Immunol 125(2 Suppl 2):S3–S23. https://doi.org/10.1016/j.jaci.2009.12.980
3. Li Z, Cheng Y, Wu F, Wu L, Cao H, Wang Q, Tang W (2020) The emerging landscape of circular RNAs in immunity: breakthroughs and challenges. Biomark Res 8:25. https://doi.org/10.1186/s40364-020-00204-5
4. Yang L, Han X, Zhang C, Sun C, Huang S, Xiao W, Gao Y, Liang Q, Luo F, Lu W, Fu J, Zhou Y (2020) Hsa_circ_0060450 negatively regulates type I interferon-induced inflammation by serving as miR-199a-5p sponge in type 1 diabetes mellitus. Front Immunol 11:576903. https://doi.org/10.3389/fimmu.2020.576903
5. Zhao C-X, Yan Z-X, Wen J-J, Fu D, Xu P-P, Wang L, Cheng S, Hu J, Zhao W-L (2021) CircEAF2 counteracts Epstein-Barr virus-positive diffuse large B-cell lymphoma progression via miR-BART19-3p/APC/β-catenin axis. Mol Cancer 20(1):153. https://doi.org/10.1186/s12943-021-01458-9
6. Luo Q, Zhang L, Li X, Fu B, Guo Y, Huang Z, Li J (2019) Identification of circular RNAs hsa_circ_0044235 and hsa_circ_0068367 as novel biomarkers for systemic lupus erythematosus. Int J Mol Med 44(4):1462–1472. https://doi.org/10.3892/ijmm.2019.4302
7. Zhao R-J, Zhang W-Y, Fan X-X (2024) Circular RNAs: potential biomarkers and therapeutic targets for autoimmune diseases. Heliyon 10(1):e23694. https://doi.org/10.1016/j.heliyon.2023.e23694
8. Williams BR (1997) Role of the double-stranded RNA-activated protein kinase (PKR) in cell regula-

tion. Biochem Soc Trans 25(2):509–513. https://doi.org/10.1042/bst0250509
9. Liu C-X, Li X, Nan F, Jiang S, Gao X, Guo S-K, Xue W, Cui Y, Dong K, Ding H, Qu B, Zhou Z, Shen N, Yang L, Chen L-L (2019) Structure and degradation of circular RNAs regulate PKR activation in innate immunity. Cell 177(4):865–880.e21. https://doi.org/10.1016/j.cell.2019.03.046
10. Wang B, Shen W, Yan L, Li X, Zhang L, Zhao S, Jin X (2023) Reveal the potential molecular mechanism of circRNA regulating immune-related mRNA through sponge miRNA in the occurrence and immune regulation of papillary thyroid cancer. Ann Med 55(2):2244515. https://doi.org/10.1080/07853890.2023.2244515
11. Wang X, Lu Z, Gomez A, Hon GC, Yue Y, Han D, Fu Y, Parisien M, Dai Q, Jia G, Ren B, Pan T, He C (2014) N6-methyladenosine-dependent regulation of messenger RNA stability. Nature 505(7481):117–120. https://doi.org/10.1038/nature12730
12. Chen YG, Chen R, Ahmad S, Verma R, Kasturi SP, Amaya L, Broughton JP, Kim J, Cadena C, Pulendran B, Hur S, Chang HY (2019) N6-Methyladenosine modification controls circular RNA immunity. Mol Cell 76(1):96–109.e9. https://doi.org/10.1016/j.molcel.2019.07.016
13. Ng WL, Marinov GK, Liau ES, Lam YL, Lim Y-Y, Ea C-K (2016) Inducible RasGEF1B circular RNA is a positive regulator of ICAM-1 in the TLR4/LPS pathway. RNA Biol 13(9):861–871. https://doi.org/10.1080/15476286.2016.1207036
14. Ng WL, Marinov GK, Chin Y-M, Lim Y-Y, Ea C-K (2017) Transcriptomic analysis of the role of RasGEF1B circular RNA in the TLR4/LPS pathway. Sci Rep 7(1):12227. https://doi.org/10.1038/s41598-017-12550-w
15. Zhang C, Han X, Yang L, Fu J, Sun C, Huang S, Xiao W, Gao Y, Liang Q, Wang X, Luo F, Lu W, Zhou Y (2020) Circular RNA circPPM1F modulates M1 macrophage activation and pancreatic islet inflammation in type 1 diabetes mellitus. Theranostics 10(24):10908–10924. https://doi.org/10.7150/thno.48264
16. Wang X, Sheng W, Xu T, Xu J, Gao R, Zhang Z (2021) CircRNA hsa_circ_0110102 inhibited macrophage activation and hepatocellular carcinoma progression via miR-580-5p/PPARα/CCL2 pathway. Aging (Albany NY) 13(8):11969–11987. https://doi.org/10.18632/aging.202900
17. Zhang J, Cheng F, Rong G, Tang Z, Gui B (2021) Circular RNA hsa_circ_0005567 overexpression promotes M2 type macrophage polarization through miR-492/SOCS2 axis to inhibit osteoarthritis progression. Bioengineered 12(1):8920–8930. https://doi.org/10.1080/21655979.2021.1989999
18. Deng Q, Huang J, Yan J, Mao E, Chen H, Wang C (2021) Circ_0001490/miR-579-3p/FSTL1 axis modulates the survival of mycobacteria and the viability, apoptosis and inflammatory response in *Mycobacterium tuberculosis*-infected macrophages. Tuberculosis (Edinb) 131:102123. https://doi.org/10.1016/j.tube.2021.102123
19. Katopodi T, Petanidis S, Domvri K, Zarogoulidis P, Anestakis D, Charalampidis C, Tsavlis D, Bai C, Huang H, Freitag L, Hohenforst-Schmidt W, Matthaios D, Porpodis K (2021) Kras-driven intratumoral heterogeneity triggers infiltration of M2 polarized macrophages via the circHIPK3/PTK2 immunosuppressive circuit. Sci Rep 11(1):15455. https://doi.org/10.1038/s41598-021-94671-x
20. Song H, Yang Y, Sun Y, Wei G, Zheng H, Chen Y, Cai D, Li C, Ma Y, Lin Z, Shi X, Liao W, Liao Y, Zhong L, Bin J (2022) Circular RNA Cdyl promotes abdominal aortic aneurysm formation by inducing M1 macrophage polarization and M1-type inflammation. Mol Ther 30(2):915–931. https://doi.org/10.1016/j.ymthe.2021.09.017
21. Zhou R-M, Shi Z-H, Shan K, Zhang S-J, Zhang Y-H, Liang Y, Yan B, Zhao C (2022) Comparative analysis of differentially expressed circular RNAs in polarized macrophages. Front Genet 13:823517. https://doi.org/10.3389/fgene.2022.823517
22. Jiang F, Liu X, Cui X, Hu J, Wang L, Xue F, Guo S, Wang X (2022) Circ_0000518 promotes macrophage/microglia M1 polarization via the FUS/CaMKKβ/AMPK pathway to aggravate multiple sclerosis. Neuroscience 490:131–143. https://doi.org/10.1016/j.neuroscience.2021.12.012
23. Huang Z, Su R, Deng Z, Xu J, Peng Y, Luo Q, Li J (2017) Identification of differentially expressed circular RNAs in human monocyte derived macrophages response to *Mycobacterium tuberculosis* infection. Sci Rep 7(1):13673. https://doi.org/10.1038/s41598-017-13885-0
24. Wang K, Bai X, Mei L, Miao Y, Jin F (2022) CircRNA_0050486 promotes cell apoptosis and inflammation by targeting miR-1270 in atherosclerosis. Ann Transl Med 10(16):905. https://doi.org/10.21037/atm-22-3745
25. Chen X, Xie X, Zhou W (2020) CircCFL1/MiR-107 axis targeting HMGB1 promotes the malignant progression of diffuse large B-cell lymphoma tumors. Cancer Manag Res 12:9351–9362. https://doi.org/10.2147/CMAR.S263222
26. Liu W, Lei L, Liu X, Ye S (2021) CircRNA_OTUD7A upregulates FOXP1 expression to facilitate the progression of diffuse large B-cell lymphoma via acting as a sponge of miR-431-5p. Genes Genomics 43(6):653–667. https://doi.org/10.1007/s13258-021-01094-z
27. Wu W, Wu Z, Xia Y, Qin S, Li Y, Wu J, Liang J, Wang L, Zhu H, Fan L, Fu J, Xu W, Jin H, Li J (2019) Downregulation of circ_0132266 in chronic lymphocytic leukemia promoted cell viability through miR-337-3p/PML axis. Aging (Albany NY) 11(11):3561–3573. https://doi.org/10.18632/aging.101997
28. Li L-J, Zhu Z-W, Zhao W, Tao S-S, Li B-Z, Xu S-Z, Wang J-B, Zhang M-Y, Wu J, Leng R-X, Fan Y-G, Pan H-F, Ye D-Q (2018) Circular RNA expression

profile and potential function of hsa_circ_0045272 in systemic lupus erythematosus. Immunology 155(1):137–149. https://doi.org/10.1111/imm.12940
29. Zhang L, Zhang K, Fang W, Li H, Li Y, Jiang W, Hu D, Coelho C, Liu X, Cai L, Liao W, Pan W (2020) CircRNA-1806 decreases T cell apoptosis and prolongs survival of mice after cryptococcal infection by sponging miRNA-126. Front Microbiol 11:596440. https://doi.org/10.3389/fmicb.2020.596440
30. Zhang C, Wang X, Chen Y, Wu Z, Zhang C, Shi W (2018) The down-regulation of hsa_circ_0012919, the sponge for miR-125a-3p, contributes to DNA methylation of CD11a and CD70 in CD4(+) T cells of systemic lupus erythematous. Clin Sci (Lond) 132(21):2285–2298. https://doi.org/10.1042/CS20180403
31. Huang Z, Cao Y, Zhou M, Qi X, Fu B, Mou Y, Wu G, Xie J, Zhao J, Xiong W (2019) Hsa_circ_0005519 increases IL-13/IL-6 by regulating hsa-let-7a-5p in CD4(+) T cells to affect asthma. Clin Exp Allergy 49(8):1116–1127. https://doi.org/10.1111/cea.13445
32. Zhou J, Xu M, Chen Z, Huang L, Wu Z, Huang Z, Liu L (2023) circ_SPEF2 regulates the balance of Treg cells by regulating miR-16-5p/BACH2 in lymphoma and participates in the immune response. Tissue Eng Regen Med 20(7):1145–1159. https://doi.org/10.1007/s13770-023-00585-2
33. Liu Z, Wang T, She Y, Wu K, Gu S, Li L, Dong C, Chen C, Zhou Y (2021) N(6)-methyladenosine-modified circIGF2BP3 inhibits CD8(+) T-cell responses to facilitate tumor immune evasion by promoting the deubiquitination of PD-L1 in non-small cell lung cancer. Mol Cancer 20(1):105. https://doi.org/10.1186/s12943-021-01398-4
34. Yang C, Wu S, Mou Z, Zhou Q, Dai X, Ou Y, Chen X, Chen Y, Xu C, Hu Y, Zhang L, Zou L, Jin S, Hu J, Mao S, Jiang H (2022) Exosome-derived circTRPS1 promotes malignant phenotype and CD8+ T cell exhaustion in bladder cancer microenvironments. Mol Ther 30(3):1054–1070. https://doi.org/10.1016/j.ymthe.2022.01.022
35. Wang Y-H, Yu X-H, Luo S-S, Han H (2015) Comprehensive circular RNA profiling reveals that circular RNA100783 is involved in chronic CD28-associated CD8(+)T cell ageing. Immun Ageing 12:17. https://doi.org/10.1186/s12979-015-0042-z
36. Chen S, Saeed AFUH, Liu Q, Jiang Q, Xu H, Xiao GG, Rao L, Duo Y (2023) Macrophages in immunoregulation and therapeutics. Signal Transduct Target Ther 8(1):207. https://doi.org/10.1038/s41392-023-01452-1
37. LeBien TW, Tedder TF (2008) B lymphocytes: how they develop and function. Blood 112(5):1570–1580. https://doi.org/10.1182/blood-2008-02-078071
38. Pan J, Hu S, Ren X, Hu H, Deng X, Yu B, Cobos I, Chen X, Zhang W (2022) Whole-transcriptome profiling and circRNA-miRNA-mRNA regulatory networks in B-cell development. Front Immunol 13:812924. https://doi.org/10.3389/fimmu.2022.812924
39. Zurawska AE, Mycko MP, Selmaj I, Raine CS, Selmaj KW (2021) Multiple sclerosis: circRNA profile defined reveals links to B-cell function. Neurol Neuroimmunol Neuroinflamm 8(5). https://doi.org/10.1212/NXI.0000000000001041
40. Huang Z, Fu B, Qi X, Xu Y, Mou Y, Zhou M, Cao Y, Wu G, Xie J, Zhao J, Wang Y, Xiong W (2021) Diagnostic and therapeutic value of Hsa_circ_0002594 for T helper 2-mediated allergic asthma. Int Arch Allergy Immunol 182(5):388–398. https://doi.org/10.1159/000511612
41. Lee Y, Choe J, Park OH, Kim YK (2020) Molecular mechanisms driving mRNA degradation by m(6)A modification. Trends Genet 36(3):177–188. https://doi.org/10.1016/j.tig.2019.12.007
42. Iparraguirre L, Muñoz-Culla M, Prada-Luengo I, Castillo-Triviño T, Olascoaga J, Otaegui D (2017) Circular RNA profiling reveals that circular RNAs from ANXA2 can be used as new biomarkers for multiple sclerosis. Hum Mol Genet 26(11):3564–3572. https://doi.org/10.1093/hmg/ddx243
43. Iparraguirre L, Alberro A, Hansen TB, Castillo-Triviño T, Muñoz-Culla M, Otaegui D (2021) Profiling of plasma extracellular vesicle transcriptome reveals that circRNAs are prevalent and differ between multiple sclerosis patients and healthy controls. Biomedicines 9(12). https://doi.org/10.3390/biomedicines9121850
44. He J, Ren M, Li H, Yang L, Wang X, Yang Q (2019) Exosomal circular RNA as a biomarker platform for the early diagnosis of immune-mediated demyelinating disease. Front Genet 10:860. https://doi.org/10.3389/fgene.2019.00860
45. Paraboschi EM, Cardamone G, Soldà G, Duga S, Asselta R (2018) Interpreting non-coding genetic variation in multiple sclerosis genome-wide associated regions. Front Genet 9:647. https://doi.org/10.3389/fgene.2018.00647
46. Yang XO, Panopoulos AD, Nurieva R, Chang SH, Wang D, Watowich SS, Dong C (2007) STAT3 regulates cytokine-mediated generation of inflammatory helper T cells. J Biol Chem 282(13):9358–9363. https://doi.org/10.1074/jbc.C600321200
47. Han J, Zhuang W, Feng W, Dong F, Hua F, Yao R, Qu X (2021) The circular RNA circINPP4B acts as a sponge of miR-30a to regulate Th17 cell differentiation during progression of experimental autoimmune encephalomyelitis. Cell Mol Immunol 18(9):2177–2187. https://doi.org/10.1038/s41423-021-00748-y
48. Yang J, Cheng M, Gu B, Wang J, Yan S, Xu D (2020) CircRNA_09505 aggravates inflammation and joint damage in collagen-induced arthritis mice via miR-6089/AKT1/NF-κB axis. Cell Death Dis 11(10):833. https://doi.org/10.1038/s41419-020-03038-z
49. Peng H, Xing J, Wang X, Ding X, Tang X, Zou J, Wang S, Liu Y (2022) Circular RNA circNUP214 modulates the T helper 17 cell response in patients

with rheumatoid arthritis. Front Immunol 13:885896. https://doi.org/10.3389/fimmu.2022.885896
50. Xin P, Tan Z, Wang Z, Chen Y, Zhuang Y (2023) Circular RNA hsa_circ_0000175 serves as a potential biomarker for rheumatoid arthritis via miR-31-5p/GSDME Axis. Biochem Genet 62:2522. https://doi.org/10.1007/s10528-023-10576-6
51. Zhong S, Ouyang Q, Zhu D, Huang Q, Zhao J, Fan M, Cai Y, Yang M (2020) Hsa_circ_0088036 promotes the proliferation and migration of fibroblast-like synoviocytes by sponging miR-140-3p and upregulating SIRT 1 expression in rheumatoid arthritis. Mol Immunol 125:131–139. https://doi.org/10.1016/j.molimm.2020.07.004
52. Wang X, Liu D, Cui G, Shen H (2022) Circ_0088036 mediated progression and inflammation in fibroblast-like synoviocytes of rheumatoid arthritis by miR-1263/REL-activated NF-κB pathway. Transpl Immunol 73:101604. https://doi.org/10.1016/j.trim.2022.101604
53. Zhao P, Ma G, Ma L (2023) Circ_0000479 promotes proliferation, invasion, migration and inflammation and inhibits apoptosis of rheumatoid arthritis fibroblast-like synoviocytes via miR-766/FKBP5 axis. J Orthop Surg Res 18(1):220. https://doi.org/10.1186/s13018-023-03700-0
54. Yu H, Yang J, Chen K, Kang W, Zhu F (2023) Circ_0000396 suppresses the proliferation and inflammation of rheumatoid arthritis synovial fibroblasts by targeting miR-574-5p/RSPO1 axis. J Orthop Surg Res 18(1):718. https://doi.org/10.1186/s13018-023-04117-5
55. Wen J, Liu J, Wang X, Wang J (2021) Expression and clinical significance of circular RNAs related to immunity and inflammation in patients with rheumatoid arthritis. Int Immunopharmacol 92:107366. https://doi.org/10.1016/j.intimp.2021.107366
56. Chen Y, Xu X, Li X, Zhong J, Wu B, Shang J, Jiang N, Wang B, Zhong S, Lu H (2021) Identification of circular RNAs hsa_circ_0140271 in peripheral blood mononuclear cells as a novel diagnostic biomarker for female rheumatoid arthritis. J Orthop Surg Res 16(1):647. https://doi.org/10.1186/s13018-021-02794-8
57. Miao X, Tian Y, Wu L, Zhao H, Liu J, Gao F, Zhang W, Liu Q, Guo H, Yang L, Yang R, Feng X, Liu S (2022) CircRTN4 aggravates mesangial cell dysfunction by activating the miR-513a-5p/FN axis in lupus nephritis. Lab Investig 102(9):966–978. https://doi.org/10.1038/s41374-022-00788-6
58. Jiang Z, Li S, Jia Y, Wu Q, Chen X, Zhang M, Miao Q, Zhong Z, Zhai Z, Ni B, Xiao J, Tang J (2023) CircPTPN22 modulates T-cell activation by sponging miR-4689 to regulate S1PR1 expression in patients with systemic lupus erythematosus. Arthritis Res Ther 25(1):206. https://doi.org/10.1186/s13075-023-03150-3
59. Zhao X, Dong R, Tang Z, Wang J, Wang C, Song Z, Ni B, Zhang L, He X, You Y (2022) Circular RNA circLOC101928570 suppresses systemic lupus erythematosus progression by targeting the miR-150--5p/c-myb axis. J Transl Med 20(1):547. https://doi.org/10.1186/s12967-022-03748-2
60. Cavanillas ML, Alcina A, Núñez C, de las Heras V, Fernández-Arquero M, Bartolomé M, de la Concha EG, Fernández O, Arroyo R, Matesanz F, Urcelay E (2010) Polymorphisms in the IL2, IL2RA and IL2RB genes in multiple sclerosis risk. Eur J Hum Genet 18(7):794–799. https://doi.org/10.1038/ejhg.2010.15
61. Guo G, Wang H, Ye L, Shi X, Yan K, Lin K, Huang Q, Li B, Lin Q, Zhu L, Xue X, Zhang H (2019) Hsa_circ_0000479 as a novel diagnostic biomarker of systemic lupus erythematosus. Front Immunol 10:2281. https://doi.org/10.3389/fimmu.2019.02281
62. Firoozi Z, Mohammadisoleimani E, Shahi A, Naghizadeh MM, Mirzaei E, Asad AG, Salmanpour Z, Javad Nouri SM, Mansoori Y (2022) Hsa_circ_0000479/Hsa-miR-149-5p/RIG-I, IL-6 Axis: a potential novel pathway to regulate immune response against COVID-19. Can J Infect Dis Med Microbiol 2022:2762582. https://doi.org/10.1155/2022/2762582
63. Arora S, Singh P, Dohare R, Jha R, Ali Syed M (2020) Unravelling host-pathogen interactions: ceRNA network in SARS-CoV-2 infection (COVID-19). Gene 762:145057. https://doi.org/10.1016/j.gene.2020.145057
64. Barbagallo D, Palermo CI, Barbagallo C, Battaglia R, Caponnetto A, Spina V, Ragusa M, Di Pietro C, Scalia G, Purrello M (2022) Competing endogenous RNA network mediated by circ_3205 in SARS-CoV-2 infected cells. Cell Mol Life Sci 79(2):75. https://doi.org/10.1007/s00018-021-04119-8
65. Zhuang Z-G, Zhang J-A, Luo H-L, Liu G-B, Lu Y-B, Ge N-H, Zheng B-Y, Li RX, Chen C, Wang X, Liu Y-Q, Liu F-H, Zhou Y, Cai X-Z, Chen ZW, Xu J-F (2017) The circular RNA of peripheral blood mononuclear cells: Hsa_circ_0005836 as a new diagnostic biomarker and therapeutic target of active pulmonary tuberculosis. Mol Immunol 90:264–272. https://doi.org/10.1016/j.molimm.2017.08.008
66. Yang R, Chen J, Xu B, Yang B, Fu J, Xiao S, Tan C, Chen H, Wang X (2020) circ_2858 helps blood-brain barrier disruption by increasing VEGFA via sponging miR-93-5p during *E. Coli* meningitis. Mol Ther Nucleic Acids 22:708–721. https://doi.org/10.1016/j.omtn.2020.09.034
67. Axisa P-P, Hafler DA (2016) Multiple sclerosis: genetics, biomarkers, treatments. Curr Opin Neurol 29(3):345–353. https://doi.org/10.1097/WCO.0000000000000319
68. Olsson T, Barcellos LF, Alfredsson L (2017) Interactions between genetic, lifestyle and environmental risk factors for multiple sclerosis. Nat Rev Neurol 13(1):25–36. https://doi.org/10.1038/nrneurol.2016.187
69. International Multiple Sclerosis Genetics Consortium (2019) Multiple sclerosis genomic map implicates peripheral immune cells and microglia

in susceptibility. Science 365(6460). https://doi.org/10.1126/science.aav7188

70. Cotsapas C, Mitrovic M (2018) Genome-wide association studies of multiple sclerosis. Clin Transl Immunol 7(6):e1018. https://doi.org/10.1002/cti2.1018
71. Steri M, Orrù V, Idda ML, Pitzalis M, Pala M, Zara I, Sidore C, Faà V, Floris M, Deiana M, Asunis I, Porcu E, Mulas A, Piras MG, Lobina M, Lai S, Marongiu M, Serra V, Marongiu M, Sole G, Busonero F, Maschio A, Cusano R, Cuccuru G, Deidda F, Poddie F, Farina G, Dei M, Virdis F, Olla S, Satta MA, Pani M, Delitala A, Cocco E, Frau J, Coghe G, Lorefice L, Fenu G, Ferrigno P, Ban M, Barizzone N, Leone M, Guerini FR, Piga M, Firinu D, Kockum I, Lima Bomfim I, Olsson T, Alfredsson L, Suarez A, Carreira PE, Castillo-Palma MJ, Marcus JH, Congia M, Angius A, Melis M, Gonzalez A, Alarcón Riquelme ME, da Silva BM, Marchini M, Danieli MG, Del Giacco S, Mathieu A, Pani A, Montgomery SB, Rosati G, Hillert J, Sawcer S, D'Alfonso S, Todd JA, Novembre J, Abecasis GR, Whalen MB, Marrosu MG, Meloni A, Sanna S, Gorospe M, Schlessinger D, Fiorillo E, Zoledziewska M, Cucca F (2017) Overexpression of the cytokine BAFF and autoimmunity risk. N Engl J Med 376(17):1615–1626. https://doi.org/10.1056/NEJMoa1610528
72. Mycko MP, Zurawska AE, Selmaj I, Selmaj KW (2022) Impact of diminished expression of circRNA on multiple sclerosis pathomechanisms. Front Immunol 13:875994. https://doi.org/10.3389/fimmu.2022.875994
73. Torkamandi S, Bahrami S, Ghorashi T, Dehani M, Bayat H, Hoseini SM, Rezaei S, Soosanabadi M (2021) Dysregulation of long noncoding RNA MEG3 and NLRC5 expressions in patients with relapsing-remitting multiple sclerosis: is there any correlation? Genes Immun 22(7–8):322–326. https://doi.org/10.1038/s41435-021-00154-4
74. Iwakawa H-O, Tomari Y (2015) The functions of MicroRNAs: mRNA decay and translational repression. Trends Cell Biol 25(11):651–665. https://doi.org/10.1016/j.tcb.2015.07.011
75. Cardamone G, Paraboschi EM, Soldà G, Liberatore G, Rimoldi V, Cibella J, Airi F, Tisato V, Cantoni C, Gallia F, Gemmati D, Piccio L, Duga S, Nobile-Orazio E, Asselta R (2023) The circular RNA landscape in multiple sclerosis: disease-specific associated variants and exon methylation shape circular RNA expression profile. Mult Scler Relat Disord 69:104426. https://doi.org/10.1016/j.msard.2022.104426
76. Huang Y, Bu T, Wu L, Sun H, Gao S, Huang D, Zhang X. Identification of the circRNA/miRNA/mRNA network and hub mRNAs involved in multiple sclerosis through bioinformatics analysis. https://doi.org/10.21203/rs.3.rs-3007890/v1
77. Keller A, Leidinger P, Lange J, Borries A, Schroers H, Scheffler M, Lenhof H-P, Ruprecht K, Meese E (2009) Multiple sclerosis: microRNA expression profiles accurately differentiate patients with relapsing-remitting disease from healthy controls. PLoS One 4(10):e7440. https://doi.org/10.1371/journal.pone.0007440
78. Di Matteo A, Bathon JM, Emery P (2023) Rheumatoid arthritis. Lancet 402(10416):2019–2033. https://doi.org/10.1016/S0140-6736(23)01525-8
79. GBD 2021 Rheumatoid Arthritis Collaborators (2023) Global, regional, and national burden of rheumatoid arthritis, 1990–2020, and projections to 2050: a systematic analysis of the Global Burden of Disease Study 2021. Lancet Rheumatol 5(10):e594–e610. https://doi.org/10.1016/S2665-9913(23)00211-4
80. van Delft MAM, Huizinga TWJ (2020) An overview of autoantibodies in rheumatoid arthritis. J Autoimmun 110:102392. https://doi.org/10.1016/j.jaut.2019.102392
81. Padyukov L (2022) Genetics of rheumatoid arthritis. Semin Immunopathol 44(1):47–62. https://doi.org/10.1007/s00281-022-00912-0
82. Huang Y, Xue Q, Cheng C, Wang Y, Wang X, Chang J, Miao C (2023) Circular RNA in autoimmune diseases: special emphasis on regulation mechanism in RA and SLE. J Pharm Pharmacol 75(3):370–384. https://doi.org/10.1093/jpp/rgac096
83. Li Z, Wang J, Lin Y, Fang J, Xie K, Guan Z, Ma H, Yuan L (2022b) Newly discovered circRNAs in rheumatoid arthritis, with special emphasis on functional roles in inflammatory immunity. Front Pharmacol 13:983744. https://doi.org/10.3389/fphar.2022.983744
84. Dema B, Charles N (2016) Autoantibodies in SLE: specificities, isotypes and receptors. Antibodies (Basel) 5(1). https://doi.org/10.3390/antib5010002
85. Satoh M, Vázquez-Del Mercado M, Chan EKL (2009) Clinical interpretation of antinuclear antibody tests in systemic rheumatic diseases. Mod Rheumatol 19(3):219–228. https://doi.org/10.1007/s10165-009-0155-3
86. Lang KS, Burow A, Kurrer M, Lang PA, Recher M (2007) The role of the innate immune response in autoimmune disease. J Autoimmun 29(4):206–212. https://doi.org/10.1016/j.jaut.2007.07.018
87. Luo Q, Li X, Fu B, Zhang L, Fang L, Qing C, Guo Y, Huang Z, Li J (2021) Expression profile and diagnostic value of circRNAs in peripheral blood from patients with systemic lupus erythematosus. Mol Med Rep 23(1):1. https://doi.org/10.3892/mmr.2020.11639
88. Miao Q, Zhong Z, Jiang Z, Lin Y, Ni B, Yang W, Tang J (2019) RNA-seq of circular RNAs identified circPTPN22 as a potential new activity indicator in systemic lupus erythematosus. Lupus 28(4):520–528. https://doi.org/10.1177/0961203319830493
89. Zhao X, Dong R, Zhang L, Guo J, Shi Y, Ge L, Wang J, Song Z, Ni B, You Y (2022) N6-methyladenosine-dependent modification of circGARS acts as a new player that promotes SLE progression through the NF-κB/A20 axis. Arthritis Res Ther 24(1):37. https://doi.org/10.1186/s13075-022-02732-x

90. Gu J, Su C, Huang F, Zhao Y, Li J (2022) Past, present and future: the relationship between circular RNA and immunity. Front Immunol 13:894707. https://doi.org/10.3389/fimmu.2022.894707
91. Chen YG, Kim MV, Chen X, Batista PJ, Aoyama S, Wilusz JE, Iwasaki A, Chang HY (2017) Sensing self and foreign circular RNAs by intron identity. Mol Cell 67(2):228–238.e5. https://doi.org/10.1016/j.molcel.2017.05.022
92. Lodde V, Floris M, Beerman I, Munk R, Guha R, Steri M, Orrù V, Abdelmohsen K, Crompton PD, Gorospe M, Idda ML, Cucca F (2020) Evolutionarily selected overexpression of the cytokine BAFF enhances mucosal immune response against P. falciparum. Front Immunol 11:575103. https://doi.org/10.3389/fimmu.2020.575103
93. Cadena C, Hur S (2017) Antiviral immunity and circular RNA: no end in sight. Mol Cell 67(2):163–164. https://doi.org/10.1016/j.molcel.2017.07.005
94. Idda ML, Pitzalis M, Lodde V, Loizedda A, Frau J, Lobina M, Zoledziewska M, Virdis F, Delogu G, Marini MG, Mingoia M, Masala M, Lorefice L, Fronza M, Carmagnini D, Carta E, Pilotto S, Castiglia P, Chessa P, Uzzau S, Farina G, Solla P, Steri M, Devoto M, Fiorillo E, Floris M, Zarbo RI, Cocco E, Cucca F (2022) Cross-sectional analysis of the humoral response after SARS-CoV-2 vaccination in Sardinian multiple sclerosis patients, a follow-up study. Front Immunol 13:946356. https://doi.org/10.3389/fimmu.2022.946356
95. Yang M, Qi M, Xu L, Huang P, Wang X, Sun J, Shi J, Hu Y (2021) Differential host circRNA expression profiles in human lung epithelial cells infected with SARS-CoV-2. Infect Genet Evol 93:104923. https://doi.org/10.1016/j.meegid.2021.104923
96. Hu Y, Xu Y, Deng X, Wang R, Li R, You L, Song J, Zhang Y (2021) Comprehensive analysis of the circRNA expression profile and circRNA-miRNA-mRNA network in the pathogenesis of EV-A71 infection. Virus Res 303:198502. https://doi.org/10.1016/j.virusres.2021.198502
97. Hu Y, Yang R, Zhao W, Liu C, Tan Y, Pu D, Song J, Zhang Y (2021) circRNA expression patterns and circRNA-miRNA-mRNA networks during CV-A16 infection of SH-SY5Y cells. Arch Virol 166(11):3023–3035. https://doi.org/10.1007/s00705-021-05190-z
98. Zhang Y, Wang C, Zhu C, Ye W, Gu Q, Shu C, Feng X, Chen X, Zhang W, Shan T (2022) Redondoviridae infection regulates circRNAome in periodontitis. J Med Virol 94(6):2537–2547. https://doi.org/10.1002/jmv.27624
99. Liu Y, Hu C, Sun Y, Wu H, Chen X, Liu Q (2019) Identification of differentially expressed circular RNAs in HeLa cells infected with chlamydia trachomatis. Pathog Dis 77(7):ftz062. https://doi.org/10.1093/femspd/ftz062
100. Broadbent KM, Broadbent JC, Ribacke U, Wirth D, Rinn JL, Sabeti PC (2015) Strand-specific RNA sequencing in Plasmodium falciparum malaria identifies developmentally regulated long non-coding RNA and circular RNA. BMC Genomics 16(1):454. https://doi.org/10.1186/s12864-015-1603-4
101. Wesselhoeft RA, Kowalski PS, Parker-Hale FC, Huang Y, Bisaria N, Anderson DG (2019) RNA circularization diminishes immunogenicity and can extend translation duration in vivo. Mol Cell 74(3):508–520.e4. https://doi.org/10.1016/j.molcel.2019.02.015
102. Liu C-X, Chen L-L (2022) Circular RNAs: characterization, cellular roles, and applications. Cell 185(13):2390. https://doi.org/10.1016/j.cell.2022.06.001
103. Li H, Peng K, Yang K, Ma W, Qi S, Yu X, He J, Lin X, Yu G (2022) Circular RNA cancer vaccines drive immunity in hard-to-treat malignancies. Theranostics 12(14):6422–6436. https://doi.org/10.7150/thno.77350
104. Huang K, Li N, Li Y, Zhu J, Fan Q, Yang J, Gao Y, Liu Y, Hou Q, Gao S, Wei K, Deng C, Zuo C, Sun Z (2022) Delivery of circular mRNA via degradable lipid nanoparticles against SARS-CoV-2 delta variant. bioRxiv 2022.05.12.491597. https://doi.org/10.1101/2022.05.12.491597
105. Pfafenrot C, Schneider T, Müller C, Hung L-H, Schreiner S, Ziebuhr J, Bindereif A (2021) Inhibition of SARS-CoV-2 coronavirus proliferation by designer antisense-circRNAs. Nucleic Acids Res 49(21):12502–12516. https://doi.org/10.1093/nar/gkab1096

CircularRNA and Musculoskeletal Diseases

25

Pujiao Yu, Yuan Qi, Hui Huang, Hairong Wang, and Jiahong Xu

Abstract

The skeletal muscle is the largest organ in the human body and plays a crucial role in the locomotion, metabolism, and homeostasis. Myogenesis is regulated by a complex molecular network and is kept finely balanced under physiological conditions. The disorder of muscular homeostasis results in musculoskeletal diseases, including myositis, dystrophy, sarcopenia, atrophy, and cachexia. These diseases severely affect the life quality of humans. In recent years, with the progress of high-throughput sequencing technology and the development of bioinformatics, a large number of noncoding RNAs (ncRNAs) have been identified in the skeletal muscle, and numerous experiments to date have shown that those ncRNAs are widely involved in multiple pathophysiological process of skeletal muscle. Circular RNAs (circRNA) are a class of non-coding RNAs formed by covalently closed loops through back-splicing and exon skipping. circRNAs have been confirmed to play a vital role in various biological functions, acting as microRNA sponges and reservoirs, as well as combining with RNA-binding proteins during the progression of myogenesis. In this chapter, we will summarize the progress of circRNA in regulating skeletal muscle and provide new information to better understand muscle diseases development and therapeutic strategy.

Keywords

Circular RNA · Skeletal muscle · Muscular disease · Myogenesis · Duchenne muscular dystrophy · Muscle atrophy · Myotonic Dystrophy · Myasthenia gravis · Cachexia

1 Background

The skeletal muscle is the largest organ in the human body, which contributes to over 40% of the body mass in adults, and plays a crucial role in the locomotion, metabolism, and homeostasis [1–4]. Skeletal muscle is composed of myofibers with multinucleated contractile which are generated by fusion of multiple mononucleated myoblasts through a process known as myogenesis. Proper growth and homeostasis of skeletal muscle are the decisive factors of human motor performance and organic health. Conversely, musculoskeletal diseases due to the disorder of muscular homeostasis, such as myositis, dystrophy, sarcopenia, atrophy, and cachexia, severely affect the life quality of humans. In recent years, with the progress of high-throughput sequencing

P. Yu · Y. Qi · H. Huang · H. Wang · J. Xu (✉)
Department of Cardiology, Gongli Hospital, School of Medicine, Shanghai University, Shanghai, China

J. Xiao (ed.), *Advances in Circular RNAs*, Advances in Experimental Medicine and Biology 1485,
https://doi.org/10.1007/978-981-96-9428-0_25

technology and the development of bioinformatics, a large number of noncoding RNAs (ncRNAs) have been identified in the skeletal muscle, and numerous experiments to date have shown the those ncRNAs are widely involved in multiple muscle formation steps, including myoblast proliferation, cell cycle exit, muscle-specific gene expression, myoblast fusion, and sarcomere assembly [5–8]. Circular RNA (circRNA) is a large class of noncoding RNAs with high stability and evolutionarily conserved which have covalently closed, single-stranded transcripts. Currently, studies indicate that circRNAs shape gene expression by sponge miRNAs and proteins, regulating transcription and interfering with splicing, or translating into proteins or polypeptides [9, 10]. Recent studies also reveal that circRNAs are abundant in skeletal muscles and dynamically expressed in the progress of myogenesis [11–13]. In this chapter, we will summarize the progress of circRNA in regulating skeletal muscle and provide new information to better understand muscle diseases' development and therapeutic strategy.

1.1 The Development of Muscle

Skeletal muscle development in vertebrates is initiated throughout the organism. The process of myogenesis is divided into multiple phases. Those that establish muscles prenatally are founder stem cells (FSC1, FSC2, FSC3). FSC1 will be largely exhausted in the embryo, while FSC2 and FSC3 will give rise to the majority of adult skeletal muscle stem cells. All the founding stem cell progenitor cells from mid-embryonic stages onward marked with paired box genes, including Pax3 and Pax7, are committed to generate myoblasts. Then, many transcription factors are involved in regulating the proliferation and differentiation of myoblasts, including Myod, Mrf4, Myf5, and myogenein. At last, committed myoblasts fuse with each other to form multiple nuclear myotubes after proliferation, growth, and elongation and finally fuse to form the multinucleated myotubes [2, 14, 15].

1.2 The Regeneration of Muscle

Adult skeletal muscle, like all renewable organs, relies on a mechanism that compensates for the turnover of terminally differentiated cells to maintain tissue homeostasis. Founder stem cells will give rise to satellite cells, a subpopulation of adult muscle stem cells, reside between the sarcolemma and basal lamina of muscle fiber, and contribute to postdamage muscle regeneration in adult muscle [16]. Satellite cells express Pax7 and remain in the quiescent state. They enter the cycle immediately after muscle damage and divide into proliferating myoblasts marked by coexpression of Pax7 and MyoD, then fuse to form nascent multinucleate myofibers or fuse with regenerating myofibers end fragments leading to repair of the injured muscles. Many signal molecules that are involved in the progress of muscle regeneration include fibroblast growth factor [17], hepatocyte growth factor (Hgf) [18], insulin-like growth factor (IGF-I and IGF-II) [19, 20], interleukins [21], and members of the transforming growth-β (TGF-β) family [22]. During repairment, a portion of the activated satellite cells will eliminate MyoD expression and return to the quiescent state to maintain the satellite cell population in the muscle [23, 24].

2 CircRNAs

CircRNAs were first discovered in plant viroids by electron microscopy in the late 1970s and then in hepatitis delta virus (HDV) and eukaryotic cells in the 1990s [25–27]. CircRNAs are covalently closed-loop structure with no 50-cap and 30-poly (A) termini. Although CircRNAs have been discovered for decades, very little is known about their mechanism so far. They were just considered as byproducts of splicing errors without biological function at very early stage, and now, it is widely proved in many species that circRNAs could serve as microRNA sponge or RNA-binding proteins interactor to regulate physiology and pathology progresses, including development, differentiation, aging, inflammation, apoptosis, and so on [9]. They have also

Table 25.1 The CircRNA expression of profiling studies in skeletal muscles

Sample	Experimental conditions	Detects circRNAs	Method	Species	References
C2c12 cell line	Growth and differentiation	1600	RNA-seq	Mouse	[28]
C2c12 cell line	Growth and differentiation	11,000	MicroArray	Mouse	[29]
C2c12 cell line	Growth and differentiation	37,700	RNA-seq	Mouse	[30]
Primary myoblasts	Growth and differentiation	2100	RNA-seq	Human	[28]
Longissimus dorsi	Aging	4400	RNA-seq	Pig	[31]
Vastus lateralis	Aging	12,000	RNA-seq	Monkey	[32]
Longissimus dorsi	Adult skeletal muscle	13,000	RNA-seq	Goat	[33]
Longissimus dorsi	Adult skeletal muscle	886	RNA-seq	Bovine	[34]
Longissimus dorsi	Adult skeletal muscle	6000	RNA-seq	Bovine	[35]
Leg muscles	Embryo development	13,400	RNA-seq	Chicken	[36]

been demonstrated to have abilities of coding and translating into polypeptides. CircRNAs are abundant in skeletal muscles. With the advanced high-throughput RNA sequencing (RNA-seq) and related bioinformatics tools, several of circRNAs have been identified and have been proved to be key regulators in the muscle growth and development [11] (Table 25.1).

2.1 CircRNAs in Myogenesis

2.1.1 CircRNAs Serve as Sponge in Regulating Myogenesis

In skeletal muscle, several circRNAs have been found to act as miRNA sponges that regulate myogenesis-related genes at posttranscriptional level. Table 25.2 summarizes some of the recently discovered mechanisms by which circRNAs regulate the growth and development of skeletal muscle in animals. circSNX29 could serve as an endogenous miR-774 sponge to activate noncanonical Wnt5a/Ca^{2+} pathway, resulting in promoting bovine myoblast differentiation progression [37]. circSVIL, previously shown to be highly expressed in leg muscle during late embryonic development in chicken, is highly expressed in differentiated myotubes compared to growing C2C12 myoblasts. Regulation of circSVIL in myoblast showed that it could promote both myoblast proliferation and differentiation by sponging miR-203 and increasing expression of its targets c-JUN and MEF2C [36]. circMYL1 could regulate the proliferation and differentiation of bovine myoblasts by acting as an miR-2400 sponge to activate the MYOG gene [38]. In pigs, circIGF1R was found to regulate myoblast differentiation of porcine skeletal muscle satellite cells by acting as a miR-16 sponge [39]. circFUT10 contains three miR-133a-binding sites and hence promotes apoptosis in myoblast by sponging miR-133a, which regulates Bcl-2. Overexpression of circFUT10 inhibits myoblast proliferation by arresting cell cycle [40]. CircLMO7 inhibits myoblast differentiation by sponging microRNA miR-378a-3p, which targets the 3′ UTR of HDAC4 mRNA to inhibit its expression. HDAC4 silencing promotes expression of myogenic transcription factors, including MyoD and MEF2C promoting myogenesis [33]. circUSP13 could significantly promote differentiation and inhibit apoptosis in goat primary myoblasts by sponging miR-29c. Mechanically, miR-29c could promote IGF1 expression that upregulated the expression of MyoG and MyHC [41]. CircZNF609 acts as a decoy for miR-194-5p to promote expression of BCLAF1 and thereby suppresses myoblast differentiation [42]. circITSN2 was verified to promote primary myoblast proliferation and differentiation via targeting miR-218-5p, which could inhibit the expression of LIM domain 7 (LMO7) [43]. circFGFR4 could promote myoblast differentiation through the upregulation of Wnt3a by sponge miR-107 [44]. An analysis of the circRNA–miRNA–mRNA regulatory network revealed that conserved circNfix might regulate myogenesis by associating with miR-204-5p, a suppressor of myocyte enhancer factor 2c (Mef2c) expression. Silencing circNfix moderately reduced Mef2c

Table 25.2 Functions and regulatory mechanisms of circRNAs in skeletal muscle development

circRNA	Regulation method	Biological function	Species	References
circSNX29	miR-774 sponge	Promote myoblast differentiation	Bovine	[37]
circSVIL	miR-203 sponge/interact with protein (STAT)	Promote myoblast proliferation and differentiation/promote proliferation and inhibit apoptosis	Chicken/bovine	[36, 61]
circMYL1	miR-2400 sponge	Inhibit differentiation of myogenic cells	Mouse	[38]
circIGF1R	miR-16 sponge	Promote myogenic cell differentiation	Pig	[39]
circLMO7	miR-378a-3p sponge	Regulate myoblast differentiation	Bovine	[40]
circFUT10	miR-133a sponge	Regulate apoptosis in myoblast	Mouse	[33]
CircUSP13	miR-29c sponge	Inhibit myogenic cells apoptosis	Goat	[41]
circZNF609	miR-194-5p sponge/code protein	Suppress myoblast differentiation	Human	[42, 28]
circITSN2	miR-218-5p sponge	Promote skeletal muscle development in embryos	Chicken	[43]
circFGFR4	miR-107 sponge	Promote myoblast differentiation	Bovine	[44]
circNfix	miR-204-5p sponge	Promote myoblast differentiation	Mouse	[45]
circTTN	miR-432 sponge/recruit protein (PURB)	Promote primary myoblasts proliferation and differentiation	Bovine	[46, 60]
circCCDC91	miR-15 family sponge	Promote myogenic proliferation and differentiation	Chicken	[47]
circUBE3A	miR-28-5p sponge	Regulate myoblast proliferation and differentiation	Goat	[48]
circUBE3C	miR-191 sponge	Inhibit proliferation and promote apoptosis in primary myoblasts	Bovine	[49]
circNDST1	miR-411a sponge	Promote proliferation and inhibit myoblast differentiation	Bovine	[50]
circFgfr2	miR-133 sponge	Regulate myogenesis and muscle regeneration	Pig	[51]
circCDR1as	miR-7 sponge	Activates myogenesis and myotube formation	Goat	[52]
circSmad4	Interact with protein (PURB)	Promote myogenesis	Bovine	
CircMYBPC1	Binds protein (MyHC)	Promotes myoblast differentiation and skeletal muscle regeneration	Bovine	[62]
circHipk2	Binds protein(Rpl7)	Regulate myoblast proliferation and promote myotube formation	Mouse	[63]
CircFAM188B	Code protein	Regulate proliferation and differentiation of skeletal muscle satellite cells	Chicken	[68]

mRNA expression and inhibited C2C12 differentiation [45]. circTTN was reported that it could promote proliferation and differentiation of bovine primary myoblasts via competitively combining with miR-432 to activate the IGF2/phosphatidylinositol 3-kinase (PI3K)/AKT signaling pathway [46]. circCCDC91 promotes the proliferation and differentiation of chicken myoblasts and rescues muscle atrophy by suppressing the expression of miR-15 family via activating IGF1/PI3K/AKT signaling pathway [47]. circUBE3A specifically binds to and inhibits miR-28-5p, boosting the expression of Hydroxyacyl Coenzyme A Dehydrogenase Beta (HADHB) and contributing to goat myoblast proliferation and differentiation [48]. circUBE3C interacts with miR-191 and alleviates the suppression of p27 through cytoplasmic separation, thereby inhibiting proliferation and promoting apoptosis in bovine primary myoblasts [49]. Sequestration of miR-411a by circNDST1 attenuates the down-regulated effect on its target gene Smad4, thereby promoting proliferation and inhibiting cell differentiation in bovine myoblasts [50]. In pig, circFgfr2 can regulate myogenesis and muscle regeneration via acting as a ceRNA sponge of miR-133 family members to activate the expression of Map 3 k20 and the downstream JNK/MAPK signaling pathway [51]. circCDR1as can greatly relieve IGF1R levels through sponging miR-7 and subsequently activates myogenesis and myotube formation [52]. These studies highlight a potential regulatory role of circRNAs in the myogenesis via the inhibition of miRNAs.

2.1.2 CircRNAs Interact with Proteins in Regulating Myogenesis

Another one of the classical mechanisms of circRNAs is to interact with RNA-binding proteins (RBPs). The RBPs' binding sites of circRNAs are less than that of the corresponding linear mRNA, and the distinct circularization mechanism of circRNA makes it reasonable to form more complex secondary and tertiary structures [53, 54]. However, there are still many results shown that RBPs can interact with circRNAs and involved in the multiple pathophysiological progress. By comparing protein sequences, 500 RBPs have been identified in humans and mice [55–57]. With the development of high-throughput sequencing technology and molecular biology experimental technology, such as mass spectrometry, more than 1900 different RBPs have been identified in the human genome [58]. circRNAs can bind various RBPs just like molecular sponges and function as endogenous competitive RNA. They can also act as protein scaffolds to promote proteins contacting with each other and act as protein decoys of RBPs to promote their transport between the nucleus and cytoplasm to maintain their specific functions [59]. Previously, we have mentioned that circTTN can act as miR-432 sponge in regulating myogenesis. A recent study also showed that circTTN could recruits the purine-rich binding proteins β (PURB) on the Titin (TTN) promoter to inhibit the expression of the TTN gene while PURB inhibits myoblast proliferation and differentiation, which is consistent with circTTN function [60]. In another study, PURB was also proved to be combined with circ-Smad4 located in the nucleus to participate in the transcription regulation of MHC, thus promoting myogenesis [61]. circSVIL can interact with STAT1 to promote proliferation and inhibit apoptosis in bovine myoblasts [62]. CircMYBPC1 directly binds MyHC proteins and promotes myoblast differentiation and skeletal muscle regeneration [63]. circHipk2 was proved to be mostly distributed in the cytoplasm, and dynamically and differentially expressed in various myogenesis systems in vitro and in vivo. It was also found that circHipk2 can regulate myoblast proliferation and promote myotube formation in C2C12 cells though directly binding to ribosomal protein Rpl7, an essential 60S preribosomal assembly factor, to inhibit ribosome translation [64].

2.1.3 CircRNAs Convert into Proteins in Regulating Myogenesis

CircRNAs were believed to be noncoding in the past due to its lacks of 5′ cap and poly(A) tail which are required for translation. However, most of the circRNAs are originated form exons and localized in the cytoplasm, which hints a possibility of circRNA translation [65, 66]. Some circRNAs with the presence of an internal ribosomal entry site (IRES) are predicted to have abilities of translation [67, 68]. CircZNF609 is one of the first endogenous circRNAs with the IRES reported to have an ability to translate. It was firstly identified in muscle differentiation by the high-content functional genomic screen and proved to specifically control myoblast proliferation. Further research showed that circZNF609 is associated with heavy polysomes, and it is translated into a protein in a splicing-dependent and cap-independent manner. It is notable that the phenotype of circZNF609 to its protein coding capacity was not clarified in the study [28]. A recent study found in chicken that circFAM188B can regulate proliferation and differentiation of skeletal muscle satellite cells. Bioinformatic analysis revealed that circFAM188B contains an opening reading frame that translates into circFAM188B-103aa, and IRES analysis further confirmed the coding potential of circ-FAM188B. Western blot assay detected a flag-tagged circFAM188B-103aa, and several peptides of circFAM188B-103aa were detected by LC-MS/MS analysis [69].

2.2 CircRNA in Muscular Disease

2.2.1 Duchenne Muscular Dystrophy

DMD is a severe muscular disorder caused by frame-shifting deletions or nonsense mutations in the DMD gene resulting in the absence or reduced production of the dystrophin protein [70–72]. Hitoshi et al. focused on the characterization of circRNA production in the region spanning exon 45 to exon 55 of the DMD gene that represents a deletion hotspot in 63% of DMD patients and identified the preferential splice sites involved in both circRNA generation and in multiple exon skipping of exons 45 to 55. These results suggested that artificial and specific increase the expression levels of these circRNAs by exon skipping might have the possibility to improve or cure DMD patients [73]. The protein-coding circ-ZNF609 was found expressed at elevated levels in differentiated DMD myoblasts [28]. circ-QKI and circ-BNC2 were upregulated during in vitro differentiation of normal myoblasts and downregulated in the DMD conditions, well correlating with the notion that dystrophic cells have altered progression into the differentiation process [74]. Song et al. performed a microarray analysis to explore the expression patterns of circRNAs in the gastrocnemius muscles in mdx mice, a DMD animal model, and C57 mice. A total of 197 differentially expressed circRNAs were identified between mdx mice and C57 mice. A total of 189 of them were predicted to have protein-coding potential, while 98 of them could potentially be translated into polypeptides with 150 or more amino acids, which indicated that circRNAs may play pivotal roles in the pathophysiological mechanisms of DMD [75].

2.2.2 Muscle Atrophy

Skeletal muscle atrophy is characterized by weakening, shrinking, and decreasing muscle mass and fiber cross-sectional area at the histological level [76, 77]. Several pathophysiological processes are involved in skeletal muscle atrophy, including inflammation, mitochondrial dysfunction, oxidative stress, activated proteolytic system, autophagy, and mTOR pathway [78–82]. circTmeff1 can directly interact with TAR DNA-binding protein 43 (TDP-43) and promotes aggregation of TDP-43 in mitochondria, thus rescuing muscle atrophy in vitro and in vivo. The researchers also identified a novel protein termed TMEFF1-339aa coded by circTMEFF1 that was responsible for the proatrophic effects of circTMEFF1 [83]. Spermine oxidase (SMOX) is a polyamine catabolic enzyme that plays a critical role in muscle differentiation. circSMOX, which has been identified recently arising from *SMOX* gene, was found significantly increased compared to the physiological state, in both in vitro and in vivo models of muscle atrophy. Oxidative stress-related protein SOD1 and RNA processing-related protein FUS were the potential downstream candidates of circSMOX [84]. Moreover, it has been reported that circCCDC91 was differentially expressed in the chicken breast muscle among four different embryonic developmental time points. And circCCDC91 could rescue skeletal muscle atrophy by modulating the expression of insulin receptor substrate 1 (IRS1), as well as activate insulin-like growth factor 1-phosphatidylinositol 3-kinase/AKT (IGF1-PI3K/AKT) signaling pathway via absorbing miR-15a, miR-15b-5p, and miR-15c-5p [47].

2.2.3 Myotonic Dystrophy

Myotonic dystrophy is a multisystemic disorder caused by microsatellite expansion mutations in two unrelated genes leading to similar, yet distinct, diseases [85, 86]. Myotonic dystrophy type 1 (DM1), also known as Steinert disease, is the most prevalent muscular dystrophy in adults and has a wide phenotypic spectrum of clinical manifestations that includes myotonia, reduced muscle strength, cardiac arrhythmia, insulin resistance, cataracts, hypogonadism, and cognitive defects [87]. Myotonic dystrophy type 2 (DM2) tends to cause a milder phenotype with later onset of symptoms and is less common than myotonic dystrophy type 1 [88]. By using publicly available RNA-Seq datasets deposited in the DMseq database, Karol et al. found that the global circRNA level is indeed increased in DM1. It is believed that this increase may simply be a side effect of splicing aberrations or secondary effect of the chronic pathological state of DM1 [89]. And another RNA-sequencing datasets identified four transcripts from circularRNAs upregulated in DM1 skeletal muscles, including CDYL, HIPK3, RTN4_03, and ZNF609, while receiver operating characteristics curves of these four circRNAs discriminated DM1 patients from controls. They were negatively correlated with skeletal muscle strength and with splicing biomarkers of disease severity, and displayed higher values in more severely affected patients, wherein increased circRTN4_03 and circZNF609 were observed in differentiated myogenic cell lines derived from DM1 patients [90].

2.2.4 Myasthenia Gravis

Myasthenia gravis is multifactorial disorder that is characterized by skeletal muscle weakness and fatigue. It is associated with antibodies directed against the acetylcholine receptor, muscle-specific kinase (MUSK), lipoprotein-related protein 4 (LRP4), or agrin in the postsynaptic membrane at the neuromuscular junction [91, 92]. Recently, next-generation sequencing of circRNAs showed some circRNAs were abnormally expressed in MG patients, which hint theirs potential in the regulation of myasthenia gravis or as biomarkers in the diagnosis and prognosis of myasthenia gravis patients [93]. Lv et al. conducted an expression profile of circRNAs in peripheral bold samples of patients with myasthenia gravis and found that hsa-circRNA5333-4 represents a novel biomarker for the diagnosis and monitoring of myasthenia gravis [94]. Circ-0076490 was found to regulate proliferation and apoptosis of Jurkat cells via targeting the miR-144-3p/MAPK1 axis in the process of myasthenia gravis [95]. circFBL promoted myogenic proliferation by negative regulation of miR-133/PAX7 axis in patients with myasthenia gravis [96].

2.2.5 Cachexia

Cachexia is a devastating multifactorial syndrome defined as skeletal muscle wasting and fat storage depletion, in the context of chronic systemic inflammation and metabolic disorders, and cachexia can't be reversed by nutritional support [97]. Cancer-associated cachexia occurs in 50–80% of advanced cancer patients (80–90% for gastric and pancreatic cancers) and eventually accounts for more than 20% of all cancer-related deaths [98]. circANAPC7 could inhibit ZIP4-mediated cell proliferation through miR-373–PHLPP2–AKT signaling axis and inhibit ZIP4-mediated cancer cachexia by inhibiting TGF-β secretion [99]. Moreover, the conditioned media from miR-373 overexpressed cells increased the protein level of muscle atrophy markers in C2C12 myotubes, suggesting that miR-373 also plays an important role in cancer cachexia. In another hands, disorder of adipogenesis also facilitates the loss of skeletal muscle. circPTK2 was reported to serve as a ceRNA to contribute to cachexia progression through the miR-182-5p/JAZF1 axis. JAZF1 was previously reported to be a potential target for regulating lipid metabolism. Overexpression of JAZF1 in adipocytes and hepatocytes could reduce lipid synthesis and increase lipolysis [100]

3 Perspective

CircRNAs are abundant and stable transcripts, emerging as the key regulators of gene expression. The emergence of circRNA provides a new insight to understand the myogenesis and the pathogenesis of skeletal muscle diseases. The application and development of RNA sequencing technology and bioinformatic technology provide more systematic and professional theoretical support for the analysis of those circRNAs. Despite recent discoveries of circRNA biogenesis and function in regulating myogenesis and skeletal muscle diseases, many challenges remain to be conquered. For example, their functions in sarcopenia and inflammatory myopathy remain unknown. Currently, the mechanisms of being molecular sponges of miRNAs are the most common research strategies for studying circRNAs' downstream regulation mechanism on pathophysiological process. Most of circRNAs express at very low levels in the human tissues, and their function as molecular sponges of miRNA was doubted to be overestimated. Much remains unknown about the specific mechanisms of the network of interactions between circRNAs and other molecules, the protein coding by circRNAs, and their functions. A recent study showed that N6-methyladenosine (m6A) also facilitates circRNA translation and was found to be very common in circRNAs as judged by circRNA-m6A-seq [101]. These results suggest that translatable circRNAs may be common in human transcriptome. However, knowledge about the functions and mechanisms of circRNA-coded proteins in the pathological and physiological progresses of skeletal muscle is limited, and more work should be done in the future. In addition, circRNA's potential as novel biomarkers for diagnosis, prognosis, and therapeutic response in other diseases has acquired many evidence-based medical proofs, including cancers, cardiovascular diseases, renal diseases, etc. However, their clinical values in skeletal diseases still await further cohort study.

References

1. Frontera WR, Ochala J (2015) Skeletal muscle: a brief review of structure and function. Calcif Tissue Int 96(3):183–195. https://doi.org/10.1007/s00223-014-9915-y
2. Bentzinger CF, Wang YX, Rudnicki MA (2012) Building muscle: molecular regulation of myogenesis. Cold Spring Harb Perspect Biol 4(2). https://doi.org/10.1101/cshperspect.a008342
3. Mukund K, Subramaniam S (2020) Skeletal muscle: a review of molecular structure and function, in health and disease. Wiley Interdiscip Rev Syst Biol Med 12(1):e1462. https://doi.org/10.1002/wsbm.1462
4. Morgan J, Partridge T (2020) Skeletal muscle in health and disease. Dis Model Mech 13(2). https://doi.org/10.1242/dmm.042192
5. Liu L, Gu T, Bao X, Zheng S, Zhao J, Zhang L (2019) Microarray profiling of circular RNA identifies hsa_circ_0126991 as a potential risk factor for essential hypertension. Cytogenet Genome Res 157(4):203–212. https://doi.org/10.1159/000500063
6. Yang Y, Wu J, Liu W, Zhao Y, Chen H (2023) The function and regulation mechanism of non-coding RNAs in muscle development. Int J Mol Sci 24(19). https://doi.org/10.3390/ijms241914534
7. Wohlwend M, Laurila PP, Williams K, Romani M, Lima T, Pattawaran P, Benegiamo G, Salonen M, Schneider BL, Lahti J, Eriksson JG, Barrès R, Wisløff U, Moreira JBN, Auwerx J (2021) The exercise-induced long noncoding RNA CYTOR promotes fast-twitch myogenesis in aging. Sci Transl Med 13(623):eabc7367. https://doi.org/10.1126/scitranslmed.abc7367
8. Ballarino M, Morlando M, Fatica A, Bozzoni I (2016) Non-coding RNAs in muscle differentiation and musculoskeletal disease. J Clin Invest 126(6):2021–2030. https://doi.org/10.1172/jci84419
9. Li X, Yang L, Chen LL (2018) The biogenesis, functions, and challenges of circular RNAs. Mol Cell 71(3):428–442. https://doi.org/10.1016/j.molcel.2018.06.034
10. Zhang Z, Yang T, Xiao J (2018) Circular RNAs: promising biomarkers for human diseases. EBioMedicine 34:267–274. https://doi.org/10.1016/j.ebiom.2018.07.036
11. Greco S, Cardinali B, Falcone G, Martelli F (2018) Circular RNAs in muscle function and disease. Int J Mol Sci 19(11). https://doi.org/10.3390/ijms19113454
12. Chen R, Lei S, Jiang T, Zeng J, Zhou S, She Y (2020) Roles of lncRNAs and circRNAs in regulating skeletal muscle development. Acta Physiol (Oxf) 228(2):e13356. https://doi.org/10.1111/apha.13356
13. Das A, Das A, Das D, Abdelmohsen K, Panda AC (2020) Circular RNAs in myogenesis. Biochim Biophys Acta Gene Regul Mech 1863(4):194372. https://doi.org/10.1016/j.bbagrm.2019.02.011
14. Fan CM, Li L, Rozo ME, Lepper C (2012) Making skeletal muscle from progenitor and stem cells: development versus regeneration. Wiley Interdiscip Rev Dev Biol 1(3):315–327. https://doi.org/10.1002/wdev.30
15. Chargé SB, Rudnicki MA (2004) Cellular and molecular regulation of muscle regeneration. Physiol Rev 84(1):209–238. https://doi.org/10.1152/physrev.00019.2003
16. Campion DR (1984) The muscle satellite cell: a review. Int Rev Cytol 87:225–251. https://doi.org/10.1016/s0074-7696(08)62444-4
17. Pawlikowski B, Vogler TO, Gadek K, Olwin BB (2017) Regulation of skeletal muscle stem cells by fibroblast growth factors. Dev Dyn 246(5):359–367. https://doi.org/10.1002/dvdy.24495
18. Anderson JE (2016) Hepatocyte growth factor and satellite cell activation. Adv Exp Med Biol 900:1–25. https://doi.org/10.1007/978-3-319-27511-6_1
19. Philippou A, Halapas A, Maridaki M, Koutsilieris M (2007) Type I insulin-like growth factor receptor signaling in skeletal muscle regeneration and hypertrophy. J Musculoskelet Neuronal Interact 7(3):208–218

20. MacGregor J, Parkhouse WS (1996) The potential role of insulin-like growth factors in skeletal muscle regeneration. Can J Appl Physiol 21(4):236–250. https://doi.org/10.1139/h96-021
21. Schiaffino S, Pereira MG, Ciciliot S, Rovere-Querini P (2017) Regulatory T cells and skeletal muscle regeneration. FEBS J 284(4):517–524. https://doi.org/10.1111/febs.13827
22. Chen JL, Colgan TD, Walton KL, Gregorevic P, Harrison CA (2016) The TGF-β signalling network in muscle development, adaptation and disease. Adv Exp Med Biol 900:97–131. https://doi.org/10.1007/978-3-319-27511-6_5
23. Yin H, Price F, Rudnicki MA (2013) Satellite cells and the muscle stem cell niche. Physiol Rev 93(1):23–67. https://doi.org/10.1152/physrev.00043.2011
24. Ten Broek RW, Grefte S, Von den Hoff JW (2010) Regulatory factors and cell populations involved in skeletal muscle regeneration. J Cell Physiol 224(1):7–16. https://doi.org/10.1002/jcp.22127
25. Hansen TB, Jensen TI, Clausen BH, Bramsen JB, Finsen B, Damgaard CK, Kjems J (2013) Natural RNA circles function as efficient microRNA sponges. Nature 495(7441):384–388. https://doi.org/10.1038/nature11993
26. Sanger HL, Klotz G, Riesner D, Gross HJ, Kleinschmidt AK (1976) Viroids are single-stranded covalently closed circular RNA molecules existing as highly base-paired rod-like structures. Proc Natl Acad Sci USA 73(11):3852–3856. https://doi.org/10.1073/pnas.73.11.3852
27. Kos A, Dijkema R, Arnberg AC, van der Meide PH, Schellekens H (1986) The hepatitis delta (delta) virus possesses a circular RNA. Nature 323(6088):558–560. https://doi.org/10.1038/323558a0
28. Legnini I, Di Timoteo G, Rossi F, Morlando M, Briganti F, Sthandier O, Fatica A, Santini T, Andronache A, Wade M, Laneve P, Rajewsky N, Bozzoni I (2017) Circ-ZNF609 is a circular RNA that can be translated and functions in myogenesis. Mol Cell 66(1):22–37.e29. https://doi.org/10.1016/j.molcel.2017.02.017
29. Chen R, Jiang T, Lei S, She Y, Shi H, Zhou S, Ou J, Liu Y (2018) Expression of circular RNAs during C2C12 myoblast differentiation and prediction of coding potential based on the number of open reading frames and N6-methyladenosine motifs. Cell Cycle 17(14):1832–1845. https://doi.org/10.1080/15384101.2018.1502575
30. Zhang P, Xu H, Li R, Wu W, Chao Z, Li C, Xia W, Wang L, Yang J, Xu Y (2018) Assessment of myoblast circular RNA dynamics and its correlation with miRNA during myogenic differentiation. Int J Biochem Cell Biol 99:211–218. https://doi.org/10.1016/j.biocel.2018.04.016
31. Chen J, Zou Q, Lv D, Wei Y, Raza MA, Chen Y, Li P, Xi X, Xu H, Wen A, Zhu L, Tang G, Li M, Jiang A, Liu Y, Fu Y, Jiang Y, Li X (2018) Comprehensive transcriptional landscape of porcine cardiac and skeletal muscles reveals differences of aging. Oncotarget 9(2):1524–1541. https://doi.org/10.18632/oncotarget.23290
32. Abdelmohsen K, Panda AC, De S, Grammatikakis I, Kim J, Ding J, Noh JH, Kim KM, Mattison JA, de Cabo R, Gorospe M (2015) Circular RNAs in monkey muscle: age-dependent changes. Aging 7(11):903–910. https://doi.org/10.18632/aging.100834
33. Wei X, Li H, Yang J, Hao D, Dong D, Huang Y, Lan X, Plath M, Lei C, Lin F, Bai Y, Chen H (2017) Circular RNA profiling reveals an abundant circLMO7 that regulates myoblasts differentiation and survival by sponging miR-378a-3p. Cell Death Dis 8(10):e3153. https://doi.org/10.1038/cddis.2017.541
34. Cao Y, You S, Yao Y, Liu ZJ, Hazi W, Li CY, Zhang XY, Hou XX, Wei JC, Li XY, Wang DW, Chen CF, Zhang YF, Ni W, Hu SW (2018) Expression profiles of circular RNAs in sheep skeletal muscle. Asian Australas J Anim Sci 31(10):1550–1557. https://doi.org/10.5713/ajas.17.0563
35. Li C, Li X, Yao Y, Ma Q, Ni W, Zhang X, Cao Y, Hazi W, Wang D, Quan R, Hou X, Liu Z, Zhan Q, Liu L, Zhang M, Yu S, Hu S (2017) Genome-wide analysis of circular RNAs in prenatal and postnatal muscle of sheep. Oncotarget 8(57):97165–97177. https://doi.org/10.18632/oncotarget.21835
36. Ouyang H, Chen X, Li W, Li Z, Nie Q, Zhang X (2018) Circular RNA circSVIL promotes myoblast proliferation and differentiation by sponging miR-203 in chicken. Front Genet 9:172. https://doi.org/10.3389/fgene.2018.00172
37. Peng S, Song C, Li H, Cao X, Ma Y, Wang X, Huang Y, Lan X, Lei C, Chaogetu B, Chen H (2019) Circular RNA SNX29 sponges miR-744 to regulate proliferation and differentiation of myoblasts by activating the Wnt5a/Ca(2+) Signaling pathway. Mol Ther Nucleic Acids 16:481–493. https://doi.org/10.1016/j.omtn.2019.03.009
38. Elnour IE, Wang X, Zhansaya T, Akhatayeva Z, Khan R, Cheng J, Hung Y, Lan X, Lei C, Chen H (2021) Circular RNA circMYL1 inhibit proliferation and promote differentiation of myoblasts by sponging miR-2400. Cells 10(1). https://doi.org/10.3390/cells10010176
39. Xie B, Lin J, Chen X, Zhou X, Zhang Y, Fan M, Xiang J, He N, Hu Z, Wang F (2023) CircXRN2 suppresses tumor progression driven by histone lactylation through activating the Hippo pathway in human bladder cancer. Mol Cancer 22(1):151. https://doi.org/10.1186/s12943-023-01856-1
40. Li H, Yang J, Wei X, Song C, Dong D, Huang Y, Lan X, Plath M, Lei C, Ma Y, Qi X, Bai Y, Chen H (2018) CircFUT10 reduces proliferation and facilitates differentiation of myoblasts by sponging miR-133a. J Cell Physiol 233(6):4643–4651. https://doi.org/10.1002/jcp.26230
41. Zhang Z, Fan Y, Deng K, Liang Y, Zhang G, Gao X, El-Samahy MA, Zhang Y, Deng M, Wang F (2022) Circular RNA circUSP13 sponges miR-29c to pro-

mote differentiation and inhibit apoptosis of goat myoblasts by targeting IGF1. FASEB J 36(1):e22097. https://doi.org/10.1096/fj.202101317R
42. Wang Y, Li M, Wang Y, Liu J, Zhang M, Fang X, Chen H, Zhang C (2019) A Zfp609 circular RNA regulates myoblast differentiation by sponging miR-194-5p. Int J Biol Macromol 121:1308–1313. https://doi.org/10.1016/j.ijbiomac.2018.09.039
43. Shen X, Wei Y, Liu W, You G, Tang S, Su Z, Du M, He J, Zhao J, Tian Y, Zhang Y, Ma M, Zhu Q, Yin H (2021) A novel circular RNA circITSN2 targets the miR-218-5p/LMO7 Axis to promote chicken embryonic myoblast proliferation and differentiation. Front Ccell Dev Biol 9:748844. https://doi.org/10.3389/fcell.2021.748844
44. Li H, Wei X, Yang J, Dong D, Hao D, Huang Y, Lan X, Plath M, Lei C, Ma Y, Lin F, Bai Y, Chen H (2018) circFGFR4 promotes differentiation of myoblasts via binding miR-107 to relieve its inhibition of Wnt3a. Mol Ther Nucleic Acids 11:272–283. https://doi.org/10.1016/j.omtn.2018.02.012
45. Das A, Shyamal S, Sinha T, Mishra SS, Panda AC (2021) Identification of potential circRNA-microRNA-mRNA regulatory network in skeletal muscle. Front Mol Biosci 8:762185. https://doi.org/10.3389/fmolb.2021.762185
46. Wang X, Cao X, Dong D, Shen X, Cheng J, Jiang R, Yang Z, Peng S, Huang Y, Lan X, Elnour IE, Lei C, Chen H (2019) Circular RNA TTN acts as a miR-432 sponge to facilitate proliferation and differentiation of myoblasts via the IGF2/PI3K/AKT signaling pathway. Mol Ther Nucleic Acids 18:966–980. https://doi.org/10.1016/j.omtn.2019.10.019
47. Zhao J, Zhao X, Shen X, Zhang Y, Zhang Y, Ye L, Li D, Zhu Q, Yin H (2022) CircCCDC91 regulates chicken skeletal muscle development by sponging miR-15 family via activating IGF1-PI3K/AKT signaling pathway. Poult Sci 101(5):101803. https://doi.org/10.1016/j.psj.2022.101803
48. Fan Y, Zhang Z, Deng K, Kang Z, Guo J, Zhang G, Zhang Y, Wang F (2023) CircUBE3A promotes myoblasts proliferation and differentiation by sponging miR-28-5p to enhance expression. Int J Biol Macromol 226:730–745. https://doi.org/10.1016/j.ijbiomac.2022.12.064
49. Yang H, Yue B, Yang S, Qi A, Yang Y, Tang J, Ren G, Jiang X, Lan X, Pan C, Chen H (2024) circUBE3C modulates myoblast development by binding to miR-191 and upregulating the expression of p27. J Cell Physiol 239(2):e31159. https://doi.org/10.1002/jcp.31159
50. Yang Z, Song C, Jiang R, Huang Y, Lan X, Lei C, Qi X, Zhang C, Huang B, Chen H (2022) CircNDST1 regulates bovine myoblasts proliferation and differentiation via the miR-411a/Smad4 Axis. J Agric Food Chem 70(32):10044–10057. https://doi.org/10.1021/acs.jafc.1c08167
51. Chen X, Ouyang H, Wang Z, Chen B, Nie Q (2018) A novel circular RNA generated by FGFR2 gene promotes myoblast proliferation and differentiation by sponging miR-133a-5p and miR-29b-1-5p. Cells 7(11). https://doi.org/10.3390/cells7110199
52. Li L, Chen Y, Nie L, Ding X, Zhang X, Zhao W, Xu X, Kyei B, Dai D, Zhan S, Guo J, Zhong T, Wang L (1862) Zhang H (2019) MyoD-induced circular RNA CDR1as promotes myogenic differentiation of skeletal muscle satellite cells. Biochim Biophys Acta Gene Regul Mech 8:807–821. https://doi.org/10.1016/j.bbagrm.2019.07.001
53. Glisovic T, Bachorik JL, Yong J, Dreyfuss G (2008) RNA-binding proteins and post-transcriptional gene regulation. FEBS Lett 582(14):1977–1986. https://doi.org/10.1016/j.febslet.2008.03.004
54. Lunde BM, Moore C, Varani G (2007) RNA-binding proteins: modular design for efficient function. Nat Rev Mol Cell Biol 8(6):479–490. https://doi.org/10.1038/nrm2178
55. McKee AE, Minet E, Stern C, Riahi S, Stiles CD, Silver PA (2005) A genome-wide in situ hybridization map of RNA-binding proteins reveals anatomically restricted expression in the developing mouse brain. BMC Dev Biol 5:14. https://doi.org/10.1186/1471-213x-5-14
56. Cook KB, Kazan H, Zuberi K, Morris Q, Hughes TR (2011) RBPDB: a database of RNA-binding specificities. Nucleic Acids Res 39(Database issue):D301–D308. https://doi.org/10.1093/nar/gkq1069
57. Galante PA, Sandhu D, de Sousa AR, Gradassi M, Slager N, Vogel C, de Souza SJ, Penalva LO (2009) A comprehensive in silico expression analysis of RNA binding proteins in normal and tumor tissue: identification of potential players in tumor formation. RNA Biol 6(4):426–433. https://doi.org/10.4161/rna.6.4.8841
58. Ashburner M, Ball CA, Blake JA, Botstein D, Butler H, Cherry JM, Davis AP, Dolinski K, Dwight SS, Eppig JT, Harris MA, Hill DP, Issel-Tarver L, Kasarskis A, Lewis S, Matese JC, Richardson JE, Ringwald M, Rubin GM, Sherlock G (2000) Gene ontology: tool for the unification of biology. The Gene Ontology Consortium. Nat Genet 25(1):25–29. https://doi.org/10.1038/75556
59. Zheng S, Zhang X, Odame E, Xu X, Chen Y, Ye J, Zhou H, Dai D, Kyei B, Zhan S, Cao J, Guo J, Zhong T, Wang L, Li L, Zhang H (2021) CircRNA-protein interactions in muscle development and diseases. Int J Mol Sci 22(6). https://doi.org/10.3390/ijms22063262
60. Ai N, Yu Z, Xu X, Liufu S, Wang K, Huang S, Li X, Liu X, Chen B, Ma H, Yin Y (2023) Circular intronic RNA circTTN inhibits host gene transcription and myogenesis by recruiting PURB proteins to form heterotypic complexes. Int J Mol Sci 24(12). https://doi.org/10.3390/ijms24129859
61. Yue B, Yang H, Wu J, Wang J, Ru W, Cheng J, Huang Y, Lan X, Lei C, Chen H (2022) circSVIL regulates bovine myoblast development by inhibiting STAT1 phosphorylation. Sci China Life Sci 65(2):376–386. https://doi.org/10.1007/s11427-020-1908-2

62. Chen M, Wei X, Song M, Jiang R, Huang K, Deng Y, Liu Q, Shi D, Li H (2021) Circular RNA circMYBPC1 promotes skeletal muscle differentiation by targeting MyHC. Mol Ther Nucleic Acids 24:352–368. https://doi.org/10.1016/j.omtn.2021.03.004
63. Yan J, Yang Y, Fan X, Tang Y, Tang Z (2021) Sp1-mediated circRNA circHipk2 regulates myogenesis by targeting ribosomal protein Rpl7. Genes 12(5). https://doi.org/10.3390/genes12050696
64. Wen SY, Qadir J, Yang BB (2022) Circular RNA translation: novel protein isoforms and clinical significance. Trends Mol Med 28(5):405–420. https://doi.org/10.1016/j.molmed.2022.03.003
65. Hwang HJ, Kim YK (2024) Molecular mechanisms of circular RNA translation. Exp Mol Med 56(6):1272–1280. https://doi.org/10.1038/s12276-024-01220-3
66. Chen CK, Cheng R, Demeter J, Chen J, Weingarten-Gabbay S, Jiang L, Snyder MP, Weissman JS, Segal E, Jackson PK, Chang HY (2021) Structured elements drive extensive circular RNA translation. Mol Cell 81(20):4300–4318.e4313. https://doi.org/10.1016/j.molcel.2021.07.042
67. Wang Y, Wang Z (2015) Efficient backsplicing produces translatable circular mRNAs. RNA (New York, NY) 21(2):172–179. https://doi.org/10.1261/rna.048272.114
68. Yin H, Shen X, Zhao J, Cao X, He H, Han S, Chen Y, Cui C, Wei Y, Wang Y, Li D, Zhu Q (2020) Circular RNA CircFAM188B encodes a protein that regulates proliferation and differentiation of chicken skeletal muscle satellite cells. Front Cell Dev Biol 8:522588. https://doi.org/10.3389/fcell.2020.522588
69. Duchenne (1867) The pathology of paralysis with muscular degeneration (Paralysie Myosclerotique), or paralysis with apparent hypertrophy. Br Med J 2(363):541–542. https://doi.org/10.1136/bmj.2.363.541
70. Falzarano MS, Scotton C, Passarelli C, Ferlini A (2015) Duchenne muscular dystrophy: from diagnosis to therapy. Molecules (Basel, Switzerland) 20(10):18168–18184. https://doi.org/10.3390/molecules201018168
71. Duan D, Goemans N, Takeda S, Mercuri E, Aartsma-Rus A (2021) Duchenne muscular dystrophy. Nat Rev Dis Prim 7(1):13. https://doi.org/10.1038/s41572-021-00248-3
72. Suzuki H, Aoki Y, Kameyama T, Saito T, Masuda S, Tanihata J, Nagata T, Mayeda A, Takeda S, Tsukahara T (2016) Endogenous multiple exon skipping and back-splicing at the DMD mutation hotspot. Int J Mol Sci 17(10). https://doi.org/10.3390/ijms17101722
73. Cazzella V, Martone J, Pinnarò C, Santini T, Twayana SS, Sthandier O, D'Amico A, Ricotti V, Bertini E, Muntoni F, Bozzoni I (2012) Exon 45 skipping through U1-snRNA antisense molecules recovers the Dys-nNOS pathway and muscle differentiation in human DMD myoblasts. Mol Ther 20(11):2134–2142. https://doi.org/10.1038/mt.2012.178
74. Song Z, Liu Y, Fang X, Xie M, Ma Z, Zhong Z, Feng X, Zhang W (2020) Comprehensive analysis of the expression profile of circRNAs and their predicted protein-coding ability in the muscle of mdx mice. Funct Integr Genomics 20(3):397–407. https://doi.org/10.1007/s10142-019-00724-w
75. Sartori R, Romanello V, Sandri M (2021) Mechanisms of muscle atrophy and hypertrophy: implications in health and disease. Nat Commun 12(1):330. https://doi.org/10.1038/s41467-020-20123-1
76. Liu Q, Deng J, Qiu Y, Gao J, Li J, Guan L, Lee H, Zhou Q, Xiao J (2021) Non-coding RNA basis of muscle atrophy. Mol Ther Nucleic Acids 26:1066–1078. https://doi.org/10.1016/j.omtn.2021.10.010
77. Schiaffino S, Dyar KA, Ciciliot S, Blaauw B, Sandri M (2013) Mechanisms regulating skeletal muscle growth and atrophy. FEBS J 280(17):4294–4314. https://doi.org/10.1111/febs.12253
78. Yin L, Li N, Jia W, Wang N, Liang M, Yang X, Du G (2021) Skeletal muscle atrophy: from mechanisms to treatments. Pharmacol Res 172:105807. https://doi.org/10.1016/j.phrs.2021.105807
79. Zhang H, Qi G, Wang K, Yang J, Shen Y, Yang X, Chen X, Yao X, Gu X, Qi L, Zhou C, Sun H (2023) Oxidative stress: roles in skeletal muscle atrophy. Biochem Pharmacol 214:115664. https://doi.org/10.1016/j.bcp.2023.115664
80. Yoshida T, Delafontaine P (2020) Mechanisms of IGF-1-mediated regulation of skeletal muscle hypertrophy and atrophy. Cells 9(9). https://doi.org/10.3390/cells9091970
81. Chen X, Ji Y, Liu R, Zhu X, Wang K, Yang X, Liu B, Gao Z, Huang Y, Shen Y, Liu H, Sun H (2023) Mitochondrial dysfunction: roles in skeletal muscle atrophy. J Transl Med 21(1):503. https://doi.org/10.1186/s12967-023-04369-z
82. Chen R, Yang T, Jin B, Xu W, Yan Y, Wood N, Lehmann HI, Wang S, Zhu X, Yuan W, Chen H, Liu Z, Li G, Bowen TS, Li J, Xiao J (2023) CircTmeff1 promotes muscle atrophy by interacting with TDP-43 and encoding a novel TMEFF1-339aa protein. Adv Sci (Weinh) 10(17):e2206732. https://doi.org/10.1002/advs.202206732
83. Reinoso-Sánchez JF, Baroli G, Duranti G, Scaricamazza S, Sabatini S, Valle C, Morlando M, Casero RA Jr, Bozzoni I, Mariottini P, Ceci R, Cervelli M (2020) Emerging role for linear and circular spermine oxidase RNAs in skeletal muscle physiopathology. Int J Mol Sci 21(21). https://doi.org/10.3390/ijms21218227
84. Soltanzadeh P (2022) Myotonic dystrophies: a genetic overview. Genes 13(2). https://doi.org/10.3390/genes13020367
85. Thornton CA (2014) Myotonic dystrophy. Neurol Clin 32(3):705–719, viii. https://doi.org/10.1016/j.ncl.2014.04.011
86. Hartman J, Patki T, Johnson NE (2024) Diagnosis and management of myotonic dystrophy type 1.

JAMA 331(14):1227–1228. https://doi.org/10.1001/jama.2024.2511
87. Finsterer J (2002) Myotonic dystrophy type 2. Eur J Neurol 9(5):441–447. https://doi.org/10.1046/j.1468-1331.2002.00453.x
88. Koscianska E, Kozlowska E, Fiszer A (2021) Regulatory potential of competing endogenous RNAs in myotonic dystrophies. Int J Mol Sci 22(11). https://doi.org/10.3390/ijms22116089
89. Voellenkle C, Perfetti A, Carrara M, Fuschi P, Renna LV, Longo M, Sain SB, Cardani R, Valaperta R, Silvestri G, Legnini I, Bozzoni I, Furling D, Gaetano C, Falcone G, Meola G, Martelli F (2019) Dysregulation of circular RNAs in myotonic dystrophy type 1. Int J Mol Sci 20(8). https://doi.org/10.3390/ijms20081938
90. Sieb JP (2014) Myasthenia gravis: an update for the clinician. Clin Exp Immunol 175(3):408–418. https://doi.org/10.1111/cei.12217
91. Gilhus NE, Verschuuren JJ (2015) Myasthenia gravis: subgroup classification and therapeutic strategies. Lancet Neurol 14(10):1023–1036. https://doi.org/10.1016/s1474-4422(15)00145-3
92. Kong X, Wu T, Cai H, Chen Z, Wang Y, He P, Liu P, Li L, Peng S, Xu F, Wang J, Zhang H, Wang L (2024) Construction of ceRNA network mediated by circRNAs screening from microarray and identification of novel biomarkers for myasthenia gravis. Gene 918:148463. https://doi.org/10.1016/j.gene.2024.148463
93. Lv J, Ren L, Han S, Zhang J, Zhao X, Zhang Y, Fang H, Zhang L, Yang H, Wang S, Yang J, Cui X, Zhang Q, Zhang Y, Gao F (2021) Peripheral blood hsa-circRNA5333-4: a novel biomarker for myasthenia gravis. Clin Immunol (Orlando, Fla) 224:108676. https://doi.org/10.1016/j.clim.2021.108676
94. Ye Q, Gu C, Yan W (2024) Circ_0076490 silencing inhibits MAPK1 expression to decrease the proliferation and increase apoptosis of Jurkat cells by regulating miR-144-3p in myasthenia gravis. Neurol Res:1–11. https://doi.org/10.1080/01616412.2024.2394324
95. Lai X, Bi Z, Yang X, Hu R, Wang L, Jin M, Li L, Bu B (2021) Upregulation of circ-FBL promotes myogenic proliferation in myasthenia gravis by regulation of miR-133/PAX7. Cell Biol Int 45(11):2287–2293. https://doi.org/10.1002/cbin.11676
96. Argilés JM, Busquets S, Stemmler B, López-Soriano FJ (2014) Cancer cachexia: understanding the molecular basis. Nat Rev Cancer 14(11):754–762. https://doi.org/10.1038/nrc3829
97. Ferrer M, Anthony TG, Ayres JS, Biffi G, Brown JC, Caan BJ, Cespedes Feliciano EM, Coll AP, Dunne RF, Goncalves MD, Grethlein J, Heymsfield SB, Hui S, Jamal-Hanjani M, Lam JM, Lewis DY, McCandlish D, Mustian KM, O'Rahilly S, Perrimon N, White EP, Janowitz T (2023) Cachexia: a systemic consequence of progressive, unresolved disease. Cell 186(9):1824–1845. https://doi.org/10.1016/j.cell.2023.03.028
98. Sadeghi M, Keshavarz-Fathi M, Baracos V, Arends J, Mahmoudi M, Rezaei N (2018) Cancer cachexia: diagnosis, assessment, and treatment. Crit Rev Oncol Hematol 127:91–104. https://doi.org/10.1016/j.critrevonc.2018.05.006
99. Shi X, Yang J, Liu M, Zhang Y, Zhou Z, Luo W, Fung KM, Xu C, Bronze MS, Houchen CW, Li M (2022) Circular RNA ANAPC7 inhibits tumor growth and muscle wasting via PHLPP2-AKT-TGF-β signaling axis in pancreatic cancer. Gastroenterology 162(7):2004–2017.e2002. https://doi.org/10.1053/j.gastro.2022.02.017
100. Ding Z, Sun D, Han J, Shen L, Yang F, Sah S, Sui X, Wu G (2021) Novel noncoding RNA CircPTK2 regulates lipolysis and adipogenesis in cachexia. Mol Metab 53:101310. https://doi.org/10.1016/j.molmet.2021.101310
101. Yang Y, Fan X, Mao M, Song X, Wu P, Zhang Y, Jin Y, Yang Y, Chen LL, Wang Y, Wong CC, Xiao X, Wang Z (2017) Extensive translation of circular RNAs driven by N(6)-methyladenosine. Cell Res 27(5):626–641. https://doi.org/10.1038/cr.2017.31

Part VII
Circular RNAs in Archaea

26 Circular RNAs in Archaea

Hubert F. Becker
and Sébastien Ferreira-Cerca

Abstract

Circular RNA molecules were first described more than 40 years ago. However, it is only recently that the broadness of their phylogenetic distribution has been revealed. Since their discoveries, numerous studies have characterized the molecular mechanisms and function underlying these peculiar molecules, mostly in diverse eukaryotic species. In contrast, studies focusing on the biology of circular RNAs in archaea remain relatively scarce.

In this chapter, we provide an overview of the discovery of circular RNAs in archaea and summarize our knowledge of their biology, with an emphasis on circular preribosomal RNAs and Box C/D RNAs.

Keywords

Archaea · Circular RNA · rRNA · snoRNA · RNA ligase · Box C/D RNA

H. F. Becker (✉)
Laboratoire d'Optique et Biosciences (CNRS UMR7645, INSERM U1182), École Polytechnique, Institut polytechnique de Paris, Palaiseau, France

Faculté des Sciences et Ingénierie, Sorbonne Université, Paris, France
e-mail: hubert.becker@polytechnique.edu

S. Ferreira-Cerca (✉)
Laboratoire de Biologie Structurale de la Cellule (BIOC—CNRS UMR 7654), École Polytechnique, Institut Polytechnique de Paris, Palaiseau, France
e-mail: sebastien.ferreira-cerca@polytechnique.edu

1 Introduction

Circular RNAs (circRNAs) were first discovered, more than 40 years ago, in a few viroids or viruses using a single-stranded closed RNA molecule as genetic material ([1, 2] for a recent review). Since the 1990s, other circRNAs, produced from noncanonical splicing, were described in cytoplasmic fractions of eukaryotic cells with a function as microRNA or protein inhibitors (miRNA sponges) for these scrambled exons [3]. In the beginning of the 2010s, the discovery of circRNAs has seen an explosion with the advancement of RNA-sequencing technologies (RNA-seq) and the development of circRNA-specific bioinformatics algorithms and enrichment methodologies. Thousands of new circRNAs were identified across all domains of life, with an emphasis on eukaryotic species, with different expression levels from low levels for most of the circRNAs to highly abundant for a few others [4, 5]. CircRNA expressions were also found to have tissue and developmental stage-specific expression patterns and have been shown to be involved in a variety of human diseases and plant stress responses. CircRNAs were also described as localized to the cytoplasm, as remarkably stable RNA; however, most natural

J. Xiao (ed.), *Advances in Circular RNAs*, Advances in Experimental Medicine and Biology 1485,
https://doi.org/10.1007/978-981-96-9428-0_26

circRNAs do not appear to be translated contrary to engineered circRNAs [6].

In bacteria, circRNAs have been less studied, and their presence and biological functionality are unclear. However, recently, a regulatory noncoding RNA (DucS) was revealed in both linear and circular conformation, suggesting that other types of circular RNAs probably exist in diverse bacteria [7].

In contrast to the wealth of information obtained in diverse eukaryotic species, our knowledge of circ-RNAs in archaea remains relatively modest.

At the end of the 1970s, Carl Woese and collaborators described a new group of organisms, now collectively known as archaea [8–11]. Based on their unique biological properties, archaea were defined as a new domain of life next to bacteria and eukarya [8–11]. Many representatives of the archaeal domains were initially isolated in extreme environmental niches, even though it is now clear that not all archaea are extremophiles, and that they can colonize diverse types of environments, including the human gut or skin [12–14]. It is also now clear that archaea play critical roles in diverse biogeochemical cycles and that our understanding of their biology may contribute to biotechnological development [12, 15, 16]. Moreover, in the past years, several lines of evidence suggest a common origin of archaea and eukaryotes [17, 18]. Accordingly, studying archaea also provides critical insight into the origin of eukaryotes and their biology.

Even though the molecular biology of archaea is still poorly characterized, it is interesting to note that a few years before the rise of high-throughput sequencing approaches, circular RNAs belonging to several different functional RNA groups had already been identified in several archaea. For example, the presence of circular RNA molecules was revealed within the excised tRNA-introns in the mesophilic halophile *Haloferax volcanii* [19, 20]. Once cleaved from its pre-tRNA-Trp and circularized, the resulting Box C/D RNA contained in the circular intron can guide the methylation of nucleotides of the mature tRNA-Trp [21]. In addition, several large circular introns derived from the 23S and 16S rRNAs were described in hyperthermophilic archaea [22, 23] and proposed as rRNAs processing intermediates [24] (see below for details). Moreover, the unexpected discovery of significant amounts of circular Box C/D RNAs, in addition to linear forms, in native complexes with proteins required for Box C/D function was very intriguing [25]. In this case, the circular form of *Pyrococcus furiosus* Box C/D RNAs could function as a guide for 2′-O-methylation on specific nucleotides in rRNAs and tRNAs, by forming an RNP complex with three specific proteins (L7ae, Nop5, and fibrillarin) [25, 26]. Still in the 2010s, thanks to the development of high-throughput RNA-sequencing technology, the list of circular RNAs was highly completed in the "third" domain of life. Interestingly, the results on circular transcriptomes for three archaea (*Pyrococcus abyssi*, *Saccharalobus solfataricus*, and *Nanoarchaeum equitans*) converged in terms of the functional categorization of circRNAs. Circular RNAs were mainly described for different classes of noncoding RNA, but also in coding RNAs and nonannotated RNAs. These findings will be discussed in the following sections of this review.

CircRNAs differ structurally from other types of RNAs as their 3′ and 5′ ends are not free but covalently joined to form a circular structure. The lack of free ends confers to circular RNAs resistance toward exonucleases and therefore enhances their stability, but it remains unclear whether their circularization affects or changes their function. The biogenesis of circRNAs seems to use different mechanisms between eukaryotic and archaeal cells. In eukaryotes, circRNAs are mostly formed by back-splicing events, wherein a downstream 5′ donor splice site is joined to an upstream acceptor splice site. Circularization and splicing processes seem to compete against each other, leading to the coexistence of linear and circular RNA forms resulting from the splicing of the same RNA precursor. This suggests that a possible function of some protein-coding genes may be to generate circular RNAs rather than linear mRNAs or proteins. The efficiency of back-splicing reactions is often facilitated by intronic repeat sequences or RNA-binding proteins, which bring the 3′ and 5′ splicing sites into proximity for the circularization reaction [3].

In archaea, circRNA formation was described in the cases of rRNAs maturation, tRNA intron excision, and Box C/D RNAs formation. Both ribosomal and tRNA maturation depend on the tRNA splicing machinery and the presence of a bulge-helix-bulge motif. For Box C/D RNAs, the circularization events result from an enzymatic ligation step of the splicing reaction involving a new family of ATP-dependent RNA ligase (Rnl3). Members of the Rnl3 family are found in several archaea and present an atypical homodimeric quaternary structure and circularize single-stranded RNA by joining RNA extremities via three nucleotidyltransferase steps similar to those described for ATP-dependent DNA ligases [27] (see below for details). Overall, these examples suggest that the circularization reactions could involve intron-dependent (bhb-motif-dependent) ligation step of the splicing reaction (rRNAs processing, tRNA intron excision), but also intron-independent (bhb-motif-independent) circularization of RNAs (most Box C/D RNAs). Noteworthy, this raises the question of how motif-independent circularization occurs on a mature RNA with already processed mature extremities.

Although the circular form of RNAs makes their detection and quantification challenging, recent advances in circRNA-specific computational tools have helped to develop novel approaches to identify circRNAs. Over the past decade, the vast majority of circRNA studies have been conducted in human and mammalian cells, due to their high involvement in neurodegenerative disorders, cardiovascular diseases, and cancers [28, 29]. All eukaryotic circRNA predictors are based on finding back-splicing junctions in RNA-seq data since the formation of these back-spliced junctions actually requires the classical spliceosome machinery and splicing signals. These methods are all based on the identification of back-spliced junctions, with reads that do not align continuously, but rather discontinuously, and in the different orientations with respect to the reference genome. These results, in so-called inverted reads, are detected computationally. Many algorithms exist for circRNA identification, including `CIRI`, `Segemehl`, `Mapsplice`, `CircSeq`, and `find_circ` [30, 31]. Identified circRNAs were categorized into three broad classes including exonic circRNAs (EcircRNAs), intronic circRNAs (ciRNAs), and exonic–intronic circRNAs (EIciRNAs).

In archaea, RNA splicing is a rare phenomenon and is mostly observed in pre-tRNA and pre-rRNA with a uniform splicing mechanism [32]. Introns are spliced by an apparently archaeal-specific mechanism in which the tRNA splicing endoribonuclease (endA) cuts a "bulge-helix-bulge" (bhb) motif that forms at exon–intron junctions, before recruiting RNA ligases [33]. The different RNA splicing scenario in archaea required the development of an adapted computational program for the identification of circRNAs, and in particular the circularization of mature RNA extremities. To date, the successful identification of a plethora of circular RNAs has been performed in several archaeal organisms by a combination of biochemical, RNA-sequencing, and computational approaches. In this review, we will present and compare archaeal circRNA transcriptomes obtained for different species, taking care to develop the junction signature-based algorithm for high-throughput detection of circRNAs, the diversity of circRNAs molecules identified, and their biogenesis and function in archaea.

2 Methodology for Circular RNA Identification by RNA-seq Deep Sequencing

The first archaeal genome-wide discovery of circular transcripts was described for the archaeon *Saccharalobus solfataricus* (previously *Sulfolobus solfataricus*) (acidophile and hyperthermophile archaeon) by sequencing the total RNA fraction by the Illumina approach [34]. RNA circularization junctions were identified for cases where reads align to the reference genome in a permuted order, i.e., with a downstream part of the sequence aligning with an upstream position in the genome. To attain higher-confidence data on circRNAs, results' analysis has been optimized by suppression of redundancy, depletion of template switching, and addition of an exo-ribonucleolytic enzyme treatment with the ribonuclease RNase R. This exonuclease degrades linear RNA molecules from the 3′-end, but leaves circRNAs intact, leading to an enrichment in the number of reads supporting circularization junctions. This RNA-seq approach yielded 26.94 million reads mapped to the genome, with at least 33 bp indicating a high quality of sequencing data. After the application of all computational criteria, the set of circRNAs identified in *S. solfataricus* contained 37 genes, divided into several functional groups [34].

For the hyperthermophilic archaeon *Pyrococcus abyssi*, a combination of RNA immunoprecipitation assays and RNA-seq following the Ion Torrent PGM protocol was used [35]. Each sequencing run generated ~400,000 reads with a read size of 80 to 90 base pairs, and all the reads were mapped to the *P. abyssi* reference genome with the unique criterion of reads splitting for the inversion of two matches within the same locus. To enrich for circRNAs, we used RNase R, which specifically degrades linear RNA molecules in a 3′–5′ direction. After analysis of the resulting 285,000 reads, 11–15% of them were classified as circular, and this pipeline finally led to the identification of 133 individual circRNA loci in *P. abyssi*.

RNA deep sequencing was also performed in the hyperthermophilic archaeon *Nanoarchaeum equitans*, a small-sized archaeon positioned in the phylum termed "Nanoarchaeota." This organism has a minimal, compact genome of only 490 kilobases and an extremely high gene density with very little noncoding DNA. The analysis of the abundant small RNA population revealed two distinct circular junctions in the Box C/D RNA family [36]. Other Illumina Hiseq2000 RNA-seq studies on *Thermoproteus* species, with "missing" canonical SRP (signal recognition particles) RNA genes, revealed that *Thermoproteus* species contain permuted SRP RNA sequences, indicating that mature SRP RNAs exist as circular molecules [37].

Advances in high-throughput sequencing and associated bioinformatics tools will now facilitate the discovery of circRNAs for other archaea to validate the extent of circRNAs in the "third" domain of life. As it has been reported for RNA-seq studies in eukaryotic cells, it is also essential to apply, in archaea, multiple and restrictive constraints to allow attaining highly selective circRNAs' identification. In addition to the developed computational criteria, we can highlight methodological improvements that have highly simplified circRNAs identification in *P. abyssi*, such as the circRNAs enrichment in the RNA sample by exonuclease treatment and the intersection of the circRNAs candidates obtained from independent and complementary experiments [35].

3 Functional Categorization of Archaeal CircRNA Transcriptome

Interestingly, the results on circular transcriptomes for *P. abyssi* and *S. solfataricus* converged in terms of functional categorization of circRNAs, with a vast majority of noncoding RNA (~80%) (see Fig. 26.1). Three noncoding RNA classes are frequently circular in all these archaeal species (Box C/D RNAs, tRNA-intron, and precursor rRNAs), but a few circRNAs were also identified in protein-coding RNAs and nonfunctionally annotated RNAs. Most circRNAs have been identified in these studies, since their sizes were compatible with the size of the sequencing reads generated by RNA fragmentation during the RNA library preparation (~200 nucleotides). For longer RNA genes (rRNA and coding RNAs), circRNAs only represent a part of the whole transcript, which may result from technical identification problems due to the more complex structure of "long" RNAs, or simply because circRNAs contain only part of the initial RNA sequence.

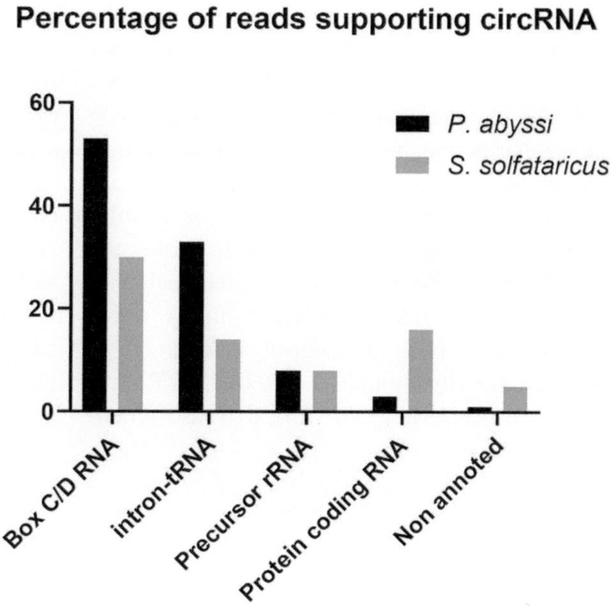

Fig. 26.1 Functional repartition of circRNAs in archaea. Functional RNA classes identified in a circular form, based on RNA-seq data obtained from two independent studies on *P. abyssi* and *S. solfataricus*. The percentage of reads supporting circular RNA is given for each RNA functional class

3.1 Circular Box C/D RNA

Results of RNA-seq for two thermophile organisms, *P. abyssi* and *S. solfataricus,* show that the most common circRNA corresponds to the Box C/D RNAs, a guide RNA for rRNA methylation [34, 35, 38, 39]. These results were confirmed when enrichment of circRNAs was done by RNase R pretreatment, to degrade linear RNA and overcome possible artifacts. For instance, 17 and 11 circular Box C/D RNA loci were identified for *P. abyssi* and *S. solfataricus*, respectively [34, 35]. But, with a relaxation of selection criteria (taking into account only the inversion of reads for the genome mapping step), the majority of the Box C/D RNA sequences have been identified as circular: 38 out of 56 in *P. abyssi* cells and 11/13 in *S. solfataricus* cells [34, 35]. Based on these data, RNA circularization in thermophiles could provide increased thermostability by limiting thermal denaturation of the stem structures formed between the 5′ and 3′ termini of Box C/D RNAs. This idea has also been proposed for the hyperthermophilic archaeon *Pyrococcus furiosus* for which it has been observed that most of the Box C/D RNAs can exist in circular forms in cells [25]. However, the proportion between circular and linear forms seems to vary between different Box C/D RNAs, as well as between hyperthermophilic archaea [25, 34, 35]. This could raise the idea that circular forms represent processing intermediates of Box C/D RNAs maturation by a processing mechanism yet uncharacterized, as it was previously demonstrated for tRNA and rRNA processing (see section below).

To date, for the second family of guide RNA, box H/ACA RNA, implicated in pseudouridinylation of RNA [38, 40], the circular form was only identified for one of them in *S. solfataricus* [34], while only linear forms have been reported in *P. abyssi* [35]. This result does not confirm the hypothesis, issued for Box C/D RNA, that RNA circularization could also provide thermostability of RNA molecules in hyperthermophiles organisms, by limiting thermal denaturation of stem structures formed between the 5′ and 3′ extremities of H/ACA RNA.

3.2 Circular RNA Generated During tRNA Maturation

The excised intron from tRNA-Trp (65 to 72 nucleotides) was also detected as an abundant and stable circRNA found in *S. solfataricus, H. volcanii, and P. abyssi* cells [19, 21, 34, 35].

Intron from tRNA-Trp does not form a new circRNA category since the intron sequence also encodes for a Box C/D RNA [19, 21]. It seems that the circularization occurs simultaneously during the splicing process, and linear intermediates are not expected to occur. This has been suggested by the absence of enrichment in circular reads after exoribonuclease treatment during the library preparation [35] and in agreement with previous results that described a highly stable intron-tRNA-Trp in *Haloferax volcanii* [19]. Among the five circular tRNA (tRNATrp, tRNALys, tRNAMet, tRNAPro, tRNASer) identified in *S. solfataricus* cells [34], tRNA-Trp is the only example overlapping with *P. abyssi* results [35]. Both pre-tRNA-Trp contain a "bhb" motif, which will allow the circularization of the cleaved-out intron during the archaeal splicing mechanisms. Furthermore, the cleaved intron sequence carries a Box C/D motif that guides the 2′-O-methylation on nucleotides in the anticodon loop of mature tRNA-Trp [21], classifying the archaeal intron of tRNA-Trp in the Box C/D RNA family.

3.3 Circular Precursor Ribosomal RNA Precursors (Circ-pre-rRNA)

3.3.1 Discovery of Circular Pre-rRNAs in Archaea

At the dawn of the RNomics era, an initial study of the Hüttenhofer laboratory and his collaborators suggested, for the first time, the possible presence of circular-pre-rRNA in two phylogenetically distant archaea representative of two different phyla [24]. Even though the method used in this study could not directly address the existence of circular-pre-rRNA in archaea, it demonstrated the presence of chimeric byproducts resulting from such a specific maturation step; hence, the indirect presence of this type of maturation intermediate has set the ground to further study this intriguing peculiarity of rRNA maturation in archaea.

However, it is only 10 years later, with the advance of next-generation sequencing technologies and the avalanche of RNAse R-treated transcriptomics, that the group of Rotem Sorek formally confirmed the existence of circular-pre-rRNA in archaea [34]. This second landmark study took advantage of the selective enrichment of circular RNA upon treatment with the exonuclease RNase R to demonstrate the presence of circular pre-rRNA in phylogenetically distant archaea [34].

Accordingly, these two pioneering studies suggested the presence of circular pre-rRNA in four different organisms, *Archaeoglobus fulgidus*, *S. solfataricus*, *Sulfolobus acidocaldarius*, and *Halobacterium* sp. NRC1, respectively. Since then, the presence of circular pre-rRNA has been confirmed in at least four additional archaea, namely, *H. volcanii*, *P. furiosus*, *P. abyssi*, *Marinobacter psychrophilus,* and *Methanococcus maripaludis* [35, 41–45].

Together, the broad phylogenetic repartition of the presence of the circular-pre-rRNA intermediates suggests that this unconventional maturation step unique to archaea is likely a widespread phenomenon across this domain of life.

3.3.2 Formation and Maturation of Circular-pre-rRNA

Early on it has been suggested that circular pre-16S rRNA and pre-23S rRNA synthesis depends on the tRNA splicing machinery in archaea, thereby functionally linking tRNA and rRNA maturation in archaea [24].

In archaea, the small and large ribosomal subunits (SSU and LSU, respectively) are composed of several r-proteins that assemble with three rRNAs, the SSU 16S rRNA, and the LSU 23S and 5S rRNA. Most of these rRNA are contained in operons and transcribed as a primary precursor containing the future mature rRNA species flanked by external and internal spacer sequences that will be trimmed during the rRNA maturation process [44, 46]. In archaea, the spacers flanking the 16S and 23S rRNA can form an antiparallel region, building a double-stranded RNA sequence, known as the processing stem. Both 16S and 23S processing stems are characterized by the presence of bulge-helix-bulge motifs, a

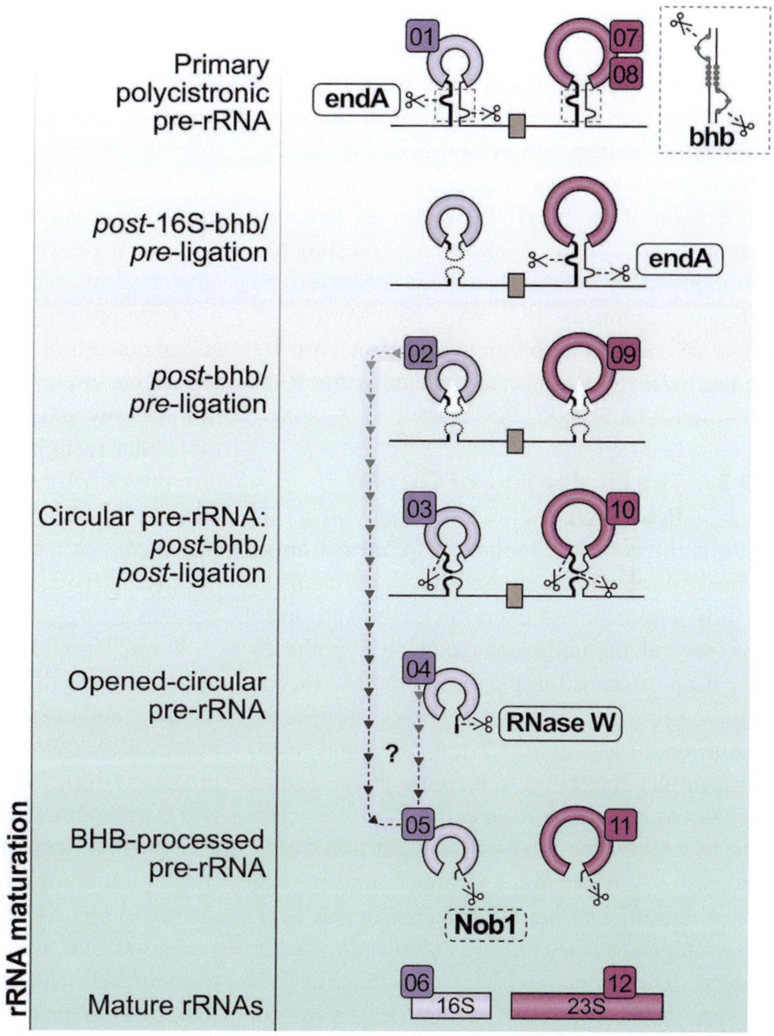

Fig. 26.2 Ribosomal RNA processing of an archetype archaeon, *H. volcanii.* Ribosomal RNA processing pathway and known RNase activities (indicated by scissors) in *H. volcanii* are depicted (based on [44]). Ribonucleases for which in vivo and in vitro are available are boxed, and ribonucleases for which only in vitro data are available are indicated by a dashed box. The canonical bulge-helix-bulge motif with three nucleotide bulges separated by a 4 bp helix, substate of the tRNA splicing endoribonuclease, endA, is depicted. Note the presence of circ-pre-16S rRNA and circ-pre-23S rRNA intermediates (intermediates 3 and 10, respectively). These circular pre-rRNA intermediates were observed in all archaeal organisms representative of different phyla analyzed thus far

motif commonly found in intron-containing precursor tRNA (see Fig. 26.2) [24, 41, 46, 47].

It is now known that the bhb motifs are required for the formation of circular-pre-16S rRNA and circular-pre-23S rRNA and the formation of mature 16S and 23S rRNA in archaea [41]. Moreover, in vitro and in vivo studies in archaea have confirmed the implication of the endonuclease *endA* for the formation circular-pre-rRNA [43, 45, 47]. In analogy to tRNA splicing downstream of the bhb-motif-dependent maturation step, the generated 5′/3′ ends of the pre-rRNA intermediates are ligated, presumably by the tRNA splicing ligase. However, there are so far no in vitro/in vivo evidence is yet supporting this model [24, 41, 44, 46].

The generated circular pre-rRNAs are then linearized. In most cases analyzed thus far, linearization generates the mature 5′ ends of the mature rRNA (see Fig. 26.2) [44, 45]. This processing order is in good agreement with the 5′ end maturation first rule observed in bacteria and eukarya [48–50]. In contrast to what seemed to be a shared order of maturation, there are some exceptions known in archaea. For instance, and in contrast to the order described above, it has been suggested that linearization of the circular pre-16S rRNA in *M. psychrophilus* generates the 16S mature 3′ end first [43]. Furthermore, an alternative linearization event has been observed in *P. furiosus*, where the circular-pre-rRNA is opened by excision of an rRNA helix from the mature 23S rRNA thereby generating a circularly permutated mature 23S rRNA [42, 44]. So far there is no evidence for bhb-motif-dependent circularization of pre-5S rRNA. However, circularization of 5S rRNA (presumably bhb-motif-independent) has been observed in different organisms [34, 35]. Noteworthy, *P. abyssi* 5S rRNA can be circularized in vitro by the pab1020 RNA ligase, a member of the Rnl3 RNA ligase family [35] (see below).

3.3.3 The Elusive Role of Circular Pre-rRNA

What is the role of circular rRNA formation in archaea? A clear and definitive answer remains to be fully demonstrated experimentally, and up to now, several mutually nonexclusive hypotheses have been discussed in the literature [24, 41, 51, 52].

The synthesis of circular-pre-rRNA has been, for example, suggested to increase protection of the nascent pre-rRNA from cellular RNase activities, as well as stabilization against extreme conditions (temperature, intracellular salt concentration) that are encountered across various ecological niches that are colonized by many archaea. The connection with extremophily and circ-pre-rRNA formation in archaea does not seem to be a major determinant for the synthesis of circular pre-rRNA. For example, mesophilic archaea, like the methanogens described above, do form circular-pre-rRNA [43]. The presence of circular pre-rRNA in thermophilic bacteria that may overcome similar thermal denaturation encountered by thermophilic archaea remains to be explored. Finally, whereas increased exposition to ribonuclease activities in archaea, which might expose the nascent pre-rRNA, is more deleterious in comparison to what we know in bacteria, and contributed to the selection of this type of maturation is yet unknown.

Another possibility to explain the "widespread" distribution of circular pre-rRNA in archaea might reflect the evolutionary constraints to which the common archaeal ancestor has been exposed. In other words, circular pre-rRNA synthesis could be an inherited molecular relic that was essentially advantageous in the archaeal common ancestor (e.g., thermostability protection against ribonucleases). Today, this feature may not necessarily provide a growth advantage in all or any modern archaea. However, the loss of circular pre-rRNA synthesis may now be disadvantageous within a maturation pathway that has been optimized around this ancient and beneficial constraint. In this regard, archaea that may not utilize circular-pre-rRNA-dependent maturation (if any exist) may (1) shed light on the molecular reengineering and conditions enabling circ-pre-rRNA independent maturation pathway and (2) provide insights into the loss of such intermediates, a necessary step for the emergence of the eukaryotic lineage form within the archaeal domain as suggested recently [14].

In any case, an emerging picture of commonalities during ribosome biogenesis in bacteria, archaea, and eukarya is that the 5′/3′ ends of the mature rRNA are structurally positioned in close proximity [46, 51, 52]. As such, circularization and pseudo-circularization emerge as a necessary intermediate step for the formation of mature, translation-competent, ribosomal subunits (discussed in [46, 51, 52]). Interestingly, these circularization and pseudo-circularization, however, are achieved with the help of orthologous molecular machineries, suggesting evolutionary constraints that have imposed a certain degree of reengineering while maintaining analogous molecular mechanisms mimicry.

3.3.4 Outstanding Remaining Questions Related to Circular Pre-rRNA

Many open questions that might improve our understanding of the role of circular pre-rRNA in archaea remain. Among those, it will be crucial to understand how widespread pre-circular-rRNA in archaea are, and whether some archaea utilize circ-pre-rRNA-independent maturation pathways, and how much those pathways diverge from the recently described "canonical" archaeal pre-rRNA maturation pathway shared by at least three phylogenetically divergent archaea [44]. In addition, structural characterization of these preribosomal intermediates across divergent archaea will also provide a critical detailed understanding of the molecular features of these archaea-specific preribosomal intermediates and how they correlate structurally and functionally to known maturation steps of bacterial or eukaryotic ribosome biogenesis.

These approaches might contribute to better deducing the role and evolutionary constraint at the origin of circular-pre-rRNA synthesis and its maintenance across phylogenetically diverse organisms.

3.4 Other Circular RNA in Archaea

Several additional circRNAs were identified in *P. abyssi* and *S. solfataricus* within protein-coding genes or nonannotated, but the number of circular reads was generally lower than linear reads for the same locus, complicating thus their analyses. It is not clear if these circRNAs represent intron excision products or functional fragments of RNA transcripts distributed all across the corresponding gene. Nevertheless, we cannot exclude an undescribed regulatory function for circRNAs detected within protein-coding genes or the possibility that these "coding" circRNAs represent degradation products of RNA, based on their variability, small size, and supposed random distribution between ORFs.

4 CircRNAs Are Connected to the Archaeal Exosome

The archaeal exosome is a multiprotein complex devoted to the phosphorolytic exoribonucleolytic degradation of single-stranded RNA in 3′ to 5′ direction. This complex is homologous to the trimeric bacterial polynucleotide phosphorylase (PNPase) and the eukaryotic exosome. Crystallographic analyses revealed for each complex a ring-shape formed by a homotrimeric complex, with the hexameric RNase PH-domains surrounding a central channel that can accommodate a single-stranded RNA molecule [53, 54].

In hyperthermophilic and methanogenic archaea, the protein complex is composed of two proteins (aRrp41 and aRrp42) homologous to RNase PH and one protein containing a KH/S1 domain characterized in RNA-binding proteins [55]. Three copies of the three proteins form a nine-subunit complex termed the archaeal exosome that, as with PNPase, is highly implicated in RNA turnover and degradation of RNA.

A recent iCLIP analysis, consisting of UV crosslinking and immunoprecipitation using antibodies directed against aRrp41 or Rrp4 proteins, was used to detect RNAs bound to the archaeal exosome in *S. solfataricus* [56]. Results showed that the exosome complex interacts with mRNAs, tRNA-derived introns, noncoding RNAs, antisense RNA, posttranscriptionally added adenine-rich RNA tails, and circRNAs. For this last example, analysis of iCLIP sequencing data presented many cDNA reads that could not be mapped to the genome since they contain permuted sequences, corresponding to the main feature of circRNA detection. Numerous circRNA candidates were identified, with only very low read numbers, and correspond to 16S/23S processing intermediates and putative new circRNAs containing 5′ regions of protein-coding genes [56]. After confirmation of the interaction between these circRNAS with the exosome, it was suggested that the exosome could be involved in degradation of circRNA intermediates occurring during the maturation of pre-rRNA and during the degradation of coding mRNAs. More precisely, circRNAs could be converted in a linear form by an unknown endoribonuclease, making them accessible for exoribonucleolytic degradation.

5 Many Archaea Species Encode Rnl3 RNA Ligase with Circularization Activity

ATP-dependent RNA ligases are extensively identified in the three domains of life (bacteria, archaea, and eukarya) and classified into six different families (Rnl1-Rnl6) based on their structure and substrate specificity [57]. Rnl3 family members, identified from the Pfam domain database, include ATP-dependent RNA ligases that join together free RNA ends from a single RNA molecule, for generating a circular RNA molecule (circRNA), but prevent the joining of two RNA molecules into multimer linear RNA. This circularization reaction is realized without a complementary bridging polynucleotide but necessarily involves bringing the 3′ and 5′ RNA extremities together, via, for example, a stem-loop structure. The structure of Rnl3 ligases is unique among polynucleotide ligases in that it forms a homodimeric quaternary structure only found in the Rnl3 family members (see below for details and for review see [58]). Thereby, suggesting a direct correlation between the atypical dimeric structure and the RNA circularization activity. This result was confirmed by genetic and functional analyses of an Rnl3 member in the hyperthermophilic archaeon *Thermococcus kodakarensis*. It was observed, by RNA-seq, that an allelic knockout of the Tko-Rnl3 gene mostly dissipates circular box C/D RNAs in the transcriptome but also produces aberrant rRNA fragments, suggesting that Tko-Rnl3 participates in RNA circularization and rRNA maturation processes [59].

The majority of Rnl3 ligases are found within the archaeal domain and are especially prominent in thermophilic and hyperthermophilic archaea, with structural and functional studies in *Methanobacterium thermoautotrophicum* [60–62], *Pyrococcus abyssi* [35, 63], and *Palaeococcus pacificum* [64] (see Fig. 26.3a). An Rnl3 structure for a thermophilic bacterial homolog was also resolved and deposited on the PDB website for an RNA ligase from *Aquifex aeolicus* (PDB ID 3QWU) (Osipiuk J. et al., not published). A perfect superimposition of all the Rnl3 high-resolution structures was obtained and revealed two additional C-terminal domains to the nucleotidyltransferase domain: a dimerization domain containing a swapping motif described for a role in oligomerization mechanisms, and a C-terminal domain without known structural homologs (see Fig. 26.3b) [58].

By bioinformatic approaches using Pfam and Uniprot [65, 66], Rnl3 members were frequently found in thermophiles and hyperthermophiles with 97 and 31 sequences in archaea and bacteria, respectively [58]. Thermophilic proteins tend to have a prominent hydrophobic core and an organization in dimer or higher oligomeric states to increase the rigidity of the individual subunits. Therefore, the higher-order quaternary structure of Rnl3 proteins can be advantageous for the stability of thermophilic proteins and may be a common feature of thermoadaptation [67].

Concerning enzymatic activities, Rnl3 ligases can adenylate both RNA and DNA substrates, with primarily intra-molecular ligation on RNA molecules, leading to RNA circles. The reaction requires a catalytic lysine residue in the active site and the conservation of the dimeric structure of Rnl3 ligases since mutation or deletion of the dimerization domain abolishes the circular reaction and triggers a partial aggregation of the proteins. Adenylation of RNA molecules by wild-type or engineered Rnl3 ligases could be highly applicable in the modification of 5′ RNA or DNA extremities for use in biotechnology, in addition to the ability for thermophilic Rnl3 to function at elevated temperatures [60–62, 64].

The formation of circular RNA has not been elucidated in all archaea, since some archaea contain circular RNA molecules but without the identification of Rnl3 member (e.g., *Haloferax volcanii*, *Nanoarchaeum equitans*, *Saccharalobus solfataricus*, *Sulfolobus acidocaldarius*). This implies, that other proteins would be able to take in charge the circularization reaction, and potential candidates are mis-annotated DNA ligases since Rnl3 proteins are closely related to the CDC9 domain found in ATP-dependent DNA ligase.

It cannot be ruled out that another family of ligase may carry out the circularization reaction if we refer to the identification of a new human RNA ligase (C12ofr29) able to self-cyclized RNA as well as 5′-adenylation of RNA [68]. The enzyme exhibits profound selectivity for intramolecular-single-strand RNA ligation, suggesting that human C12ofr29 RNA ligase may be involved in the RNA ligation machinery.

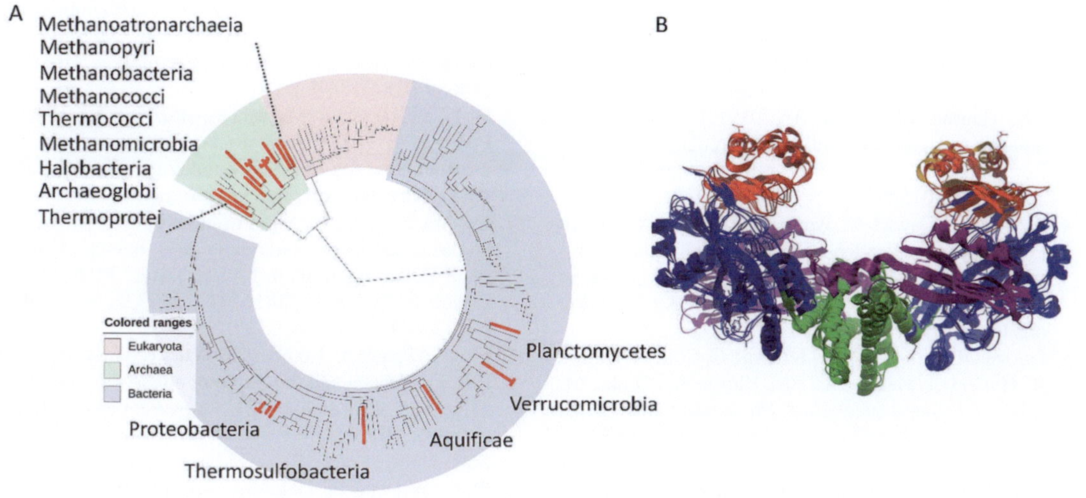

Fig. 26.3 Phylogenetic distribution and structural features of the RNA ligase Rnl3. (**a**) Phylogenetic distribution of Rnl3 RNA ligase family in archaea and bacteria. The tree was built using the online tool Interactive Tree Of Life (/iTOL.embl.de) to display Rnl3 family members by red branches on the tree. The names of archaeal and bacterial classes are given as soon as a member of Rnl3 ligase has been identified in this class through the Pfam protein families' database. (**b**) Superimposition of the structures of four Rnl3 RNA ligases, including three archaea *P. abyssi* Pab1020 (PDB ID 2VUG), *M. thermoautotrophicum* MTH1221 (PDB ID 5D1P), *P. pacificas* (PBD ID 8UCE) and one bacterium *Aquifex aeolicus* (PDB ID 3QWU). The conserved domains of Rnl3 proteins are colored as follows: the N-terminal domain in red, the catalytic nucleotidyltransferase domain in blue, the dimerization domain in green, and the C-terminal domain in purple

6 Conclusions/Future Perspectives

Many open questions regarding the role of circular RNAs in archaea remain. Among those, it will be important to define how widespread circular RNAs in archaea are, and whether there are additional molecular factors involved in this intriguing process. Moreover, it will be important to understand whether circRNA formation is regulated and how circularization contributes to modulating the function or stability of these RNAs. This also means studying the mechanism of RNA circularization and the involvement of the atypic homodimeric RNA ligase (Rnl3 family) mainly present in thermophilic organisms. The direct correlation between dimeric quaternary structure and RNA circularization has not yet been established.

In addition, structural characterization of these circRNAs may also provide a critical detailed understanding of the molecular features. And the particular high stability of circRNA and their resistance to most RNA-degrading machinery could provide them special advantages in regulating gene expression intra- or extracellularly.

All these approaches might contribute to better elucidating the function and evolutionary constraint at the origin of circular RNA synthesis in archaea.

Acknowledgments The authors wish to thank all past and present members of their laboratories, in particular Dr. Hannu Myllykallio and Maria Imezar (École Polytechnique), for fruitful scientific discussions.

References

1. Sanger HL, Klotz G, Riesner D, Gross HJ, Kleinschmidt AK (1976) Viroids are single-stranded covalently closed circular RNA molecules existing as highly base-paired rod-like structures. Proc Natl Acad Sci 73(11):3852–3856. https://doi.org/10.1073/pnas.73.11.3852

2. Patop IL, Wüst S, Kadener S (2019) Past, present, and future of circRNAs. EMBO J 38(16):e100836. https://doi.org/10.15252/embj.2018100836
3. Kristensen LS, Andersen MS, Stagsted LVW, Ebbesen KK, Hansen TB, Kjems J (2019) The biogenesis, biology and characterization of circular RNAs. Nat Rev Genet 20(11):675–691. https://doi.org/10.1038/s41576-019-0158-7
4. Salzman J, Gawad C, Wang PL, Lacayo N, Brown PO (2012) Circular RNAs are the predominant transcript isoform from hundreds of human genes in diverse cell types. PLoS One 7(2):e30733. https://doi.org/10.1371/journal.pone.0030733
5. Izuogu OG, Alhasan AA, Mellough C, Collin J, Gallon R, Hyslop J, Mastrorosa FK, Ehrmann I, Lako M, Elliott DJ, Santibanez-Koref M, Jackson MS (2018) Analysis of human ES cell differentiation establishes that the dominant isoforms of the lncRNAs RMST and FIRRE are circular. BMC Genomics 19(1):276. https://doi.org/10.1186/s12864-018-4660-7
6. Barrett SP, Salzman J (2016) Circular RNAs: analysis, expression and potential functions. Development 143(11):1838–1847. https://doi.org/10.1242/dev.128074
7. He T-T, Xu Y-F, Li X, Wang X, Li J-Y, Ou-Yang D, Cheng H-S, Li H-Y, Qin J, Huang Y, Wang H-Y (2023) A linear and circular dual-conformation noncoding RNA involved in oxidative stress tolerance in Bacillus altitudinis. Nat Commun 14(1):5722. https://doi.org/10.1038/s41467-023-41491-4
8. Fox GE, Magrum LJ, Balch WE, Wolfe RS, Woese CR (1977) Classification of methanogenic bacteria by 16S ribosomal RNA characterization. Proc Natl Acad Sci 74:4537–4541. https://doi.org/10.1073/pnas.74.10.4537
9. Woese CR, Fox GE (1977) Phylogenetic structure of the prokaryotic domain: the primary kingdoms. Proc Natl Acad Sci 74(11):5088–5090. https://doi.org/10.1073/pnas.74.11.5088
10. Woese CR, Kandler O, Wheelis ML (1990) Towards a natural system of organisms: proposal for the domains archaea, bacteria, and eucarya. Proc Natl Acad Sci 87(12):4576–4579
11. Albers S-V, Forterre P, Prangishvili D, Schleper C (2013) The legacy of Carl Woese and Wolfram Zillig: from phylogeny to landmark discoveries. Nat Rev Microbiol 11:713–719
12. Offre P, Spang A, Schleper C (2013) Archaea in biogeochemical cycles. Ann Rev Microbiol 67:437–457. https://doi.org/10.1146/annurev-micro-092412-155614
13. Adam PS, Borrel G, Brochier-Armanet C, Gribaldo S (2017) The growing tree of archaea: new perspectives on their diversity, evolution and ecology. ISME J 11(11):2407–2425. https://doi.org/10.1038/ismej.2017.122
14. Tahon G, Geesink P, Ettema TJG (2021) Expanding archaeal diversity and phylogeny: past, present, and future. Annu Rev of Microbiol 75:359. https://doi.org/10.1146/annurev-micro-040921-050212
15. Cabrera MÁ, Blamey JM (2018) Biotechnological applications of archaeal enzymes from extreme environments. Biol Res 51:37–37. https://doi.org/10.1186/s40659-018-0186-3
16. De Lise F, Iacono R, Moracci M, Strazzulli A, Cobucci-Ponzano B (2023) Archaea as a model system for molecular biology and biotechnology. Biomolecules 13(1). https://doi.org/10.3390/biom13010114
17. Koonin EV (2015) Origin of eukaryotes from within archaea, archaeal eukaryome and bursts of gene gain: Eukaryogenesis just made easier? Philos Trans R Soc B Biol Sci 370(1678):20140333. https://doi.org/10.1098/rstb.2014.0333
18. Eme L, Spang A, Lombard J, Stairs CW, Ettema TJG (2017) Archaea and the origin of eukaryotes. Nat Rev Microbiol 15(12):711–723. https://doi.org/10.1038/nrmicro.2017.133
19. Salgia SR, Singh SK, Gurha P, Gupta R (2003) Two reactions of Haloferax volcanii RNA splicing enzymes: joining of exons and circularization of introns. RNA 9(3):319–330. https://doi.org/10.1261/rna.2118203
20. Singh SK, Gurha P, Tran EJ, Maxwell ES, Gupta R (2004) Sequential 2′-O-methylation of archaeal pre-tRNATrp nucleotides is guided by the intron-encoded but trans-acting box C/D ribonucleoprotein of pre-tRNA. J Biol Chem 279(46):47661–47671. https://doi.org/10.1074/jbc.M408868200
21. Clouet d'Orval B, Bortolin M-L, Gaspin C, Bachellerie J-P (2001) Box C/D RNA guides for the ribose methylation of archaeal tRNAs. The tRNATrp intron guides the formation of two ribose-methylated nucleosides in the mature tRNATrp. Nucleic Acids Res 29(22):4518–4529. https://doi.org/10.1093/nar/29.22.4518
22. Dalgaard JZ, Garrett RA, Belfort M (1994) Purification and characterization of two forms of I-DmoI, a thermophilic site-specific endonuclease encoded by an archaeal intron. J Biol Chem 269(46):28885–28892. https://doi.org/10.1016/S0021-9258(19)61990-3
23. Lykke-Andersen J, Garrett RA (1994) Structural characteristics of the stable RNA introns of archaeal hyperthermophiles and their splicing junctions. J Mol Biol 243(5):846–855. https://doi.org/10.1006/jmbi.1994.1687
24. Tang TH, Rozhdestvensky TS, d'Orval BC, Bortolin M-L, Huber H, Charpentier B, Branlant C, Bachellerie J-P, Brosius J, Hüttenhofer A (2002) RNomics in archaea reveals a further link between splicing of archaeal introns and rRNA processing. Nucleic Acids Res 30(4):921–930
25. Starostina NG, Marshburn S, Johnson LS, Eddy SR, Terns RM, Terns MP (2004) Circular box C/D RNAs in Pyrococcus furiosus. Proc Natl Acad Sci 101(39):14097–14101. https://doi.org/10.1073/pnas.0403520101
26. Clouet-d'Orval B, Gaspin C, Mougin A (2005) Two different mechanisms for tRNA ribose methylation in archaea: a short survey. Biochimie 87(9):889–895. https://doi.org/10.1016/j.biochi.2005.02.004

27. Pascal JM (2008) DNA and RNA ligases: structural variations and shared mechanisms. Curr Opin Struct Biol 18(1):96–105. https://doi.org/10.1016/j.sbi.2007.12.008
28. Haddad G, Lorenzen JM (2019) Biogenesis and function of circular RNAs in health and in disease. Front Pharmacol 10:428. https://doi.org/10.3389/fphar.2019.00428
29. Vo JN, Cieslik M, Zhang Y, Shukla S, Xiao L, Zhang Y, Wu Y-M, Dhanasekaran SM, Engelke CG, Cao X, Robinson DR, Nesvizhskii AI, Chinnaiyan AM (2019) The landscape of circular RNA in cancer. Cell 176(4):869–881.e13. https://doi.org/10.1016/j.cell.2018.12.021
30. Zeng X, Lin W, Guo M, Zou Q (2017) A comprehensive overview and evaluation of circular RNA detection tools. PLoS Comput Biol 13(6):e1005420. https://doi.org/10.1371/journal.pcbi.1005420
31. Jakobi T, Dieterich C (2019) Computational approaches for circular RNA analysis. WIREs RNA 10(3):e1528. https://doi.org/10.1002/wrna.1528
32. Lykke-Andersen J, Aagaard C, Semionenkov M, Garrett RA (1997) Archaeal introns: splicing, intercellular mobility and evolution. Trends Biochem Sci 22(9):326–331. https://doi.org/10.1016/S0968-0004(97)01113-4
33. Englert M, Sheppard K, Aslanian A, Yates JR, Söll D (2011) Archaeal 3′-phosphate RNA splicing ligase characterization identifies the missing component in tRNA maturation. Proc Natl Acad Sci 108(4):1290–1295. https://doi.org/10.1073/pnas.1018307108
34. Danan M, Schwartz S, Edelheit S, Sorek R (2012) Transcriptome-wide discovery of circular RNAs in archaea. Nucleic Acids Res 40(7):3131–3142. https://doi.org/10.1093/nar/gkr1009
35. Becker HF, Héliou A, Djaout K, Lestini R, Regnier M, Myllykallio H (2017) High-throughput sequencing reveals circular substrates for an archaeal RNA ligase. RNA Biol 14:1075–1085. https://doi.org/10.1080/15476286.2017.1302640
36. Randau L (2012) RNA processing in the minimal organism Nanoarchaeum equitans. Genome Biol 13(7):R63. https://doi.org/10.1186/gb-2012-13-7-r63
37. Plagens A, Daume M, Wiegel J, Randau L (2015) Circularization restores signal recognition particle RNA functionality in Thermoproteus. elife 4:e11623. https://doi.org/10.7554/eLife.11623
38. Watkins NJ, Bohnsack MT (2012) The box C/D and H/ACA snoRNPs: key players in the modification, processing and the dynamic folding of ribosomal RNA. WIREs RNA 3(3):397–414. https://doi.org/10.1002/wrna.117
39. Breuer R, Gomes-Filho J-V, Randau L (2021) Conservation of archaeal C/D box sRNA-guided RNA modifications. Front Microbiol 12:496. https://doi.org/10.3389/fmicb.2021.654029
40. Czekay DP, Kothe U (2021) H/ACA small ribonucleoproteins: structural and functional comparison between archaea and eukaryotes. Front Microbiol 12:488. https://doi.org/10.3389/fmicb.2021.654370
41. Jüttner M, Weiß M, Ostheimer N, Reglin C, Kern M, Knüppel R, Ferreira-Cerca S (2019) A versatile cis-acting element reporter system to study the function, maturation and stability of ribosomal RNA mutants in archaea. Nucleic Acids Res:gkz1156. https://doi.org/10.1093/nar/gkz1156
42. Birkedal U, Beckert B, Wilson DN, Nielsen H (2020) The 23S ribosomal RNA from Pyrococcus furiosus is circularly permuted. Front Microbiol 11:3114. https://doi.org/10.3389/fmicb.2020.582022
43. Qi L, Li J, Jia J, Yue L, Dong X (2020) Comprehensive analysis of the pre-ribosomal RNA maturation pathway in a methanoarchaeon exposes the conserved circularization and linearization mode in archaea. RNA Biol:1–15. https://doi.org/10.1080/15476286.2020.1771946
44. Grünberger F, Jüttner M, Knüppel R, Ferreira-Cerca S, Grohmann D (2023) Nanopore-based RNA sequencing deciphers the formation, processing, and modification steps of rRNA intermediates in archaea. RNA 29(8):1255–1273. https://doi.org/10.1261/rna.079636.123
45. Qi L, Liu H, Gao J, Deng K, Wang X, Dong X, Li J (2023) Endonucleolytic processing plays a critical role in the maturation of ribosomal RNA in Methanococcus maripaludis. RNA Biol 20(1):760–773. https://doi.org/10.1080/15476286.2023.2258035
46. Londei P, Ferreira-Cerca S (2021) Ribosome biogenesis in archaea. Front Microbiol 12:1476. https://doi.org/10.3389/fmicb.2021.686977
47. Schwarz TS, Berkemer SJ, Bernhart SH, Weiß M, Ferreira-Cerca S, Stadler PF, Marchfelder A (2020) Splicing endonuclease is an important player in rRNA and tRNA maturation in archaea. Front Microbiol 11:2856. https://doi.org/10.3389/fmicb.2020.594838
48. Shajani Z, Sykes MT, Williamson JR (2011) Assembly of bacterial ribosomes. Annu Rev Biochem 80:501–526. https://doi.org/10.1146/annurev-biochem-062608-160432
49. Henras AK, Plisson-Chastang C, O'Donohue M-F, Chakraborty A, Gleizes P-E (2015) An overview of pre-ribosomal RNA processing in eukaryotes. Wiley Interdiscip Rev RNA 6(2):225–242. https://doi.org/10.1002/wrna.1269
50. Davis JH, Williamson JR (2017) Structure and dynamics of bacterial ribosome biogenesis. Philos Trans R Soc B Biol Sci 372:20160181. https://doi.org/10.1098/rstb.2016.0181
51. Ferreira-Cerca S (2017) Life and death of ribosomes in archaea. In: Clouet-d'Orval B (ed) RNA metabolism and gene expression in archaea. Springer, pp 129–158. https://doi.org/10.1007/978-3-319-65795-0_6
52. Bohnsack KE, Henras AK, Nielsen H, Bohnsack MT (2023) Making ends meet: a universal driver of large ribosomal subunit biogenesis. Trends Biochem Sci 48(3):213–215. https://doi.org/10.1016/j.tibs.2022.09.003

53. Lorentzen E, Walter P, Fribourg S, Evguenieva-Hackenberg E, Klug G, Conti E (2005) The archaeal exosome core is a hexameric ring structure with three catalytic subunits. Nat Struct Mol Biol 12(7):575–581. https://doi.org/10.1038/nsmb952
54. Evguenieva-Hackenberg E, Hou L, Glaeser S, Klug G (2014) Structure and function of the archaeal exosome. WIREs RNA 5(5):623–635. https://doi.org/10.1002/wrna.1234
55. Evguenieva-Hackenberg E, Walter P, Hochleitner E, Lottspeich F, Klug G (2003) An exosome-like complex in Sulfolobus solfataricus. EMBO Rep 4(9):889–893. https://doi.org/10.1038/sj.embor.embor929
56. Bathke J, Gauernack AS, Rupp O, Weber L, Preusser C, Lechner M, Rossbach O, Goesmann A, Evguenieva-Hackenberg E, Klug G (2020) iCLIP analysis of RNA substrates of the archaeal exosome. BMC Genomics 21(1):797. https://doi.org/10.1186/s12864-020-07200-x
57. Unciuleac M-C, Goldgur Y, Shuman S (2015) Structure and two-metal mechanism of a eukaryal nick-sealing RNA ligase. Proc Natl Acad Sci 112(45):13868–13873. https://doi.org/10.1073/pnas.1516536112
58. Becker HF, L'Hermitte-Stead C, Myllykallio H (2019) Diversity of circular RNAs and RNA ligases in archaeal cells. Biochimie 164:37. https://doi.org/10.1016/j.biochi.2019.06.011
59. Liu Y, Takagi Y, Sugijanto M, Nguyen KDM, Hirata A, Hori H, Ho CK (2022) Genetic and functional analyses of archaeal ATP-dependent RNA ligase in C/D box sRNA circularization and ribosomal RNA processing. Front Mol Biosci 9:811548. https://doi.org/10.3389/fmolb.2022.811548
60. Torchia C, Takagi Y, Ho CK (2008) Archaeal RNA ligase is a homodimeric protein that catalyzes intramolecular ligation of single-stranded RNA and DNA. 36(19):6218–6227. https://doi.org/10.1093/nar/gkn602
61. Zhelkovsky AM, McReynolds LA (2012) Structure-function analysis of Methanobacterium thermoautotrophicum RNA ligase—engineering a thermostable ATP independent enzyme. BMC Mol Biol 13(1):24. https://doi.org/10.1186/1471-2199-13-24
62. Gu H, Yoshinari S, Ghosh R, Ignatochkina AV, Gollnick PD, Murakami KS, Ho CK (2016) Structural and mutational analysis of archaeal ATP-dependent RNA ligase identifies amino acids required for RNA binding and catalysis. Nucleic Acids Res 44(5):2337–2347. https://doi.org/10.1093/nar/gkw094
63. Brooks MA, Meslet-Cladiére L, Graille M, Kuhn J, Blondeau K, Myllykallio H, van Tilbeurgh H (2008) The structure of an archaeal homodimeric ligase which has RNA circularization activity. Protein Sci 17(8):1336–1345. https://doi.org/10.1110/ps.035493.108
64. Rousseau M, Oulavallickal T, Williamson A, Arcus V, Patrick WM, Hicks J (2024) Characterisation and engineering of a thermophilic RNA ligase from Palaeococcus pacificus. Nucleic Acids Res 52(7):3924–3937. https://doi.org/10.1093/nar/gkae149
65. Mistry J, Chuguransky S, Williams L, Qureshi M, Salazar GA, Sonnhammer ELL, Tosatto SCE, Paladin L, Raj S, Richardson LJ, Finn RD, Bateman A (2021) Pfam: the protein families database in 2021. Nucleic Acids Res 49(D1):D412–D419. https://doi.org/10.1093/nar/gkaa913
66. The UniProt Consortium (2023) UniProt: the universal protein knowledgebase in 2023. Nucleic Acids Res 51(D1):D523–D531. https://doi.org/10.1093/nar/gkac1052
67. Reed CJ, Lewis H, Trejo E, Winston V, Evilia C (2013) Protein adaptations in archaeal extremophiles. Archaea 2013(1):373275. https://doi.org/10.1155/2013/373275
68. Yuan Y, Stumpf FM, Schlor LA, Schmidt OP, Saumer P, Huber LB, Frese M, Höllmüller E, Scheffner M, Stengel F, Diederichs K, Marx A (2023) Chemoproteomic discovery of a human RNA ligase. Nat Commun 14(1):842. https://doi.org/10.1038/s41467-023-36451-x

Part VIII

Future Prospects

27 Challenges and Future Prospectives in Circular RNA Investigation

Danielle Hiam, Matthew J. Gedye, Glenn D. Wadley, and Bianca C. Bernardo

Abstract

Circular RNAs (circRNAs) are a novel class of highly stable noncoding RNAs that play significant biological roles in the regulation of many cellular processes including proliferation, apoptosis, angiogenesis and cell cycle progression. CircRNAs belong to a diverse class of stable RNAs expressed in most species and cell types and are highly tissue specific. Their high stability and tissue-specific expression makes them ideal biomarkers for disease. Indeed, their differential expression is well described in various diseases including several cancers, cardiovascular disease, diabetes and neurological disorders. However, their specific role in associated molecular pathways is yet to be properly determined. Furthermore, although studies demonstrate the potential of blood circRNAs as biomarkers in several diseases, the challenge remains to identify consistent and reproducible data across independent studies. Whilst the high stability and lower immunogenicity of circRNAs make them advantageous over traditional mRNA as an RNA therapeutic, more work is needed to develop efficient, cost-effective production systems that support scalable, commercial circRNA development.

Keywords

Noncoding RNAs · Circular RNAs · Sequencing technologies · Biomarkers · Therapeutic applications · Bioinformatics

1 Introduction

Circular RNAs (CircRNAs) are a novel class of highly stable noncoding RNAs that, unlike linear RNA, form a covalently closed circular molecule. CircRNAs belong to a diverse class of stable RNAs expressed in most species and cell types and are highly tissue-specific [1–3]. For example, over 9000 candidate circRNAs have been detected in the hearts of each species of mouse, rat and humans [4]. They are well recognised as playing significant biological roles in the regulation of many cellular processes including proliferation, apoptosis, angiogenesis and cell cycle progression (reviewed in [5]). Their most well-established biological function is to act as

D. Hiam · M. J. Gedye · G. D. Wadley
Institute for Physical Activity and Nutrition (IPAN), School of Exercise and Nutrition Sciences, Deakin University, Burwood, VIC, Australia

B. C. Bernardo (✉)
Faculty of Health, School of Exercise and Nutrition Science, Deakin University, Burwood, VIC, Australia

Department of Cardiometabolic Healths, The University of Melbourne, Parkville, VIC, Australia
e-mail: bianca.bernardo@deakin.edu.au

J. Xiao (ed.), *Advances in Circular RNAs*, Advances in Experimental Medicine and Biology 1485,
https://doi.org/10.1007/978-981-96-9428-0_27

microRNA sponges [6], but they can also act as protein scaffolds and regulate transcription. Although typically classed as ncRNAs, there is evidence that some circRNAs can be translated into functional proteins such as circ-ZNF609, which has been shown to control myoblast proliferation [7]. Their high stability and tissue-specific expression make them ideal biomarkers for disease. Indeed, their aberrant expression is well described in various cancers, diabetes and in neurological disorders such as Alzheimer's disease [5, 6, 8–10]. CircRNAs are also involved in cardiovascular disease, with some circRNAs having a cardioprotective role [11], whilst others play a detrimental role in pathological remodelling [12]. Therefore, the focus of this chapter will be to discuss the major challenges that lie ahead for circRNA to be analysed at the computation level and utilised as biomarkers for diseases and their role in potential future therapies.

2 CircRNA Biology: Biogenesis and Basic Characteristics

Most circRNAs originate from protein-coding genes, but may also derive from noncoding, antisense, intronic, 3′ untranslated region (UTR), 5′ UTR or intergenic genomic regions [13, 14]. Precise knowledge of circRNA biogenesis is still in its infancy, but unlike the splicing of linear RNA, circRNAs are largely accepted to be generated from back-splicing of exons of pre-mRNA [15]. The phenomenon of back-splicing is the product of the spliceosome ligating a downstream donor (5′ splice site) to its own upstream acceptor (3′ splice site) or the 3′ exon of another molecule, bound by a phosphodiester bond [16, 17]. Canonical splicing sites and spliceosomal machinery are therefore still necessary for the back-splicing process, whereby intronic repeats and exonic sequences collaborate with one another, implying that circularisation occurs post-transcriptionally by *cis* and *trans* regulators [14, 18]. The resulting circular covalently closed RNA molecule is highly stable, exhibiting a much greater half-life when compared to linear RNA [19]. CircRNAs are therefore better protected against degradation, likely as a result of resistance to debranching enzymes and RNA exonucleases [20], making them ideal for long-term protein expression or target regulation. While back-splicing is less efficient than canonical splicing, resulting in circRNAs being expressed at low levels compared to linear RNA [21], they are ubiquitously expressed and highly conserved across mammalian species [3, 19]. Early algorithms for detecting circRNAs from RNA-sequencing data exclusively identified novel circRNA expression [22, 23]. More recently, however, circRNA expression has been observed to be highly tissue and cell specific, showing the most abundant expression in human brain, heart and liver where they are thought to be largely associated with tissue development and differentiation [24]. Among the best characterised functions of circRNAs are as microRNA (miRNA) sponges, in which miRNA activity is negatively regulated through the interference of circRNAs with miRNA and mRNA interactions [25]. For example, circRNA CDR1 was revealed to contain 63 conserved binding sites for miR-7 in mouse neural tissue, implying post-transcriptional regulation of miR-7 target genes [20]. Similarly, circRNAs also function as RNA-binding protein (RBP) sponges, such as circFOXO3, which binds to stress and senescence proteins inhibiting their translocation to the nucleus, thereby preventing their function in the heart cells of aged mice [26]. In addition, circRNAs have also been found to regulate gene transcription, with circZNF827 recruiting regulator proteins to the promoter region in the nucleus [27]. Further in-depth exploratory work is required to validate these relationships and better understand the mechanistic interactions in these pathways.

3 CircRNA Sequencing, Detection and Associated Challenges

Advancements in sequencing technologies and bioinformatic tools have enabled researchers to detect, quantify and characterise circRNAs with a greater accuracy [28–30]. CircRNAs can be identified using several methods including RNA-sequencing, microarray and RT-PCR. Many of these techniques and analysis tools rely on the back-splice junction (BSJ) to detect circRNAs. The BSJ acts as a molecular signature for circRNA and distinguishes them from the linear RNA transcripts [29, 31]. Despite circRNAs representing an increasingly popular area of research, there are still many challenges to accurately determine circRNA both at the experimental and the computational levels [28]. This section summarises the current circRNA methods and bioinformatic tools.

Short-read RNA sequencing of rRNA-depleted total RNA is a popular choice for profiling of circRNA due to high-throughput and cost-effectiveness. It also has the added advantage of being able to simultaneously analyse the expression of coding and noncoding RNAs [28, 31]. Preparation of the libraries is critical for sequencing circRNAs, and there are many variations of library preparations that ultimately affect the circRNA detection [28, 30]. Typically, RNA libraries for sequencing are prepared by either selection of RNA based on poly(A) tails or depleted of rRNA before library preparation. Libraries prepared using selection of poly(A) tail cannot be used as circRNAs lack the poly(A) tail [29]. rRNA-depleted libraries can be used; however, there are issues with the low abundance of circRNA for accurate detection and analysis that bioinformatic pipelines can only partially rectify [28]. Therefore, protocols have been adapted to enrich the circRNA pool with the RNase R treatment followed by polyadenylation and poly(A) + RNA Depletion (RPAD) protocol being shown to yield the highest number of circRNA [32].

One of the biggest challenges in analysis of circRNA using RNA sequencing is the annotation of the circRNA. There are many bioinformatic tools that have been created to annotate circRNAs, and these have been extensively reviewed elsewhere (see comprehensive reviews [28, 33, 34]). CircRNA annotation is generally based on the identification of the BSJ region using one of two methods, segmented-based circRNA detection or pseudo-reference-based circRNA detection [28]. However, relying solely on BSJ for detection of circRNA can be a limitation. This is because most circRNAs are derived from exons and retain the same sequence elements as the linear RNA from which they were derived, making it difficult to distinguish between the two. Further, isoforms of circRNA can have an identical BSJ region but differ in the internal sequences and therefore biologically may interact differently or have different biological functions [29, 31]. Long-read RNA sequencing developed by Oxford Nanopore Technology and Pac Bio allows for the full circle reconstruction of circRNA, circumventing some of issues associated with short-read sequencing [28, 35, 36]. For example, it has greater accuracy in identifying circRNAs isoforms and can distinguish circRNAs from the overlapping regions of corresponding linear RNA transcripts more accurately [28, 37]. Full-length annotation of the circRNA gives additional context into the biological function as well as its interaction with other biological pathways [29, 38]. Finally, it has been shown to have enhanced sensitivity in detecting circRNAs of relatively low abundance [36]. While long-read sequencing is associated with greater cost, it provides a promising new avenue for circRNA research.

Microarrays are particularly useful for analysis and comparison of *known* circRNAs, making them an invaluable resource for diagnostic and biomarkers applications [29]. The microarray probes are designed based on the BSJ sequence and are specific to a circRNA. A recent study found that microarrays can detect 10 times more circRNAs using microarray compared to RNA sequencing read at a depth of 20 million reads [39]. A drawback of microarray is that they cannot identify novel circRNAs or isoforms of circRNAs [31]. However, depending on the research question, microarrays may be a suitable technique due to efficiency, sensitivity and ease of analysis when compared to the more complex RNA sequencing and bioinformatic analysis techniques.

Advancements in library preparations methods, sequencing technologies and computational approaches have facilitated an exponential growth in circRNA research. Future development of circRNA-specific techniques will result in a more advanced understanding of circRNAs and lead to potential applications in diagnostics and therapeutics.

4 Possibilities and Challenges of CircRNAs as Biomarkers

Growing evidence supports a role of circRNAs in normal physiology and several diseases, including cancer, cardiovascular disease, diabetes and neurological disorders [5]. Although circRNAs have been found to be either up- or down-regulated in various disease states, the challenge remains to determine their specific role in associated molecular pathways.

The stability and long half-life of circRNAs make them ideal candidates as biomarkers. They have been recognised to exist at high levels in body fluids, including human whole blood, where over 2400 different circRNAs have been reproducibly detected [20]. To develop novel biomarkers for disease, it is important to analyse the expression of circRNA levels in samples that are easy to collect, such as blood. Here, we provide a summary of several studies that have measured circRNAs in the blood of patients with disease and explored the possibility of using them as disease markers.

Several published reports have implicated circulating circRNAs as predictors and biomarkers for cardiovascular disease (reviewed in [40–43]). The myocardial infarction-associated circRNA (MICRA) was found to be significantly lower in the peripheral blood of patients with myocardial infarction compared to healthy controls, predicting the presence of cardiac dysfunction [44]. Elevated circulating levels of hsa_circ_025016 were able to predict the onset of post-operative atrial fibrillation, a common complication in patients who have undergone coronary artery bypass grafting surgery [45]. The ratio of serum circR-284:miR-221 shows promise to be a diagnostic biomarker of carotid plaque rupture and stroke; however, longitudinal studies are required, and the functional relationship between circR-284 and miR-221 currently remains unknown [46].

There are distinct differences in the expression profile of circRNAs between patients with cancer and healthy controls, suggesting a clinical application for cancer diagnosis and prognosis. An atlas of Blood-Based Biomarkers for Early Diagnosis of Cancers (BBcancer) has recently been established [47]. This is a web-accessible and comprehensive open resource providing the expression landscape of six types of RNAs in blood samples from over 5000 patients across 15 cancer types [47]. This platform may allow the development of new blood-based biomarkers for early detection of cancer.

Not only do circRNAs have potential as biomarkers, but they can also differentiate Alzheimer's disease (AD) from cognitively normal individuals and other types of dementia. AD is a neurogenerative disease that is difficult to diagnose and distinguish from other types of dementia. Current diagnostic methods for AD are invasive, expensive and time consuming. Studies have shown that it is possible to differentiate patients with AD and healthy controls by testing the expression of blood circRNAs [48, 49]; however, sample cohorts were small (five and eight patients with AD respectively). In a larger population (485 subjects), a panel of six blood circRNAs (hsa_circ_0077001, hsa_circ_0022417, hsa_circ_0014356, hsa_circ_0014353, hsa_circ_0074533, hsa_circ_0089894) demonstrated superior efficacy to differentiate AD from healthy controls, as well as from other types of dementias [50]. Separately, circ_0003391 may be an important biomarker for AD. A significant decrease in the expression of circ_0003391 was reported in the peripheral blood of AD patients compared to patients with other types of dementia, which may allow a more accurate diagnosis of AD [51].

Many studies demonstrate the potential of blood circRNAs as biomarkers in a number of different diseases. However, the challenge remains to identify consistent and reproducible data across independent studies. Significant differences in circRNAs detected between independent studies may be due to methods of sample preparation and different types of samples used (e.g. whole blood, serum or plasma). Often, detailed descriptions of methodology are lacking in published reports; thus, identifying the optimal method for reliable measurement of blood circRNA biomarkers is challenging. In addition, larger longitudinal studies need to be performed to validate reliability and sensitivity of circRNA as biomarkers.

5 Considerations for the Development of CircRNA Therapeutics

Advances in deep sequencing, computational analysis and identification algorithms have revealed a huge number (more than 30,000) of different circRNAs expressed in a wide range of mammalian tissues [52]. The expression of circRNAs has been reported to change in association with disease, and many studies have described functional roles for circRNAs in promoting or providing protection against disease (reviewed in detail elsewhere [5, 42, 53–58]). However, further functional and population validation studies are required.

Two key features of circRNAs make them advantageous over traditional mRNA as an RNA therapeutic: (1) high stability and (2) less immunogenic. Increased stability of circRNA results in longer duration of expression, which is ideal for long-term protein expression or target regulation. This characteristic could mean that lower doses are required for therapeutic impact, which in turn reduces manufacturing costs and may improve patient safety [59]. Lower immunogenicity will help avoid activation of innate immune response, limiting adverse effects. Despite these clear advantages, there are still some challenges in the process of moving from bench to bedside [60]. The pharmacokinetics and biodistribution of circRNA therapeutics in the body are unknown, making it challenging to predict the most effective dosage and administration regimen. This can make translation to the clinic difficult. CircRNAs often participate in more than one molecular mechanism, and some circRNAs are present in more than one tissue or cell type [61]. Thus, targeting these circRNAs therapeutically may lead to off-target effects in other cells/tissues. Although circRNAs are generally considered to be less immunogenic, this is largely dependent on production strategy, purification method and purity of the final artificial circRNA [62–64]. Impure circRNA may induce strong immune responses, increasing the safety risk to patients. The question whether artificial circRNAs can escape the cellular innate immune response was addressed in a recent paper [65]. In this study, the authors demonstrate that the immunogenic potential of circRNAs is dependent on the purification method. Completely purified circRNAs escape immune-surveillance and are less immunogenic compared with unpurified or incompletely purified circRNAs [65]. Despite the recent progress, several factors remain important to consider when evaluating the immunogenic potential of circRNAs including (i) design of circRNA (i.e. length and structure), (ii) modifications, (iii) cell line model, (iv) choice of reagents, (v) production and purification strategies and (vi) delivery method [66, 67]. Pioneering work by Chen et al. [64] and Wesselhoeft et al. [62] can serve as a useful starting point for strategies to alter these key variables.

In order for circRNA to serve as successful therapeutic targets in disease, two main obstacles need to be addressed: (i) efficient and cost-effective production and purification methods need to be developed and (ii) highly efficient delivery systems need to be optimised. Currently, synthesis of very long (>5000 nucleotides) strands remains difficult, and purification and quality control at scale is challenging and expensive for biomedical circRNA production. Delivery systems need to protect the RNA cargo that is easily damaged by RNAses in blood and tissues and are currently limited to systemic delivery via intravenous or intraperitoneal injection. How each step of the process of exogenous circRNA synthesis can be modified to be more efficient and suitable for circRNA synthesis has been eloquently reviewed in detail by Choi and Nam [68], so here we will focus on scalable technologies and delivery vehicles of circRNAs.

Recent efforts have developed new and scalable technologies that produce highly stable, low immunogenicity and prolonged protein expression. Chen and colleagues [69] based their design on the self-splicing group II intron to achieve co-transcriptional circularisation of circRNAs, that when delivered to mice, produced robust protein expression. Therapeutically, circRNA encoding the receptor-binding domain of the SARS-CoV-2 S protein was able to induce a strong antibody production in mice [69]. This technology could be suitable for industry-level applications soon. A new in vitro circRNA engineering method using the *Tetrahymena* group I intron ribozyme to develop circRNA was shown to produce efficient protein expression and stability with negligible innate immune responses [70, 71]. This method could provide an alternative option over current methods for circRNA preparation and could be suitable for large-scale industrial production of circRNA.

To effectively deliver circRNAs to target organs for disease treatment, a number of techniques have been employed including adeno-associated virus (AAV), nanoparticles and exosomes (see review [58]). These methods are also popular delivery tools for linear mRNA and other noncoding RNAs (reviewed in [72, 73]). Demonstrating proof of concept, Maganck and colleagues [74] described the use of recombinant AAV vectors to deliver circRNA-producing transgenes in vivo. Others have shown improvements in cardiac outcomes following AAV-circRNA-mediated delivery in preclinical models of heart disease [75, 76]. However, several challenges remain with the use of AAV as a delivery vehicle (see review [77]): (i) packaging size: circRNA transcripts larger than 4.7 kb are difficult to package into AAV cassettes, limiting the number of circRNAs that can be targeted with the use of an AAV-based approach; (ii) potential immune response, as up to 70% of healthy individuals have anti-AAV neutralising antibodies that are built up through natural exposure to AAVs, which can limit the effectiveness of an AAV-based therapy [40, 78]; (iii) they are expensive to manufacture [79] and (iv) potentially poor uptake in particular disease settings (e.g. preclinical model of diabetic cardiomyopathy [80]). Nanoparticles have also been used to successfully deliver several circRNA plasmids and vaccines into mice [81–83] and have the potential to improve targeting to specific cells. However, it remains unclear how safe some types of nanoparticles are for clinical use (i.e. gold nanoparticles) [58]. Exosomes are proving to be a popular delivery vehicle of choice for circRNA therapeutics. They naturally carry a wide range of molecules (including circRNAs, long noncoding RNAs, microRNAs, DNA fragments, lipids, proteins), can protect RNAs from degradation, promote cellular uptake without triggering an immune response and can be used to directly deliver circRNA-expressing vectors [58]. The use of nanoparticles as a delivery system has recently been demonstrated in nonhuman primates [84]. Delivery of circSCMH1 extracellular vesicles into the brain after stroke improved stroke recovery outcomes in both mice and monkeys [84]. Like most delivery systems, the use of exosomes is not without manufacturing challenges (e.g. purification methods, upstream cell culture, quality control measures) [85].

6 Future Prospectives

Many studies have contributed to our understanding of circRNA biology, function, detection and their important roles in disease, but we still need to learn more about the regulation of circRNAs in healthy and diseased tissues. Bioinformatic technologies have advanced, and circRNAs have promising prospects as protein replacement therapies, diagnostic biomarkers and vaccines. A number of biotechnology companies are also focusing their research and development on circRNA therapeutics. Here, we discuss recent developments and future prospects in these areas.

6.1 Emerging Bioinformatics Technologies

Limited attempts have been made to characterise circRNA expression patterns at single-cell resolution [86, 87]. More recently, the single-cell landscape of circRNAs in human and mouse tissues was reported, where a total of 139,643 and 214,747 circRNAs in human and mouse single cells, respectively, were identified [88]. With developments in single-cell or spatial level sequencing technologies specifically for detecting circRNA, single-cell RNA-sequencing will likely to be used more in the future.

Sequence information is vital to study the function of circRNAs. Complete sequence information is required to analyse potential interactions and mechanism of action of circRNAs. A major limitation of standard next-generation sequencing techniques is that although circRNAs are detected, their full sequence is not revealed. Recently, technologies such as Oxford Nanopore MinION technology [89] have been developed, which is capable of generating longer read lengths, between 5000 and 30,000 base pairs. This has made direct sequencing of the full-length sequence of circRNAs possible. Research groups have reported new protocols for experimental library preparation that enrich and sequence full-length circRNAs for Nanopore Technology, as well as developed new algorithms to reconstruct the sequence of circRNAs [36, 37, 90–94]. Very recently, a long-read-based database, called FL-circAS, was developed [95]. This interactive database provides identifications and related analyses of full-length circRNAs (over two million) at the isoform level, from 20 cell lines/tissues from human and mouse [95]. Over time, this database will be expanded to other organisms and more cell lines/tissues. This will be a valuable resource when investigating functional and regulatory roles of circRNAs.

6.2 Protein Replacement Therapy

CircRNAs are highly stable, allowing for prolonged protein expression. Synthetic circRNAs and some endogenous circRNAs can encode proteins and have been recently engineered for robust stable protein production in eukaryotic cells [96]. Further, purified circRNA is more stable than unpurified circRNA and results in longer duration of protein expression following injection in mouse adipose tissue [62]. This in turn would lead to less frequent dosing of a circRNA therapeutic, where repeated dosing is often required for the treatment of chronic disease. In an effort to build better circRNA therapeutics and enhance translation, Chen and colleagues [97] developed a modular high-throughput circRNA platform to make and test synthetic circRNAs. Five elements (vector topology, 5′ and 3′ untranslated regions, internal ribosome entry sites and synthetic aptamers recruiting translation initiation machinery) effecting circRNA translation were tested, that when optimised, increased circRNA protein yields by several hundred-fold, and provided greater duration of protein translation in vivo [97]. It is anticipated that this platform could be used more widely in the future.

6.3 The Potential of circRNAs as Novel Diagnostic or Prognostics Markers

The high tissue/cell specificity of circRNAs, coupled with their relatively high stability in body fluids and ease of detection using simpler technologies (e.g. quantitative PCR), make them ideal candidates as novel diagnostic or prognostic markers. However, there is a need to detect circRNAs in body fluids more efficiently. Recently, a group has proposed an electrochemical ultrasensitive point-of-care testing platform for circRNA detection with a smartphone [98]. This technology is based on Au nanoflowers/peptide nucleic acid-modified carbon-fibre microelectrode [98]. It was shown to exhibit excellent sensitivity and stability towards circRNA detection, and analytical performance was verified in human serum samples [98]. This 'plug-and-play' diagnostic system could be a promising system for future clinical analyses of circRNAs in disease.

6.4 Vaccine Development

Intrinsic features of circRNA (e.g. increased stability, longer shelf life, extended durability and superior potency) make it a more suitable and superior RNA format of choice for vaccine development over traditional mRNA [99]. These features bring the possibility of requiring less genetic material of the circRNA to be as effective and a single-dose option. This in turn means it is likely to have a better safety profile as well as being cheaper to manufacture. Recently, circRNA vaccines were shown to provide robust protection against SARS-CoV-2 in both mice and rhesus macaques, enabling a higher and more durable antigen production than the traditional modified mRNA vaccine [100, 101]. When combined with protein antigens, circRNA induced potent and persistent T cell responses in mice following immunisation, demonstrating efficacy as a therapeutic cancer vaccine [102]. Wan and colleagues [82] recently designed a novel lyophilised mannose-lipid nanoparticle-encapsulated circRNA vaccine against Rabies virus and SARS-CoV-2. This design and modification was able to elicit a robust immune response in mice, was stable for at least 24 weeks of storage at 4 °C and maintained immunogenicity [82]. The improvement in shelf life and storage has the potential to make circRNA vaccines more easily accessible, especially in developing countries. However, circRNA vaccines are yet to be tested in the clinic, and it will take time to test their safety profile and become available on a larger manufacturing scale.

6.5 Biotechnology Companies

Collectively, these promising applications of circRNAs have been noted by industry. Several biotech companies are investigating circRNA therapeutics for a range of indications, including (i) Orna™, specialising in cirRNA products [103] to deliver a gene therapy for Duchenne Muscular Dystrophy; (ii) Circio, who are developing circRNA medications for cancer and rare diseases; and (iii) Ginkgo Bioworks and Esperovax, a joint venture to develop a circRNA therapy to specifically target colorectal cancer cells. In the future, we anticipate more biotech companies to focus on the design and development of circRNA therapeutics, especially as more academic and medical institutions devote time and research to overcome associated production, delivery and purification challenges.

7 Conclusion

CircRNAs are covalently closed loop structures that are highly stable, conserved, abundant and often exhibit tissue- or cell-specific expression. It is for these reasons that circRNAs could be the next revolution in RNA therapeutics. However, despite the potential of blood circRNAs to act as biomarkers in several different diseases, many challenges remain in identifying consistent and reproducible data across independent studies.

The high stability and lower immunogenicity of circRNAs make them advantageous over traditional mRNA as an RNA therapeutic, but significant challenges remain before they can be moved from the benchtop to the bedside. The pharmacokinetics and biodistribution of circRNA therapeutics in the body are unknown, making it challenging to predict the most effective dosage and administration regimen. Also, although circRNAs are generally considered to be less immunogenic, this is largely dependent on production strategy, purification method and purity of the final artificial circRNA. Currently, synthesis of very long (>5000 nucleotides) strands remains difficult, and purification and quality control at scale are challenging and expensive for biomedical circRNA production. For circRNA to serve as successful therapeutic targets in disease, two main obstacles need to be addressed: (i) efficient and cost-effective production and purification methods need to be developed and (ii) highly efficient delivery systems need to be optimised. Thus, more work needs to be done to develop efficient, cost-effective production systems that support scalable and commercial circRNA development.

Acknowledgements Not applicable.

CRediT Authorship Contribution Statement Danielle Hiam, Matthew Gedye: Investigation, Writing—original draft, Writing—review & editing. **Glenn D. Wadley:** Writing—original draft, Writing—review & editing. **Bianca C. Bernardo**: Conceptualisation, Investigation, Writing—original draft, Writing—review & editing. All authors approved the final version of manuscript.

Competing Financial Interests The authors declare no competing financial interests.

Funding D.H. was supported by an Executive Dean's Postdoctoral Research Fellowship from Deakin University. M.J.G is supported by a Deakin University Postgraduate Research Scholarship.

References

1. Memczak S, Jens M, Elefsinioti A, Torti F, Krueger J, Rybak A, Maier L, Mackowiak SD, Gregersen LH, Munschauer M, Loewer A, Ziebold U, Landthaler M, Kocks C, le Noble F, Rajewsky N (2013) Circular RNAs are a large class of animal RNAs with regulatory potency. Nature 495(7441):333–338. https://doi.org/10.1038/nature11928
2. Maass PG, Glažar P, Memczak S, Dittmar G, Hollfinger I, Schreyer L, Sauer AV, Toka O, Aiuti A, Luft FC, Rajewsky N (2017) A map of human circular RNAs in clinically relevant tissues. J Mol Med (Berl) 95(11):1179–1189. https://doi.org/10.1007/s00109-017-1582-9
3. Salzman J, Chen RE, Olsen MN, Wang PL, Brown PO (2013) Cell-type specific features of circular RNA expression. PLoS Genet 9(9):e1003777. https://doi.org/10.1371/journal.pgen.1003777
4. Werfel S, Nothjunge S, Schwarzmayr T, Strom TM, Meitinger T, Engelhardt S (2016) Characterization of circular RNAs in human, mouse and rat hearts. J Mol Cell Cardiol 98:103–107. https://doi.org/10.1016/j.yjmcc.2016.07.007
5. Verduci L, Tarcitano E, Strano S, Yarden Y, Blandino G (2021) CircRNAs: role in human diseases and potential use as biomarkers. Cell Death Dis 12(5):468. https://doi.org/10.1038/s41419-021-03743-3
6. Kristensen LS, Andersen MS, Stagsted LVW, Ebbesen KK, Hansen TB, Kjems J (2019) The biogenesis, biology and characterization of circular RNAs. Nat Rev Genet 20(11):675–691. https://doi.org/10.1038/s41576-019-0158-7
7. Legnini I, Di Timoteo G, Rossi F, Morlando M, Briganti F, Sthandier O, Fatica A, Santini T, Andronache A, Wade M, Laneve P, Rajewsky N, Bozzoni I (2017) Circ-ZNF609 is a circular RNA that can be translated and functions in myogenesis. Mol Cell 66(1):22–37.e29. https://doi.org/10.1016/j.molcel.2017.02.017
8. Dube U, Del-Aguila JL, Li Z, Budde JP, Jiang S, Hsu S, Ibanez L, Fernandez MV, Farias F, Norton J, Gentsch J, Wang F, Salloway S, Masters CL, Lee JH, Graff-Radford NR, Chhatwal JP, Bateman RJ, Morris JC, Karch CM, Harari O, Cruchaga C (2019) An atlas of cortical circular RNA expression in Alzheimer disease brains demonstrates clinical and pathological associations. Nat Neurosci 22(11):1903–1912. https://doi.org/10.1038/s41593-019-0501-5
9. Dehghanbanadaki H, Asili P, Haji Ghadery A, Mirahmad M, Dehnavi AZ, Parsaei A, Baradaran HR, Azami M, Jose Justo da Silva G, Parvan R, Moradi Y (2023) Diagnostic accuracy of circular RNA for diabetes mellitus: a systematic review and diagnostic meta-analysis. BMC Med Genet 16(1):48. https://doi.org/10.1186/s12920-023-01476-0
10. Xu H, Guo S, Li W, Yu P (2015) The circular RNA Cdr1as, via miR-7 and its targets, regulates insulin transcription and secretion in islet cells. Sci Rep 5:12453. https://doi.org/10.1038/srep12453
11. Wang K, Long B, Liu F, Wang JX, Liu CY, Zhao B, Zhou LY, Sun T, Wang M, Yu T, Gong Y, Liu J, Dong YH, Li N, Li PF (2016) A circular RNA protects the heart from pathological hypertrophy and heart failure by targeting miR-223. Eur Heart J 37(33):2602–2611. https://doi.org/10.1093/eurheartj/ehv713
12. Bibi A, Bartekova M, Gandhi S, Greco S, Madè A, Sarkar M, Stopa V, Tastsoglou S, de Gonzalo-Calvo D, Devaux Y, Emanueli C, Hatzigeorgiou AG, Nossent AY, Zhou Z, Martelli F (2024) Circular RNA regulatory role in pathological cardiac remodelling. Br J Pharmacol 182:316. https://doi.org/10.1111/bph.16434
13. Pamudurti NR, Bartok O, Jens M, Ashwal-Fluss R, Stottmeister C, Ruhe L, Hanan M, Wyler E, Perez-Hernandez D, Ramberger E, Shenzis S, Samson M, Dittmar G, Landthaler M, Chekulaeva M, Rajewsky N, Kadener S (2017) Translation of CircRNAs. Mol Cell 66(1):9–21.e27. https://doi.org/10.1016/j.molcel.2017.02.021
14. Sharma AR, Bhattacharya M, Bhakta S, Saha A, Lee SS, Chakraborty C (2021) Recent research progress on circular RNAs: biogenesis, properties, functions, and therapeutic potential. Mol Ther Nucleic Acids 25:355–371. https://doi.org/10.1016/j.omtn.2021.05.022
15. Greene J, Baird AM, Brady L, Lim M, Gray SG, McDermott R, Finn SP (2017) Circular RNAs: biogenesis, function and role in human diseases. Front Mol Biosci 4:38. https://doi.org/10.3389/fmolb.2017.00038
16. Liu CX, Chen LL (2022) Circular RNAs: characterization, cellular roles, and applications. Cell 185(12):2016–2034. https://doi.org/10.1016/j.cell.2022.04.021
17. Zhou WY, Cai ZR, Liu J, Wang DS, Ju HQ, Xu RH (2020) Circular RNA: metabolism, functions and interactions with proteins. Mol Cancer 19(1):172. https://doi.org/10.1186/s12943-020-01286-3
18. Liang D, Wilusz JE (2014) Short intronic repeat sequences facilitate circular RNA production. Genes

Dev 28(20):2233–2247. https://doi.org/10.1101/gad.251926.114
19. Jeck WR, Sorrentino JA, Wang K, Slevin MK, Burd CE, Liu J, Marzluff WF, Sharpless NE (2013) Circular RNAs are abundant, conserved, and associated with ALU repeats. RNA 19(2):141–157. https://doi.org/10.1261/rna.035667.112
20. Memczak S, Papavasileiou P, Peters O, Rajewsky N (2015) Identification and characterization of circular RNAs as a new class of putative biomarkers in human blood. PLoS One 10(10):e0141214. https://doi.org/10.1371/journal.pone.0141214
21. Chen LL (2016) The biogenesis and emerging roles of circular RNAs. Nat Rev Mol Cell Biol 17(4):205–211. https://doi.org/10.1038/nrm.2015.32
22. Glazar P, Papavasileiou P, Rajewsky N (2014) circBase: a database for circular RNAs. RNA 20(11):1666–1670. https://doi.org/10.1261/rna.043687.113
23. Liu YC, Li JR, Sun CH, Andrews E, Chao RF, Lin FM, Weng SL, Hsu SD, Huang CC, Cheng C, Liu CC, Huang HD (2016) CircNet: a database of circular RNAs derived from transcriptome sequencing data. Nucleic Acids Res 44(D1):D209–D215. https://doi.org/10.1093/nar/gkv940
24. Xia S, Feng J, Lei L, Hu J, Xia L, Wang J, Xiang Y, Liu L, Zhong S, Han L, He C (2017) Comprehensive characterization of tissue-specific circular RNAs in the human and mouse genomes. Brief Bioinform 18(6):984–992. https://doi.org/10.1093/bib/bbw081
25. Hansen TB, Jensen TI, Clausen BH, Bramsen JB, Finsen B, Damgaard CK, Kjems J (2013) Natural RNA circles function as efficient microRNA sponges. Nature 495(7441):384–388. https://doi.org/10.1038/nature11993
26. Du WW, Yang W, Chen Y, Wu ZK, Foster FS, Yang Z, Li X, Yang BB (2017) Foxo3 circular RNA promotes cardiac senescence by modulating multiple factors associated with stress and senescence responses. Eur Heart J 38(18):1402–1412. https://doi.org/10.1093/eurheartj/ehw001
27. Hollensen AK, Thomsen HS, Lloret-Llinares M, Kamstrup AB, Jensen JM, Luckmann M, Birkmose N, Palmfeldt J, Jensen TH, Hansen TB, Damgaard CK (2020) circZNF827 nucleates a transcription inhibitory complex to balance neuronal differentiation. elife 9. https://doi.org/10.7554/eLife.58478
28. Digby B, Finn S, Ó Broin P (2024) Computational approaches and challenges in the analysis of circRNA data. BMC Genomics 25(1):527. https://doi.org/10.1186/s12864-024-10420-0
29. Drula R, Braicu C, Berindan-Neagoe I (2024) Current advances in circular RNA detection and investigation methods: are we running in circles? WIREs RNA 15(3):e1850. https://doi.org/10.1002/wrna.1850
30. Szabo L, Salzman J (2016) Detecting circular RNAs: bioinformatic and experimental challenges. Nat Rev Genet 17(11):679–692. https://doi.org/10.1038/nrg.2016.114
31. Pandey PR, Munk R, Kundu G, De S, Abdelmohsen K, Gorospe M (2020) Methods for analysis of circular RNAs. WIREs RNA 11(1):e1566. https://doi.org/10.1002/wrna.1566
32. Shi H, Zhou Y, Jia E, Liu Z, Pan M, Bai Y, Zhao X, Ge Q (2022) Comparative analysis of circular RNA enrichment methods. RNA Biol 19(1):55–67. https://doi.org/10.1080/15476286.2021.2012632
33. Zeng X, Lin W, Guo M, Zou Q (2017) A comprehensive overview and evaluation of circular RNA detection tools. PLoS Comput Biol 13(6):e1005420. https://doi.org/10.1371/journal.pcbi.1005420
34. Chen L, Wang C, Sun H, Wang J, Liang Y, Wang Y, Wong G (2021) The bioinformatics toolbox for circRNA discovery and analysis. Brief Bioinform 22(2):1706–1728. https://doi.org/10.1093/bib/bbaa001
35. Zheng Y, Ji P, Chen S, Hou L, Zhao F (2019) Reconstruction of full-length circular RNAs enables isoform-level quantification. Genome Med 11(1):2. https://doi.org/10.1186/s13073-019-0614-1
36. Zhang J, Hou L, Zuo Z, Ji P, Zhang X, Xue Y, Zhao F (2021) Comprehensive profiling of circular RNAs with nanopore sequencing and CIRI-long. Nat Biotechnol 39(7):836–845. https://doi.org/10.1038/s41587-021-00842-6
37. Rahimi K, Færch Nielsen A, Venø MT, Kjems J (2021) Nanopore long-read sequencing of circRNAs. Methods 196:23–29. https://doi.org/10.1016/j.ymeth.2021.09.010
38. Ma XK, Zhai SN, Yang L (2023) Approaches and challenges in genome-wide circular RNA identification and quantification. Trends Genet 39(12):897–907. https://doi.org/10.1016/j.tig.2023.09.006
39. Li S, Teng S, Xu J, Su G, Zhang Y, Zhao J, Zhang S, Wang H, Qin W, Lu ZJ, Guo Y, Zhu Q, Wang D (2019) Microarray is an efficient tool for circRNA profiling. Brief Bioinform 20(4):1420–1433. https://doi.org/10.1093/bib/bby006
40. Lu D, Thum T (2019) RNA-based diagnostic and therapeutic strategies for cardiovascular disease. Nat Rev Cardiol 16(11):661–674. https://doi.org/10.1038/s41569-019-0218-x
41. Aufiero S, Reckman YJ, Pinto YM, Creemers EE (2019) Circular RNAs open a new chapter in cardiovascular biology. Nat Rev Cardiol 16(8):503–514. https://doi.org/10.1038/s41569-019-0185-2
42. Long Q, Lv B, Jiang S, Lin J (2023) The landscape of circular RNAs in cardiovascular diseases. Int J Mol Sci 24(5). https://doi.org/10.3390/ijms24054571
43. Kim Y-K (2022) Circular RNAs as a promising biomarker for heart disease. Biomed Pharmacother 156:113935. https://doi.org/10.1016/j.biopha.2022.113935
44. Vausort M, Salgado-Somoza A, Zhang L, Leszek P, Scholz M, Teren A, Burkhardt R, Thiery J, Wagner DR, Devaux Y (2016) Myocardial infarction-associated circular RNA predicting left ventricular

dysfunction. J Am Coll Cardiol 68(11):1247–1248. https://doi.org/10.1016/j.jacc.2016.06.040
45. Zhang J, Xu Y, Xu S, Liu Y, Yu L, Li Z, Xue X, Wang H (2018) Plasma circular RNAs, Hsa_circRNA_025016, predict postoperative atrial fibrillation after isolated off-pump coronary artery bypass grafting. J Am Heart Assoc 7(2):e006642. https://doi.org/10.1161/JAHA.117.006642
46. Bazan HA, Hatfield SA, Brug A, Brooks AJ, Lightell DJ, Woods TC (2017) Carotid plaque rupture is accompanied by an increase in the ratio of serum circR-284 to miR-221 levels. Circ Cardiovasc Genet 10(4):e001720. https://doi.org/10.1161/CIRCGENETICS.117.001720
47. Zuo Z, Hu H, Xu Q, Luo X, Peng D, Zhu K, Zhao Q, Xie Y, Ren J (2019) BBCancer: an expression atlas of blood-based biomarkers in the early diagnosis of cancers. Nucleic Acids Res 48(D1):D789–D796. https://doi.org/10.1093/nar/gkz942
48. Li Y, Fan H, Sun J, Ni M, Zhang L, Chen C, Hong X, Fang F, Zhang W, Ma P (2020) Circular RNA expression profile of Alzheimer's disease and its clinical significance as biomarkers for the disease risk and progression. Int J Biochem Cell Biol 123:105747. https://doi.org/10.1016/j.biocel.2020.105747
49. Li Y, Lv Z, Zhang J, Ma Q, Li Q, Song L, Gong L, Zhu Y, Li X, Hao Y, Yang Y (2020) Profiling of differentially expressed circular RNAs in peripheral blood mononuclear cells from Alzheimer's disease patients. Metab Brain Dis 35(1):201–213. https://doi.org/10.1007/s11011-019-00497-y
50. Ren Z, Chu C, Pang Y, Cai H, Jia L (2022) A circular RNA blood panel that differentiates Alzheimer's disease from other dementia types. Biomark Res 10(1):63. https://doi.org/10.1186/s40364-022-00405-0
51. Liu L, Chen X, Chen YH, Zhang K (2020) Identification of circular RNA hsa_Circ_0003391 in peripheral blood is potentially associated with Alzheimer's disease. Front Aging Neurosci 12:601965. https://doi.org/10.3389/fnagi.2020.601965
52. Chen X, Han P, Zhou T, Guo X, Song X, Li Y (2016) circRNADb: a comprehensive database for human circular RNAs with protein-coding annotations. Sci Rep 6:34985. https://doi.org/10.1038/srep34985
53. Haque S, Harries LW (2017) Circular RNAs (circRNAs) in health and disease. Genes 8(12):353
54. Pisignano G, Michael DC, Visal TH, Pirlog R, Ladomery M, Calin GA (2023) Going circular: history, present, and future of circRNAs in cancer. Oncogene 42(38):2783–2800. https://doi.org/10.1038/s41388-023-02780-w
55. Gu A, Jaijyan DK, Yang S, Zeng M, Pei S, Zhu H (2023) Functions of circular RNA in human diseases and illnesses. Noncoding RNA 9(4). https://doi.org/10.3390/ncrna9040038
56. Ward Z, Pearson J, Schmeier S, Cameron V, Pilbrow A (2021) Insights into circular RNAs: their biogenesis, detection, and emerging role in cardiovascular disease. RNA Biol 18(12):2055–2072. https://doi.org/10.1080/15476286.2021.1891393
57. Holdt LM, Kohlmaier A, Teupser D (2018) Circular RNAs as therapeutic agents and targets. Front Physiol 9:1262. https://doi.org/10.3389/fphys.2018.01262
58. He AT, Liu J, Li F, Yang BB (2021) Targeting circular RNAs as a therapeutic approach: current strategies and challenges. Signal Transduct Target Ther 6(1):185. https://doi.org/10.1038/s41392-021-00569-5
59. Liu X, Zhang Y, Zhou S, Dain L, Mei L, Zhu G (2022) Circular RNA: an emerging frontier in RNA therapeutic targets, RNA therapeutics, and mRNA vaccines. J Control Release 348:84–94. https://doi.org/10.1016/j.jconrel.2022.05.043
60. Young TL, Wang KC, Varley AJ, Li B (2023) Clinical delivery of circular RNA: lessons learned from RNA drug development. Adv Drug Deliv Rev 197:114826. https://doi.org/10.1016/j.addr.2023.114826
61. Ji P, Wu W, Chen S, Zheng Y, Zhou L, Zhang J, Cheng H, Yan J, Zhang S, Yang P, Zhao F (2019) Expanded expression landscape and prioritization of circular RNAs in mammals. Cell Rep 26(12):3444–3460.e3445. https://doi.org/10.1016/j.celrep.2019.02.078
62. Wesselhoeft RA, Kowalski PS, Parker-Hale FC, Huang Y, Bisaria N, Anderson DG (2019) RNA circularization diminishes immunogenicity and can extend translation duration in vivo. Mol Cell 74(3):508–520.e504. https://doi.org/10.1016/j.molcel.2019.02.015
63. Chen YG, Kim MV, Chen X, Batista PJ, Aoyama S, Wilusz JE, Iwasaki A, Chang HY (2017) Sensing self and foreign circular RNAs by intron identity. Mol Cell 67(2):228–238.e225. https://doi.org/10.1016/j.molcel.2017.05.022
64. Chen YG, Chen R, Ahmad S, Verma R, Kasturi SP, Amaya L, Broughton JP, Kim J, Cadena C, Pulendran B, Hur S, Chang HY (2019) N6-methyladenosine modification controls circular RNA immunity. Mol Cell 76(1):96–109.e109. https://doi.org/10.1016/j.molcel.2019.07.016
65. Breuer J, Barth P, Noe Y, Shalamova L, Goesmann A, Weber F, Rossbach O (2022) What goes around comes around: artificial circular RNAs bypass cellular antiviral responses. Mol Ther Nucleic Acids 28:623–635. https://doi.org/10.1016/j.omtn.2022.04.017
66. Awan FM (2022) The HOPE for Pandora's Box of artificial circular RNA immunogenicity. Mol Ther Nucleic Acids 28:640–642. https://doi.org/10.1016/j.omtn.2022.04.029
67. Obi P, Chen YG (2021) The design and synthesis of circular RNAs. Methods 196:85–103. https://doi.org/10.1016/j.ymeth.2021.02.020
68. Choi S-W, Nam J-W (2024) Optimal design of synthetic circular RNAs. Exp Mol Med 56:1281. https://doi.org/10.1038/s12276-024-01251-w
69. Chen C, Wei H, Zhang K, Li Z, Wei T, Tang C, Yang Y, Wang Z (2022) A flexible, efficient, and scalable platform to produce circular RNAs as new

therapeutics. bioRxiv:2022.2005.2031.494115. https://doi.org/10.1101/2022.05.31.494115
70. Lee KH, Kim S, Song J, Han SR, Kim JH, Lee SW (2023) Efficient circular RNA engineering by end-to-end self-targeting and splicing reaction using Tetrahymena group I intron ribozyme. Mol Ther Nucleic Acids 33:587–598. https://doi.org/10.1016/j.omtn.2023.07.034
71. Cui J, Zhang L, Zhang Z, Luo X, Liu Y, Li C, Huang W, Zou L, Yu X, Xiao F (2023) A precise and efficient circular RNA synthesis system based on a ribozyme derived from Tetrahymena thermophila. Nucleic Acids Res 51(14):e78. https://doi.org/10.1093/nar/gkad554
72. Ooi JYY, Bernardo BC (2020) Translational potential of non-coding RNAs for cardiovascular disease. Adv Exp Med Biol 1229:343–354. https://doi.org/10.1007/978-981-15-1671-9_21
73. Wadley GD, Lamon S, Alexander SE, McMullen JR, Bernardo BC (2019) Noncoding RNAs regulating cardiac muscle mass. J Appl Physiol (1985) 127(2):633–644. https://doi.org/10.1152/japplphysiol.00904.2018
74. Meganck RM, Borchardt EK, Castellanos Rivera RM, Scalabrino ML, Wilusz JE, Marzluff WF, Asokan A (2018) Tissue-dependent expression and translation of circular RNAs with recombinant AAV vectors in vivo. Mol Ther Nucleic Acids 13:89–98. https://doi.org/10.1016/j.omtn.2018.08.008
75. Lavenniah A, Luu TDA, Li YP, Lim TB, Jiang J, Ackers-Johnson M, Foo RS (2020) Engineered circular RNA sponges act as miRNA inhibitors to attenuate pressure overload-induced cardiac hypertrophy. Mol Ther 28(6):1506–1517. https://doi.org/10.1016/j.ymthe.2020.04.006
76. Garikipati VNS, Verma SK, Cheng Z, Liang D, Truongcao MM, Cimini M, Yue Y, Huang G, Wang C, Benedict C, Tang Y, Mallaredy V, Ibetti J, Grisanti L, Schumacher SM, Gao E, Rajan S, Wilusz JE, Goukassian D, Houser SR, Koch WJ, Kishore R (2019) Circular RNA CircFndc3b modulates cardiac repair after myocardial infarction via FUS/VEGF-A axis. Nat Commun 10(1):4317. https://doi.org/10.1038/s41467-019-11777-7
77. Bass-Stringer S, Bernardo BC, May CN, Thomas CJ, Weeks KL, McMullen JR (2018) Adeno-associated virus gene therapy: translational progress and future prospects in the treatment of heart failure. Heart Lung Circ 27(11):1285–1300. https://doi.org/10.1016/j.hlc.2018.03.005
78. Santer L, Bar C, Thum T (2019) Circular RNAs: a novel class of functional RNA molecules with a therapeutic perspective. Mol Ther 27(8):1350–1363. https://doi.org/10.1016/j.ymthe.2019.07.001
79. Tanne JH (2022) FDA approves $3.5m gene therapy for adults with haemophilia B. BMJ 379:o2858. https://doi.org/10.1136/bmj.o2858
80. Weeks KL, Kiriazis H, Wadley GD, Masterman EI, Sergienko NM, Raaijmakers AJA, Trewin AJ, Harmawan CA, Yildiz GS, Liu Y, Drew BG, Gregorevic P, Delbridge LMD, McMullen JR, Bernardo BC (2024) A gene therapy targeting medium-chain acyl-CoA dehydrogenase (MCAD) did not protect against diabetes-induced cardiac pathology. J Mol Med (Berl) 102(1):95–111. https://doi.org/10.1007/s00109-023-02397-2
81. Wu N, Xu J, Du WW, Li X, Awan FM, Li F, Misir S, Eshaghi E, Lyu J, Zhou L, Zeng K, Adil A, Wang S, Yang BB (2021) YAP circular RNA, circYap, attenuates cardiac fibrosis via binding with tropomyosin-4 and gamma-actin decreasing actin polymerization. Mol Ther 29(3):1138–1150. https://doi.org/10.1016/j.ymthe.2020.12.004
82. Wan J, Wang Z, Wang L, Wu L, Zhang C, Zhou M, Fu ZF, Zhao L (2024) Circular RNA vaccines with long-term lymph node-targeting delivery stability after lyophilization induce potent and persistent immune responses. MBio 15(1):e01775-01723. https://doi.org/10.1128/mbio.01775-23
83. Du WW, Yang W, Li X, Awan FM, Yang Z, Fang L, Lyu J, Li F, Peng C, Krylov SN, Xie Y, Zhang Y, He C, Wu N, Zhang C, Sdiri M, Dong J, Ma J, Gao C, Hibberd S, Yang BB (2018) A circular RNA circ-DNMT1 enhances breast cancer progression by activating autophagy. Oncogene 37(44):5829–5842. https://doi.org/10.1038/s41388-018-0369-y
84. Yang L, Han B, Zhang Z, Wang S, Bai Y, Zhang Y, Tang Y, Du L, Xu L, Wu F, Zuo L, Chen X, Lin Y, Liu K, Ye Q, Chen B, Li B, Tang T, Wang Y, Shen L, Wang G, Ju M, Yuan M, Jiang W, Zhang JH, Hu G, Wang J, Yao H (2020) Extracellular vesicle-mediated delivery of circular RNA SCMH1 promotes functional recovery in rodent and nonhuman primate ischemic stroke models. Circulation 142(6):556–574. https://doi.org/10.1161/CIRCULATIONAHA.120.045765
85. Ahn SH, Ryu SW, Choi H, You S, Park J, Choi C (2022) Manufacturing therapeutic exosomes: from bench to industry. Mol Cells 45(5):284–290. https://doi.org/10.14348/molcells.2022.2033
86. Fan X, Zhang X, Wu X, Guo H, Hu Y, Tang F, Huang Y (2015) Single-cell RNA-seq transcriptome analysis of linear and circular RNAs in mouse preimplantation embryos. Genome Biol 16(1):148. https://doi.org/10.1186/s13059-015-0706-1
87. Zhong C, Yu S, Han M, Chen J, Ning K (2017) Heterogeneous circRNA expression profiles and regulatory functions among HEK293T single cells. Sci Rep 7(1):14393. https://doi.org/10.1038/s41598-017-14807-w
88. Wu W, Zhang J, Cao X, Cai Z, Zhao F (2022) Exploring the cellular landscape of circular RNAs using full-length single-cell RNA sequencing. Nat Commun 13(1):3242. https://doi.org/10.1038/s41467-022-30963-8
89. Jain M, Koren S, Miga KH, Quick J, Rand AC, Sasani TA, Tyson JR, Beggs AD, Dilthey AT, Fiddes IT, Malla S, Marriott H, Nieto T, O'Grady J, Olsen HE, Pedersen BS, Rhie A, Richardson H, Quinlan AR, Snutch TP, Tee L, Paten B, Phillippy AM,

Simpson JT, Loman NJ, Loose M (2018) Nanopore sequencing and assembly of a human genome with ultra-long reads. Nat Biotechnol 36(4):338–345. https://doi.org/10.1038/nbt.4060

90. Fuchs S, Babin L, Andraos E, Bessiere C, Willier S, Schulte JH, Gaspin C, Meggetto F (2022) Generation of full-length circular RNA libraries for Oxford Nanopore long-read sequencing. PLoS One 17(9):e0273253. https://doi.org/10.1371/journal.pone.0273253
91. Xin R, Gao Y, Gao Y, Wang R, Kadash-Edmondson KE, Liu B, Wang Y, Lin L, Xing Y (2021) isoCirc catalogs full-length circular RNA isoforms in human transcriptomes. Nat Commun 12(1):266. https://doi.org/10.1038/s41467-020-20459-8
92. Rahimi K, Venø MT, Dupont DM, Kjems J (2021) Nanopore sequencing of brain-derived full-length circRNAs reveals circRNA-specific exon usage, intron retention and microexons. Nat Commun 12(1):4825. https://doi.org/10.1038/s41467-021-24975-z
93. Hou L, Zhang J, Zhao F (2023) Full-length circular RNA profiling by nanopore sequencing with CIRI-long. Nat Protoc 18(6):1795–1813. https://doi.org/10.1038/s41596-023-00815-w
94. Zhang J, Hossain MT, Liu W, Peng Y, Pan Y, Wei Y (2022) Evaluation of CircRNA sequence assembly methods using long reads. Front Genet 13:816825. https://doi.org/10.3389/fgene.2022.816825
95. Chiang T-W, Jhong S-E, Chen Y-C, Chen C-Y, Wu W-S, Chuang T-J (2023) FL-circAS: an integrative resource and analysis for full-length sequences and alternative splicing of circular RNAs with nanopore sequencing. Nucleic Acids Res 52(D1):D115–D123. https://doi.org/10.1093/nar/gkad829
96. Wesselhoeft RA, Kowalski PS, Anderson DG (2018) Engineering circular RNA for potent and stable translation in eukaryotic cells. Nat Commun 9(1):2629. https://doi.org/10.1038/s41467-018-05096-6
97. Chen R, Wang SK, Belk JA, Amaya L, Li Z, Cardenas A, Abe BT, Chen C-K, Wender PA, Chang HY (2023) Engineering circular RNA for enhanced protein production. Nat Biotechnol 41(2):262–272. https://doi.org/10.1038/s41587-022-01393-0
98. Zhang B, Chen M, Cao J, Liang Y, Tu T, Hu J, Li T, Cai Y, Li S, Liu B, Xu J, Liang B, Ye X, Cai X (2021) An integrated electrochemical POCT platform for ultrasensitive circRNA detection towards hepatocellular carcinoma diagnosis. Biosens Bioelectron 192:113500. https://doi.org/10.1016/j.bios.2021.113500
99. Niu D, Wu Y, Lian J (2023) Circular RNA vaccine in disease prevention and treatment. Signal Transduct Target Ther 8(1):341. https://doi.org/10.1038/s41392-023-01561-x
100. Qu L, Yi Z, Shen Y, Lin L, Chen F, Xu Y, Wu Z, Tang H, Zhang X, Tian F, Wang C, Xiao X, Dong X, Guo L, Lu S, Yang C, Tang C, Yang Y, Yu W, Wang J, Zhou Y, Huang Q, Yisimayi A, Liu S, Huang W, Cao Y, Wang Y, Zhou Z, Peng X, Wang J, Xie XS, Wei W (2022) Circular RNA vaccines against SARS-CoV-2 and emerging variants. Cell 185(10):1728–1744.e1716. https://doi.org/10.1016/j.cell.2022.03.044
101. Seephetdee C, Bhukhai K, Buasri N, Leelukkanaveera P, Lerdwattanasombat P, Manopwisedjaroen S, Phueakphud N, Kuhaudomlarp S, Olmedillas E, Saphire EO, Thitithanyanont A, Hongeng S, Wongtrakoongate P (2022) A circular mRNA vaccine prototype producing VFLIP-X spike confers a broad neutralization of SARS-CoV-2 variants by mouse sera. Antivir Res 204:105370. https://doi.org/10.1016/j.antiviral.2022.105370
102. Amaya L, Grigoryan L, Li Z, Lee A, Wender PA, Pulendran B, Chang HY (2023) Circular RNA vaccine induces potent T cell responses. Proc Natl Acad Sci USA 120(20):e2302191120. https://doi.org/10.1073/pnas.2302191120
103. Shen H, Liu B, Xu J, Zhang B, Wang Y, Shi L, Cai X (2021) Circular RNAs: characteristics, biogenesis, mechanisms and functions in liver cancer. J Hematol Oncol 14(1):134. https://doi.org/10.1186/s13045-021-01145-8

MIX
Papier aus verantwortungsvollen Quellen
Paper from responsible sources
FSC® C105338

If you have any concerns about our products,
you can contact us on
ProductSafety@springernature.com

In case Publisher is established outside the EU,
the EU authorized representative is:
Springer Nature Customer Service Center GmbH
Europaplatz 3, 69115 Heidelberg, Germany

Printed by Libri Plureos GmbH
in Hamburg, Germany